ENGINEERING MECHANICS

STATICS

COMPUTATIONAL EDITION
SI EDITION

Robert W. Soutas-Little
Michigan State University

Daniel J. Inman
Virginia Polytechnic Institute and State University

Daniel S. Balint
Imperial College London

THOMSON™

Australia · Canada · Mexico · Singapore · Spain · United Kingdom · United States

Engineering Mechanics: Statics, **Computational Edition,** ***SI Edition***

by Robert W. Soutas-Little, Daniel J. Inman, and Daniel S. Balint

Associate Vice President and Editorial Director:
Evelyn Veitch

Publisher:
Chris Carson

Developmental Editor:
Hilda Gowans

Permissions Coordinator:
Vicki Gould

Production Services:
RPK Editorial Services

Copy Editor:
Harlan James

Proofreader:
Erin Wagner

Indexer:
Shelly Gerger-Knechtl

Production Manager:
Renate McCloy

Creative Director:
Angela Cluer

Interior Design:
Carmela Pereira

Cover Design:
Andrew Adams

Compositor:
International Typesetting and Composition

Printer:
Courier–Kendalville

Cover Image Credit:
© IML Images Group Ltd/Alamy

Printed and bound in the United States
1 2 3 4 07

For more information contact Nelson, 1120 Birchmount Road, Toronto, Ontario, Canada, MIK 5G4. Or you can visit our Internet site at http://www.nelson.com

Library Congress Control Number: 2007932743

ISBN-10: 0-495-43811-1

ISBN-13: 978-0-495-43811-3

North America
Nelson
1120 Birchmount Road
Toronto, Ontario MIK 5G4
Canada

Asia
Thomson Learning
5 Shenton Way #01-01
UIC Building
Singapore 068808

Australia/New Zealand
Thomson Learning
102 Dodds Street
Southbank, Victoria
Australia 3006

Europe/Middle East/Africa
Thomson Learning
High Holborn House
50/51 Bedford Row
London WCIR 4LR
United Kingdom

Latin America
Thomson Learning
Seneca, 53
Colonia Polanco
11560 Mexico D.F.
Mexico

Spain
Paraninfo
Calle/Magallanes, 25
28015 Madrid, Spain

CONTENTS

PREFACE

Most basic texts in Statics have changed little since the 1960s. Over this period, one of the most popular texts for Statics was by Beer and Johnston, first published in 1962 in a non-vector form, and then in vector editions in the early 1970s. Another very popular text was by J. L. Meriam, who pioneered the use of SI units in mechanics. Meriam also held the strong view that vector operations should be conceptual whenever possible, and placed strong emphasis on drawing vector polygons and solving most two-dimensional problems using trigonometry. With the increase in computer usage, computer problems were added to some books in the early 1990s, but only in a limited manner as most students were not familiar with programming languages. In addition, there was limited time in the course schedule to cover solution of problems of this type. Basically, there have been few changes in the material presented and the mathematical treatment of introductory mechanics since the introduction of vectors.

Why then would we spend years writing yet another book on Statics? The main topics covered in any Statics course have not changed and, indeed, have survived the test of time. The course has always served to present the concepts of equilibrium both of particles and rigid bodies and to introduce concepts needed in Strength of Materials and Dynamics courses. The main reason for the development of this text was that we did not feel that the current texts were fully in phase with the students' computer sophistication and therefore were limited in their treatment. This book represents a totally new presentation of Statics and has been developed over the last 10 years. The approach used in the book is embodied by its title "Engineering Mechanics: Statics—*Computational Edition*." We have incorporated the use of computational software into the solution of some problems in special manuals for MATLAB, Mathcad, Maple, and Mathematica. Homework problems are marked in such a manner that the instructor and student will know if a particular problem can and in some cases must be solved with the aid of software or if it could be easily solved "by hand."

We found that we could greatly expand the scope of problems that could be solved by using computational software. Indeed, the nonlinear nature of some of these problems made the use of the computational software necessary. Computational methods were separated in the text so that they can be omitted if the instructor chooses. These methods would still be available as a reference for the student for later courses. We have introduced new solution methods, such as, a direct vector solution where the vector equilibrium equation is solved without expanding it into scalar equations. We have guarded against a "black box" approach and have tried to enhance the students' physical understanding of basic mechanics. The computational software may be thought of as a sophisticated graphical calculator. This approach, the use of computational software, is simply the next step in the evolution from slide-rule and trig tables, to calculators and now to computers.

We and other instructors have been testing this approach for the last 10 years and have found widespread acceptance and enthusiasm for the material. A greater emphasis has been placed on modeling (free body diagrams) and mathematically forming the vector equations of equilibrium and the required geometrical constraints. Many of the problems are formulated in general parametric terms allowing the student to see the effect of each parameter, that is, the effect of the change of an angle, length, load placement, spring constant, coefficient of friction, etc. This allows the student to consider design implications as well as developing the skill of analysis.

The biggest change in courses taught in this manner was in the type of examinations that were given. Since a greater emphasis has been placed on modeling and expressing the equations of equilibrium while deemphasizing the numerical work, examinations needed to reflect this change. Therefore, many exam questions ask that problems be solved to a point where only numerical calculations remain. Typically, one take-home problem that required full numerical solution was included in the exams, allowing the student to use computational software. With the increased capabilities of laptop computers and graphical calculators, examination techniques need to change. As one author said, "My grandchildren are far more computer literate than I am and I'm the one who has to catch up."

All books in Statics introduce the concepts of forces, moments, and equilibrium. In addition, the basic concepts of centroids, second moments of area, internal forces, and moments in beams are introduced as a bridge to Mechanics of Materials (deformable bodies) courses. In addition, the concept of dry friction is introduced with applications to wedges, belts, and common mechanisms. We tested introducing mathematical methods on a "need-to-know" basis but we found this to be disjointed in some cases and therefore presented all vector analysis in Chapter 2. We have also introduced systems of linear equations and matrices in this chapter. Equilibrium is expressed as a system of linear equations that, in the past, students have spent much of their time solving numerically. Most calculators and all software programs solve these using a matrix approach. It was not the purpose to present a complete treatment of linear algebra in the text, but to add enough material so that the student could use the computational software in an informed manner. Chapter 3 discusses particle equilibrium with emphasis on modeling and writing the vector equation of equilibrium. One can alter the presentation by introducing vectors from the first three sections in Chapter 2 and then cover the first two sections of Chapter 3 followed by methods of solution of these equations from Chapter 2.

Springs and statically indeterminate problems have been introduced in Chapter 3 that provide a smooth bridge to the study of deformable bodies. This material is of a "stand alone" nature and may be omitted without loss in continuity of the course. Two special sections were added to this chapter; introduction to static friction and the keystone of an arch. Friction is covered in detail in Chapter 9 following the standard approach, but a brief introduction in particle equilibrium emphasizes that the friction force will equal only that required for equilibrium and that the coefficient of static friction only specifies a maximum value before movement. The friction force would never be equal to the coefficient of static friction multiplied by the normal force as the system would be completely unstable. The discussion of arches is mainly for historical purposes and may be treated as a single homework problem.

Chapters 4 through 6 follow the traditional presentation with some additional vector methods. A section entitled "Method of Joints using Matrix Techniques" was added to Chapter 7 to introduce modern structural analysis. This section may be covered or omitted without loss of continuity. A section on discontinuity functions was added to Chapter 8 to provide a smooth transition to bending of beams in Mechanics of Materials courses and to introduce the Heaviside step function for use in Dynamics. A complete coverage of friction is given in Chapter 9 and duplicates some of the material in Chapter 3 on particle equilibrium. Chapter 10 presents moments of inertia in the standard manner but adds a discussion of the eigenvalue problem. Chapter 11 covers the topic of virtual work . This topic is usually omitted in the standard course but is included for completeness and as a reference.

A limited number of biomechanics problems have been added to introduce students to the principals of mechanics as applied to biomedical engineering. Biomechanics is becoming a major tool for clinical understanding of orthopaedic and rehabilitation problems as well as applications in sports medicine. For easy reference, a "Statics Index Dictionary" is included allowing students to find a quick definition of a topic and its location in the text.

Separate manuals in MATLAB, Mathcad, Maple, and Mathematica present details on each computational software package and how it can be used in the solution of problems in Statics. The applicable sample problems in the text are solved in these manuals and they should be used to guide the student in using of a specific software package to solve Statics problems. We have found little or no additional class time is required.

Many individuals have contributed to the development of this text. We cannot individually thank the many students who have given us 'feed back' and made many useful suggestions, but we can thank the many faculty who have helped. We apologize in advance for any we have missed, but here in alphabetical order is a partial list.

Dean Nicholas Altiero, College of Engineering, Tulane University
Professor Greg Heiner, Utah State University
Dr. K. Jimmy Hsia, University of Illinois Urbana-Champaign
Dr. Dallas Kingsbury, Arizona State University
Professor Carl Knowlen, University of Washington
Professor John B. Ligon, Michigan Technical University
Dr. Jun Nogami, Michigan State University
Dr. K. Papadakis, University of Patras
Professor Roger Haut, Michigan State University
Professor Robert Hubbard, Michigan State University
Dr. Tamara Reed-Bush, Michigan State University
Dr. Wendy Reffeor, Grand Valley State University
Professor Henry Scanton, RPI
Dr. Joe Slater, Wright State University
Professor Arnold E. Somers, Jr, Valdosta State University
Patricia Soutas-Little, retired; Michigan State University
Professor Bill Spencer, University of Illinois, Urbana-Champaign (formerly at Notre Dame)
Bill Stenquist, Editor, McGraw Hill
Dr. Wayne Whitman, Georgia Tech (formally at West Point)

and from Thomson Engineering

Chris Carson
Hilda Gowans
Rose Kernan, RPK Editorial Services.

Chapter 1

INTRODUCTION

The 300-meter-high Eiffel Tower was designed by the French engineer Alexandre Gustave Eiffel (1832–1923) and erected for the Paris Exposition of 1889. Eiffel also designed the framework of the Statue of Liberty, which is found in New York Harbor. A miniature model of this statue stands in the middle of the River Seine on a small island near the Eiffel Tower. (Photo courtesy of ErickN/Shutterstock)

1.1 MECHANICS

Mechanics is the oldest area of study in physics, originating with the Greek culture in 400 to 300 BC. The earliest record of the study of mechanical problems was the work of Aristotle (384–322 BC). Although historical questions remain as to the authorship of some of these early works, considerations of the equilibrium of vertical forces and the action of levers were evident in these writings. Archimedes of Syracuse (287–212 BC) introduced the formal study of levers and the concept of the center of gravity. He also examined concepts of geometry and wrote about the theory of buoyancy and the equilibrium of floating bodies. Many later scientists and mathematicians, including Kepler and Galileo, contributed to the development of mechanics, but the formal presentation of the principles of mechanics was made by Sir Isaac Newton (1642–1727).

Mechanics is a basic engineering science that lays the foundation for machine design, structural analysis, stress analysis, vibrations, electromagnetic field theory, and fluid and solid mechanics. In particular, courses in mechanics use the methods of modeling, the techniques of vector algebra and calculus, and computational methods. Mechanics is the branch of physical science that deals with motion and the effects of forces on gases, liquids, and solid bodies. For ease of study, mechanics has been divided into the study of deformable solid bodies, the study of fluids and gases, and the study of rigid bodies. To further simplify the approach, the mechanics of rigid bodies is subdivided into statics, which deals with rigid bodies at rest or moving at a constant velocity, and dynamics, which is the study of rigid bodies undergoing acceleration. Although statics may be considered a special case of dynamics, it is helpful to study this special case first in order to obtain a firm understanding of the concepts of forces and moments, learn the methods of modeling physical situations, and become skilled at using the mathematical tools necessary to describe these ideas.

Statics, which is based upon surprisingly few basic principles, employs a field of mathematics called vector algebra. We will introduce another field of mathematics, called matrix algebra, that proves useful when one is working with common computational software. It is not necessary to use matrix notation to solve problems in statics, but most presently available commercial software uses this notation.

1.2 BASIC CONCEPTS

The starting point of the study of mechanics is an examination of the basic concepts upon which Newton based his three laws of motion. These concepts are as follows.

Space is a boundless expanse in which objects and events occur and have relative position and orientation. Space allows the measurement of length, area, and volume. The measurement of length is made by comparison with another object of known or standard length, such as a ruler. Area is defined as the product of two perpendicular lengths and allows the measurement of a two-dimensional space. Volume is three-dimensional space and is the product of three lengths. Newton considered space to be infinite, homogeneous, isotropic, and absolute. The last property, absoluteness, allowed Newton to assume the existence of a primary inertial frame of reference, which is not moving relative to the "fixed" stars and has an origin located at the center of mass of the universe. If space is isotropic, the properties of a closed system at any point in space are unaffected by the orientation of the system. "Homogeneous" means that the space is the same at every point and does not change from point to point. Our present-day understanding of space is somewhat different from Newton's, owing to Einstein's theory of relativity; however, in this book, we will assume the same idea of space that Newton did and neglect any small deviations owing to relativistic effects.

Time is the concept used to order the flow of events. Newton assumed time is absolute—that is, time is the same for all observers and is independent of all objects in the world. Physicists consider time to be an abstraction arrived at by changes in the physical world; in other words, the flow of a series of events *defines* time, and time depends upon that flow of events. We use Newton's definition of absolute time, and time is measured by comparison with some repeatable event, such as the rotation of the Earth, the oscillation of a pendulum, or atomic vibrations (i.e., a transition frequency in the element cesium).

Newton defined ***mass*** as a "quantity of matter" that related an object's volume to its density. He stated that *gravitational mass,* as defined by the law of gravitational attraction, is equivalent to *inertial mass,* which measures an object's resistance to being accelerated. In current times, this statement is called the "principle of equivalence" and is a postulate of Einstein's theory of relativity. In this book, we assume that the mass of a body is independent of its motion, neglecting any relativistic effects.

Force is defined as the action of one body on another. This action may be the result of direct contact between the two bodies, or it may arise from gravitational, magnetic, or electrical effects between two bodies separated by a distance. Newton postulated that forces always occur in pairs, equal and opposite, each acting on one of the two bodies. Force is not measured directly; only the effect a force produces can be measured. For example, the force required to stretch a spring is measured by determining the distance the spring stretches.

Newton formulated four axioms, or laws, that are the basis of the study of rigid-body statics and dynamics. The first three laws, known as ***Newton's laws of motion***, may be stated as follows:

1. *Every body or particle continues in a state of rest or uniform motion (constant velocity) in a straight line, unless it is compelled to change that state by forces acting upon it.* That is, the body will remain at rest or continue to move with the same speed and direction if no net external forces act on it.
2. *The change of motion of a body is proportional to the net force imposed on the body and is in the direction of the net force.* That is, the net force is equal to the change in the product of the mass and the velocity. The mass is the resistance of the body to acceleration.
3. *If one body exerts a force on a second body, then the second body exerts a force on the first that is equal in magnitude, opposite in direction, and collinear.*

Newton's final law is the law of universal gravitational attraction:

4. *Any two particles are attracted to each other with a force whose magnitude is proportional to the product of their gravitational masses and inversely proportional to the square of the distance between them.* An example of a gravitational attraction force is shown in Figure 1.1.

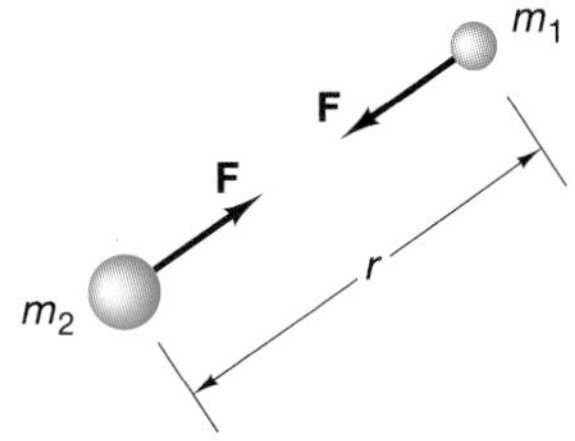

m_1 F F r m_2

Figure 1.1 The gravitational force F between two bodies of masses m_1 and m_2 separated by a distance r between their centers.

The magnitude of the gravitational force F is stated mathematically as

$$F = \frac{Gm_1m_2}{r^2} \tag{1.1}$$

where

$$G = 66.73 \times 10^{-12}\ \text{m}^3/\text{kg}\cdot\text{s}^2$$

is the ***universal gravitational constant***.

The ***weight*** of a body on the Earth's surface is the force due to the gravitational attraction of an object to the mass of the Earth. Weight is expressed as

$$W = \frac{GmM}{R^2} \tag{1.2}$$

where G is the universal gravitational constant, m is the mass of the object (in kilograms), M is the mass of the Earth (in kilograms), and R is the radius of the Earth (in meters). For the gravitational attraction of an object on the surface of the Earth, the ratio

$$g = (GM)/R^2$$

is taken to be a constant, and the value of g is 9.807 m/s^2 or 32.17 ft/s^2. This value differs at different points on the Earth, as the Earth is not a perfect sphere of radius R, and elevations vary. However, for most engineering problems, g is considered to be a constant. The weight of an object near the surface of the Earth is caused by the gravitational attraction and is equal to

$$W = mg \tag{1.3}$$

The constant g has the units of acceleration and is sometimes called the ***gravitational acceleration*** or *the acceleration due to gravity.* This concept will be discussed in detail when the subject of particle dynamics is presented. For the present, consider g to be a constant that relates the mass of an object to its weight on the surface of the Earth.

Problems

Some problems will be conceptual in nature, not requiring the solution of equations or any quantitative effort. These problems are based on ideas from Eric Mazur, Peer Instruction, Prentice Hall, Upper Saddle River, N. J., 1997.

1.1 A collision occurs between a bus and a small sports car. During the impact:
a. There is no force between the bus and the car.
b. The bus exerts a force on the car, but the car does not exert any force on the bus.
c. The bus exerts a much larger force on the car than the car does on the bus.
d. The force that the bus exerts on the car is equal to the force the car exerts on the bus.
e. The car exerts a larger force on the bus than the bus does on the car.

The resulting deceleration (negative acceleration) would be
a. the same for the bus and the car.
b. greater for the bus than the car.
c. greater for the car than the bus.

1.2 An elevator ascends at a constant rate (constant speed). The net force acting on the elevator is
a. equal to zero.
b. equal to the weight of the elevator and its occupants.
c. equal to the difference between the weight of the elevator and the tension in the cables lifting the elevator.

1.3 If you moved to a planet of the same radius, but a greater mass, your weight would
a. decrease.
b. increase.
c. remain the same.

1.4 If you moved to a planet of the same mass, but a larger radius, your mass would
a. decrease.
b. increase.
c. remain the same.

1.3 UNITS

It is important to establish units of measure for length, mass, time, and force before proceeding further with the discussion of statics. Although most engineers and scientists use an ***international absolute system of units*** (SI units, or Systeme International d'Unites), metric units are used in many countries and U.S. customary units are still used in the United States. The common metric system and the U.S. customary system of units are based upon weight and is therefore called a ***gravitational system of units***. In these systems, force is taken as a basic unit. In the absolute system, mass is taken as a basic unit and force is a derived system. Confusion still exists in the use of kilogram force (kgf) in the metric system and kilogram mass in the SI system. Kilogram force can only be used in the earth's gravitation, but the conversion to force in newtons is the same as in the SI system, that is, multiplying by 9.81. All problems in the text will use SI units, but we will show conversion tables in the section.

1.3.1 SI UNITS

The base units in the SI system are those of length, mass, and time, while force is recognized as a derived quantity. The ***SI unit*** of length is the ***meter*** (m), mass is measured in ***kilograms*** (kg), and time is expressed in ***seconds*** (s). The unit of force is called a ***newton*** (N) and is defined as the force required to accelerate 1 kilogram of mass by 1 meter per second per second, or, symbolically,

$$1\ \text{N} = (1\ \text{kg}) \cdot (1\ \text{m/s}^2)$$

Larger and smaller units in the SI system are decimal multiples of the base units, as indicated by a prefix. The SI prefixes are given in Table 1.1. When pronouncing these units, one places the accent upon the prefix for emphasis. "Kilometer" is an excellent example that is frequently pronounced differently in nontechnical conversations. The use of "hecto-", "deka-", "deci-", and "centi-" is avoided, except for measurements of areas and volumes and in the nontechnical common use of "centimeter".

One SI prefix that has come into common usage during the last few years is nano or 10^{-9}. A major scientific and engineering initiative of the 21st century is nanotechnology. Nanotechnology is a new research and technology at the atomic, molecular, or macromolecular levels in the length scale of 1 to 100 nm. A nanometer (nm) is one billionth of a meter and nanotechnology is new hybrid science combining engineering, chemistry, and biology. This technology would build nanomachines one atom or molecule at a time. Nanogears would be about a nanometer across and could rotate at speeds of 6 trillion revolutions per minute (rpm). In 1990, researchers at IBM showed that it is possible to manipulate a single atom. Research now is looking into creating nanoscopic machines called *assemblers* that are programmed to manipulate atoms and molecules. *Replicators* would be

Table 1.1
SI Prefixes

Multiplication Factor	Prefix Name	Symbol
10^{12}	tera	T
10^{9}	giga	G
10^{6}	mega	M
10^{3}	kilo	k
10^{2}	hecto	h
10	deka	da
10^{-1}	deci	d
10^{-2}	centi	c
10^{-3}	milli	m
10^{-6}	micro	μ
10^{-9}	nano	n
10^{-12}	pico	p
10^{-15}	femto	f
10^{-18}	atto	a

programmed to build the trillions of assemblers needed to build the nanomachines. Trillions of assemblers and replicators will fill a space smaller than a cubic millimeter. The first product to be made by nanomachines will be stronger fibers; one hundred times stronger and four times lighter than steel. In the medical field, nanorobots or nanobots could change the molecular structures of cancer cells or viruses. Molecular computers could store trillions of bytes of information and be billions times faster than silicon-based computers. The United States' federal government is allocating millions of dollars for research in nanotechnology.

One other unit of measurement that should be presented at this time is the measure of an angle. Although we will frequently specify the value of an angle in degrees and will have no difficulty in finding trigonometric functions (sine, cosine, etc.) of that angle in degrees, when the angle appears in an equation as an independent unit, it is measured in radians:

$$\pi \text{ radians} = 180^\circ$$

Radians are the basic measure of angles, and degrees are used only because students are familiar with these units. When angular velocity (rate of rotation) is introduced, it will be expressed in radians per second (rad/s). Note that angular measurement is dimensionless; that is, it has no specific unit. Instead, the angle is considered as the ratio of the circular arc length to the radius of the arc. One complete revolution is 2π radians, or 360°. Most computational software packages require that angles be expressed in radians.

1.3.2 U.S. CUSTOMARY UNITS

Many U.S. engineers still use a system of units based on the old English units for length, force, and time. This system is not only a different system of units, but also a different concept, since force instead of mass is taken as a fundamental unit. No problems in these units will be included in this text. This system will be briefly reviewed to aid students when they encounter them. The standard units of length is the *foot* (ft), the unit of force is the *pound* (lb), and the unit of time is the *second* (s). Defining the pound as a standard unit makes the system dependent upon the gravitational attraction or weight. The constant g has a value of 32.17 ft/s^2 in U.S. units. Mass in U.S. units is measured in slugs, expressed as pound-second squared divided by feet. The U.S. system is not a decimal system, and conversion within the system from one unit to another is cumbersome. For example:

$$\begin{aligned} 1 \text{ foot} &= 12 \text{ inches} \\ 1 \text{ mile} &= 5280 \text{ ft} \\ 1 \text{ yard} &= 3 \text{ ft} \\ 1 \text{ nautical mile} &= 6080.2 \text{ ft} = 1853.248 \text{ meters} \end{aligned}$$

Astronomers must measure very large distances and therefore use different units. For example, an astronomical unit (AU) is the distance from the Earth to the Sun; 150 million kilometers [1 AU $= 150 \times 10^6$ km]. For these distances, miles or kilometers become useless for scaling distances and even astronomical units soon become cumbersome. For tabulating interstellar distances, astronomers use light years, the distance that light travels in a year at its constant velocity of 7.2 AU per hour (299,792 kilometers per second). Our Milky Way Galaxy is 90,000 light-years in diameter. To enclose the known universe, a cube of billions of light-years on each side is needed. According to the widely accepted

Big Bang theory, the universe is expanding at near light speed since coming into existence about 15 billion years ago in a colossal genesis explosion.

1.3.3 CONVERSION BETWEEN SYSTEMS OF UNITS

It is frequently necessary to convert from one system of units to another. Some of these conversions are shown in Table 1.2.

Sample Problem 1.1

Suppose you are driving your car in the U.S. along a road with a posted speed limit of 60 mph (miles per hour), what is the speed limit in km/hr?

Solution

The conversion from mph to km/hr can be obtained from Table 1.2:

$$60 \text{ mph} = 60 \text{ mi/h} \times 1.609 \text{ km/mi} = 96.54 \text{ km/hr}$$

For the calculation of dynamic parameters such as acceleration, time lapsed, of distance traveled, the speed must be in meters per second. Conversion from km/hr to m/s is accomplished by multiplying by 1000 m/km and dividing by 3600 s/hr;

$$1 \text{ km/hr} = 1 \times \frac{1000 \text{ m/km}}{3600 \text{ s/hr}} \times \text{km/hr} = 0.278 \text{ m/s}$$

Table 1.2
Units Conversion

LENGTH Meter (m) is the Basic SI Unit	
1 ft → 0.3048 m	1 m → 3.281 ft
1 in → 2.54 cm	1 cm → 0.3937 in
1 yd → 0.9144 m	1 m → 1.094 yd
1 mi → 1.609 km	1 km → 0.6214 mile
1 angstrom → 10^{-10} m	→ 3.937×10^{-9}
1 furlong → 201.2 m	→ 660 ft
1 light year → 9.461×10^{15} m	→ 5.879×10^{12} mi
1 micron → 10^{-6} m	1 m → 10^{6} microns μ
1 nanometer (nm) → 10^{-9} m	1 m → 10^{9} nm
AREA Square Meter (m^2) is the Basic SI Unit	
1 ft² → 9.290×10^{-2} m²	1 m² → 10.764 ft² → 1.550×10^{3} in²
1 hectare → 1×10^{4} m²	1 hectare → 2.471 acre
1 sq mile → 640 acres	1 sq mile → 2.590×10^{6} m²
1 sq in → 6.452×10^{-4} m²	1 sq in → 6.452 cm²
VOLUME Cubic Meter (m^3) is the Basic SI Unit	
1 m³ → 3.541 cubic ft → 6.102×10^{4} in³	
1 cubic ft → 2.832×10^{-2} m³	
1 cubic in → 1.639×10^{-5} m³	
1 liter → 1.00×10^{-3} m³ → 0.2642 gal (U.S.) → 2.113 pints (U.S.)	
1 gallon (U.S.) → 3.785 liter	
1 gallon (U.K.) → 1.201 gallon (U.S.)	

Continued

FORCE Newton (N) is the Basic SI Unit

1 lb force → 4.448 N	1 N → 0.2248 lb force
1 lb force → 4.534 × 10^{-1} kg force	1 kgf → 2.208 lb force
1 kg force → 9.807 N	1 kgf → 0.1020 N
1 dyne → 1.00 × 10^{-5} N	1 N → 1.000 10^{5} dynes
1 U.S. ton → 2000 lb	

kgf (kilo) is used as a basic measure of weight in the gravitational metric system

MASS Kilogram Mass (kg) is the Basic SI Unit

1 slug → 32.17 lb force gravitational weight (g = 32.17 ft/s^2)
1 slug → 14.59 kg mass
1 kg · 1.0 m/s^2 = 1.00 N
Weight on Earth 1 kg · g = 9.807 N (g = 9.807 m/s^2)

VELOCITY Meter/Second (m/s) is the Basic SI Unit

1 ft/s → 0.3048 m/s	1 m/s → 3.281 ft/s
1 km/hr → 0.6214 mph	1 mph → 1.609 km/hr
1 km/hr → 0.5396 knot	1 knot → 1.852 km/hr → 1.152 mph

PRESSURE Pascal (Pa) N m^{-2} is the Basic SI Unit

1 psi → 6.895 × 10^{3} Pa	1 Pa → 1.450 × 10^{-4} psi
1 Mm Hg → 1.333 × 10^{2} Pa	1 Pa → 7.501 × 10^{-3} mm Hg
1 atmosphere → 1.013 × 10^{5} Pa	1 Pa → 9.869 × 10^{-6} atmosphere (atm)
1 atmosphere → 14.7 psi	1 psi → 6.804 × 10^{-2} atm

ENERGY Joule (J) kg m^{2}s^{-2} is the Basic SI Unit

1 ft-lb → 1.356 J	1 J → 0.7377 ft-lb
1 erg → 1.00 × 10^{-7} J	1 J → 1.00 × 10^{7} erg

POWER Watt (W) J s^{-1} is the Basic SI Unit

1 ft-lb/s → 1.356 W	1 W → 0.7376 ft-lb/s
1 HP (British) → 7.457 × 10^{2} W	1 W → 1.341 × 10^{-3} HP

Sample Problem 1.2

In engineering practice, data often appear in mixed and sometimes incompatible units. In such cases, it is extremely important to convert all the units to a consistent system before doing any calculations. For example, suppose you want to compute the volume of air in an automobile tire from the data commonly given on its sidewall. The designation P215/65R15 gives the width of the tire in millimeters (215 mm), the ratio of the sidewall height to the tire width in percent (65%), and the wheel diameter in inches (15 in.). Calculate the approximate volume of the air in the tire, assuming that the sidewalls and width are straight and that the tire can be treated as a hollow cylindrical shape. Express the result in cubic meters.

Solution

We want the answer to be in cubic meters, so we will convert the data into units of meters. The diameter of the wheel is in inches, and this is converted first to millimeters and then to meters:

$$D = (15 \text{ in.})(25.40 \text{ mm/in.})(0.001 \text{ m/mm}) = 0.381 \text{ m}$$

The inner radius of the tire is

$$R_i = D/2 = 0.190 \text{ m}$$

The tire width is

$$w = 215 \text{ mm} = 0.215 \text{ m}$$

The tire sidewall height is

$$h = 0.65\, w = 0.65\, (0.215 \text{ m}) = 0.140 \text{ m}$$

Therefore, the outer radius is

$$R_o = R_i + h = 0.190 + 0.140 = 0.330 \text{ m}$$

The volume of the air in the tire is

$$V = \pi\, [R_o^2 - R_i^2]\, w$$

$$V = \pi\, [(0.330)^2 - (0.190)^2] 0.215 = 0.049 \text{ m}^3$$

Breaking the tire code

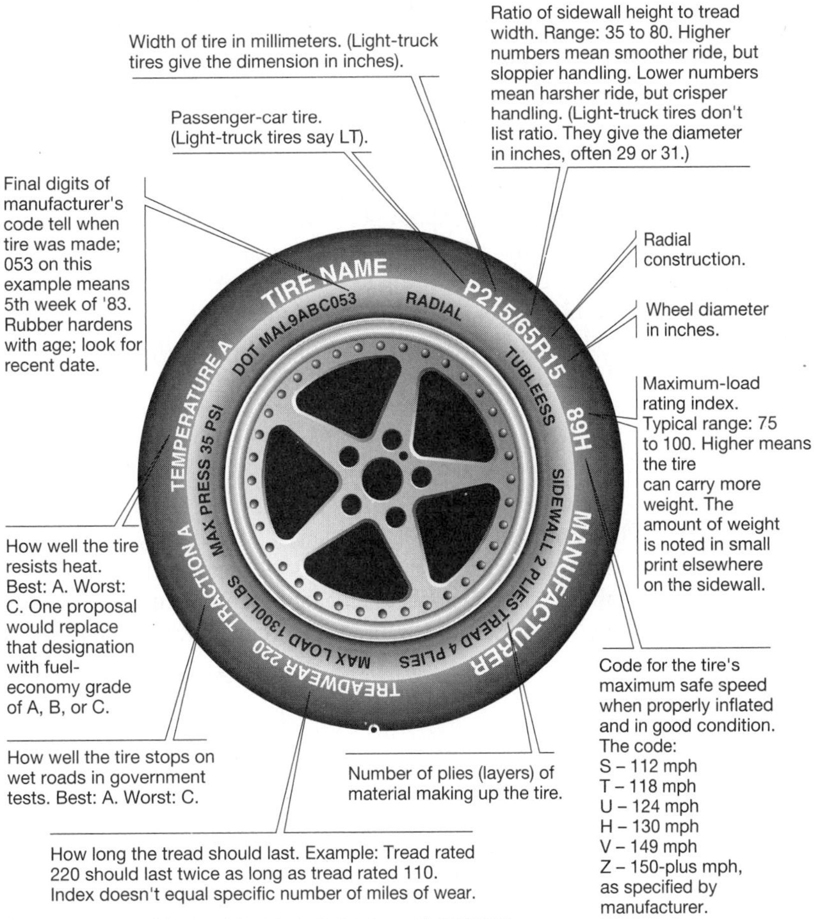

Width of tire in millimeters. (Light-truck tires give the dimension in inches).

Ratio of sidewall height to tread width. Range: 35 to 80. Higher numbers mean smoother ride, but sloppier handling. Lower numbers mean harsher ride, but crisper handling. (Light-truck tires don't list ratio. They give the diameter in inches, often 29 or 31.)

Passenger-car tire. (Light-truck tires say LT).

Final digits of manufacturer's code tell when tire was made; 053 on this example means 5th week of '83. Rubber hardens with age; look for recent date.

Radial construction.

Wheel diameter in inches.

Maximum-load rating index. Typical range: 75 to 100. Higher means the tire can carry more weight. The amount of weight is noted in small print elsewhere on the sidewall.

TIRE NAME
DOT MAL9ABC053
RADIAL
P215/65R15
TUBLESS
89H
SIDEWALL 2 PLIES
MANUFACTURER
TREAD 4 PLIES
MAX LOAD 1300LBS
TREADWEAR 220
TRACTION A
TEMPERATURE A
MAX PRESS 35 PSI

How well the tire resists heat. Best: A. Worst: C. One proposal would replace that designation with fuel-economy grade of A, B, or C.

Code for the tire's maximum safe speed when properly inflated and in good condition. The code:
S – 112 mph
T – 118 mph
U – 124 mph
H – 130 mph
V – 149 mph
Z – 150-plus mph, as specified by manufacturer.

How well the tire stops on wet roads in government tests. Best: A. Worst: C.

Number of plies (layers) of material making up the tire.

How long the tread should last. Example: Tread rated 220 should last twice as long as tread rated 110. Index doesn't equal specific number of miles of wear.

Source: Bridgestone/Firestone, Tire Industry Safety Council USA TODAY

P215/65R15 tire. (Courtesy of Bridgestone/Firestone and Tire Industry Safety Council. Reprinted by permission)

Problems

1.5 Is the average NBA basketball player taller than 2 m?

1.6 If your height is 1.83 m, how tall are you in feet?

1.7 Determine your weight in newtons.

1.8 Determine your mass in kilograms.

1.9 Determine your mass in slugs.

1.4 NUMERICAL CALCULATIONS

The precision of any number is designated by the number of significant figures it contains. A significant figure is any digit, including zero, that is not used to specify the location of the decimal point. For example, the number 2701 has four significant figures. However, it is difficult to know whether the number 2700 has two, three, or four significant figures. To overcome this difficulty, and also to represent large and small numbers in a convenient way, a *scientific notation* using powers of 10 has been adopted. In this notation, the number 2701 is written 2.701×10^3. The number 2700 would be written as 2.700×10^3 if all of its digits were significant, or as 2.7×10^3 if it had only two significant figures.

In statics problems, it may be difficult to determine the number of significant figures for data not given in scientific notation. For example, suppose two forces are given with magnitudes of 10 lb and 8 lb. Examining the number 10, are there two significant figures or one? If it were given as 1.0×10^1, then it would be clear that there were two significant figure in the data. However, if the number 8 is examined and it is noted that it has not been rounded up to 10, one would also have to conclude that the data is accurate to the nearest pound, and that both figures in 10 are significant. One could reach the same conclusion if the forces were 10 lb and 32 lb. If the load was 30,000 lb and no other loads were given, there would be no way to determine the number of significant figures. Therefore, in these problems, the load will be followed by the scientific notation in parentheses; 30,000 lb (3.00×10^4), indicating that the data is accurate only to the nearest 100 lb.

Determining the number of significant figures in mechanics problems is more difficult due to the fact that these problems always involve vectors as will be shown in Chapter 2. Vector mathematics uses trigonometric functions such as the sine and cosine of angles. It might be reasonable to assume that an angle might be measured only to the nearest degree so that a reported angle of 25° might vary from 24.5° to 25.5° and still maintain two significant figures. However, many calculations yield the sine or cosine of the angle and the inverse sine or cosine is used to determine the angle. Suppose in a problem, it is determined that using two significant figures that the sine of an unknown angle is 0.97. It would not be possible to determine the angle to the accuracy of one degree as:

$$\begin{aligned} \sin(75^\circ) &= 0.966 \approx 0.97 \\ \sin(76^\circ) &= 0.970 \approx 0.97 \\ \sin(77^\circ) &= 0.974 \approx 0.97 \end{aligned}$$

Therefore, unless three significant figures are carried in the trigonometric calculations, the angle cannot be determined to the accuracy of 1°. The answer should still be reported to the nearest degree and not include tenths of a degree.

The number of significant figures is determined by the accuracy of the data. The solution cannot be more accurate than the least accurate of any of the data. For example, if a length of 1.000 m can be measured to an accuracy of 2 mm, then the degree of accuracy of the data is

$$\frac{0.002}{1.000} = 0.002 = 0.2 \text{ percent}$$

Suppose that this measure of length is used in the solution of a problem, resulting in the determination that a force of 4008 N was acting on a particle. It would be a misrepresentation of the accuracy of the calculation to report this force as 4008 N. The largest error is 0.2 percent, so the confidence in the magnitude of the force varies from 4000 to 4016 N. Thus, the solution should be reported as 4.01×10^3 N.

Any computing machine—from calculator to large supercomputer—has a finite amount of space to represent a number. Unfortunately, many of the numbers that arise in engineering calculations are ***irrational numbers*** (numbers that cannot be represented by integers or fractions)—for example, π, e, and $\sqrt{2}$. The designer of any computing device must decide the most efficient method of representing numbers in a finite record length; that is, each number must be represented by a limited number of zeros and ones in a binary base. This can sometimes be varied by using double- or triple-precision calculations. The finite record length then determines the number of digits displayed in the computer results. For example, an answer may appear as 1.497532×10^2. The data, however, may limit the number of significant figures to four, so the answer should be reported as 1.498×10^2. Rounding off is frequently ignored in large calculations on computers, but you must always realize that the accuracy is limited to the accuracy of the data used in the calculation. Engineering answers should be rounded off, so that they cannot be misinterpreted in later calculations.

1.5 PROBLEM-SOLVING STRATEGY

The solution of a problem in mechanics may be considered to consist of four major parts.

1. **Modeling of the physical problem.** Modeling can be accomplished by inspecting the object to be analyzed or by working with layout drawings, photographs, or a scale model from which physical dimensions can be obtained and each individual part of the object examined. The model should be as simple as possible to allow full investigation of the phenomena involved. It is useful to make a sketch of the object and decide whether it can be modeled as a two-dimensional or a three-dimensional object. A powerful tool for this modeling is the ***free-body diagram***. This diagram isolates the object from its surrounding environment and accurately represents its interaction with that environment. Unless an accurate model and the corresponding free-body diagram are created, the balance of the solution is wasted. An example of a free-body diagram is shown in Figure 1.2, in which a ladder is leaned against a smooth wall. The forces acting on the ladder include the weight of the person, the force between the ladder and the wall, and the normal and friction forces between the ladder and the ground. Internal forces are neglected because of Newton's third law. Therefore, the forces between the ladder rungs and the rails are not included in the free-body diagram.

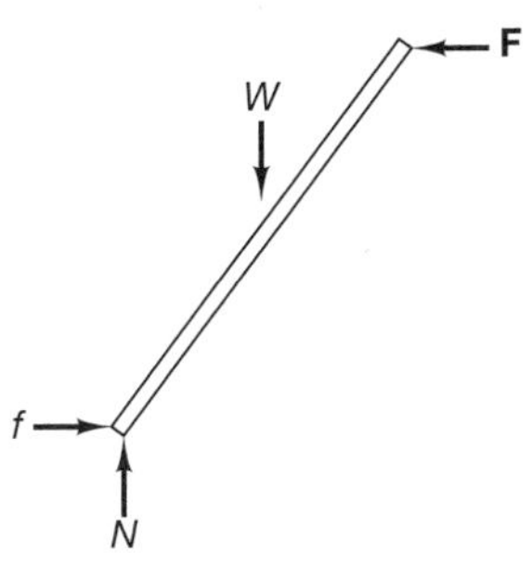

Figure 1.2 Free-body diagram.

2. **Expressing the governing physical laws in mathematical form.** In statics, this involves writing the equations of equilibrium and any necessary constraint equations. In dynamics, the equations of motion and constraints on the motion will be expressed in vector form according to Newton's laws. There are different ways to express these mathematical equations; for example, the equations are different in different coordinate systems. Correctly formulated, each of these approaches will yield the same solution, but some may require more computations than others. It is important at this point to determine what is known, what is not known, and what is being asked.
3. **Solution of the equations.** Modern computational tools should be used, when available, to reduce the labor involved in solving the problem and to allow the investigation of what happens with a change of parameters. Most statics problems

require the solution of a system of simultaneous linear equations, but some involve nonlinear equations. Dynamics problems require the solution of a system of linear or nonlinear differential equations. Computational tools also allow the presentation of results in the form of x–y plots, surface plots, bar or pie diagrams, etc.

4. **Interpretation of results.** This element is fundamental to the concept of design. The results may indicate that the design is flawed or overly complex for the application. Careful examination of the results also reveals errors that may have occurred in the modeling or during the analysis. For example, if you calculated the speed of a runner and the result of the calculation was 100 m/s, you should suspect an error because this speed would be almost 10 times the world record. If your calculations for the load on the Golden Gate Bridge during rush-hour traffic yield a result of 50 lb, that implausible result clearly indicates that an error was made in one of the previous steps.

1.6 COMPUTATIONAL SOFTWARE

Undergraduate engineering courses are employing commercial ***computational software*** in greater frequency as more students gain access to computers and as the capabilities of the computational software increase. Therefore, most engineering companies expect students to be familiar with computational software. At this time, there are student editions of Mathematica®, MATLAB®, Mathcad®, Maple®, and other packages. Many computational methods are also available on hand-held calculators. While many problems in this text can be done without the use of computational software, these tools can greatly enhance your understanding of the problem and, particularly, increase your insight into the design process. Changes in geometry or loading can be investigated simply by changing input parameters and the results visually displayed in the form of graphs. ***Computational methods will be shown in the supplemental manuals***, and any of the available programs can be adopted to do these calculations. You can graph the results, which may then be transferred to a word-processing document for preparing reports in a professional manner. Your ability to use tools of numerical calculation and graphing will be very useful in this course, in later courses, and throughout your professional career. In addition, many problems presented in this book cannot be solved by hand, as they involve large systems of linear or nonlinear equations. However, these are the types of problems that engineers encounter during their professional careers.

Chapter 2

VECTOR ANALYSIS

The German mathematician Hermann Gunter Grassmann (1809–1877) developed a system of space geometry that led to modern vector analysis. (Photo courtesy of Stockholm University, Department of Linguistics)

2.1 INTRODUCTION

Previously a force has been defined as the action of one body upon another body. This action may be due to the contact of one body with another, called a ***surface force***, or the action may arise from a gravitational or electromagnetic effect between separated bodies, called a ***body force***. As the names imply, surface forces act on the surface of a body, and body forces act on each element of mass within the body. The effects of the force can change the motion of the body, deform the body, or change the forces restraining the body at its supports (or all three at once). These effects are studied separately. Statics examines the restraints and supports on a body, along with the characteristics of forces and how their actions can be combined or equivalent actions defined. Dynamics is the study of the changes in motion caused by forces. Mechanics of materials or strength of materials is the study of the deformations of bodies subjected to forces.

A force is described mathematically as a ***vector***, a quantity with both magnitude and direction. Many physical quantities are vectors, including velocity, acceleration, position, moments, and forces. This chapter presents the basics of vector algebra. These will be expanded in the study of dynamics to include vector calculus. Vector algebra is used in this chapter to study the actions of forces on a ***particle***. Modeling an object as a particle implies not that the object is of microscopic size, but that the object's size and shape are such that the effects of forces acting on it do not depend upon its size and shape. Only in a limited number of engineering problems can an object be modeled as a particle, and, in general, size and shape must be considered. In these cases, when size and shape—but not deformation—are important, the object is classified as a ***rigid body***. An air-traffic controller will consider an airplane as a particle as the plane is tracked in flight, but the pilot of that plane must consider the plane as a rigid body and react to forces acting on different points on the plane. A more general consideration of rigid bodies will be introduced in Chapter 4.

In this chapter, the basic mathematics of vector analysis and matrix methods will be discussed. Until the 1950s and 1960s, the only use of vectors in undergraduate mechanics texts was simple vector addition in a plane using trigonometric analysis. That approach will be introduced first in the discussion in this text. The concept of adding lines (vector addition) dates back to Archimedes and Hero of Alexandria who considered the parallelogram of velocities. The concept of parallelogram of forces was common in the 16th and 17th centuries. It was not until the 1800s, that mathematicians discussed the need for a system of space analysis that allowed for the use of coordinates.

The origin of vector analysis, can be traced back to the work by Sir William Rowan Hamilton (1805–1865). In 1831, Hamilton studied the geometrical representations of complex numbers (numbers involving real and imaginary $\sqrt{-1}$ parts). He introduced the *term quaternion* that was equal to $w + \mathbf{i}x + \mathbf{j}y + \mathbf{k}z$, where **i**, **j**, and **k** were unit vectors and introduced relations between these vectors. Although he made suggestions that this mathematics might have some use in space analysis, there is little evidence that he pursued the topic.

The German mathematician Hermann Gunter Grassmann (1809–1877), although aware of Hamilton's work independently began what many consider the origin of vector analysis. He developed a system of space geometry that could have led to modern vector analysis. He introduced his system in a long and complicated book. The system was so broad and complicated that others had difficulty understanding it. He did introduce the concept of vector multiplication in his space. He realized that once geometry is viewed in an algebraic form, it could be easily expanded into three-dimensional space. His book was mostly ignored. One reviewer of his work wrote that it was *"commendably good material expressed in a deficient form."* A historian wrote: *"It seems to be Grassmann's fate to be rediscovered from time to time, each time as if he had been virtually forgotten since his death in 1879."* Fearnly-Sander wrote in a paper in the American Mathematical Monthly in 1979: *"All mathematicians stand, as Newton said he did, on the shoulders of giants, but few have come closer than Hermann Grassmann to creating, single-handedly, a new subject."*

Josiah Willard Gibbs, who was the first individual to receive a doctorate in Engineering in the United States (Yale University 1868), published "**Elements of Vector Analysis**" in 1881. Gibbs relied highly on Grassmann's work and carefully organized the concepts of vector algebra. He can be considered with Oliver Heaviside (1850–1915), the founders of modern vector analysis. The two men's work were independent and their backgrounds totally different. Gibbs was Professor of mathematical physics at Yale University and Heaviside's formal education in England ended when he was 16 in 1866. Two years later, he took a job as a telegraph operator. He became interested in electrical problems and published his first paper in 1878. In 1874, he went to live with his parents to pursue independent study and research. He did not become aware of Gibbs' work until 1888, and his own work independently organized the field of vector calculus. Gibbs and Heaviside organized the full subject of vector analysis composed of vector algebra and vector calculus.

The 1890s were a time of serious debate between the "vectorists" and those that saw limited use of the vector analysis. By the turn of the century, vector analysis was becoming a standard subject in physics and engineering. As stated earlier, it did not find its way into engineering undergraduate mechanics texts until the late 1950s and early 1960s. The title of the texts added the term "Vector Edition."

In previous texts, this material was introduced only on "the need to know" basis. In the presentation that follows, the simplest mathematical vector algebra; using trigonometric solutions based in part on the law of cosines and the law of sines, will be presented. Although most students will be familiar with this approach, these methods are limited to two-dimensional problems and are more labor intensive. Full vector algebra will be presented using component notation. Vector equations will be solved using component notation to generate a system of simultaneous equations. These equations may be solved by hand or by matrix methods. Two types of vector multiplication will be presented and applications of each discussed. Finally, a direct solution of vector equations will be presented. Most calculators and all computer computational programs support these vector and matrix methods. Mastery of these vector methods will aid greatly in allowing the student to focus on modeling of problems of statics and writing the vector equations of equilibrium. Once these equations are written, they may be solved by any of the mathematical methods presented in this chapter.

2.2 VECTORS

2.2.1 DEFINITION OF A SCALAR AND A VECTOR

Physical quantities that have magnitude only are called scalars. Examples of scalars are mass, temperature, volume, and energy. Scalars can be represented by a number with an appropriate sign and a unit scale of measure—for example, kg,°F, or ft^3—or as a function, called a scalar function. Thus, the temperature of a point on a stove may be given as 50°C, representing the temperature of that point at that time. Or, during heating of the stove's burner, the temperature may be expressed as a function of time, such as $T(t) = 37 + 10t$, where t represents time in seconds. In this case, the burner was initially at 37°C and the burner temperature is increasing at a rate of 10°/s. The temperature inside an oven may vary with position in the oven, and temperature as a function of position might be $T = 120 + 3x - 4y$, where x and y are coordinates inside the oven. The mathematical treatment of these scalar quantities depends upon the standard rules of arithmetic, algebra, and calculus.

Many physical quantities have not only magnitude, but direction, and both must be known to fully describe the quantity. Examples of these quantities are forces, displacements, velocities, accelerations, and momenta. ***These quantities are called vectors if they have magnitude and direction, and if they satisfy the laws of vector addition.*** The requirement that such quantities obey the laws of vector addition (presented later in this section) is fundamental to vector mathematics. Quantities that have magnitude and direction, but that do not satisfy the laws of vector addition, cannot be treated meaningfully as vectors. Vectors are defined over a field of real numbers and form a three-dimensional space. The algebra of vectors is therefore founded upon two operations: addition of vectors and multiplication of a vector by a field element or scalar multiplication.

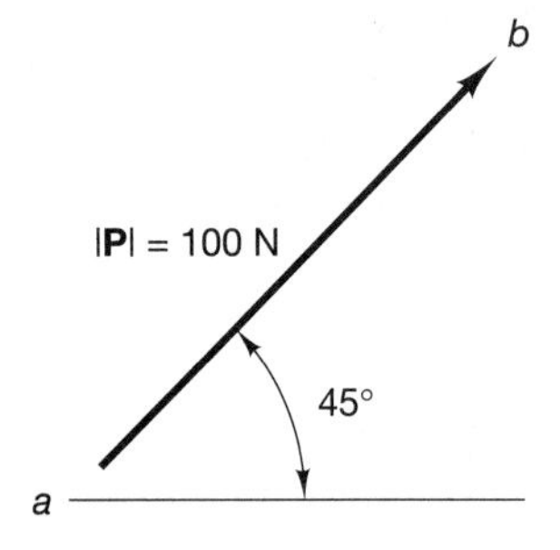

Figure 2.1

Vectors may be represented as *directed line segments having magnitude and direction, but not a specific location in space.* The specification of magnitude and direction is necessary for complete knowledge of a vector. For example, if you moved a total "distance" of 10 meters from a certain point, your current location would not be known, except that you would be somewhere inside a sphere of 10-meters radius centered at the original position. In fact, you could have moved 5 meters to the right and then 5 meters to the left, returning to your starting point. Therefore, displacement magnitudes are of limited use without the specification of directions.

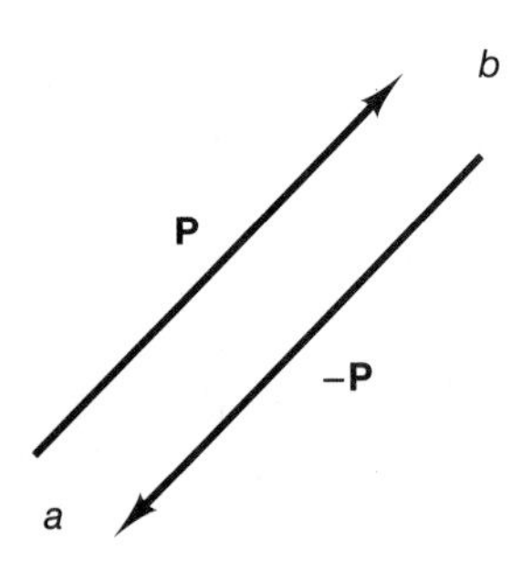

Figure 2.2

A vector may be represented by an arrow, as shown in Figure 2.1. This type of representation is useful conceptually, and historically it has been used to analyze vectors by employing graphical methods. The direction of the arrow indicates the direction of the vector in space, such as northeast, or 45° above the horizontal. The length of the arrow represents the magnitude of the vector, scaled in appropriate units. Note that a vector can have a variety of units, depending on what physical quantity it represents, and the magnitude of the vector is expressed in these appropriate units. The vector is denoted by a bold letter **P**, and the magnitude of the vector is designated by P or $|\mathbf{P}|$. For example, if **P** in Figure 2.1 were a force vector, its units would be newtons or pounds, and its magnitude might, for example, be 100 N. If the vector represented a displacement, it would have units of length, and its magnitude might be 100 mm. The magnitude of the vector is a scalar that is always positive. The ***sense*** of the vector is indicated by the positive direction of the arrow. The vector with the opposite sense, but equal magnitude, is, by definition, designated $-\mathbf{P}$, as shown in Figure 2.2. The vectors **P** and $-\mathbf{P}$ are *equal and opposite* vectors.

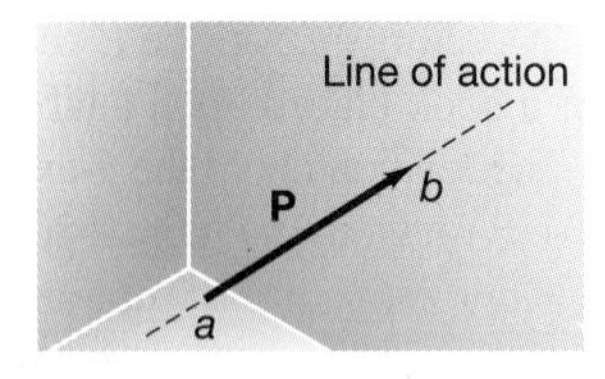

Figure 2.3

Although a vector is completely specified by its magnitude and direction, some applications require that the location, or point of application, of the vector in space also be specified. A straight line parallel to the vector **P**, shown in Figure 2.3, passing through the ***origin*** (tail) of the vector (point a) and the ***terminus*** (head) of the vector (point b) is called the ***line of action*** of the vector. The line of action may be thought of as a line in space of infinite length along which the vector acts. Specifying any point on the line of action gives the location of the vector in space.

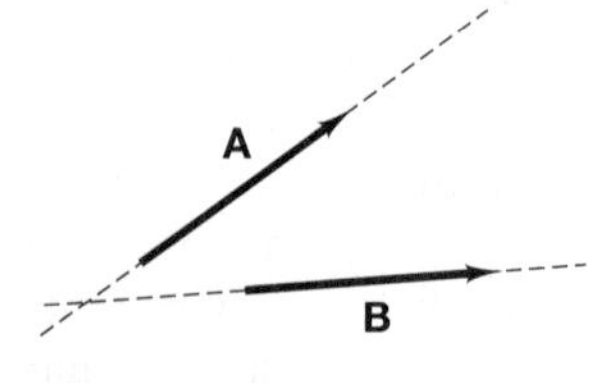

Figure 2.4

2.2.2 VECTOR ADDITION

Consider two vectors **A** and **B** that are not parallel and have no specific line of action in space. Maintaining their magnitude and direction, they may be considered to act at any point of application in space. Therefore, they may be treated as acting in a plane, as shown in Figure 2.4. Vectors that have specified locations in space and lines of action that intersect

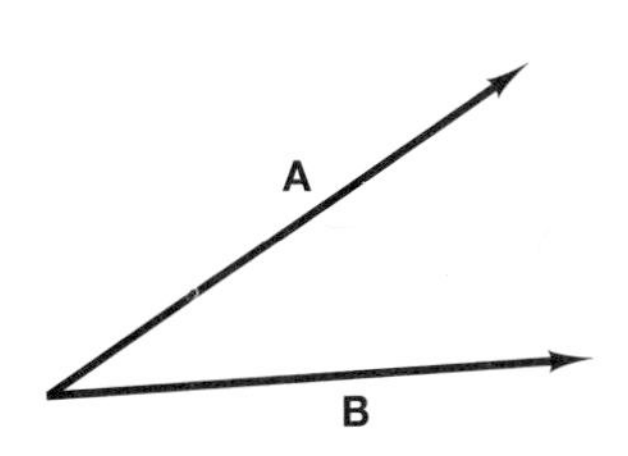

Figure 2.5

lie in a common plane and are called ***coplanar*** vectors. If only the lines of action of the two vectors are specified, the origins of the vectors can be moved along the lines of action and placed at the intersection of their lines of action in the plane. If two vectors have specified spatial locations and have a common origin, as shown in Figure 2.5, they are called ***concurrent*** vectors. If the lines of action or the points at which the vectors act are of no physical importance, any two vectors can be mathematically considered to be coplanar and concurrent.

Now consider how the effects of the two vectors **A** and **B** might be added. For example, if **A** and **B** were both force vectors with units in newtons, then their sum would represent their combined effect at the point of intersection of their lines of action. The sum of the two vectors is a vector **R**, called the ***resultant***, which is obtained by putting the origin of vector **B** at the terminus of vector **A** and taking the vector **R** from the origin of **A** to the terminus of **B**:

$$\mathbf{R} = \mathbf{A} + \mathbf{B} \tag{2.1}$$

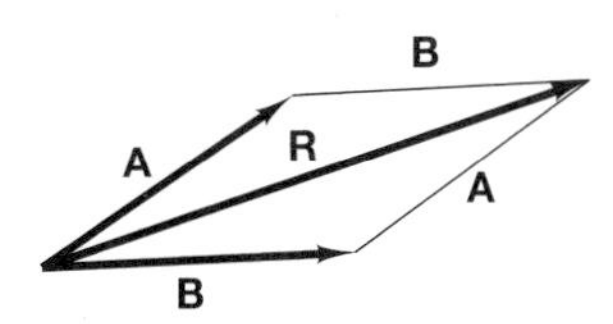

Figure 2.6

Conceptually, this operation illustrates the ***parallelogram law*** of vector addition. In Figure 2.6, the parallelogram for vector addition is formed by putting the origin of vector **B** at the terminus of vector **A** and the origin of vector **A** at the terminus of vector **B**. Vectors **A** and **B** then form the sides of a parallelogram. The resultant or sum **R** is the diagonal of the parallelogram. If these were displacements, **R** would be equivalent to displacing an object by a distance and direction described by **A** followed by **B**, or in the reverse order. The resultant is equivalent to the sum of **A** and **B**, regardless of the order in which they were selected to form the parallelogram. We can conclude that vector addition is ***commutative***; that is,

$$\mathbf{A} + \mathbf{B} = \mathbf{B} + \mathbf{A} \tag{2.2}$$

The definition of a vector requires that it not only have magnitude and direction, but also satisfy the law of vector addition. Vector addition is commutative, a fact taken for granted in ordinary scalar addition. This is an important property of vectors; even if a physical quantity has both magnitude and direction, it is not a vector if it violates the commutative rule of addition.

An interesting example of a quantity that has magnitude and direction, but is not a vector, is a finite rotation. A rotation may be specified as being about a line or axis in space (direction) with a sense, generally chosen by what is called the right-hand rule. This rule states that if you point the thumb of the right hand in the direction of the axis of rotation, the fingers will curve in the positive direction of rotation. The magnitude of the rotation is specified in degrees or radians. A representation of such a rotation is shown in Figure 2.7.

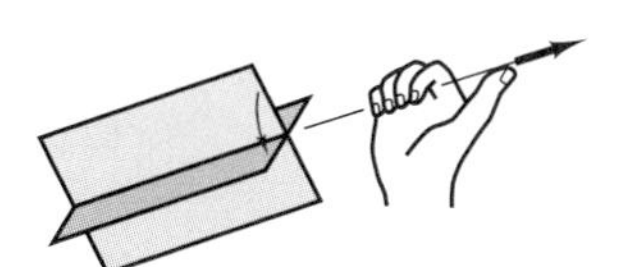

Figure 2.7

Now, by all appearances, this rotation should be treated as a vector; for it to be a vector, however, it must satisfy the law of vector addition. Consider a book placed upon a table and subjected to 90° rotations about the x- and y-axes in space. The rotation about the x-axis will be designated the **A** rotation and the rotation about the y-axis the **B** rotation. In Figure 2.8(a) rotation **A** is performed first, followed by rotation **B**. Figure 2.8(b) shows these two rotations in the inverse order—that is, first **B** and then **A**. It is apparent that the order of rotations is important and that rotations are sequence dependent. Finite rotations do not satisfy the law of vector addition, as the addition of these rotations is *not* commutative:

$$\mathbf{A} + \mathbf{B} \neq \mathbf{B} + \mathbf{A}$$

Noncommutativity also means that if the original and final positions of the book were given, unique rotations about orthogonal axes could not be determined that would move the book from the original to the final position. For example, the final position of the book would be the same if we rotated it first +90° about the x-axis and then +90° about the z-axis or if we rotated it +90° about the y-axis followed by +90° about the x-axis. Thus, finite rotations are not vectors, because they do not satisfy the commutative law of vector

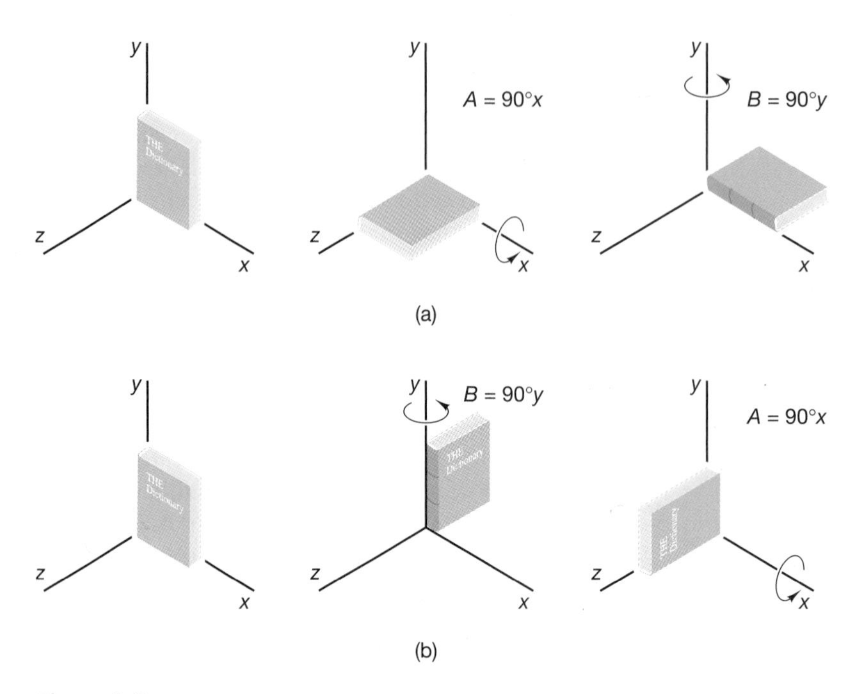

Figure 2.8

addition. Finite rotations, therefore, offer special difficulties when we consider a body rotating in space. These difficulties will be examined in detail in our study of the dynamics of rigid bodies.

Examining the parallelogram in Figure 2.6, we can determine the sum of the vectors by using the trigonometric laws for triangles. Consider the lower triangle formed by the three vectors, as shown in Figure 2.9. The magnitude of the vector **R** in the figure can be found by using the law of cosines (see Mathematics Window 2.1 for a review of the laws of sines and cosines):

$$R^2 = A^2 + B^2 - 2AB\cos\gamma = A^2 + B^2 + 2AB\cos\theta_{AB} \tag{2.3}$$

where A, B, R, and θ_{AB} are defined in Figure 2.9.

The orientation of the resultant **R** with respect to the vector **B** is given by the angle α and can be found by using the law of sines:

$$\sin\alpha = \frac{A}{R}\sin\gamma \tag{2.4}$$

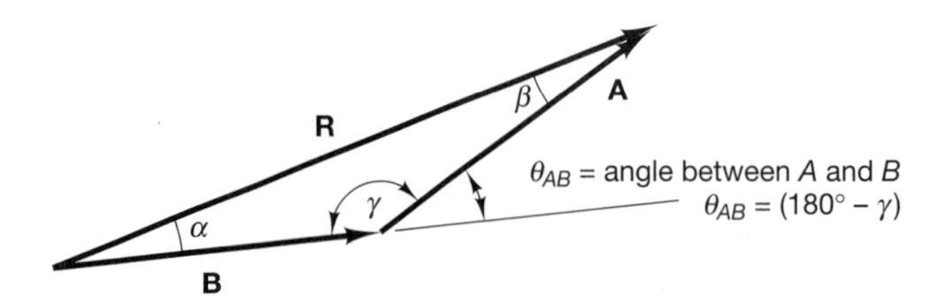

Figure 2.9

MATHEMATICS WINDOW 2.1

Review of the law of sines and law of cosines For any plane triangle $\alpha + \beta + \gamma = 180°$.

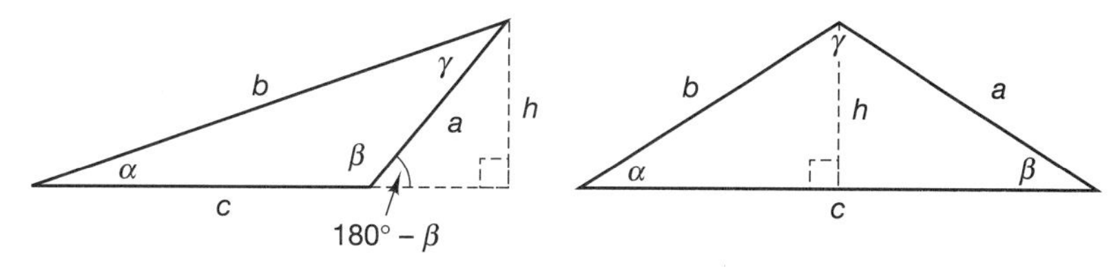

$$\textbf{Law of sines: } \frac{a}{\sin\alpha} = \frac{b}{\sin\beta} = \frac{c}{\sin\gamma}$$

$$\textbf{Law of cosines: } a^2 = b^2 + c^2 - 2bc\cos\alpha$$

$$b^2 = a^2 + c^2 - 2ac\cos\beta$$

$$c^2 = a^2 + b^2 - 2ab\cos\gamma$$

The height of a plane triangle $= h = b\sin\alpha = a\sin\beta$.

The lengths and directions of the vectors also could have been laid out with drafting tools and the magnitude and orientation of the resultant determined graphically. Graphical techniques of this nature were common in the past and still have some conceptual value, but computers and calculators now allow for more accurate and systematic solutions of such problems.

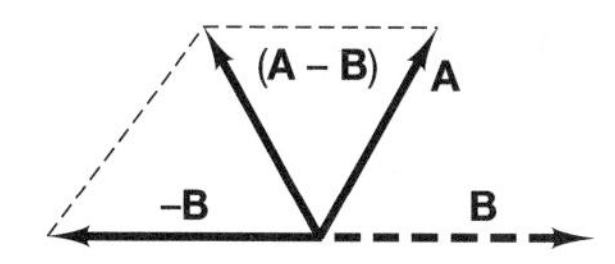

Figure 2.10

Subtraction of **B** from **A** is defined as the addition of vector **A** to $(-\mathbf{B})$, where $-\mathbf{B}$ is the equal and opposite vector of **B**. The subtraction

$$\mathbf{A} - \mathbf{B} = \mathbf{A} + (-\mathbf{B}) \tag{2.5}$$

is shown in Figure 2.10. Clearly, the addition or subtraction of vectors is not equivalent to the scalar addition or subtraction of the magnitudes of the vectors.

Consider the addition of three vectors, as illustrated in Figure 2.11. Vectors **A**, **B**, and **C** can be added together by successive uses of the law of vector addition, either by first forming the vector **A** + **B** and then adding **C**, or by adding **A** to the vector **B** + **C**. In this example, all three vectors lie in the same plane, and the addition may be done graphically or with the use of the law of sines and the law of cosines.

Vector addition satisfies an ***associative*** law, as illustrated in Figure 2.11; that is:

$$(\mathbf{A} + \mathbf{B}) + \mathbf{C} = \mathbf{A} + (\mathbf{B} + \mathbf{C}) \tag{2.6}$$

Therefore, **vectors satisfy both the commutative law of addition and the associative law of addition**, as do scalars. The vectors need not be coplanar, but three-dimensional perspectives are difficult to represent graphically.

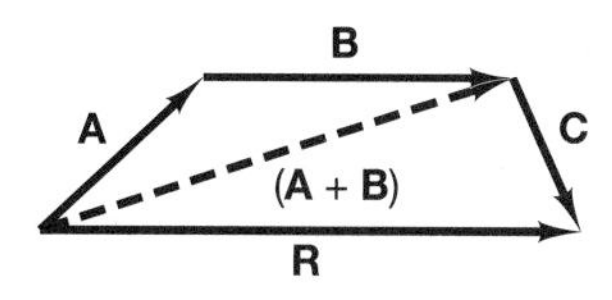

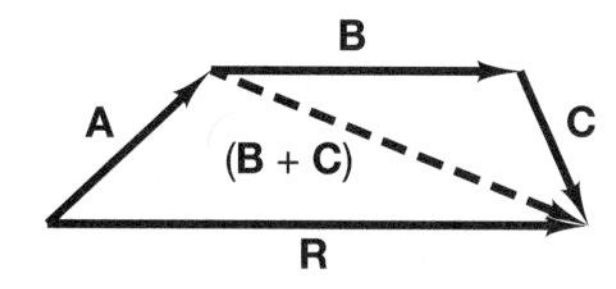

Figure 2.11

Figures 2.10 and 2.11 are called vector diagrams and have been used in the past to add vectors together graphically. They are also the basis for understanding the repeated use of the parallelogram law when all the vectors lie in a single plane. If vector diagrams are used as a graphical or conceptual tool in vector addition, care must be taken to represent each vector in the proper orientation and to a common scale.

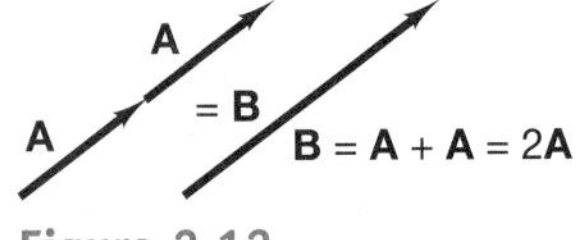

Figure 2.12

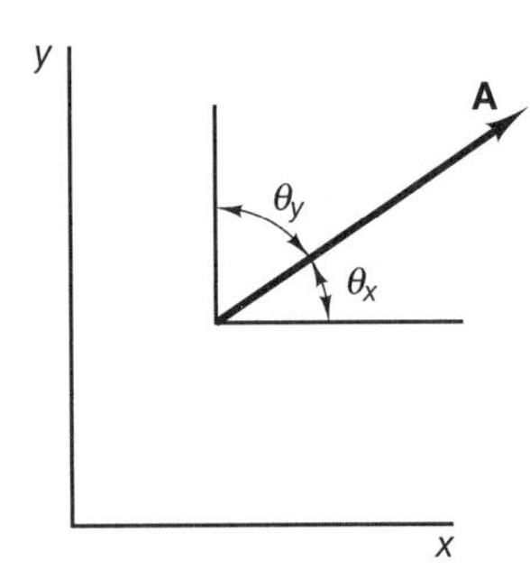

Figure 2.13

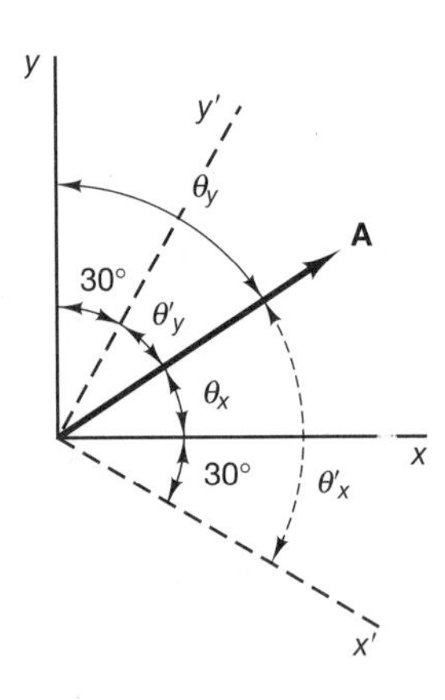

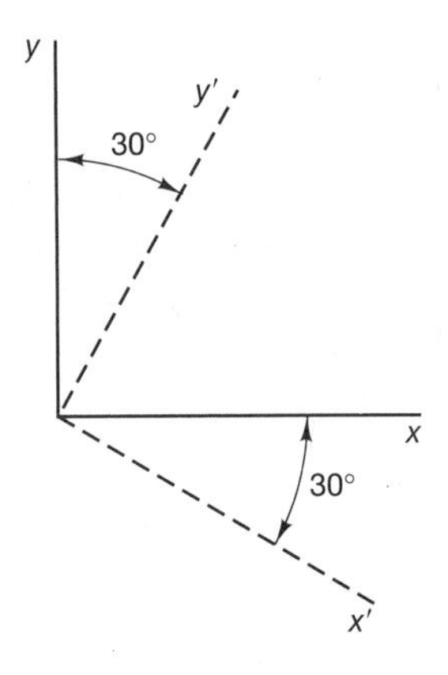

Figure 2.14 The magnitude and direction of a vector do not change, even though the vector's description depends upon the choice of coordinate system.

2.2.3 MULTIPLICATION OF A VECTOR BY A SCALAR

A vector added to itself is equal to a vector of twice the magnitude in the same direction, as shown in Figure 2.12. The vector **B** is equal to 2**A**, or equal to a vector whose magnitude is twice the magnitude of **A**. We can generalize to say that any vector may be multiplied by a scalar, and the resulting vector will have its magnitude increased by scalar multiplication with no change in direction.

2.2.4 VECTOR COMPONENTS

To specify the direction and magnitude of a vector in space, an appropriate reference system must be established. The brilliant French mathematician and philosopher René Descartes (1596–1650) suggested that, for two-dimensional problems (all the vectors lie in one plane), two perpendicular lines could be drawn in the plane. Then all directions and orientations could be referenced to these two lines, or coordinate axes. Following Descartes' suggestion, the vector **A** can be drawn as shown in Figure 2.13, and the familiar rectangular or Cartesian coordinates, x and y, are used to define the plane in which the vector lies. Since the x- and y-axes are perpendicular, they are called ***orthogonal*** axes. Coordinate axes are not initially specified in real engineering problems, and the choice of orientation is left to the engineer. The coordinate directions are often chosen to correlate with physical directions, such as horizontal and vertical, north–south and east–west, or parallel to a slope and perpendicular to it.

In Figure 2.13, the horizontal line is the x-axis (the *abscissa*), and the vertical line is the y-axis (the *ordinate*). The orientation of the vector relative to these two orthogonal lines is known if the angle θ_x or θ_y between the vector **A** and the x-axis or the y-axis, respectively, is specified. In two dimensions, if θ_x is known, θ_y is its complement: $\theta_y = 90° - \theta_x$. Note that this gives the *orientation of the vector relative to a specific choice of coordinate axes x and y.* If the reference axes had been chosen in a different manner, the angles that describe the orientation of the vector would be different, even though the vector itself remains in the same direction in space. This is illustrated in Figure 2.14, where the vector **A** is referenced to two different sets of orthogonal axes, $x-y$ and x'–y'. This dependence upon reference coordinates should be considered at all times, since the choice of coordinate systems is arbitrary, although a particular choice may simplify the mathematics involved. Remember that *the vector's magnitude and orientation are independent of the choice of coordinates; only its description relative to a coordinate system changes with a change of system.*

Note that, in Figure 2.14, the second coordinate system (x'–y') is rotated 30° clockwise relative to the first system. The angles are related as follows:

$$\begin{aligned} \theta_x' &= \theta_x + 30° \\ \theta_y' &= \theta_y - 30° \\ \theta_x' + \theta_y' &= \theta_x + \theta_y = 90° \end{aligned} \tag{2.7}$$

The last condition is true because the coordinates are orthogonal. Note again that the coordinates are called orthogonal coordinates if they are mutually perpendicular; only in certain special applications are nonorthogonal coordinates used.

In the above discussion of coordinates, no mention was made of dimensions or units. Instead, x and y have been used only to indicate the directions of two perpendicular lines in two-dimensional space. If graphical methods are to be used, some scale relating a length along an axis to the magnitude of the vector must be given. This is similar to scale drawings, where, say, 1 cm might represent 1 m of length. The

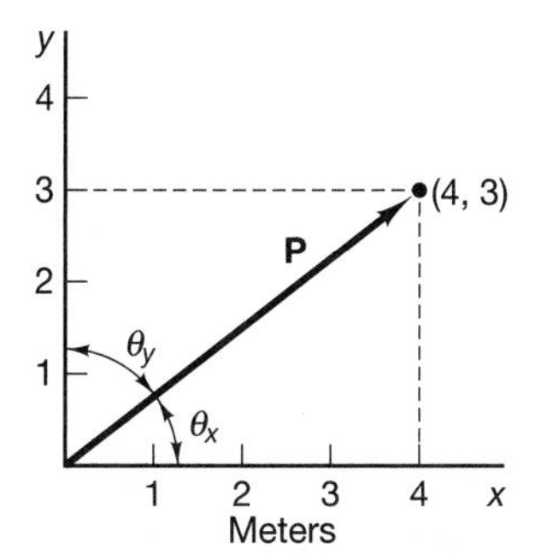

Figure 2.15 The units of a vector's magnitude must be consistent with the scale of the coordinate axes. If the axes indicate meters, the magnitude of the vector **P** is measured in meters.

difference between the two types of scaling is that in a force vector diagram 1 cm might represent 10 N, whereas in a velocity vector diagram 1 cm might represent 10 m/s.

If the length of a vector is scaled to correspond to its magnitude, some simple concepts of trigonometry can be used to describe the vector. Consider a displacement vector from the origin to a point in space having coordinates $(x,y) = (4,3)$. The vector's length might be given in meters, as shown in Figure 2.15. The length, or magnitude, of **P** is 5 meters, as found by the Pythagorean theorem. The angle between **P** and the x-axis is $36.9° = 0.644$ radians. The magnitude and angle of the vector are determined by the trigonometry of a right triangle (see Mathematics Window 2.2 for a review).

MATHEMATICS WINDOW 2.2

Some common formulas from trigonometry used frequently in mechanics For right triangles:

$$r = \sqrt{x^2 + y^2} \quad \text{Pythagorean theorem}$$

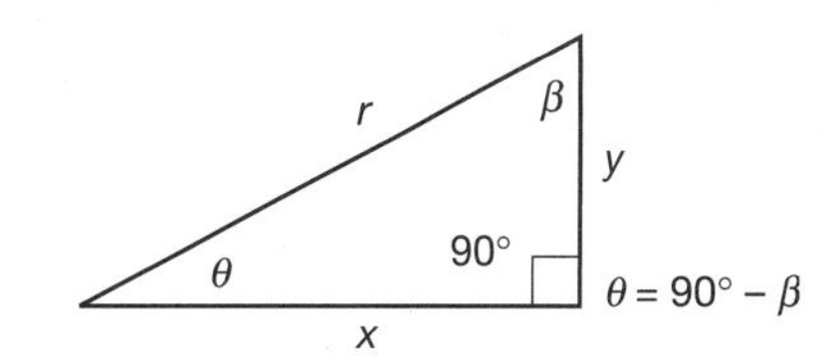

$$\sin\theta = \tfrac{y}{r} \qquad \theta = \sin^{-1}\left(\tfrac{y}{r}\right)$$

$$\cos\theta = \tfrac{x}{r} \qquad \theta = \cos^{-1}\left(\tfrac{x}{r}\right)$$

$$\tan\theta = \tfrac{y}{x} \qquad \theta = \tan^{-1}\left(\tfrac{y}{x}\right)$$

$$\sin\beta = \tfrac{x}{r} \qquad \cos\beta = \tfrac{y}{r} \qquad \tan\beta = \tfrac{x}{y}$$

$$y = r\sin\theta = r\cos\beta \qquad x = r\cos\theta = r\sin\beta$$

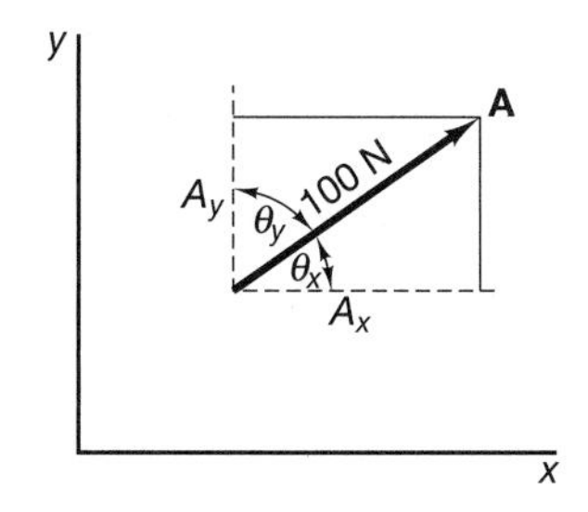

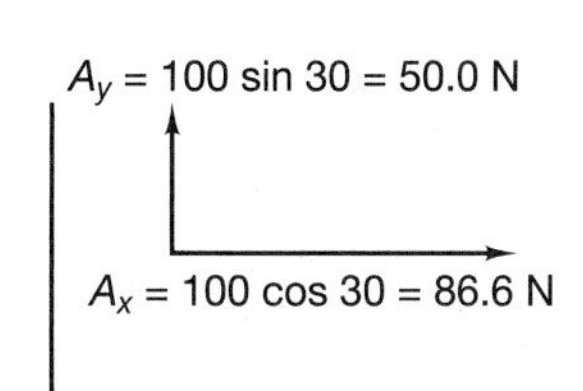

Figure 2.16

Now consider the force vector shown in Figure 2.16. The magnitude of this force vector is given in newtons. A scale—for example, 1 cm = 25 N—can be used to represent the length of the force vector and thereby allow graphical measurements. (Historically, many vector operations were carried out in some graphical manner, avoiding the need for tedious trigonometric calculations, before the availability of electronic calculators and computers.) The magnitudes of the lines A_x and A_y may also be expressed in newtons. If the angle between **A** and the x-axis is 30°, then the trigonometric relationships show that **A** has an effect of 86.6 N in the x-direction and 50.0 N in the y-direction. If A_x and A_y are written as vectors, they are called the ***Cartesian components*** of the vector **A** in the x- and y-directions. A general definition of vector components will be given later in this section. However, the components of the vector must satisfy the vector equation

$$\mathbf{A} = \mathbf{A}_x + \mathbf{A}_y \tag{2.8}$$

The magnitudes of the components can be written

$$A_x = |\mathbf{A}|\cos\theta_x = 100\,[\cos 30°] = 86.6\text{ N}$$
$$A_y = |\mathbf{A}|\cos\theta_y = |\mathbf{A}|\sin\theta_x = 100\,[\sin 30°] = 50.0\text{ N} \tag{2.9}$$

where $|\mathbf{A}|$ is the magnitude of the vector **A**. Both of these equations have been written with a positive sign, as the vector **A** is oriented in the positive x- and y-directions. Note that, for a particular coordinate system, the magnitude and direction of the vector **A** are completely specified by its components. From the Pythagorean theorem, the magnitude of **A** may be calculated in terms of its Cartesian components as

$$|\mathbf{A}| = \sqrt{A_x^2 + A_y^2} \tag{2.10}$$

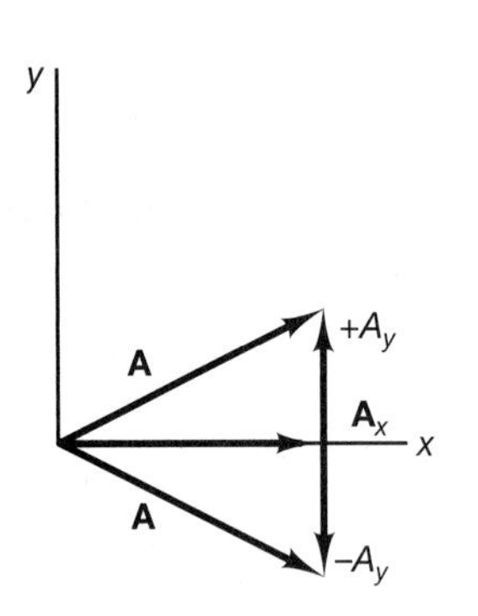

Figure 2.17

Note that $|100\text{ N}| = \sqrt{(86.6)^2 + (50.0)^2}$ in the example shown in Figure 2.16. The angle the vector **A** makes with either the x-axis or the y-axis may be determined by taking the inverse tangent of the ratio of the components; for example, $\theta_x = \tan^{-1}\left(\frac{A_y}{A_x}\right)$. Therefore, if the components are given, the magnitude and direction of the vector are easily obtained. In many engineering applications, instruments are used to measure the components of a force in perpendicular directions so that the magnitude and direction of the force may be computed.

If the magnitude and only one component of a vector are given, the second component cannot be uniquely determined. Consider that $|\mathbf{A}|$ and $\mathbf{A}_x$ are known, only the magnitude of $\mathbf{A}_y$ can be determined, as shown in Figure 2.17, and calculated by

$$\mathbf{A}_y = \pm\sqrt{|\mathbf{A}|^2 - |\mathbf{A}_x|^2}$$

An alternative method for referencing a vector to a coordinate system is to specify the magnitude of the vector and the angles it makes with the x- and y-axes. Note that the cosines of the angles between vector **A** and its components are given by

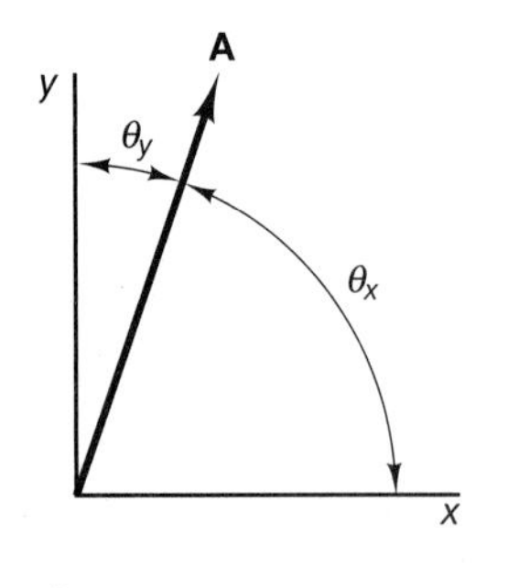

Figure 2.18

$$\begin{aligned} \cos\theta_x &= \frac{A_x}{|\mathbf{A}|} \qquad -180° < \theta_x \leqslant 180° \\ \cos\theta_y &= \frac{A_y}{|\mathbf{A}|} \qquad -180° < \theta_y \leqslant 180° \end{aligned} \tag{2.11}$$

The cosines of the two angles θ_x and θ_y (to the x- and y-axes) specify the components of the vector by "projecting" the vector onto the x- and y-axes, respectively. Projecting the vector onto an axis is accomplished by drawing a line from the head of the vector perpendicular to the axis. The cosines are called ***direction cosines*** and are of great importance in mechanics. For simplicity, we denote the direction cosines as

$$\begin{aligned} \lambda_x &= \cos\theta_x \\ \lambda_y &= \cos\theta_y \end{aligned} \tag{2.12}$$

These equations specify the direction cosines of **A** in two dimensions. Note that $\cos\theta_y = \sin\theta_x$ and that the sum of the squares of the direction cosines is always equal to unity:

$$\lambda_x^2 + \lambda_y^2 = 1 \tag{2.13}$$

From Figure 2.18, it may also be noted that

$$\cos\theta_x = \pm\sin\theta_y\text{:} \quad + \text{ if } \theta_x \text{ is} < 90° \text{ and } - \text{ if } \theta_x \text{ is} > 90° \tag{2.14}$$

In the case of Cartesian coordinates, the component A_x may be thought of as the *projection* of **A** onto the x-axis and A_y as the projection of **A** onto the y-axis.

Often, a vector will be specified by giving its magnitude and the direction cosines referenced to a coordinate system. Direction cosines will be developed further when vectors in three dimensions are discussed.

Up until now, the x- and y-axes have been used only to specify the orientation of the vector and have made no reference to the vector's point of application in space. For

Figure 2.19

example, in Figure 2.19, the vectors **A** and **A**′ are in the same direction in the plane. Their origins are located at two different points in space, and, since the vectors are parallel, their lines of action do not intersect. However, if their magnitudes and directions are the same, their components are the same. Vectors **A** and **A**′ are described by their components or by their magnitudes and direction cosines; since these are equal, we conclude that **A** = **A**′. The two vectors are equal if they have the same magnitude and the same spatial direction, regardless of their points of application.

The effects of some vectors do not depend on their location in space; that is, the effects are independent of the location of their origin. This means that the effects of these vectors depend only on their direction and magnitude. If **A** and **A**′ were this type of vector, then for all applications, the vectors would be equal and equivalent: **A** = **A**′. Such vectors are called ***free vectors***. Moments or torques and angular velocities are vectors of this type. Other vectors act along a line in space (the line of action), but do not have a unique point of application along that line. The origin and terminus can be arbitrarily chosen along the line of action. These are called ***sliding vectors***. Forces may be treated as sliding vectors if their local effects, such as deformations and internal forces, are not considered. Finally, there are vector applications where the point at which the vector acts is of great importance. In these cases, the vectors are called ***fixed vectors***. The velocity of a point on a solid body rotating in space is an example of a fixed vector.

Sample Problem 2.1

A pit crew member is pushing on a race car with a force ***F*** of magnitude 500 N at an angle of 20° with the horizontal. Choose an x–y coordinate system along the vertical and horizontal directions, and compute the components F_x and F_y of the vector, as well as the direction cosines.

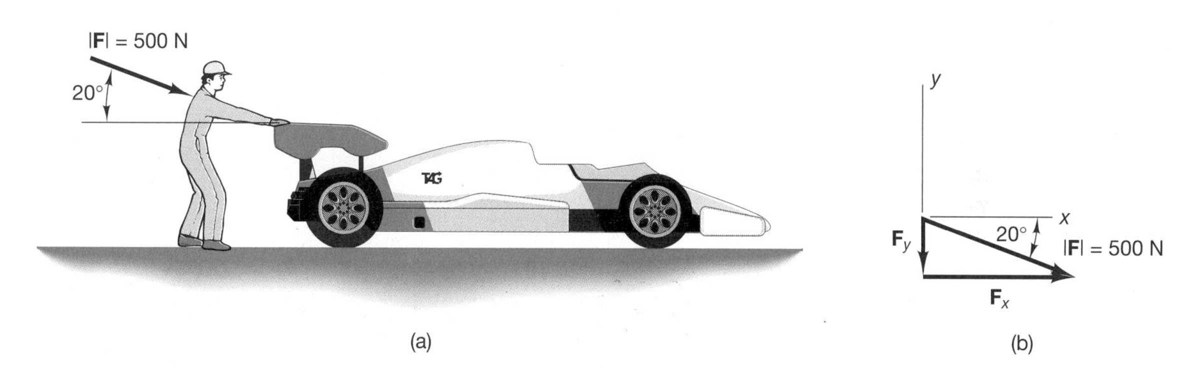

Solution

The first step in solving this problem is to draw a figure showing the force ***F***, the x–y axes, and the components of ***F*** along the axes. Using the definition of components of ***F*** and equations from Mathematics Window 2.1, we can calculate the magnitudes of the components when the angle the force makes with the x-axis is $-20°$:

$$F_x = 500\text{ N} \times \cos(-20°) = 470\text{ N}$$
$$F_y = 500\text{ N} \times \sin(-20°) = -171\text{ N}$$

Note that, although the pit crew member is exerting a 500-N force, he is putting only 470 N in the direction of the car's motion. As shown later, it is only this component of the 500-N force that does any work on the car to move it onto the track.

The direction cosines for this force vector are

$$\lambda_x = \cos\theta_x = \cos(-20°) = 0.940$$
$$\lambda_y = \cos\theta_y = \cos(110°) = -0.342$$

Note that

$$\lambda_x^2 + \lambda_y^2 = (0.940)^2 + (-0.342)^2 = 1.00$$

as it should.

Sample Problem 2.2

For the vector shown, determine its components and direction cosines in both the x–y coordinate system and the x'–y' coordinate system.

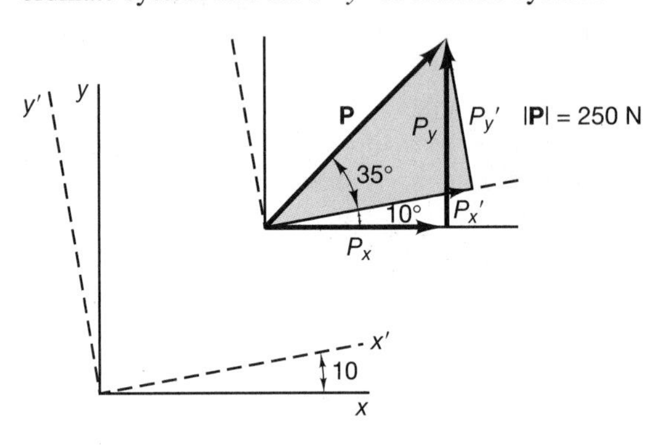

Solution In the x–y system,

$$\theta_x = 45° \quad \theta_y = 45°$$

In the x'–y' system,

$$\theta_x' = 35° \quad \theta_y' = 55°$$

Therefore, the direction cosines are

$$\lambda_x = \cos\theta_x = 0.707 \qquad \lambda_x' = 0.819$$
$$\lambda_y = \cos\theta_y = 0.707 \qquad \lambda_y' = 0.574$$

and the components are

$$P_x = \lambda_x|\mathbf{P}| = 177\text{ N} \qquad P_x' = 205\text{ N}$$
$$P_y = \lambda_y|\mathbf{P}| = 177\text{ N} \qquad P_y' = 144\text{ N}$$

Note that, in each coordinate system, the sum of the squares of the direction cosines is equal to unity, and the magnitude of the force **P** is 250 N.

Small changes in a vector's angle of orientation often can have counterintuitive effects on components. This occurs because of the nonlinear nature of the trigonometric functions used to compute a vector's component values. It is important to understand and recognize this behavior in engineering measurement and design. The nature of the trigonometric functions is such that changes in the direction of a vector are not reflected equally in each component. This is best illustrated by an example. Suppose that a protractor is used to measure the 10° angle between the two coordinate systems in Sample Problem 2.2, and suppose that the measurement is off by 1°, so that you measure 9° instead of 10°. Then the x'-component of the force **P** becomes $P_x' = 250 \cos 36 = 202$ N instead of 205 N, as it should be. Likewise, $P_y' = 250 \cos 54 = 147$ N instead of 144 N, as it should be. Note that the percent error in the x-component is $(205 - 202)/205 \times 100 = 1.3\%$, while that in the y-component is $(147 - 144)/144 \times 100 = 2.1\%$. In other words, one and half as much change occurs in the y-component as in the x-component.

Figure 2.20

As a more dramatic example of the effects of small changes of direction, consider the error a baseball pitcher faces when trying to throw a strike. The pitcher must throw the ball a distance of about 18.29 m (60 ft) such that the ball passes over 0.43-m (17 in) wide home plate. Suppose that the pitcher's aim in the horizontal direction is off by 2°, as illustrated in Figure 2.20. How much error results in the horizontal component? From the definition of the tangent function, the horizontal component of the pitch is $18.29 \tan(2) = 0.64$ m Thus, just a 2° error to the left or right in the pitcher's throw results in missing the center of home plate by more than 0.3 m!

2.2.5 RESOLUTION OF A VECTOR INTO COMPONENTS

We have seen that two or more vectors can be added to give a single vector that is equivalent to the original vectors. Conversely, a single vector **A** may be replaced by two coplanar vectors whose sum is equivalent to **A**. When two noncollinear coplanar vectors can be added so that their sum is equal to the original vector **A**, they are called the ***components of the vector* A**. The process of determining these components is called "resolving the vector into components."

Webster's dictionary defines the word "component" as "serving as one of the parts of the whole," from the Latin *com-*, "together" + *ponere*, "put." The mathematical and technical definition of the components of a vector is more precise than this lay definition and is dependent upon the ability to uniquely resolve a vector into components in any coordinate directions.

Any vector can be resolved into two, and only two, noncollinear coplanar components. In addition, in three-dimensional space, any vector can be resolved into three, and only three, noncollinear and noncoplanar components.

The limit of two vectors in a coplanar resolution and three vectors in a three-dimensional resolution is based upon the fact that this is the number of vectors required to ***span*** two-dimensional and three-dimensional spaces, respectively. The specification of the minimum number and type of vectors that are necessary to represent any vector in that space is called the requirement to span the space. This means that we can find a vector **A** that is equivalent to the sum of two coplanar vectors **B** and **C**, or, given the vector **A** and the directions of **B** and **C**, we can find the magnitudes of **B** and **C**. However, if we know that **A** is equal to the sum of three *coplanar* vectors **B**, **C**, and **D**, we cannot determine the magnitudes of **B**, **C**, and **D** even if their directions are known and the magnitude and direction of **A** is known. We will discuss this in greater detail when the topic of statically indeterminate problems is presented.

Earlier in this section, we introduced the idea of components of a vector, relative to a planar Cartesian coordinate system containing the vector. The resolution of a vector into its two Cartesian components is equivalent to determining the two vectors, $\boldsymbol{A}_x$ and $\boldsymbol{A}_y$ that, when added together, will give the vector **A**:

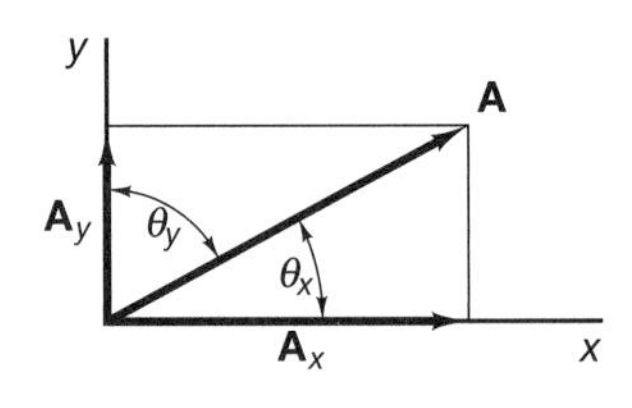

Figure 2.21

$$\mathbf{A} = \boldsymbol{A}_x + \boldsymbol{A}_y$$

The magnitudes of $\mathbf{A}_x$ and $\mathbf{A}_y$ are sometimes called the scalar components of the vector **A** in the x–y coordinate system, and the vectors themselves are the vector components, as in Figure 2.21. Assuming that the vector **A** is described in an x–y coordinate system, determining the two vectors $\mathbf{A}_x$ and $\mathbf{A}_y$ is a simple matter of examining a right triangle. The magnitudes of the two vectors, as given in Eq. (2.9), are

$$\begin{aligned} A_x &= |\mathbf{A}| \cos \theta_x \\ A_y &= |\mathbf{A}| \cos \theta_y \end{aligned} \quad \text{where } |\mathbf{A}| \text{ is the magnitude of the vector } \mathbf{A}. \tag{2.15}$$

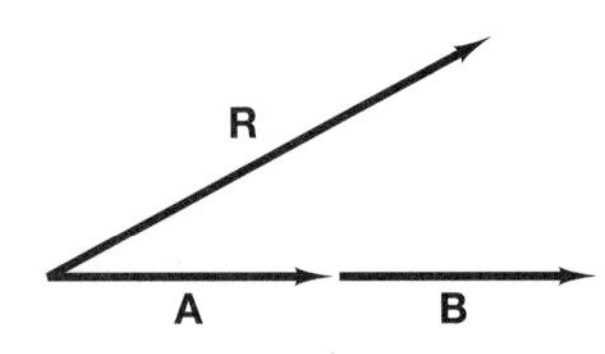

Figure 2.22

This approach may be generalized to determine the components of a vector in a nonorthogonal coordinate system. A vector can be resolved into components that are not orthogonal, but the component vectors cannot be ***collinear***. This may be seen by the example in Figure 2.22. No combination of the vectors **A** and **B** will sum to the vector **R**. If **R** were collinear with **A** and **B**, there would be an infinite number of combinations of **A** and **B** that would equal **R**.

In Figure 2.23, the vector **A** is shown resolved into its components $\mathbf{A}_u$ and $\mathbf{A}_v$ along the lines u and v, respectively, where u and v are not orthogonal. The axis u is located α degrees counterclockwise from the x-axis, and the axis v is located β degrees clockwise from the y-axis, as shown. Remembering that $\theta_x + \theta_y = 90°$, we can form the triangle for vector addition as shown in Figure 2.24. The two components of **A** in the u and v directions can be determined by applying the law of sines to the triangle shown in Figure 2.24. This yields

$$\frac{\sin(90 + \alpha + \beta)}{|\mathbf{A}|} = \frac{\sin(\theta_y - \beta)}{A_u} = \frac{\sin(\theta_x - \alpha)}{A_v}$$

Solving for A_u and A_v yields

$$\begin{aligned} A_u &= \frac{|\mathbf{A}| \sin(\theta_y - \beta)}{\sin(90 + \alpha + \beta)} \\ A_v &= \frac{|\mathbf{A}| \sin(\theta_x - \alpha)}{\sin(90 + \alpha + \beta)} \end{aligned} \tag{2.16}$$

where A_u and A_v are the scalar components of the vector **A** in the u and v directions, respectively.

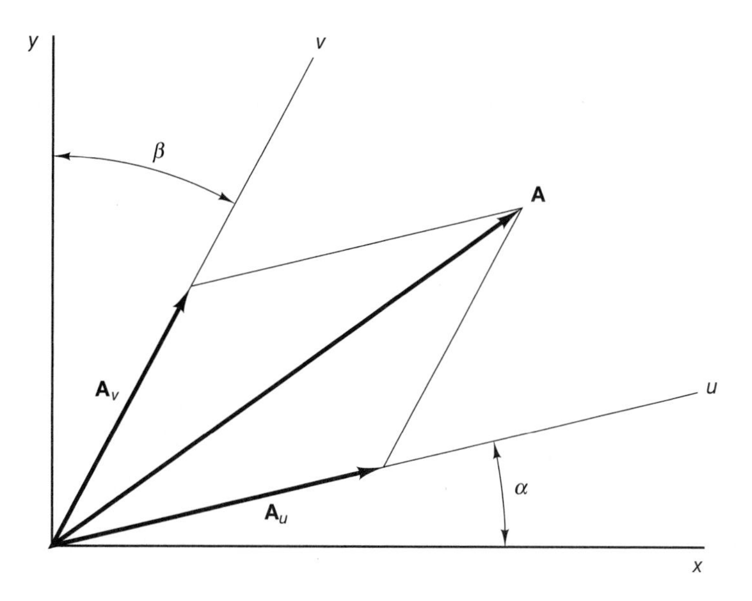

Figure 2.23 The vector **A** resolved into components $\mathbf{A}_u$ and $\mathbf{A}_v$ along nonorthogonal axes.

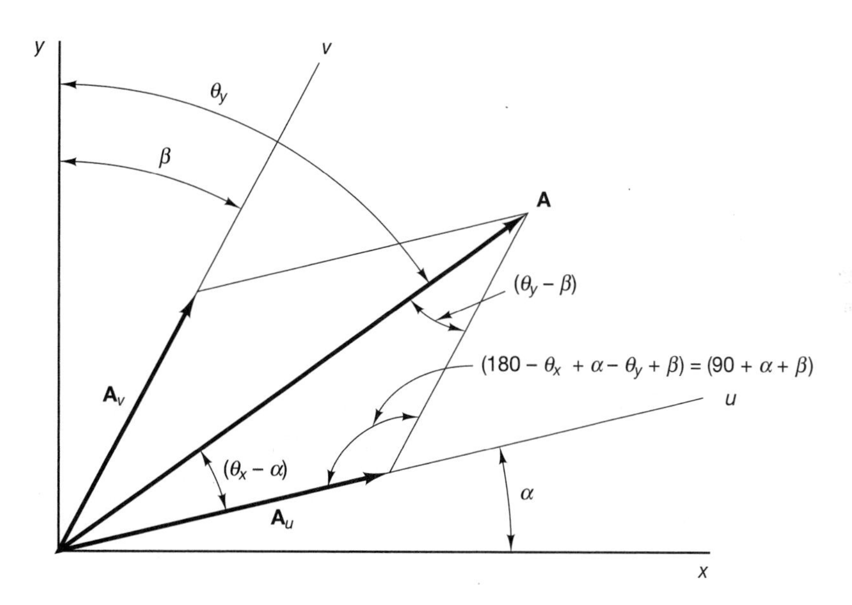

Figure 2.24

This process may be considered either as obtaining the components of a vector in nonorthogonal coordinate systems or as decomposing a vector into the sum of two vectors having given lines of action.

Clearly, vector addition is fundamental to all aspects of vector mathematics, and the resolution of a vector into components, whether orthogonal or not, is an extremely important concept in mathematics and mechanics. Many scientific calculators can obtain the components of a vector in orthogonal coordinates in a single operation and can be programmed to determine the components in nonorthogonal directions.

Sample Problem 2.3

An ocean liner is being towed by two tugboats. If the total force applied to the ocean liner is 130,000 N (1.3×10^5 N) in the direction N45 °E, and tug **A** is pulling in a northern direction while tug **B** pulls N70 °E, determine the forces applied by each of the tugboats.

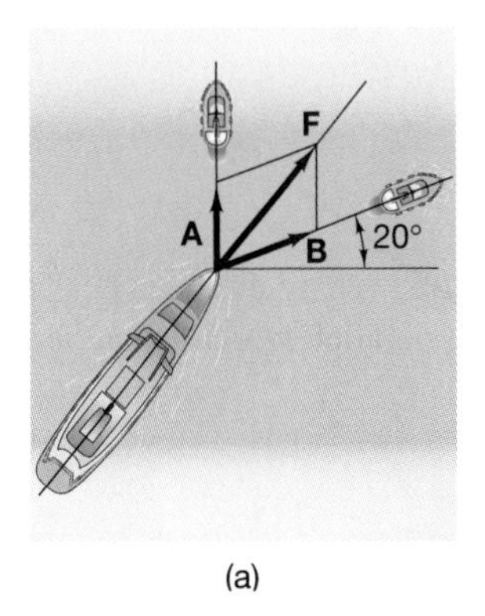

(a)

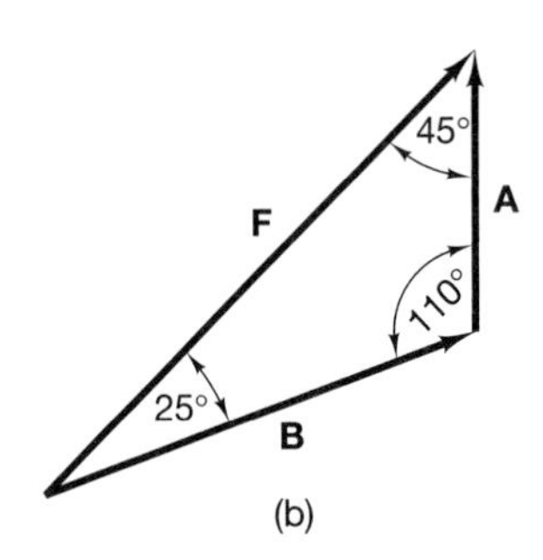

(b)

Solution The forces applied by the two tugs must add to equal the total force on the ocean liner:

$$\mathbf{B} + \mathbf{A} = \mathbf{F}$$

We draw the vector diagram to represent this, as shown in the accompanying diagram. Now, by the law of sines,

$$\frac{A}{\sin 25°} = \frac{B}{\sin 45°} = \frac{F}{\sin 110°}$$
$$F = 130{,}000 \text{ N}$$

Therefore,

$$A = 58{,}500 \text{ N and } B = 97{,}900 \text{ N}$$

Note that, since this is not a right triangle, the Pythagorean theorem does not apply.

Sample Problem 2.4

A support frame consists of two members A and B arranged as illustrated in the diagram below. The force $\boldsymbol{F}$ applied to the bracket can be resolved into components **A** and **B** along the members A and B, respectively. For the case where the angle β is fixed, design the frame by choosing the angle θ such that the magnitude of the force in each member is the same when a force is applied at the end.

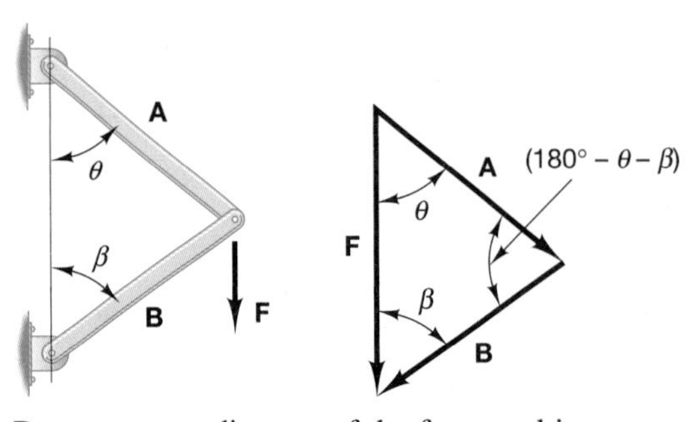

Solution Draw a vector diagram of the force and its components, as shown in the figure above on the right. The components can be determined by use of the law of sines; note that

$$\sin(180 - \theta - \beta) = \sin(\theta + \beta)$$
$$\frac{F}{\sin(\theta + \beta)} = \frac{A}{\sin \beta} = \frac{B}{\sin \theta}$$

Thus, the law of sines requires that

$$A = B\frac{\sin\ \beta}{\sin\ \theta}$$

The magnitudes of **A** and **B** are to be equal, therefore, the sine of θ must equal the sine of β, and the two solutions are

$$\theta = \beta \qquad \theta = \pi - \beta$$

The first solution is correct, and the two angles are equal. The second solution yields a configuration such that **A** and **B** are parallel, which is mathematically correct but physically incorrect.

Sample Problem 2.5

A 500-N (5.0×10^2 N) force is to be resolved into components along lines a–a' and b–b'. Determine the angle β and the component along b–b' if it is known that the component along a–a' is 320 N.

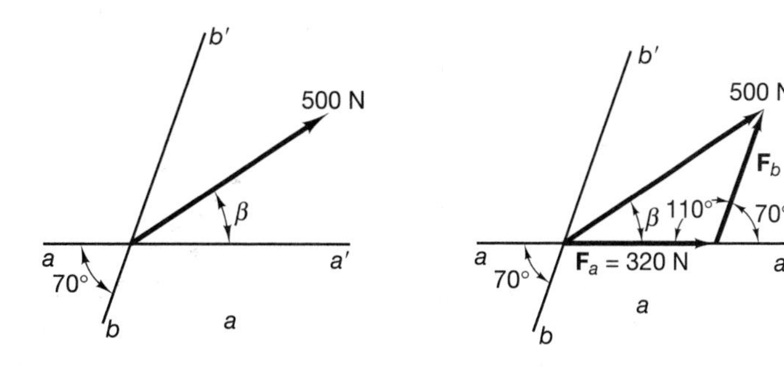

Solution Draw a force vector diagram of the force and the two components. The angle between F_a and F_b is $(180 - 70) = 110°$. Therefore, by the law of sines

$$\frac{500}{\sin(110°)} = \frac{F_b}{\sin(\beta)} \quad \Rightarrow \quad 500 \sin\beta = F_b \sin(110°)$$

$$\frac{500}{\sin(110°)} = \frac{320}{\sin(180 - 110 - \beta)} \quad \Rightarrow \quad 500 \sin(70 - \beta) = 320 \sin(110°)$$

Now, $\sin(110°) = \sin(70°)$ and $\sin(70 - \beta) = \sin(70)\cos\beta - \cos(70)\sin\beta$; therefore, the second equation may be written as

$$500 \sin(70)\cos\beta - 500\cos(70)\sin\beta = 320\sin(70)$$

This is a transcendental equation for β and can be solved by taking the $\sin\beta$ term to the right side and squaring both sides of the equation. The trignometric identity $\cos^2\beta = 1 - \sin^2\beta$ can then be used to form a quadratic equation in $\cos\beta$. This equations has two solutions, $\beta = 33°$ and $\beta = -73°$, but only the positive angle is physically correct. Once the angle has been determined, the magnitude F_b is obtained from the first equation, yielding

$$\beta = 33° \qquad F_b = 290 \text{ N}$$

Many calculators and most computational software packages have graphing and root functions that will solve transcendental equations. (See the Computational Supplement.)

2.3 FORCES AND THEIR CHARACTERISTICS

In Chapter 1, a force was defined as the action of one body upon another body. The characteristics of a force are:

1. Its magnitude.
2. Its direction (orientation and sense).
3. Its point of application (origin or terminus).

A force is a vector, and it obeys all the laws of vector mathematics, including vector addition, as described in Section 2.2. The units used to specify the magnitude of a force are given in either the system SI—such as newtons (N), kilonewtons (kN), etc.—or the U.S. customary system—such as pounds (lb), kilopounds (kip), tons, ounces, etc. In two-dimensional problems, the force vector is characterized by specifying its components in a coordinate system or by giving its magnitude and direction by the use of angles or direction cosines (defined by Eq. 2.12). The sense of the force may be represented by the direction of the vector along its line of action. The sense is also given in terms describing the effect of the force, such as a *tensile* force on a cable, or a *compressive* force on a support structure. When a force is applied on a body, it will produce both internal and external effects. The external effects are changes in the forces that support the body. The internal effects are the deformation of the body and the distributed internal forces. The intensities of these internal forces are called ***stresses*** and have units of Pascals (Pa), newtons per square meter (N/m^2) or pounds per square inch ($lb/in.^2$, or psi). These stresses can try to pull the body apart—called tension—as shown in Figure 2.25(a). Stresses can also compress the body, as shown in Figure 2.25(b).

Directions of forces may also be given in descriptive terms, such as *normal to a surface* (perpendicular) or *tangential to a surface*. The forces that are tangential to a surface are called *shear forces*, and the intensities of these internal forces are called *shear stresses*, as shown in Figure 2.26. Internal forces in structural members are discussed in detail in Chapter 7.

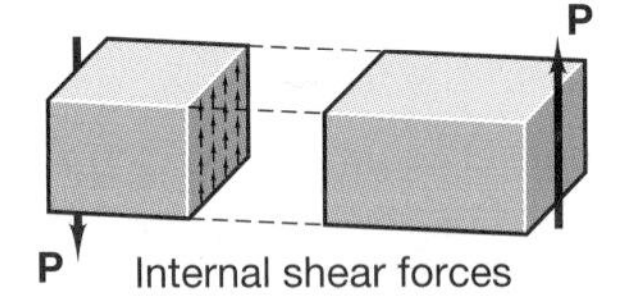

Figure 2.26 Shear forces **P** tangent to the surface of the block.

In this section, we consider the action of forces upon a particle. All the forces will be assumed to have a common point of application on the particle, as the particle occupies only a point in space. These forces form a concurrent force system, and initially we examine only coplanar and concurrent force systems. The combined effect of two or more coplanar forces is obtained by the laws of vector addition. This addition may be accomplished by use of either the parallelogram law or vector components.

One force that acts on most particles is the force of gravitational attraction to the Earth. This is termed the weight of the particle and is always directed toward the center of

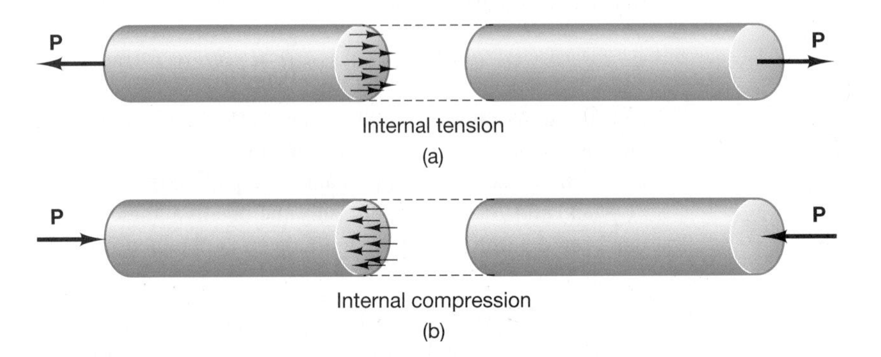

Figure 2.25 (a) Tensile forces on a bar; (b) compressive forces on a bar.

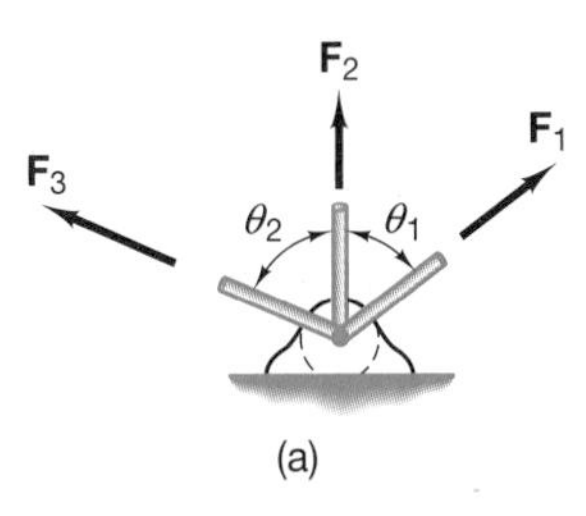

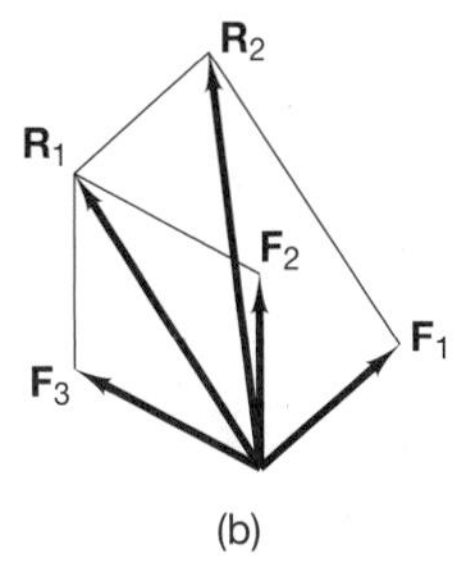

Figure 2.27 (a) Three concurrent forces acting at a point. (b) Determining the resultant of the three forces by using the parallelogram law twice.

the Earth. (See Section 1.2.) In SI units, the mass of the object is specified in kilograms, and its weight is obtained by multiplying by g:

$$W = m\,g \tag{2.17}$$

In U.S. customary units, the weight of an object is measured directly in pounds, and its mass is obtained by division by g:

$$m = W/g \tag{2.18}$$

The unit of weight, as for any other force in SI units, is the newton. In some dynamic applications, the force is expressed in g's but this is confusing, as the actual force is the product of the value in g's and the mass of the object.

2.3.1 CONCURRENT COPLANAR FORCES

The ***resultant*** of a system of concurrent forces is defined to be the a force equal to vector sum of all the forces. This resultant force produces the same effect as that produced by the sum of the effects of each of the forces acting independently. Therefore, the resultant is equivalent to the concurrent forces. **If the body is modeled as a particle, all forces acting upon it are concurrent**. The relevant equation may be written mathematically as

$$\text{Resultant } \mathbf{R} = \sum_{i} \mathbf{F}_i \tag{2.19}$$

If the forces are coplanar and noncollinear, this summation may be done by use of the law of sines and the law of cosines. An example of such a force system is shown in Figure 2.27. Forces $\mathbf{F}_2$ and $\mathbf{F}_3$ can be added using the parallelogram law to obtain $\mathbf{R}_1$. Then $\mathbf{R}_1$ can be added to $\mathbf{F}_1$ to obtain $\mathbf{R}_2$, which is the resultant or the sum of the three forces $\mathbf{F}_1$, $\mathbf{F}_2$, and $\mathbf{F}_3$. All forces lie in the same plane, and the orientation and magnitude of the vector resultant $\mathbf{R}_2$ can be found by using trigonometry. This procedure could be continued for any number of concurrent forces lying in a plane. Note that the definition of the resultant of a concurrent coplanar force system depends upon the definition of vector addition and, indeed, is an example of the repeated use of this definition.

In theory, there is no limitation to using this method when the concurrent forces do not lie in the same plane; however, determining the necessary angles in three-dimensional space becomes difficult. Three-dimensional vector diagrams are extremely difficult to construct, and, once done, the required angles necessary for the analysis are not easily obtained. An example of a three-dimensional vector diagram is shown in Figure 2.28. Vectors **A** and **B** form a plane in space, so their resultant $\mathbf{R}_1$ can be determined by use of the parallelogram law. The vectors $\mathbf{R}_1$ and **C** determine another plane in space, and the desired vector sum $\mathbf{R}_2$ lies in this plane. Note that determining the angle between $\mathbf{R}_1$ and **C**, which is necessary to solve the problem, could prove to be difficult. A pure graphical solution to this problem is very difficult. It is desirable to seek a means of adding vectors in space using coordinate reference systems and vector components. This alternative, more efficient method is discussed in the next section.

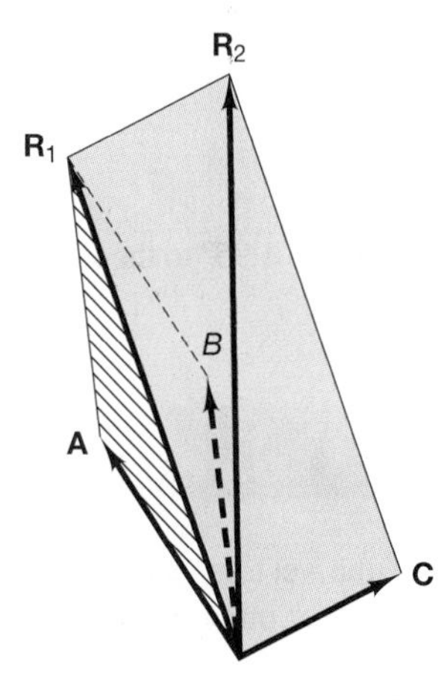

Figure 2.28 Adding three concurrent vectors not in the same plane.

Sample Problem 2.6

Determine the resultant of the three forces acting concurrently on the ring shown in Figure 2.27. The force $\mathbf{F}_2$ acts in a vertical direction, and $\theta_1 = 30°$ and $\theta_2 = 45°$. The magnitudes of the forces are

$$F_1 = 100\text{ N}$$
$$F_2 = 50\text{ N}$$
$$F_3 = 80\text{ N}$$

Solution Let us find the intermediate resultant $\mathbf{R}_1$ between $\mathbf{F}_2$ and $\mathbf{F}_3$. By the law of cosines,

$$R_1^2 = 50^2 + 80^2 - 2(50)(80)\cos 135°$$

where R_1 is the magnitude of the vector $\mathbf{R}_1$. Taking the square root of both sides, we obtain

$$R_1 = 120.7 \text{ N}$$

The angle $\mathbf{R}_1$ makes with the vertical may be found by the law of sines:

$$\sin\beta = \frac{80}{120.7}\sin 135°$$

$$\beta = 27.95°$$

The resultant of the system of three forces may now be found by adding $\mathbf{F}_1$ and $\mathbf{R}_1$, as illustrated in the accompanying diagram. The angle between $\mathbf{R}_1$ and $\mathbf{F}_1$ is

$$\gamma = 180° - 30° - 27.95° = 122.05°$$

Again, by the law of cosines,

$$R_2^2 = 120.7^2 + 100^2 - 2(120.7)(100)\cos 122.05°$$

$$R_2 = 193.34 \text{ N}$$

The angle between $\mathbf{R}_1$ and $\mathbf{R}_2$ may be found by the law of sines:

$$\sin\phi = \frac{100}{193.34}\sin 122.05°$$

$$\phi = 26°$$

The resultant of the system is 193.34 N, acting at an angle of $(27.95 - 26) = 1.95°$ to the left of the vertical.

Problems

2.1 Add the vectors **A** and **B** in the order **C** = **B** + **A**, that is, place the tail of **A** to the head of **B**. Determine the magnitude of **C** and the angle between **C** and **B**.

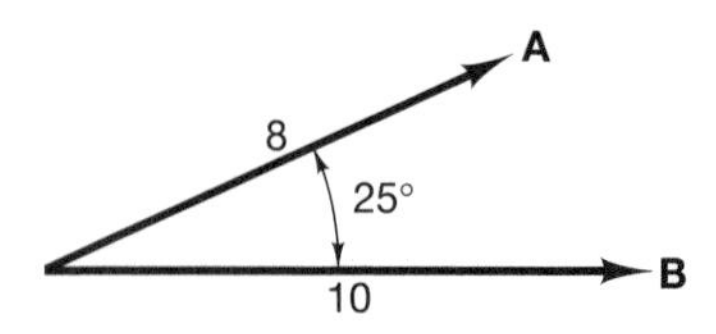

Figure P2.1

2.2 In Problem 2.1, form **D** = **A** + **B**; that is, start with **A** and form the triangle by adding **B**. Determine the magnitude of **D** and the angle between **D** and **B**.

2.3 Determine the difference of the vectors **A** and **B** in Figure P2.1, in the order **D** = **B** − **A**. Compute the magnitude of **D** and the angle to the horizontal.

2.4 Determine the difference of the vectors **A** and **B** of Figure P2.1, in the order **C** = **A** − **B**. Compute the magnitude of **C** and the angle to the horizontal.

2.5 Calculate the resultant of adding the forces $\mathbf{F}_1$ and $\mathbf{F}_2$ in Figure P2.5.

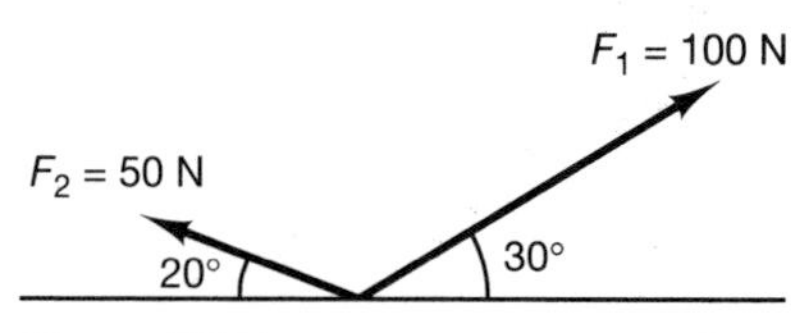

Figure P2.5

2.6 Two water skiers are being pulled by a boat that provides a net 500-N (5.00×10^2) force along the x-axis in Figure P2.6. This force causes a tension in each of the ropes, which in turn pulls the water skier. Calculate the tension in the ropes using the law of sines.

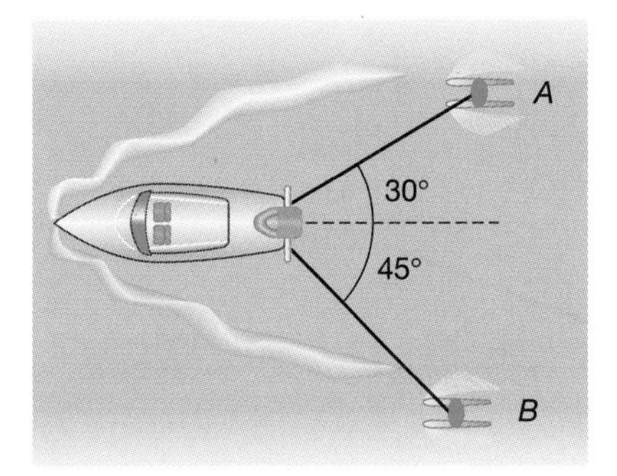

Figure P2.6

2.7 Calculate the resultant of the two velocity vectors acting on the glider shown in Figure P2.7. One of these vectors represents the wind's effect and the other the effect of the towing aircraft.

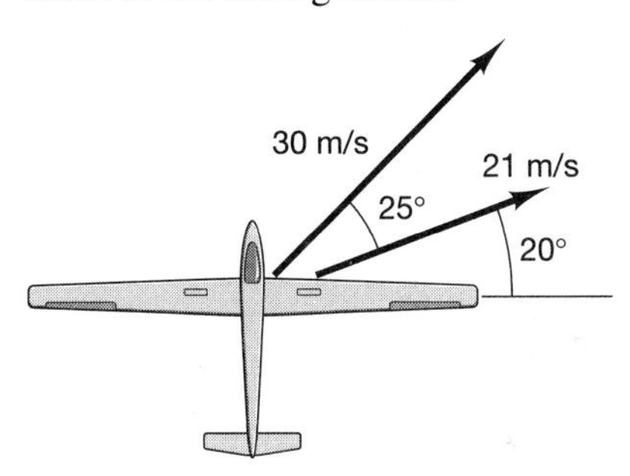

Figure P2.7

2.8 A barge is pulled by two tug boats. To move the barge along in the water, the tug boats must exert a resultant force (the sum of $\mathbf{T}_A$ and $\mathbf{T}_B$) of 20,000 N (20.0×10^3) along the direction of motion of the barge. (a) Determine the tension in each rope if the position of tug B is such that $\beta = 45°$ (b) Suppose that tug B can move anywhere such that $0 < \beta < 90°$, determine that angle β where the tension in tug B's rope has a minimum value still maintaining the resultant of 20,000 N (20.0×10^3) along the x-direction on the barge.

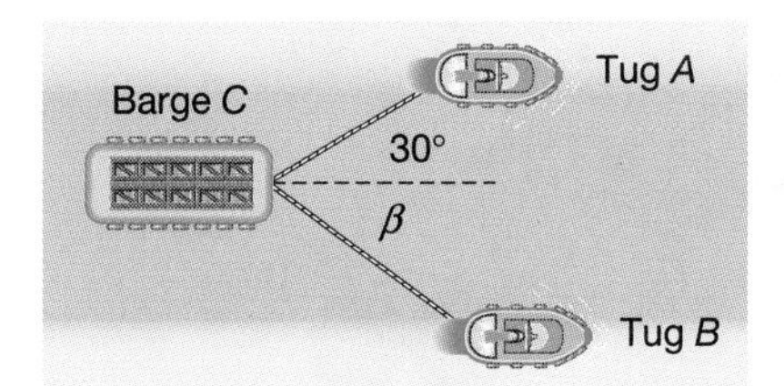

Figure P2.8

2.9 A 500-N (5.00×10^2) force $\mathbf{F}$ acts on the frame shown in Figure P2.9. Calculate the components of $\mathbf{F}$ along the struts AC and AB for the case $\theta = 25°$.

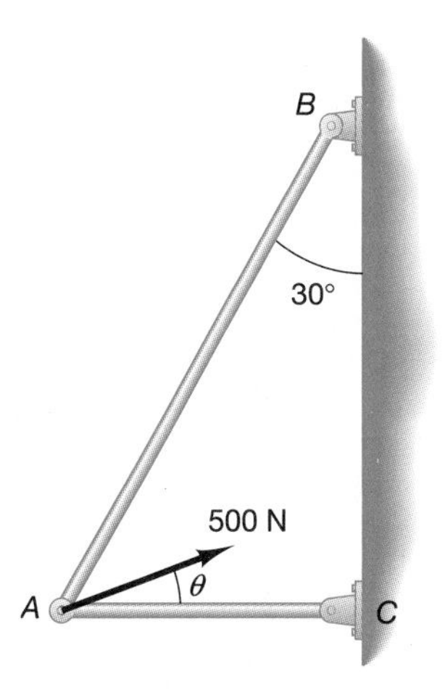

Figure P2.9

2.10 In order to design the frame in Problem 2.9 properly, it is necessary to calculate the angle θ such that the force component along AC has a magnitude of 400 N (4.00×10^2) when the applied force $\mathbf{F}$ is 500 N. Find the required, angle θ and the resulting force along AB.

2.11 Compute the components of the 250-N (2.50×10^2) force acting along the two struts AB and CB in Figure P2.11.

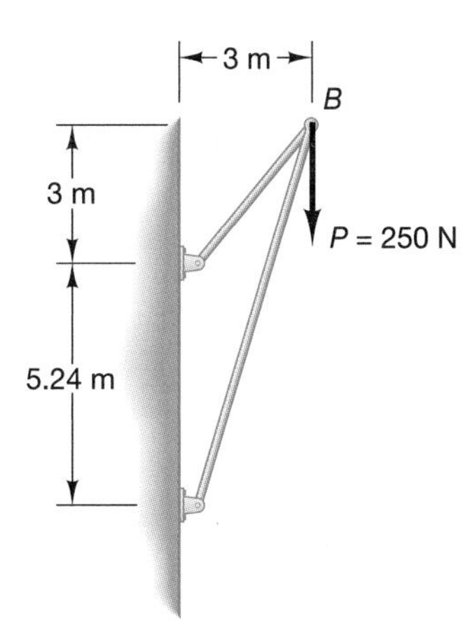

Figure P2.11

2.12 In a two-member roof frame element, the angle θ between the members, is 30°. A force $\mathbf{F} = 4000$ N (4.000×10^3) at an angle of $\beta = 20°$ is applied to the system. (See Figure P2.12.) (a) Calculate the components of this force along each of the two members A and B. (b) Let F be fixed at 20° ($\beta = 20°$). Calculate the forces A and B as the frame angle θ ranges from 5° to 90°, and hence, investigate various roof designs for the given load. Use your calculations to find the best or optimal design, which, in this case, is defined as the angle θ causing the loads in members A and B to be of equal magnitude. What happens to your calculations for $\theta = 0°$?

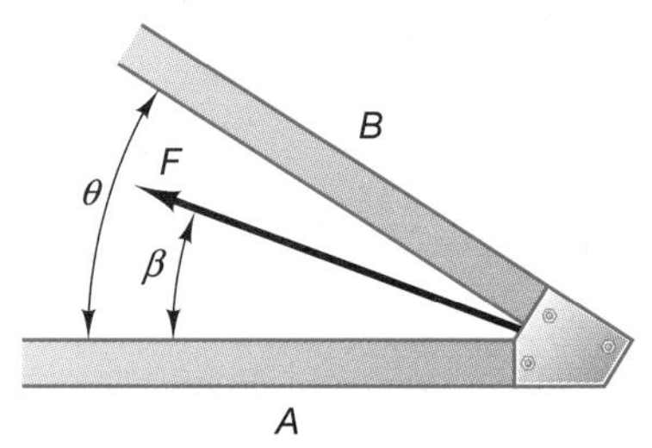

Figure P2.12

2.13 Resolve the 1000-N (1.000×10^3) force into components in the x- and y-directions. Determine the direction cosines.

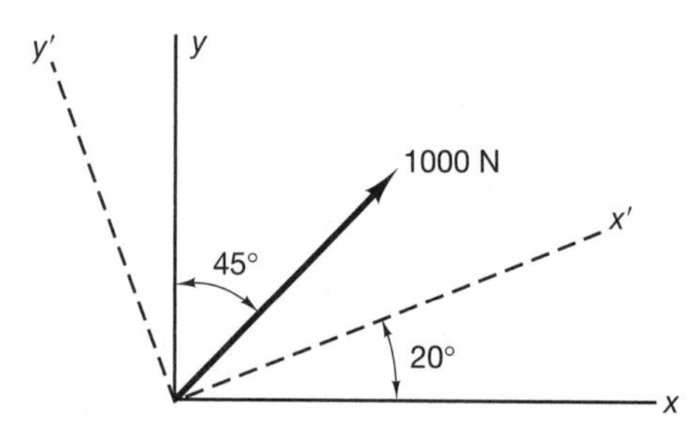

Figure P2.13

2.14 In Problem 2.13, resolve the force into components in the x'- and y'-directions. Determine the direction cosines.

2.15 Given the components of a vector in the x and y coordinates, determine the components in the x' and y' coordinates.

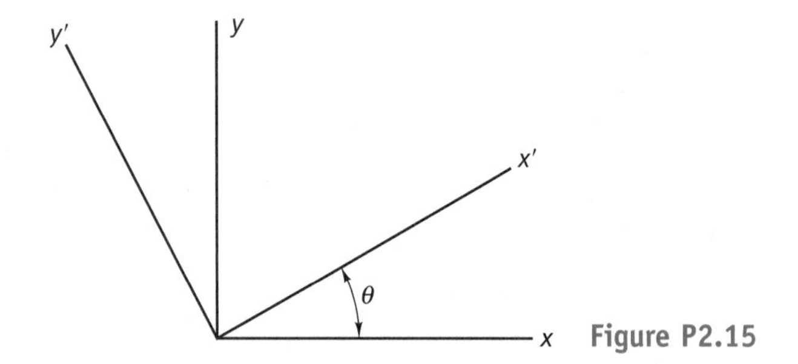

Figure P2.15

2.16 Resolve the 1.00×10^2 m displacement vector into components in the u and v directions.

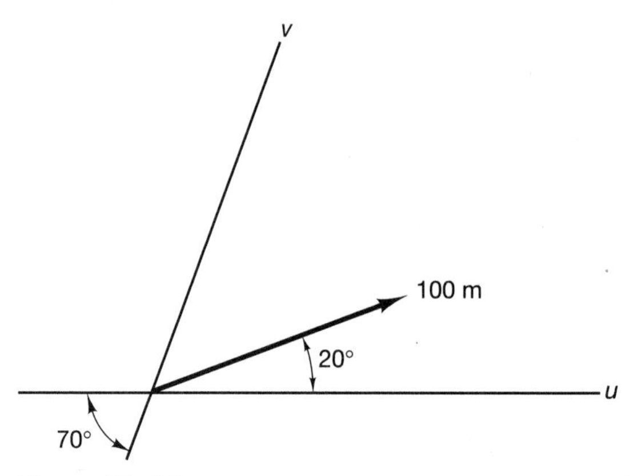

Figure P2.16

2.17 A 120-N (1.20×10^2) force is to be resolved into components along lines a–a' and b–b'. Determine the angle β and the component along b–b' if the component along a–a' is 80 N.

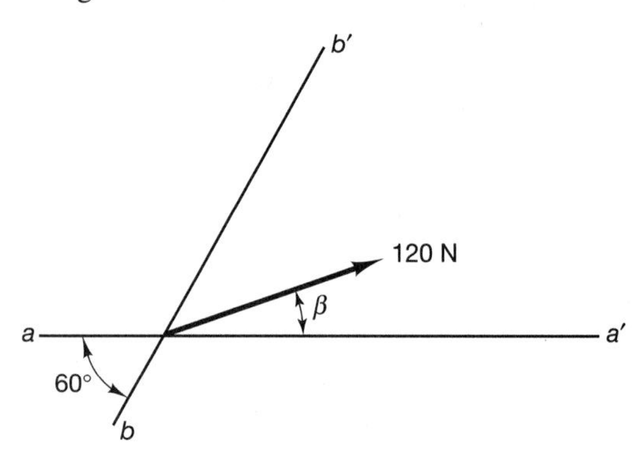

Figure P2.17

2.18 For the 120-N force in Problem 2.17, determine the angle β and the component along a–a' if the component along b–b' is 58 N.

2.19 Figure P2.19 shows three forces acting concurrently in a plane at a single point (an eye bolt on a truck bed). Determine the resultant force acting on the bolt.

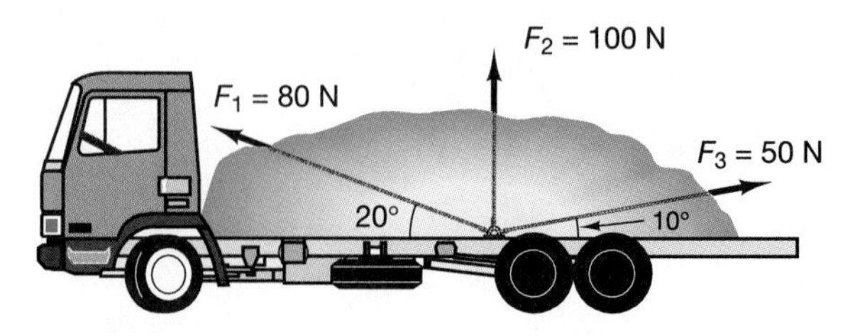

Figure P2.19

2.20 The post on a speedboat is used to tow three water skiers. Each water skier has a slightly different drag force in the water; hence, the tensions on each are different: $T_1 = 1.20 \times 10^2$ N, $T_2 = 1.50 \times 10^2$ N and $T_3 = 1.00 \times 10^2$ N acting at the angles shown in Figure P2.20. Calculate the resultant force acting on the boat and its angle relative to an axis perpendicular to the boat's axis.

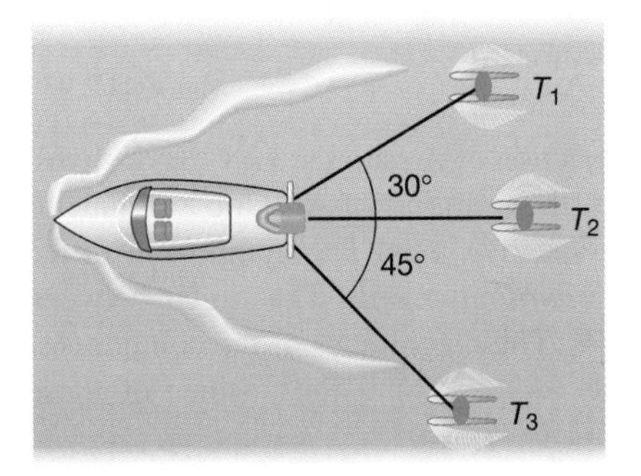

Figure P2.20

2.21 Four concurrent forces act on the center of mass of a landing airplane, as shown in Figure P2.21. Calculate the resultant force and the angle it makes with the horizontal axis.

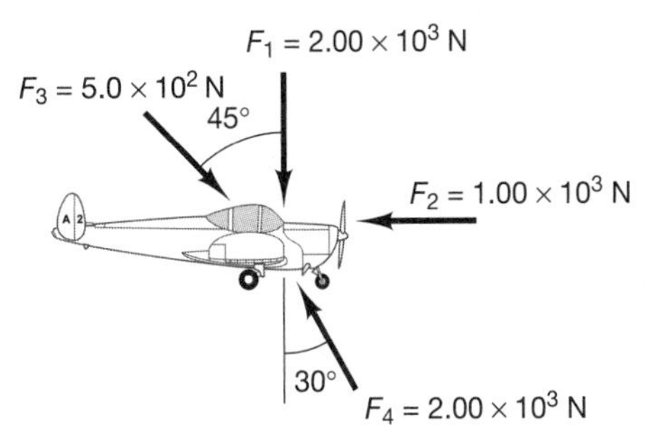

Figure P2.21

2.4 THREE-DIMENSIONAL CARTESIAN COORDINATES AND UNIT BASE VECTORS

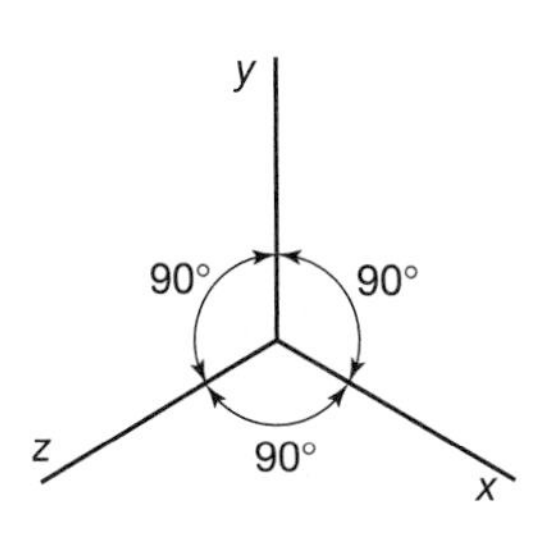

Figure 2.29 A Cartesian coordinate system of three mutually perpendicular axes.

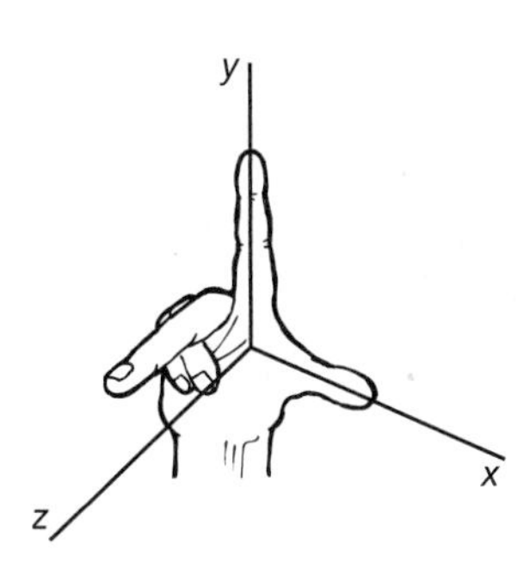

Figure 2.30

In Section 2.2, a two-dimensional orthogonal x–y coordinate system called the two-dimensional Cartesian coordinates was constructed. This system can be extended to three dimensions by forming three mutually perpendicular reference axes that have their origin at the common point of intersection, as shown in Figure 2.29. The coordinate axes are designated the x-, y-, and z-axes, and the plane formed by any two coordinate axes is called a ***coordinate plane***—that is, the xy-plane, the yz-plane, and the zx-plane. The x-axis is perpendicular to the yz-plane, the y-axis is perpendicular to the zx-plane, and the z-axis is perpendicular to the xy-plane. The coordinate system shown is a ***right-handed coordinate system***. This means that the x-, y-, and z-coordinates may be aligned with the thumb, index, and middle fingers of the right hand, respectively, as shown in Figure 2.30. Vector mathematics is based upon right-handed coordinate systems, and when you establish a coordinate system, you should make sure that it is right handed. As has been stated before, the choice of the coordinate system is arbitrary, but that choice must be a right-handed coordinate system. Just as the right hand may be rotated so that the thumb points in any direction, a right-handed coordinate system may be rotated in any manner, and it remains right handed, as shown in Figure 2.31. The opposite of a right-handed coordinate system is a *left-handed coordinate system.* The only way to change a right-handed coordinate system to a left-handed coordinate system is to pull one axis back through itself while keeping the others unchanged, as shown in Figure 2.32. This is equivalent to turning a right-hand glove inside out so that it will fit the left hand. Mathematically, such an operation is called an ***inversion of coordinates***, compared to a ***rotation*** of the coordinate axes. A rotation preserves the right-handedness of the system, whereas an inversion does not.

In a manner similar to the two-dimensional cases, the vector **A** may be considered to be the sum of three vectors, $\mathbf{A}_x$, $\mathbf{A}_y$, and $\mathbf{A}_z$, as shown in Figure 2.33. Vectors $\mathbf{A}_x$ and $\mathbf{A}_y$ may be added by using the Pythagorean theorem, as they are at right angles in the vector parallelogram. The vector $\mathbf{A}_z$ is perpendicular to the vector $(\mathbf{A}_x + \mathbf{A}_y)$ and could be added

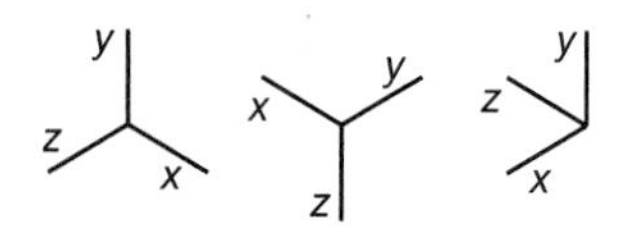

Figure 2.31 Right-handed coordinate systems in various orientations.

to this sum by again using the Pythagorean theorem. The magnitude of the vector **A** is, then,

$$|\mathbf{A}| = \sqrt{A_x^2 + A_y^2 + A_z^2} \tag{2.20}$$

As in two dimensions, the magnitudes $|\mathbf{A}_x|$, $|\mathbf{A}_y|$, and $|\mathbf{A}_z|$ are the scalar components of the vector **A**, which is equal to the unique sum of its components:

$$\mathbf{A} = \mathbf{A}_x + \mathbf{A}_y + \mathbf{A}_z \tag{2.21}$$

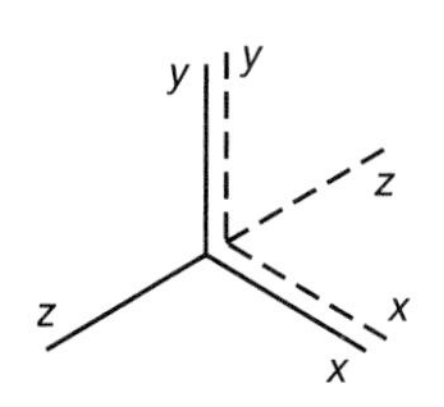

Figure 2.32 The solid coordinate system is right-handed, and the dashed-line system is left-handed.

The vectors $\mathbf{A}_x$, $\mathbf{A}_y$, and $\mathbf{A}_z$ may be viewed as the projection of **A** onto the x-, y-, and z-axes, respectively, as shown in Figure 2.34. The line from the terminus of **A** to the terminus of each of the components is perpendicular to the respective axis. The component is the projection of the vector on the axis *for orthogonal coordinates only.* The angles between the vector **A** and the axes are designated θ_x, θ_y, and θ_z. The magnitudes of the components may be written in terms of the magnitude of the vector **A** and the cosines of these angles as:

$$|\mathbf{A}_x| = |\mathbf{A}| \cos\theta_x, \quad |\mathbf{A}_y| = |\mathbf{A}| \cos\theta_y, \quad |\mathbf{A}_z| = |\mathbf{A}| \cos\theta_z \tag{2.22}$$

In a manner consistent with the two-dimensional case, the direction cosines of the vector **A** are denoted and defined by

$$\lambda_x = \cos\theta_x = \frac{|\mathbf{A}_x|}{|\mathbf{A}|}, \quad \lambda_y = \cos\theta_y = \frac{|\mathbf{A}_y|}{|\mathbf{A}|}, \quad \lambda_z = \cos\theta_z = \frac{|\mathbf{A}_z|}{|\mathbf{A}|} \tag{2.23}$$

Note that, the *sum of the squares of the direction cosines is equal to unity.*

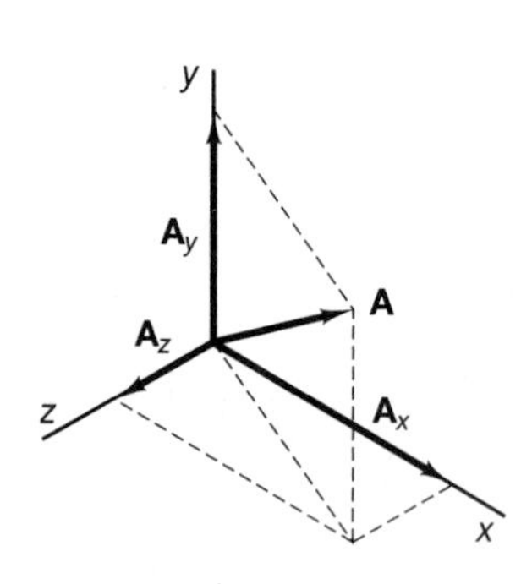

Figure 2.33

2.4.1 UNIT BASE VECTORS

We have referenced a vector **A** to the Cartesian coordinate system by specifying its components in that system or by specifying the magnitude of **A** and its direction cosines in the coordinate system. However, it is difficult to add vectors in space using these specifications. In this section, we will simplify vector algebra by introducing unit vectors.

A ***unit vector*** is defined as a vector having a magnitude of unity and a specific direction. A unit vector does not have units; that is, its magnitude is not given in meters, newtons, pounds, etc. (Note that here we are using the word "unit" to mean two different things. A unit vector is a vector of magnitude unity. A unit of measure is the smallest quantity of a standard basic measure—for example, a m/s, a meter, a newton, etc.)

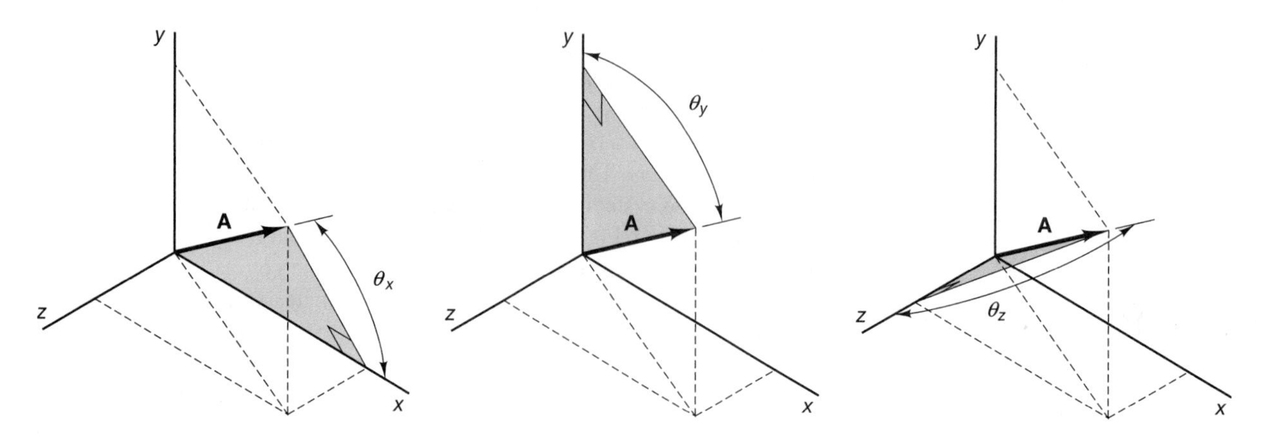

Figure 2.34 The vector components of vector **A** are the projections of **A** onto the three Cartesian coordinate axes.

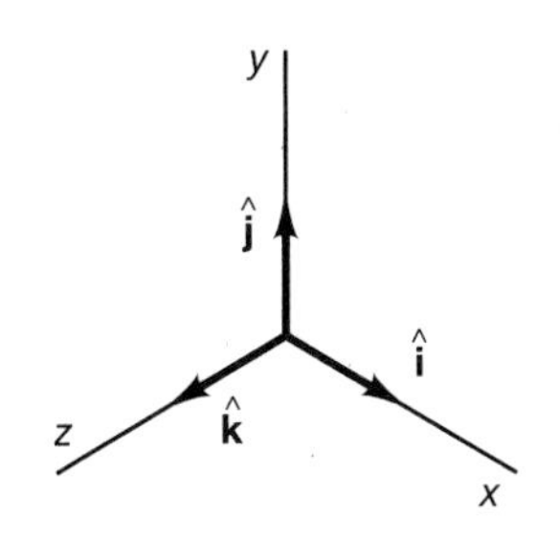

Figure 2.35 The $\hat{\mathbf{i}}, \hat{\mathbf{j}}$, and $\hat{\mathbf{k}}$ unit vectors along the x, y, z axes, respectively.

A unit vector may be thought of as a *pointer* in space, having a magnitude of unity and a specific direction, and not expressed in any unit of measure. It is used mathematically to specify a particular direction. Unit vectors are denoted by a caret over the vector.

Special unit vectors are used to point in the Cartesian coordinate directions. Common engineering notation uses $\hat{\mathbf{i}}$, $\hat{\mathbf{j}}$, and $\hat{\mathbf{k}}$ to point in the x-, y-, and z-directions, respectively, as shown in Figure 2.35. Note that $\hat{\mathbf{i}}$, $\hat{\mathbf{j}}$, and $\hat{\mathbf{k}}$ are vectors and, therefore, have components in the coordinate system. These components are

$$\begin{array}{lll} i_x = 1 & j_x = 0 & k_x = 0 \\ i_y = 0 & j_y = 1 & k_y = 0 \\ i_z = 0 & j_z = 0 & k_z = 1 \end{array} \tag{2.24}$$

The unit vectors $\hat{\mathbf{i}}$, $\hat{\mathbf{j}}$, and $\hat{\mathbf{k}}$ form the basis of the coordinate system and are called the ***unit base vectors*** of the system. Any vector **A** may be written as a sum of its components by the use of unit vectors. In particular, a general three-dimensional vector **A** may be written in terms of the unit vectors $\hat{\mathbf{i}}$, $\hat{\mathbf{j}}$, and $\hat{\mathbf{k}}$ as

$$\mathbf{A} = A_x\hat{\mathbf{i}} + A_y\hat{\mathbf{j}} + A_z\hat{\mathbf{k}} \tag{2.25}$$

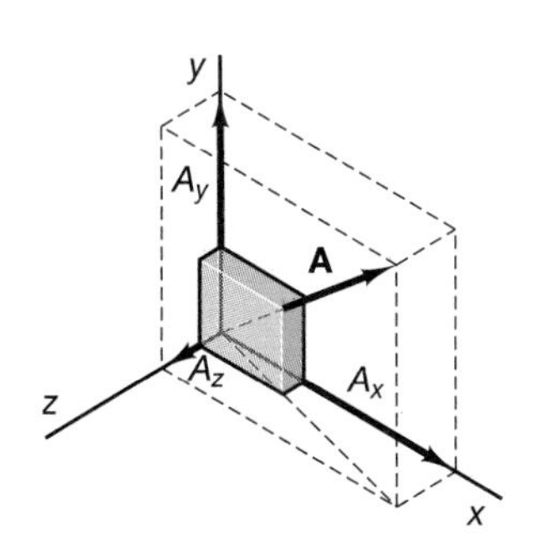

Figure 2.36

This arrangement is illustrated in Figure 2.36. The component of the vector **A** can be determined even when the origin of the vector does not coincide with the origin of the coordinate system. The magnitudes of the components of the vectors $\mathbf{A}_x$, $\mathbf{A}_y$, and $\mathbf{A}_z$ are the scalar components. Writing a vector in this form is called writing the vector in ***component notation***. Note that since the coordinate system is orthogonal, $\mathbf{A}_x$ is independent of $\mathbf{A}_y$ and $\mathbf{A}_z$; that is, $\mathbf{A}_x$ makes no contribution in the y- or z- directions. The vector **A**, with components $\mathbf{A}_x$, $\mathbf{A}_y$, and $\mathbf{A}_z$, is completely specified in a given coordinate system x, y, z with unit base vectors $\hat{\mathbf{i}}$, $\hat{\mathbf{j}}$, $\hat{\mathbf{k}}$ by Eq. (2.25).

The magnitude of the vector, as shown before, may be written as

$$|\mathbf{A}| = \sqrt{A_x^2 + A_y^2 + A_z^2} \tag{2.26}$$

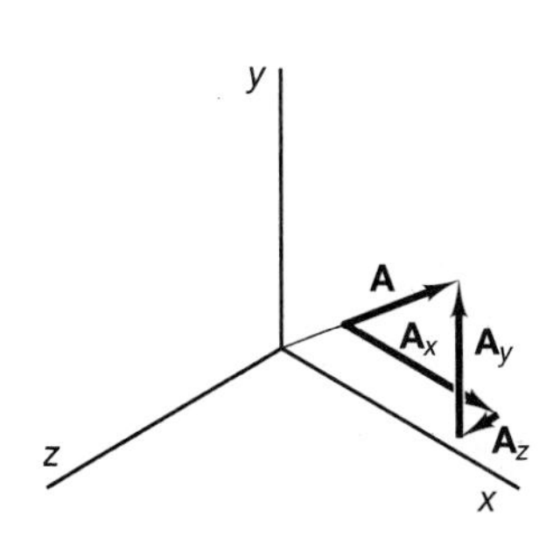

Figure 2.37

This equation results from repeated use of the Pythagorean theorem, as illustrated in Figure 2.37. The Pythagorean theorem was used for the right triangle in the plane parallel to the x–z plane and then a second time for the right triangle in the plane perpendicular to the x–z plane.

A unit vector $\hat{\mathbf{a}}$ in the **A** direction is given by

$$\begin{aligned} \hat{\mathbf{a}} &= \frac{\mathbf{A}}{|\mathbf{A}|} = \frac{A_x}{|\mathbf{A}|}\hat{\mathbf{i}} + \frac{A_y}{|\mathbf{A}|}\hat{\mathbf{j}} + \frac{A_z}{|\mathbf{A}|}\hat{\mathbf{k}} \\ \hat{\mathbf{a}} &= a_x\hat{\mathbf{i}} + a_y\hat{\mathbf{j}} + a_z\hat{\mathbf{k}} \end{aligned} \tag{2.27}$$

Using the definition of the direction cosines from Eq. (2.23) yields

$$\hat{\mathbf{a}} = \lambda_x\hat{\mathbf{i}} + \lambda_y\hat{\mathbf{j}} + \lambda_z\hat{\mathbf{k}}$$

2.4.2 VECTOR EQUALITY IN COMPONENT NOTATION

Two vectors are defined to be equal to one another, denoted $\mathbf{A} = \mathbf{B}$, if the magnitudes of their respective components are equal—that is, if

$$A_x\hat{\mathbf{i}} + A_y\hat{\mathbf{j}} + A_z\hat{\mathbf{k}} = B_x\hat{\mathbf{i}} + B_y\hat{\mathbf{j}} + B_z\hat{\mathbf{k}}$$

Equating the magnitudes of the $\hat{\mathbf{i}}$, $\hat{\mathbf{j}}$, and $\hat{\mathbf{k}}$ components yields

$$A_x = B_x, \ A_y = B_y, \ A_z = B_z \tag{2.28}$$

Because the unit vectors are orthogonal, any vector equation that equates one vector to another vector must treat their respective x-, y-, and z-component magnitudes as equal independently. This treatment is equivalent to stating three scalar equations. Note that the equality of two vectors depends only upon the fact that their components are equal and does not refer to where the two vectors may be in space. That is, two vectors may be mathematically equal even if they do not have common origins or lines of action, as was shown in Figure 2.19. The physical equivalence of the effects of equal vectors depends upon whether they are fixed, sliding, or free vectors—that is, upon the physical quantities represented.

2.4.3 VECTOR ADDITION BY COMPONENTS

Two vectors may be added by adding their respective components. That is, the vector sum

$$\mathbf{C} = \mathbf{A} + \mathbf{B} \tag{2.29}$$

This equation may be written in component form

$$\begin{aligned} C_x\hat{\mathbf{i}} + C_y\hat{\mathbf{j}} + C_z\hat{\mathbf{k}} &= A_x\hat{\mathbf{i}} + A_y\hat{\mathbf{j}} + A_z\hat{\mathbf{k}} + B_x\hat{\mathbf{i}} + B_y\hat{\mathbf{j}} + B_z\hat{\mathbf{k}} \\ &= (A_x + B_x)\hat{\mathbf{i}} + (A_y + B_y)\hat{\mathbf{j}} + (A_z + B_z)\hat{\mathbf{k}} \end{aligned} \tag{2.30}$$

Equating the $\hat{\mathbf{i}}$, $\hat{\mathbf{j}}$, and $\hat{\mathbf{k}}$ components yields the three scalar equations

$$\begin{aligned} C_x &= A_x + B_x \\ C_y &= A_y + B_y \\ C_z &= A_z + B_z \end{aligned} \tag{2.31}$$

The addition of two vector is commutative, as has been shown before:

$$\mathbf{C} = \mathbf{A} + \mathbf{B} = \mathbf{B} + \mathbf{A} \tag{2.32}$$

Since scalar addition is commutative and the component method reduces vector addition to addition of scalar components, vector addition is also commutative. We noted this fact when we discussed vector addition by the parallelogram law. Vector addition is clearly associative also, as it may be accomplished by scalar addition of components, and scalar addition is associative and independent of sequence. That is,

$$\mathbf{A} + \mathbf{B} + \mathbf{C} = (\mathbf{A} + \mathbf{B}) + \mathbf{C} = \mathbf{A} + (\mathbf{B} + \mathbf{C}) \tag{2.33}$$

If vectors are expressed as components in the same coordinate system, any number of vectors may be added by simply forming the sum of their components. The use of vector components simplifies vector algebra, as it is not necessary to determine the angles between vectors, as is done in trigonometry. We shall see in Section 2.5 that the use of components is compatible with computational software.

2.4.4 MULTIPLICATION OF A VECTOR BY A SCALAR

As was discussed earlier, a vector added to itself is equal to a vector whose components are each twice the original vector:

$$\mathbf{B} = \mathbf{A} + \mathbf{A} = 2\mathbf{A} \tag{2.34}$$

Generalizing from this equation, we can say that multiplication of a vector by a scalar is equivalent to multiplication of each component by the scalar. Thus,

$$\alpha\mathbf{A} = \alpha A_x\hat{\mathbf{i}} + \alpha A_y\hat{\mathbf{j}} + \alpha A_z\hat{\mathbf{k}} \tag{2.35}$$

where α denotes any scalar value. Note that this definition is consistent with the development of the multiplication of numbers from addition, as in elementary arithmetic. This type of multiplication is the basis for expressing a vector **A** as the sum of its scalar components times their respective unit base vectors.

2.4.5 VECTOR SUBTRACTION

The subtraction of a vector from another vector is accomplished by noting that the *negative of a vector* is equivalent to multiplying the vector by the scalar (-1). The resulting vector will have the same magnitude as the original vector but will be opposite in direction to it. Therefore, subtraction may be represented by

$$\mathbf{D} = \mathbf{A} - \mathbf{B} = \mathbf{A} + (-\mathbf{B}) \tag{2.36}$$

In component notation, subtraction results in the three scalar equations

$$\begin{aligned} D_x &= A_x - B_x \\ D_y &= A_y - B_y \\ D_z &= A_z - B_z \end{aligned} \tag{2.37}$$

This allows vector subtraction to be defined in a manner similar to the way it is defined in scalar arithmetic.

2.4.6 GENERAL UNIT VECTORS

Unit base vectors may be used to represent vectors in terms of their coordinate components, but unit vectors may also be used in directions other than coordinate directions. In particular, if a vector lies along a desired direction, a unit vector may be formed in that direction. Consider a vector **B** that has a specific direction in space. The vector may be a force vector, a velocity vector, a displacement vector, or any other physical quantity that can be treated as a vector. A unit vector may be defined by dividing vector **B** by a scalar that is equal to the vector's magnitude. This yields a unit vector $\hat{\mathbf{b}}$ pointing in the direction of **B**. Division by a scalar may be thought of as multiplication of the vector by the inverse of the scalar $(1/\alpha)$ and may be treated using the laws of multiplication by a scalar. The unit vector $\hat{\mathbf{b}}$ in the **B** direction is thus formed by

$$\begin{aligned} \hat{\mathbf{b}} &= \frac{\mathbf{B}}{|\mathbf{B}|} \quad \text{where } |\mathbf{B}| = \sqrt{B_x^2 + B_y^2 + B_z^2} \\ \hat{\mathbf{b}} &= \frac{B_x}{|\mathbf{B}|}\hat{\mathbf{i}} + \frac{B_y}{|\mathbf{B}|}\hat{\mathbf{j}} + \frac{B_z}{|\mathbf{B}|}\hat{\mathbf{k}} \end{aligned} \tag{2.38}$$

Note that the components of the unit vector $\hat{\mathbf{b}}$ are the direction cosines of the vector **B**, as defined in Eq. (2.23). It was shown that in two dimensions the direction of a vector could be specified by giving its magnitude and direction cosines. Therefore, it is reasonable that the components of a unit vector have only directional characteristics, and these are direction cosines. The vector **B** may be written simply as $\mathbf{B} = |\mathbf{B}|\hat{\mathbf{b}}$.

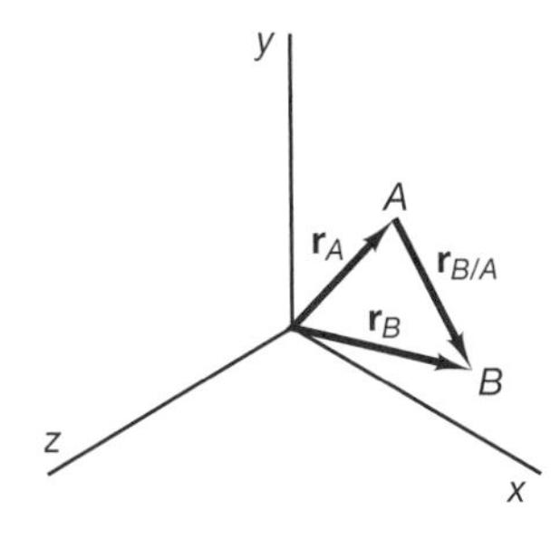

Figure 2.38 Position vectors for points A and B and the position vector of B relative to A.

2.4.7 VECTOR DIRECTIONS IN SPACE

A point in space may be located by the use of a ***position vector***. The magnitudes of the components of a position vector are the coordinates (x,y,z) of the point. Consider the two points A and B shown in Figure 2.38. The position vectors to point A and B are:

$$\begin{aligned} \mathbf{r}_\mathrm{A} &= x_A\hat{\mathbf{i}} + y_A\hat{\mathbf{j}} + z_A\hat{\mathbf{k}} \\ \mathbf{r}_\mathrm{B} &= x_B\hat{\mathbf{i}} + y_B\hat{\mathbf{j}} + z_B\hat{\mathbf{k}} \end{aligned} \tag{2.39}$$

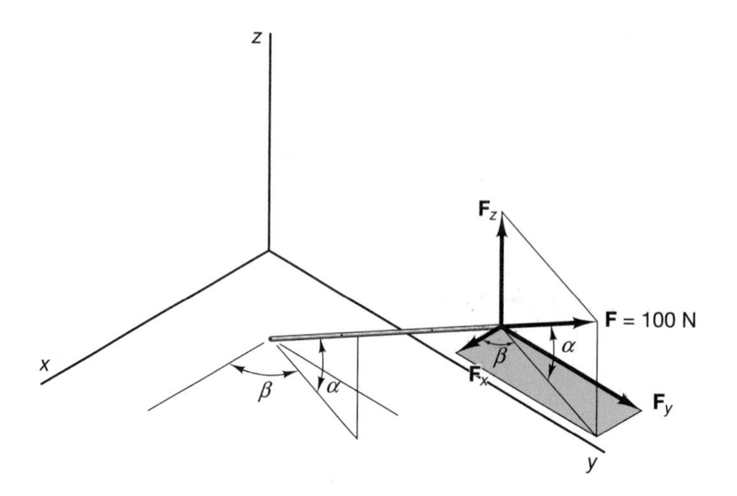

Figure 2.39

It is useful in many applications to locate point B relative to point A. This is done by the use of a ***relative position vector*** $\mathbf{r}_{B/A}$ (B relative to A), defined as

$$\mathbf{r_{B/A}} = \mathbf{r_B} - \mathbf{r_A} = (x_B - x_A)\hat{\mathbf{i}} + (y_B - y_A)\hat{\mathbf{j}} + (z_B - z_A)\hat{\mathbf{k}} \tag{2.40}$$

The position vector of A relative to B is the negative of the position vector of B relative to A.

$$\mathbf{r_{B/A}} = -\mathbf{r_{A/B}} \tag{2.41}$$

The directions of vectors in space are usually specified by giving geometric information about their line of action in space. This information may be given in the form of the angles that the line makes with the coordinate axes or coordinate planes, or by specifying the coordinates of two points on the line. Consider a force of 100 N acting along a cable, as shown in Figure 2.39. The location of the cable is given by the angle α between the cable and its projection on the x–y plane and the angle β between the projection on the x–y plane and the x-axis. The component of the force in the z-direction is

$$F_z = F \sin \alpha \tag{2.42}$$

The projection of the force onto the x–y plane is

$$F_{xy} = F \cos \alpha \tag{2.43}$$

The components of $\mathbf{F}$ in the x- and y-directions can now be found and are

$$F_x = F_{xy} \cos \beta = F \cos \alpha \cos \beta \tag{2.44}$$

$$F_y = F_{xy} \sin \beta = F \cos \alpha \sin \beta \tag{2.45}$$

The force vector may now be written in terms of a unit vector $\hat{\mathbf{f}}$ in the $\mathbf{F}$ direction, where

$$\hat{\mathbf{f}} = \cos \alpha \cos \beta \hat{\mathbf{i}} + \cos \alpha \sin \beta \hat{\mathbf{j}} + \sin \alpha \hat{\mathbf{k}} \tag{2.46}$$

and the force becomes

$$\mathbf{F} = 100\hat{\mathbf{f}} = 100(\cos \alpha \cos \beta \hat{\mathbf{i}} + \cos \alpha \sin \beta \hat{\mathbf{j}} + \sin \alpha \hat{\mathbf{k}}) \tag{2.47}$$

The second manner in which the direction of a vector may be given is to specify the coordinates of any two points on its line of action, as illustrated in Figure 2.40. A position vector from the origin to a point A on the line of action of $\mathbf{F}$ is given by

$$\mathbf{r_A} = x_A\hat{\mathbf{i}} + y_A\hat{\mathbf{j}} + z_A\hat{\mathbf{k}} \tag{2.48}$$

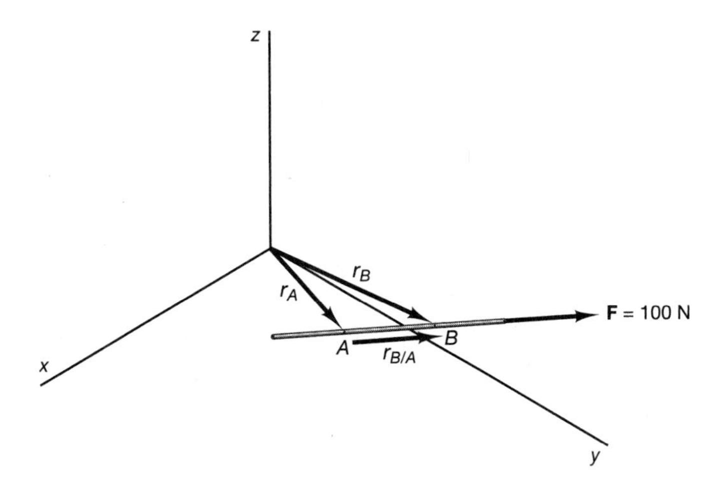

Figure 2.40

and the position vector to a second point ***B*** on the line of action is

$$\mathbf{r_B} = x_B\hat{\mathbf{i}} + y_B\hat{\mathbf{j}} + z_B\hat{\mathbf{k}} \tag{2.49}$$

A relative position vector $\mathbf{r}_{B/A}$ is

$$\mathbf{r}_B = \mathbf{r}_A + \mathbf{r}_{B/A} \tag{2.50a}$$

Therefore,

$$\mathbf{r}_{B/A} = \mathbf{r}_B - \mathbf{r}_A = (x_B - x_A)\hat{\mathbf{i}} + (y_B - y_A)\hat{\mathbf{j}} + (z_B - z_A)\hat{\mathbf{k}} \tag{2.50b}$$

Now a unit vector in the direction of the force can be written as

$$\hat{\mathbf{f}} = \frac{\mathbf{r}_{B/A}}{|\mathbf{r}_{B/A}|} \tag{2.51}$$

The force can be written in terms of this unit vector as

$$\mathbf{F} = 100\,\hat{\mathbf{f}}\,\mathrm{N} \tag{2.52}$$

Vectors in space are generally indicated by their magnitude times a unit vector specifying their direction.

2.4.8 MATRIX NOTATION FOR VECTORS

A second form of component notation for vectors is introduced to prepare for the use of computational software for vector calculations. The vector **A** is written as a column of numbers—that is, in matrix notation. Matrices will be discussed in detail in Section 2.7. We have

$$\mathbf{A} = \begin{pmatrix} A_x \\ A_y \\ A_z \end{pmatrix} \tag{2.25a}$$

(The equation number refers back to the number of the equation in component notation.) A ***matrix*** is an array of numbers presented in rows and columns. The ***elements*** of the matrix are the individual entries in designated rows and columns. For example, a_{21} is the element in the second row, first column. The first subscript designates the element's row, and the second subscript designates the element's column. Therefore, a_{ij} is the element in the ith row and the jth column. Details of the use of matrix notation for vectors are discussed in Section 2.5, where computational software is introduced. The unit base vectors are written in matrix notation as

$$\hat{\mathbf{i}} = \begin{pmatrix} 1 \\ 0 \\ 0 \end{pmatrix} \quad \hat{\mathbf{j}} = \begin{pmatrix} 0 \\ 1 \\ 0 \end{pmatrix} \quad \hat{\mathbf{k}} = \begin{pmatrix} 0 \\ 0 \\ 1 \end{pmatrix} \tag{2.24a}$$

This form of notation always lists all three components of the vector, that is, if the vector lies in the x–y plane, the third component is listed as zero in the third row of the array of numbers.

The unit vector $\hat{\mathbf{a}}$ in the $\mathbf{A}$ direction is written in terms of the direction cosines of $\mathbf{A}$ as

$$\hat{\mathbf{a}} = \begin{pmatrix} \lambda_x \\ \lambda_y \\ \lambda_z \end{pmatrix} \tag{2.27a}$$

In matrix notation, vector addition is

$$\mathbf{C} = \mathbf{A} + \mathbf{B}$$

$$\begin{pmatrix} C_x \\ C_y \\ C_z \end{pmatrix} = \begin{pmatrix} A_x \\ A_y \\ A_z \end{pmatrix} + \begin{pmatrix} B_x \\ B_y \\ B_z \end{pmatrix}$$

Therefore, Eq. (2.30) is written as

$$\begin{pmatrix} C_x \\ C_y \\ C_z \end{pmatrix} = \begin{pmatrix} A_x + B_x \\ A_y + B_y \\ A_z + B_z \end{pmatrix} \tag{2.30a}$$

The multiplication of a vector $\mathbf{A}$ by a scalar α, given in Eq. (2.35), in matrix notation is:

$$\alpha\mathbf{A} = \alpha \begin{pmatrix} A_x \\ A_y \\ A_z \end{pmatrix} = \begin{pmatrix} \alpha A_x \\ \alpha A_y \\ \alpha A_z \end{pmatrix} \tag{2.35a}$$

The position vector from the origin to a point A in space can be written in matrix notation as

$$\mathbf{r}_A = \begin{pmatrix} x_A \\ y_A \\ z_A \end{pmatrix} \tag{2.39a}$$

Matrix notation has no advantage over the standard component notation, in particular, but is useful in manipulation of vectors using computational tools such as calculators and computer codes used by most practicing engineers.

2.5 COMPUTATION OF VECTOR OPERATIONS

Three-dimensional vectors are described by their three components, and the computation of vector operations requires that the three scalar equivalent equations be numerically evaluated. Several common commercial software packages and calculators can do these calculations in the same manner that an ordinary calculator does arithmetic calculations. Many of these software programs have been released in inexpensive student editions or are available through university computing facilities. Such software programs can be used directly as a ***vector calculator***, or they can be used by identifying the vectors as algebraic symbols and indicating the operations to be performed by keying in the appropriate symbol. A third method is to specify the vectors and perform symbolic operations. These types of operations are shown in the Computational Windows in the supplement manuals. In all cases, the vector is written as a column or row of three numbers or symbols in matrix notation. The general subject of matrices is discussed in Section 2.7, but for the present, let us consider a matrix to be an array of numbers consisting of a specified number of rows and columns. The use of software packages to perform vector operations is not necessary, but it reduces the labor of numerical calculations and eliminates many careless errors.

Sample Problem 2.7

Find the resultant **C** of the two vectors: **C** = **A** + **B**

$$\mathbf{A} = 6\hat{\mathbf{i}} + 2\hat{\mathbf{j}} + 0\hat{\mathbf{k}}$$
$$\mathbf{B} = 0\hat{\mathbf{i}} + 7\hat{\mathbf{j}} + 5\hat{\mathbf{k}}$$

Write vector **C** as the product of a magnitude and a unit vector in the direction of **C**.

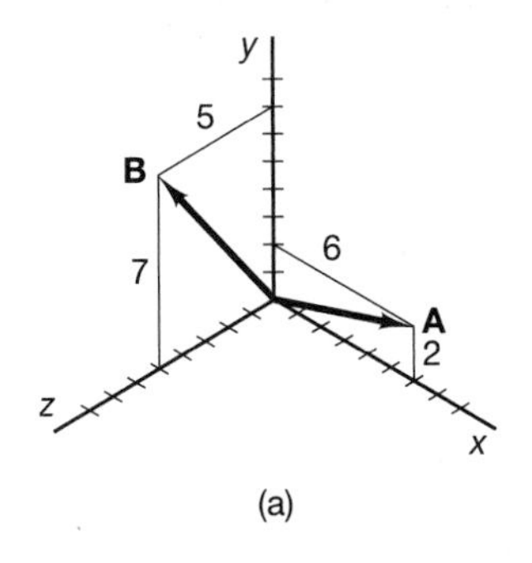

(a)

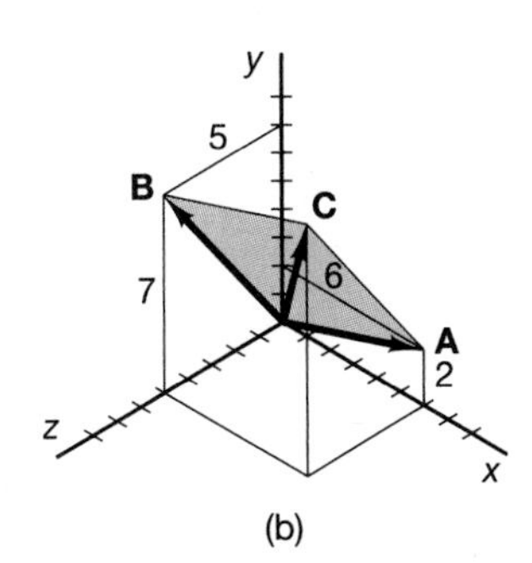

(b)

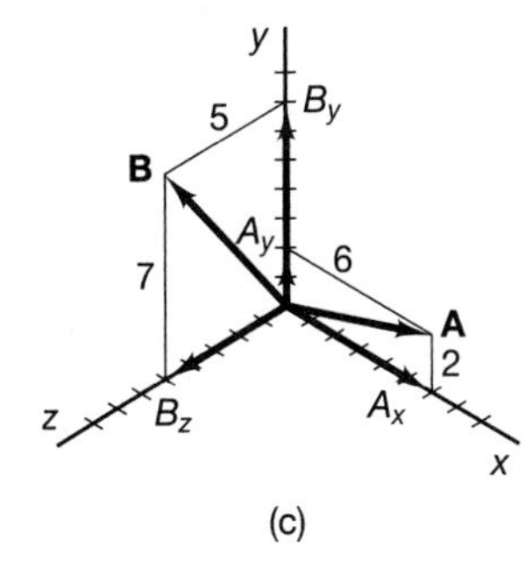

(c)

Solution The two vectors may be sketched in a reference coordinate system. Note that **A** and **B** form a plane in space and that **C** lies in this plane. The vectors **A** and **B** can be broken into components in the coordinate directions.

Although the methods of trigonometry might be attempted in this example, it is difficult to determine the angle between the two vectors. It is apparent that the use of the components greatly simplifies the calculation. We have

$$\mathbf{C} = \mathbf{A} + \mathbf{B} = (6 + 0)\,\hat{\mathbf{i}} + (2 + 7)\,\hat{\mathbf{j}} + (0 + 5)\,\hat{\mathbf{k}} = 6\hat{\mathbf{i}} + 9\hat{\mathbf{j}} + 5\hat{\mathbf{k}}$$

The magnitude of **C** is

$$|C| = \sqrt{C_x^2 + C_y^2 + C_z^2} = 11.92$$

Then the directional cosines of **C** are

$$\lambda_x = \cos\theta_x = \frac{C_x}{|\mathbf{C}|} = 0.503$$

$$\lambda_y = \cos\theta_y = \frac{C_x}{|\mathbf{C}|} = 0.755$$

$$\lambda_z = \cos\theta_z = \frac{C_z}{|\mathbf{C}|} = 0.419$$

Thus, the unit vector in the $\mathbf{C}$ direction is

$$\hat{\mathbf{c}} = 0.503\hat{\mathbf{i}} + 0.755\hat{\mathbf{j}} + 0.419\hat{\mathbf{k}}$$

The vector $\mathbf{C}$ could have been written as

$$\mathbf{C} = |\mathbf{C}|\hat{\mathbf{c}} = 11.92(0.503\hat{\mathbf{i}} + 0.755\hat{\mathbf{j}} + 0.419\hat{\mathbf{k}})$$

Sample Problem 2.8

An electronic scoreboard in a gymnasium is supported by three cables as shown in the diagram. The tension in each cable is

$$T_A = 368 \text{ N}$$
$$T_B = 259 \text{ N}$$
$$T_C = 482 \text{ N}$$

Specify the force vector for each cable in terms of the magnitude of the tension and a unit vector along the cable. Then determine the resultant of the tensions.

Solution

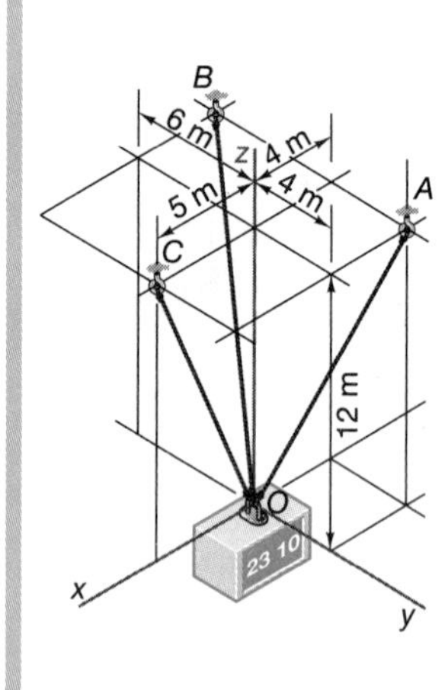

The unit vectors along the cables may be obtained by finding position vectors from the origin O to the cable attachments at A, B, and C. A unit vector along each cable can then be determined. Since each cable must be in tension, the unit vectors must be directed from point O to the point of attachment.

A vector from point O to point A along the cable A would be

$$\mathbf{A} = -4\hat{\mathbf{i}} + 4\hat{\mathbf{j}} + 12\hat{\mathbf{k}} \text{ (m)}$$

Therefore, a unit vector $\hat{\mathbf{a}}$ in that direction may be obtained by dividing $\mathbf{A}$ by its magnitude.

$$|\mathbf{A}| = \sqrt{(-4)^2 + (4)^2 + (12)^2}$$
$$|\mathbf{A}| = 13.27 \text{ (m)}$$
$$\hat{\mathbf{a}} = -0.302\hat{\mathbf{i}} + 0.302\hat{\mathbf{j}} + 0.905\hat{\mathbf{k}}$$

In a similar manner, vectors from O to B and O to C can be determined:

$$\mathbf{B} = -4\hat{\mathbf{i}} - 6\hat{\mathbf{j}} + 12\hat{\mathbf{k}} \text{ (m)}$$
$$\mathbf{C} = 5\hat{\mathbf{i}} + 12\hat{\mathbf{k}} \text{ (m)}$$

The unit vectors in these two directions become

$$|\mathbf{B}| = 14.0 \qquad \hat{\mathbf{b}} = -0.286\hat{\mathbf{i}} - 0.429\hat{\mathbf{j}} + 0.857\hat{\mathbf{k}}$$
$$|\mathbf{C}| = 13.0 \qquad \hat{\mathbf{c}} = 0.385\hat{\mathbf{i}} + 0.923\hat{\mathbf{k}}$$

The vector force in each cable can be written in Cartesian notation as

$$\mathbf{T}_A = 368\hat{\mathbf{a}} = -111\hat{\mathbf{i}} + 111\hat{\mathbf{j}} + 333\hat{\mathbf{k}} \text{ (N)}$$
$$\mathbf{T}_B = 259\hat{\mathbf{b}} = -74\hat{\mathbf{i}} - 111\hat{\mathbf{j}} + 222\hat{\mathbf{k}} \text{ (N)}$$
$$\mathbf{T}_C = 482\hat{\mathbf{c}} = 185\hat{\mathbf{i}} + 445\hat{\mathbf{k}} \text{ (N)}$$

If we represent the total force acting on the scoreboard from the cables by $\mathbf{R}$, then

$$\mathbf{R} = \mathbf{T}_A + \mathbf{T}_B + \mathbf{T}_C = 1000\hat{\mathbf{k}} \text{ (N)}$$

Physically, the tensions in the cables are stabilizing the scoreboard in the x- and y-directions (holding it in place). The z-force of 1000 N suggests that the scoreboard weighs 1000 N. This intuitive reasoning will be verified using formal arguments in Chapter 3.

Since the components of the unit vectors are the direction cosines to the coordinate axes, the angles the cables make with the vertical are easily obtained. For cable A;

$$\cos \theta_z^A = 0.905, \text{ and the angle to the vertical is } 25.2°$$

For the other two cables, the angles are

$$\cos \theta_z^B = 0.857 \qquad \theta_z^B = 31°$$

$$\cos \theta_z^C = 0.923 \qquad \theta_z^C = 22.6°$$

The angles to the other coordinate axes may be found in the same manner.

The numerical work can be reduced by use of computational software, as shown in the supplements.

Problems

2.22 A force vector $\mathbf{F} = 300\hat{\mathbf{i}} - 700\hat{\mathbf{j}} + 200\hat{\mathbf{k}}(\text{N})$. Determine the magnitude of the force and a unit vector $\hat{\mathbf{f}}$ acting along it.

2.23 Calculate the magnitude of the vector $\mathbf{A}$ defined by $\mathbf{A} = 3\hat{\mathbf{i}} + 2\hat{\mathbf{j}} + 6\hat{\mathbf{k}}$. Determine a unit vector along $\mathbf{A}$.

2.24 A vector makes a 30° angle with the x axis and a 65° angle with the y-axis. Determine the angle the vector makes with the z-axis.

2.25 A force vector $\mathbf{F} = 30\hat{\mathbf{i}} + F_y\hat{\mathbf{j}} - 40\hat{\mathbf{k}}$ (N). If the magnitude of the force is 130 N, determine the component F_y.

2.26 A vector $\mathbf{A} = 10\hat{\mathbf{i}} - 20\hat{\mathbf{j}} + 5\hat{\mathbf{k}}$ and a vector $\mathbf{B} = 15\hat{\mathbf{i}} + 5\hat{\mathbf{j}} - 20\hat{\mathbf{k}}$. Determine the magnitude of a vector $\mathbf{C} = \mathbf{A} + \mathbf{2B}$ and a unit vector along $\mathbf{C}$.

2.27 Two vectors $\mathbf{U} = 2\hat{\mathbf{i}} - \hat{\mathbf{j}} + 5\hat{\mathbf{k}}$ and $\mathbf{V} = -3\hat{\mathbf{i}} + 2\hat{\mathbf{j}}$. Determine the magnitude of the vector $\mathbf{2U} - \mathbf{V}$ and the angles it makes with the x-, y-, and z-axes.

2.28 A rope tied to an eye bolt on a ship makes an angle of $\theta_x = 30°$ and $\theta_y = 85°$ relative to the three-dimensional $\hat{\mathbf{i}}, \hat{\mathbf{j}}, \hat{\mathbf{k}}$ coordinate system shown in Figure P2.28. The tension in the rope is measured to be 200 N. Calculate the force component in the z-direction.

2.29 A rope is holding the mast of a sailboat. (See Figure P2.29.) The tension in the rope is 1000 N. Calculate the angles θ_x, θ_y, and θ_z the rope makes with the coordinate system shown. Determine the forces F_x, F_y, and F_z that the rope applies to the eye bolt.

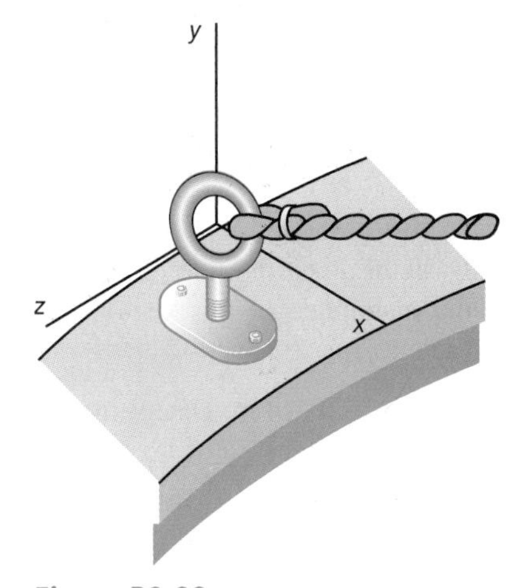

Figure P2.28

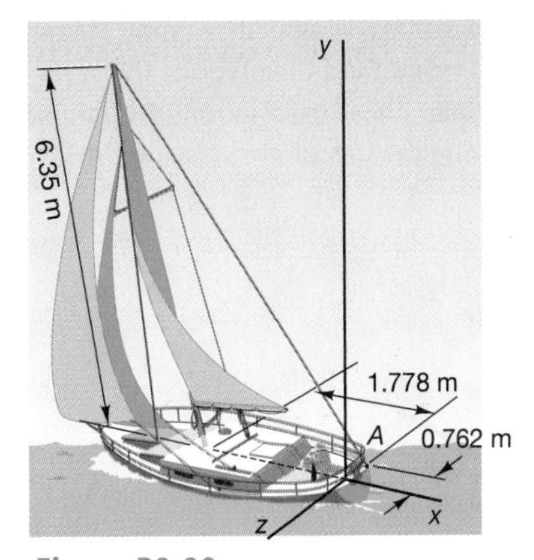

Figure P2.29

2.30 A tow truck is attempting to pull a car out of a ditch, as shown in Figure P2.30. The tension in the cable is 9600 N, and the geometry is as indicated in the figure. Calculate the components of the force exerted by the towing cable on the car.

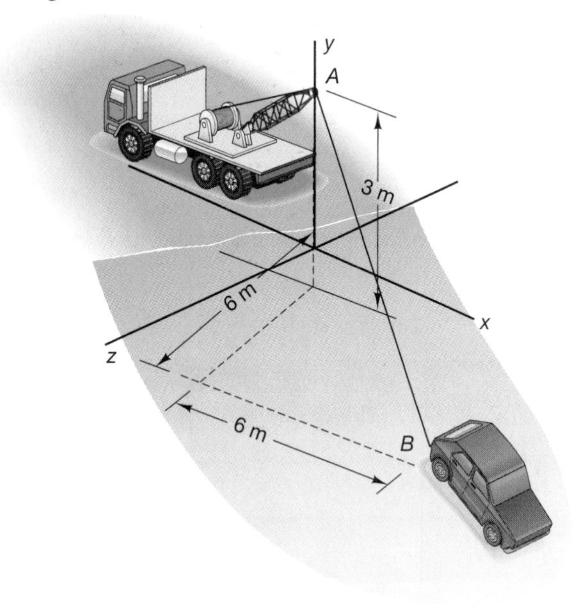

Figure P2.30

2.31 Calculate the unit vector along the line AB in Figure P2.30. Calculate the magnitude of this vector, and verify that it is a unit vector.

2.32 Point A has coordinates (5, 0, 6) m. Determine the coordinate z_B of the point B, (9, 12, z_B) m, if it is known that the distance from A to B is 13 m.

2.33 Point A has coordinates (2, 1, 4) m and point B has coordinates (1, −3, 2) m. Determine the relative position vector $\mathbf{r}_{B/A}$ (B relative to A) and the unit vector $\hat{\mathbf{e}}_{B/A}$.

2.34 Consider the two forces shown in Figure P2.34. Use the vector component method to solve for the resultant force $\mathbf{F}_1 + \mathbf{F}_2$. Compare these calculations with those made in Problem 2.5 using the law of cosines and the law of sines.

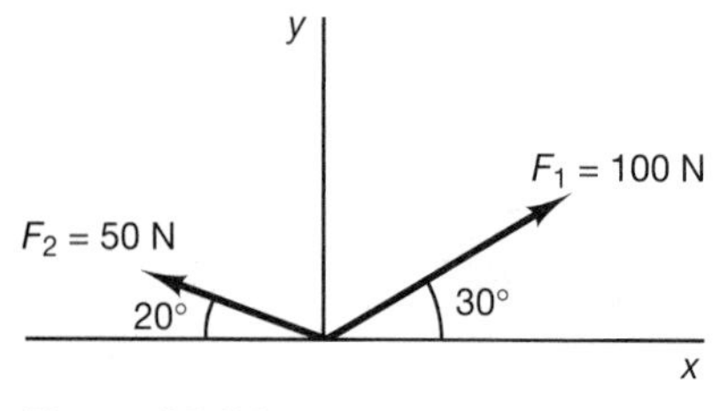

Figure P2.34

2.35 Consider the water skiers shown in Figure P2.35. the two tensions T_1 and T_2 must add up to 500 N in the x-direction. Determine the magnitudes of T_1 and T_2 using vector components.

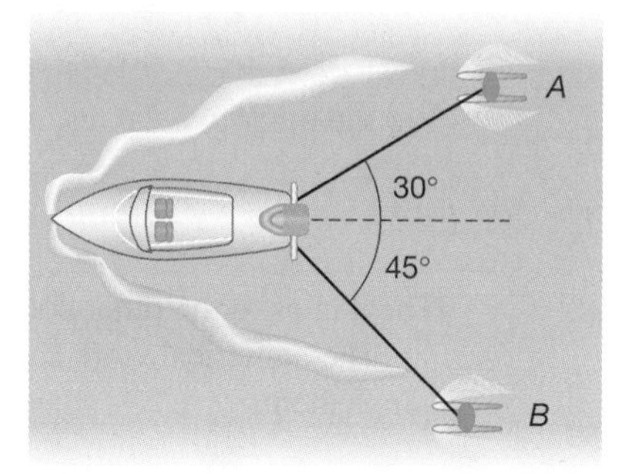

Figure P2.35

2.36 Calculate the resultant force acting on the eye bolt of the truck shown in Figure P2.36.

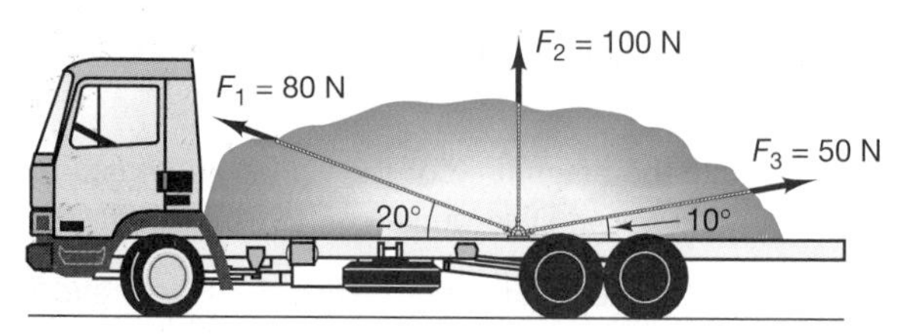

Figure P2.36

2.37 Consider the three force vectors acting on the speedboat shown in Figure P2.37 and calculate their resultant using the vector component method; T_1 = 120 N, T_2 = 150 N and T_3 = 100 N.

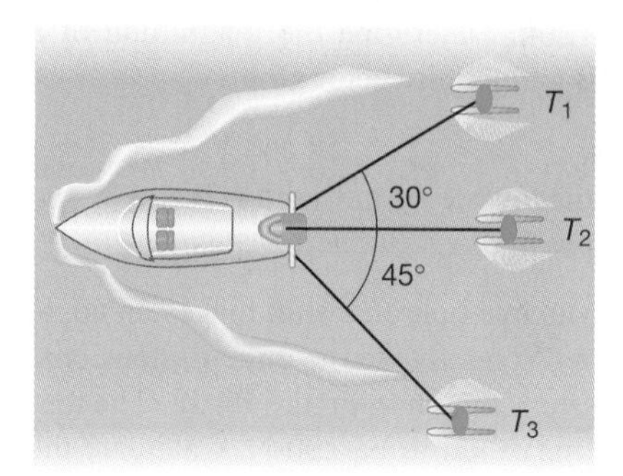

Figure P2.37

2.38 A vector $\mathbf{A} = 100\hat{\mathbf{i}}$ is equal to the sum of two vectors **B** and **C**. If the magnitude of $|\mathbf{B}|$ is 70 and the magnitude of $|\mathbf{C}|$ is 80, determine the x and y components of **B** and **C**.

2.39 Calculate the resultant force of the four forces acting on an airplane shown in Figure P2.39 as it lands.

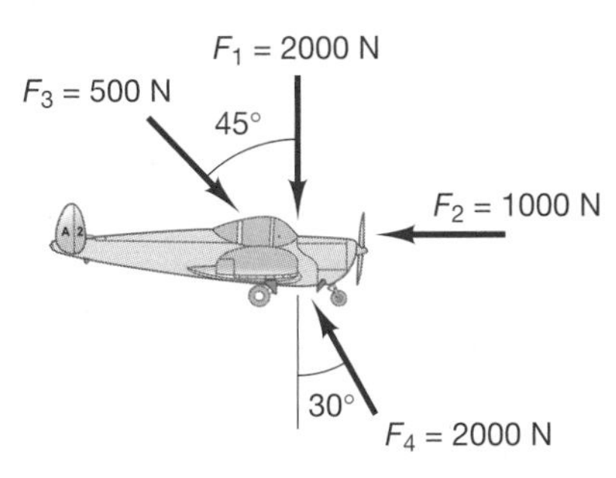

Figure P2.39

2.40 Three coplaner forces are applied to a ring. Calculate the angles θ and β so that the three forces add up to zero. (See Figure P2.40.)

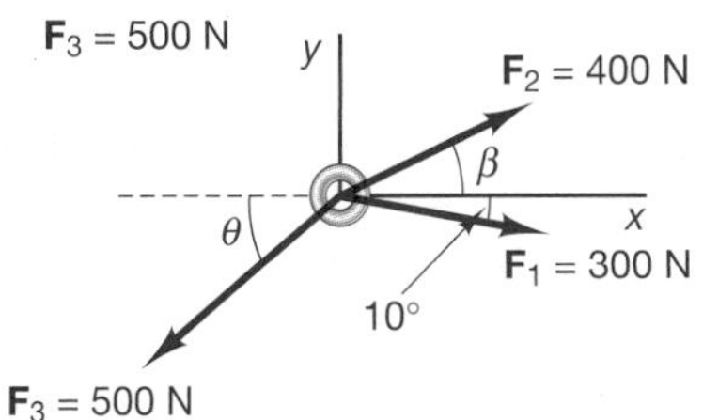

Figure P2.40

2.41 Solve the vector equation $\mathbf{A} + \mathbf{B} = \mathbf{D}$ for the vector $\mathbf{B}$, where $\mathbf{A} = 2\hat{\mathbf{i}} - 3\hat{\mathbf{j}} + 2\hat{\mathbf{k}}$ and $\mathbf{D} = \hat{\mathbf{i}} - \hat{\mathbf{j}}$. Also, compute the direction cosines of **B**.

2.42 Solve the vector equation $\mathbf{A} + \mathbf{B} = \mathbf{D}$ for the vector **A**, where $\mathbf{D} = \hat{\mathbf{i}} - \hat{\mathbf{j}} + \hat{\mathbf{k}}$ and $\mathbf{B} = 10\hat{\mathbf{i}} + 10\hat{\mathbf{j}} - 10\hat{\mathbf{k}}$. Also, compute the direction cosines of **A**.

2.43 Solve the vector equation $\mathbf{A} + \mathbf{B} + \mathbf{C} = \mathbf{D}$ for **B**, where $\mathbf{A} = 3\hat{\mathbf{i}} + 2\hat{\mathbf{j}}$, $\mathbf{C} = \hat{\mathbf{i}} + 15\hat{\mathbf{j}} + 3\hat{\mathbf{k}}$, and $\mathbf{D} = \hat{\mathbf{i}} + \hat{\mathbf{j}} + \hat{\mathbf{k}}$. Calculate the magnitude and the direction cosines of **B**.

2.44 Consider the two position vectors **A** and **B** shown in Figure P2.44. Calculate the sum $\mathbf{R} = \mathbf{A} + \mathbf{B}$ and determine its direction cosines.

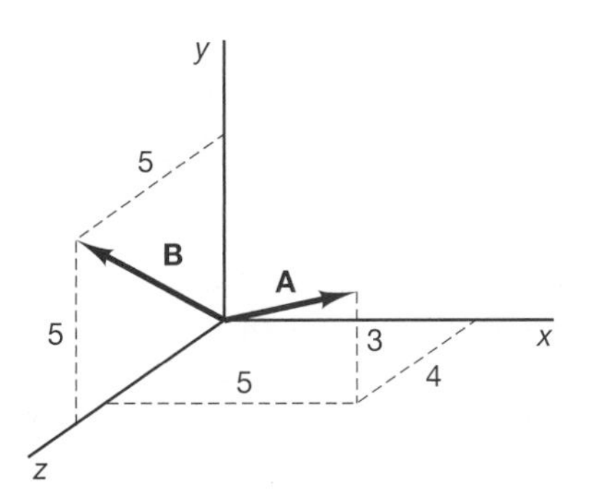

Figure P2.44

2.45 A force F of 300 N is applied at the origin along the line of action defined by the vector **A** in Problem 2.44. Calculate the components of the force in each coordinate direction.

2.46 A 500-N force is applied along the line of action determined by the resultant vector **R** made up of the sum of the two vectors **A** and **B** in Problem 2.44. Calculate the components of this vector along the coordinate directions x, y, and z, and the angle that this force makes with each coordinate.

2.47 A force $\boldsymbol{F}$ is applied to an eye by a rope used to hold down a circus tent. The angles made by the rope are measured, and the values are indicated in Figure P2.47. The magnitude of the force is to be a maximum of 1500 N. In choosing the bolt fixture, it is important to know the maximum value of force that will be exerted on the bolt in each of the three coordinate directions. Calculate the x-, y-, and z-components of $\boldsymbol{F}$, and write $\boldsymbol{F}$ in component form.

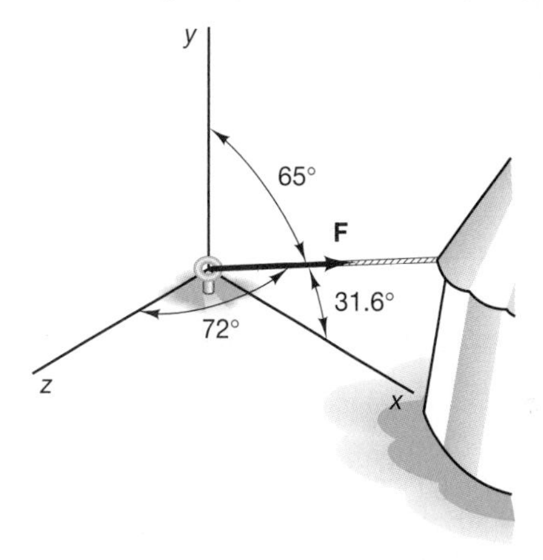

Figure P2.47

2.48 Three forces act along the directions illustrated by the given coordinates in Figure P2.48. Calculate the resultant of these three forces, as well as the angles made by the resultant and each of the three coordinate directions. Here, $F_1 = 100$ N, $F_2 = 50$ N, and $F_3 = 75$ N.

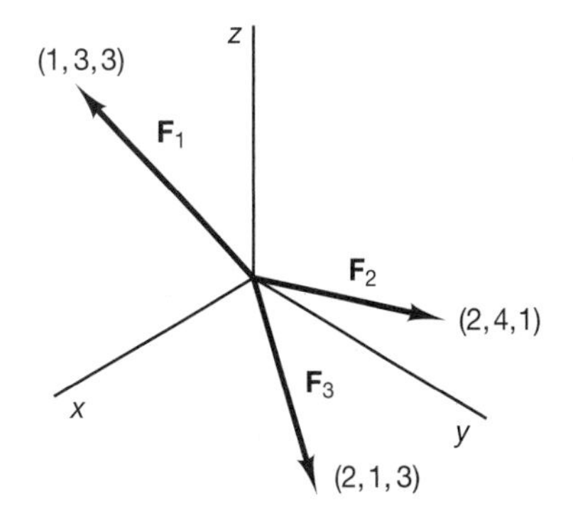

Figure P2.48

2.49 Determine the components of the 800-N (8.00×10^2 N) force, F in the x–y–z coordinate system. (See Figure P2.49.)

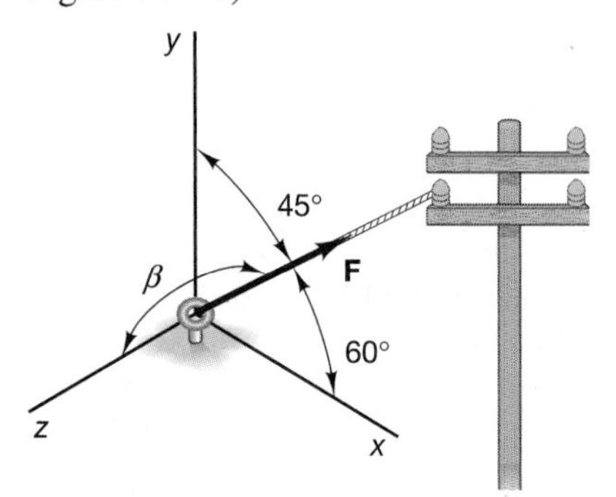

Figure P2.49

2.50 Calculate the components of the force F of magnitude 2000 N, that makes an angle of $\alpha = 40°$ with the x–z plane and whose component in the x–z plane makes an angle $\beta = 60°$ with the z-axis. (See Figure P2.50.)

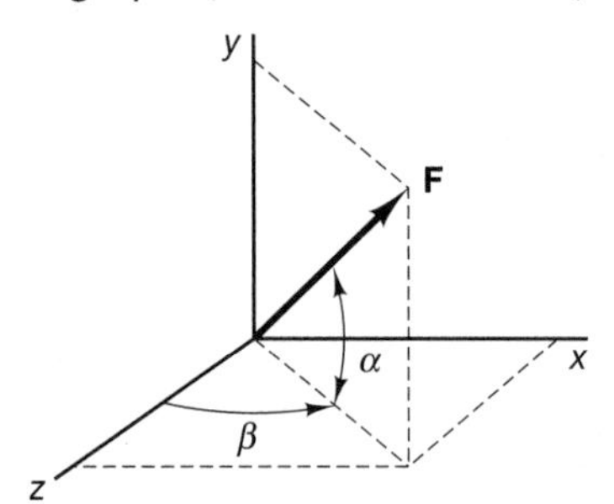

Figure P2.50

2.51 Calculate the components of the 1000-N force that forms an angle $\alpha = 50°$ with the x–y plane and an angle $\beta = 70°$ between the x–y component and the y-axis. (See Figure P2.51.)

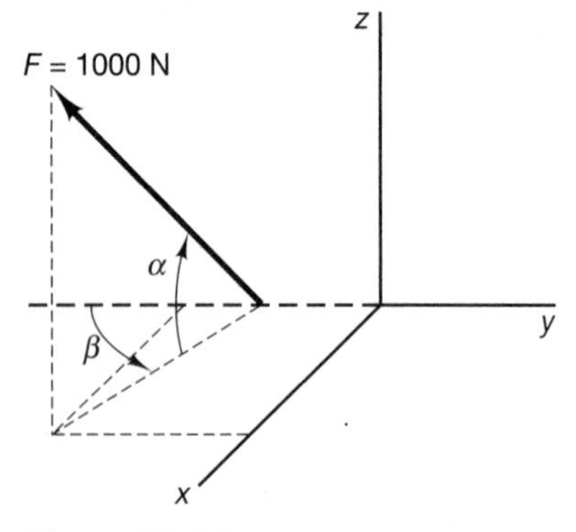

Figure P2.51

2.52 The force F of Problem 2.50 is given in component form as $\mathbf{F} = 250\hat{\mathbf{i}} + 230\hat{\mathbf{j}} + 125\hat{\mathbf{k}}$ N. Calculate the angle α between the vector $\mathbf{F}$ and the x–z plane and the angle β between the projection F_{xz} and the z-axis.

2.53 Consider the vector $\mathbf{P}$ equal to the sum of $\mathbf{A}$ and $\mathbf{B}$ making the angles indicated in Figure P2.53. The vector $\mathbf{P}$ is known to be $35\hat{\mathbf{i}} + 35\hat{\mathbf{j}}$ in newtons, $\beta = 15°$, and $\alpha = 20°$. Calculate the components $\mathbf{A}$ and $\mathbf{B}$.

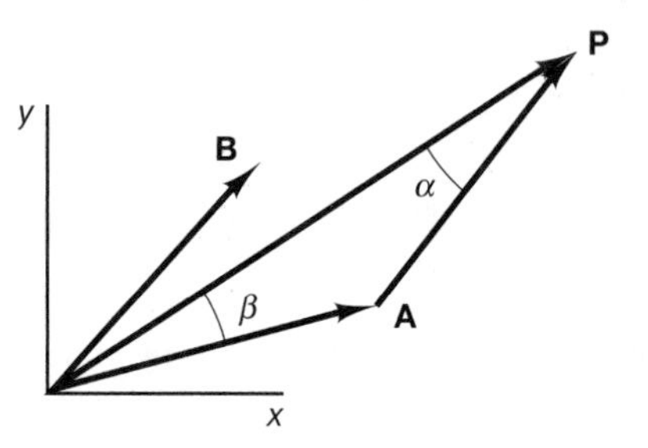

Figure P2.53

2.54 If the resultant of the tensions in cables A, B, and C is $\mathbf{R} = 250\hat{\mathbf{k}}$ (N) and $y_C = 6$ m, determine the tension in each cable.

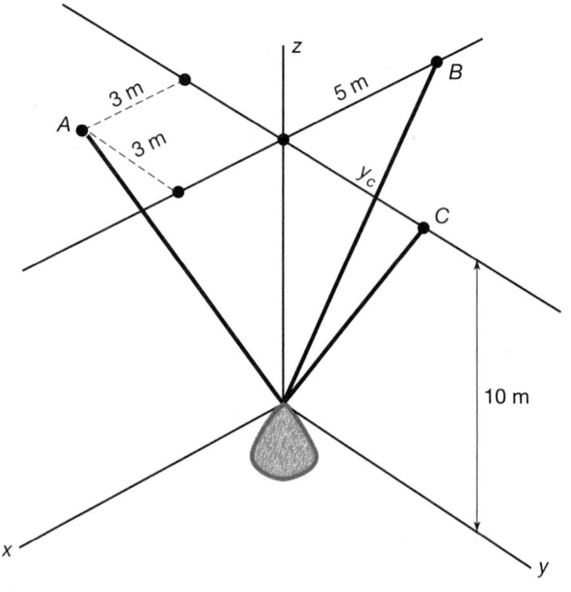

Figure P2.54

2.55 Determine the tensions in each cable if $y_C = 8$ m and $\mathbf{R} = 300\hat{\mathbf{k}}$ (N) in Problem 2.54.

2.56 In Problem 2.54, construct a parametric solution for the cable tensions as a function of y_C if the coordinates of point C are (0, y_C, 10) m. Examine the solution for $1 \leq y_C \leq 10$ m and determine the value of y_C where $T_B = T_C$.

2.57 The following vectors are written in matrix notation:

$$\mathbf{A} = \begin{pmatrix} 2 \\ -3 \\ 1 \end{pmatrix} \quad \mathbf{B} = \begin{pmatrix} 1 \\ 1 \\ 1 \end{pmatrix} \quad \mathbf{C} = \begin{pmatrix} 3 \\ 0 \\ -2 \end{pmatrix}$$

Determine the angles each vector makes with the x-, y-, and z-axes.

2.58 Two tugboats pull on a barge as illustrated in Figure P2.58. From the resulting motion of the barge, it is determined that the resultant force **P** is $15{,}000\hat{\mathbf{i}}$ N. It is desired to know what forces are transmitted through the lines by the tugs. The angles the tow lines make are indicated in the figure. Calculate the magnitude of the forces applied by tugs A and B.

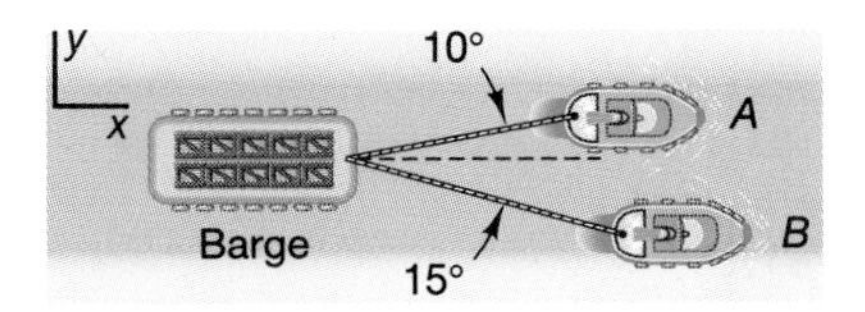

Figure P2.58

2.59 Two tow trucks attempt to pull a wrecked truck out of a ditch. (See Figure P2.59.) The drivers need the damaged truck to move along the path OP, but because of trees and other features of the terrain, they must position their trucks at the angles illustrated. The drivers know from experience (and the weight of the wrecked truck) that it will take about 20,000 N of force to move the truck out of the ditch. How much force must each tow truck apply so that the wrecked vehicle moves along OP?

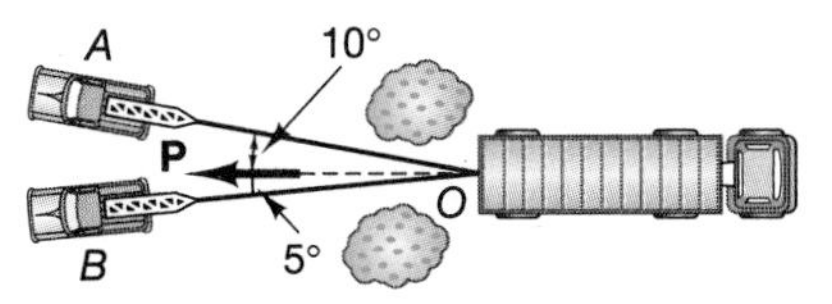

Figure P2.59

2.6 COMPONENTS OF A VECTOR IN NONORTHOGONAL DIRECTIONS

We showed in Section 2.2 that any vector can be resolved into two, and only two, noncollinear coplanar components. Frequently, it is desirable to represent a vector as the sum of vectors that are not orthogonal. These vectors are still called components, but they are not Cartesian orthogonal components.

Consider first the two-dimensional case, with a vector **P** expressed in nonorthogonal components **A** and **B** in a plane, as shown in Figure 2.41. We examined this problem in Section 2.2 using trigonometric methods; now we will solve it using a reference system of orthogonal unit base vectors.

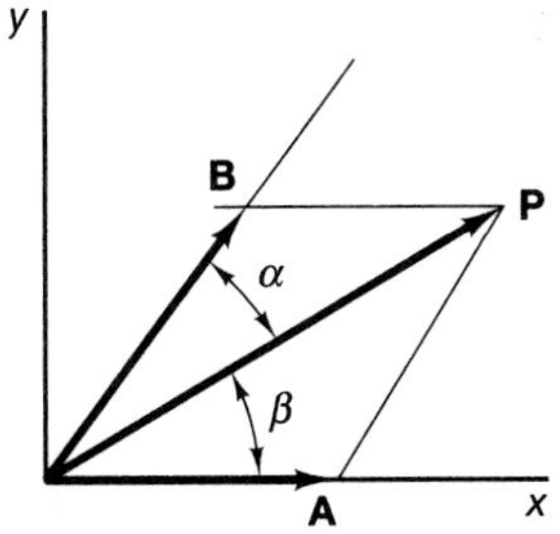

Figure 2.41

For the two-dimensional case, three types of problems arise:

1. One of the two components is known—for example, **A**. The second component, **B**, may be obtained by solving the vector equation

$$\mathbf{A} + \mathbf{B} = \mathbf{P} \tag{2.53}$$

which becomes

$$\mathbf{B} = \mathbf{P} - \mathbf{A} \tag{2.54}$$

This yields the desired component vector **B**. Eq. (2.54) represents two scalar equations involving six variables, namely, the two orthogonal Cartesian components of the three vectors. Thus, at least four of the variables must be known in order to solve the equation.

2. The two vectors **A** and **B** have known directions, but unknown magnitudes, and their sum is given as equal to **P**:

$$\mathbf{A} + \mathbf{B} = \mathbf{P} \tag{2.55}$$

That is, the directions of the three vectors are known, but only the magnitude of **P** is known, and we want to find the magnitudes of **A** and **B**. This, again, is a system of two equations in two unknowns.

The components of **P** in the nonorthogonal directions **A** and **B** may be obtained using Cartesian component notation. Unit vectors in the **A** and **B** directions may be written as

$$\begin{aligned}\hat{\mathbf{a}} &= \hat{\mathbf{i}} \\ \hat{\mathbf{b}} &= \cos(\alpha + \beta)\hat{\mathbf{i}} + \sin(\alpha + \beta)\hat{\mathbf{j}}\end{aligned} \tag{2.56}$$

If computational software is used, Eq. (2.56) is written in matrix notation as

$$\hat{\mathbf{a}} = \begin{pmatrix} 1 \\ 0 \\ 0 \end{pmatrix} \qquad \hat{\mathbf{b}} = \begin{pmatrix} \cos(\alpha + \beta) \\ \sin(\alpha + \beta) \\ 0 \end{pmatrix} \tag{2.56a}$$

The vector **P** is known and may be written in component notation as

$$\mathbf{P} = |\mathbf{P}| \cos \beta\hat{\mathbf{i}} + |\mathbf{P}| \sin \beta\hat{\mathbf{j}} \tag{2.57}$$

Written in matrix notation,

$$\mathbf{P} = |\mathbf{P}|\begin{pmatrix} \cos\beta \\ \sin\beta \\ 0 \end{pmatrix} \tag{2.57a}$$

If **A** and **B** are the components of **P**, then

$$\mathbf{P} = |\mathbf{P}| \cos \beta\hat{\mathbf{i}} + |\mathbf{P}| \sin \beta\hat{\mathbf{j}} = |\mathbf{A}|\hat{\mathbf{a}} + |\mathbf{B}|\hat{\mathbf{b}} \tag{2.58}$$

In matrix notation, we have

$$\mathbf{P} = \begin{pmatrix} |\mathbf{P}| \cos\beta \\ |\mathbf{P}| \sin\beta \\ 0 \end{pmatrix} = \begin{pmatrix} |\mathbf{A}| + |\mathbf{B}| \cos(\alpha + \beta) \\ |\mathbf{B}| \sin(\alpha + \beta) \\ 0 \end{pmatrix} \tag{2.58a}$$

The scalar equations are obtained by equating the components of the vectors, yielding

$$\begin{aligned}|\mathbf{P}| \cos\beta &= |\mathbf{A}| + |\mathbf{B}| \cos(\alpha + \beta) \\ |\mathbf{P}| \sin\beta &= |\mathbf{B}| \sin(\alpha + \beta)\end{aligned} \tag{2.59}$$

The two scalar equations may be solved for the magnitudes of the components in the **A** and **B** directions.

3. Another type of vector problem arises in some design situations in which the magnitude of one vector is unknown and the direction of another vector is unknown. Therefore, in the vector equation $\mathbf{A} + \mathbf{B} = \mathbf{C}$, $\hat{\mathbf{a}}$ and $|\mathbf{B}|$ and **C** are known, and $\hat{\mathbf{b}}$ and **A** are to be determined.

   A coordinate system should be established such that **A** lies along one of the coordinate axes, for example the *x*-axis. The *y*-axis will be perpendicular to the

x-axis and the known vector **C** can be expressed in that coordinate system. The vectors may be written as:

$$\begin{aligned}\mathbf{A} &= |\mathbf{A}|\hat{\mathbf{i}} \\ \mathbf{B} &= |\mathbf{B}|(\cos\lambda_x\hat{\mathbf{i}} + \cos\lambda_y\hat{\mathbf{j}}) \\ \mathbf{C} &= C_x\hat{\mathbf{i}} + C_y\hat{\mathbf{j}}\end{aligned} \tag{2.60}$$

Equating the x and y components yields two equations with three unknowns; λ_x, λ_y and $|\mathbf{A}|$. The sum of the squares of the direction cosines is one:

$$\lambda_x^2 + \lambda_y^2 = 1 \tag{2.61}$$

There will be two solutions to this equation and therefore to the problem

Sample Problem 2.9 If $\hat{\mathbf{a}}$ and $|\mathbf{B}|$ are known, determine $\hat{\mathbf{b}}$ and **A** such that $\mathbf{A} + \mathbf{B} = \mathbf{C}$.

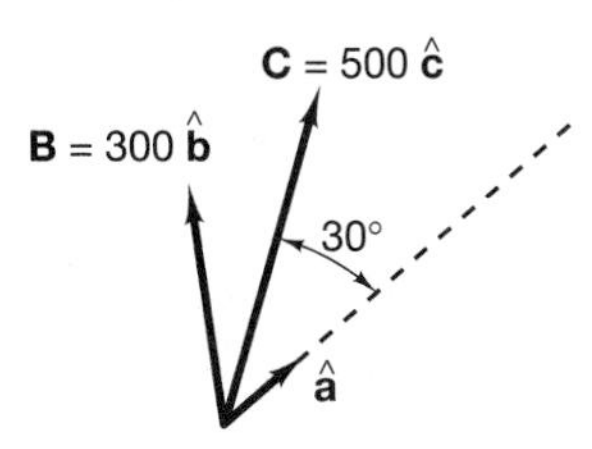

Solution Establish a coordinate system such that $\hat{\mathbf{i}} = \hat{\mathbf{a}}$

The vectors may be written as:

$$\begin{aligned}\mathbf{C} &= 500(\cos 30\hat{\mathbf{i}} + \sin 30\hat{\mathbf{j}}) \\ \mathbf{A} &= A\hat{\mathbf{i}} \\ \mathbf{B} &= 300(\lambda_x\hat{\mathbf{i}} + \lambda_y\hat{\mathbf{j}})\end{aligned}$$

Equating x and y components yields

$$\begin{aligned}&\hat{\mathbf{i}}: A + 300\lambda_x = 500\cos 30 \\ &\hat{\mathbf{j}}: 300\lambda_y = 500\sin 30\end{aligned}$$

A third equation is the property of the direction cosines.

$$\lambda_x^2 + \lambda_y^2 = 1$$

We have three equations for the three unknowns. Solving these equations, yields:

$$\begin{aligned}\lambda_y &= 0.834 \\ \lambda_x &= \sqrt{1 - \lambda_y^2} = \pm 0.553 \\ A_+ &= 267 \quad A_- = 599\end{aligned}$$

Note that there are two solutions. The two solutions are shown on the following pages.

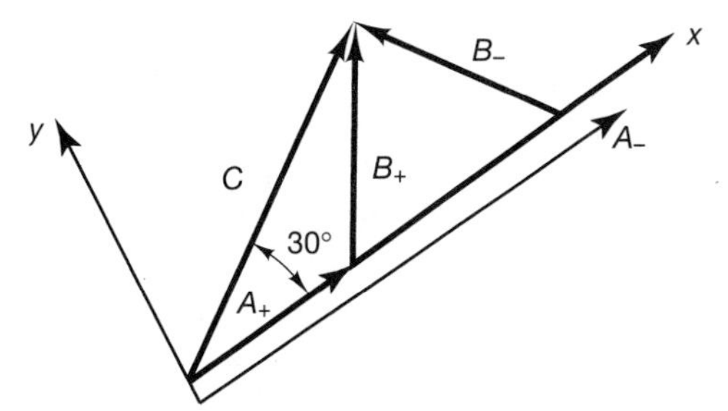

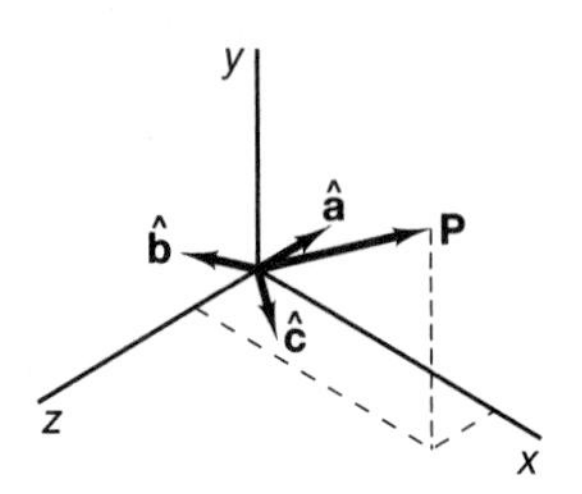

Figure 2.42

Three-dimensional nonorthogonal components are best approached by first expressing the vectors in Cartesian coordinates. Consider the vector **P** specified by its components P_x, P_y, and P_z. Three noncoplanar directions are specified by three unit vectors $\hat{\mathbf{a}}$, $\hat{\mathbf{b}}$, and $\hat{\mathbf{c}}$ as shown in Figure 2.42. The Cartesian components of these unit vectors are known. Written in component form, the vectors are

$$\begin{aligned}\hat{\mathbf{a}} &= a_x\hat{\mathbf{i}} + a_y\hat{\mathbf{j}} + a_z\hat{\mathbf{k}}\\ \hat{\mathbf{b}} &= b_x\hat{\mathbf{i}} + b_y\hat{\mathbf{j}} + b_z\hat{\mathbf{k}}\\ \hat{\mathbf{c}} &= c_x\hat{\mathbf{i}} + c_y\hat{\mathbf{j}} + c_z\hat{\mathbf{k}}\\ \mathbf{P} &= P_x\hat{\mathbf{i}} + P_y\hat{\mathbf{j}} + P_z\hat{\mathbf{k}}\end{aligned} \tag{2.62}$$

Again, these three equations may be expressed as one vector equation in the form

$$|\mathbf{A}|\hat{\mathbf{a}} + |\mathbf{B}|\hat{\mathbf{b}} + |\mathbf{C}|\hat{\mathbf{c}} = \mathbf{P} \tag{2.63}$$

The unknowns are the three magnitudes of the components—that is, $|\mathbf{A}|$, $|\mathbf{B}|$, and $|\mathbf{C}|$. Expanding the vector equation into equivalent scalar equations by equating the x-, y-, and z-components yields three equations for the three unknowns:

$$\begin{aligned}a_xA + b_xB + c_xC &= P_x\\ a_yA + b_yB + c_yC &= P_y\\ a_zA + b_zB + c_zC &= P_z\end{aligned} \tag{2.64}$$

The problem has now been reduced to the solution of a system of three linear equations. If $\hat{\mathbf{a}}$, $\hat{\mathbf{b}}$, and $\hat{\mathbf{c}}$ were coplanar, it would not be possible to express a vector **P** that had a component perpendicular to that plane in components A, B, and C. If **P** is coplanar with $\hat{\mathbf{a}}$, $\hat{\mathbf{b}}$, and $\hat{\mathbf{c}}$, only two non-collinear coplanar vectors would be needed.

As in the two-dimensional case, design problems arise where in Eq. (2.63) the vectors **B** and **P** are known but only the magnitude of **C** and the direction of **A** are known. These cases arise when there are design load limits on **C** but it can be placed in any direction. The unknowns are the magnitude of **A** and the unit vector $\hat{\mathbf{c}}$ The vector equation can be expressed as three scalar equations but there are four unknowns. The final equation is

$$c_x^2 + c_y^2 + c_z^2 = 1 \tag{2.65}$$

The choice of the coordinate system is left to the problem solver and problems of this type are easier if one of the coordinate axes is aligned with the unit vector in the **A** direction. If the x axis is aligned with the unit vector $\hat{\mathbf{a}}$, the three scalar equations become:

$$\begin{aligned}A + Cc_x &= P_x - Bb_x\\ Cc_y &= P_y - Bb_y\\ Cc_z &= P_z - Bb_z\end{aligned} \tag{2.66}$$

Therefore, the two components of $\hat{\mathbf{c}}$ are:

$$\begin{aligned}c_y &= \frac{P_y - Bb_y}{C}\\ c_z &= \frac{P_z - Bb_z}{C}\end{aligned} \tag{2.67}$$

The two values of the remaining component of $\hat{\mathbf{c}}$ are:

$$c_x = \pm\sqrt{1 - c_y^2 - c_z^2} \tag{2.68}$$

The two values of A may now be determined.

Sample Problem 2.10 A tripod supports a 10-kg movie camera. (See figure left.) Because this weight is transmitted down the legs of the tripod, the resultant of the three leg forces is 98.1 N downward. Determine the force on each leg of the tripod by considering the legs as nonorthogonal components of the resultant vertical force.

Solution We first draw the vector diagram of the problem, as shown. The resultant vector may be written as

$$\mathbf{R} = -98.1\hat{\mathbf{j}}$$

The problem is equivalent to resolving **R** into components in the directions of each leg. The unit vectors directed down each leg may be found from the spatial dimensions as follows:

Tripod

$$\hat{\mathbf{a}} = \frac{1}{\sqrt{60^2 + 100^2}}(-60\hat{\mathbf{i}} - 100\hat{\mathbf{j}}) = -0514\hat{\mathbf{i}} - 0.857\hat{\mathbf{j}}$$

$$\hat{\mathbf{b}} = \frac{1}{\sqrt{52.5^2 + 100^2 + 30^2}}(52.5\hat{\mathbf{i}} - 100\hat{\mathbf{j}} + 30\hat{\mathbf{k}}) = 0.449\hat{\mathbf{i}} - 0.856\hat{\mathbf{j}} + 0.257\hat{\mathbf{k}}$$

$$\hat{\mathbf{c}} = \frac{1}{\sqrt{52.5^2 + 100^2 + 30^2}}(52.5\hat{\mathbf{i}} - 100\hat{\mathbf{j}} - 30\hat{\mathbf{k}}) = 0.449\hat{\mathbf{i}} - 0.856\hat{\mathbf{j}} - 0.257\hat{\mathbf{k}}$$

To determine the magnitudes of the tripod leg forces, we solve the vector equation

$$F_A\hat{\mathbf{a}} + F_B\hat{\mathbf{b}} + F_C\hat{\mathbf{c}} = \mathbf{R}$$

Expanding this into scalar equations by equating the coefficients of the $\hat{\mathbf{i}}$, $\hat{\mathbf{j}}$, and $\hat{\mathbf{k}}$ unit vectors, respectively, yields

$$-0.514F_A + 0.449F_B + 0.449F_C = 0$$
$$-0.857F_A - 0.856F_B - 0.856F_C = -98.1$$
$$+0.257F_B - 0.257F_C = 0$$

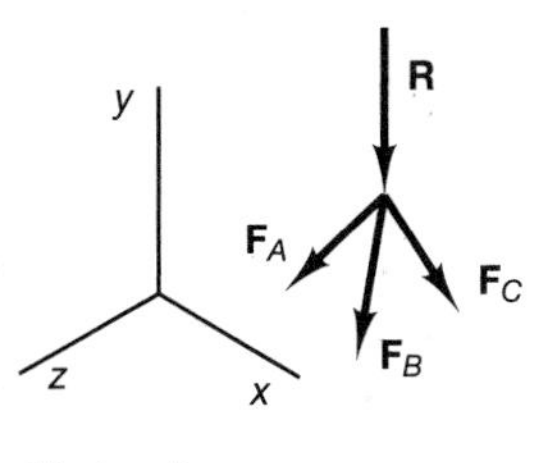

Vector diagram

This system of equations may be solved by algebraic means or by the use of a calculator or computer routine for linear systems of equations, as shown in Section 2.7. The results are

$$F_A = 53.4 \text{ N}$$
$$F_B = 30.6 \text{ N}$$
$$F_C = 30.6 \text{ N}$$

The compressive forces in the tripod's legs are not equal.

2.7 SYSTEMS OF LINEAR EQUATIONS

Most problems in statics result in a system of linear equations that must be solved to determine the unknown quantities. Modern computational methods are available to solve such systems. Consider the system

$$\begin{aligned} a_{11}x_1 + a_{12}x_2 + \ldots + a_{1n}x_n &= c_1 \\ a_{21}x_1 + a_{22}x_2 + \ldots + a_{2n}x_n &= c_2 \\ &\ldots \\ a_{m1}x_1 + a_{m2}x_2 + \ldots + a_{mn}x_n &= c_m \end{aligned} \quad (2.69)$$

If $m = n$, there are as many equations as unknowns, and the unknowns can usually be determined if the equations are linearly independent. If $m < n$, the system of equations is *underdetermined,* or *indeterminate,* and there are too few equations to determine the unknowns. If $m > n$, the system is called an *overdetermined* system of equations, and the unknowns are chosen to approximately satisfy the equations. The overdetermined system is solved using a least squares method. (See any introductory text on numerical analysis.)

Elementary methods for solving a system of linear equations with $m = n$ require the elimination of unknowns by adding or subtracting equations until an equation with only one unknown is obtained. That equation is then solved by ordinary algebra, and the value obtained is substituted back into the other equations to determine another unknown. This method is formalized as the ***Gauss–Jordan reduction*** see Appendix A. As the method is time consuming, not surprisingly, software routines have been developed to solve systems of linear equations. These routines use a special mathematical notation, called matrix notation, and involve rules for using matrices, forming a basis for matrix algebra. Matrix rules will be discussed later in this section; first we examine some other characteristics of systems of linear equations.

Consider the simple system consisting of the following two equations in two unknowns x and y:

$$\begin{aligned} 2.0x + 3.0y &= 10.0 \\ 1.0x - 1.0y &= 2.0 \end{aligned}$$

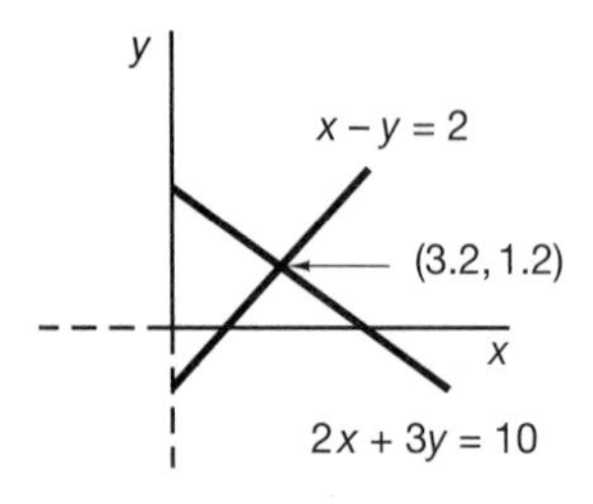

Figure 2.43 Graphical representation of a system of two simultaneous equations.

Multiplying the second equation by 3 and adding to the first yields

$$x = 3.2$$

Substituting back into the second equation yields

$$y = 1.2$$

Each equation represents a line in x–y space and may be drawn as shown in Figure 2.43. The solution of the system of equations is the intersection of the lines in x–y space, as shown in Figure 2.43. If these two equations were not *linearly independent,* the two lines would coincide, and there would have been no intersection. Occasionally, two equations are almost the same, and their two lines almost coincide, making the intersection difficult to discern. Mathematically, this system would then be called *ill conditioned,* and increased numerical accuracy would be required to determine the intersection point. An example of an ill-conditioned system is

$$\begin{aligned} 1.000x - 1.000y &= 1.000 \\ 1.999x - 2.001y &= 1.988 \end{aligned}$$

Multiplying the first equation by 1.999 and subtracting it from the second equation yields

$$-0.002y = -0.011$$

so that $x = 6.5$ and $y = 5.5$. Note that all four significant figures were necessary to obtain the solution, which is accurate to only two significant figures. A small variation in even the last digit of any of the coefficients would have made large differences in the answer. Hence, the equations are called ill-conditioned.

In formulating the solution to problems, errors are possible, leading to a system of equations that has no solution. Consider the two equations

$$\begin{aligned} x - y &= 1.00 \\ -2x + 2y &= 1.00 \end{aligned}$$

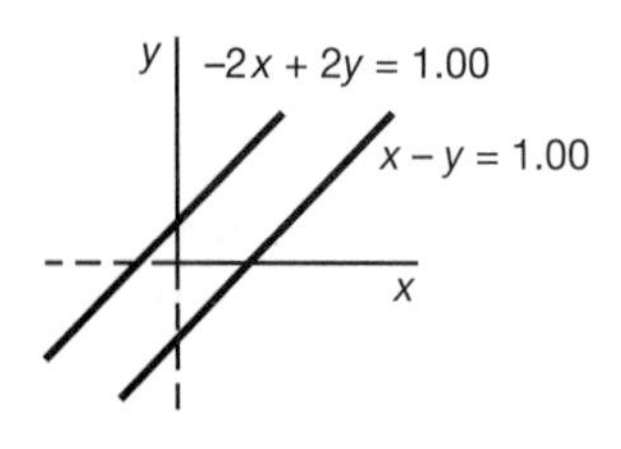

Figure 2.44 Graph of a system of two equations with no solution.

These equations would appear as the two lines in x–y space shown in Figure 2.44. These two equations form parallel lines and have no intersection. If we attempted an algebraic solution by multiplying the first equation by 2 and adding it to the second, the resulting reduced equation would be

$$0 = 3.00$$

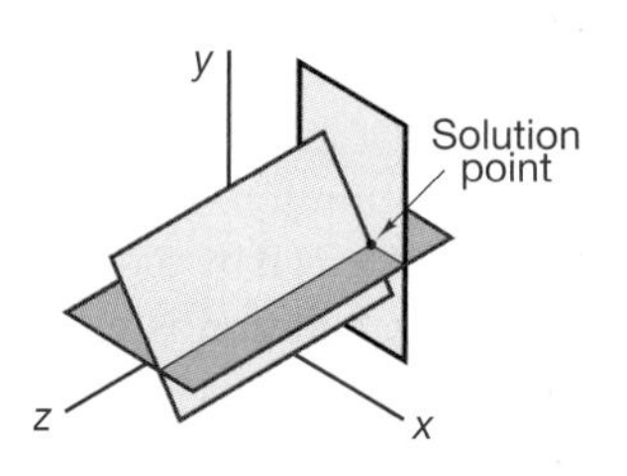

Figure 2.45 Graphical representation of a system of three simultaneous equations.

This system of equations is said to be *incompatible* or *inconsistent.* Usually, when a system of equations is incompatible, an error has been made in setting up the problem (mechanics principles used incorrectly) or in the mathematics leading to the system of equations.

If a system of linear equations consists of three equations in the three unknowns x, y, and z, each equation would represent a plane in x–y–z space. The solution to this system would be the intercept of the three planes, as shown in Figure 2.45. If any two planes almost coincide, the system is ill conditioned, and if two planes are parallel, the system is incompatible. Conceptually, these concepts can be extended into a multidimensional space for systems of equations greater than three, but graphs of such systems cannot be drawn.

2.7.1 MATRICES

The known coefficients a_{ij} in Eq. (2.69) may be written in an ordered rectangular array as

$$\begin{bmatrix} a_{11} & a_{12} & a_{13} & \dots & a_{1n} \\ a_{21} & a_{22} & a_{23} & \dots & a_{2n} \\ & & & \dots & \\ a_{m1} & a_{m2} & a_{m3} & \dots & a_{mn} \end{bmatrix} \tag{2.70}$$

The array consists of m rows and n columns. The notation a_{ij} means that this element of the array is in the ith row and the jth column. When suitable laws of equality, addition, subtraction, and multiplication are associated with such arrays, the arrays are called *matrices.* If the unknowns x_i and the constants on the right side of the system of equations (2.69) are written in column arrays, three matrices are formed:

$$[A] = \begin{bmatrix} a_{11} & a_{12} & a_{13} & \dots & a_{1n} \\ a_{21} & a_{22} & a_{23} & \dots & a_{2n} \\ & & & \dots & \\ a_{m1} & a_{m2} & a_{m3} & \dots & a_{mn} \end{bmatrix} \tag{2.71}$$

$$[x] = \begin{bmatrix} x_1 \\ x_2 \\ \dots \\ x_m \end{bmatrix} \tag{2.72}$$

$$[C] = \begin{bmatrix} c_1 \\ c_2 \\ \dots \\ c_m \end{bmatrix} \tag{2.73}$$

The system of equations (2.69) may now be written in matrix notation as

$$[A]\,[x] = [C] \tag{2.74}$$

Equation (2.74) implies multiplication between $[A]$ and $[x]$ equaling $[C]$. Before discussing matrix multiplication, we must define the equality of two matrices. If matrix $[A]$ is an $(m \times n)$ array consisting of m rows and n columns, with elements a_{ij} (ith row and jth column), then matrix $[\boldsymbol{B}]$ is equal to $[\boldsymbol{A}]$ if and only if

$$b_{ij} = a_{ij} \tag{2.75}$$

Note particularly that $[A]$ and $[B]$ must be the same size $(m \times n)$. We define the sum of two matrices $[A]$ and $[B]$ (that is, $[C] = [A] + [B]$), if $[A]$ and $[B]$ are of the same size $(m \times n)$, by

$$c_{ij} = a_{ij} + b_{ij} \tag{2.76}$$

for each value of i from 1 to m and each value of j from 1 to n. This definition of matrix addition simply creates a new matrix $[C]$ of the same size with elements formed by the sum of the corresponding elements of the two matrices $[A]$ and $[B]$. Matrix addition is commutative; that is,

$$[A] + [B] = [B] + [A] \tag{2.77}$$

The ith equation in the system of linear equations (2.69) can be written as

$$\sum_{k=1}^{n} a_{ik}x_k = c_i \qquad (i = 1,2, \ldots m) \tag{2.78}$$

Since this is equivalent to the matrix equation (2.68), matrix multiplication is defined as

$$[A][x] = [a_{ik}][x_k] = \left[\sum_{k=1}^{n} a_{ik}x_k\right] \tag{2.79}$$

Note that there are n coefficients in each row, corresponding to the n unknowns. This implies that in Eq. (2.74), $[A]$ must have *n columns* and $[x]$ must have *n rows*. The formal definition of matrix multiplication is

$$\begin{matrix} [A] & [B] & = [C] \\ (m \times k) & (k \times n) & = (m \times n) \end{matrix} \tag{2.80}$$

where the elements of $[C]$ are

$$c_{ij} = \sum_{k} a_{ik}b_{kj} \tag{2.81}$$

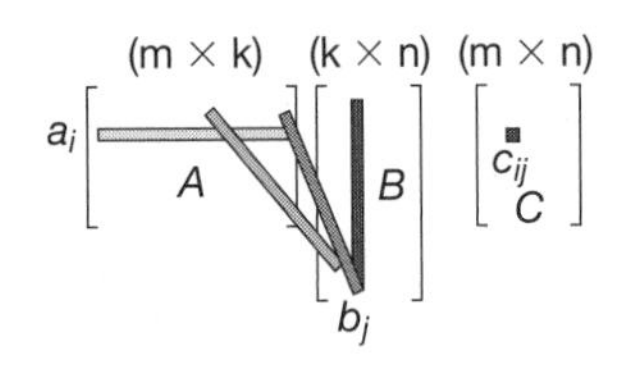

Figure 2.46 In matrix multiplication, each element c_{ij} of the product matrix $[C]$ is formed by multiplying each element a_{ik} in the ith row of $[A]$ by the corresponding element b_{kj} in the jth column of matrix $[B]$ and summing the element products. It is as if you folded the ith row of $[A]$ onto the jth column of $[B]$ and added the results.

The product $[C]$ is an $(m \times n)$ matrix having the same number of rows as $[A]$ and the same number of columns as $[B]$. The number of columns of $[A]$ must equal the number of rows of $[B]$ for the matrix multiplication to be defined. The value of an element of $[C]$ is obtained by "folding" the row of $[A]$ onto the column of $[B]$ and summing the products of corresponding elements, as diagrammed in Figure 2.46. Matrix multiplication is not commutative; that is,

$$[A][B] \neq [B][A] \tag{2.82}$$

Furthermore, if the number of columns of $[B]$ does not equal the number of rows of $[A]$, the product $[B][A]$ is not even defined. Even if both the matrices are square matrices (having an equal number of rows and columns), matrix multiplication is generally not commutative. However, matrix multiplication is associative; so that

$$[A]\,([B][C]) = ([A][B])\,[C] \tag{2.83}$$

and distributive, so that

$$\begin{aligned} [A]\,([B] + [C]) &= [A][B] + [A][C] \\ ([B] + [C])\,[A] &= [B][A] + [C][A] \end{aligned} \tag{2.84}$$

Multiplication by the ***inverse*** or reciprocal of a square nonsingular matrix is equivalent to division by a matrix and is fundamental to the solution of systems of linear equations. A singular matrix is one whose ***determinant*** is equal to zero. The inverse of the matrix $[A]$ is written as

$$\text{Inverse } [A] = [A]^{-1} \tag{2.85}$$

A singular matrix does not possess an inverse, and a nonsingular matrix can have only one unique inverse. (See Appendix.) If the determinant of $[A]$ is equal to zero, the system of equations $[A][X] = [C]$ is not linearly independent and cannot be solved. If the determinant of $[A]$ is not zero, the system of equations is linearly independent and a solution exists.

The special square matrix equivalent to unity is called a ***unit*** or ***identity matrix*** and has diagonal elements equal to unity and all off-diagonal elements equal to zero. The (3×3) identity matrix is

$$[I] = \begin{bmatrix} 1 & 0 & 0 \\ 0 & 1 & 0 \\ 0 & 0 & 1 \end{bmatrix} \tag{2.86}$$

Any matrix multiplied by the unit matrix is equal to itself:

$$[A][I] = [I][A] = [A] \tag{2.87}$$

The product of a matrix with its inverse is equal to the identity matrix:

$$[A][A]^{-1} = [A]^{-1}[A] = [I] \tag{2.88}$$

Modern computational software uses the concept of a matrix inverse to solve systems of linear equations. All the computational software programs have the capability to solve systems of equations using matrix techniques.

Formally, the solution of the system of linear equations

$$[A][x] = [C] \tag{2.89}$$

would be accomplished by multiplying both sides of the equation by the inverse of $[A]$ as follows:

$$[A]^{-1}[A][x] = [A]^{-1}[C] \tag{2.90}$$

The product of a matrix with its inverse is equal to the unit matrix, and consequently, the product of the unit matrix with another matrix is equal to that matrix. Therefore, Eq. (2.90) becomes

$$[x] = [A]^{-1}[C] \tag{2.91}$$

Most problems in statics reduce to the solution of a series of simultaneous equations, and although a little effort is required to learn the language, you will find that the use of computer programs will greatly reduce the time spent doing the calculations. In the past, students solved these equations by hand—that is, by adding or subtracting the equations in such a manner that the number of unknowns could be reduced. It is important to know how to solve these equations by hand, but as you progress in your education, you should use the available modern computing tools. At different points in the text, equations will be solved using computer software such as MATLAB, Maple, Mathematica, or Mathcad, the details of the solutions are shown in the computational supplements.

Sample Problem 2.11

Solve the following system of three equations by (a) a type of Gauss–Jordan reduction and (b) matrix inversion:

$$\begin{aligned} 2x - 3y + z &= 100 \\ 2y - z &= -50 \\ 3x + y + z &= 35 \end{aligned}$$

Solution **(a)** Add the first and second equations to eliminate z, yielding

$$2x - y = 50$$

Add the second and third equations to eliminate z again:

$$3x + 3y = -15$$

Now we have two equations involving the two unknowns x and y. Multiply the first of the reduced equations by 3, and add it to the second reduced equation, yielding

$$9x = 135 \qquad \Rightarrow\ x = 15$$

Substituting this into the first reduced equation yields

$$30 - y = 50 \qquad \Rightarrow\ y = -20$$

Substituting the values for x and y into any of the original equations yields

$$z = 10$$

(b) The system can be written in matrix notation as

$$[C]\,[X] = [R]$$

Where

$$[C] = \begin{bmatrix} 2 & -3 & 1 \\ 0 & 2 & -1 \\ 3 & 1 & 1 \end{bmatrix} \quad [X] = \begin{bmatrix} x \\ y \\ z \end{bmatrix} \quad [R] = \begin{bmatrix} 100 \\ -50 \\ 35 \end{bmatrix}$$

The values for the $[X]$ matrix can be found by matrix inversion:

$$[X] = [C]^{-1}\,[R] \rightarrow \begin{bmatrix} x \\ y \\ z \end{bmatrix} = \begin{bmatrix} 15 \\ -20 \\ 10 \end{bmatrix}$$

The actual matrix inversion is obtained by using a calculator or computational software.

Sample Problem 2.12

For the two matrices

$$[A] = \begin{bmatrix} 2 & -3 & 1 \\ 0 & 2 & -1 \\ 3 & 1 & 1 \end{bmatrix} \quad [B] = \begin{bmatrix} 5 & 3 & 2 \\ -2 & 1 & 4 \\ -1 & 0 & -1 \end{bmatrix}$$

determine: (a) $[A] + [B]$
(b) $[A]\,[B]$
(c) $[B]\,[A]$

Solution Using a calculator or computational software; we obtain the following:

$$\mathbf{A} = \begin{pmatrix} 2 & -3 & 1 \\ 0 & 2 & -1 \\ 3 & 1 & 1 \end{pmatrix} \qquad \mathbf{B} = \begin{pmatrix} 5 & 3 & 2 \\ -2 & 1 & 4 \\ -1 & 0 & -1 \end{pmatrix}$$

$$\text{(a)}\ \mathbf{A} + \mathbf{B} = \begin{pmatrix} 7 & 0 & 3 \\ -2 & 3 & 3 \\ 2 & 1 & 0 \end{pmatrix}$$

$$\text{(b)}\quad \mathbf{A}\cdot\mathbf{B} = \begin{pmatrix} 15 & 3 & -9 \\ -3 & 2 & 9 \\ 12 & 10 & 9 \end{pmatrix}$$

$$\text{(c)}\quad \mathbf{B}\cdot\mathbf{A} = \begin{pmatrix} 16 & -7 & 4 \\ 8 & 12 & 1 \\ -5 & 2 & -2 \end{pmatrix}$$

Problems

2.60 A vector $\mathbf{R} = 100\hat{\mathbf{i}} + 50\hat{\mathbf{j}}$ is resolved into components $\mathbf{A} = 30\hat{\mathbf{i}} + 60\hat{\mathbf{k}}$, $\mathbf{B} = 50\hat{\mathbf{j}} - 50\hat{\mathbf{k}}$ and $\mathbf{C}$, Determine $\mathbf{C}$.

2.61 A Vector $\mathbf{R} = 30\hat{\mathbf{i}} - 40\hat{\mathbf{j}} + 60\hat{\mathbf{k}}$ is resolved into components along lines defined by

$$\mathbf{a} = \hat{\mathbf{i}} + 2\hat{\mathbf{j}} - \hat{\mathbf{k}}$$
$$\mathbf{b} = 2\hat{\mathbf{i}} - \hat{\mathbf{j}} + \hat{\mathbf{k}}$$
$$\mathbf{c} = -\hat{\mathbf{i}} + \hat{\mathbf{j}} + 2\hat{\mathbf{k}}$$

Determine the magnitudes of the three components.

2.62 It is known that $\mathbf{R} = -\hat{\mathbf{i}} + 3\hat{\mathbf{j}}$, as viewed in the x–y coordinate system illustrated in Figure P2.62. Use orthogonal base vectors along x and y to compute the components of $\mathbf{R}$ along the two indicated directions u and v.

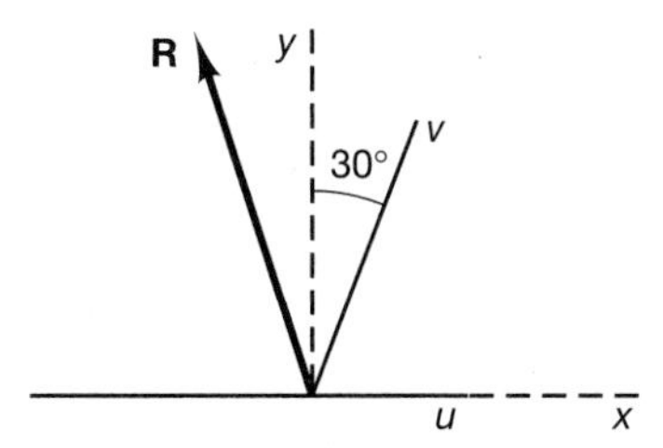

Figure P2.62

2.63 Resolve the vector $\mathbf{R} = 100\hat{\mathbf{i}}$ into components in the directions of the three unit vectors

$$\hat{\mathbf{a}} = 0.707\hat{\mathbf{i}} + 0.707\hat{\mathbf{j}}$$
$$\hat{\mathbf{b}} = -0.612\hat{\mathbf{i}} + 0.612\hat{\mathbf{j}} + 0.5\hat{\mathbf{k}}$$
$$\hat{\mathbf{c}} = 0.353\hat{\mathbf{i}} + 0.353\hat{\mathbf{j}} + 0.866\hat{\mathbf{k}}$$

2.64 Calculate the components of the 200-N force (applied along a line $3\hat{\mathbf{i}} + 2\hat{\mathbf{j}} + 3\hat{\mathbf{k}}$) along the three ropes $\mathbf{A}$, $\mathbf{B}$, and $\mathbf{C}$ acting along lines (See Figure P2.64):

$$\mathbf{A} = \hat{\mathbf{i}} + \hat{\mathbf{j}}$$
$$\mathbf{B} = \hat{\mathbf{j}} + \hat{\mathbf{k}}$$
$$\mathbf{C} = \hat{\mathbf{i}} + \hat{\mathbf{k}}$$

Note the vectors designating these directions are not unit vectors.

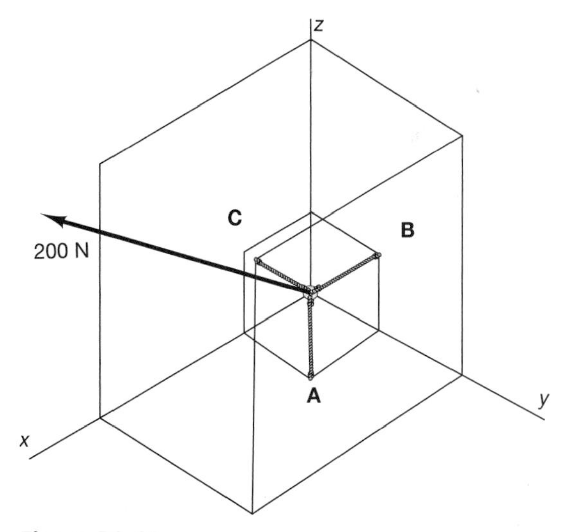

Figure P2.64

2.65 Find the components of the 100-N force vector $\mathbf{P}$ having direction cosines $\lambda_x = 0.2617$, $\lambda_y = -0.2094$, and $\lambda_z = 0.9422$ along the three vectors given by $\mathbf{A} = \hat{\mathbf{i}} + \hat{\mathbf{j}} + \hat{\mathbf{k}}$, $\mathbf{B} = -\hat{\mathbf{j}} - \hat{\mathbf{k}}$, and $\mathbf{C} = \hat{\mathbf{i}} - \hat{\mathbf{k}}$. (Hint: Note that $\mathbf{A}$, $\mathbf{B}$, and $\mathbf{C}$ are not unit vectors. To resolve $\mathbf{P}$ into components along these lines, you must use unit vectors.)

2.66 Determine the components of the vector $\mathbf{R} = 50\hat{\mathbf{i}} + 50\hat{\mathbf{j}} + 50\hat{\mathbf{k}}$ along lines

$$\mathbf{A} = \hat{\mathbf{i}} + 3\hat{\mathbf{j}} + \hat{\mathbf{k}}$$
$$\mathbf{B} = 5\hat{\mathbf{i}} + 5\hat{\mathbf{j}} + 2\hat{\mathbf{k}}$$
$$\mathbf{C} = \hat{\mathbf{i}} + 2\hat{\mathbf{j}} + 3\hat{\mathbf{k}}$$

2.67 The 500-N force vector **P** in Figure P2.67 is resolved into components such that: **P** = **A** + **B** + **C** where the magnitude of **A** is unknown, the magnitude of **B** is 300 N in the direction shown and the magnitude of **C** is 400 N in an unknown direction. Determine the magnitude of **A** and the unit vector $\hat{\mathbf{c}}$

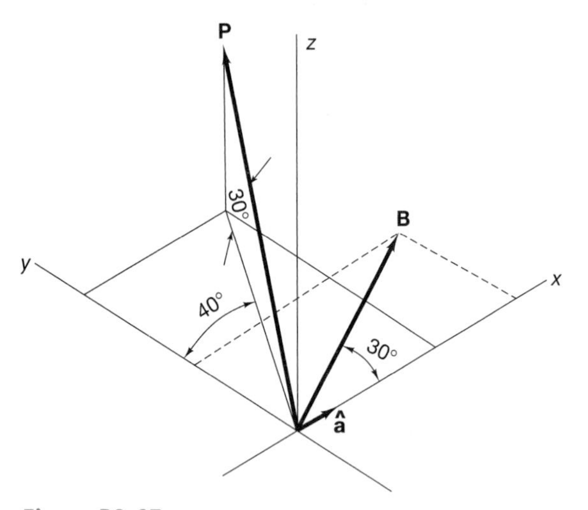

Figure P2.67

2.68 Solve the following set of linear equations by a type of Gauss–Jordan reduction:

$$3F_A + 6F_B - 2F_C = 1000$$
$$-F_A + 2F_B = 500$$
$$2.5F_A - F_B + 3F_C = -200$$

2.69 Solve the system of linear equations in Problem 2.68 using matrix inversion. Examine the determinant to show that the system is nonsingular.

2.70 Show that the following equations are linearly dependent:

$$3x + 6y - 2z = 20$$
$$-2x + 2z = -30$$
$$x + 6y = -10$$

2.71 Solve the following system of equations:

$$T_1 - T_2 + 3T_3 - 2T_4 = 100$$
$$6T_2 - 2T_3 + T_4 - T_5 = -60$$
$$-2T_1 + 5T_2 + T_3 + 6T_5 = 200$$
$$-T_1 - 4T_3 - 3T_4 + T_5 = -40$$
$$-T_1 - 3T_2 + 2T_3 + T_4 - 3T_5 = 150$$

2.72 For the matrices $[A] = \begin{bmatrix} 4 & -2 & 0 \\ -1 & 3 & 2 \\ 1 & 0 & 2 \end{bmatrix}$ and $[B] = \begin{bmatrix} 1 & 3 & 1 \\ 0 & 2 & -2 \\ -3 & -1 & 2 \end{bmatrix}$, determine (a) $[A] + [B]$ and (b) $[B] - [A]$.

2.73 For the matrices in Problem 2.72, determine (a) $[A]\,[B]$ and (b) $[B]\,[A]$.

2.74 Find the determinant of $[A]$ and $[B]$ in Problem 2.72.

2.75 Find the inverse of the matrices in Problem 2.72.

2.76 Solve the following system of linear equations:

$$F_1 - F_2 + 2F_3 - F_4 = 100$$
$$3F_1 - 2F_2 - F_3 + 3F_4 = -250$$
$$-2F_1 + 4F_2 - 5F_4 = 0$$
$$-F_1 + 6F_2 + F_3 + F_4 = 50$$

2.77 Determine the difference in the solution of the system of equations in Problem 2.76 if the right side of the second equation is +250 instead of −250. Note how sensitive the solution is to a change in sign.

2.78 Determine the difference in the solution of the system of equations in Problem 2.76 if the coefficient of F_3 in the fourth equation is −1 instead of +1. Note the sensitivity of the system of equations to the change in sign.

2.8 SCALAR PRODUCT OF TWO VECTORS

The need to compute certain physical quantities requires the development of special mathematical operations. One form of vector multiplication is required to calculate the work done by a force, and a different form of vector multiplication is required to determine the turning effect of a force. These two types of vector multiplication arise in mechanics and need to be defined. We will examine each one independently and show applications of each. The two multiplications are named for the type of mathematical quantity that results from the multiplication. The first type of multiplication of two vectors results in a scalar and is called the ***scalar product***. The second type of multiplication results in another vector, and is called a ***vector product***. The vector product will be discussed in detail in Section 2.9. It is interesting to note that the inverse of multiplication, division between two vectors, is not defined.

The scalar product is represented symbolically by placing a "dot" between the two vectors. For this reason, the scalar product is also called the ***dot product***. The symbolic notation for the scalar product between **A** and **B** is

$$\mathbf{A} \cdot \mathbf{B} \tag{2.92}$$

The terms "scalar product" and "dot product" will be used interchangeably in this text.

Consider the two vectors in Figure 2.47, which have been shown with their origins coinciding. The angle θ is the smaller of the two possible nonnegative angles between the two vectors, $0 \leq \theta \leq 180°$. The dot product of **A** and **B** is defined to be equal to the product of the magnitude of **A**, the magnitude of **B**, and the cosine of the angle between them:

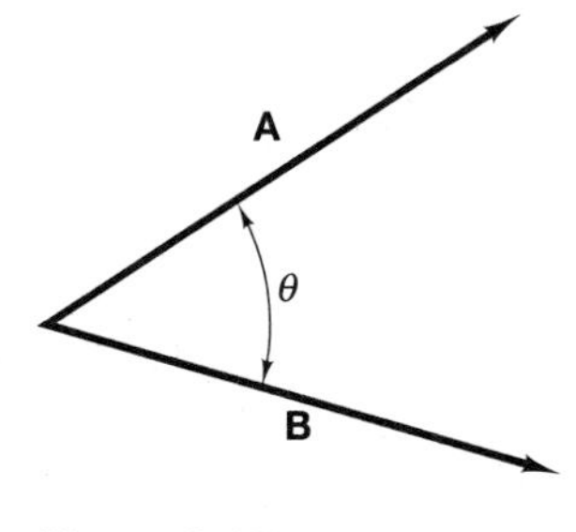

Figure 2.47

$$\mathbf{A} \cdot \mathbf{B} = |\mathbf{A}|\,|\mathbf{B}| \cos\theta \tag{2.93}$$

This multiplication may be thought of conceptually as a projection of one vector onto the other, followed by the multiplication of the magnitude of this projection by the magnitude of the vector upon which it is projected. (See Figure 2.48) For example, the dot product projects **A** upon **B** or **B** upon **A** and may be considered as the contribution of **A** in the **B** direction or vice versa. If **A** and **B** are perpendicular, their dot product is zero.

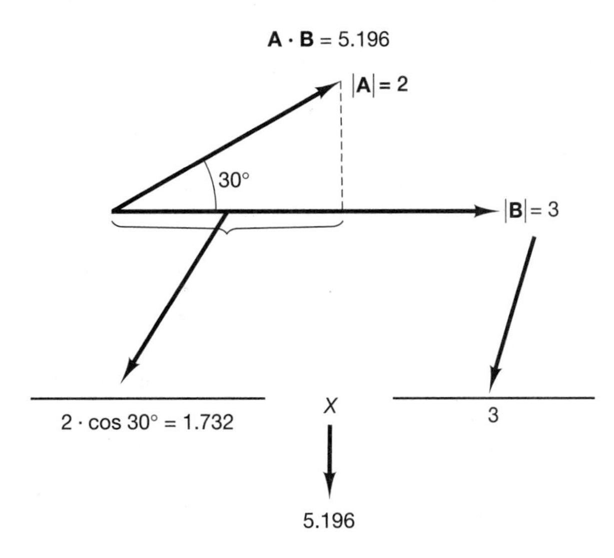

Figure 2.48

If two vectors are written in terms of orthogonal base vectors, the scalar product will involve these base vectors. Since the base vectors are unit vectors and perpendicular to each other, the scalar products between the base vectors are

$$\begin{aligned} \hat{\mathbf{i}} \cdot \hat{\mathbf{i}} &= 1 \quad \hat{\mathbf{j}} \cdot \hat{\mathbf{j}} = 1 \quad \hat{\mathbf{k}} \cdot \hat{\mathbf{k}} = 1 \\ \hat{\mathbf{i}} \cdot \hat{\mathbf{j}} &= 0 \quad \hat{\mathbf{i}} \cdot \hat{\mathbf{k}} = 0 \quad \hat{\mathbf{j}} \cdot \hat{\mathbf{k}} = 0 \end{aligned} \tag{2.94}$$

From the definition, it can be shown that the dot product is commutative and distributive:

$$\begin{aligned} \mathbf{A}\cdot\mathbf{B} &= \mathbf{B}\cdot\mathbf{A} = \alpha \\ \mathbf{A}\cdot(\mathbf{B}+\mathbf{C}) &= \mathbf{A}\cdot\mathbf{B} + \mathbf{A}\cdot\mathbf{C} \end{aligned} \tag{2.95}$$

The dot product between two vectors may now be written using the base unit vectors and the distributive nature of the dot product:

$$\begin{aligned} \mathbf{A}\cdot\mathbf{B} &= (A_x\hat{\mathbf{i}} + A_y\hat{\mathbf{j}} + A_z\hat{\mathbf{k}})\cdot(B_x\hat{\mathbf{i}} + B_y\hat{\mathbf{j}} + B_z\hat{\mathbf{k}}) \\ \mathbf{A}\cdot\mathbf{B} &= A_xB_x + A_yB_y + A_zB_z \end{aligned} \tag{2.96}$$

If the scalar product is taken between a vector **A** and a unit base vector, the projection of **A** upon that coordinate axis or the component of **A** in Cartesian coordinates is obtained:

$$\begin{aligned} \mathbf{A}\cdot\hat{\mathbf{i}} &= A_x \\ \mathbf{A}\cdot\hat{\mathbf{i}} &= |\mathbf{A}|\cos\theta_x \end{aligned} \tag{2.97}$$

where the term θ_x is the cosine of the angle between the x-axis and the vector **A** or the direction cosine.

Note that the dot product of any vector with itself is equal to the square of the vector's magnitude:

$$\mathbf{A}\cdot\mathbf{A} = |\mathbf{A}|^2 \tag{2.98}$$

2.8.1 APPLICATIONS OF THE SCALAR PRODUCT

Two important applications of the dot product will be examined fully. First, *the dot product may be used to determine the angle between any two vectors or intersecting lines.* Consider the two vectors **A** and **B** shown in Figure 2.47. The angle θ between the vectors is

$$\theta = \cos^{-1}\left(\frac{\mathbf{A}\cdot\mathbf{B}}{|\mathbf{A}|\,|\mathbf{B}|}\right) \qquad 0 \leq \theta \leq 180^\circ \tag{2.99}$$

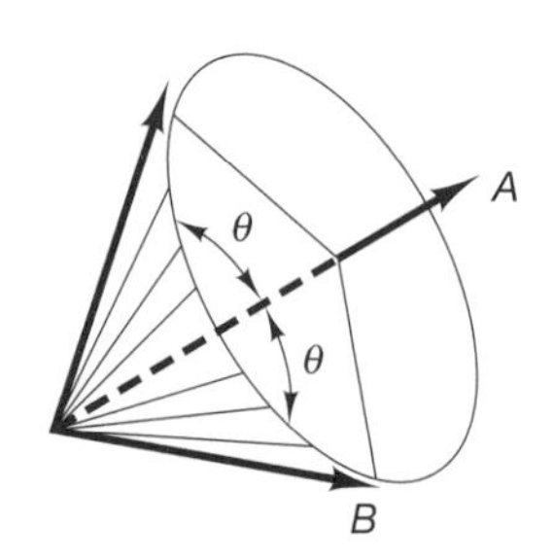

Figure 2.49 The dot product can be used to determine the angle θ between two vectors **A** and **B**, but not their orientation in space.

Care must be taken to understand that the dot product may be used to determine the angle θ, but not the plane in which **A** and **B** lie. That is, **B** may lie anywhere on a cone with cone angle θ relative to **A**, as shown in Figure 2.49. The sign of the angle θ will always be positive, since $\cos\theta = \cos(-\theta)$.

The second application of the dot product is *to resolve a vector into components parallel and perpendicular to a line in space.* To resolve a vector into the perpendicular and parallel components to a line, a unit vector must be determined along the line. Consider the vector and line shown in Figure 2.50, where $\hat{\mathbf{u}}$ is a unit vector along that line. The component parallel to the line a is found by the projection of the vector **A** onto the line. The parallel component may be written as

$$\mathbf{A}_{\parallel} = (\mathbf{A}\cdot\hat{\mathbf{u}})\hat{\mathbf{u}} = |\mathbf{A}|\cos\theta\,\hat{\mathbf{u}} \tag{2.100}$$

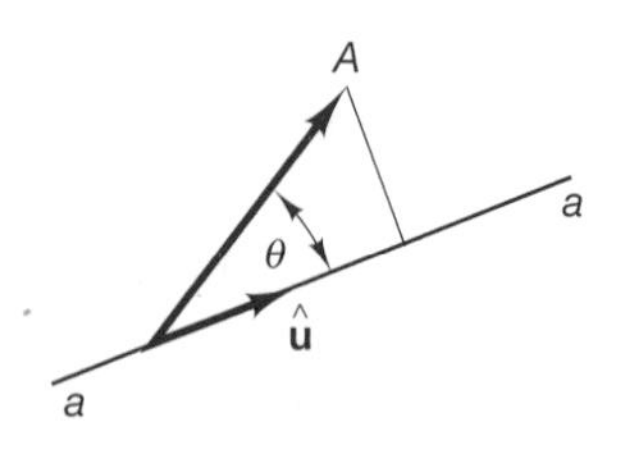

Figure 2.50

The component perpendicular to the line may be found by using the fact that the sum of the components must always add to the original vector:

$$\begin{aligned} \mathbf{A} &= \mathbf{A}_{\parallel} + \mathbf{A}_{\perp} \\ \mathbf{A}_{\perp} &= \mathbf{A} - \mathbf{A}_{\parallel} = \mathbf{A} - (\mathbf{A}\cdot\hat{\mathbf{u}})\hat{\mathbf{u}} \end{aligned} \tag{2.101}$$

When the projection or the dot product is used to determine the component of a vector along a line, the second component is *always* perpendicular to that line. Therefore, the dot product may not be used to determine components along two nonorthogonal lines or

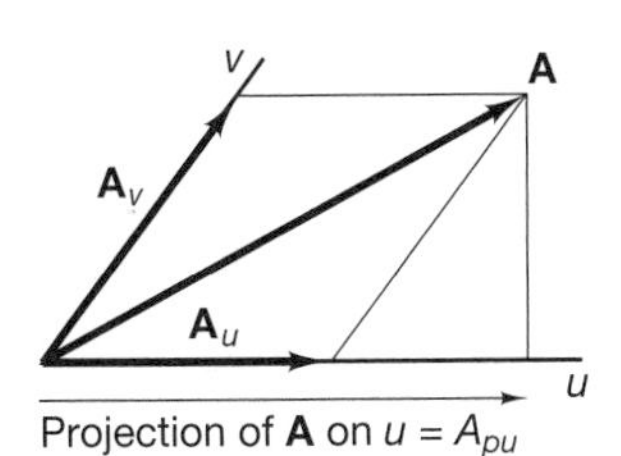

Projection of **A** on $u = A_{pu}$

Figure 2.51

coordinate axes. The definition of vector components states that the vector must always equal the vector sum of its components. The difference between the projection of a vector along a line and the component of the vector along that line may be seen in Figure 2.51. Let $\mathbf{A}_u$ and $\mathbf{A}_v$ be the components of the vector **A** in the u and v directions, respectively, and let $\hat{\mathbf{u}}$ and $\hat{\mathbf{v}}$ be unit vectors in these directions. Then A may be written

$$\mathbf{A} = \mathbf{A}_u + \mathbf{A}_v = A_u\hat{\mathbf{u}} + A_v\hat{\mathbf{v}} \tag{2.102}$$

Let us designate the projection of **A** in the $\hat{\mathbf{u}}$ direction by A_{pu}. This projection is

$$\begin{aligned} A_{\text{pu}} &= \mathbf{A}\cdot\hat{\mathbf{u}} = A_u\,\hat{\mathbf{u}}\cdot\hat{\mathbf{u}} + A_v\,\hat{\mathbf{v}}\cdot\hat{\mathbf{u}} \\ A_{\text{pu}} &= A_u + A_v\,\hat{\mathbf{v}}\cdot\hat{\mathbf{u}} \end{aligned} \tag{2.103}$$

Therefore, the projection A_{pu} equals the component A_u in the u direction only if the dot product $\hat{\mathbf{v}}\cdot\hat{\mathbf{u}} = 0$ or if the lines u and v are orthogonal. *The dot product can be used only to form components of a vector that are parallel and perpendicular to a given line in space.*

Most computational software packages and calculators will compute the scalar product of two vectors automatically, either numerically or symbolically. Consult the Computational Supplements for details.

Sample Problem 2.13

Two vectors, shown in the diagram at the left, are given as:

$$\mathbf{A} = 3\hat{\mathbf{i}} - 2\hat{\mathbf{j}} - 2\hat{\mathbf{k}}$$
$$\mathbf{B} = 4\hat{\mathbf{i}} + 4\hat{\mathbf{j}}$$

Sketch the two vectors in space and determine the angle between them.

Solution We have drawn the two vectors in the coordinate system as shown.
The magnitudes of the two vectors are

$$|\mathbf{A}| = \sqrt{\mathbf{A}\cdot\mathbf{A}} = \sqrt{3^2 + 2^2 + 2^2} = 4.123$$
$$|\mathbf{B}| = \sqrt{\mathbf{B}\cdot\mathbf{B}} = \sqrt{4^2 + 4^2} = 5.657$$

The dot product of **A** and **B** is

$$\mathbf{A}\cdot\mathbf{B} = 3(4) - 2(4) = 4$$

Therefore, the angle θ between the vectors is

$$\cos\theta = \frac{\mathbf{A}\cdot\mathbf{B}}{|\mathbf{A}||\mathbf{B}|} = \frac{4}{4.123(5.657)} = 0.172 \Rightarrow \theta = 80.1^\circ$$

Since the two vectors have been sketched in three dimensions, the angle in space may be "seen." The use of the scalar product gives only the magnitude of the smaller of the two angles between the two vectors when their origins coincide.

Sample Problem 2.14

A force $\mathbf{F} = 10\hat{\mathbf{i}} - 10\hat{\mathbf{j}} + 5\hat{\mathbf{k}}$ N is applied to a bar, whose position is shown in the diagram on the next page. Determine the component of force that is transmitted along the bar and the component that is perpendicular to the bar.

Solution First form a unit vector $\hat{\mathbf{n}}$ along the bar AB directed toward B from the position vector $\mathbf{r}_{B/A}$, point B relative to point A:

$$\mathbf{r}_{B/A} = 12\hat{\mathbf{i}} + 3\hat{\mathbf{j}} + 4\hat{\mathbf{k}}$$

The magnitude of this vector is

$$|\mathbf{r}_{B/A}| = 13$$

The unit vector $\hat{\mathbf{n}}$ may be obtained by dividing the vector $\mathbf{r}_{B/A}$ by its magnitude:

$$\hat{\mathbf{n}} = \frac{\mathbf{r}_{B/A}}{\mathbf{r}_{B/A}} = 0.923\hat{\mathbf{i}} + 0.231\hat{\mathbf{j}} + 0.308\hat{\mathbf{k}}$$

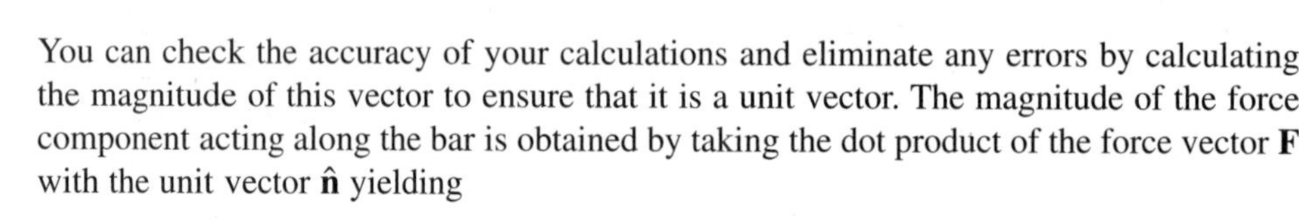

You can check the accuracy of your calculations and eliminate any errors by calculating the magnitude of this vector to ensure that it is a unit vector. The magnitude of the force component acting along the bar is obtained by taking the dot product of the force vector **F** with the unit vector $\hat{\mathbf{n}}$ yielding

$$\mathbf{F}\cdot\hat{\mathbf{n}} = 10(0.923) - 10(0.231) + 5(0.308) = 8.46$$

This is the magnitude of the force component along AB, and it can be written as a vector, $\mathbf{F}_{AB}$, by multiplying this magnitude by the unit vector $\hat{\mathbf{n}}$:

$$\mathbf{F}_{AB} = (\mathbf{F}\cdot\hat{\mathbf{n}})\hat{\mathbf{n}} = 8.46(0.923\hat{\mathbf{i}} + 0.231\hat{\mathbf{j}} + 0.308\hat{\mathbf{k}})$$
$$= 7.81\hat{\mathbf{i}} + 1.95\hat{\mathbf{j}} + 2.61\hat{\mathbf{k}}$$

The force component perpendicular to the bar, $\mathbf{F}_{\perp}$, can be found by subtracting the force parallel to the bar from the total force:

$$\mathbf{F}_{\perp} = \mathbf{F} - \mathbf{F}_{AB} = 2.19\hat{\mathbf{i}} - 11.95\hat{\mathbf{j}} + 2.39\hat{\mathbf{k}}$$

The dot product may be used in other ways in this example. For instance, suppose that we want to know the angle the force makes with the bar. A unit vector along the direction of the force is

$$\hat{\mathbf{f}} = \mathbf{F}/|\mathbf{F}| = 0.667\hat{\mathbf{i}} - 0.667\hat{\mathbf{j}} + 0.333\hat{\mathbf{k}}$$

Since the dot product equals the product of the magnitudes of the vectors and the cosine of the angle between them, the angle between the bar and the force is

$$\cos^{-1}(\hat{\mathbf{f}}\cdot\hat{\mathbf{n}}) = 55.7°$$

A check for errors can be made by seeing whether $\mathbf{F}_{AB}$ is perpendicular to $\mathbf{F}_{\perp}$, as was desired. If they are perpendicular, the dot product between them is zero, as the cosine of 90° is zero. We have

$$\mathbf{F}_{AB}\cdot\mathbf{F}_{\perp} = [7.81(2.19) - (1.95)(11.95) + 2.61\,(2.39)] = 0.04$$

Round-off error makes this dot product slightly different from zero.

Sample Problem 2.15 Derive the Law of Cosines using vector addition and the scalar product from the vector equations $\mathbf{C} = \mathbf{A} + \mathbf{B}$.

Solution Take the scalar product of the equation of vector addition with itself.

$$\mathbf{C}\cdot\mathbf{C} = (\mathbf{A} + \mathbf{B})\cdot(\mathbf{A} + \mathbf{B})$$
$$= \mathbf{A}\cdot\mathbf{A} + \mathbf{A}\cdot\mathbf{B} + \mathbf{B}\cdot\mathbf{A} + \mathbf{B}\cdot\mathbf{B}$$

Evaluating the dot products and combining terms, yields

$$C^2 = A^2 + B^2 + 2AB\cos\theta$$

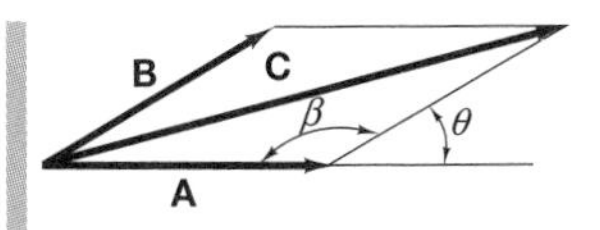

This equation is the result of use of the distributive and associative laws of the dot product. From the figure we can relate the dot product to the law of cosines by examining the relationship between the angles.

$$\beta = (180 - \theta)$$
$$\cos(180 - \theta) = -\cos\theta$$

Using the relationship between β and θ, we obtain the law of cosines.

$$C^2 = A^2 + B^2 - 2AB\cos\beta$$

This is the law of cosines.

Problems

2.79 Consider the two vectors $\mathbf{A} = 3\hat{\mathbf{i}} - 2\hat{\mathbf{j}} + \hat{\mathbf{k}}$ and $\mathbf{B} = \hat{\mathbf{i}} + \hat{\mathbf{j}} + \hat{\mathbf{k}}$. Calculate $\mathbf{A}\cdot\mathbf{B}$, $\mathbf{A}\cdot\hat{\mathbf{i}}$, $\mathbf{A}\cdot\hat{\mathbf{j}}$, and $\mathbf{A}\cdot\hat{\mathbf{k}}$.

2.80 Calculate the angle between the vectors **A** and **B** of Problem 2.79.

2.81 Calculate the unit vector $\hat{\mathbf{n}}$ along the line ON, and determine the projection of vector **F** onto this line. (See Figure P2.81.)

2.82 Let $\mathbf{A} = 4\hat{\mathbf{i}} - \hat{\mathbf{j}} + \hat{\mathbf{k}}$ and $\mathbf{B} = -\hat{\mathbf{i}} + B_y\hat{\mathbf{j}} + 3\hat{\mathbf{k}}$, where B_y is unknown. Determine B_y such that the vector **B** is orthogonal to **A**.

2.83 Consider the vector $\mathbf{R} = -\hat{\mathbf{i}} + 3\hat{\mathbf{j}}$, and compute the component of **R** in the direction of the line V and a second component perpendicular to the line V. (See Figure P2.83.)

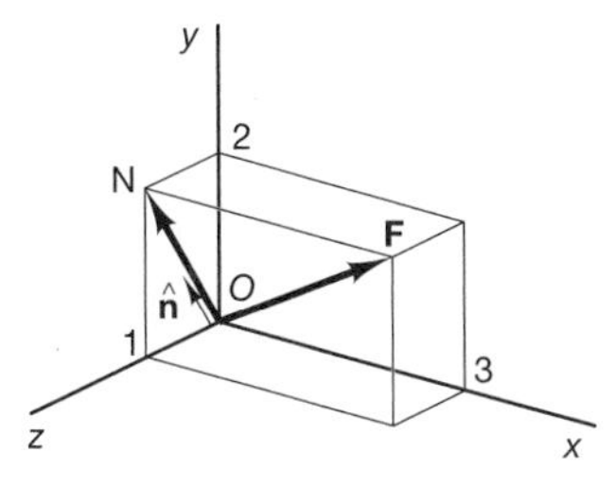

Figure P2.81

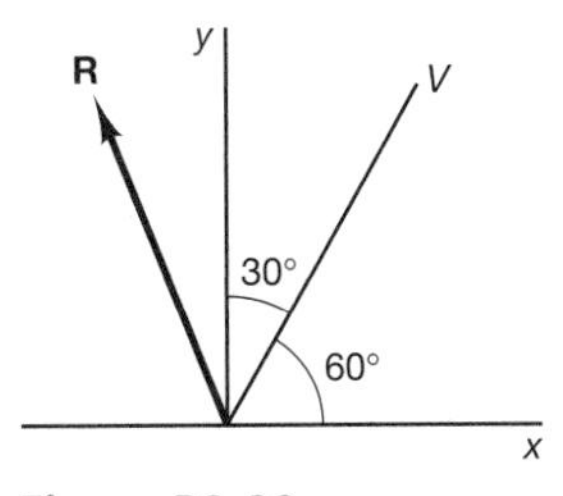

Figure P2.83

2.84 Consider the force **R** of Problem 2.83. Compare the projection of $\mathbf{R} = -\hat{\mathbf{i}} + 3\hat{\mathbf{j}}$ along the line at V to the component of **R** along V, where **R** is resolved into components along V and along the positive x-axis.

2.85 Consider the vector $\mathbf{A} = 3\hat{\mathbf{i}} + 2\hat{\mathbf{j}} - \hat{\mathbf{k}}$ and the vector $\mathbf{B} = \hat{\mathbf{i}} - \hat{\mathbf{j}} + B_z\hat{\mathbf{k}}$. Calculate B_z such that **A** and **B** are perpendicular.

2.86 A force $\mathbf{F} = 30\hat{\mathbf{i}} + 30\hat{\mathbf{j}}$ N is applied to the handle of a crank. (See Figure P2.86.) Determine the components of **F** parallel and perpendicular to the handle. Determine the two unit vectors, one parallel and one perpendicular to the handle.

2.87 Resolve the vector **F** shown in Figure P2.87 into the sum of a component along a and one along b; and compare these components to the projection of **F** onto **a** and the projection of **F** onto **b**.

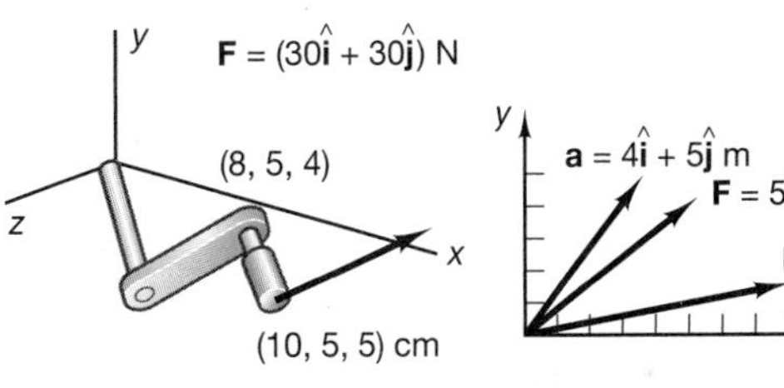

Figure P2.86 Figure P2.87

2.88 Repeat Problem 2.87 for the system shown in Figure P2.88.

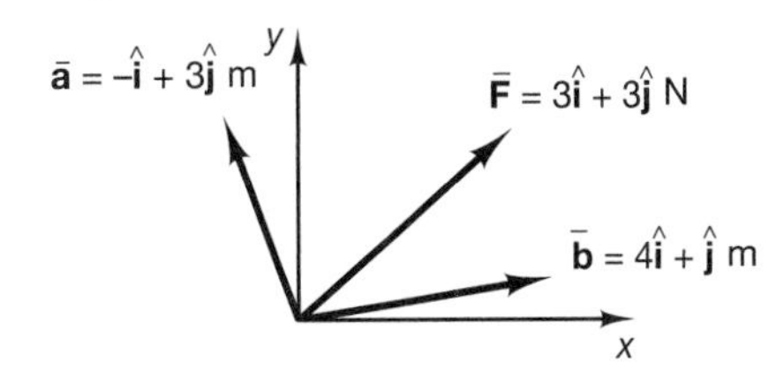

Figure P2.88

2.89 Repeat Problem 2.87 for the vectors illustrated in Figure P2.89.

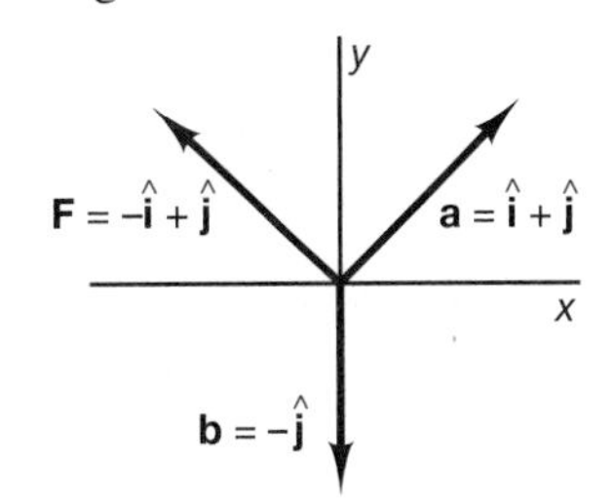

Figure P2.89

2.90 Consider a force vector $\mathbf{F} = (210\hat{\mathbf{i}} + 305\hat{\mathbf{j}} - 400\hat{\mathbf{k}})$N and resolve it into components along lines $\mathbf{A} = 2\hat{\mathbf{i}} + 3\hat{\mathbf{j}} + \hat{\mathbf{k}}$, $\mathbf{B} = -3\hat{\mathbf{i}} - 2\hat{\mathbf{j}} + 3\hat{\mathbf{k}}$ and $\mathbf{C} = -\hat{\mathbf{i}} - \hat{\mathbf{j}} - \hat{\mathbf{k}}$ Determine the projection of **F** on the three lines and show that the sum of the components equals the vector **F** and the sum of the projection does not equal the vector **F**.

2.91 A force $\mathbf{F} = 100\hat{\mathbf{i}} + 50\hat{\mathbf{j}}$, is applied to the joint of two frame elements 45° apart. (See Figure P2.91.) Resolve the vector into components along the two frame members. Then compare these components to the projection of **F** along *a* and the projection of **F** along *b*.

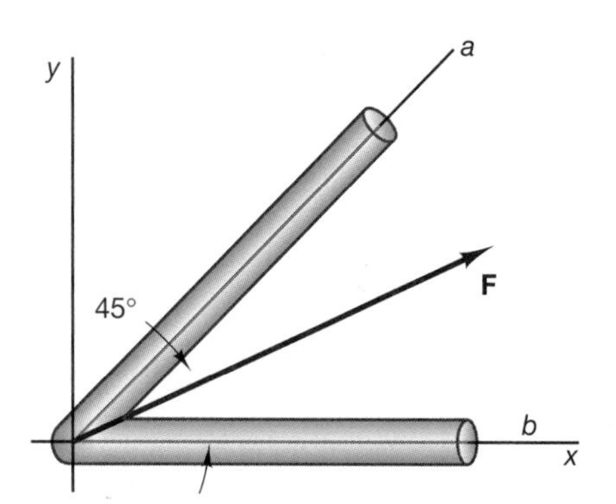

Figure P2.91

2.92 A 200-N force is applied to the joint of a frame. Resolve the force **F** into components along the frame elements, *a* and *b*. Compute the projections of **F** along *a* and *b* using both the dot product and the relationship between projections and components.

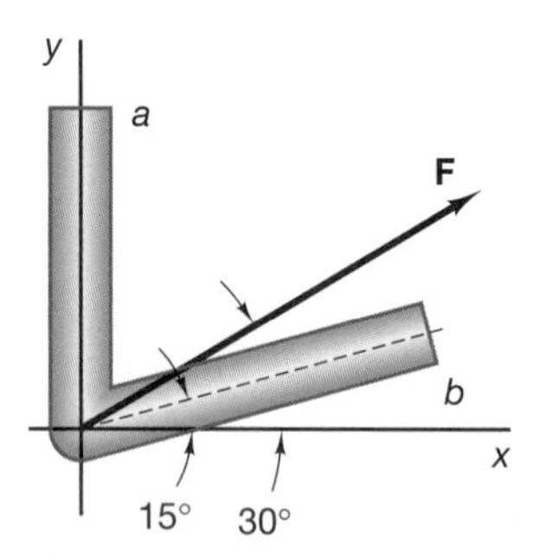

Figure P2.92

2.93 A 300-N force is applied to the joint of a frame. (See Figure P2.93) Resolve the force **F** into components along the frame elements *a* and *b*. Also, compute the projections of **F** along *a* and *b*, using both the dot product and the relationship between the projection and the components of a vector.

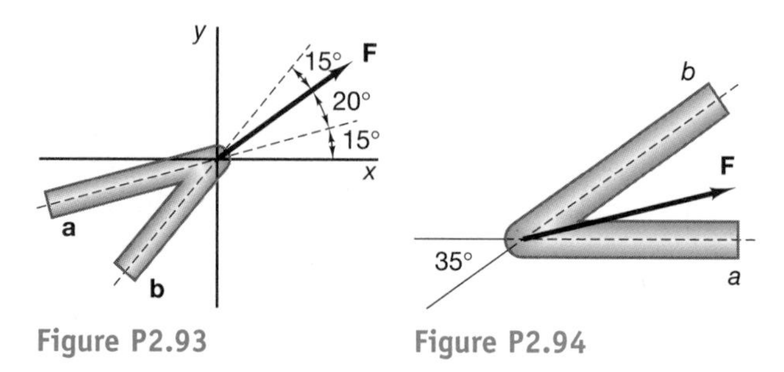

Figure P2.93

Figure P2.94

2.94 From measurements, it is known that $\mathbf{F} = 200\hat{\mathbf{a}} + 100\hat{\mathbf{b}}$ N, where $\hat{\mathbf{a}}$ and $\hat{\mathbf{b}}$ are unit vectors along the frame members *a* and *b*. (See Figure P2.94.) The frame members make an angle of 35° with each other. Calculate the magnitude of the force **F**, the angle that **F** makes with the frame (say, with $\hat{\mathbf{a}}$), and the projection of **F** along $\hat{\mathbf{b}}$.

2.95 The positions of the markers on a gymnast's leg while landing after an aerial maneuver are measured relative to a coordinate system in the laboratory. As is known from watching the Olympics, the angle the knee makes upon landing is one of the factors used by judges. In training athletes, markers are often used to determine these angles by measuring the locations relative to the laboratory frame of reference. Some sample measurements are shown in Figure P2.95. Determine (a) the length of the gymnast's leg and (b) the angle of flexion of the knee—that is, the angle between the lower leg and the thigh.

	x (meters)	*y* (meters)
Hip	2.10	1.0
Knee	2.30	0.6
Ankle	2.15	0.15

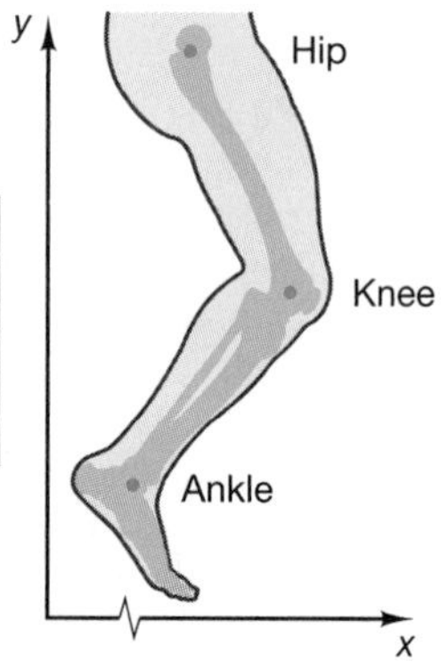

Figure P2.95

2.9 VECTOR PRODUCT OR CROSS PRODUCT

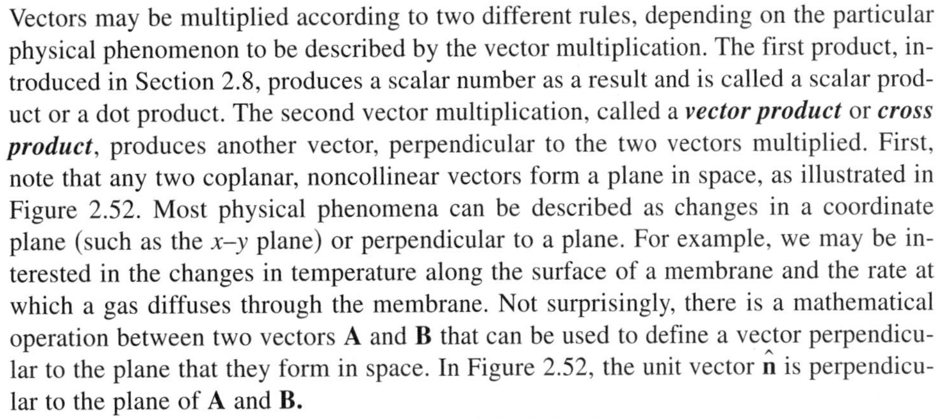

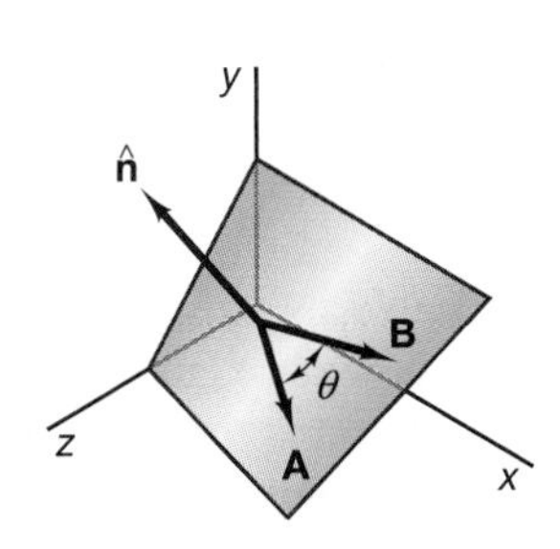

Figure 2.52 Two vectors **A** and **B** form a plane in space with $\hat{\mathbf{n}}$ normal to the plane.

Vectors may be multiplied according to two different rules, depending on the particular physical phenomenon to be described by the vector multiplication. The first product, introduced in Section 2.8, produces a scalar number as a result and is called a scalar product or a dot product. The second vector multiplication, called a ***vector product*** or ***cross product***, produces another vector, perpendicular to the two vectors multiplied. First, note that any two coplanar, noncollinear vectors form a plane in space, as illustrated in Figure 2.52. Most physical phenomena can be described as changes in a coordinate plane (such as the x–y plane) or perpendicular to a plane. For example, we may be interested in the changes in temperature along the surface of a membrane and the rate at which a gas diffuses through the membrane. Not surprisingly, there is a mathematical operation between two vectors **A** and **B** that can be used to define a vector perpendicular to the plane that they form in space. In Figure 2.52, the unit vector $\hat{\mathbf{n}}$ is perpendicular to the plane of **A** and **B.**

The cross product of two vectors **A** and **B** is defined as

$$\mathbf{A} \times \mathbf{B} = |\mathbf{A}||\mathbf{B}| \sin\theta\, \hat{\mathbf{n}} \tag{2.104}$$

where the sense of the unit vector $\hat{\mathbf{n}}$ is specified by the right-hand rule illustrated in Figure 2.53. The positive sense of $\mathbf{A} \times \mathbf{B}$, or $\hat{\mathbf{n}}$, is obtained by placing the right hand with the fingers aligned with the first vector, **A**, and curving them toward the second vector, **B**. As shown in the figure, the thumb of the right hand will point in the positive direction of $\hat{\mathbf{n}}$. Most threaded screws and nuts follow a right-hand rule. Turned in the right-hand direction, the screw moves in and the nuts tightens.

The angle between **A** and **B** is denoted by θ. If $\theta = 90°$, then $\sin\theta = 1$; if $\theta = 0°$, then $\sin\theta = 0$. Therefore, if **A** and **B** are parallel vectors, $\mathbf{A} \times \mathbf{B} = 0$. These vectors could be considered collinear and would not define a plane in space.

Note that, in Figure 2.53, if the fingers of the right hand are first aligned with **B** and then curved into **A**, forming the cross product $\mathbf{B} \times \mathbf{A}$, the thumb points in the opposite direction. Thus, the cross product is *not* commutative; that is, $\mathbf{A} \times \mathbf{B} \neq \mathbf{B} \times \mathbf{A}$, but instead,

$$\mathbf{A} \times \mathbf{B} = -(\mathbf{B} \times \mathbf{A}) \tag{2.105}$$

However, the cross product is distributive, and

$$\mathbf{A} \times (\mathbf{B} + \mathbf{C}) = \mathbf{A} \times \mathbf{B} + \mathbf{A} \times \mathbf{C} \tag{2.106}$$

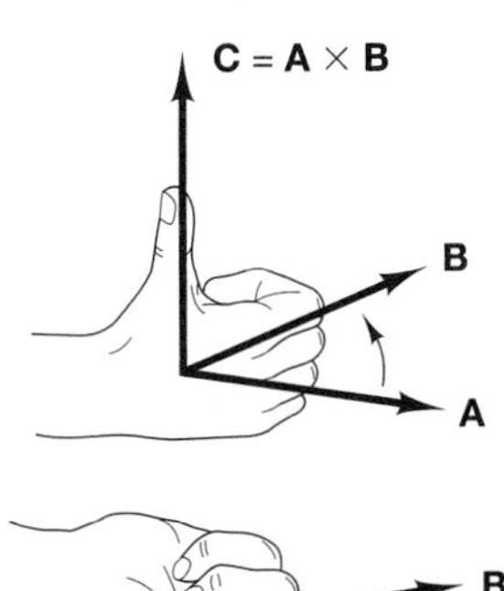

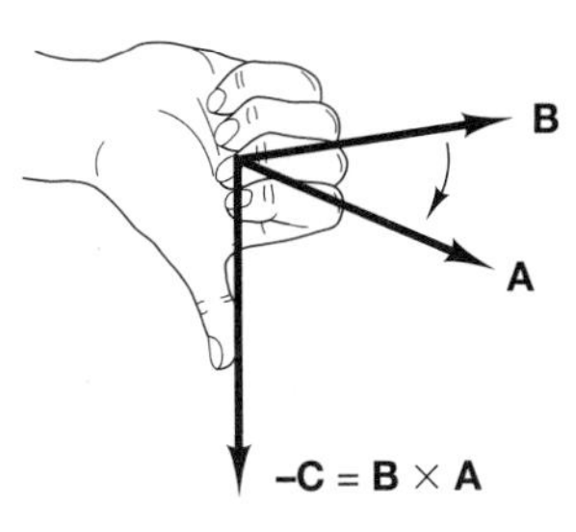

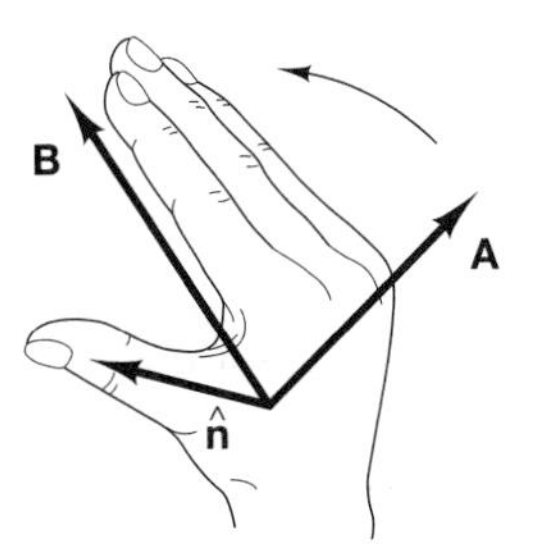

Figure 2.53 The direction of the vector product $\mathbf{A} \times \mathbf{B}$ is the direction in which your right thumb points when you move the fingers of your right hand from **A** to **B**.

Referring back to Figure 2.52, note that the cross product may be used to define a direction perpendicular to the plane of **A** and **B**—that is, the unit vector $\hat{\mathbf{n}}$**.** This can be written mathematically as

$$\hat{\mathbf{n}} = \frac{\mathbf{A} \times \mathbf{B}}{|\mathbf{A} \times \mathbf{B}|} \tag{2.107}$$

Recall that a unit vector in a given direction is formed by dividing the vector in that direction by its magnitude.

The cross product arises in mechanics when rotational effects are considered. Note that if **A** is a position vector **r** and **B** is a force vector **F**, the cross product $\mathbf{r} \times \mathbf{F}$ has the units of a moment. The connection between the cross product and the moment of a force will be formalized in the Chapter 4. Note also that the right-hand rule coincides with the direction required to tighten most screws (those with right-hand threads).

It is useful to examine the cross products between the unit base vectors of a rectangular coordinate system, since then the cross product can be formally computed using component notation. The cross products of the base unit vectors $\hat{\mathbf{i}}$, $\hat{\mathbf{j}}$, and $\hat{\mathbf{k}}$ are

$$\begin{array}{lll} \hat{\mathbf{i}} \times \hat{\mathbf{i}} = \mathbf{0} & \hat{\mathbf{j}} \times \hat{\mathbf{i}} = -\hat{\mathbf{k}} & \hat{\mathbf{k}} \times \hat{\mathbf{i}} = \hat{\mathbf{j}} \\ \hat{\mathbf{i}} \times \hat{\mathbf{j}} = \hat{\mathbf{k}} & \hat{\mathbf{j}} \times \hat{\mathbf{j}} = \mathbf{0} & \hat{\mathbf{k}} \times \hat{\mathbf{j}} = -\hat{\mathbf{i}} \\ \hat{\mathbf{i}} \times \hat{\mathbf{k}} = -\hat{\mathbf{j}} & \hat{\mathbf{j}} \times \hat{\mathbf{k}} = \hat{\mathbf{i}} & \hat{\mathbf{k}} \times \hat{\mathbf{k}} = \mathbf{0} \end{array} \tag{2.108}$$

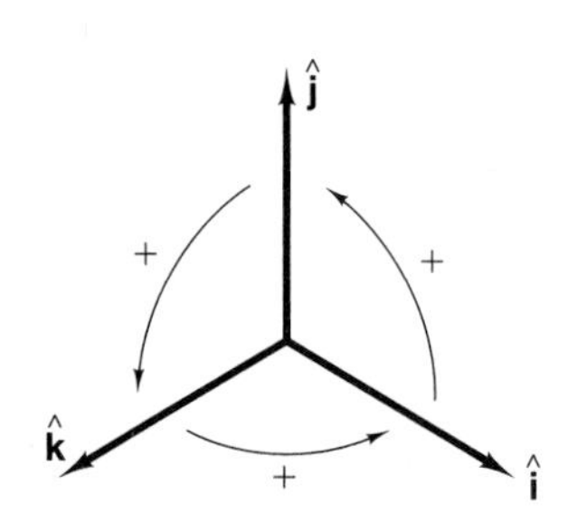

Figure 2.54 To find the cross product of two base unit vectors, go around the circle from the first vector in the product to the second. The remaining vector is the cross product. If you went around the circle clockwise, the sign is negative; if you went around the circle counter-clockwise, the sign is positive.

A small diagram representing the unit vectors as coordinate base vectors may help to remember the various cross products between unit vectors. This is shown in Figure 2.54. Unit base vectors can be used to examine formally the calculation of the cross product between two vectors **A** and **B**. Let the two vectors be written in component notation as

$$\mathbf{A} = A_x\hat{\mathbf{i}} + A_y\hat{\mathbf{j}} + A_z\hat{\mathbf{k}} \text{ and } \mathbf{B} = B_x\hat{\mathbf{i}} + B_y\hat{\mathbf{j}} + B_z\hat{\mathbf{k}} \tag{2.109}$$

Then $\mathbf{A} \times \mathbf{B}$ becomes

$$\begin{aligned} &(A_x\hat{\mathbf{i}} + A_y\hat{\mathbf{j}} + A_z\hat{\mathbf{k}}) \times (B_x\hat{\mathbf{i}} + B_y\hat{\mathbf{j}} + B_z\hat{\mathbf{k}}) \\ &\quad = A_xB_y\hat{\mathbf{k}} - A_xB_z\hat{\mathbf{j}} - A_yB_x\hat{\mathbf{k}} + A_yB_z\hat{\mathbf{i}} + A_zB_x\hat{\mathbf{j}} - A_zB_y\hat{\mathbf{i}} \end{aligned}$$

The right-hand side of this equation is obtained by repeated application of the distributive law and the cross product between the unit vectors. The components of the cross product may be grouped according to unit vectors as

$$\mathbf{A} \times \mathbf{B} = (A_yB_z - A_zB_y)\hat{\mathbf{i}} + (A_zB_x - A_xB_z)\hat{\mathbf{j}} + (A_xB_y - A_yB_x)\hat{\mathbf{k}} \tag{2.110}$$

Let the vector **C** denote the result of the cross product; that is, let $\mathbf{C} = \mathbf{A} \times \mathbf{B}$. Then the cross product may also be written in the form of the determinant of the array

$$\mathbf{C} = \begin{vmatrix} \hat{\mathbf{i}} & \hat{\mathbf{j}} & \hat{\mathbf{k}} \\ A_x & A_y & A_z \\ B_x & B_y & B_z \end{vmatrix} \tag{2.111}$$

The determinant of this three-by-three array may be evaluated as the sum of the products along the diagonals from top left to lower right, minus the sum of the products of the diagonals from lower left to top right:

$$\begin{aligned} \mathbf{C} &= \begin{vmatrix} \hat{\mathbf{i}} & \hat{\mathbf{j}} & \hat{\mathbf{k}} \\ A_x & A_y & A_z \\ B_x & B_y & B_z \end{vmatrix} \\ &= (\hat{\mathbf{i}}A_yB_z + \hat{\mathbf{j}}A_zB_x + \hat{\mathbf{k}}A_xB_y) - (B_xA_y\hat{\mathbf{k}} + B_yA_z\hat{\mathbf{i}} + B_zA_x\hat{\mathbf{j}}) \end{aligned} \tag{2.112}$$

Again, the cross product can be written by grouping terms as components multiplying the base unit vectors, so that

$$\mathbf{C} = \mathbf{A} \times \mathbf{B} = (A_yB_z - A_zB_y)\hat{\mathbf{i}} + (A_zB_x - A_xB_z)\hat{\mathbf{j}} + (A_xB_y - A_yB_x)\hat{\mathbf{k}}$$

This is the same form as eq. (2.110). the determinant calculation may be easier to remember than the formulas for the cross products between the unit vectors. the cross product $\mathbf{C} = \mathbf{A} \times \mathbf{B}$ may be written in terms of scalar components as

$$\begin{aligned} C_x &= A_yB_z - A_zB_y \\ C_y &= A_zB_x - A_xB_z \\ C_z &= A_xB_y - A_yB_x \end{aligned} \tag{2.113}$$

Note that the x-component of the vector **C** is independent of the x-components of the vectors **A** and **B**, and involves only their y- and z-components. This is due to the fact that the **C** vector is perpendicular to the **A** and **B** vectors. Such a relationship can be used to advantage in the analysis of problems of mechanics and is also useful in remembering the formulas.

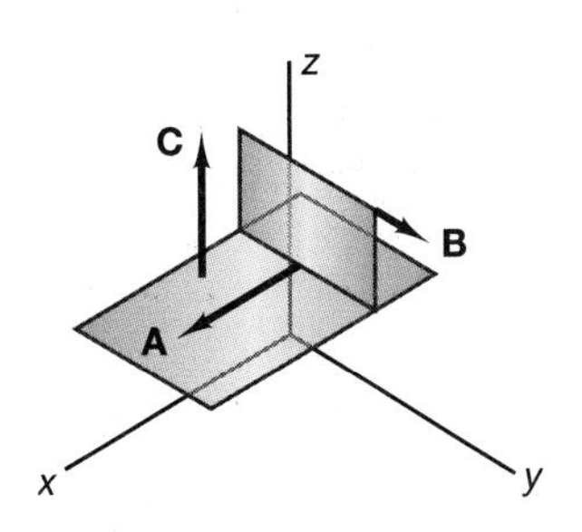

Figure 2.55 Two non-coplanar vectors **A** and **B** can still be multiplied to form a cross product **C**.

Any vector is fully specified by giving its magnitude and direction, and its position in space is given only for particular applications or physical interpretations. Therefore, it is interesting to note that, mathematically, the cross product can be formed between two vectors in space that are not coplanar, as shown in Figure 2.55.

If we let $\mathbf{A} = A_x\hat{\mathbf{i}}$, $\mathbf{B} = B_y\hat{\mathbf{j}}$, and $\mathbf{C} = \mathbf{A} \times \mathbf{B}$, then $\mathbf{C} = A_xB_y\hat{\mathbf{k}}$. Although **C** is perpendicular to both **A** and **B**, the vectors **A** and **B** do not lie in a single plane. This concept is very useful in many mathematical developments and physical applications.

Two types of vector multiplication have been presented, but it should be noted that division between vectors is not defined. This means that there is no inverse to either the scalar (dot) or vector (cross) product of vectors. In other words, given the equation of the form

$$\mathbf{A} \cdot \mathbf{B} = \alpha \tag{2.114}$$

where **A** and α are known, there is no direct vector operation to determine an unknown vector **B**, as is the case in ordinary scalar algebra. If Eq. (2.110) is expanded using the rectangular components of the vector, the scalar form is

$$A_xB_x + A_yB_y + A_zB_z = \alpha \tag{2.115}$$

Even with the components of **A** known and α known, there are an infinite number of values of the components of **B** that would satisfy this equation. Therefore, there is no unique method to invert the dot product of two vectors.

Consider the vector product

$$\mathbf{A} \times \mathbf{B} = \mathbf{C} \tag{2.116}$$

where the vectors **A** and **C** are known.

Eq. (2.116), when expanded into scalar form, becomes Eq. (2.113), which may be written in matrix notation as

$$\begin{bmatrix} 0 & -A_z & A_y \\ A_z & 0 & -A_x \\ -A_y & A_x & 0 \end{bmatrix} \begin{bmatrix} B_x \\ B_y \\ B_z \end{bmatrix} = \begin{bmatrix} C_x \\ C_y \\ C_z \end{bmatrix} \tag{2.117}$$

Attempts to solve these three equations for the components of **B** will fail, because the three equations are linearly dependent rendering the matrix of the coefficients singular. This may be noted by examining the determinant of the coefficient matrix involving the components of **A**. The determinant of this matrix is zero, and the equations cannot be solved. Thus there is no unique solution for the vector **B** in Eq. (2.116). A better understanding of why there are an infinite number of solutions for the vector **B** can be obtained by expressing, in a general manner, the vector into two components, one parallel to **A** and one perpendicular to **A**. The vector **A** may be written as

$$\mathbf{A} = A\hat{\mathbf{a}}$$

where $\hat{\mathbf{a}}$ is a unit vector in the direction of **A**. The vector **B** can be written as

$$\mathbf{B} = \mathbf{B}_{\parallel} + \mathbf{B}_{\perp} \tag{2.118}$$

Eq. (2.109) can be written in the form:

$$\begin{aligned} \mathbf{A} \times (\mathbf{B}_{\parallel} + \mathbf{B}_{\perp}) &= \mathbf{C} \\ \mathbf{A} \times \mathbf{B}_{\parallel} &= \mathbf{0} \quad \text{and hence} \\ \mathbf{A} \times \mathbf{B}_{\perp} &= \mathbf{C} \end{aligned} \tag{2.119}$$

Since $\mathbf{B}_{\parallel}$ could have any value and Eq. 2.119 would still be satisfied, the vector $\boldsymbol{B}$ is not uniquely determined.

However, a unique value of $\mathbf{B}_{\perp}$ can be determined. Note that the vectors **A**, $\mathbf{B}_{\perp}$ and **C** are mutually perpendicular or orthogonal.

$$\mathbf{A} \times \mathbf{B}_{\perp} = |\mathbf{A}||\mathbf{B}_{\perp}|\hat{\mathbf{c}} = |\mathbf{C}|\hat{\mathbf{c}}$$

where

$$|\mathbf{A}||\mathbf{B}_{\perp}| = |\mathbf{C}| \tag{2.120}$$

and

$$|\mathbf{B}_{\perp}| = \frac{|\mathbf{C}|}{|\mathbf{A}|}$$

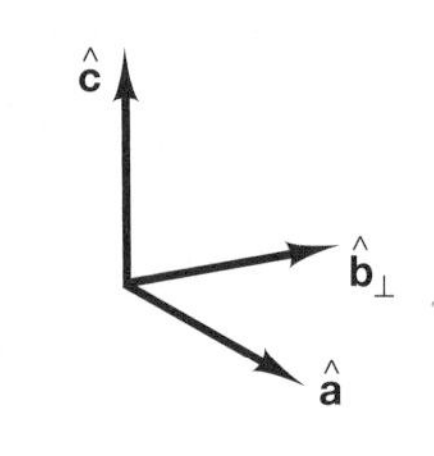

Figure 2.56

Therefore, the magnitude of the portion of **B** perpendicular to **A** is uniquely determined. Examining the vector diagram shown in Figure 2.56 shows the direction of **B** perpendicular.

These three unit vectors form an orthogonal set

$$\hat{\mathbf{b}}_{\perp} = \hat{\mathbf{c}} \times \hat{\mathbf{a}}$$

so that

$$\mathbf{B}_{\perp} = \frac{|\mathbf{C}|}{|\mathbf{A}|}(\hat{\mathbf{c}} \times \hat{\mathbf{a}}) \tag{2.121}$$

2.9.1 MULTIPLE PRODUCTS OF VECTORS

There are advantages to examining multiple products of vectors, as these will arise in many applications in mechanics. Let **A**, **B**, and **C** be any three vectors. The outcome of the operation

$$\mathbf{A} \cdot (\mathbf{B} \times \mathbf{C}) \tag{2.122}$$

is a scalar, called the **scalar triple product** (or mixed product) of **A**, **B**, and **C**. The cross product between **B** and **C** is separated by parentheses, indicating that this operation must be performed before the dot product is taken. The order is necessary because, if the dot product between **A** and **B** were taken first, yielding a scalar, then the cross product could not be performed. In general, the parentheses are used to avoid any confusion in the order of operations. Note that this use of parentheses is consistent with most computer programming languages and commercial software packages, as well as programmable calculators.

If the rectangular components of the three vectors are denoted in the usual manner, then the scalar triple product may be written in component form as

$$\mathbf{A} \cdot (\mathbf{B} \times \mathbf{C}) = A_x(B_yC_z - B_zC_y) + A_y(B_zC_x - B_xC_z) + A_z(B_xC_y - B_yC_x) \tag{2.123}$$

Or it may be written in determinant form as

$$\mathbf{A} \cdot (\mathbf{B} \times \mathbf{C}) = \begin{vmatrix} A_x & A_y & A_z \\ B_x & B_y & B_z \\ C_x & C_y & C_z \end{vmatrix} \tag{2.124}$$

A property of determinants is that if any two rows of a determinant are interchanged, only the sign of the determinant changes. This leads directly to the permutation theorem for the scalar triple product. A *permutation* is a change in the order of an ordered set. A permutation of the vectors in a scalar triple product is defined as the interchange of any two vectors in the product. The result of a permutation of the product will be the negative of the original scalar triple product. If the vectors in a scalar triple product are subjected to an odd number of permutations, the value of this product is changed in sign only; if the number of permutations is even, the value of the product is not changed. Therefore, the following vector identities are true:

$$\mathbf{A}\cdot(\mathbf{B}\times\mathbf{C}) = \mathbf{B}\cdot(\mathbf{C}\times\mathbf{A}) = \mathbf{C}\cdot(\mathbf{A}\times\mathbf{B})$$
$$\mathbf{A}\cdot(\mathbf{B}\times\mathbf{C}) = -\mathbf{C}\cdot(\mathbf{B}\times\mathbf{A}) = -\mathbf{A}\cdot(\mathbf{C}\times\mathbf{B}) = -\mathbf{B}\cdot(\mathbf{A}\times\mathbf{C}) \tag{2.125}$$

One immediate application of the scalar triple product is in the calculation of the volume of a parallelepiped with vectors **A**, **B**, and **C** forming its edges, as shown in Figure 2.57. The volume of the parallelepiped is given as the scalar triple product

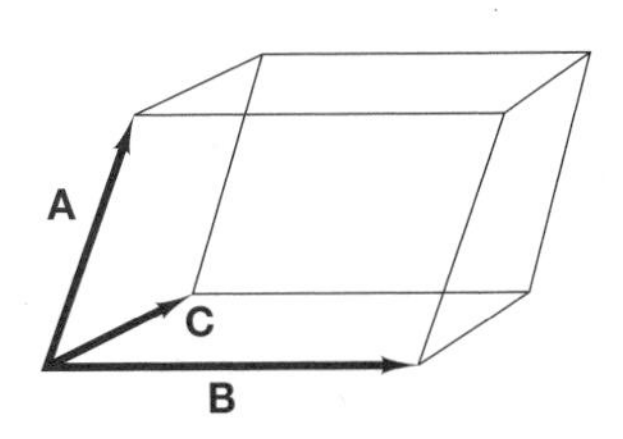

Figure 2.57

$$V = \mathbf{A}\cdot(\mathbf{B}\times\mathbf{C})$$

Other applications of the scalar triple product occur in civil, mechanical and aerospace engineering design.

Let **A**, **B**, and **C** be any three vectors. The expression

$$\mathbf{A}\times(\mathbf{B}\times\mathbf{C}) \tag{2.126}$$

is a vector, called the ***vector triple product*** of **A**, **B**, and **C**. An important vector identity, given without proof, is

$$\mathbf{A}\times(\mathbf{B}\times\mathbf{C}) = \mathbf{B}(\mathbf{A}\cdot\mathbf{C}) - \mathbf{C}(\mathbf{A}\cdot\mathbf{B}) \tag{2.127}$$

In Section 2.8, it was shown that one application of the dot product is to resolve a vector into components parallel and perpendicular to a given line in space. The direction of the line was defined by a unit vector $\hat{\mathbf{u}}$ along that line, and the component of a vector **A** parallel to that line is

$$\mathbf{A}_{\parallel} = (\mathbf{A}\cdot\hat{\mathbf{u}})\hat{\mathbf{u}} \tag{2.128}$$

The component perpendicular to the line is

$$\mathbf{A}_{\perp} = \mathbf{A} - \mathbf{A}_{\parallel} \tag{2.129}$$

Now consider the vector triple product

$$\hat{\mathbf{u}}\times(\mathbf{A}\times\hat{\mathbf{u}}) = \mathbf{A}(\hat{\mathbf{u}}\cdot\hat{\mathbf{u}}) - \hat{\mathbf{u}}(\mathbf{A}\cdot\hat{\mathbf{u}}) \tag{2.130}$$

where the vector identity in Eq. (2.127) has been used. Since the dot product of any unit vector with itself is unity, the first term on the right-hand side of Eq. (2.130) is equal to the vector **A**. Comparing Eq. (2.128) to the second term on the right-hand side of Eq. (2.130) shows that the second term is the parallel component of the vector **A**. Therefore, we may conclude that the perpendicular component of **A** is

$$\mathbf{A}_{\perp} = \hat{\mathbf{u}}\times(\mathbf{A}\times\hat{\mathbf{u}}) \tag{2.131}$$

Eqs. (2.128) and (2.131) may be used as an alternative to the method presented in Section 2.8 for finding the components of a vector parallel and perpendicular to a line in space.

Another application of the vector triple product is found in the solution of determining the perpendicular vector in the equation

$$\mathbf{A}\times\mathbf{B} = \mathbf{C} \tag{2.132}$$

when **A** and **C** are known. It was shown previously that only the component of **B** that is perpendicular to **A** can be uniquely determined. This component will be designated by the vector **P**. The three vectors **A**, **P** and **C** are mutually perpendicular and the vector equation to be solved for **P** is

$$\mathbf{A} \times \mathbf{P} = \mathbf{C} \tag{2.133}$$

The vector triple product will be formed by taking the cross product of this equation with the known vector **A**.

$$\mathbf{A} \times (\mathbf{A} \times \mathbf{P}) = \mathbf{A} \times \mathbf{C} \tag{2.134}$$

Using Eq. (2.123), the triple vector product may be written as:

$$\mathbf{A} \times (\mathbf{A} \times \mathbf{P}) = \mathbf{A}(\mathbf{A} \cdot \mathbf{P}) - \mathbf{P}(\mathbf{A} \cdot \mathbf{A}) \tag{2.135}$$

The first term is zero as **A** is perpendicular to **P**. Using the properties of the cross product, Eq. (2.131) may be written as:

$$\mathbf{P}(\mathbf{A} \cdot \mathbf{A}) = \mathbf{C} \times \mathbf{A}$$

or

$$\mathbf{P} = \frac{\mathbf{C} \times \mathbf{A}}{\mathbf{A} \cdot \mathbf{A}} = \frac{|\mathbf{C}|}{|\mathbf{A}|}\hat{\mathbf{c}} \times \hat{\mathbf{a}} \tag{2.136}$$

This agrees with Eq. (2.121).

The triple vector product can be used to determine the perpendicular distance from a point in space to a line in space. This operation will be useful in Chapter 4 when the concept of *a moment of a force* is discussed. Consider the line AB and the point C as shown in Figure 2.58. A vector from A to B can be defined as

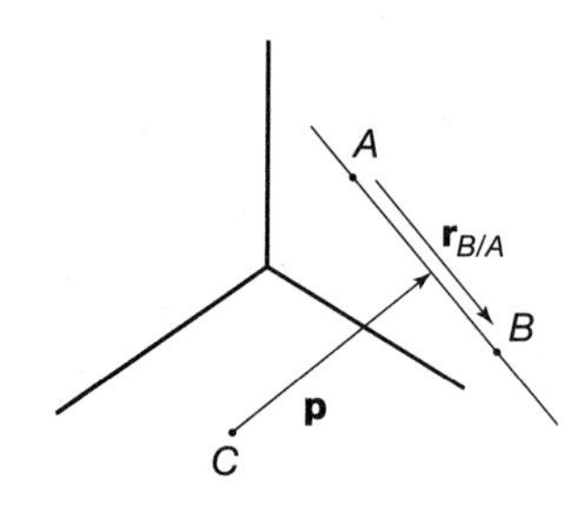

Figure 2.58

$$\mathbf{r}_{B/A} = \mathbf{r}_B - \mathbf{r}_A$$

In a similar manner, a vector can be formed from point C to a point on the line, point B.

$$\mathbf{r}_{B/C} = \mathbf{r}_B - \mathbf{r}_C$$

The cross product of these two vectors will only involve the portion of $\mathbf{r}_{B/C}$ that is perpendicular to $\mathbf{r}_{B/A}$ or the vector **p**, therefore we may write:

$$\mathbf{p} \times \mathbf{r}_{B/A} = \mathbf{r}_{B/C} \times \mathbf{r}_{B/A}$$

Form the cross-product of this equation with $\mathbf{r}_{B/A}$ yielding

$$\mathbf{r}_{B/A} \times (\mathbf{p} \times \mathbf{r}_{B/A}) = \mathbf{r}_{B/A} \times (\mathbf{r}_{B/C} \times \mathbf{r}_{B/A})$$

Using Eq. (2.123), the left side becomes

$$\mathbf{p}(\mathbf{r}_{B/A} \cdot \mathbf{r}_{B/A}) - \mathbf{r}_{B/A}(\mathbf{r}_{B/A} \cdot \mathbf{p}) = \mathbf{r}_{B/A} \times (\mathbf{r}_{B/C} \times \mathbf{r}_{B/A})$$

The second term is zero as the two vectors $\mathbf{r}_{B/A}$ and **p**, are perpendicular by definition. A unit vector along the line AB may be defined as:

$$\hat{\mathbf{n}} = \frac{\mathbf{r}_{B/A}}{|\mathbf{r}_{B/A}|}$$

The perpendicular vector **p** can now be defined as

$$\mathbf{p} = \frac{\mathbf{r}_{B/A} \times (\mathbf{r}_{B/C} \times \mathbf{r}_{B/A})}{|\mathbf{r}_{B/A}||\mathbf{r}_{B/A}|} = \hat{\mathbf{n}} \times (\mathbf{r}_{B/C} \times \hat{\mathbf{n}})$$

This agrees with Eq. (2.131)

A summary of useful vector identities involving multiple vector products is given in Mathematics Window 2.3.

MATHEMATICS WINDOW 2.3

Vector Identities The following are some useful vector identities involving combinations of dot products and cross products:

$$\mathbf{A}\cdot(\mathbf{B}\times\mathbf{C}) = \mathbf{B}\cdot(\mathbf{C}\times\mathbf{A}) = \mathbf{C}\cdot(\mathbf{A}\times\mathbf{B})$$
$$\mathbf{A}\times(\mathbf{B}\times\mathbf{C}) = \mathbf{B}(\mathbf{A}\cdot\mathbf{C}) - \mathbf{C}(\mathbf{A}\cdot\mathbf{B})$$
$$(\mathbf{A}\times\mathbf{B})\times(\mathbf{C}\times\mathbf{D}) = \mathbf{B}[\mathbf{A}\cdot(\mathbf{C}\times\mathbf{D})] - \mathbf{A}[\mathbf{B}\cdot(\mathbf{C}\times\mathbf{D})]$$
$$= \mathbf{C}[(\mathbf{A}\times\mathbf{B})\cdot\mathbf{D}] - \mathbf{D}[(\mathbf{A}\times\mathbf{B})\cdot\mathbf{C}]$$

The last form may be used to obtain

$$(\mathbf{A}\times\mathbf{B})\times(\mathbf{A}\times\mathbf{C}) = \mathbf{A}[(\mathbf{A}\times\mathbf{B})\cdot\mathbf{C}] - \mathbf{C}[(\mathbf{A}\times\mathbf{B})\cdot\mathbf{A}]$$

Examining the last term and using the first of the identities yields

$$[(\mathbf{A}\times\mathbf{B})\cdot\mathbf{A}] = \mathbf{B}\cdot(\mathbf{A}\times\mathbf{A}) = 0$$

Therefore,

$$(\mathbf{A}\times\mathbf{B})\times(\mathbf{A}\times\mathbf{C}) = \mathbf{A}[(\mathbf{A}\times\mathbf{B})\cdot\mathbf{C}]$$

Many calculators and computational software packages have incorporated an operator to perform the cross product between two vectors. This operation may be performed either numerically or analytically.

Sample Problem 2.16

Consider two vectors

$$\mathbf{A} = 5\hat{\mathbf{i}} + 3\hat{\mathbf{j}}$$

and

$$\mathbf{B} = 3\hat{\mathbf{i}} + 6\hat{\mathbf{j}}$$

Determine: (a) $\mathbf{A} + \mathbf{B}$
(b) $\mathbf{A}\cdot\mathbf{B}$
(c) the angle θ between **A** and **B**
(d) $\mathbf{A}\times\mathbf{B}$ (compare the magnitude of the cross product to the magnitude given in the definition $|\mathbf{A}||\mathbf{B}|\sin\theta$)
(e) $\mathbf{B}\times\mathbf{A}$

Solution

(a) $\mathbf{A} + \mathbf{B} = (5\hat{\mathbf{i}} + 3\hat{\mathbf{j}}) + (3\hat{\mathbf{i}} + 6\hat{\mathbf{j}}) = 8\hat{\mathbf{i}} + 9\hat{\mathbf{j}}$

(b) $\mathbf{A}\cdot\mathbf{B} = (5\hat{\mathbf{i}} + 3\mathbf{j})\cdot(3\hat{\mathbf{i}} + 6\hat{\mathbf{j}}) = 15 + 18 = 33$

(c) $\mathbf{A}\cdot\mathbf{B} = |\mathbf{A}||\mathbf{B}|\cos\theta$

$|\mathbf{A}| = 5.83 \quad |\mathbf{B}| = 6.71$

$$\frac{\mathbf{A}\cdot\mathbf{B}}{|\mathbf{A}||\mathbf{B}|} = \cos\theta = 33/(5.83\times 6.71) = 0.844$$

$$\theta = 32.48°$$

(d) $\mathbf{A}\times\mathbf{B} = (5\hat{\mathbf{i}} + 3\hat{\mathbf{j}})\times(3\hat{\mathbf{i}} + 6\hat{\mathbf{j}}) = (30 - 9)\,\hat{\mathbf{k}} = 21\hat{\mathbf{k}}$

$|\mathbf{A}||\mathbf{B}|\sin\theta = (5.83)(6.71)\sin 32.48 = 21$

(e) $\mathbf{B}\times\mathbf{A} = (3\hat{\mathbf{i}} + 6\hat{\mathbf{j}})\times(5\hat{\mathbf{i}} + 3\hat{\mathbf{j}}) = (9 - 30)\,\hat{\mathbf{k}} = -21\hat{\mathbf{k}}$

Solutions using computational software are shown in the Computational Supplements.

Sample Problem 2.17 Determine the components of the 50-kg mass as a weight vector perpendicular and tangent to the plane formed by points A (4,0,2) m, B (0,2,4) m, and C (1,3,0) m. The x–y plane is horizontal and the z-axis is vertical.

Solution Any three non-collinear points in space form a plane. First sketch the three-dimensional axes and locate the points A, B, and C. Sketch the plane formed by these three points to obtain a conceptual feel for the problem.

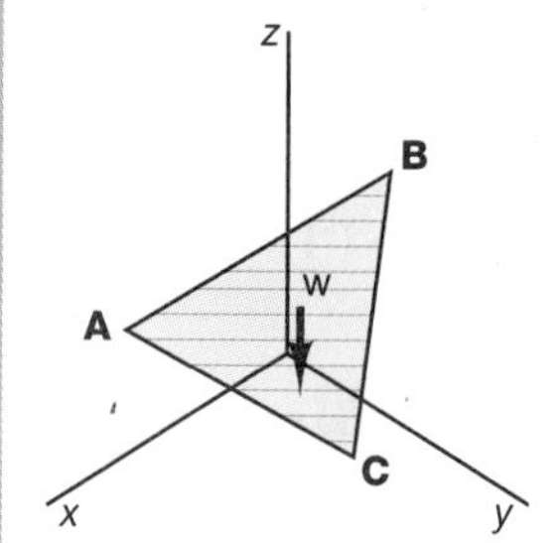

Form vectors from B to A and from B to C.

$$\mathbf{r}_{A/B} = \mathbf{r}_A - \mathbf{r}_B = 4\hat{\mathbf{i}} - 2\hat{\mathbf{j}} - 2\hat{\mathbf{k}}$$
$$\mathbf{r}_{C/B} = \mathbf{r}_C - \mathbf{r}_B = \hat{\mathbf{i}} + \hat{\mathbf{j}} - 4\hat{\mathbf{k}}$$

A unit vector normal to the plane is:

$$\hat{\mathbf{n}} = \frac{\mathbf{r}_{A/B} \times \mathbf{r}_{C/B}}{|\mathbf{r}_{A/B} \times \mathbf{r}_{C/B}|} = 0.549\hat{\mathbf{i}} + 0.768\hat{\mathbf{j}} + 0.329\hat{\mathbf{k}}$$

The weight vector is:

$$\mathbf{W} = -50 * 9.81\hat{\mathbf{k}}$$

The normal and tangential components of this vector are:

$$\mathbf{W}_n = (\mathbf{W}\cdot\hat{\mathbf{n}})\hat{\mathbf{n}} = -88.6\hat{\mathbf{i}} - 124.1\hat{\mathbf{j}} - 53.2\hat{\mathbf{k}}$$
$$\mathbf{W}_t = \mathbf{W} - \mathbf{W}_n = 88.6\hat{\mathbf{i}} + 124.1\hat{\mathbf{j}} - 437.3\hat{\mathbf{k}}$$

The normal component is obtained by use of the dot product with the normal vector and the remaining part of the weight vector must be tangential to the plane.

This vector operation is important when considering objects on inclined planes supported by friction.

Problems

2.96 Given the two vectors $\mathbf{r} = 3\hat{\mathbf{i}} + 2\hat{\mathbf{j}} - \hat{\mathbf{k}}$ and $\mathbf{F} = \hat{\mathbf{i}} + 2\hat{\mathbf{j}} + \hat{\mathbf{k}}$, compute the cross product $\mathbf{r} \times \mathbf{F}$.

2.97 Two vectors are given by $\mathbf{A} = \hat{\mathbf{i}} + \hat{\mathbf{j}}$ and $\mathbf{B} = 3\hat{\mathbf{i}} + 5\hat{\mathbf{j}}$. Calculate the angle between them.

2.98 Calculate the cross product $\mathbf{A} \times \mathbf{B}$ of the two vectors $\mathbf{A} = 3\hat{\mathbf{i}} - \hat{\mathbf{j}} - \hat{\mathbf{k}}$ and $\mathbf{B} = \hat{\mathbf{i}} + \hat{\mathbf{j}}$.

2.99 Consider the three vectors $\mathbf{A} = \hat{\mathbf{i}}$, $\mathbf{B} = 2\hat{\mathbf{i}} + 3\hat{\mathbf{j}} + \hat{\mathbf{k}}$, and $\mathbf{C} = -\hat{\mathbf{i}} - \hat{\mathbf{j}} + \mathbf{k}$. Compute $\mathbf{A} \times (\mathbf{B} + \mathbf{C})$ and $\mathbf{A} \times \mathbf{B} + \mathbf{A} \times \mathbf{C}$ to convince yourself that the distributive law is valid. Use the determinant rule.

2.100 Consider the two vectors $\mathbf{A} = A_x\hat{\mathbf{i}} + 3\hat{\mathbf{j}} + \hat{\mathbf{k}}$ and $\mathbf{B} = \hat{\mathbf{i}} + 2\hat{\mathbf{j}} + \hat{\mathbf{k}}$. Calculate the component A_x so that the vector $\mathbf{C} = \mathbf{A} \times \mathbf{B}$ lies only in the x–y plane. Check to see that the solution $\mathbf{C}$ is perpendicular to both $\mathbf{A}$ and $\mathbf{B}$. Note that, in general, the equation $\mathbf{C} = \mathbf{A} \times \mathbf{B}$ cannot be solved directly for $\mathbf{A}$, given values of $\mathbf{C}$ and $\mathbf{B}$, as discussed in the text.

2.101 Consider the vector $\mathbf{F} = 3\hat{\mathbf{i}} + 3\hat{\mathbf{j}} + \hat{\mathbf{k}}$. Find an x-component of the vector $\mathbf{r} = r_x\hat{\mathbf{i}} + \hat{\mathbf{j}} - \hat{\mathbf{k}}$ such that the vector $\mathbf{M} = \mathbf{r} \times \mathbf{F}$ lies in the x–y plane. Show by direct calculation that $\mathbf{M}$ is perpendicular to both $\mathbf{F}$ and $\mathbf{r}$, as it must be. Sketch the plane made by the vectors $\mathbf{r}$ and $\mathbf{F}$; it should be perpendicular to the x–y plane. Note that, in general, the equation $\mathbf{M} = \mathbf{r} \times \mathbf{F}$ cannot be solved for $\mathbf{r}$, given values of $\mathbf{F}$ and $\mathbf{M}$, as discussed in the text.

2.102 In both statics and dynamics, the multiplication of three vectors together, using both the dot and cross products, occurs repeatedly. This product, called the mixed triple product, or scalar triple product, is $\mathbf{A} \cdot (\mathbf{B} \times \mathbf{C})$. (a) Given $\mathbf{A} = 3\hat{\mathbf{i}} + 2\hat{\mathbf{j}}$, $\mathbf{B} = \hat{\mathbf{i}} - \hat{\mathbf{j}} + \hat{\mathbf{k}}$, and $\mathbf{C} = \hat{\mathbf{i}} + 2\hat{\mathbf{j}} + 3\hat{\mathbf{k}}$, calculate the number (scalar) $\mathbf{A} \cdot (\mathbf{B} \times \mathbf{C})$. (b) Also, calculate $(\mathbf{A} \times \mathbf{B}) \cdot \mathbf{C}$.

2.103 Let $\mathbf{F} = 3\hat{\mathbf{i}} + 10\hat{\mathbf{j}}$ kN, and let $\mathbf{r}_1 = 4\hat{\mathbf{i}} + 5\hat{\mathbf{j}} - 3\hat{\mathbf{k}}$ and $\mathbf{r}_2 = 3\hat{\mathbf{i}} - 2\hat{\mathbf{j}}$, both in meters. Calculate $(\mathbf{r}_1 + \mathbf{r}_2) \times \mathbf{F}$ and $(\mathbf{r}_1 \times \mathbf{F}) + (\mathbf{r}_2 \times \mathbf{F})$, showing that the distributive law holds.

2.104 Use the distributive law of vector cross products and knowledge of the unit vector cross products to compute $\mathbf{r} \times \mathbf{F}$, where $\mathbf{r} = 30\hat{\mathbf{i}} + 10\hat{\mathbf{j}} - 100\hat{\mathbf{k}}$ and $\mathbf{F} = -100\hat{\mathbf{i}} + 200\hat{\mathbf{j}} + 325\hat{\mathbf{k}}$.

2.105 The two vectors $\mathbf{A} = \hat{\mathbf{i}} + \hat{\mathbf{j}} + 3\hat{\mathbf{k}}$ and $\mathbf{B} = -\hat{\mathbf{i}} - \hat{\mathbf{j}} + 3\hat{\mathbf{k}}$ form a plane. Calculate the unit normal vector to this plane.

2.106 Determine the perpendicular part of $\mathbf{B}$ that will satisfy the equation $\mathbf{A} \times \mathbf{B} = \mathbf{C}$ where $\mathbf{A} = 5\hat{\mathbf{i}} - 4\hat{\mathbf{j}} + 2\hat{\mathbf{k}}$ and $\mathbf{C} = -28\hat{\mathbf{i}} - 24\hat{\mathbf{j}} + 22\hat{\mathbf{k}}$.
Check to be sure $\mathbf{A}$ is perpendicular to $\mathbf{C}$ so that a solution exists.

2.107 Determine the perpendicular part of $\mathbf{B}$ that will satisfy the equation $\mathbf{A} \times \mathbf{B} = \mathbf{C}$ where $\mathbf{A} = 2\hat{\mathbf{i}} - 3\hat{\mathbf{j}} + 2\hat{\mathbf{k}}$ and $\mathbf{C} = -16\hat{\mathbf{i}} - 10\hat{\mathbf{j}} + \hat{\mathbf{k}}$. Check to be sure $\mathbf{A}$ is perpendicular to $\mathbf{C}$ so that a solution exists.

2.108 Determine the perpendicular part of the vector $\mathbf{A}$ that satisfies the equation $\mathbf{A} \times \mathbf{B} = \mathbf{C}$ where $\mathbf{B} = -\hat{\mathbf{i}} + 2\hat{\mathbf{j}} + 4\hat{\mathbf{k}}$ and $\mathbf{C} = -20\hat{\mathbf{i}} - 10\hat{\mathbf{j}}$.

2.109 A plane is defined by the three points; $A = (3,1,6)$, $B = (0,5,2)$ and $C = (1,1,8)$. Define the tangential and normal components of a 200-N weight resting on the plane if the vertical direction is in the z direction.

2.110 A ball is placed on a slope and released as shown in Figure P2.110. If two other points on the slope have coordinates $A\,(3,0.5,-0.2)$ m and $B(1,3,-0.3)$ m relative to the ball's initial position, determine a unit vector in the direction the ball will roll.

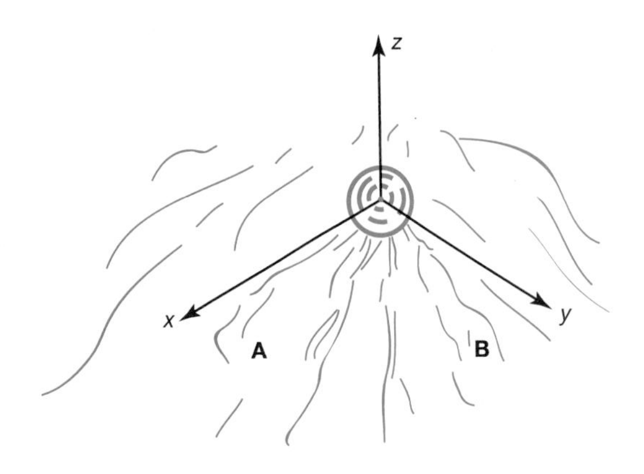

Figure P2.110

2.111 It is desired to track an object moving through space establishing both the position and orientation of the object. Three non-collinear points (A,B,C) are marked on the object and tracked with a three-dimensional motion analysis system. Establish a local coordinate system (x,y,z) attached to the object in terms of the global coordinate system (X,Y,Z) used by the motion analysis system. The x-axis in the local coordinate system should lie along the vector $\mathbf{r}_{B/A}$ and the z-axis should be perpendicular to the plane formed by the points A, B, and C.

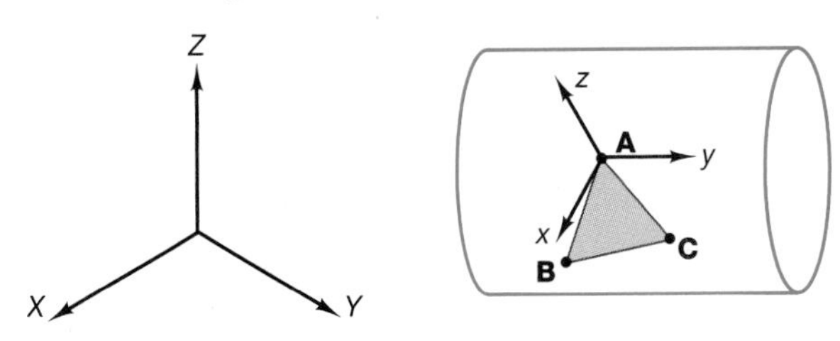

Figure P2.111

a) Develop the general vector equations to relate the unit base vectors in the local system in terms of the global coordinates of the three points and the global unit base vectors. See Figure P2.111.
b) Solve to determine the 3×3 matrix relating the unit vectors in the local system to the unit vectors in the global system, if:

$$\mathbf{r}_A = 1.632\hat{\mathbf{I}} + 2.010\hat{\mathbf{J}} + 4.356\hat{\mathbf{K}}$$
$$\mathbf{r}_B = 1.420\hat{\mathbf{I}} + 2.516\hat{\mathbf{J}} + 3.140\hat{\mathbf{K}}$$
$$\mathbf{r}_C = 1.711\hat{\mathbf{I}} + 3.220\hat{\mathbf{J}} + 3.241\hat{\mathbf{K}}$$

2.10 DIRECT VECTOR SOLUTIONS

A ***direct vector solution*** is the solution of a vector equation that does not expand the vector equation into scalar form to solve the equation. Examples of this have been shown in Eq. (2.131) and Eq. (2.136). In this Chapter, operations of vector algebra have been presented and methods for the solution of a vector equation:

$$A\hat{\mathbf{a}} + B\hat{\mathbf{b}} + C\hat{\mathbf{c}} = \mathbf{D} \tag{2.137}$$

when the magnitudes A, B, and C are unknown have been examined. The two-dimensional form where C was zero and **D** was coplanar with the vectors **A** and **B** was solved using trigonometry first and then using components in an orthogonal coordinate system. The three-dimensional problem was solved by expressing the vectors in the x,y,z coordinate system and then solving the three equations for the three unknowns. Matrix methods were introduced to solve this system of equations and to prepare for the solution of larger systems of equations that are encountered in complex statics problems.

The simplest form of a direct vector solution of Eq. (2.137) occurs when only one vector is unknown, for example, **C**. The vector equation for this case can be written as:

$$\mathbf{C} = \mathbf{D} - \mathbf{A} - \mathbf{B} \tag{2.138}$$

The unknown vector is then obtained by simple vector addition and subtraction. Although the vector **C** may be written as $C\hat{\mathbf{c}}$ involving four unknowns, the magnitude **C** and the three components of the unit vector, these are easily obtained as:

$$C = |\mathbf{C}| \qquad \hat{\mathbf{c}} = \frac{\mathbf{C}}{|\mathbf{C}|} \tag{2.139}$$

In Sections 2.8 and 2.9, two types of multiplication between two vectors, the scalar and vector products, were introduced. It was shown that if two vectors are perpendicular, the dot product between them is zero as the cosine of 90° is zero. The vector product between two vectors that are not parallel yields a vector that is perpendicular to those two vectors. These product definitions can be used to solve Eq. (2.137) without solving a system of linear equations. However, to solve any vector equation, the vectors must be expressed in an orthogonal x,y,z coordinate system.

In the general three-dimensional problem, the unit vectors $\hat{\mathbf{a}}$, $\hat{\mathbf{b}}$, and $\hat{\mathbf{c}}$ cannot be coplanar and in general are not orthogonal. The three linear equations can be formed by taking the dot product of Eq. (2.137) with the three base unit vectors.

$$\begin{aligned} A\hat{\mathbf{a}}\cdot\hat{\mathbf{i}} + B\hat{\mathbf{b}}\cdot\hat{\mathbf{i}} + C\hat{\mathbf{c}}\cdot\hat{\mathbf{i}} &= \mathbf{D}\cdot\hat{\mathbf{i}} \\ A\hat{\mathbf{a}}\cdot\hat{\mathbf{j}} + B\hat{\mathbf{b}}\cdot\hat{\mathbf{j}} + C\hat{\mathbf{c}}\cdot\hat{\mathbf{j}} &= \mathbf{D}\cdot\hat{\mathbf{j}} \\ A\hat{\mathbf{a}}\cdot\hat{\mathbf{k}} + B\hat{\mathbf{b}}\cdot\hat{\mathbf{k}} + C\hat{\mathbf{c}}\cdot\hat{\mathbf{k}} &= \mathbf{D}\cdot\hat{\mathbf{k}} \end{aligned} \tag{2.140}$$

Eq. (2.140) is in the same form as Eq. (2.64) and this system may be solved by hand or using matrix techniques.

A different approach is to form a system of biorthogonal unit vectors. Each of these three biorthogonal vectors will be perpendicular to two of the vectors $\hat{\mathbf{a}}$, $\hat{\mathbf{b}}$, and $\hat{\mathbf{c}}$. These unit vectors are formed using the cross product as follows:

$$\hat{\mathbf{r}} = \frac{\hat{\mathbf{b}} \times \hat{\mathbf{c}}}{|\hat{\mathbf{b}} \times \hat{\mathbf{c}}|} \qquad \hat{\mathbf{s}} = \frac{\hat{\mathbf{c}} \times \hat{\mathbf{a}}}{|\hat{\mathbf{c}} \times \hat{\mathbf{a}}|} \qquad \hat{\mathbf{t}} = \frac{\hat{\mathbf{a}} \times \hat{\mathbf{b}}}{|\hat{\mathbf{a}} \times \hat{\mathbf{b}}|} \tag{2.141}$$

Note, since the original unit vectors $\hat{\mathbf{a}}$, $\hat{\mathbf{b}}$, and $\hat{\mathbf{c}}$ were not mutually perpendicular, the biorthogonal vectors would not be unit vectors unless divided by the magnitude of the cross product. The value of these biorthogonal vectors will be obvious now in a direct solution of the vector equation Eq. (2.137). First, take the dot product of Eq. (2.137) with the unit vector $\hat{\mathbf{r}}$ yielding;

$$A\hat{\mathbf{a}}\cdot\hat{\mathbf{r}} + B\hat{\mathbf{b}}\cdot\hat{\mathbf{r}} + C\hat{\mathbf{c}}\cdot\hat{\mathbf{r}} = \mathbf{D}\cdot\hat{\mathbf{r}}$$

$\hat{\mathbf{r}}$ is perpendicular to both $\hat{\mathbf{b}}$ and $\hat{\mathbf{c}}$. Therefore,

$$A\hat{\mathbf{a}}\cdot\hat{\mathbf{r}} = \mathbf{D}\cdot\hat{\mathbf{r}} \qquad (2.142)$$

$$A = \frac{\mathbf{D}\cdot\hat{\mathbf{r}}}{\hat{\mathbf{a}}\cdot\hat{\mathbf{r}}}$$

In a similar manner, the values of B and C may be obtained:

$$B = \frac{\mathbf{D}\cdot\hat{\mathbf{s}}}{\hat{\mathbf{b}}\cdot\hat{\mathbf{s}}} \qquad C = \frac{\mathbf{D}\cdot\hat{\mathbf{t}}}{\hat{\mathbf{c}}\cdot\hat{\mathbf{t}}} \qquad (2.143)$$

This solution was obtained using only the properties of the cross product and the dot product.

The two-dimensional vector equation:

$$A\hat{\mathbf{a}} + B\hat{\mathbf{b}} = \mathbf{D} \qquad (2.144)$$

where $\hat{\mathbf{a}}$, $\hat{\mathbf{b}}$, and $\mathbf{D}$ lie in the same plane, the *x–y* plane.

This equation may be solved by a direct vector method but the component solution may be as simple so the choice is left to the individual solving the problem. The two component equations are:

$$\begin{aligned} Aa_x + Bb_x &= D_x \\ Aa_y + Bb_x &= D_y \end{aligned} \qquad (2.145)$$

The direct vector solution will treat the unit vector in the z-direction, $\hat{\mathbf{k}}$, as the third unit vector. The biorthogonal vectors to $\hat{\mathbf{a}}$ and $\hat{\mathbf{b}}$ are:

$$\hat{\mathbf{r}} = \hat{\mathbf{k}} \times \hat{\mathbf{b}} \qquad \hat{\mathbf{s}} = \hat{\mathbf{k}} \times \hat{\mathbf{a}} \qquad (2.146)$$

The values of A and B are:

$$A = \frac{\mathbf{D}\cdot\hat{\mathbf{r}}}{\hat{\mathbf{a}}\cdot\hat{\mathbf{r}}} \qquad B = \frac{\mathbf{D}\cdot\hat{\mathbf{s}}}{\hat{\mathbf{b}}\cdot\hat{\mathbf{s}}} \qquad (2.147)$$

The direct vector solution has been obtained by forming a biorthogonal set of unit vectors. This was done for conformance with other advanced mathematical techniques but the solution to Eq. (2.137) could have been written as:

$$A = \frac{\mathbf{D}\cdot(\hat{\mathbf{b}} \times \hat{\mathbf{c}})}{\hat{\mathbf{a}}\cdot(\hat{\mathbf{b}} \times \hat{\mathbf{c}})} \qquad B = \frac{\mathbf{D}\cdot(\hat{\mathbf{c}} \times \hat{\mathbf{a}})}{\hat{\mathbf{b}}\cdot(\hat{\mathbf{c}} \times \hat{\mathbf{a}})} \qquad C = \frac{\mathbf{D}\cdot(\hat{\mathbf{a}} \times \hat{\mathbf{b}})}{\hat{\mathbf{c}}\cdot(\hat{\mathbf{a}} \times \hat{\mathbf{b}})}$$

Sample Problem 2.18 Solve the following vector equation for the magnitudes A, B, and C.

$$A\hat{\mathbf{a}} + B\hat{\mathbf{b}} + C\hat{\mathbf{c}} = \mathbf{D} \text{ where}$$

$$\hat{\mathbf{a}} = 0.231\hat{\mathbf{i}} + 0.308\hat{\mathbf{j}} + 0.923\hat{\mathbf{k}}$$

$$\hat{\mathbf{b}} = -0.231\hat{\mathbf{i}} + 0.308\hat{\mathbf{j}} + 0.923\hat{\mathbf{k}}$$

$$\hat{\mathbf{c}} = -0.385\hat{\mathbf{i}} + 0.923\hat{\mathbf{k}}$$

$$\mathbf{D} = 100\hat{\mathbf{k}}$$

Solution The three biorthogonal unit vectors can be formed:

$$\mathbf{R} = \hat{\mathbf{b}} \times \hat{\mathbf{c}} = -0.284\hat{\mathbf{i}} - 0.568\hat{\mathbf{j}} - 0.118\hat{\mathbf{k}} \quad |\mathbf{R}| = 0.646$$

$$\hat{\mathbf{r}} = \frac{\mathbf{R}}{|\mathbf{R}|} = -0.440\hat{\mathbf{i}} - 0.879\hat{\mathbf{j}} - 0.183\hat{\mathbf{k}}$$

$$\mathbf{S} = \hat{\mathbf{c}} \times \hat{\mathbf{a}} = -0.284\hat{\mathbf{i}} + 0.568\hat{\mathbf{j}} - 0.118\hat{\mathbf{k}} \quad |\mathbf{S}| = 0.646$$

$$\hat{\mathbf{s}} = \frac{\mathbf{S}}{|\mathbf{S}|} = -0.440\hat{\mathbf{i}} + 0.879\hat{\mathbf{j}} - 0.183\hat{\mathbf{k}}$$

$$\mathbf{T} = \hat{\mathbf{a}} \times \hat{\mathbf{b}} = 0.568\hat{\mathbf{i}} - 0.142\hat{\mathbf{k}} \quad |\mathbf{T}| = 0.586$$

$$\hat{\mathbf{t}} = \frac{\mathbf{T}}{|\mathbf{T}|} = 0.970\hat{\mathbf{i}} - 0.243\hat{\mathbf{k}}$$

The values of the vector magnitudes may now be obtained.

$$A = \frac{\hat{\mathbf{r}} \cdot \mathbf{D}}{\hat{\mathbf{r}} \cdot \hat{\mathbf{a}}} = 33.854$$

$$B = \frac{\hat{\mathbf{s}} \cdot \mathbf{D}}{\hat{\mathbf{s}} \cdot \hat{\mathbf{b}}} = 33.854$$

$$C = \frac{\hat{\mathbf{t}} \cdot \mathbf{D}}{\hat{\mathbf{t}} \cdot \hat{\mathbf{c}}} = 40.625$$

These calculations are time consuming but most computational software programs will perform them quickly.

Problems

2.112 Given a vector $\mathbf{A} = 8\hat{\mathbf{i}} - 3\hat{\mathbf{j}} + 2\hat{\mathbf{k}}$ and a unit vector $\hat{\mathbf{u}} = 0.577\hat{\mathbf{i}} + 0.577\hat{\mathbf{j}} + 0.577\hat{\mathbf{k}}$, determine the components of **A** parallel and perpendicular to $\hat{\mathbf{u}}$.

2.113 Using the direct vector method, determine the magnitudes of **A** and **B** to satisfy the equation $\mathbf{A} + \mathbf{B} = \mathbf{C}$, if

$$\hat{\mathbf{a}} = 0.707\hat{\mathbf{i}} + 0.707\hat{\mathbf{j}}$$

$$\hat{\mathbf{b}} = \hat{\mathbf{j}}$$

$$\mathbf{C} = 4\hat{\mathbf{i}} - 3\hat{\mathbf{j}}$$

2.114 Verify the solution of 2.113 by solving the scalar equations from the vector equation.

2.115 Using the direct vector method, determine the magnitudes of **A**, **B**, and **C** to satisfy the equation $\mathbf{A} + \mathbf{B} + \mathbf{C} = \mathbf{D}$ if **A**, **B**, and **C** lie along lines defined by the vectors:

$$\mathbf{A}l = 3\hat{\mathbf{i}} + 4\hat{\mathbf{j}} + 10\hat{\mathbf{k}}$$

$$\mathbf{B}l = -4\hat{\mathbf{i}} - 3\hat{\mathbf{j}} + 12\hat{\mathbf{k}}$$

$$\mathbf{C}l = -3\hat{\mathbf{i}} + 4\hat{\mathbf{j}} + 10\hat{\mathbf{k}}$$

and

$$\mathbf{D} = 100\hat{\mathbf{k}}$$

2.116 Verify the solution of 2.115 by expanding the vector equation into a system of equations.

2.117 Using the direct vector method, determine the magnitudes of the unknown vectors, **A**, **B**, and **C** if their directions are known for solution of the vector equation:

$$A\hat{\mathbf{a}} + B\hat{\mathbf{b}} + C\hat{\mathbf{c}} = 250\hat{\mathbf{k}}$$

$$\hat{\mathbf{a}} = 0.318\hat{\mathbf{i}} + 0.424\hat{\mathbf{j}} + 0.848\hat{\mathbf{k}}$$

$$\hat{\mathbf{b}} = -0.566\hat{\mathbf{i}} + 0.226\hat{\mathbf{j}} + 0.793\hat{\mathbf{k}}$$

$$\hat{\mathbf{c}} = 0.324\hat{\mathbf{i}} - 0.487\hat{\mathbf{j}} + 0.811\hat{\mathbf{k}}$$

Verity this solution by expanding the equation into three scalar equations and solving them.

2.118 Using the direct vector method, solve the vector equation:

$A\hat{\mathbf{a}} + B\hat{\mathbf{b}} + C\hat{\mathbf{c}} = 191\hat{\mathbf{i}} + 217\hat{\mathbf{j}} - 68\hat{\mathbf{k}}$ where

$$\hat{\mathbf{a}} = 0.6\hat{\mathbf{i}} + 0.8\hat{\mathbf{j}}$$

$$\hat{\mathbf{b}} = 0.707\hat{\mathbf{i}} + 0.707\hat{\mathbf{k}}$$

$$\hat{\mathbf{c}} = 0.385\hat{\mathbf{j}} + 0.923\hat{\mathbf{k}}$$

Verify the solution by solving the three scalar component equations.

2.119 Using the direct vector method, solve the vector equation

$A\hat{\mathbf{a}} + B\hat{\mathbf{b}} + C\hat{\mathbf{c}} = 25\hat{\mathbf{i}} + 135\hat{\mathbf{j}} + 90\hat{\mathbf{k}}$ where

$$\hat{\mathbf{a}} = 0.6\hat{\mathbf{i}} + 0.8\hat{\mathbf{j}}$$

$$\hat{\mathbf{b}} = -0.707\hat{\mathbf{i}} + 0.707\hat{\mathbf{k}}$$

$$\hat{\mathbf{c}} = 0.707\hat{\mathbf{j}} + 0.707\hat{\mathbf{k}}$$

Verify the solution by solving the three scalar equations.

2.120 Using the direct vector method, solve the vector equation: $\mathbf{A} + \mathbf{B} + \mathbf{C} = \mathbf{D}$ where

$$\hat{\mathbf{a}} = 0.577\hat{\mathbf{i}} + 0.577\hat{\mathbf{j}} + 0.577\hat{\mathbf{k}}$$

$$\hat{\mathbf{b}} = -0.707\hat{\mathbf{i}} + 0.707\hat{\mathbf{k}}$$

$$\hat{\mathbf{c}} = 0.577\hat{\mathbf{i}} - 0.577\hat{\mathbf{j}} - 0.577\hat{\mathbf{k}}$$

$$\mathbf{D} = -6\hat{\mathbf{i}} + 87\hat{\mathbf{j}} + 120\hat{\mathbf{k}}$$

Verify the solution by solving the three scalar equations.

2.121 Using the direct vector method, solve the vector equation: $\mathbf{A} + \mathbf{B} + \mathbf{C} = \mathbf{D}$ where

$$\hat{\mathbf{a}} = 0.302\hat{\mathbf{i}} - 0.905\hat{\mathbf{j}} + 0.302\hat{\mathbf{k}}$$
$$\hat{\mathbf{b}} = -0.371\hat{\mathbf{i}} - 0.928\hat{\mathbf{k}}$$
$$\hat{\mathbf{c}} = -0.707\hat{\mathbf{j}} - 0.707\hat{\mathbf{k}}$$
$$\mathbf{D} = 12\hat{\mathbf{i}} - 55\hat{\mathbf{j}} + 10\hat{\mathbf{k}}$$

Verify the solution by solving the three scalar equations.

2.122 Using the direct vector method, solve the vector equation: $\mathbf{A} + \mathbf{B} + \mathbf{C} = \mathbf{D}$ where

$$\hat{\mathbf{a}} = 0.302\hat{\mathbf{i}} + 0.905\hat{\mathbf{j}} + 0.302\hat{\mathbf{k}}$$
$$\hat{\mathbf{b}} = -0.371\hat{\mathbf{i}} - 0.928\hat{\mathbf{k}}$$
$$\hat{\mathbf{c}} = -0.707\hat{\mathbf{j}} + 0.707\hat{\mathbf{k}}$$
$$\mathbf{D} = 12\hat{\mathbf{i}} + 50\hat{\mathbf{j}} + 10\hat{\mathbf{k}}$$

Verify the solution by solving the three scalar equations.

Chapter Summary

This chapter has focused on vector algebra and methods of solution of vector equations. Two-dimensional equations may be solved using trigonometry after forming vector diagrams. However, the trigonometric method does not work well for three-dimensional vector equations. For three-dimensional problems, a reference coordinate system is required. The coordinate system is established by the individual working the problem. In some cases, specific choices of the coordinate system will reduce the labor in solving the problem. The vector equation is expressed in terms of components in the chosen coordinate system. The equation can then be expanded into three scalar equations that can be solved as a system of equations either by hand or use of linear algebra. Matrix methods have the advantage of reduction of errors and determining situations where errors have produced a system of linearly dependent equations. For some vector equations, a direct vector solution is available that does not require expansion of the equation into scalar components.

The remaining chapters will concentrate on modeling equilibrium conditions and forming the required vector equations. These equations can be solved by one or more of the methods introduced in this chapter.

Chapter 3

PARTICLE EQUILIBRIUM

The rock climber can be modeled as a particle with all the forces concurrent with her center of gravity. (Photo courtesy of Joe Gough/Shutterstock)

3.1 FREE-BODY DIAGRAMS OF A PARTICLE

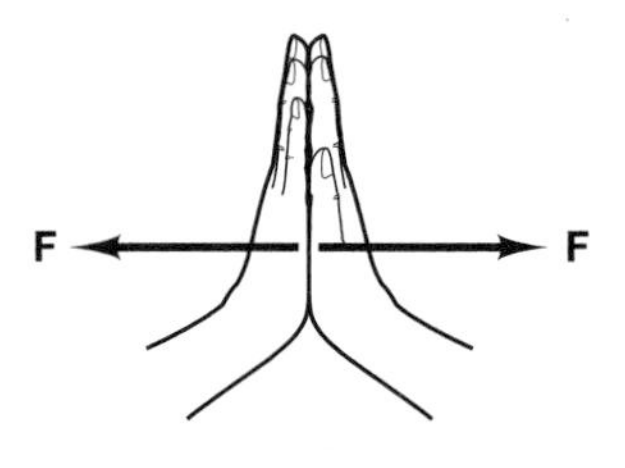

Figure 3.1 When two people press their hands together, each person exerts a force in one direction and feels an equal force in the opposite direction.

Before examining concepts of modeling, let us review Newton's third law: "*If one body exerts a force on a second body, then the second body exerts a force on the first body that is equal in magnitude, opposite in direction and collinear.*" This is the concept of action and reaction. Suppose that two people press their right hands together, as shown in Figure 3.1. The person on the right is pushing to the left (action), but senses the force pushing to the right (reaction) from the person on the left. Who is generating the force and who is resisting cannot be determined. The action and reaction forces are equal, opposite, and collinear.

Newton's third law can be used to isolate a particle from its environment and represent all the forces that act upon the particle, including gravitational attraction when appropriate. This isolation of the particle and representation of the forces acting upon it produces a free-body diagram. This is one of the most useful skills that an engineer can obtain. Drawing a free-body diagram is an art, and can be learned only by practice. If a correct free-body diagram is constructed, then the balance of the solution can be accomplished in a very systematic manner.

A formal definition of a free-body diagram is as follows:

A free-body diagram is a sketch of an object or a connected group of objects, that isolates the body from its environment or surrounding bodies and represents the interactions of its environment by appropriate external forces.

If all the forces acting on an object can be considered to act on a single point (a concurrent force system), the object can be modeled as a particle. If the forces cannot be modeled as concurrent forces, the object cannot be modeled as a particle and the geometry of the body must be considered. Concurrent forces produce no turning effect (moments) about the point of application.

There may be—and usually are—internal forces acting within the isolated body, but, by Newton's third law, these forces do not contribute to the resultant of the external forces acting on the body. The external forces may be due to contact with other bodies or due to the attraction of separated bodies. If all of the external forces can be treated as concurrent forces, the body can be modeled as a *particle*. In that case, only the magnitudes and directions of the external forces need to be shown. In section 4.2, we will use free-body diagrams to model objects that cannot be treated as particles; that is, the forces acting on them are not concurrent. In these cases, the point of application of each external force is important and must be accurately shown in the free body diagram.

The construction of a free-body diagram is an essential step in solving any problem in mechanics. Before such a diagram can be developed, however, a *space diagram* must be available. The space diagram shows the actual object within its environment. The diagram may be a photograph, a blueprint, a scale model, a sketch, or any other means that allows the determination of dimensions and interactions between objects. Space diagrams are presented as figures in most of the homework problems in this text.

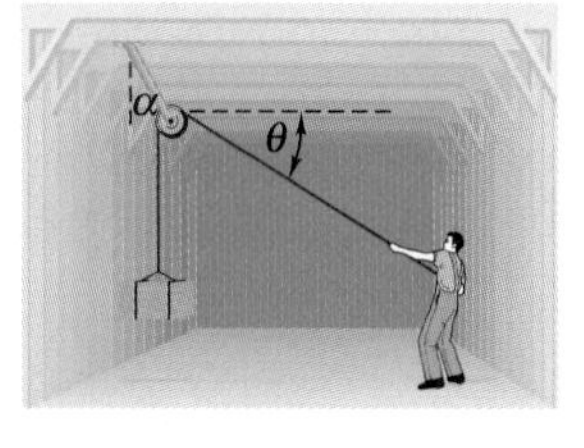

Figure 3.2 Space diagram of a rancher holding a bale of hay with a rope and pulley.

Figure 3.2 shows a rancher holding a bale of hay by means of a pulley hung from the ceiling of the barn. This space diagram is the first step to modeling a real problem, and it must be examined before a free-body diagram can be prepared. Suppose that you walked into a barn and observed someone supporting a bale of hay by the system shown. If one wanted to be sure that the cable between the pulley and the barn rafter was strong enough, you would need to know or measure two things. First, one would need to know the weight of the bale of hay; second, one would need to determine the angle θ that the rope makes with the horizontal. This angle might be calculated by measuring the length of the rope between the pulley and the rancher's hands and the horizontal distance from the hands to the vertical part of the rope and then using trigonometry. In general, with a little experience, a very good estimate of an angle can be made by observation. In Section 2, it will be shown that if we know the angle α the cable makes with the ceiling, we can determine the weight of the bale of hay. The weight and the angle are related and cannot be specified independently of each other.

The space diagram can be used to draw a free-body diagram of the pulley, showing all the forces acting upon it. The pulley is then isolated and treated as a particle. The weight of the bale of hay is designated by a force vector **W**. The rope transmits this force to the pulley. If there is no friction present in the pulley, the tension in the rope is constant. Therefore, the rancher must exert a force **F** on the other end of the rope that is equal to the weight of the bale or the magnitude of the **W** vector. Neglecting the pulley's weight, we find that the only other force acting on the pulley is the tension **T**, acting along the cable to the ceiling. The completed free-body diagram of the pulley is shown in Figure 3.3.

Figure 3.3

As a free-body diagram is developed, it is necessary to specify what is known or can be measured, and what must be calculated by the equations of statics. If the weight of the bale of hay is known, that weight can be shown on the free-body diagram or written in equation form, such as

$$\mathbf{W} = -200\hat{\mathbf{j}}\ \text{N}$$

Note that since the weight is known, the magnitude of the vector **F** is also known to be 200 N, based upon the assumptions made earlier.

The space sketch frequently contains more dimensions than may be required to solve the problem, but it is always better to have too much information than not enough. The construction of a free-body diagram is essential to the correct solution of any problem in mechanics.

The force due to friction between two bodies will be discussed in detail in special section 3.5A and Chapter 9. This force always opposes motion and has a maximum value which is dependent on the materials of the contacting bodies and the normal force between the contacting surfaces.

Sample Problem 3.1

A man is pushing his sports car along level ground as shown in the figure below. In the photo, the man's arms appear to make an angle of approximately 40° with the horizontal. If we assume that the reaction force between the man's legs and the ground acts parallel to his right leg (farthest from the car), then the angle this force makes with the vertical is approximately 30°. Construct a free-body diagram of the man, modeled as a particle, showing the forces acting on him.

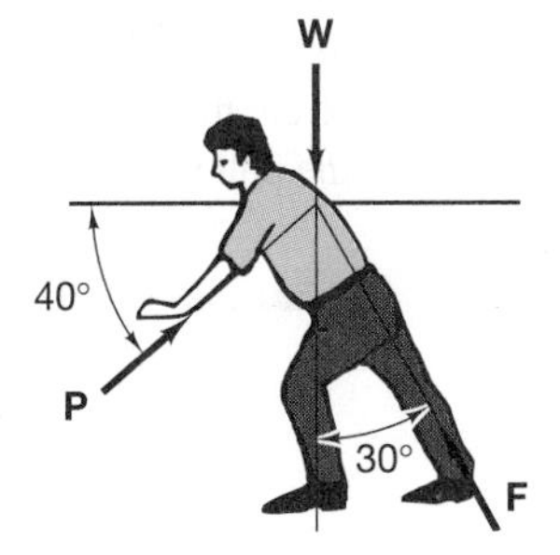

Solution The photo serves as a space diagram for this situation. We construct the free-body diagram at the right, assuming that the weight of the man is known. Notice that, in this case, the man has been modeled as a particle under the influence of coplanar, concurrent forces. The force **P** is the reaction force from the car, the force **F** is the ground reaction force, and the weight **W** is due to gravitational attraction. The force on the man's left leg is neglected, and **F** is parallel to his right leg. Note that the force between the ground and the feet is composed of a normal force to the ground and a friction force. If the man is modeled as a particle, the three forces **W**, **P**, and **F** form a concurrent force system.

Sample Problem 3.2

The tripod shown at the left supports a movie camera. Each leg is in compression and exerts a force on the camera. Based upon the space diagram of the camera and tripod shown, construct a free-body diagram, taking the camera to be a particle.

Solution

First, examine the space drawing to determine whether all the necessary measurements were taken to allow a proper free-body diagram to be drawn. Note that the lengths of the legs of the tripod are not given, but these may be found from the dimensions that are given. It is also necessary to know the locations of the bottom of the tripod legs in order to know their orientation in space. To analyze the problem, these orientations expressed by the use of unit vectors acting along each leg. Thus, the direction of the compressive force on each leg is known. These forces would have to be compressive, or the legs would lift off the floor. The direction and magnitude of the weight of the camera is known, and so a free-body diagram can be drawn in which $\hat{\mathbf{a}}$, $\hat{\mathbf{b}}$, and $\hat{\mathbf{c}}$ are unit vectors acting on the legs and A, B, and C represent the magnitudes of the compressive forces on the legs. (See diagram below.)

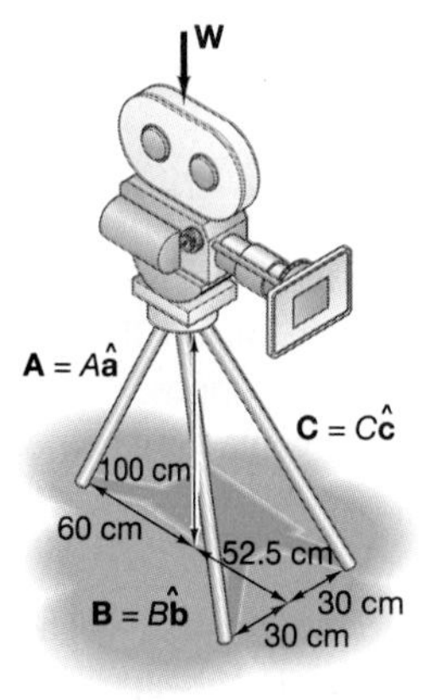

Sample Problem 3.3

For the pulley system shown below, construct a free-body diagram to be used to determine the tension in the cable if friction is neglected in the pulleys.

Solution

If friction is neglected, the tension T in the cable is constant. The block and the pulley attached to the cable will be considered a single particle, and the cable will be cut in three places by the section line a–a. Note that if only the block were used for the free-body diagram, two different tensions would act on the block, and a second free-body diagram would be needed for the pulley. Therefore, the simplest free-body diagram is that shown below.

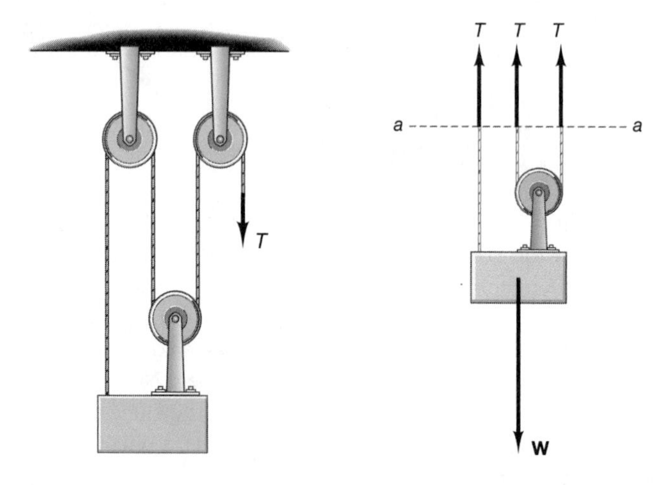

Problems

3.1 A box of mass 20 kg remains at rest on a 30° incline. (See Figure P3.1.) Knowing that frictional resistance holds the box in place, construct a free-body diagram of the box, modeled as a particle.

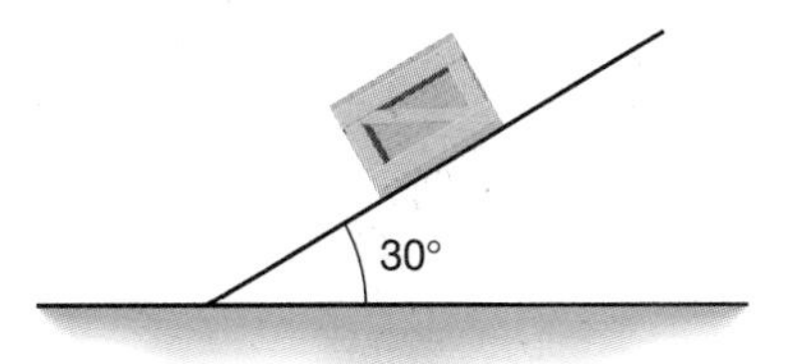

Figure P3.1

3.2 A climber is rappelling down a cliff, as shown in Figure P3.2. Construct a free-body diagram of the climber.

Figure P3.2

3.3 A 100-kg box is supported as shown in the space diagram of Figure P3.3. Construct a free-body diagram of the mass.

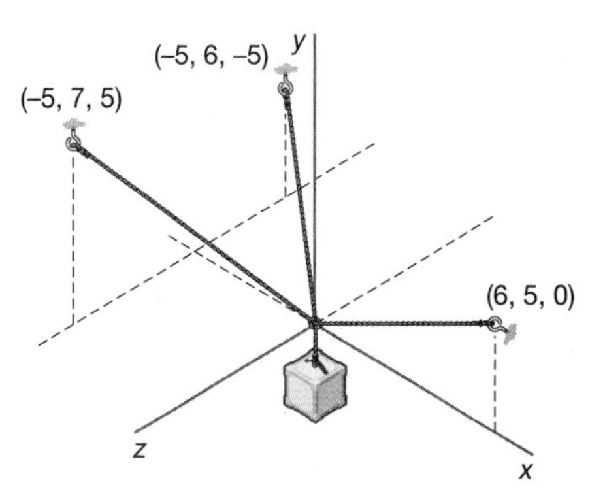

Figure P3.3

3.4 Two cylinders of weight W and radius R rest against smooth planes, as shown in Figure P3.4. Construct a free-body diagram of each cylinder.

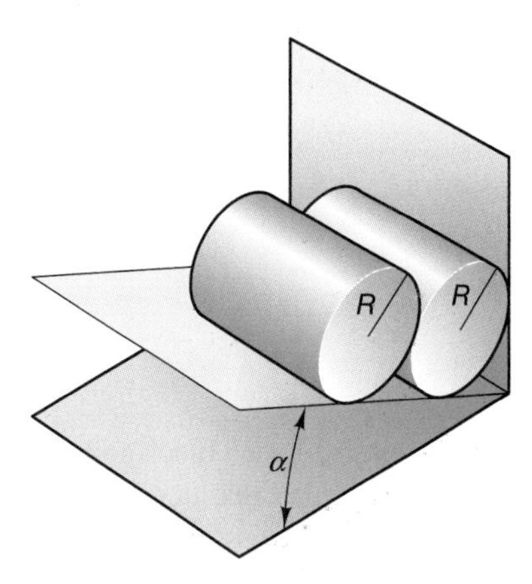

Figure P3.4

3.5 A 100-kg block is suspended from a pulley so that it remains at rest. (See Figure P3.5.) Construct a free-body diagram of the block m.

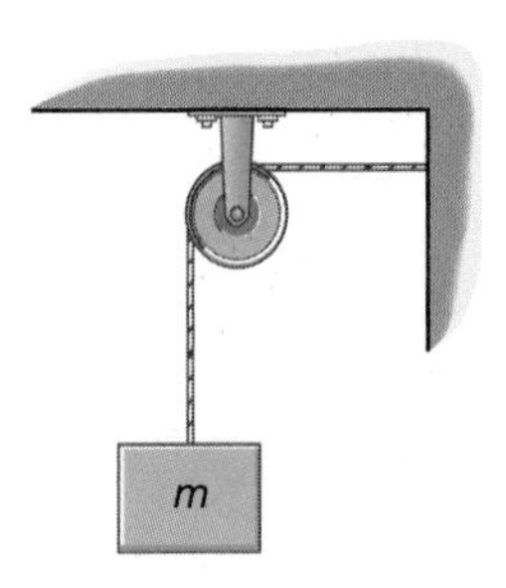

Figure P3.5

3.6 A pulley and two blocks are held in place by frictional resistance on mass m_1 as shown in Figure P3.6. Construct a free-body diagram of each block, modeled as a particle.

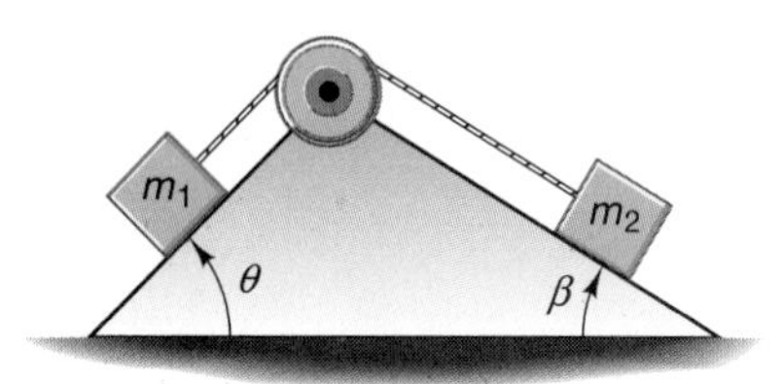

Figure P3.6

3.7 In Figure P3.7, the block m_2 is held in place by a resistive friction force. Construct a free-body diagram of m_2, modeled as a particle.

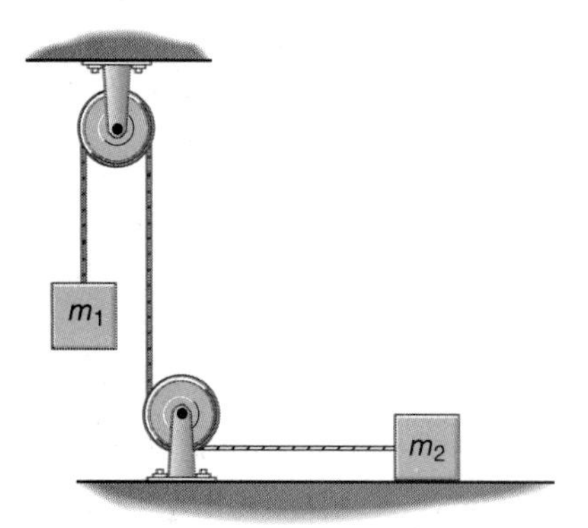

Figure P3.7

3.8 Construct a free-body diagram of each block and pulley as shown in Figure P3.8, assuming that the system is "balanced," not moving, and frictionless. (That is, the tension in the continuous cable is the same on either side of a pulley.) Treat each component as a particle by neglecting the radius of the pulley.

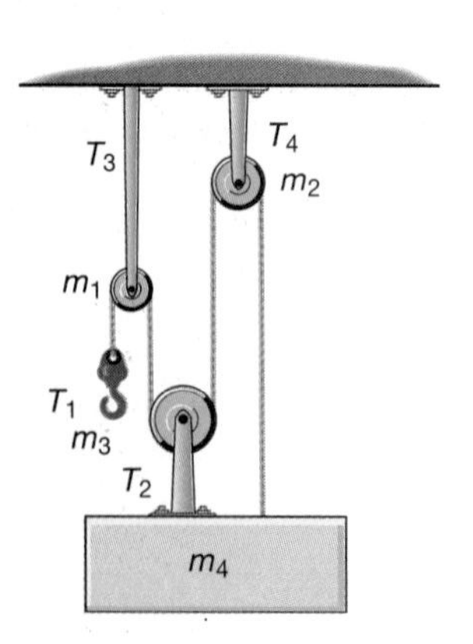

Figure P3.8

3.9 A crate of mass m is held in place by three ropes, as indicated in Figure P3.9. The position of the end of each rope is indicated by coordinates, given in meters. The magnitude of each force is given symbolically by T_1, T_2, and T_3. Construct the free-body diagram of the crate, and then write each vector acting on the mass in $\hat{\mathbf{i}}, \hat{\mathbf{j}}, \hat{\mathbf{k}}$ coordinates. Model the crate as a point mass.

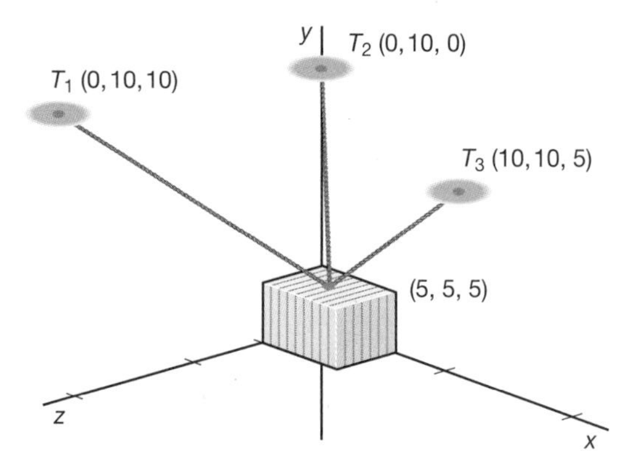

Figure P3.9

3.10 A crate of mass m is held in place by three ropes, as indicated in Figure P3.10. The position of the end of each rope is indicated by coordinates, given in feet. The magnitude of each force is given symbolically by T_1, T_2, and T_3. Construct the free-body diagram of the crate, and then write each vector acting on the mass in $\hat{\mathbf{i}}, \hat{\mathbf{j}}, \hat{\mathbf{k}}$ coordinates. Model the crate as a point mass.

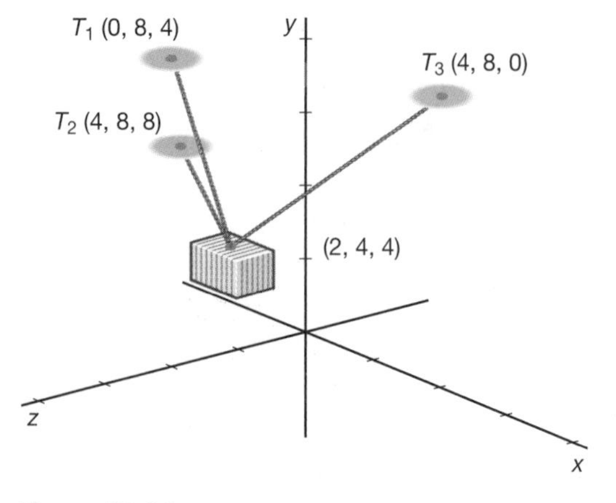

Figure P3.10

3.2 EQUILIBRIUM OF A PARTICLE

In the preceding chapter, the vector properties of a force were discussed and the resultant of a series of concurrent forces was defined as the sum of those forces. If the resultant of all the forces acting on a particle is zero, the particle is in ***equilibrium***. Newton's second law states that "***the change in motion (or the acceleration) of a particle is proportional to the net force exerted upon it.***" The net force is equal to the resultant force. When the particle is in equilibrium, the resultant force is zero, and the particle remains at rest or continues to move at a constant speed in a given direction (constant velocity). This is a special case of dynamics and should always be considered as such.

Equilibrium problems are approached by first observing that the particle remains at rest or moves with a constant velocity. Then, by Newton's second law, there can be no net force acting on the particle. If a particle in equilibrium has m concurrent forces acting upon it, the sum of these forces is zero:

$$\mathbf{R} = \sum_{i=1}^{m} \mathbf{F}_i = 0 \tag{3.1}$$

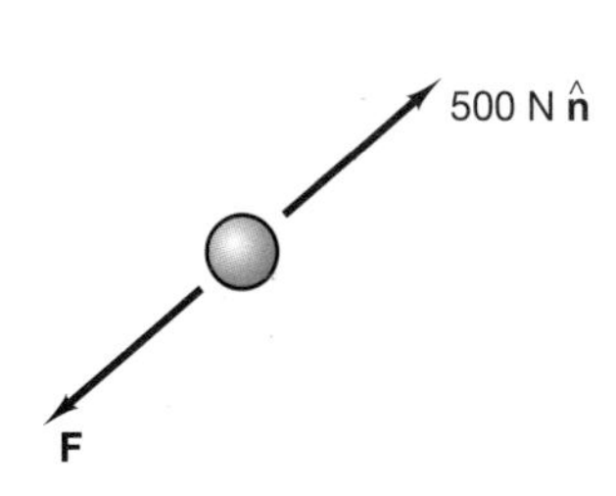

Figure 3.4

The concept of equilibrium will be used not to solve for the resultant, which is zero, but to solve for the required unknown forces that act on the particle and maintain its equilibrium.

For example, suppose that the particle shown in Figure 3.4 is observed to be in equilibrium under a known force of 500 N in the $\hat{\mathbf{n}}$ direction and another, unknown, force $\mathbf{F}$. Then the force $\mathbf{F}$ can be determined by the equation of equilibrium.

The resultant, or the sum of the forces, must be zero. Therefore;

$$\mathbf{R} = \mathbf{F} + 500\,\hat{\mathbf{n}} = 0$$
$$\mathbf{F} = -500\,\hat{\mathbf{n}}$$

The force $\mathbf{F}$ has a magnitude of 500 N and has the same line of action indicated by the unit vector $\hat{\mathbf{n}}$, but in the opposite direction $(-\hat{\mathbf{n}})$.

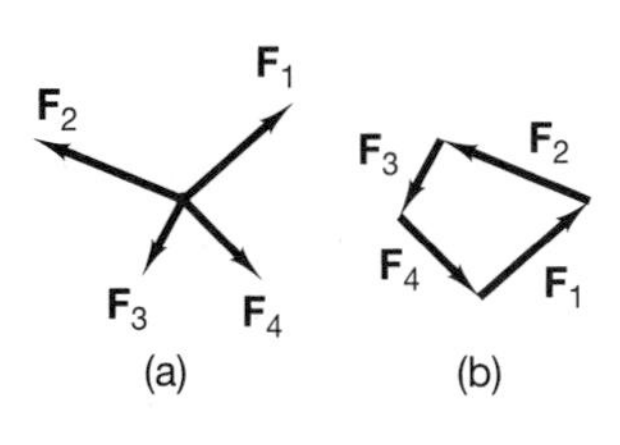

Figure 3.5 (a) A particle in equilibrium under the action of four forces. (b) The force polygon of the forces in part (a).

Now consider a particle observed to be in equilibrium with more than two forces acting upon it, as in Figure 3.5(a). If the forces are coplanar as well as concurrent, the force polygon shown in Figure 3.5(b) could be used as a graphical means to solve for unknown forces. In cases of equilibrium, these force polygons are closed; that is, the resultant is zero. This is written in vector notation as

$$\mathbf{R} = 0 = \mathbf{F}_1 + \mathbf{F}_2 + \mathbf{F}_3 + \mathbf{F}_4 \tag{3.2}$$

Equation (2.99) may be written in matrix notation as

$$\begin{pmatrix} F_{1x} \\ F_{1y} \\ F_{1z} \end{pmatrix} + \begin{pmatrix} F_{2x} \\ F_{2y} \\ F_{2z} \end{pmatrix} + \begin{pmatrix} F_{3x} \\ F_{3y} \\ F_{3z} \end{pmatrix} + \begin{pmatrix} F_{4x} \\ F_{4y} \\ F_{4z} \end{pmatrix} = 0 \tag{3.2a}$$

As has been discussed earlier, although vectors may be added by using trigonometry, the use of Cartesian components greatly simplifies the solution. For three-dimensional problems, force vector diagrams are difficult to draw, and solutions may be obtained using the scalar equations derived from the vector equilibrium equation. The three scalar equations of equilibrium for a particle are:

$$\begin{aligned} F_{1x} + F_{2x} + F_{3x} + F_{4x} &= 0 \\ F_{1y} + F_{2y} + F_{3y} + F_{4y} &= 0 \\ F_{1z} + F_{2z} + F_{3z} + F_{4z} &= 0 \end{aligned} \tag{3.2b}$$

or

$$\mathbf{R} = \sum \mathbf{F} = 0$$
$$\sum (F_x\hat{\mathbf{i}} + F_y\hat{\mathbf{j}} + F_z\hat{\mathbf{k}}) = 0$$
$$(\sum F_x)\hat{\mathbf{i}} + (\sum F_y)\hat{\mathbf{j}} + (\sum F_z)\hat{\mathbf{k}} = 0$$

The last expression requires the coefficients of the unit base vector to be equal to zero resulting in

$$\sum F_x = 0$$
$$\sum F_y = 0 \qquad (3.3)$$
$$\sum F_z = 0$$

Therefore, the sum of the forces in each orthogonal direction must be zero. The last three equations are the scalar equations of equilibrium for a particle, and each must be satisfied independently. There are three scalar equations of equilibrium in a three-dimensional problem involving a particle, and therefore only three unknowns can be determined. Occasionally, problems will be encountered in which there are more unknowns than there are equations of equilibrium. This may be the result of an error in modeling; that is, some information that was available from the space diagram was not considered. However, there are instances when the particle is overconstrained and the problem is statically indeterminate. ***Statically indeterminate*** means that the solution cannot be determined solely by using the concept of static equilibrium.

In the free-body diagrams shown in Section 3.1, the tensile and compressive forces in support members were shown acting along the center line of the support member. Therefore, the line of action of the force is along a supporting rope or cable, and the cables must be in tension, as they cannot support compressive forces. If the supports are bars such as the legs on the tripod, the bars are in compression. If the legs are intended to transmit tension, they would have to be anchored to the floor in order to transmit that tension to the floor. When solving a statics problem, we initially specify the sense of the forces on the free-body diagram. If the supports are such that they can transmit only tension or compression, this should be specified on the diagram. If the results of the analysis are negative, indicating an opposite sense, then either an error has been made or the particle has been improperly supported. The examination of functional restrictions of supports is of great importance to the solution of mechanics problems.

Sample Problem 3.4

Consider the pulley system used to hold a 200-N bale of hay in Figure 3.2. Determine the force **F** and the tension **T** in the system if the angle θ is specified.

Solution

The free-body diagram was constructed in Figure 3.3, and the solution of the static equilibrium problem is straightforward. The problem can be solved for any angle θ that the rope makes with the horizontal, from 0 to 90 degrees. (See figure at left.) If there is no friction in the pulley, the magnitude of the force **F** applied to the rope will equal the weight **W**. Treating **T** as an unknown vector, we have, for the equation of equilibrium,

$$\mathbf{R} = \sum F = 0$$
$$\mathbf{R} = \mathbf{T} + 200\,(\cos\theta\hat{\mathbf{i}} - \sin\theta\hat{\mathbf{j}}) - 200\hat{\mathbf{j}} = \mathbf{0}$$

Therefore,

$$\mathbf{T} = 200\left[-\cos\theta\hat{\mathbf{i}} + (1 + \sin\theta)\hat{\mathbf{j}}\right]$$

The magnitude of **T** is

$$|\mathbf{T}| = 200\sqrt{\cos^2\theta + (1 + \sin\theta)^2} = 200\sqrt{2 + 2\sin\theta}$$

Note that the minimum value of **T** occurs when the rancher pulls in a horizontal direction so that the angle θ is zero and the tension is $T = 283$ N. The maximum value of the tension can be found by setting the derivative of the magnitude of the tension with respect to θ equal to 0:

$$\frac{d}{d\theta}|\mathbf{T}| = \frac{d}{d\theta}200\sqrt{2 + 2\sin\theta} = \frac{-200\cos\theta}{\sqrt{2 + 2\sin\theta}} = 0$$

$$\theta = \pm 90^\circ$$

The maximum value occurs when the rancher pulls straight down so that the angle θ is 90° and $T = 400$ N. The tangent of the angle **T** makes with the vertical is

$$\tan\alpha = -\frac{T_x}{T_y} = \frac{\cos\theta}{(1 + 2\sin\theta)}$$

$$\text{for } \theta = 0, \quad \alpha = 45^\circ$$

$$\theta = 90^\circ, \quad \alpha = 0$$

If the angle θ is 30°, then

$$\mathbf{T} = 200\left[-\cos 30^\circ\hat{\mathbf{i}} + (1 + \sin 30^\circ)\hat{\mathbf{j}}\right] = -173\hat{\mathbf{i}} + 300\hat{\mathbf{j}}$$

This equation completely specifies the tension, but it also can be written in terms of the magnitude of the tension and the angle α: $|\mathbf{T}| = 346$ N, $\alpha = 30^\circ$.

Sample Problem 3.5

Consider a scoreboard of mass 200 kg supported by three cables, as shown at the left. Determine the tension in each of the cables.

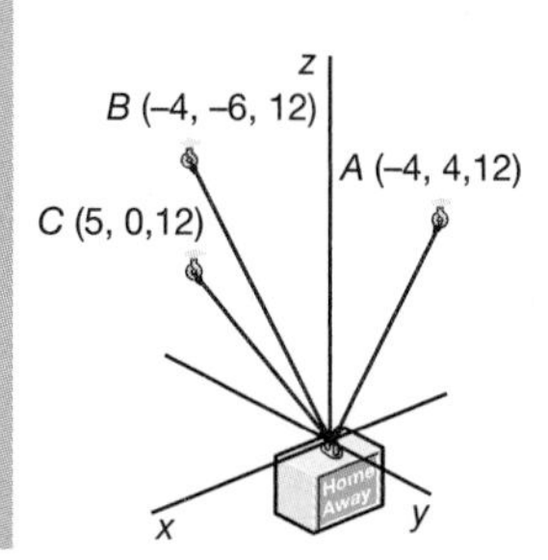

Solution Unit vectors along the cables may be determined from the space diagram. Unit vectors were obtained along each of the cables in that sample problem.

$$\hat{\mathbf{a}} = -0.302\hat{\mathbf{i}} + 0.302\hat{\mathbf{j}} + 0.905\hat{\mathbf{k}}$$

$$\hat{\mathbf{b}} = -0.286\hat{\mathbf{i}} - 0.429\hat{\mathbf{j}} + 0.857\hat{\mathbf{k}}$$

$$\hat{\mathbf{c}} = 0.385\hat{\mathbf{i}} + 0.923\hat{\mathbf{k}}$$

The vector force in each cable can be written as

$$\mathbf{T}_A = T_A\,\hat{\mathbf{a}}$$

$$\mathbf{T}_B = T_B\,\hat{\mathbf{b}}$$

$$\mathbf{T}_C = T_C\,\hat{\mathbf{c}}$$

Each cable is in tension, and the unit vectors are taken in the direction consistent with that fact. The magnitudes of the force vectors must be positive. The weight of the block is $W = mg = 200 \times 9.8 = 1962$ N and is in the $-\hat{\mathbf{k}}$ direction.

The vector equation for equilibrium is:

$$\mathbf{T_A} + \mathbf{T_B} + \mathbf{T_C} + \mathbf{W} = 0$$

The scalar equations are

$$\sum F_x = 0; \quad -0.302\,T_A - 0.286\,T_B + 0.385\,T_C = 0$$

$$\sum F_y = 0; \quad 0.302\,T_A - 0.429\,T_B = 0$$

$$\sum F_z = 0; \quad 0.905\,T_A + 0.857\,T_B + 0.923\,T_C = 1962$$

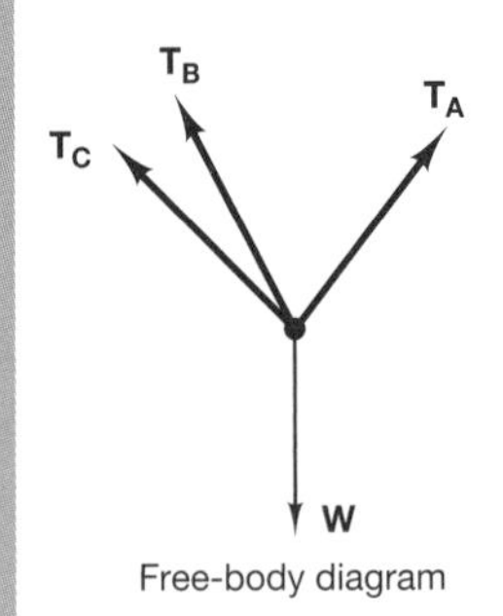

Free-body diagram

This three-dimensional problem can be solved by the direct vector method without expanding the vector equation into components. Write the vector equation as:

$$\mathbf{T_A} + \mathbf{T_B} + \mathbf{T_C} = -\mathbf{W}$$

Create three biorthogonal vectors $\hat{\mathbf{r}}$, $\hat{\mathbf{s}}$ and $\hat{\mathbf{t}}$ and use these for a direct vector solution.

$$\hat{\mathbf{r}} = \frac{\hat{\mathbf{b}} \times \hat{\mathbf{c}}}{|\hat{\mathbf{b}} \times \hat{\mathbf{c}}|} \quad \hat{\mathbf{s}} = \frac{\hat{\mathbf{c}} \times \hat{\mathbf{a}}}{|\hat{\mathbf{c}} \times \hat{\mathbf{a}}|} \quad \hat{\mathbf{t}} = \frac{\hat{\mathbf{a}} \times \hat{\mathbf{b}}}{|\hat{\mathbf{a}} \times \hat{\mathbf{b}}|}$$

$$\hat{\mathbf{r}} = -0.34\hat{\mathbf{i}} + 0.811\hat{\mathbf{j}} + 0.225\hat{\mathbf{k}}$$

$$\hat{\mathbf{S}} = -0.4\hat{\mathbf{i}} - 0.901\hat{\mathbf{j}} + 0.167\hat{\mathbf{k}}$$

$$\hat{\mathbf{t}} = 0.949\hat{\mathbf{i}} + 0.317\hat{\mathbf{k}}$$

The magnitudes of the cable tensions can now be easily obtained:

$$T_A = \frac{-\hat{\mathbf{r}} \cdot \mathrm{W}}{\hat{\mathbf{r}} \cdot \hat{\mathbf{a}}} = 723\ \mathrm{N}$$

$$T_B = \frac{-\hat{\mathbf{s}} \cdot \mathrm{W}}{\hat{\mathbf{s}} \cdot \hat{\mathbf{b}}} = 509\ \mathrm{N}$$

$$T_C = \frac{-\hat{\mathbf{t}} \cdot \mathrm{W}}{\hat{\mathbf{t}} \cdot \hat{\mathbf{c}}} = 945\ \mathrm{N}$$

This system of three equations and three unknowns may be solved by hand or by using computational software (see the Computational Supplements). The results are:

$$T_A = 723\ \mathrm{N} \qquad T_B = 509\ \mathrm{N} \qquad T_C = 945\ \mathrm{N}$$

If a fourth cable were run from the block vertically to aid in the support, there would have been four unknown tensions, and the system would have been overconstrained and would be statically indeterminate.

Sample Problem 3.6

The rock climber is shifting her weight from her left leg to her right leg as she climbs. Model her as a particle with all forces concurrent with her center of mass at her hip and determine the force on her left leg and on her right leg when she is weight bearing on each leg alone.

Solution The angle between the forces can be approximated on the free-body diagram as shown. The problem can be solved in terms of the weight of the climber. The vectors are

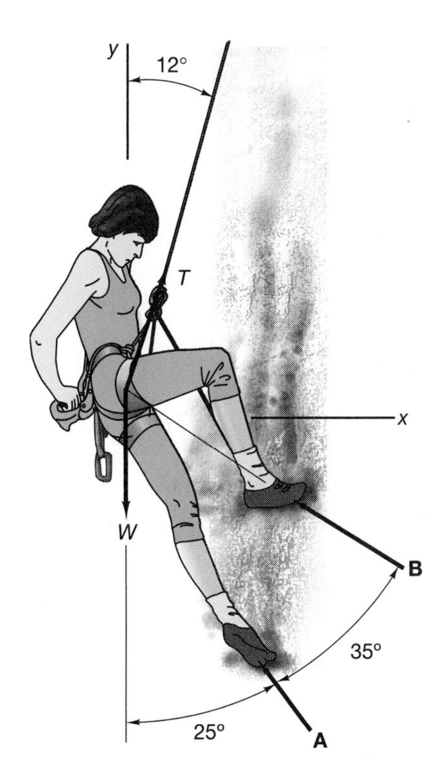

expressed in components in the coordinate system chosen as horizontal and vertical in a two-dimensional plane.

$$\mathbf{T} = T\left(\sin 12^\circ\,\hat{\mathbf{i}} + \cos 12^\circ\,\hat{\mathbf{j}}\right)$$
$$\mathbf{W} = -W\hat{\mathbf{j}}$$
$$\mathbf{A} = A\left(-\sin 25^\circ\,\hat{\mathbf{i}} + \cos 25^\circ\,\hat{\mathbf{j}}\right)$$
$$\mathbf{B} = B\left(-\sin 60^\circ\,\hat{\mathbf{i}} + \cos 60^\circ\,\hat{\mathbf{j}}\right)$$

When the left leg is the only one weight bearing, the vector equilibrium equation is:

$$\mathbf{A} + \mathbf{T} + \mathbf{W} = \mathbf{0}$$

This equation can be solved either by expanding into components or by the direct vector method, yielding:

$$A = 0.345\ \mathrm{W}$$
$$T = 0.703\ \mathrm{W}$$

When the weight is shifted totally to the right leg, the vector equilibrium equation is:

$$\mathbf{B} + \mathbf{T} + \mathbf{W} = \mathbf{0}$$

Solving the vector equation yields:

$$B = 0.219\ \mathrm{W}$$
$$T = 0.911\ \mathrm{W}$$

Sample Problem 3.7

Consider the cord of weight **W** pulled at both ends by a tension **T**, as shown at the left. For a first approximation, suppose the cord is equivalent to a particle, with its weight concentrated at the center of the cord. How large must the tension be for the cord to be completely horizontal?

Solution

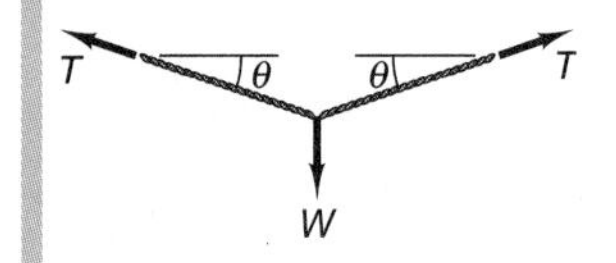

If the cord makes an angle θ with the horizontal, summation of the forces in the vertical direction yields

$$2\,T \sin\theta - \mathrm{W} = 0$$

or

$$T = \frac{W}{2\sin\theta}$$

Note that, for the cord to be stretched horizontally, $\theta = 0$ and $\sin\theta = 0$. The tension T then becomes infinitely large. Therefore:

"And so no force, however great,
Can stretch a cord, however fine,
Into a horizontal line
That shall be absolutely straight."

[Quoted as an example of accidental meter and rhyme; See *Elementary Treatise on Mechanics*, William Whewell (1794–1866)].

Sample Problem 3.8

The cable *AB* keeps the 10-kg collar at *A* in equilibrium along the smooth vertical rod. Determine the tension in the cable.

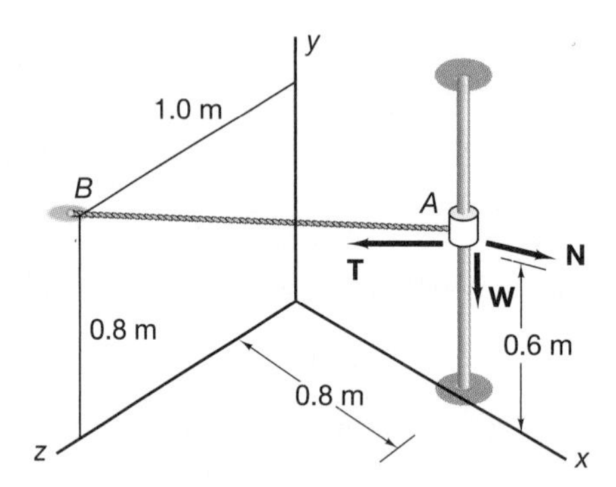

Solution

Three forces act on the collar; the tension in the cable, a force normal to the rod and the weight of the collar. These three vectors can be written as:

$$\mathbf{T} = T\hat{\mathbf{t}}$$
$$\mathbf{N} = N\hat{\mathbf{n}}$$
$$\mathbf{W} = -10*9.81\hat{\mathbf{j}}$$

The coordinates of the end points of the cable are:

$$\mathbf{A} = 0.8\hat{\mathbf{i}} + 0.6\hat{\mathbf{j}}$$
$$\mathbf{B} = 0.8\hat{\mathbf{j}} + 1.0\hat{\mathbf{k}}$$

A unit vector from A to B is:

$$\hat{\mathbf{t}} = \frac{\mathbf{B} - \mathbf{A}}{|\mathbf{B} - \mathbf{A}|}$$

$$\hat{\mathbf{t}} = -0.617\,\hat{\mathbf{i}} + 0.154\hat{\mathbf{j}} + 0.772\hat{\mathbf{k}}$$

A unit vector along the rod is $\hat{\mathbf{j}}$ and the vector equation of equilibrium is:

$$\mathbf{T} + \mathbf{N} + \mathbf{W} = \mathbf{0}$$

Taking dot product of this equation is taken with the unit vector along the rod, yields

$$\mathbf{T}\cdot\hat{\mathbf{j}} + \mathbf{N}\cdot\hat{\mathbf{j}} + \mathbf{W}\cdot\hat{\mathbf{j}} = \mathbf{0}$$

The normal force is perpendicular to the unit vector along the rod and the scalar equation yields:

$$T = \frac{-\mathbf{W}\cdot\hat{\mathbf{j}}}{\hat{\mathbf{t}}\cdot\hat{\mathbf{j}}} = 636\text{ N}$$

The normal force can be obtained from the equilibrium equation:

$$\mathbf{N} = -(\mathbf{T} + \mathbf{W})$$

$$\mathbf{N} = 393\hat{\mathbf{i}} - 491\hat{\mathbf{k}}\text{ N} \qquad |\mathbf{N}| = 629\text{ N}$$

Sample Problem 3.9

The collars hang on the vertical frame composed of two smooth rods. If the mass of collar A is 8 kg and the mass of collar B is 4 kg, determine the equilibrium angle α and the tension in the cable between the collars.

Solution

Construct free-body diagrams for the two particles.

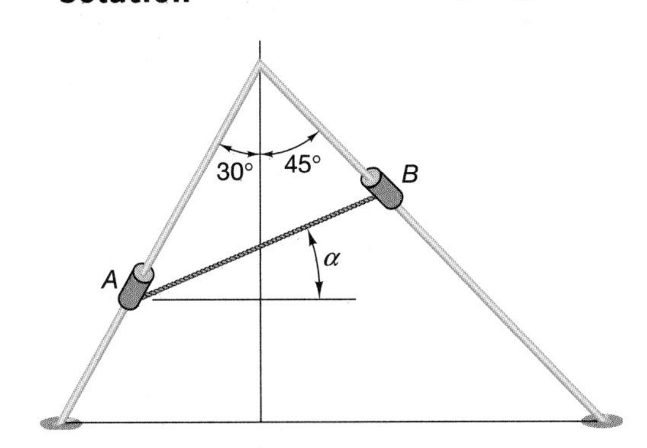

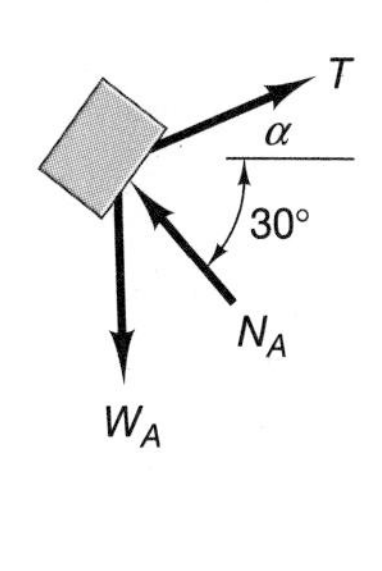

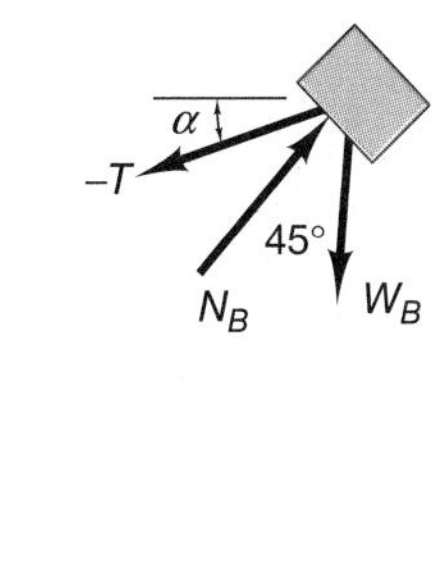

The equilibrium equation for collar A is

$$\mathbf{W_A} + \mathbf{N_A} + \mathbf{T} = \mathbf{0}$$

and for collar B is

$$\mathbf{W_B} + \mathbf{N_B} - \mathbf{T} = \mathbf{0}$$

Since all vector lie in a plane, the four scalar equations are:

$$\begin{aligned} T\cos\alpha - N_A\cos 30^\circ &= 0 \\ T\sin\alpha + N_A\sin 30^\circ - WA &= 0 \\ -T\cos\alpha + N_B\sin 45^\circ &= 0 \\ -T\sin\alpha + N_B\cos 45^\circ - W_B &= 0 \end{aligned}$$

Eliminating T and α yields

$$N_A \cos 30° = N_B \sin 45° \Rightarrow N_A = 0.816\, N_B$$
$$N_A \sin 30° + N_B \cos 45° = 117.7$$
$$N_A = 86.1\text{N} \qquad N_B = 105.6 \text{ N}$$

From the first two equations:

$$T \cos \alpha = 74.6$$
$$T \sin \alpha = 35.6$$
$$\tan \alpha = 0.476$$
$$\alpha = 25°$$
$$T = 82.3 \text{ N}$$

Sample Problem 3.10

A 10-kg collar at A and a 15-kg collar at B slide on smooth rods as shown. The collars are supported by a 2-m long cable. If the system is in equilibrium, determine the distance from the bottom of each rod to each collar and the tension in the cable.

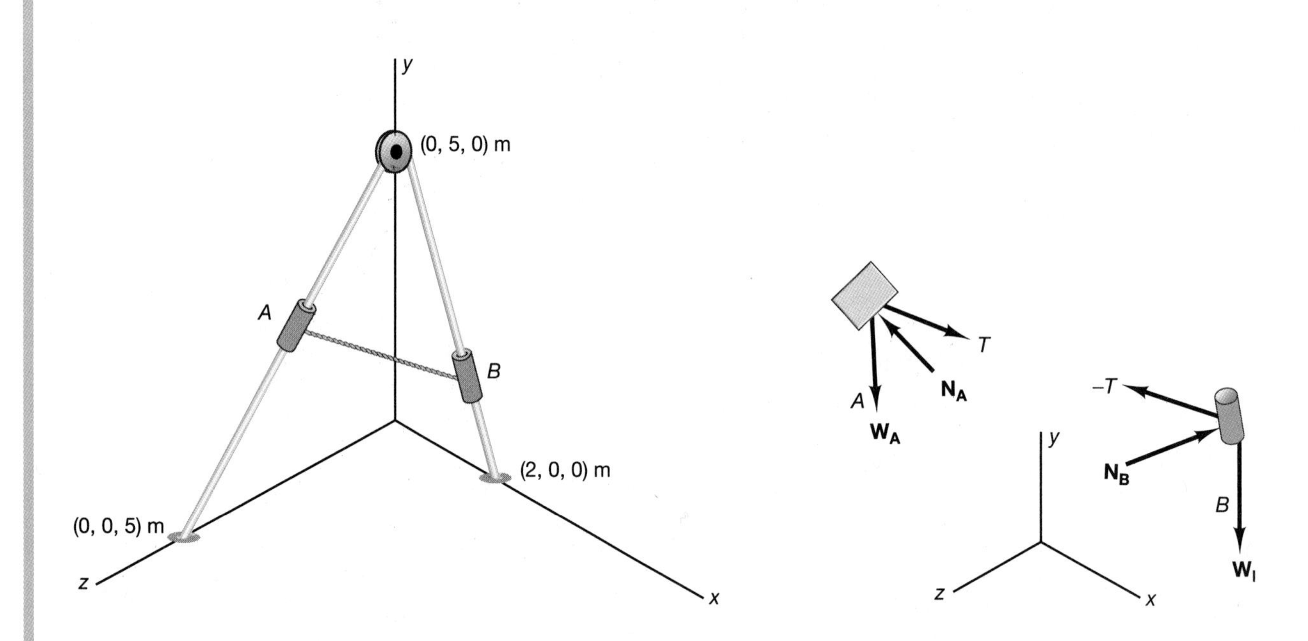

Solution

The equilibrium equation for collar A is

$$\mathbf{W_A} + \mathbf{N_A} + \mathbf{T} = \mathbf{0}$$

The equilibrium equation for collar B is

$$\mathbf{W_B} + \mathbf{N_B} - \mathbf{T} = \mathbf{0}$$

Now, construct unit vectors that run from the base of each rod directed toward the top of the rod.

$$\mathbf{L_A} = \mathbf{5\hat{j}} - 5\mathbf{\hat{k}} \qquad \mathbf{\hat{l}_A} = 0.707\mathbf{\hat{j}} - 0.707\mathbf{\hat{k}}$$
$$\mathbf{L_B} = -2\mathbf{\hat{i}} + 5\mathbf{\hat{j}} \qquad \mathbf{\hat{i}_B} = -0.371\mathbf{\hat{i}} + 0.929\mathbf{\hat{j}}$$

The points A and B are located from the origin as

$$\mathbf{A} = 5\hat{\mathbf{k}} + a\hat{\mathbf{I}}_A$$

$$\mathbf{B} = 2\hat{\mathbf{i}} + b\hat{\mathbf{I}}_B$$

where a and b are the distances from the bottom of the rod to the collar.
A vector from collar A to collar B along the cable is

$$\mathbf{B} - \mathbf{A} = (2 - 0.371b)\hat{\mathbf{i}} + (0.929b - 0.707a)\hat{\mathbf{j}} - (5 - 0.707a)\hat{\mathbf{k}}$$

Since the cable has a length of 2 meters, a unit vector along $\mathbf{B} - \mathbf{A}$ is

$$\hat{\mathbf{t}} = \frac{\mathbf{B} - \mathbf{A}}{2}$$

The dot product may now be used to eliminate the normal forces from the equilibrium equations as these forces must be perpendicular to the rods.

$$T(\hat{\mathbf{I}}_A \cdot \hat{\mathbf{t}}) - 10 \cdot 9.81(\hat{\mathbf{I}}_A \cdot \hat{\mathbf{j}}) = \mathbf{0}$$

$$-T(\hat{\mathbf{I}}_B \cdot \hat{\mathbf{t}}) - 15 \cdot 9.81(\hat{\mathbf{I}}_B \cdot \hat{\mathbf{j}}) = \mathbf{0}$$

Expanding these two scalar equations yields:

$$T[0.707*0.5*(0.929b - 0.707a) + 0.707*0.5*(5 - 0.707a)] = 10*9.81*0.707$$

$$T[0.371*0.5*(2 - 0.371b) - 0.929*0.5*(0.929b - 0.707a)] = 15*9.81*0.929$$

These are two nonlinear equations for a, b and T. The third equation is obtained by observing that

$$(\mathbf{B} - \mathbf{A}) \cdot (\mathbf{B} - \mathbf{A}) = 4$$

$$(2 - 0.371b)^2 + (0.929b - 0.707a)^2 + (5 - 0.707a)^2 = 4$$

Computational software may be used to solve the three nonlinear equations, yielding

$$a = 4.85 \text{ m}$$
$$b = 2.84 \text{ m}$$
$$T = 252 \text{ N}$$

The normal forces may be found directly from the equilibrium equations

$$\mathbf{N_A} = -\mathbf{T} - \mathbf{W_A}$$

$$\mathbf{N_B} = \mathbf{T} - \mathbf{W_B}$$

$$\mathbf{N_A} = -119\hat{\mathbf{i}} + 198\hat{\mathbf{j}} + 198\hat{\mathbf{k}} \quad |\mathbf{N_A}| = 304 \text{ N}$$

$$\mathbf{N_B} = 119\hat{\mathbf{i}} + 48\hat{\mathbf{j}} - 198\hat{\mathbf{k}} \quad |\mathbf{N_B}| = 236 \text{ N}$$

Sample Problem 3.11 Determine the forces normal and tangential to maintain equilibrium of the 50 kg mass on the inclined surface.

Solution Construct two vectors in the incline plane.

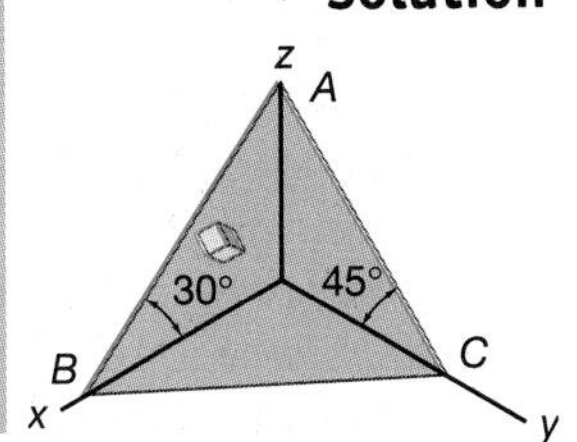

$$\hat{\mathbf{e}}_{B/A} = -\cos 30°\hat{\mathbf{i}} - \sin 30°\hat{\mathbf{k}}$$

$$= -0.866\hat{\mathbf{i}} - 0.5\hat{\mathbf{k}}$$

$$\hat{\mathbf{e}}_{C/A} = -\cos 45°\hat{\mathbf{j}} - \sin 45°\hat{\mathbf{k}}$$

$$= -0.707\hat{\mathbf{j}} - 0.707\hat{\mathbf{k}}$$

Construct a unit vector that is normal to the surface using the cross product.

$$\hat{\mathbf{n}} = \frac{\hat{\mathbf{e}}_{B/A} \times \hat{\mathbf{e}}_{C/A}}{|\hat{\mathbf{e}}_{B/A} \times \hat{\mathbf{e}}_{C/A}|}$$

$$\hat{\mathbf{n}} = -0.378\hat{\mathbf{i}} - 0.655\hat{\mathbf{j}} + 0.655\hat{\mathbf{k}}$$

The weight vector is

$$\mathbf{W} = -50g\hat{\mathbf{k}} = -490.5\hat{\mathbf{k}}$$

The normal force is opposite to the component of the weight in the normal direction.

$$\mathbf{F}_n = -(\mathbf{W} \cdot \hat{\mathbf{n}})\hat{\mathbf{n}}$$

$$\mathbf{F}_n = -121.4\hat{\mathbf{i}} - 210.2\hat{\mathbf{j}} + 210.2\hat{\mathbf{k}}$$

$$|\mathbf{F}_n| = 321.1 \text{ N}$$

The equilibrium force in the tangential direction can be obtained from the solution of the equilibrium equation.

$$\mathbf{F}_n + \mathbf{F}_t + \mathbf{W} = \mathbf{0}$$

$$\mathbf{F}_t = -(\mathbf{F}_n + \mathbf{W})$$

$$\mathbf{F}_t = 121.4\hat{\mathbf{i}} + 210.2\hat{\mathbf{j}} + 280.3\hat{\mathbf{k}}$$

$$|\mathbf{F}_t| = 370.8 \text{ N}$$

Problems

3.11 A block of mass $m = 1000$ kg is suspended from a three-cable arrangement as illustrated in Figure P3.11. For the case where $\alpha = 30°$ and $\beta = 45°$, compute the tensions T_1, T_2, and T_3.

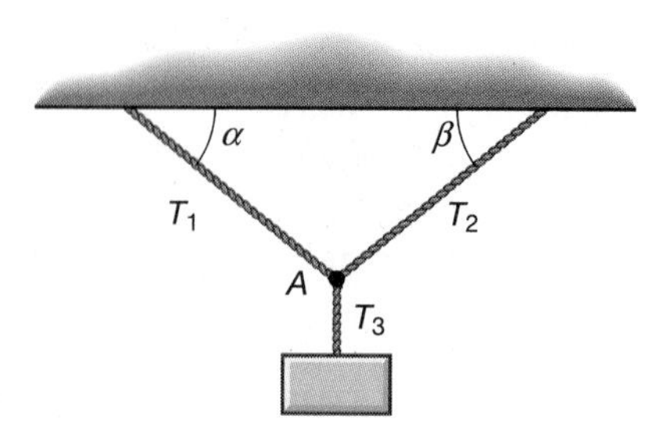

Figure P3.11

3.12 A 100-kg crate is held in equilibrium as it rests on a frictionless inclined plane making an angle θ with the horizontal. (See Figure P3.12.) (a) Compute the tension in the cable for $\theta = 30°$. (b) Compute the tension in the cable T and the normal force at the surface for θ at 5° increments, increasing from 0 to 90°. Is there some value of θ for which T equals the normal force.

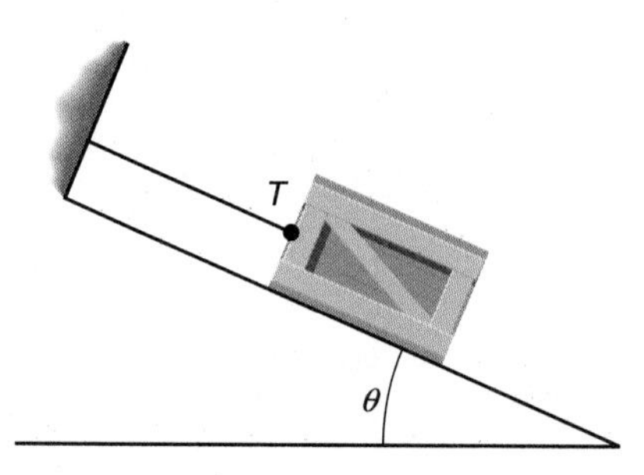

Figure P3.12

3.13 A barge is tied to a dock and held stationary against the river current by applying a force with the barge engine. (See Figure P3.13.) The net force of the river current and the barge's engine is a 4000-N force at 10°, acting through the point C. Calculate the tension in the ropes tied to points A and B.

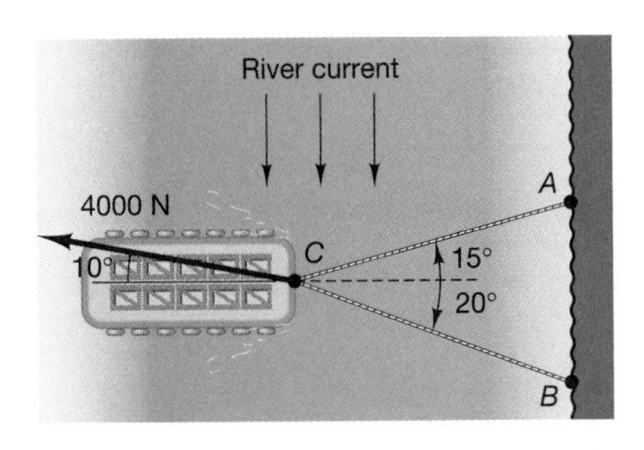

Figure P3.13

3.14 A 95-kg crate is held in place by three ropes, as shown in Figure P3.14. Calculate the tension in the ropes. Assume that T_2 is horizontal and T_1 is vertical.

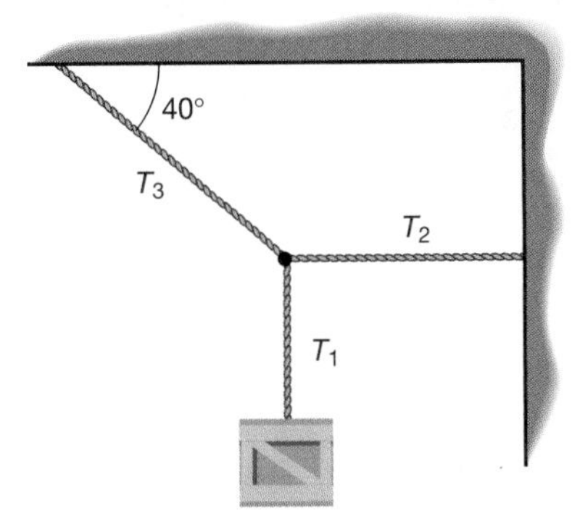

Figure P3.14

3.15 An 11-kg light fixture is suspended from a ceiling by four cords attached at a point, with the geometry given in Figure P3.15. (a) Assuming that gravity acts along the z-axis, calculate the tension in the four cords. The dimensions are given in meters. (b) Investigate the effect on the tension in the cords of moving the lamp to different heights by changing the z-coordinate of the point where the four cords meet from 1.25 m to 2.5 m and then to 0.25 m

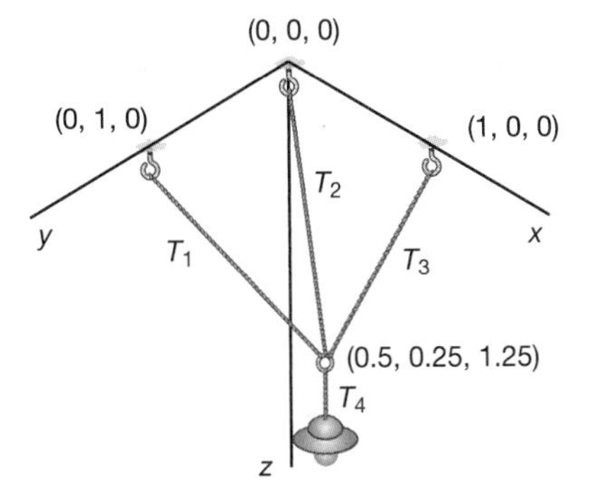

Figure P3.15

3.16 A three-cable harness is used to pull a stone block along some friction free rollers. (See Figure P3.16.) The applied force is 5000 N. (a) Calculate the tensions in the cables, T_1, T_2, and T_3, for equilibrium of point A for the values $l_1 = 2$ m and $l_2 = 2$ m. The dimensions are all in meters. (b) Calculate the tensions T_1, T_2, and T_3 for various values l_2, and then "design" the system by picking the location of point B that results in the minimum tensions in all cables. (*Hint*: Part (b) is design oriented and is intended to use a numerical solution.)

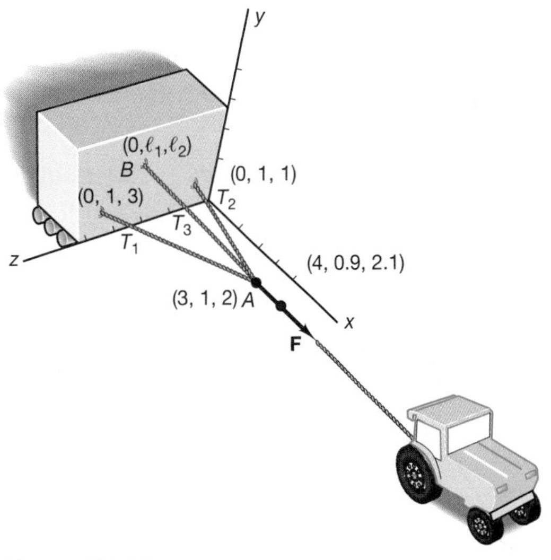

Figure P3.16

3.17 A motor is held in equilibrium by the tension in two fixed cables (T_1, T_2) and by the tension T_3 in a movable cable, provided by a counterbalance m_2. (See Figure P3.17.) Calculate the tensions T_1, T_2, T_3, and T_4 and, hence, the mass of the counterbalance for the configuration in which the mass of the motor is 350 kg. With the corner of the garage as a coordinate system, A is at A $\rightarrow$ (0,4,4) m, B is located at B $\rightarrow$ (0,4,0) m, point C is located at C $\rightarrow$ (2,2,2) m, and the pulley is located at pulley $\rightarrow$ (3.33,2,2.33) m (that is, $l_1 = 2$ m and $l_2 = 2.33$ m). Assume that the pulley–rope surface is frictionless, so that the tension on either side of the pulley is the same.

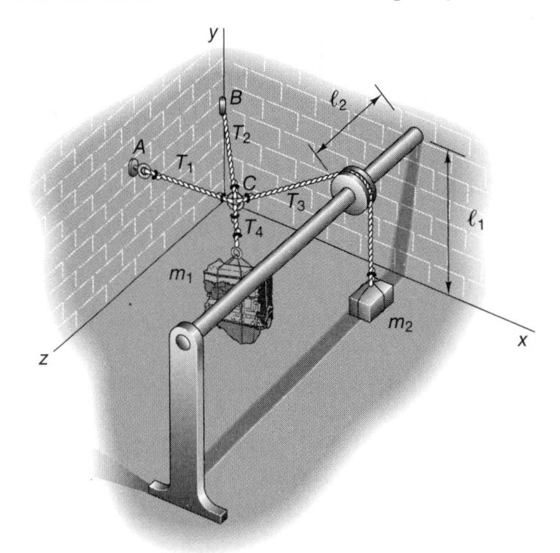

Figure P3.17

3.18 Consider again the counterbalance system of Problem 3.17. Calculate the best design for the pulley system by changing the positions l_1 and l_2 until $W_2 = m_2 g$ is at its minimum value for a given motor mass of 350 kg. Here, the "best design" refers to a configuration of the pulley (that is, the values of l_1 and l_2) resulting in the smallest value of the counterweight W_2.

3.19 A typical load on a traffic light, resulting from both gravity and wind, is modeled by $\mathbf{F} = -25\hat{\mathbf{i}} - 500\hat{\mathbf{j}} + 25\hat{\mathbf{k}}$ (N) acting at the top of the light. (See Figure P3.19.) Determine the tension in the support cables. The dimensions are given in centimeters.

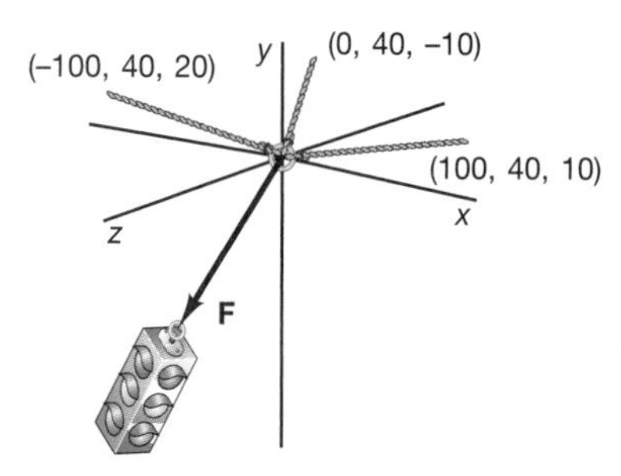

Figure P3.19

3.20 In Figure P3.20 calculate the mass m_2 required to keep the 100-kg mass m_1 in equilibrium. Assume that the pulley provides no friction.

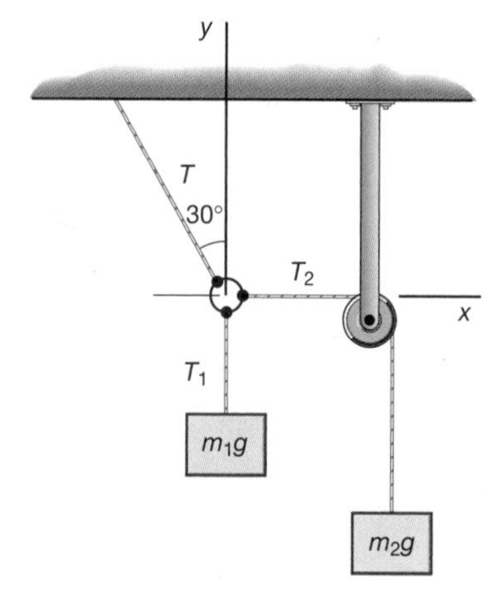

Figure P3.20

3.21 A helicopter attempts to hover in a headwind modeled by the force $\mathbf{F}_1 = 800\hat{\mathbf{i}} - 2000\hat{\mathbf{j}}$ N acting at its center of gravity. (See Figure P3.21.) The mass of the helicopter is 6350 kg. Assuming that the hover condition corresponds to equilibrium, what force $\mathbf{F}_2$ must the pilot set the controls to produce? Here the helicopter is modeled as a particle.

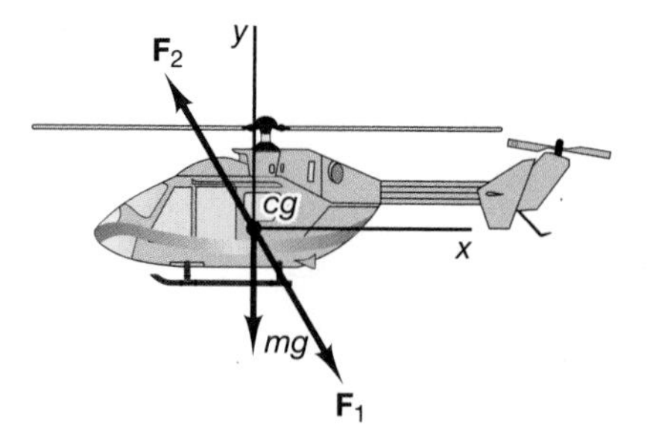

Figure P3.21

3.22 A 440-kg engine is supported by three cables, as shown in Figure P3.22. Calculate the tensions in the cables.

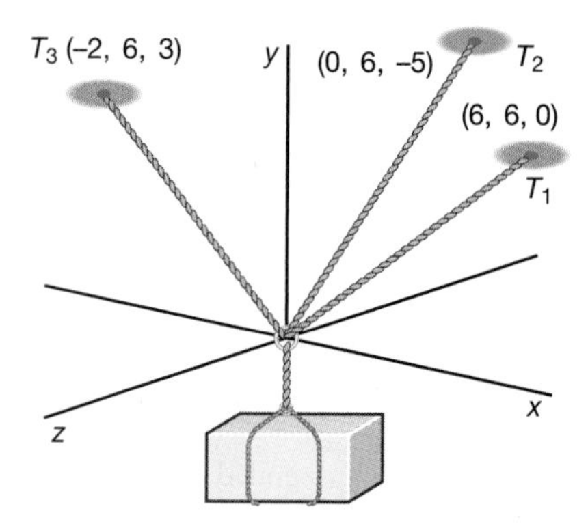

Figure P3.22

3.23 Calculate the tensions T_1, T_2, and T_3, and the corresponding vector description, for the system of Figure P3.23 to be in equilibrium. The pulleys provide no friction.

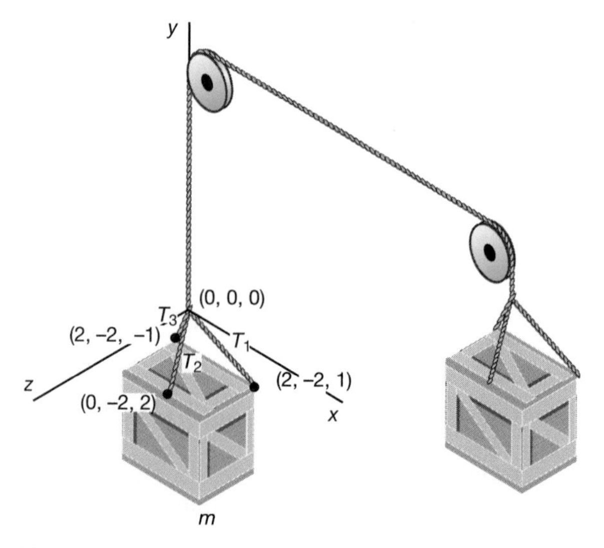

Figure P3.23

3.24 The pulley system shown in Figure P3.24 supports three masses in equilibrium. If the mass at A is 12 kg, determine the mass at B and the mass at C. Neglect the effects of friction

3.25 The mass of each pulley is m and neglecting friction, determine the force $\boldsymbol{T}$ necessary to support the large mass M for the system shown in Figure P3.25.

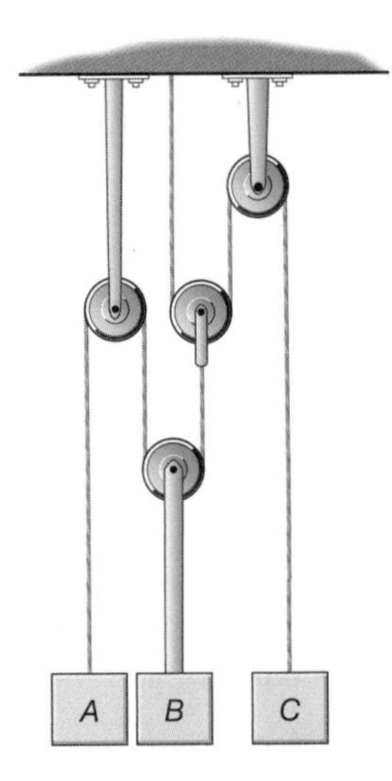

Figure P3.24

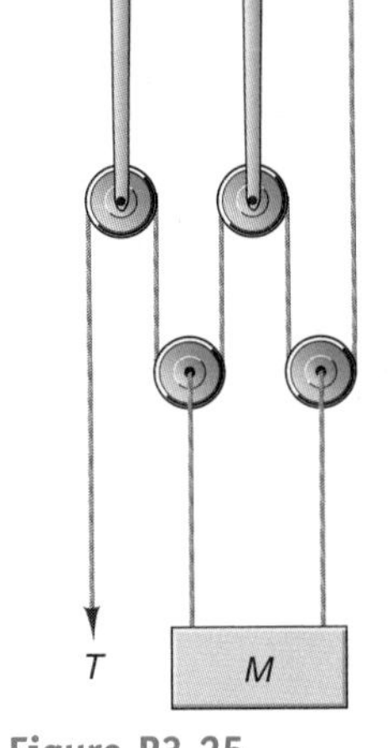

Figure P3.25

3.26 Repeat problem 3.25 for the system shown in Figure P3.26.

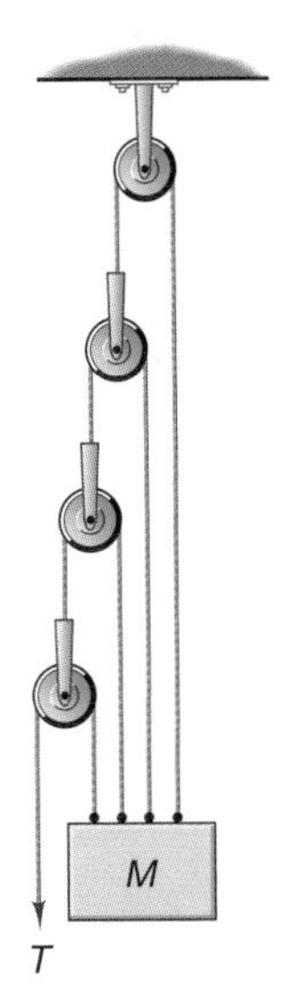

Figure P3.26

3.27 A 8-kg cylinder rests on two smooth inclined surfaces as shown in Figure P3.27. Determine the forces between the surfaces and the cylinder for values of β from 5° to 60° and graph the results.

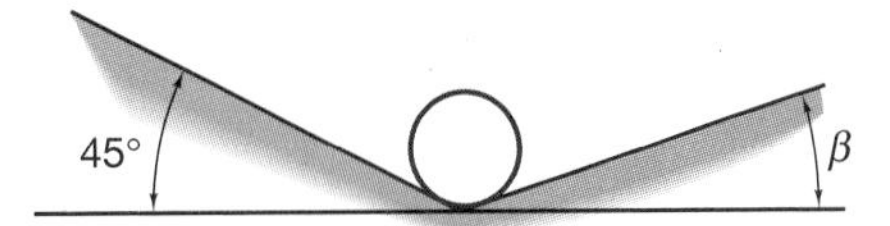

Figure P3.27

3.28 A 12-kg collar is supported on a smooth rod by a cable AD as shown in Figure P3.28. Determine the tension in the cable and the normal force between the collar and the rod.

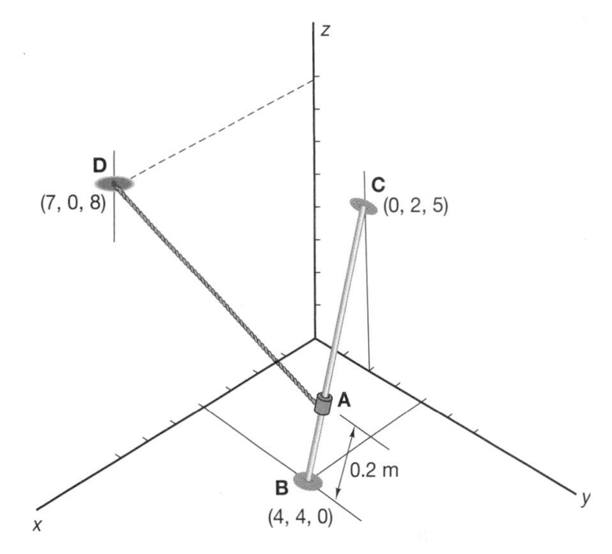

Figure P3.28

3.29 A triangular frame composed of two smooth rods is setup in the vertical plane as shown in Figure P3.29. A cable as shown connects the two collars A and B. If the angle $\theta = \beta = 30°$ and the masses of the collars are $m_A =$ 6 kg and $m_B =$ 4 kg, determine the angle α and the tension in the cable T when the system is in equilibrium.

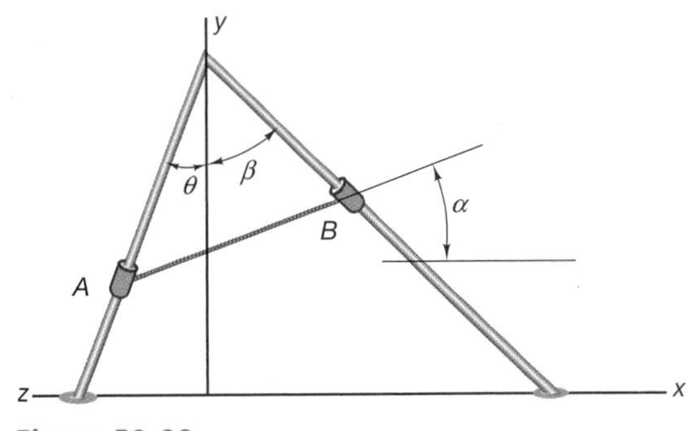

Figure P3.29

3.30 For the frame shown in Figure P3.29, determine the equilibrium angle α and the tension in the cable if the masses of the collars are equal to 6 kg each and the angles θ and β equal 45°.

3.31 For the frame shown in Figure P3.29, derive the general expression for the tension in the cable and the equilibrium angle for any angles θ and β and for any collar masses $m_A > m_B$.

3.32 For the frame shown in Sample Problem 3.10, determine the equilibrium position if the collar at A is of negligible mass. Show that the cable is normal (perpendicular) to the rod holding collar A and determine the tension in the cable.

3.33 For the frame shown in Sample Problem 3.10, determine the equilibrium position if the mass of collar B is negligible and show that the cable is perpendicular to the rod holding collar B.

3.34 The 20-kg mass A and the 30-kg mass B are free to slide on smooth rods as shown in Figure P3.24. If the collars are connected by a 3-m cable, determine the tension in the cable and the equilibrium position.

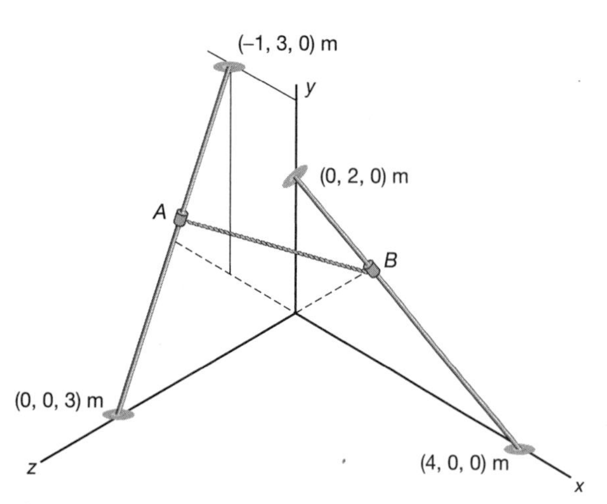

Figure P3.34

3.35 Determine the equilibrium position for the system shown in Figure P3.34 if the mass of collar A is neglible.

3.36 A 90-kg roofer works on the steep roof shown in Figure P3.36, determine the required tangential force (friction) required for the man to be in equilibrium.

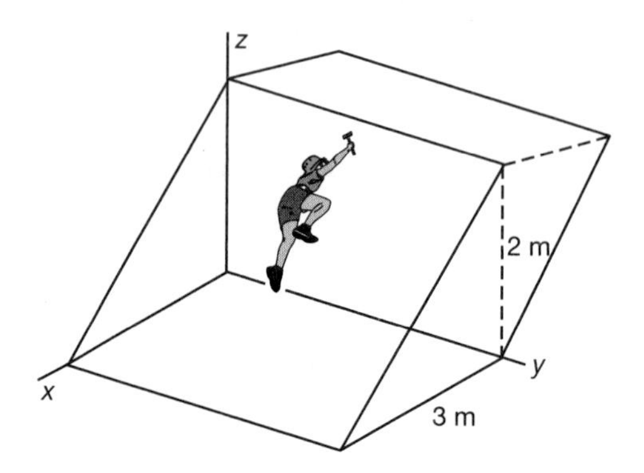

Figure P3.36

3.37 A 72-kg climber practices on the climbing wall shown in Figure P3.37. Determine the tangential force that he must maintain to keep from slipping.

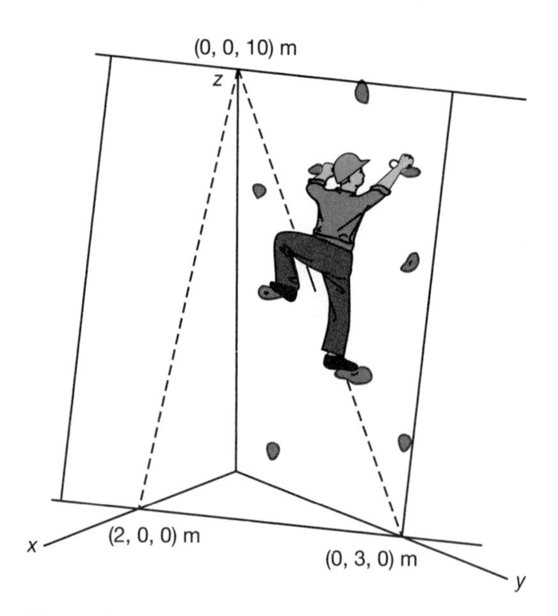

Figure P3.37

3.3 SPRINGS

Springs are used in many machines and can be classified by their geometry, such as coil springs and leaf springs, or by their function, such as linear, nonlinear and torsional springs. The principal characteristic of a spring is that the force required to deform it—that is, to stretch it or compress it—is a function of the amount of deformation of the spring. The simplest spring is a linear spring, wherein the force required to stretch the spring can be expressed as a linear function of the change in position from an "unstretched" position. Most springs can transmit either a tensile force when stretched or a compressive force when compressed. The force–deformation relation is

$$F = k(l_f - l_0) = k\delta \tag{3.4}$$

where l_f is the final length of the spring, l_0 is the unstretched length of the spring, k is the spring constant, and δ is the deformation of the spring. Note in Eq. (3.4) that the force will be positive if the final length of the spring is greater than the unstretched length and negative if the final length is less than the unstretched length of the spring. If the spring force is plotted against the amount of stretch, the curve is a straight line, as illustrated in Figure 3.6. The spring constant k is the slope of the line. The stiffer the spring, the greater is the slope. It is apparent that springs are useful in machines that need to control forces or limit the amount of movement of a part. It will be shown in Chapter 11 that a spring stores potential energy as it is stretched or compressed.

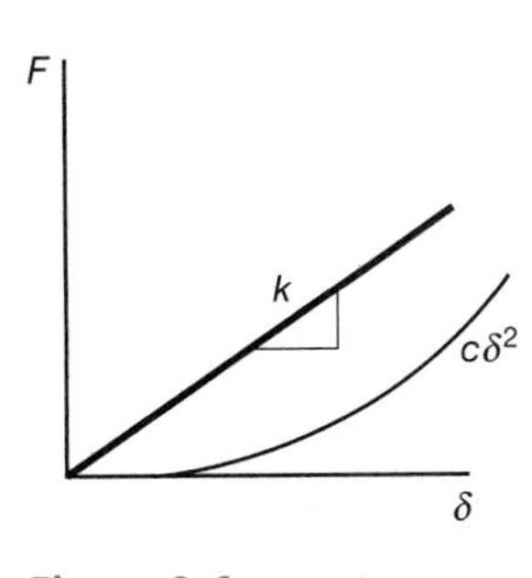

Figure 3.6

Some springs are nonlinear, and the equation relating the spring force to the displacement is therefore nonlinear:

$$F = c\delta^2 \tag{3.5}$$

The stiffness of this spring would increase with stretch, as

$$\frac{dF}{d\delta} = 2c\delta \tag{3.6}$$

The human body has "springs" that connect bone to bone (ligaments) and that connect muscle to bone (tendons). The Achilles tendon in the heel is a nonlinear spring with a force–deformation relationship similar to Eq. (3.5).

When springs are used to support an object, the free-body diagram must show the springs in their stretched position, and the geometry must be determined in that position. The geometry of the system is coupled with the determination of the unknown spring support forces, and the problem becomes nonlinear. Therefore, these problems are not easily solved without the use of computational software. The solution of a system of nonlinear equations is obtained by iterative techniques. An iterative technique starts at an initial "guessed" solution and refines that guess after each attempt (iteration), until it converges on the correct solution.

Sample Problem 3.12

A weight W, shown at the left, is supported by two springs in series. Determine (1) the total distance that the weight will stretch the system and (2) an equivalent spring constant—that is, the spring constant of a single spring with the same stiffness as the two springs.

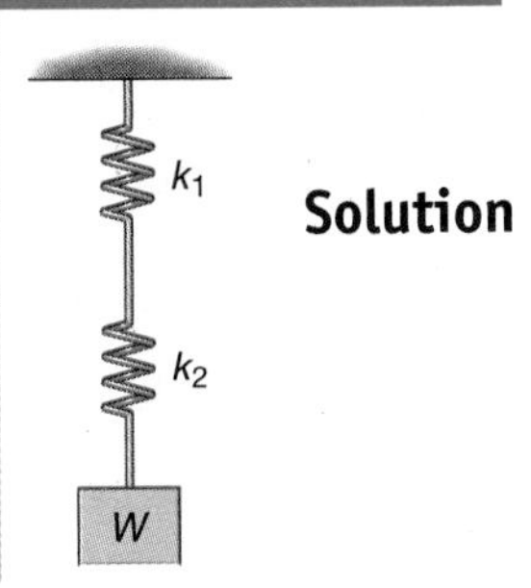

Solution The force in each spring is equal to the weight W. The total stretch is the sum of the stretches of the two springs. So we have

$$\delta_1 = \frac{W}{k_1};\ \delta_2 = \frac{W}{k_2}$$

$$\delta = \delta_1 + \delta_2 = W\left(\frac{1}{k_1} + \frac{1}{k_2}\right) = W\left(\frac{k_1 + k_2}{k_1 k_2}\right)$$

An equivalent spring constant for the two springs in series is given by

$$W = k_e\delta$$

Therefore,

$$k_e = \frac{k_1 k_2}{k_1 + k_2}$$

Sample Problem 3.13 The spring system shown in the unstretched configuration shown below at the left supports a vertical force **F** of 1000 N at point C. The lengths of the unstretched springs are $L_1 = 14$ cm and $L_2 = 11.4$ cm, and the spring constants are $k_1 = 2000$ N/cm and $k_2 = 1000$ N/cm; this is a very stiff system. Determine the tension in each spring. Solve the problem when the spring constants are reduced by a factor of 10—that is, when we have a "soft system."

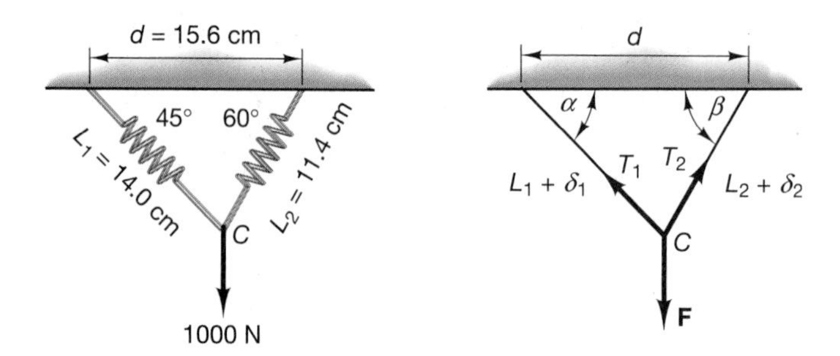

Solution This is a much more complex problem than those considered previously, because the geometry of the system changes as the force is applied. The general deformed geometry is illustrated a the right.

The angles α and β can be expressed in terms of the spring deformations by the law of cosines:

$$\cos\alpha = \frac{d^2 + (L_1 + \delta_1)^2 - (L_2 + \delta_2)^2}{2d(L_1 + \delta_1)}$$

$$\cos\beta = \frac{d^2 + (L_2 + \delta_2)^2 - (L_1 + \delta_1)^2}{2d(L_2 + \delta_2)}$$

The tensions in the springs are related to the deformations by

$$T_1 = k_1\delta_1$$

$$T_2 = k_2\delta_2$$

The equations of equilibrium are:

$$\sum F_x = T_1 \cos\alpha - T_2 \cos\beta = 0$$

$$\sum F_y = T_1 \sin\alpha + T_2 \sin\beta - F = 0$$

There are now six equations for the six unknowns α, β, δ_1, δ_2, T_1, and T_2. The difficulty in obtaining a solution is that these equations are nonlinear. The solution will be very difficult to accomplish by hand, but again, computational software may be employed to obtain the solution. Details of the solution are presented in the Computational Supplements.

Although this problem has used only simple geometric relations and the equations of equilibrium, it is actually the equilibrium solution of a complex nonlinear system that has many analogies in engineering. Frequently, a system of supports must be designed to have a particular deformed geometry. If, for example, the deformed system had to be such that the angles α and β had to be 60°, these angles would then be known, and the spring constants could be selected as design unknowns.

The three-dimensional equilibrium problem in Sample Problem 3.5 can also be solved by treating the three cables as springs.

Sample Problem 3.14

Consider a scoreboard of mass 200 kg supported by three cables, as shown in the diagram at the left. Determine the tension in each of the cables if each cable has a spring constant equal to the product of its cross-sectional area and a material property (Young's modulus), divided by its length ($k = EA/L$). Assume that EA for each cable is equal to 1.2×10^5 N/m.

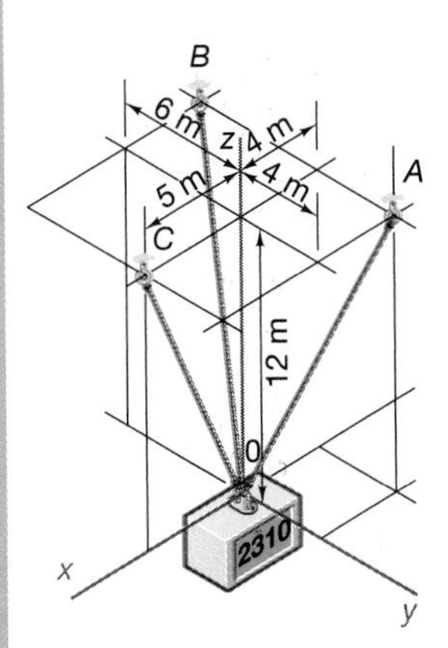

Solution The undeformed system is shown, and the origin of the coordinate system is taken at the attachment point of the cables to the scoreboard. This point will move in space as the cables deform. Let the coordinates of the final position of the point be $(x0, y0, z0)$. If the coordinates of the attachment points A, B, and C are (xA, yA, zA), (xB, yB, zB), and (xC, yC, zC), respectively, the square of the stretched lengths of each cable are

$$(xA - x0)^2 + (yA - y0)^2 + (zA - z0)^2 = LA^2$$
$$(xB - x0)^2 + (yB - y0)^2 + (zB - z0)^2 = LB^2$$
$$(xC - x0)^2 + (yC - y0)^2 + (zC - z0)^2 = LC^2$$

The unstretched lengths of each cable are

$$(xA)^2 + (yA)^2 + (zA)^2 = LA_0^{\,2}$$
$$(xB)^2 + (yB)^2 + (zB)^2 = LB_0^{\,2}$$
$$(xC)^2 + (yC)^2 + (zC)^2 = LC_0^{\,2}$$

The unit vectors along each stretched cable are

$$\hat{\mathbf{a}} = \frac{(xA - x0)}{LA}\hat{\mathbf{i}} + \frac{(yA - y0)}{LA}\hat{\mathbf{j}} + \frac{(zA - z0)}{LA}\hat{\mathbf{k}}$$
$$\hat{\mathbf{b}} = \frac{(xB - x0)}{LB}\hat{\mathbf{i}} + \frac{(yB - y0)}{LB}\hat{\mathbf{j}} + \frac{(zB - z0)}{LB}\hat{\mathbf{k}}$$
$$\hat{\mathbf{c}} = \frac{x(C - x0)}{LC}\hat{\mathbf{i}} + \frac{(yC - y0)}{LC}\hat{\mathbf{j}} + \frac{(zC - z0)}{LC}\hat{\mathbf{k}}$$

The tensions in the cables are related to the stretch of the cables by the spring constant:

$$T_A = k_A\,(LA - LA_0)$$
$$T_B = k_B\,(LB - LB_0)$$
$$T_C = k_C\,(LC - LC_0)$$

As in Sample Problem 2.18, the equilibrium equations are obtained from the scalar components of the vector equilibrium equation

$$T_A\,\hat{\mathbf{a}} + T_B\,\hat{\mathbf{b}} + T_C\,\hat{\mathbf{c}} + \mathbf{W} = 0$$

These scalar equations are

$$\frac{(xA - x0)}{LA}T_A + \frac{(xB - x0)}{LB}T_B + \frac{(xC - x0)}{LC}T_C = 0$$

$$\frac{(yA - y0)}{LA}T_A + \frac{(yB - y0)}{LB}T_B + \frac{(yC - y0)}{LC}T_C = 0$$

$$\frac{(zA - z0)}{LA}T_A + \frac{(zB - z0)}{LB}T_B + \frac{(zC - z0)}{LC}T_C = 1962$$

This nonlinear system of equations may be solved to determine the tension and stretched length of each cable and the final position of the scoreboard. The tensions in the cables and the final coordinates of the attachment to the scoreboard are

$$T_A = 723 \text{ N} \quad T_B = 516 \text{ N} \quad T_C = 935 \text{ N}$$

$$x0 = -0.039 \text{ m} \quad y0 = -0.022 \text{ m} \quad z0 = -0.924 \text{ m}$$

A design variation of this problem is to choose the spring constants of the cables so that, when deformed, the scoreboard will move only downward; that is, $x0$ and $y0$ are zero. The spring constant for one of the cables (in this example, cable A) is chosen first, and then the equations are solved for the other two spring constants and the tensions in the cables. The values of EA and the tensions in the cables are:

$$T_A = 722 \text{ N} \quad T_B = 508 \text{ N} \quad T_C = 944 \text{ N}$$

$$(EA)_A = 1.2 \times 10^5 \text{ N} \quad (EA)_B = 9.4 \times 10^4 \text{ N} \quad (EA)_C = 1.51 \times 10^5 \text{ N}$$

Notice that the stiffness of cable B was decreased, and the stiffness of cable C was increased. These adjustments can be obtained by changing either the material or the cross-sectional area of the cable. Details of solution of systems of nonlinear equations are given in the Computational Supplements.

3.4 STATICALLY INDETERMINATE PROBLEMS

Statically indeterminate problems occur when the object of interest is oversupported—that is, when there are more supports than are required to keep the object in equilibrium. Consider an object supported by two cables, as shown in Figure 3.7. The object will be modeled as a particle; therefore, the two cables are parallel and are attached to the same points on the object and at the top. *They are separated in the figure for clarity.* The cables which will stretch when the weight is released, are modeled as springs, with spring constants related to their material and size. Before the weight is released, these springs are unstretched. When the weight is released, both springs will support the weight, and will stretch an equal amount. The two spring forces $\mathbf{F}_A$ and $\mathbf{F}_B$ are collinear with the weight $\mathbf{W}$, and therefore, only one scalar equation of static equilibrium is necessary: $\Sigma \mathbf{F}_{\text{vertical}} = 0$. Thus we have

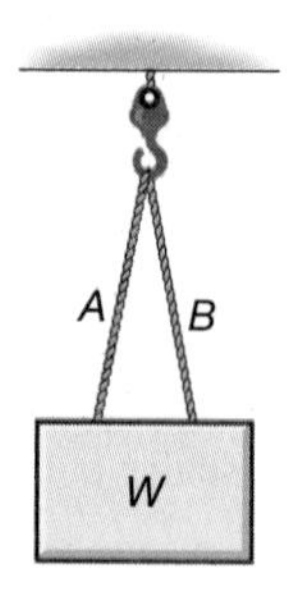

Figure 3.7

$$F_A + F_B - W = 0 \tag{3.7}$$

This scalar equation has two unknown spring forces; therefore, the force on each spring cannot be determined by equilibrium concepts alone. Accordingly, the problem is thus said to be statically indeterminate. However, if the spring constants of the two springs are known, the following relationships between the spring forces and their deformations apply:

$$F_A = k_A \delta_A$$
$$F_B = k_B \delta_B \tag{3.8}$$

Here, k and δ are the spring constant and stretch, respectively. Since the two springs are unstretched before the weight is released, the final stretches of the springs must be equal. This is called a condition of ***geometric*** or ***deformation compatibility***, and may be written mathematically as

$$\delta = \frac{F_A}{k_A} = \frac{F_B}{k_B}$$

or

$$F_A = \frac{k_A}{k_B} F_B \tag{3.9}$$

This is a second equation for the two unknown forces, and the solution is

$$F_A = \frac{k_A}{k_A + k_B} W$$
$$F_B = \frac{k_B}{k_A + k_B} W \tag{3.10}$$

If the springs are of equal stiffness, each will carry half the weight. Otherwise, the weight will be distributed proportionally to each spring's contribution to the total stiffness of the system. Statically indeterminate problems are considered in detail in the study of the mechanics of deformable bodies, structures, vibrations, and many other fields.

Sample Problem 3.15

A fourth cable D is added to the scoreboard in Sample Problem 3.5 for increased support. This cable is of the same material and cross-sectional area as the other three and is run 12 meters vertically to the ceiling in the undeformed condition. Determine the tension in the cables supporting the scoreboard.

Solution Although the system is now statically indeterminate, the solution is determined by adding another equation to the nonlinear system solved in Sample Problem 3.14 and adding the tension in this cable to the equilibrium equations. The additional equation gives the deformed length of the cable and the unit vector along the deformed cable:

$$(xD - x0)^2 + (yD - y0)^2 + (zD - z0)^2 = LD^2$$

Again, it is necessary to solve a system of nonlinear equations, and computational software must be used to obtain a numerical solution. For the case given, the tensions in the four cables and the coordinates of the scoreboard attachment are

$$T_A = 125 \text{ N} \quad T_B = 89 \text{ N} \quad T_C = 162 \text{ N} \quad T_D = 1.623 \times 10^3 \text{ N}$$
$$x_0 = -0.007 \text{ m} \quad y_0 = -0.004 \text{ m} \quad z_0 = -0.016 \text{ m}$$

Notice that the addition of the fourth cable in the position chosen causes that cable to carry almost all the weight of the scoreboard, and the other cables only stabilize the system. This could lead to a failure of the extra cable and an unsafe design. Details of the solution can be found in the Computational Supplements.

Sample Problem 3.16

An important example of a statically indeterminate problem in biomechanics occurs when a fracture of a load bearing bone is reduced using a bone plate as shown in Figure SP3.16a. This figure shows the fracture of the femur with a bone plate reduced by a plate. Bone remodels or heals best under load so the problem is how to stabilize the fracture but allow the bone to take enough load to promote healing. The simplest model is shown in the free-body diagram in Figure SP3.16b. The one equation of static equilibrium is:

$$F_B + F_P = W$$

Both the bone and the plate can be treated a linear springs with spring constants of k_B and k_P, respectively. We have already examined to the solution of this problem in the form of Eq. (3.10).

$$F_B = \frac{k_B}{k_B + k_P} W$$

$$F_P = \frac{k_P}{k_B + k_P} W$$

The biomechanist now has a method to adjust the spring constant of the bone plate to minimize the "stress shielding" on the bone and thus promote healing.

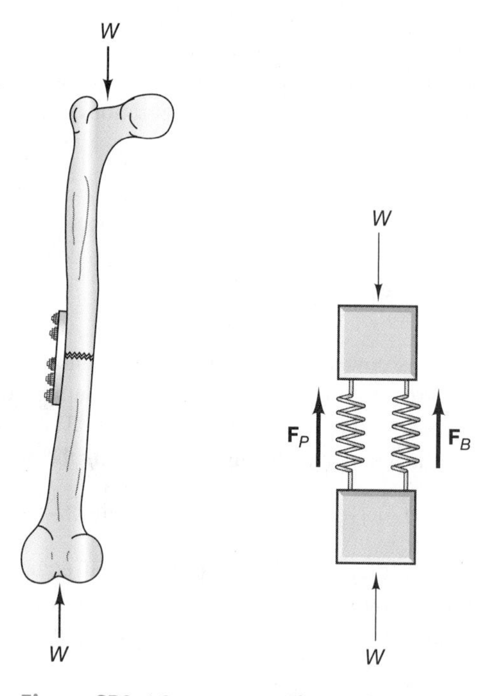

Figure SP3.16a Figure SP3.16b

In setting the bone, if the raw fracture ends have a gap between them, the spring constant for the bone would be zero and all load will be carried by the plate. Bone will ultimately fill the gap but the healing will be slow. Special plates have been developed, called compression plates, apply initial compression at the fracture site.

Problems*

3.38 Compute the forces in the spring system illustrated in Figure P3.38, using equilibrium and the linear spring deformation law.

3.39 A force is applied to the spring system shown in Figure P3.39. (a) What force is required to close the gap c, and (b) what force is required to deflect the system to twice the value of the gap?

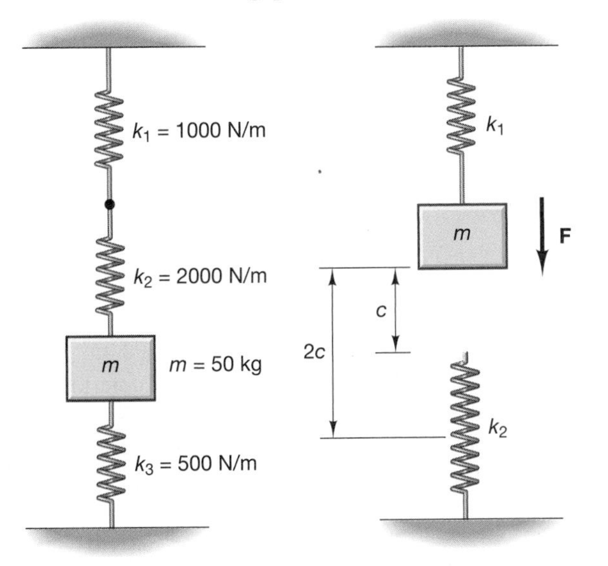

Figure P3.38 Figure P3.39

3.40 A spring and two cables, arranged as shown in Figure P3.40, support a 100-kg block. If the spring deflects $x = 0.01$ m with angular measures $\theta = 30°$ and $\beta = 20°$, calculate the spring stiffness and the tension in the cables. Neglect the deformation of the cables.

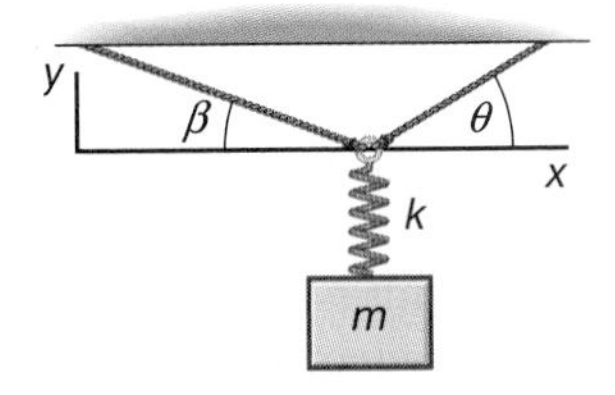

Figure P3.40

3.41 Consider the spring-and-cable arrangement of Problem 3.40. If the mass of the block is 100 kg, the stiffness is 10,000 N/m, $\theta = 15°$, and $\beta = 20°$, calculate the tensions T_1 and T_2 and the deflection of the block.

3.42 Determine the weight that causes the three springs with equal spring constant to deflect as shown in Figure P3.42. (This procedure is similar to calibrating a scale.)

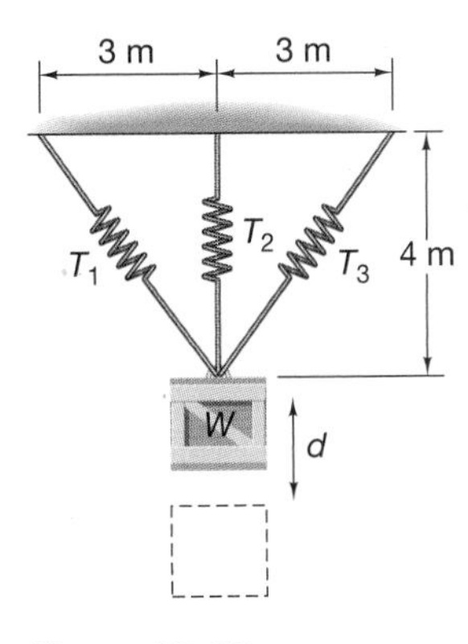

Figure P3.42

3.43 In Figure P3.43 calculate the deflection in the spring along the vector T_1 that satisfies the equilibrium condition. The spring constant is $k = 2000$ N/m, and the crate has a mass of 46.38 kg. Note that the dimensions already given account for the deflection of the spring.

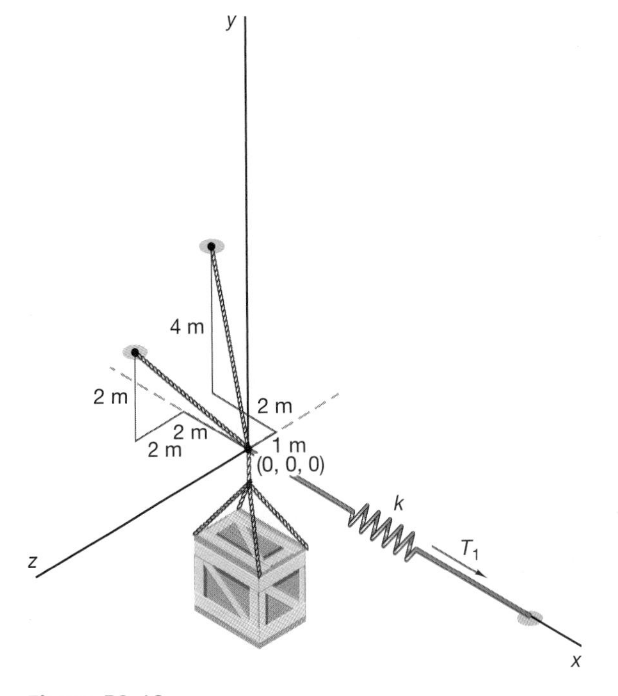

Figure P3.43

* In all problems, system shown in unloaded configuration.

3.44 A 50-kg collar is supported by a spring as it slides on a smooth rod as shown in Figure P3.44. If the spring is unstretched when the spring is in the horizontal position. If the spring constant is 2000 N/m, determine the equilibrium position of the system.

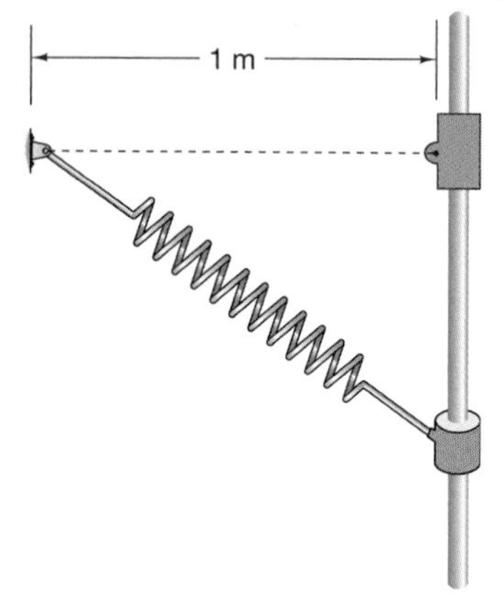

Figure P3.44

3.45 The collar system shown in Figure P3.45 consists of a 10-kg collar at A and a 15-kg collar at B. The two collars are connected by a spring with a spring constant of 500 N/m and an unstretched length of 2 m. Determine the stretch of the spring when the system is inthe equilibrium position.

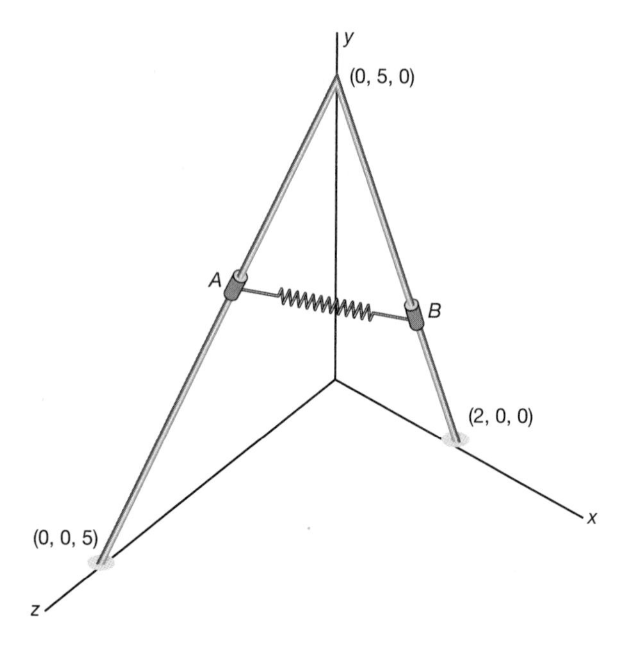

Figure P3.45

3.46 For the system shown in Figure P3.45, show that if the spring is very stiff (for example greater than 500,000 N/m), the equilibrium position is the same as that determined in Sample Problem 3.10. Determine the force in the spring

3.5 SPECIAL SECTIONS

Two special sections are added to this chapter as the principle of particle equilibrium can be used to analyze simple friction problems and simple structures. Both of these topics are covered in detail in later chapters but for some engineering studies only a brief introduction of these subjects is needed. Mechanical engineering students will need to consider machines that are driven by friction and loss of energy due to friction. Civil engineers need a detailed study of structures.

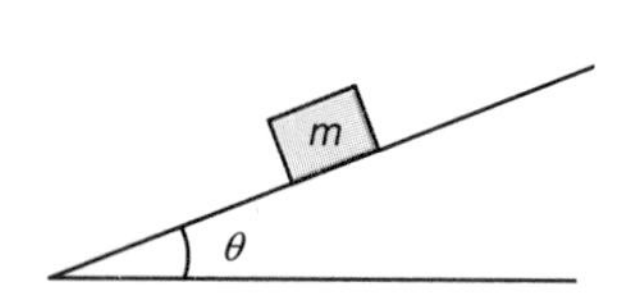

Figure 3.5A.1

3.5A INTRODUCTION TO FRICTION

Another important example of action and reaction, Newton's third law, occurs when friction between two contacting bodies is considered. The study of friction or *tribology* is a very important concept in mechanics and two surfaces in contact with no adhesive or bonding will slide relative to one another if the applied force is great enough. The resistance to sliding is caused by **friction**. Useful applications of friction, such as drive belts and wedges will be covered in detail in Chapter 9.

Consider the block of mass m on an inclined plane as shown in Figure 3.5A.1. The block is held in equilibrium by the normal force from the plane and friction between the plane and the block. The free-body diagram of the block is shown in Figure 3.5A.2. The equations of equilibrium are:

Figure 3.5A.2

$$N - mg\cos\theta = 0$$
$$mg\sin\theta - f = 0$$

The friction force for equilibrium is equal to $mg \sin\theta$. Note that the friction force would be zero if the block set on a level plane. Therefore, the friction force will only equal the force trying to slide the block. If a block on a level plane is acted upon by a force **P** acting parallel to the plane, the friction force will be equal and opposite to **P** until some maximum value is reached.

The friction between the block and the plane resists this tendency to slide until it reaches a maximum value. The **maximum friction** before slippage between any two surfaces depends upon the normal force between the two objects and the surface material of each object and is expressed by a *coefficient of static friction*, and usually designated by μ_s. The maximum static friction is

$$f_{max} = \mu_s N \tag{3.5A.2}$$

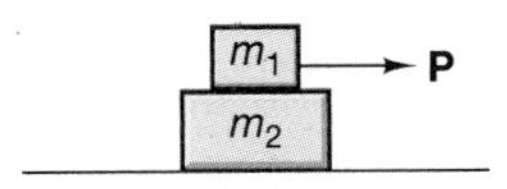

Figure 3.5A.3

where N is the normal force acting on the block. For the block on an inclined plane shown in Figure 3.5A.1, the ratio between the friction force and the normal force is

$$\frac{f}{N} = \frac{mg \sin\theta}{mg \cos\theta} = \tan\theta$$

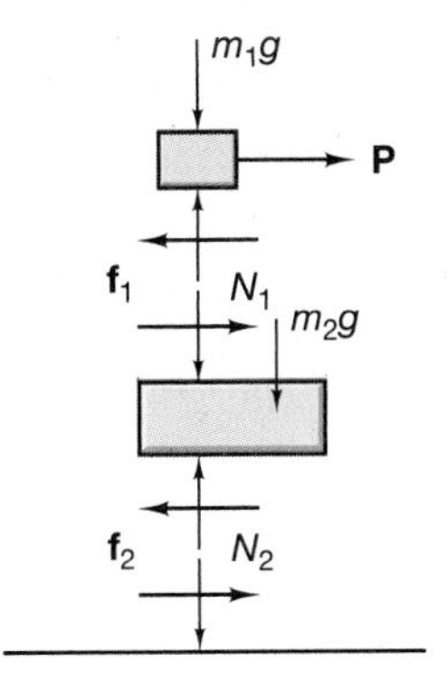

Figure 3.5A.4

If the angle of inclination is slowly increases until the block slips, the coefficient of static friction will equal the tangent of the angle of inclination when slippage occurs. When the block slips, the resisting friction decreases by as much as 25%, and the ratio of the friction force to the normal force is denoted by the coefficient of kinetic friction, μ_k, which is a unitless coefficient.

In general, the coefficient of static friction is used only to determine whether slip will occur between two surfaces. If the required friction force for equilibrium is greater than the maximum possible between two surfaces, relative motion between the surfaces will occur. See Table 3.5.1 below.

Two blocks, one on top of the other, are shown in Figure 3.5A.3. A force is applied to the top block trying to slide it across the lower block. If friction exists between all surfaces, the surface between the two blocks and the lower block and the plane, the forces acting on the blocks are shown in Figure 3.5A.4.

Table 3.5.1

Surface Materials	Coefficient of Static Friction
Steel on steel	0.75
Rubber on concrete	0.50–0.90
Rubber on ice	0.05–0.3
Metal on ice	0.04
Copper on copper	1.2
Metal on wood	0.20–0.60
Teflon® on Teflon®	0.04
Glass on glass	0.90
Copper on steel	0.50
Wood on wood	0.25–0.50
Aluminum on aluminum	1.10
Rubber tires on gravel	0.5
Rubber tires on dirt	0.3–0.5
Rubber tires on snow	0.1–0.3

The friction force on m_1 opposes the applied force **P** to prevent the top block from sliding on the lower block. The friction force will increase to counteract the applied force until the friction force reaches a maximum value and then slipping will occur. The friction force has an equal and opposite force acting on the lower block as shown. The friction force $\mathbf{f}_1$ acts to try to slide the lower block on the level plane and is resisted by the friction between the plane and the lower block, $\mathbf{f}_2$.

If no slippage occurs in the system the magnitude of the friction forces between the two blocks and the lower block and the plane are equal to the magnitude of **P**. The normal force between the two blocks is $m_1 g$ and the normal force between the lower block and the plane is $(m_1 + m_2)g$. Therefore, if the coefficient of static friction is the same between all the surfaces, the upper block will slip on the lower block.

Sample Problem 3.5A.1

Determine the range of values of the force P required to prevent the 400 N box shown from slipping up or down the inclined plane if the coefficient of static friction between the box and the plane is 0.2.

Solution

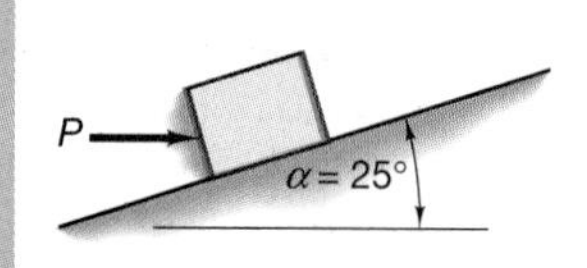

We choose a coordinate system parallel and perpendicular to the inclined plane and construct a free-body diagram of the box. (See figure at the left.) The friction force is shown acting in both directions. When P is a minimum, the tendency for slipping is down the incline, and friction will act up the incline to resist slipping. When P is a maximum, the tendency for slipping is up the plane, and friction resists by acting down the plane. Since we have specified a condition of impending slip, the magnitude of the friction force is the product of the coefficient of static friction and the normal force. Writing the equation of equilibrium, we have

$$P + W + N + f = 0$$

$$\mathbf{P} = P(\cos \mathbf{25\hat{i}} - \sin \mathbf{25\hat{j}})$$

$$\mathbf{W} = 400(-\sin \mathbf{25\hat{i}} - \cos \mathbf{25\hat{j}})$$

where

$$\mathbf{N} = N\mathbf{\hat{j}}$$

$$\mathbf{f} = \pm 0.2\, N\mathbf{\hat{i}}\ [(+)\text{ for the minimum } P \text{ and } (-) \text{ for the maximum } P]$$

Equating the $\mathbf{\hat{i}}$ and $\mathbf{\hat{j}}$ components of the vector equation gives the two scalar equations of equilibrium:

$$P_{max/min} \cos 25° - 400 \sin 25°(-/+)0.2\,N = 0$$

$$-P_{max/min} \sin 25° - 400 \cos 25° + N = 0$$

When the friction force is negative (acting down the plane), impending motion is up the plane, and P will be a maximum. The matrix equilibrium equation is

$$\begin{bmatrix} \cos 25° & -0.2 \\ -\sin 25° & 1 \end{bmatrix} \begin{bmatrix} P_{max} \\ N \end{bmatrix} = \begin{bmatrix} 300 \sin 25° \\ 400 \cos 25° \end{bmatrix}$$

Solving the matrix equation gives $P_{max} = 293.9$ N.

When the friction force is positive (acting up the plane), impending motion is down the plane, and the matrix equation is

$$\begin{bmatrix} P_{min} \\ N \end{bmatrix} = \begin{bmatrix} \cos 25° & +0.2 \\ -\sin 25° & 1 \end{bmatrix}^{-1} \begin{bmatrix} 300 \sin 25° \\ 400 \cos 25° \end{bmatrix}$$

The minimum force required to resist motion down the plane is thus $P_{min} = 97.4$ N.

Now let us solve the problem a second time with a different choice of coordinate system. The results are invariant to the choice of the coordinate system, but the equilibrium equations will have a different algebraic appearance. The free-body diagram is the same for either coordinate system. A unit vector tangent to the plane with a positive sense up the plane is

$$\hat{\mathbf{t}} = \cos 25°\hat{\mathbf{i}} + \sin 25°\hat{\mathbf{j}}$$

The unit normal vector is obtained using Eq. (5.6)—that is, the cross product of two tangential vectors:

$$\hat{\mathbf{n}} = \hat{\mathbf{k}} \times \hat{\mathbf{t}} = -\sin 25°\hat{\mathbf{i}} + \cos 25°\hat{\mathbf{j}}$$

The forces for this coordinate system are

$$\begin{aligned} \mathbf{P} &= P_{\max/\min}\hat{\mathbf{i}} \\ \mathbf{W} &= -400\hat{\mathbf{j}} \\ \mathbf{N} &= N(-\sin 25°\hat{\mathbf{i}} + \cos 25°\hat{\mathbf{J}}) \\ \mathbf{f} &= (-/+)\ 0.2N\,(\cos 25°\hat{\mathbf{i}} + \sin 25°\,\hat{\mathbf{j}}) \end{aligned}$$

The matrix equation to determine the value of P for impending motion up the plane and to resist impending motion down the plane is

$$\begin{bmatrix} P_{\max/\min} \\ N \end{bmatrix} = \begin{bmatrix} 1 & -\sin 25° \mp 0.2\cos 25° \\ 0 & \cos 25° \mp 0.2\sin 25° \end{bmatrix}^{-1} \begin{bmatrix} 0 \\ 400 \end{bmatrix}$$

The maximum and minimum magnitudes of P are

$$P_{\min} = 97.4 \quad P_{\max} = 293.9$$

Sample Problem 3.5A.2

For the block shown at the left, determine the minimum force **P** as a function of θ for any angle of incline, α, to initiate motion.

Solution

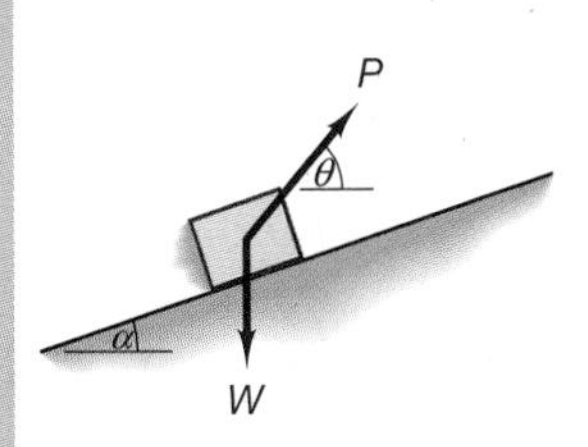

We select a coordinate system with x parallel to the incline and y perpendicular to the incline. We then construct a free-body diagram of the block, as shown on the next page and to the right. Since we are seeking the minimum value of P required for impending slip, the magnitude of the friction force is obtained using the coefficient of static friction. The equilibrium equations become

$$\begin{aligned} &P\cos(\theta - \alpha) - f - W\sin\alpha = 0 \\ &N - W\cos\alpha + P\sin(\theta - \alpha) = 0 \\ &f = \mu_s N \end{aligned}$$

$$\begin{bmatrix} P \\ N \\ f \end{bmatrix} = \begin{bmatrix} \cos(\theta - \alpha) & 0 & -1 \\ \sin(\theta - \alpha) & 1 & 0 \\ 0 & -\mu_s & 1 \end{bmatrix}^{-1} \begin{bmatrix} W\sin\alpha \\ W\cos\alpha \\ 0 \end{bmatrix}$$

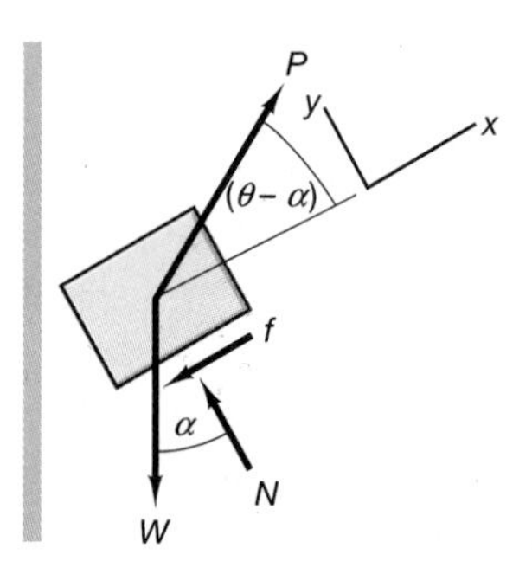

Solving these for P yields

$$P(\theta) = \frac{W(\sin\alpha + \mu_s\cos\alpha)}{\cos(\theta - \alpha) + \mu_s\sin(\theta - \alpha)}$$

It is clear that P is a function of θ, and the minimum value of P can be found by differentiating $P(\theta)$ with respect to θ and setting the differential equal to zero for a local maximum or minimum. The details of this computation are shown in the Computational Supplement.

Sample Problem 3.5A.3

A block of mass m rests on an inclined plane, as shown in the diagram at the left. Determine the minimum coefficient of static friction required to keep the mass from sliding. The plane makes an angle α with the x-axis in the x–z plane and makes an angle β with the y-axis in the y–z plane.

Solution The normal to the surface and the direction of the tangential force acting down the surface will be determined using Eq. (5.6). We construct vectors along the intercepts of the surface with the x–y plane and the y–z plane:

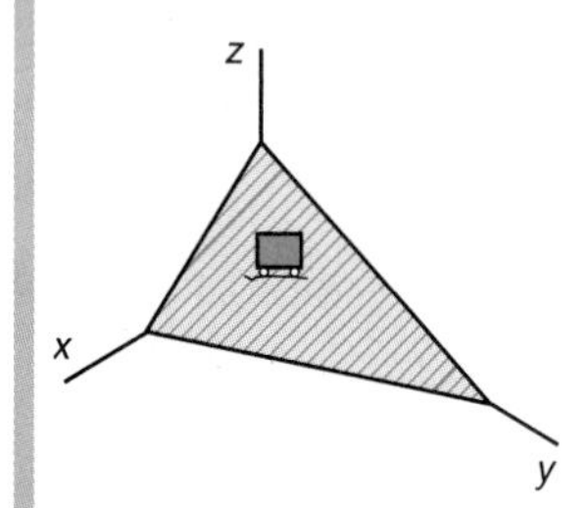

$$\hat{\mathbf{m}} = \cos\alpha\hat{\mathbf{i}} - \sin\alpha\hat{\mathbf{k}}$$

$$\hat{\mathbf{q}} = \cos\beta\hat{\mathbf{j}} - \sin\beta\hat{\mathbf{k}}$$

The unit vector normal to the plane is defined by $\hat{\mathbf{m}}$ and $\hat{\mathbf{q}}$, as shown in the diagram at the left. Since these two vectors are not orthogonal, the unit normal is the cross product divided by its magnitude:

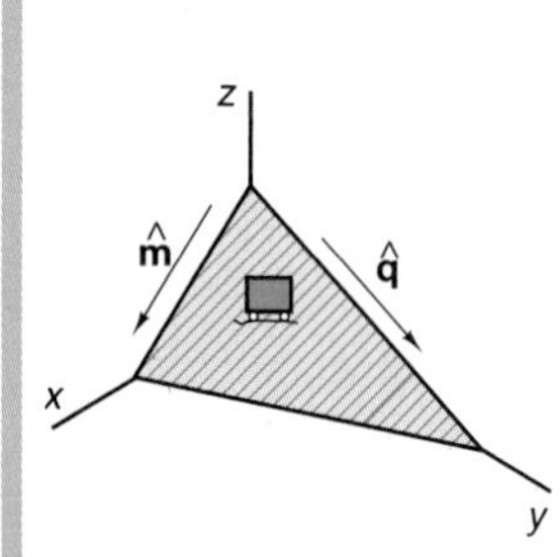

$$\hat{\mathbf{n}} = \frac{\hat{\mathbf{m}} \times \hat{\mathbf{q}}}{|\hat{\mathbf{m}} \times \hat{\mathbf{q}}|}$$

$$\hat{\mathbf{n}} = \frac{1}{\sqrt{\cos^2\alpha + \sin^2\alpha\cos^2\beta}}[\sin\alpha\cos\beta\hat{\mathbf{i}} + \cos\alpha\sin\beta\hat{\mathbf{j}} + \cos\alpha\cos\beta\hat{\mathbf{k}}]$$

The only force—other than the normal force and the friction force—acting on the mass is gravitational attraction, which results in the block's weight.

$$\mathbf{W} = -mg\hat{\mathbf{k}}$$

The components of $\mathbf{W}$ in the normal and tangential directions are, respectively,

$$\mathbf{W}_n = (\mathbf{W} \cdot \hat{\mathbf{n}})\hat{\mathbf{n}}$$

$$\mathbf{W}_t = \mathbf{W} - \mathbf{W}_n$$

The required friction force must be equal in magnitude and opposite in sense to $\mathbf{W}_t$, and the normal force is equal in magnitude and opposite in sense to $\mathbf{W}_n$ so that

$$\mathbf{N} = -\mathbf{W}_n \quad \mathbf{f} = -\mathbf{W}_t$$

The minimum coefficient of static friction is

$$\mu_{s\,\min} = \frac{|\mathbf{f}|}{|\mathbf{N}|}$$

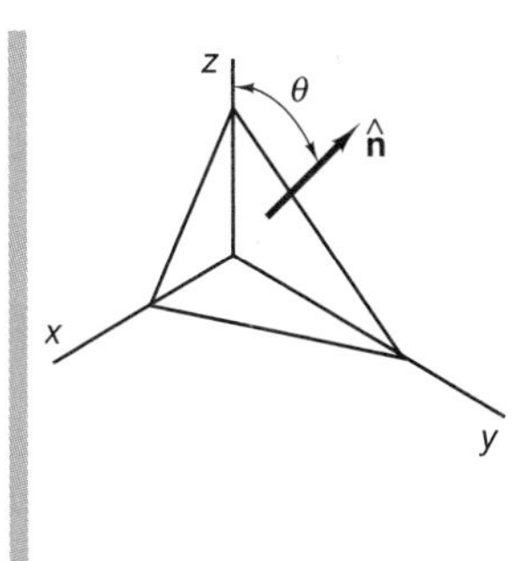

Since the unit vector $\hat{\mathbf{n}}$ is normal to the surface, the angle θ between the z-axis and the normal vector is the angle of incline of the plane. (See figure at left.) This angle is

$$\theta = \cos^{-1}(\hat{\mathbf{n}} \cdot \hat{\mathbf{k}})$$

We have shown earlier that the coefficient of static friction can be defined by the tangent of the angle of incline corresponding to impending slip. Therefore, the minimum coefficient of static friction for any inclined surface is

$$\mu_s = \tan\theta = \tan[\cos^{-1}(\hat{\mathbf{n}} \cdot \hat{\mathbf{k}})]$$

The vector calculations can be carried out for specific numerical cases, and an example is shown in the Computation Supplement.

Sample Problem 3.5A.4

Two blocks are supported by a cable as shown in the figure. Consider the case where $\alpha > \beta$ and $m_1 > m_2$. If friction occurs only between block 2 and the inclined surface, determine the friction force necessary to hold the system in equilibrium and the tension in the cable.

Solution Construct a free-body diagram of each block as shown below. As there is no friction on block 1, the equilibrium equation in the direction of the incline

$$m_1 g \sin\alpha - \mathrm{T} = 0$$

is:

$$T = m_1 g \sin\alpha$$

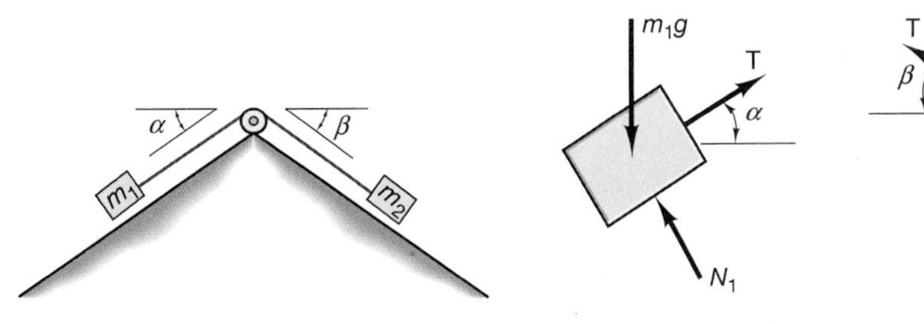

For equilibrium of the second block, we obtain:

$$T - f - m_2 \sin\beta = 0$$

Substituting for T yields

$$f = (m_1 \sin\alpha - \mathrm{m}_2 \sin\beta)g$$

Problems

3.47 A 100-N force pulls a 200-N block as shown in Figure P3.47. The coefficient of static friction between the block and the floor is $\mu_s = 0.6$, and the coefficient of kinetic friction is $\mu_k = 0.4$. (a) What is the friction force between the block and the floor? (b) Will the block move?

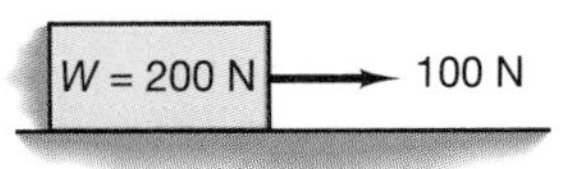

Figure P3.47

3.48 (a) A 200-N block is resting on a 30° incline. The coefficient of static friction between the block and the plane is $\mu_s = 0.8$. Will the block remain at rest? (b) A 50-kg mass is resting on a 45° incline. The coefficient of static friction between the block and the plane is $\mu_s = 0.6$. Will the block remain at rest?

3.49 What is the maximum angle of incline for a 50-kg block to rest on an inclined plane if the coefficient of friction between the block and the plane is $\mu_s = 0.35$?

3.50 A 10-N force is pushing a 40-N block, as shown in Figure P3.50. The angle of incline of the plane is $\alpha = 40°$. The coefficient of static friction between the block and the incline is $\mu_s = 0.75$, and the coefficient of kinetic friction is $\mu_k = 0.65$. Will the block slide on the plane? If it does, will it slide up or down the plane? What is the friction force between the block and the plane?

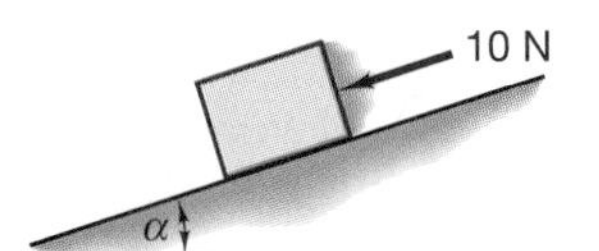

Figure P3.50

3.51 A 60-N force is pushing a 10-kg mass down the plane as shown in Figure P3.51. The angle of incline of the plane is $\alpha = 20°$. The coefficient of static friction between the block and the plane is $\mu_s = 0.6$, and the coefficient of kinetic friction is $\mu_k = 0.5$. Will the block slide on the plane? If it does, will it slide up or down the plane? What is the friction force between the block and the plane?

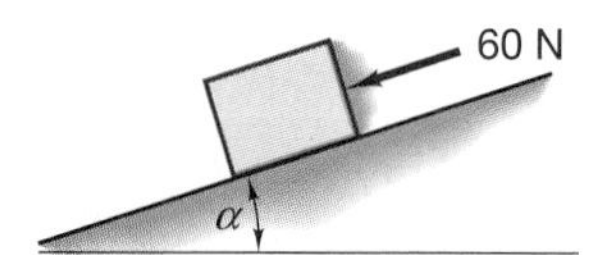

Figure P3.51

3.52 For Problem 3.50, what is the maximum angle of incline, α, if static equilibrium is to be maintained?

3.53 A 100-N force is pulling a 50-kg block as shown in Figure P3.53. The coefficient of static friction between the block and the incline is $\mu_s = 0.4$, and the coefficient of kinetic friction is $\mu_k = 0.3$. The angle of incline is $\alpha = 35°$. (a) Will the block move? (b) What is the friction force between the block and the plane?

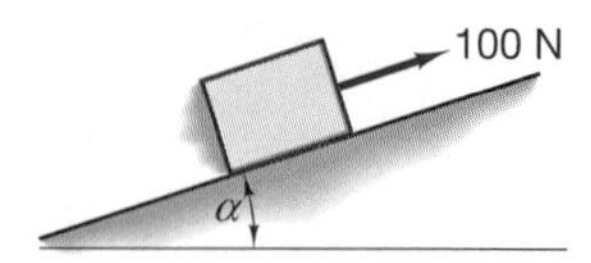

Figure P3.53

3.54 A 75-N force, inclined at an angle $\alpha = 45°$ as shown in Figure P3.54, pulls a 100-N block. The coefficient of static friction between the block and the floor is $\mu_s = 1.3$, and the coefficient of kinetic friction is $\mu_k = 1.1$. (a) Will the block move? (b) What is the friction force between the block and the plane?

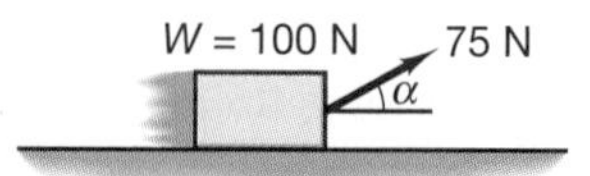

Figure P3.54

3.55 A 5-kg block rests on a rail with a 100-kg mass hanging from it. (See Figure P3.55.) (a) What force is required to begin moving the block along the rail? The coefficient of static friction for the block on the rail is $\mu_s = 0.6$, and the coefficient of kinetic friction is $\mu_k = 0.5$. The force is inclined at 60°, as shown. (b) For what angles α can the force F slide the block to the right?

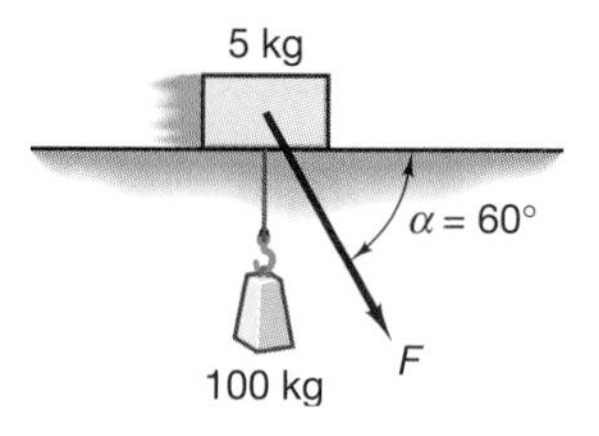

Figure P3.55

3.56 A 1000-N force, inclined at an angle $\beta = 35°$, pulls a 100-kg block. (See Figure P3.56.) The coefficient of static friction between the block and the incline is $\mu_s = 0.25$, and the coefficient of kinetic friction is $\mu_k = 0.15$. The angle of incline α, is 40°. (a) What is the friction force between the block and the plane? (b) Will the block move?

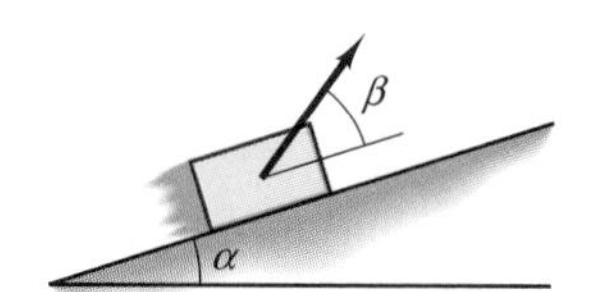

Figure P3.56

3.57 A 650-N force, inclined at an angle $\beta = 25°$, pushes a 1500-N block, as shown in Figure P3.57. The coefficient of static friction between the block and the floor is $\mu_s = 1.05$, and the coefficient of kinetic friction is $\mu_k = 0.95$. The angle of incline, α, is 30°. (a) Will the block move? (b) What is the friction force between the block and the plane?

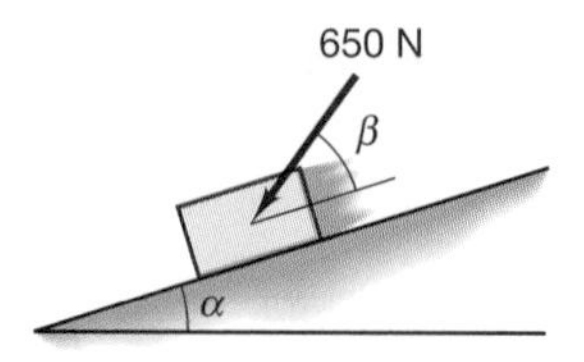

Figure P3.57

3.58 Two blocks are on an inclined plane, as shown in Figure P3.58. Block A of mass 100 kg and block B of mass 10 kg. The angle of incline is 35°. (a) What is the minimum coefficient of friction required to prevent the blocks from slipping? Assume that the coefficient of friction is the same between the plane and the block and between the two blocks. (b) What is the tension in the rope?

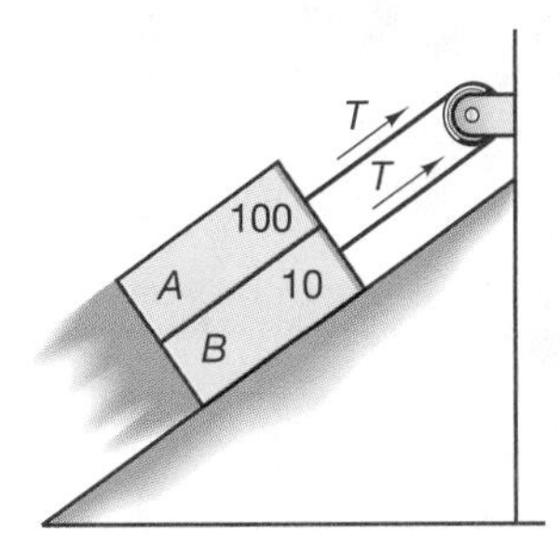

Figure P3.58

3.59 Two blocks are on an inclined plane, as shown in Figure P3.59. Block A has a mass of 10 kg and block B has a mass of 25 kg. The angle of incline is 20°. (a)What is the minimum coeffiecient of friction required to prevent the blocks from slipping? Assume that the coefficient of friction between the blocks is four times the coefficient of friction between the block and the plane. (b) What is the tension in the rope?

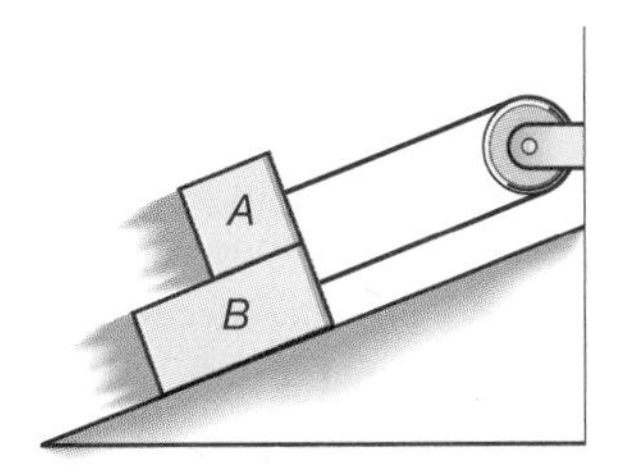

Figure P3.59

3.60 A plane is inclined 30° from the x-axis. A 50-kg mass is resting on the plane. The coefficient of static friction between the plane and the block is $\mu_s = 0.5$ and the coefficient of kinetic friction is $\mu_k = 0.4$. A force $\mathbf{F} = 20\hat{\mathbf{i}} + 40\hat{\mathbf{j}}$ N is applied to the block. Assume that gravity acts in the y-direction. Will the block slide?

3.61 A roofer is standing on a roof. The roofer has a mass of 80 kg. The coefficient of friction between the roofer's boots and the plywood is $\mu_s = 0.4$. What is the maximum incline on the roof required to prevent the roofer from slipping?

3.62 A tow truck is pulling a car out of a ditch, as shown in Figure P3.62. The car has mass of 1000 kg. What is the necessary tension in the tow cable to start moving the car out of the ditch? The coefficient of friction between the car's wheel bearings and axles is $\mu_s = 0.8$, and the coefficient of kinetic friction is $\mu_k = 0.7$.

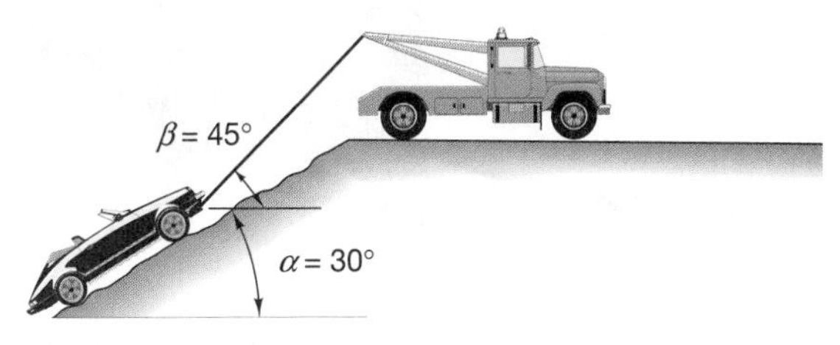

Figure P3.62

3.63 A conveyor belt is used in an airport to move baggage from the ground to the cargo hold of an airplane. (See Figure P3.63.) The angle of incline of the belt is 40°. What minimum coefficient of friction between the baggage and the belt is required to prevent the baggage from slipping?

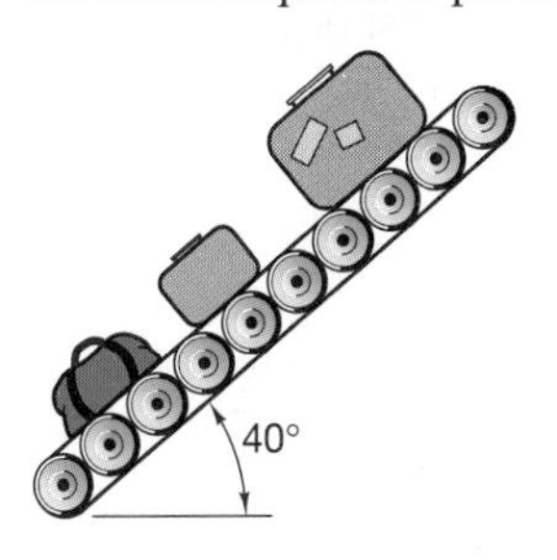

Figure P3.63

3.5B KEYSTONE OF THE ARCH

Arch bridges are one of the oldest types of bridges and have great natural strength, see Figure 3.5B.1. The weight of an arch bridge is carried outward along the curve of the arch to the supports at each end. The supports, called *abutments*, carry the load and keep the ends of the arch from spreading out. The Romans used arches to support their bridges and aqueducts. Many are still standing today and one of the most famous is the Pont du Gard aqueduct near Nimes, France. Built before the birth of Christ, this aqueduct is held together by mortar only in its top tier. The stones in the rest of the structure stay together

Figure 3.5B.1 (Photo courtesy of Jose Gil/Shutterstock)

by the sheer force of their own weight. The arch was used much earlier by the Egyptians evidenced by an arch in the tombs of Thebes dating to 1540 BC, 460 years before the Temple of Solomon was built.

The arch is composed of an odd number of total stones with the *keystone*, being the center and an even number of stones, called *voussoirs*, on either side of the keystone, as shown in Figure 3.5B.2. The keystone is the most important stone in an arch bridge, without this stone the arch would collapse. During construction of these arch structures, the arch must be supported by a frame until the keystone is in place.

The arch can be considered as a structure and the forces acting on the keystone can best be understood by considering the simple arch truss shown in Figure 3.5B.3. A free-body diagram of the pin connecting the two sides of the structure is shown in Figure 3.5B.4.

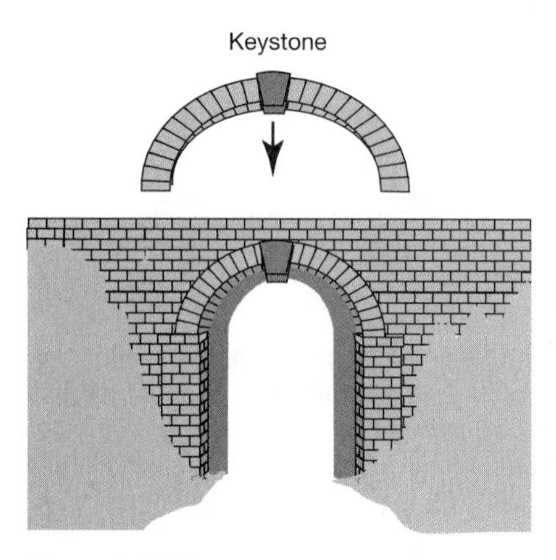

Figure 3.5B.2

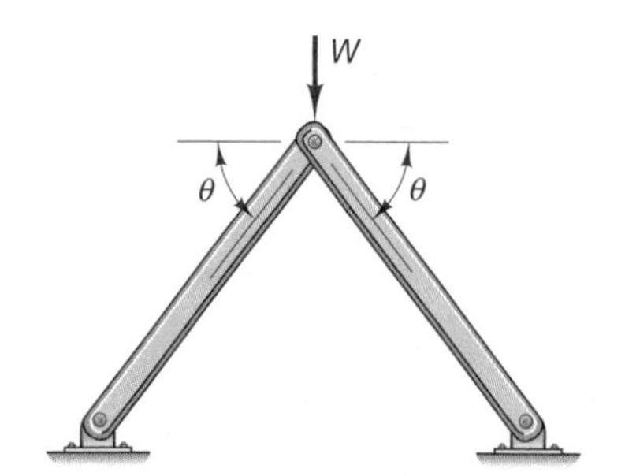

Figure 3.5B.3

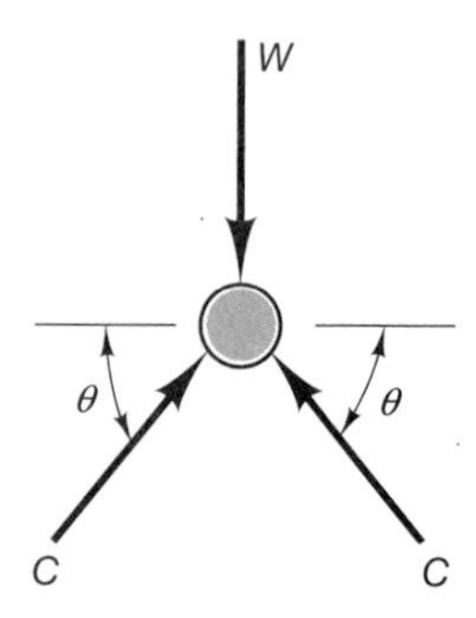

Figure 3.5B.4

Symmetry or summation of the forces in the horizontal direction shows that the compression in the left and right members are equal. Equilibrium in the vertical direction yields:

$$C = \frac{W}{2 \sin \theta} \tag{3B.1}$$

The pin in this truss is equivalent to the keystone of the arch. Note that as the angle θ decreases in value the compression in the members increases. This arch truss carries the weight downward to the abutments where vertical and horizontal forces produce equilibrium.

The analysis of the loads in truss members by examining the equilibrium of the pin is called the method of joints and will be covered in detail in Chapter 7. As can be seen, the pin is treated as a particle and the analysis follows methods outlined in this chapter.

The forces acting on the keystone of a stone arch can be determined using the concepts of particle equilibrium. Consider a semicircular arch having seventeen stones, eight on either side of the keystone. The free-body diagram of the keystone is shown in Figure 3.5B.5. The Equations of equilibrium are:

$$\begin{aligned} 2C_{10} \sin 10° - W_{ks} &= 0 \\ C_{10} = 1/2 \frac{W_{ks}}{\sin 10°} &= 2.879\, W_{ks} \end{aligned} \tag{3.5B.2}$$

W_{ks} is the total weight that is acting on the keystone plus the weight of the keystone itself. The weight acting on the keystone can be very large as it includes the structure above the stone in addition to the load crossing the structure.

Each voussoir completes ten degrees of the arch and the first 10° stone is shown in the free-body diagram in Figure 3.5B.6. The equations of equilibrium in the x and y directions, where x is taken normal to the 20° face, are:

$$\begin{aligned} -C_{20} + C_{10} \cos(20° - 10°) + W_{15} \sin 20° &= 0 \\ -S_{20} + C_{10} \sin(20° - 10°) - W_{15} \cos 20° &= 0 \end{aligned} \tag{3.5B.3}$$

where W_{15} is the total weight acting down on this stone with a center 15° off the center of the arch.

Solving for C_{20} and S_{20} yields:

$$\begin{aligned} C_{20} &= 2.835 W_{ks} + 0.342\, W_{15} \\ S_{20} &= 0.5 W_{ks} - 0.940\, W_{15} \end{aligned} \tag{3.5B.4}$$

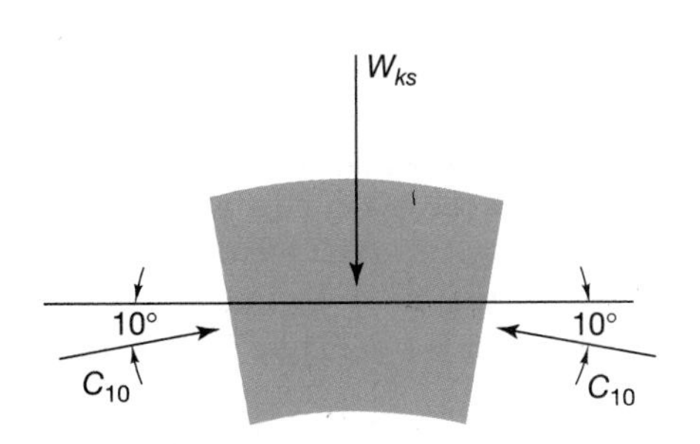

Figure 3.5B.5

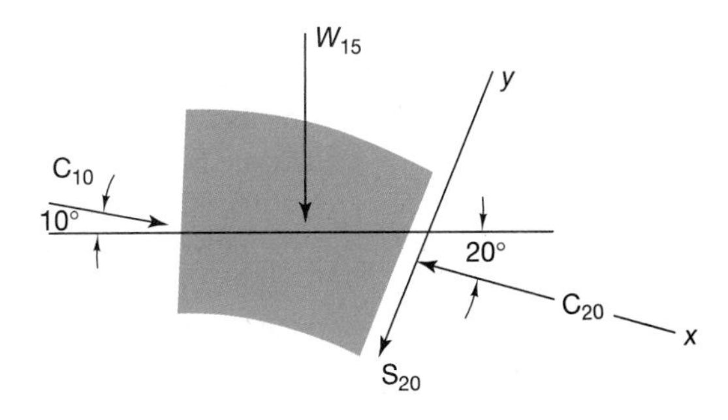

Figure 3.5B.6

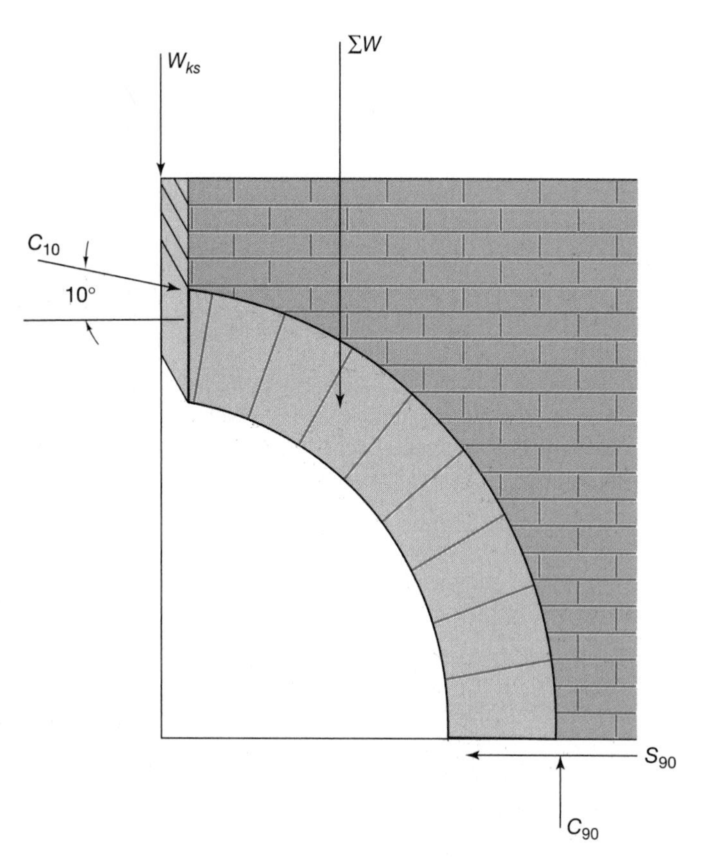

Figure 3.5B.7

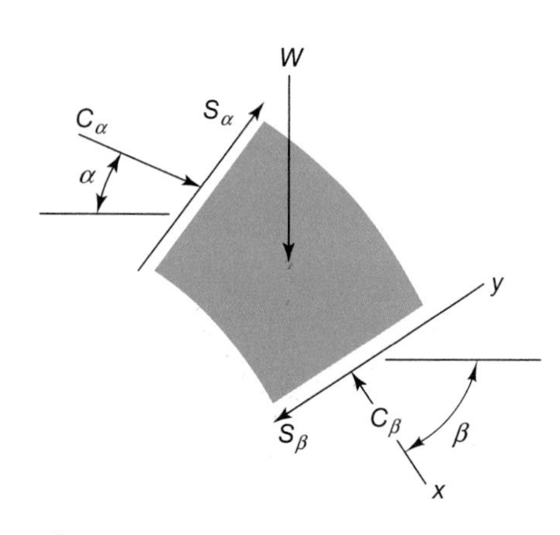

Figure 3.5B.8

A general solution for the loading on any voussoir can be obtained from the free-body diagram shown in Figure 3.5B.7.

$$C_\beta = C_\alpha \cos(\beta - \alpha) - S_\alpha \sin(\beta - \alpha) + W \sin \beta$$
$$S_\beta = C_\alpha \sin(\beta - \alpha) - S_\alpha \cos(\beta - \alpha) + W \cos \beta \qquad (3.5B.5)$$

The forces acting on the bottom voussoir can be obtained by considering the portion of the arch shown in Figure 3.5B.8. Summing forces in the horizontal and vertical directions yields:

$$C_{90} = C_{10} \sin 10° + \sum W = 0.5W_{ks} + \sum W$$
$$S_{90} = C_{10} \cos 10° = 2.835W_{ks} \qquad (3.5B.6)$$

In Figure 3.5B.6, the first voussoir was modeled with a compression and shear force on the 20° face of the stone. Some may question if the shear force can exist on the inclined face. The large compression on this face would produce adequate friction and wedge shape of the stone also resists slipping. A different model might place a horizontal force acting on the stone, necessary for equilibrium, and no shear force between the voussoirs. We can obtain the results of this model by breaking down S_{20} into nonorthogonal components in the horizontal and the C_{20} directions, as shown in Figure 3.5B.9. These components are:

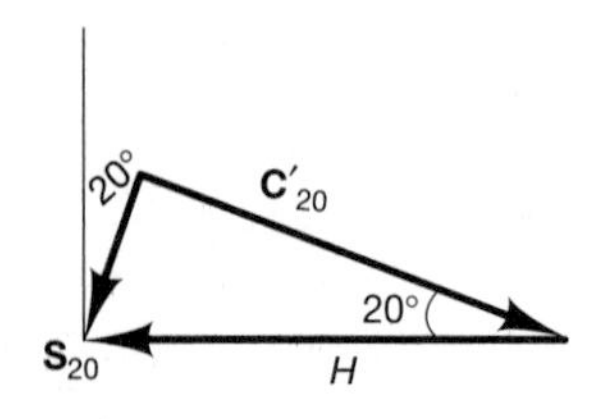

Figure 3.5B.9

$$H \sin 20° = S_{20}$$
$$C_{20}\, H \cos 20° \qquad (3.5B.7)$$

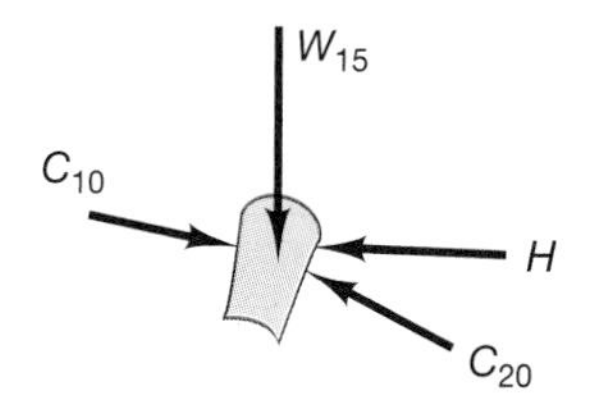

Figure 3.5B.10

Using Eq. (3.5B.4) yields

$$H = \frac{1}{\sin 20°}(0.5W_{ks} - 0.94W_{15}) = 1.462W_{ks} - 2.747W_{15} \quad (3.5B.8)$$

$$C'_{20} = H\cos 20° = 1.374\, W_{ks} - 2.581W_{15}$$

Examining Figure 3B.10, the compression on the second voussoir is:

$$C^{H}_{20} = C_{20} - C'_{20} = 1.461W_{ks} + 1.641W_{15} \quad (3.5B.9)$$

The correct model of this voussoir is some combination of the two models shown. There definitely will be some shear force between the two voussoirs, but if this is insufficient a horizontal force will develop with the stones of the rest of the structure. Note that the total forces acting on the first voussoir is equivalent in the two models. When the horizontal force at the abutment is calculated, the two models are the same.

Arches play an important part in civil engineering structures and analyzed in detail in structural design courses. St. Louis was selected in 1935 for a national monument for the 19th century expansion into the West. An inverted catenary curved arch with height and base of 630 feet was completed in the 1960s commemorating the settlers opening the western United States. Concrete and steel arch bridges are as popular now as were the stone arch bridges built in the last 2000 years.

Problems

3.64 Consider a semi-circular arch of 11 stones; the keystone and five voussoirs on each side of the arch. The sides of the keystone are at an angle of fifteen degrees with the vertical and the lower side of each voussoir has an increased angle of 15°. Calculate the required forces on the keystone and each voussoir.

3.65 Calculate the horizontal and compression force using the free-body diagram in Figure 3.5B.10 and compare results with those shown in equations. (3.5B.8) and (3.5B.9).

Chapter Summary

This chapter discussed equilibrium of a particle. An object may be modeled as a particle when it is small enough such that all the forces may be considered to act at a single point within the object. The free-body diagram will be a point in space where the concurrent forces act. The vector sum of all the forces must be zero if the particle is in a state of equilibrium. This requires the solution of a vector equation that can be accomplished by any of the methods shown in Chapter 2.

Springs are introduced to illustrate the effects of deformation on the state of equilibrium of a particle. In some cases, the problem will be classified as statically indeterminate but a solution can be obtained by considering the deformations of the supports.

Two special particle equilibrium conditions were introduced. First, the concept of friction was discussed and will be further examined in Chapter 9. Second, the oldest simple structure of an arch was analyzed using methods of particle equilibrium.

There are no particular formulae that can be memorized to solve particle equilibrium problems and the important parts are construction of an accurate free-body diagram, writing the vector equilibrium equation for all forces acting on the particle and solving this vector equation.

Chapter 4

RIGID BODIES: EQUIVALENT FORCE SYSTEMS

The equivalent forces due to gravitational attraction act at a point on the high jumper (center of mass) that passes under the bar. (Photo © Corel)

4.1 RIGID BODIES

In Chapter 3, we studied the equilibrium of a particle subjected to concurrent forces. A body modeled as a particle occupies only a single point in space, and its size and geometry can be ignored. Therefore, the forces acting on the particle are concurrent, and equilibrium of the particle is independent of the location of the particle in space. The study of bodies modeled as particles provides an understanding of the effects of the magnitude and direction of forces that form the concurrent force system, but not the effects of spatial location. These forces are completely specified by their Cartesian components, and all are considered to act on the same point of the particle.

However, in the real world, all bodies have size and shape. Modeling them as single particles must therefore be viewed as special cases. In general, objects cannot be modeled as particles, and the size and shape of the object cannot be ignored. When the size and shape of the object are considered, the point of application of the forces acting on the object must be specified. In Figure 4.1, the force exerted on the steering wheel of a car could cause the car to turn left or right, or could maintain the car in its current direction, depending upon where the force is applied to the steering wheel. In this case, it is apparent that the point of application of the force on the object is important.

All objects are composed of a system of particles. The size of these particles may be atomic, molecular, granular, or larger. The particles are bound together by ***internal forces***, which are the "glue" that hold the body together. Internal forces always occur in pairs and are equal in magnitude, opposite in direction, and collinear. Therefore, the resultant, or sum, of all of these internal forces is zero and does not affect the external motion of the body or its state of equilibrium. Figure 4.2 shows two particles, m_i and m_j, with the pair of internal forces $\mathbf{f}_{ij}$ and $\mathbf{f}_{ji}$, where, of course,

$$\mathbf{f}_{ji} = -\mathbf{f}_{ij} \tag{4.1}$$

The resultant of these forces, when the particles are considered a system, is zero. The system of particles comprising the body responds only to the forces that act on it due to contact with other bodies or attraction forces from other, separated bodies. These forces are called ***external forces*** and may be surface forces (contact forces) or body forces (such as gravitational attraction).

When the external forces acting on a body are not balanced, the body accelerates according to Newton's second law. Even when the body is in equilibrium, these external

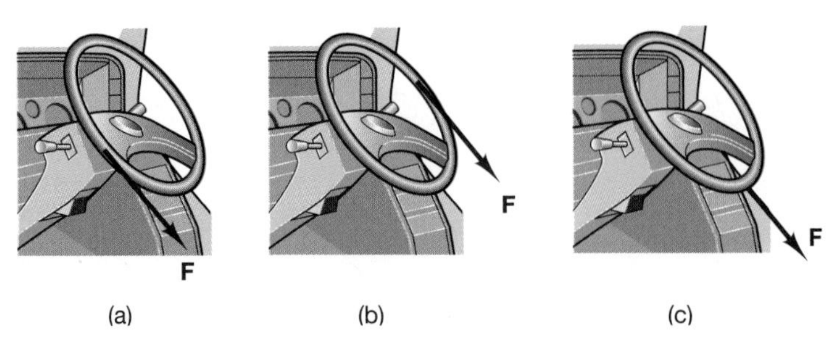

Figure 4.1 The effect of applying a force to a steering wheel depends upon the location of the force. (a) Counterclockwise rotation; (b) clockwise rotation; (c) no rotation.

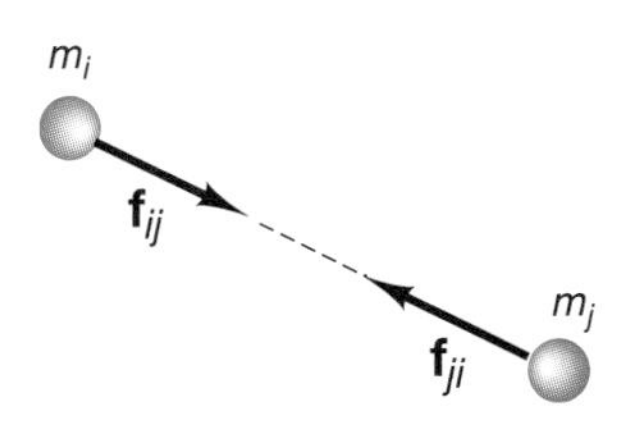

Figure 4.2 The internal force on the *i*th particle from the *j*th particle, $\mathbf{f}_{ij}$, is equal, opposite, and collinear with the internal force on the *j*th particle from the *i*th particle, $\mathbf{f}_{ij}$.

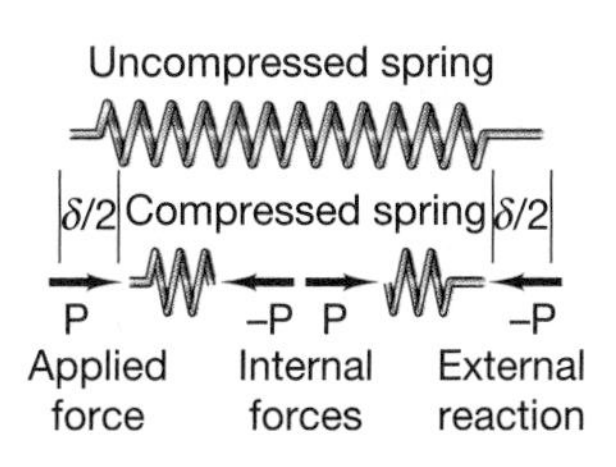

Figure 4.3 Internal forces on a compressed spring.

forces can cause the body to deform or change shape, thereby changing the internal forces within the body. For example, if two ends of a spring are pushed together, the spring will shorten in length and be subjected to internal compression. A force is applied at the left end of the spring, and it is resisted by an external reaction force at the right end. The compressed spring with the internal forces is shown in Figure 4.3.

Many engineering materials, such as concrete or steel, are stiff, and their changes in length or deformations are small in comparison with the size of the object. Therefore, the size and shape of the body do not significantly change due to the action of external forces. In these cases, the body is modeled as a *rigid body*, and the small deformations are neglected when one considers the equilibrium geometry. These deformations and internal forces are important in design, determining the selection of a body's size and material to prevent its failure due to breaking or excessive deformation. The study of the effects of internal forces and deformations is the subject of ***mechanics of materials*** or ***strength of materials***. In addition, there are cases when the deformations are large compared to the size of the body, and equilibrium conditions depend upon these deformations. An example is the fiberglass pole used in pole vaulting.

When an object is modeled as a rigid body, the point of application of each external force is specified, but the response of the rigid body is dependent only on the line of action of these forces. This allows the force to be moved along its line of action (***the principle of transmissibility***). Such movement will be discussed in more detail in Section 4.2. When the body has size and shape, the forces can cause a rotational effect by producing a ***moment*** about a point on the body or about an axis through the body. The term ***torque*** is used interchangeably with moment in many applications, but is generally associated with twisting. The term ***moment*** is the preferred terminology for the turning effect of a force. A moment is the product of a force and the perpendicular distance to the line of action of the force relative to some point. Forces can cause a body to translate; moments can cause it to rotate. The cross-product or vector product will be used to calculate moments.

The concept of a ***couple*** will be introduced in Section 4.7. A couple consists of two equal and opposite forces that are noncollinear. A couple produces a pure turning effect on a body.

Finally, we will discuss the concept of *equivalent force systems* in Sections 4.8 and 4.9. A force system can be replaced by equivalent force systems that produce the same global effects on the body. This concept is fundamental to the solution of problems in dynamics.

4.2 MODELING OF RIGID BODIES AND MOMENT OF A FORCE

When an object is modeled as a rigid body, the effects of deformation and change in shape are neglected. You might expect that when an object has significant size and shape, the point of application of a force acting on the object must be specified, and the force must be represented as a fixed vector. Indeed, if the object is not modeled as a rigid body, this would be the case. For example, consider the cardboard box shown in Figure 4.4. If the box is treated as a rigid body, the effect on the box is the same whether you push or pull on it with a force *P*. However, if deformations and internal forces are considered, the effects of the force are different, as the figure illustrates. Most modeling in engineering starts with the simple case and adds complexity as need demands. Hence, a first approach at modeling the box in Figure 4.4 is to ignore the deformation and treat the box as a rigid body. The

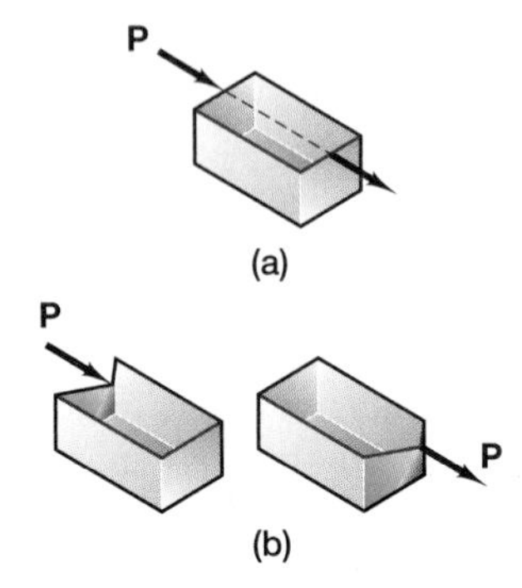

Figure 4.4 (a) A cardboard box modeled as a rigid body; (b) a cardboard box not modeled as a rigid body.

deformation can be considered later if that information is needed for design or for selecting the material out of which the box is to be made. A more sophisticated deformable-body model would be necessary to study the crush of the box, or how much force may be required before it fails.

When an object is considered as a rigid body, the force applied to it may be represented as a sliding vector. Therefore, it has a unique line of action in space, but may be considered to act anywhere along that line of action. This concept is called the ***principle of transmissibility*** and states that, for considerations of equilibrium, or for analyzing the motion of an object, the effects are the same regardless of where the force is applied on its line of action. The response of the rigid body shown in Figure 4.5 is the **same because the forces *F* and *F′*** have the same magnitude, sense, and line of action. In this case, the response of the body subjected to the force **F** is equal to the response of the body subjected to the force **F′**, or, put another way, **F** applied at *a* and **F** applied at *b* produce the same result. Note that the principle of transmissibility does *not* say that the point of application of the force is not important; it says only that the force may be considered to act anywhere along its line of action. Note also that if you push on the top of the packing crate shown in Figure 4.6, it may tip over, whereas if you push on the bottom of the crate, it may tip back upon you. In this example, the force ***P*** has two different lines of action, and the resulting behavior of the object is different.

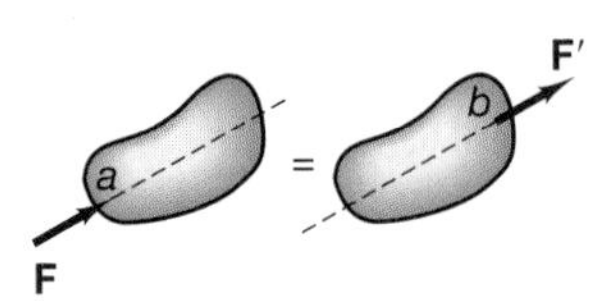

Figure 4.5 A rigid body responds in the same way to equal forces *F* and *F′* with the same line of action.

A change in the position of the line of action of a force can change the turning effects the force produces. The turning effect is called the ***moment of a force*** about a point *O* and is defined to be equal to the magnitude of the force times the perpendicular distance from the point *O* to the line of action of the force, as shown in Figure 4.7. The perpendicular distance is called the ***moment arm***. The moment of **F** about point *O* is equal in magnitude to

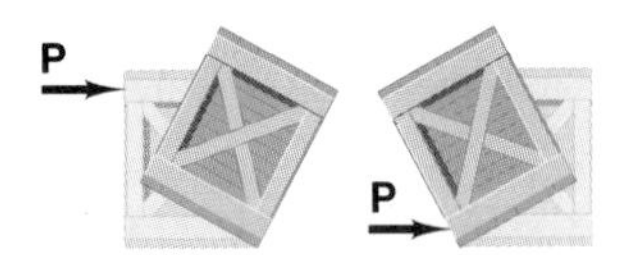

Figure 4.6 A rigid body responds differently to two equal forces with different lines of action.

$$|\mathbf{M}| = Fd \tag{4.2}$$

and causes a counterclockwise rotational effect (viewed from the top) around the point *O* in the plane shown. Note that the force **F** and the perpendicular distance *d* (the moment arm) form the plane in Figure 4.7 and that the axis of the rotating effect is perpendicular to the plane at the point *O*. It will be shown in Section 4.4 that the moment can be represented by a vector having the magnitude of *Fd* and a direction perpendicular to the plane, with a sense defined by a right-hand rule.

The magnitude of the moment can be increased by either increasing the force or increasing the moment arm. For example, the moment that can be applied to a lug nut can be increased by either increasing the force or adding an extension to the lug-nut wrench. For some objects, the perpendicular distance from the point about which the moment is to be taken to the line of action can be determined by simple mathematics. This will be illustrated in the examples that follow. However, as previously shown, for more complex problems the use of vectors can greatly simplify the task. The concept of moments (or torques) is so important and widely used that special mathematical tools have been introduced to deal with these quantities. The ***vector product*** or ***cross product***, was defined to accommodate the physical concept of a moment.

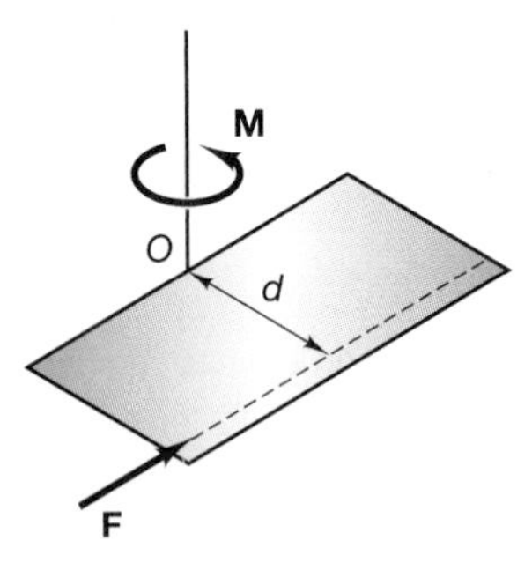

Figure 4.7 The moment **M** of a force **F** about a point *O* produces a turning effect on the object. The perpendicular distance *d* is the moment arm.

Note that the magnitude of a moment is the product of a force and a distance. Hence, common units of a moment are newton-meters (N · m) in the SI system and pound-feet (lb · ft) in the U.S. Customary system. Other choices of units are also used; for example, the torque wrench used in tightening a bolt is often calibrated in N-cm.

Sample Problem 4.1

Determine the magnitude of the moment of the force $|\mathbf{F}| = 200$ N about point A in the figure below.

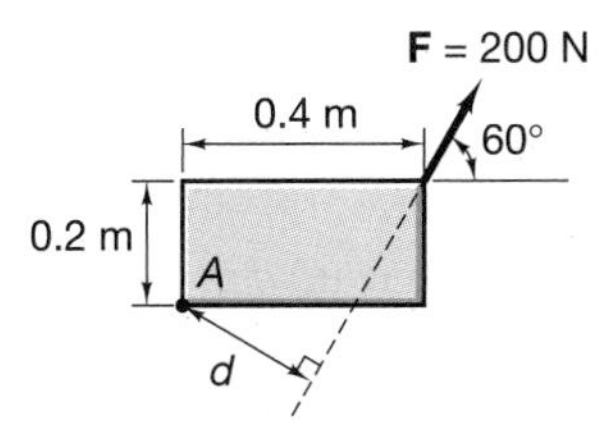

Solution

Draw a line along the line of action of **F**, and then find the perpendicular distance from A to the line of action. The moment is the product of the force and the moment arm, which in this case is simply the distance d: $M_A = Fd$. This distance can be determined by simple trigonometry:

$$x = \frac{0.2}{\tan 60°} = 0.115$$

The distance from A to the line of action along the bottom of the box is $0.4 - 0.115 = 0.285$. Therefore, the distance d can be determined from the triangle in the diagram below:

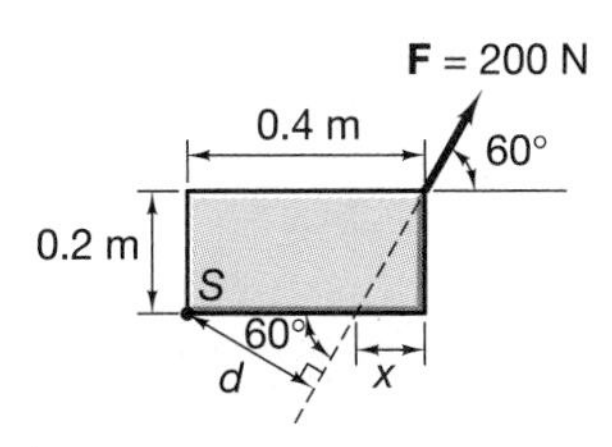

$$\sin 60° = \frac{d}{0.285}$$
$$d = 0.247 \text{ m}$$
$$M_A = Fd = 200(0.247) = 49.4 \text{ N} \cdot \text{m}$$

Sample Problem 4.2

Determine the moment of the 200-N force about points A, B, and C on the bracket shown at the left.

Solution

The moment arms for the 200-N force with respect to points A, B, and C are

$$d_A = 0$$
$$d_B = 1 \text{ m} \times \sin 45° = 0.707 \text{ m}$$
$$d_C = 3 \text{ m} \times \sin 45° = 2.121 \text{ m}$$

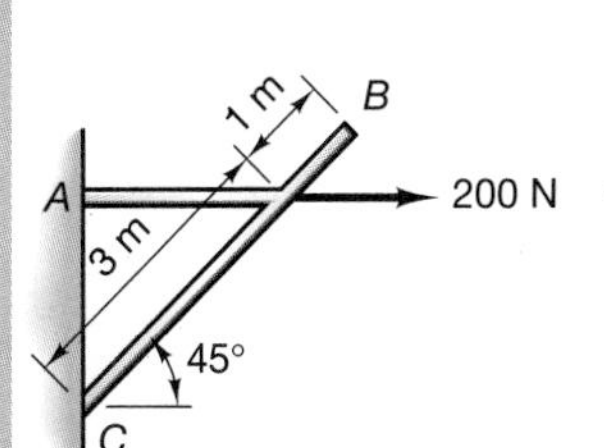

The moments are

$$M_A = 0$$
$$M_B = 141.4 \text{ N} \cdot \text{m ccw (counterclockwise)}$$
$$M_C = 424.2 \text{ N} \cdot \text{m cw (clockwise)}$$

Since the line of action of the 200-N force passes through point A, the moment about that point is zero. Point C is farther away from the line of action and has a greater moment than that about point B.

Problems

4.1 A 20-N force F is applied to the end of a 0.4-m lever, as shown in Figure P4.1. Calculate the magnitude of the moment that this force produces about the point O, and state the direction of the rotational effect.

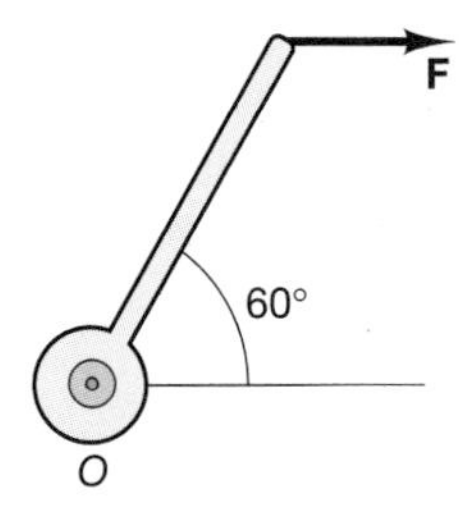

Figure P4.1

4.2 A mechanic must tighten a bolt to 45 N · cm. If the mechanic's wrench is 15-cm long, what force must be applied by the mechanic perpendicular to the end of the wrench?

4.3 The maximum force that a mechanic can exert with one arm is 500 N. If the mechanic must close a valve that requires a torque of 150 N · m, how long must the moment arm be?

4.4 Compare the moments produced at point O by the pipe wrench shown in Figure P4.4 as a 300-N force is applied at points A, B, and C along the wrench's handle. What rotational sense goes with these moments?

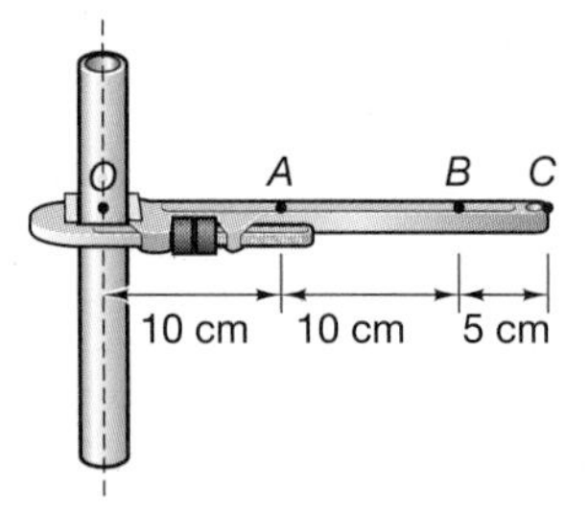

Figure P4.4

4.5 Calculate the moment of the force F about the point O and then again about the point P in Figure P4.5. What direction of rotation is produced?

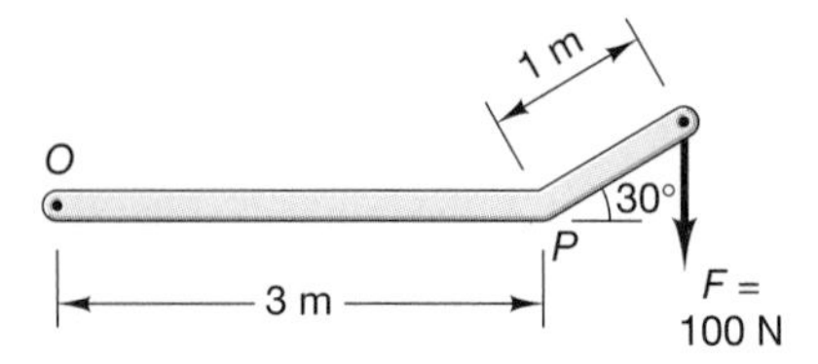

Figure P4.5

4.6 You need to produce a 40 N · cm torque in the counterclockwise direction in order to loosen a right-hand thread bolt. Choose the force **F** (magnitude and direction) for the case where (a) $d = 10$ cm.; (b) $d = 8$ cm. (See Figure P4.6.)

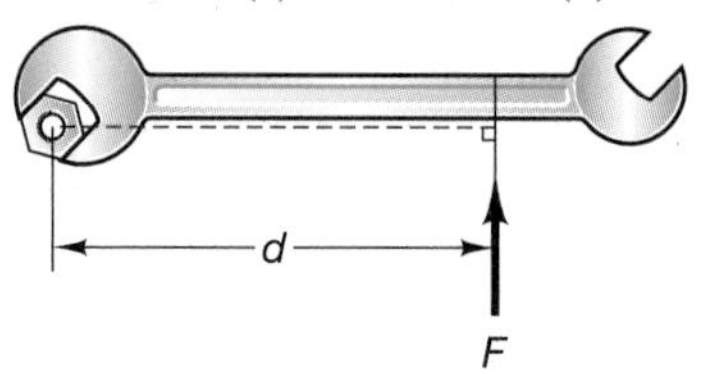

Figure P4.6

4.7 A 250-N force is applied to a wrench at an angle θ at a point 0.4 m from the origin. (See Figure P4.7.) Calculate the moment about point O as the angle θ varies from 0 to 90° in 10° increments. What are the maximum and minimum values of the moment? What sense of rotation results?

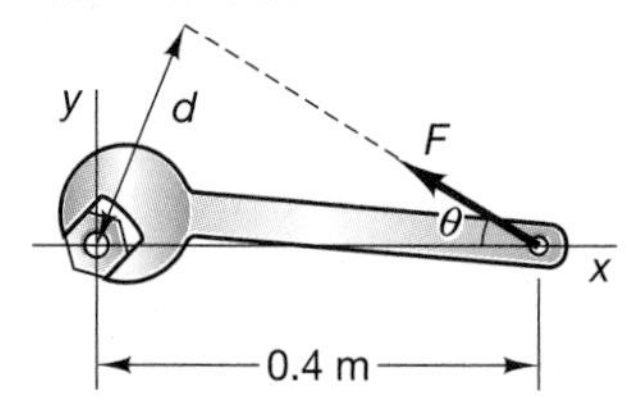

Figure P4.7

4.8 Determine the moment of the 200-N force applied as indicated in Figure P4.8 about points A, B, C, and D.

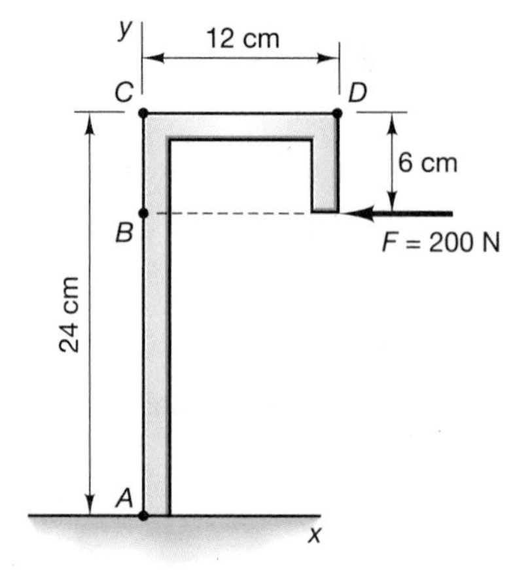

Figure P4.8

4.9 A 25-N planter is hung from a bracket. (See Figure P4.9.) Calculate the moments about the mounting points at A and B.

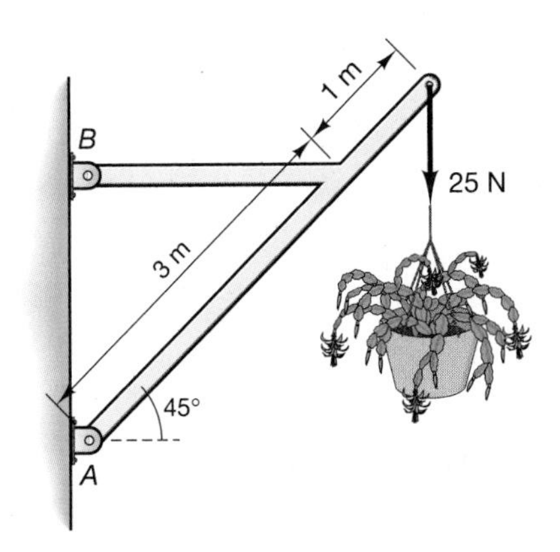

Figure P4.9

4.10 A 4000-N force is applied at the trailer hitch as illustrated in Figure P4.10. (a) Calculate the moments produced at the bumper locations A and B. (b) As the pulling car turns, the force **F** changes its direction. Calculate the moments at A and B if the line of action of **F** is changed so that it lies along the line BC.

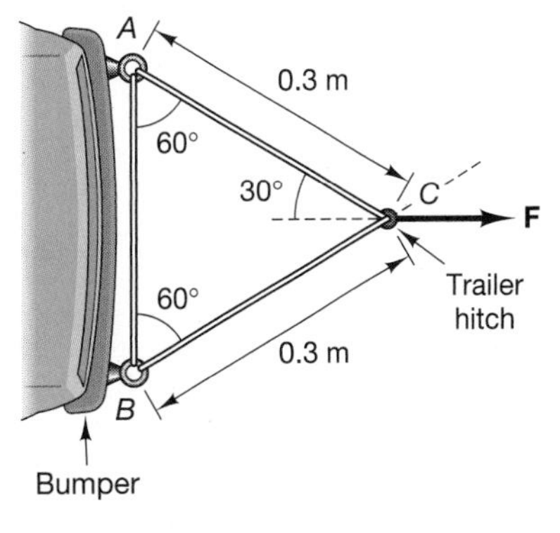

Figure P4.10

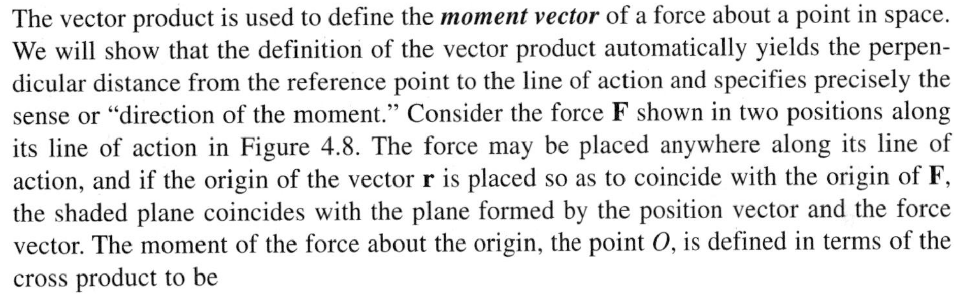

4.3 MOMENT OF A FORCE ABOUT A POINT IN SPACE

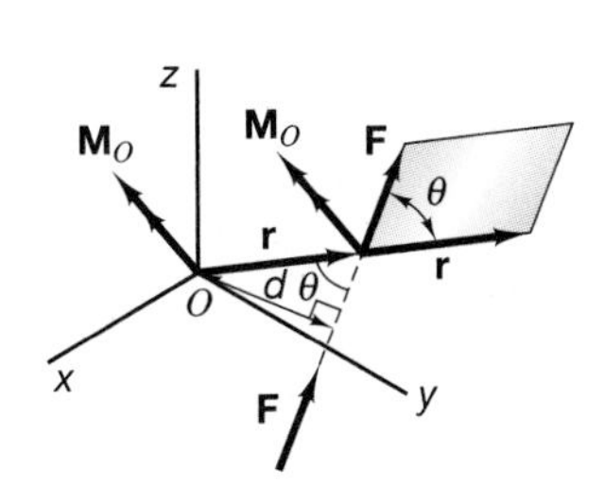

Figure 4.8 The moment vector $\mathbf{M}_O$ of a force **F** about a point O in space.

The vector product is used to define the ***moment vector*** of a force about a point in space. We will show that the definition of the vector product automatically yields the perpendicular distance from the reference point to the line of action and specifies precisely the sense or "direction of the moment." Consider the force **F** shown in two positions along its line of action in Figure 4.8. The force may be placed anywhere along its line of action, and if the origin of the vector **r** is placed so as to coincide with the origin of **F**, the shaded plane coincides with the plane formed by the position vector and the force vector. The moment of the force about the origin, the point O, is defined in terms of the cross product to be

$$\mathbf{M}_O = \mathbf{r} \times \mathbf{F} \tag{4.3}$$

where **r** is a vector from the point O to *any point* on the line of action of the force vector **F**. From the mathematical definition of the cross product, the moment vector $\mathbf{M}_O$ is perpendicular to the plane formed by the vectors **r** and **F**. In Figure 4.8, note that $|\mathbf{r}| \sin\theta$ is equal to the perpendicular distance d from 0 to the line of action, as shown in Figure 4.9. In the latter figure, the perpendicular distance to the line of action is $d = |\mathbf{r}| \sin\theta$. Since $|\mathbf{r} \times \mathbf{F}| = |\mathbf{F}|\,|\mathbf{r}| \sin\theta = |\mathbf{F}|d$, this satisfies the scalar concept of the definition of a moment presented in Section 4.2, that is, $M = Fd$. Thus, the cross-product definition is equivalent to the scalar definition. Note, however, that the cross product automatically determines the perpendicular distance from the point O to the line of action of the force and thus is easier to use than the scalar concept. In two-dimensional problems, the advantages of using vector algebra may not be apparent, but the use of vectors becomes a necessity in dealing with general three-dimensional problems. The sense of the moment vector is defined by the right-hand rule, as shown previously. The moment vector is a free vector, although it is usually shown as if it acts at the reference point O.

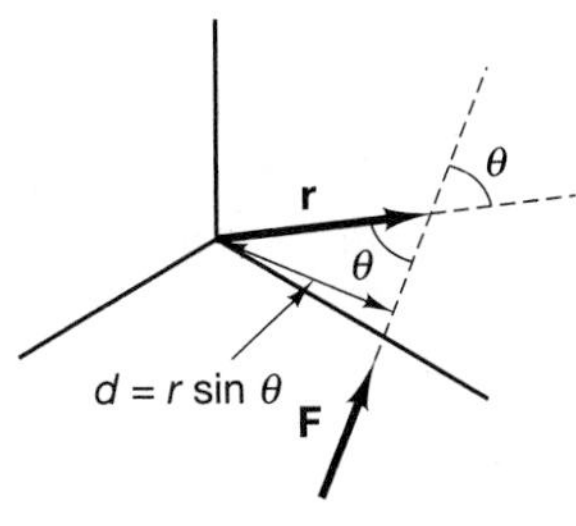

Figure 4.9

Sample Problem 4.3

Sample Problem 4.1 was solved by using trigonometry to determine the moment arm and then using the scalar definition of the moment. The same problem is solved here by using vectors. Determine the magnitude of the moment of the 200-N force about point A in the figure at the left.

Solution

A vector from point A to a point on the line of action of the force is

$$\mathbf{r} = 0.4\hat{\mathbf{i}} + 0.2\hat{\mathbf{j}} \text{ meters}$$

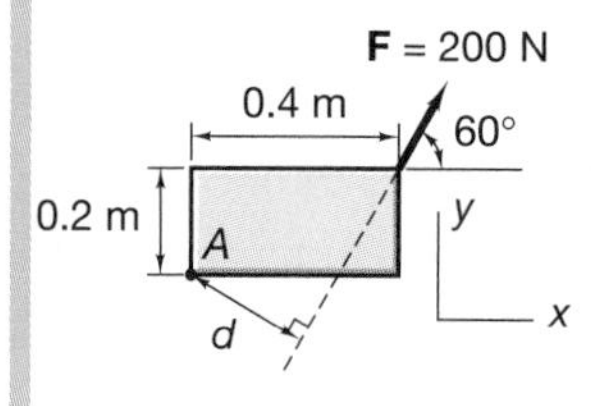

The force vector is

$$\mathbf{F} = 200\,(\cos 60°\hat{\mathbf{i}} + \sin 60°\hat{\mathbf{j}}) = 100\hat{\mathbf{i}} + 173.2\hat{\mathbf{j}} \text{ N}$$

The moment about point A is

$$\mathbf{M} = \mathbf{r} \times \mathbf{F} = (0.4\hat{\mathbf{i}} + 0.2\hat{\mathbf{j}}) \times (100\hat{\mathbf{i}} + 173.2\hat{\mathbf{j}})$$

$$= (69.28 - 20)\,\hat{\mathbf{k}} = 49.28\hat{\mathbf{k}} \text{ N}\cdot\text{m}$$

The two solutions are equal, to within numerical accuracy, when expressed to a N · m.

Sample Problem 4.4

Determine the moment of the force $P = 500$ N about each corner of the box shown in the diagram at the left.

Solution

A right-handed coordinate system has been chosen, with the origin at the corner D. The line of action of the force lies along the edge BG, so that the moments about the corners B and G are zero because their moment arms are zero.

Next, consider the moments about points A, C, and D. Let $\mathbf{r}_{B/A}$ be a vector from point A to point B (B relative to A). Point B is a point on the line of action of the force. Thus, we have

$$\mathbf{r}_{B/A} = -4\hat{\mathbf{k}}$$

$$\mathbf{M}_A = \mathbf{r}_{B/A} \times \boldsymbol{P} = -4\hat{\mathbf{k}} \times 500\hat{\mathbf{i}} = -2000\hat{\mathbf{j}} \text{ N}\cdot\text{m}$$

In a similar manner,

$$\mathbf{r}_{B/C} = 2\hat{\mathbf{j}} - 4\hat{\mathbf{k}}$$

$$\mathbf{M}_C = \mathbf{r}_{B/C} \times \mathbf{P} = (2\hat{\mathbf{j}} - 4\hat{\mathbf{k}}) \times 500\hat{\mathbf{i}} = (-2000\hat{\mathbf{j}} - 1000\hat{\mathbf{k}}) \text{ N}\cdot\text{m}$$

The moment is expressed in Cartesian components, but may be written as a magnitude times a unit vector defining its direction in space. For example, the unit vector for $\mathbf{M}_C$ is

$$\hat{\mathbf{m}}_C = \mathbf{M}_C/|\mathbf{M}_C| = (-2000\hat{\mathbf{j}} - 1000\hat{\mathbf{k}})/2236 = (-0.894\hat{\mathbf{j}} - 0.447\hat{\mathbf{k}})$$

and the moment may be written as

$$\mathbf{M}_C = 2236\,(-0.894\hat{\mathbf{j}} - 0.447\hat{\mathbf{k}}) \text{ N}\cdot\text{m}$$

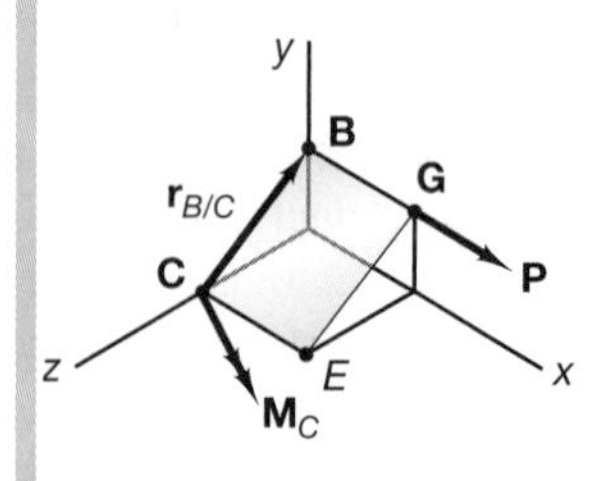

The magnitude of the moment at point C due to the force P is 2236 N · m and has a turning effect about an axis defined by the unit vector $\hat{\mathbf{m}}_c$. The direction of this turning effect is specified by the right-hand rule (the thumb of the right-hand points in the direction of the unit vector, and the fingers curl in the direction of the turning effect), as shown in the figure at the left. The moment at D can be calculated in a similar manner; hence,

$$\mathbf{r}_{B/D} = 2\hat{\mathbf{j}}$$

$$\mathbf{M}_D = \mathbf{r}_{B/D} \times \boldsymbol{P} = 2\hat{\mathbf{j}} \times 500\hat{\mathbf{i}} = -1000\hat{\mathbf{k}} \text{ N}\cdot\text{m}$$

Since the vector from the point about which the moment is desired may be taken to any point on the line of action, the following position vectors are equal and therefore the moments are equal. Thus,

$$\mathbf{r}_{G/F} = \mathbf{r}_{B/A} \text{ implies } \mathbf{M}_F = \mathbf{M}_A$$

$$\mathbf{r}_{G/E} = \mathbf{r}_{B/C} \text{ implies } \mathbf{M}_E = \mathbf{M}_C$$

$$\mathbf{r}_{G/H} = \mathbf{r}_{B/D} \text{ implies } \mathbf{M}_H = \mathbf{M}_D$$

Again, this problem may be easily solved using computational software.

4.3.1 DIRECT VECTOR SOLUTIONS

Thus far in this section, we have discussed how to calculate the moment of a force when the force and its point of application are known. Frequently in design problems, the desired moment is known, and the designer wishes to determine either where a given force should be applied—that is, the line of action of the force—or, knowing the point where the force can be applied, what should be the magnitude and direction of the force. An example might be the design of a mechanism that could exert a required torque on a shaft.

When we calculated the moment about a specific point, the equation to be solved was

$$\mathbf{M} = \mathbf{r} \times \mathbf{F} \tag{4.4}$$

when $\mathbf{r}$ and $\mathbf{F}$ were both known. This represents a *direct* solution of the vector equation. If, on the other hand, $\mathbf{M}$ and $\mathbf{F}$ are given, the *inverse* problem must be solved to determine $\mathbf{r}$. A similar situation arises if $\mathbf{M}$ and $\mathbf{r}$ are given and the problem requires determining the force $\mathbf{F}$. In Section 2.9, we noted that vector division is not defined. Therefore, there is no formal way to divide $\mathbf{M}$ by $\mathbf{F}$ or by $\mathbf{r}$ to determine the remaining vector.

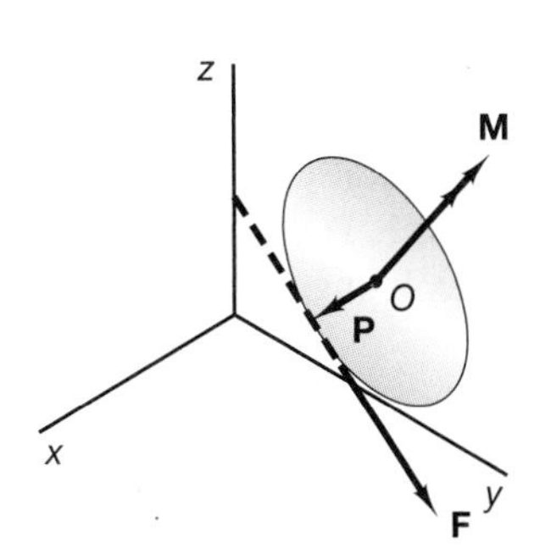

Figure 4.10 Only one position vector $\mathbf{p}$ lies on a line from O that is perpendicular to the line of action of $\mathbf{F}$.

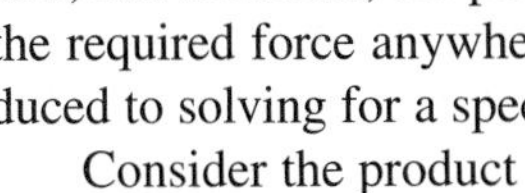

However, if we apply some conceptual understanding of the cross product, vector analysis can be used to obtain a solution to this problem. The inverse problem is illustrated in Figure 4.10. The properties of the cross product require that the moment $\mathbf{M}$ be perpendicular to both $\mathbf{r}$ and $\mathbf{F}$. In general, however, $\mathbf{r}$ and $\mathbf{F}$ are not perpendicular to one another. In fact, there are an infinite number of $\mathbf{r}$'s that can be drawn from the point O to the line of action of the force $\mathbf{F}$. Only one of these, $\mathbf{r} = \mathbf{p}$, is the perpendicular distance from O to the line of action. If $\mathbf{p}$ can be determined, then we have found a point in space on the line of action, and therefore, the position of the line of action in space is known. Then we can place the required force anywhere on this line of action. The inverse problem has now been reduced to solving for a specific vector $\mathbf{p}$ on the line of action.

Consider the product

$$\mathbf{M} = \mathbf{p} \times \mathbf{F} \tag{4.5}$$

The three vectors in Eq. (4.4) are now mutually perpendicular, or orthogonal, to one another. With this restriction on the three vectors, the inverse problem can be solved, and the vector $\mathbf{p}$ may be determined directly from the vector triple product introduced in Section 2.9. To solve the inverse problem, cross both sides of Eq. (4.5) with the force F, which yields

$$\mathbf{F} \times \mathbf{M} = \mathbf{F} \times (\mathbf{p} \times \mathbf{F}) \tag{4.6}$$

The right-hand side of Eq. (4.6) is a vector triple product, which, by the use of Eq. (2.123), can be written as

$$\mathbf{F} \times \mathbf{M} = \mathbf{F} \times (\mathbf{p} \times \mathbf{F}) = \mathbf{p}(\mathbf{F}\cdot\mathbf{F}) - \mathbf{F}(\mathbf{F}\cdot\mathbf{p}) \tag{4.7}$$

The last term is zero because **F** and **p** are perpendicular and the dot product between any two perpendicular vectors is zero. The dot product $\mathbf{F} \cdot \mathbf{F}$ is a scalar, equal to the square of the magnitude of the known force. Division by a scalar is well defined, and the vector **p** is thus

$$\mathbf{p} = \frac{\mathbf{F} \times \mathbf{M}}{\mathbf{F} \cdot \mathbf{F}} \tag{4.8}$$

This provides the solution of the inverse problem, "For a given moment and force, find the position vector." Note that if $\hat{\mathbf{f}}$ is a unit vector in the **F** direction, defined by

$$\hat{\mathbf{f}} = \mathbf{F}/|\mathbf{F}| \tag{4.9}$$

then the force may be applied at any point on the line of action at a position

$$\mathbf{r} = \mathbf{p} + d\hat{\mathbf{f}} \tag{4.10}$$

from point O, and d can be of any value, since $(d\hat{\mathbf{f}} \times \mathbf{F})$ is always zero. Thus, Eq. (4.10) can be used to construct an infinite number of solutions to this inverse problem, since

$$\mathbf{M} = \mathbf{r} \times \mathbf{F} = (\mathbf{p} + d\hat{\mathbf{f}}) \times \mathbf{F} = \mathbf{p} \times \mathbf{F} + d(\hat{\mathbf{f}} \times \mathbf{F}) = \mathbf{p} \times \mathbf{F}$$

The second type of inverse problem that often arises in design situations occurs when a desired moment at a point is given and the point at which an unknown force can be applied is known, but the magnitude and direction of the required force are unknown. Again, this problem may be solved by a direct vector solution if a sketch is drawn to conceptualize the situation.

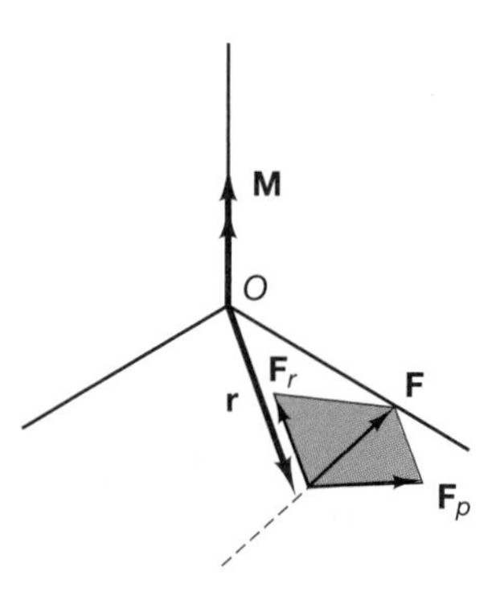

Figure 4.11 A given moment **M** caused by a force **F** applied at a given point A, but with unknown magnitude and direction.

The revelant vectors are drawn in Figure 4.11. The moment **M** is perpendicular to the plane formed by **r** and **F**, as is dictated by the definition of the moment. Vectors **M** and **r** (the position vector from O to point A) are known, and the force applied at A is to be determined such that it will produce the moment **M** at O. The force **F** is resolved into components $\mathbf{F}_p$ perpendicular to **r** and $\mathbf{F}_r$ along the direction of **r**. Since $\mathbf{F}_r$ is along the line of action of **r** and passes through O, it produces no moment about O. Using this form of the force, we find that the moment equation becomes

$$\mathbf{M} = \mathbf{r} \times \mathbf{F} = \mathbf{r} \times (\mathbf{F}_p + \mathbf{F}_r) = \mathbf{r} \times \mathbf{F}_p \tag{4.11}$$

The component of the force parallel to **r**, $\mathbf{F}_r$, cannot be uniquely determined, because it does not affect the moment at O. The component perpendicular to **r**, $\mathbf{F}_p$, can be determined in a manner similar to that used in the previous inverse solution. Crossing both sides of Eq. (4.11) with the known vector **r** yields

$$\mathbf{r} \times \mathbf{M} = \mathbf{r} \times (\mathbf{r} \times \mathbf{F}_p) \tag{4.12}$$

The vector triple product may be written as

$$\mathbf{r} \times \mathbf{M} = \mathbf{r} \times (\mathbf{r} \times \mathbf{F}_p) = \mathbf{r}(\mathbf{r} \cdot \mathbf{F}_p) - \mathbf{F}_p(\mathbf{r} \cdot \mathbf{r}) \tag{4.13}$$

The term involving the dot product between **r** and $\mathbf{F}_p$ is zero, as these two vectors are perpendicular to each other. The dot product $\mathbf{r} \cdot \mathbf{r}$ is a known scalar, and the unknown force is

$$\mathbf{F}_p = -\frac{\mathbf{r} \times \mathbf{M}}{\mathbf{r} \cdot \mathbf{r}} \tag{4.14}$$

This equation provides a solution to the second inverse problem related to the moment equation.

The solutions to these two inverse problems given by Eqs. (4.8) and (4.14) are called ***direct vector solutions*** and are useful in solving equations involving cross products when the three vectors are orthogonal.

Sample Problem 4.5

In the diagram at the left, determine the minimum force **F** that can be placed at point A, with coordinates (10, − 6, 8) m, to produce a moment at the origin given by

$$\mathbf{M}_O = 3000\hat{\mathbf{i}} + 1000\hat{\mathbf{j}} - 3000\hat{\mathbf{k}} \text{ N}\cdot\text{m}$$

Solution The moment must be perpendicular to the position vector from O to A. We have

$$\mathbf{r}_{A/O} = 10\hat{\mathbf{i}} - 6\hat{\mathbf{j}} + 8\hat{\mathbf{k}} \text{ m}$$

Since the vectors are perpendicular, their dot product must be zero:

$$\mathbf{M}_O \cdot \mathbf{r}_{A/O} = 0$$

Thus, we can obtain the inverse solution for

$$\mathbf{r} \times \mathbf{F} = \mathbf{M}$$

Using Eq. (4.14), we find that the minimum force (perpendicular to **r**) is

$$\mathbf{F}_p = -\frac{\mathbf{r} \times \mathbf{M}}{\mathbf{r}\cdot\mathbf{r}} = -(10{,}000\hat{\mathbf{i}} + 54{,}000\hat{\mathbf{j}} + 28{,}000\hat{\mathbf{k}})/200$$

$$= -50\hat{\mathbf{i}} - 270\hat{\mathbf{j}} - 140\hat{\mathbf{k}} \text{ (N)}$$

Since there is no component of the force **F** parallel to **r**, this is the minimum force that can be applied. Math errors can be avoided by checking to see that the force and position vectors are indeed perpendicular to each other and that the moment and force vectors are also perpendicular to one another.

4.4 VARIGNON'S THEOREM

In Chapter 2, we showed that the resultant of a concurrent force system is equal to the vector sum of the forces. The French mathematician Varignon (1654–1722) developed a concept called the ***principle of moments***, or ***Varignon's theorem***, which states that "the moment of a force about a point is equal to the sum of the moments of the force's components about the point." This principle is the direct result of the distributive property of the cross product, Eq. (2.102) in Section 2.9.

We know that the sum of a system of concurrent forces may be written as a single resultant. Then the moment of the resultant becomes

$$\mathbf{R} = \sum_i \mathbf{F}_i = \mathbf{F}_1 + \mathbf{F}_2 + \mathbf{F}_3 + \cdots$$

$$\mathbf{M} = \mathbf{r} \times \mathbf{R} = \mathbf{r} \times (\mathbf{F}_1 + \mathbf{F}_2 + \mathbf{F}_3 + \cdots)$$

$$= \mathbf{r} \times \mathbf{F}_1 + \mathbf{r} \times \mathbf{F}_2 + \mathbf{r} \times \mathbf{F}_3 + \cdots \quad (4.15)$$

where the last expression follows from the distributive property of the cross product. This is easily modified to take the exact form of Varignon's theorem. First, any force may be expressed as the vector sum of its components, such that

$$\mathbf{F} = F_x\hat{\mathbf{i}} + F_y\hat{\mathbf{j}} + F_z\hat{\mathbf{k}}$$

Next, the moment about any point is

$$\mathbf{M} = \mathbf{r} \times \mathbf{F} = \mathbf{r} \times F_x\hat{\mathbf{i}} + \mathbf{r} \times F_y\hat{\mathbf{j}} + \mathbf{r} \times F_z\hat{\mathbf{k}}$$

The position vector from any point may be written as

$$\mathbf{r} = x\hat{\mathbf{i}} + y\hat{\mathbf{j}} + z\hat{\mathbf{k}}$$

The moment then becomes

$$\mathbf{M} = (x\hat{\mathbf{i}} + y\hat{\mathbf{j}} + z\hat{\mathbf{k}}) \times (F_x\hat{\mathbf{i}} + F_y\hat{\mathbf{j}} + F_z\hat{\mathbf{k}})$$

or

$$\mathbf{M} = (yF_z - zF_y)\hat{\mathbf{i}} + (zF_x - xF_z)\hat{\mathbf{j}} + (xF_y - yF_x)\hat{\mathbf{k}} \tag{4.16}$$

The moment about the x-axis is the $\hat{\mathbf{i}}$-component of the moment vector as seen by the right-hand rule. This moment involves only position and force components in the yz-plane. Recalling the discussion following Eq. (2.109), we may make a similar observation about the moment about the y- and z-axes. In this manner, the moment may be considered to be the sum of three planar analyses, or triplanar in composition. In some applications, such as the bending of beams about axes of symmetry, the bending analysis about each axis is treated independently.

Varignon's theorem is very useful for obtaining the scalar components of the moment, but if the vector cross product is used, you will find that moments may be systematically obtained without the use of the theorem.

Sample Problem 4.6

A crank is pinned at point O such that it is free to rotate about that point and is subjected to a 1000-N force, as shown in the figure at the left. Determine the moment at point O.

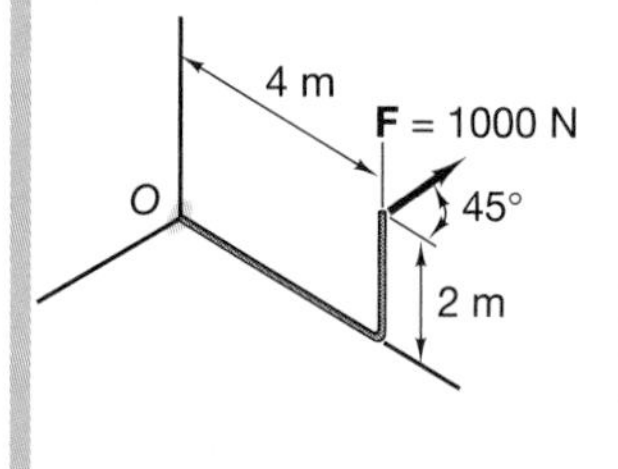

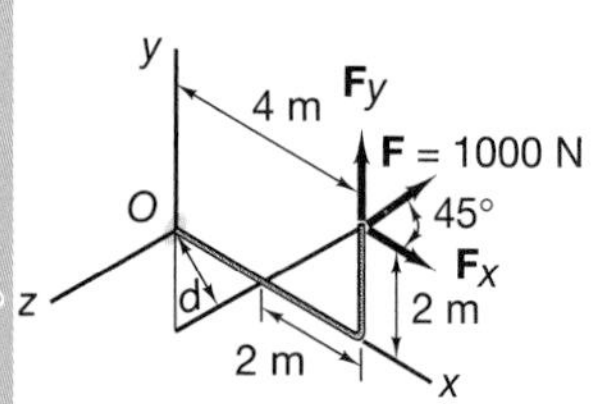

Solution We first establish a coordinate system, such that the crank lies in the xy-plane. (See diagram at left.) We can then determine the z-component of the moment at point O, using Varignon's theorem. The components of force $\mathbf{F}$ are

$$F_x = 707 \text{ N} \quad F_y = 707 \text{ N}$$

The moment arm from O to F_x is 2 m, and the moment of this component is 1414 N·m acting in a clockwise direction about the z-axis. The moment arm of the F_y component is 4 m, and the moment of this component is 2828 N·m in a counterclockwise direction. By Varignon's theorem, the moment of the force $\mathbf{F}$ is the sum of these moments, equal to 1414 N·m in a counterclockwise direction, or, by the right-hand rule, in the positive z-direction.

Using Varignon's theorem in vector notation yields

$$\begin{aligned}\mathbf{M}_O &= \mathbf{r} \times \mathbf{F} = \mathbf{r} \times (F_x\hat{\mathbf{i}} + F_y\hat{\mathbf{j}}) \\ &= (4\hat{\mathbf{i}} + 2\hat{\mathbf{j}}) \times (707\hat{\mathbf{i}} + 707\hat{\mathbf{j}}) = -1414\hat{\mathbf{k}} + 2828\hat{\mathbf{k}} = 1414\hat{\mathbf{k}} \text{ N}\cdot\text{m}\end{aligned}$$

Alternatively, since the problem is two dimensional, the distance d could have been determined directly by trigonometry:

$$d = 2 \sin 45 = 1.414 \text{ m}$$

$$\mathbf{M}_O = 1.414 \text{ m } (1000 \text{ N}) = 1414 \text{ N}\cdot\text{m}$$

However, moments are best calculated using vectors, except in simple two-dimensional cases.

Sample Problem 4.7

A cable with a tension of 300 N is attached to the bar as shown in the figure on the next page. Calculate

(i) the moment of the cable tension about point A using a position vector $\mathbf{r}_{B/A}$

(ii) the moment of the cable tension about point A using a position vector $\mathbf{r}_{C/A}$

(iii) the perpendicular position vector from A to the cable.

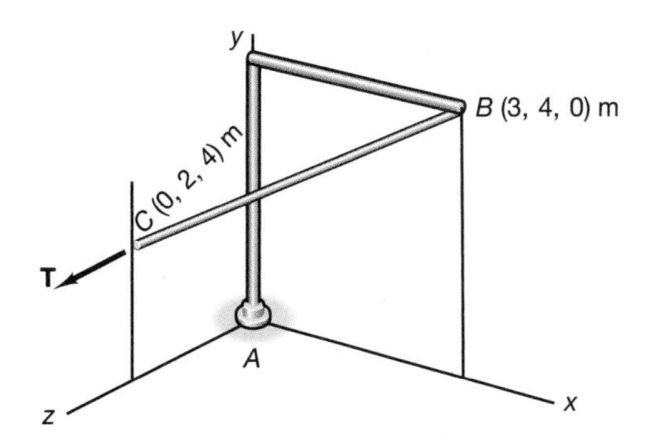

Solution The relative position vectors are:

$$\mathbf{r}_{B/A} = 3\hat{\mathbf{i}} + 4\hat{\mathbf{j}} \text{ m}$$
$$\mathbf{r}_{C/A} = 2\hat{\mathbf{j}} + 4\hat{\mathbf{k}} \text{ m}$$
$$\mathbf{r}_{C/B} = -3\hat{\mathbf{i}} - 2\hat{\mathbf{j}} + 4\hat{\mathbf{k}} \text{ m}$$

A unit vector along the cable is:

$$\hat{\mathbf{t}} = \frac{\mathbf{r}_{C/B}}{|\mathbf{r}_{C/B}|} = -0.557\hat{\mathbf{i}} - 0.371\hat{\mathbf{j}} + 0.743\hat{\mathbf{k}}$$

The tension vector in the cable is;

$$\mathbf{T} = T\hat{\mathbf{t}} = -167\hat{\mathbf{i}} - 111\hat{\mathbf{j}} + 223\hat{\mathbf{k}} \text{ N}$$

The moment of the tension about point A is:

$$\textbf{(i) } \mathbf{M}_A = \mathbf{r}_{B/A} \times \mathbf{T} = 891\hat{\mathbf{i}} - 669\hat{\mathbf{j}} + 334\hat{\mathbf{k}} \text{ Nm}$$
$$\textbf{(ii) } \mathbf{M}_B = \mathbf{r}_{C/A} \times \mathbf{T} = 891\hat{\mathbf{i}} - 669\hat{\mathbf{j}} + 334\hat{\mathbf{k}} \text{ Nm}$$

The perpendicular position vector from A to the cable is defined by:

$$\textbf{(iii)} \quad \mathbf{p} \times \mathbf{T} = \mathbf{M}_A$$
$$\mathbf{T} \times (\mathbf{p} \times \mathbf{T}) = \mathbf{T} \times \mathbf{M}_A$$
$$\mathbf{P}(\mathbf{T} * \mathbf{T}) - \mathbf{T}(\mathbf{T} * \mathbf{p}) = \mathbf{T} \times \mathbf{M}_A$$
$$\mathbf{p} = \frac{\mathbf{T} \times \mathbf{M}_A}{\mathbf{T} \cdot \mathbf{T}} = 1.24\hat{\mathbf{i}} + 2.83\hat{\mathbf{j}} \times 2.35\hat{\mathbf{k}} \text{ m}$$

$\mathbf{P}*\mathbf{T} = 0$ proves that $\mathbf{p}$ is perpendicular to $\mathbf{T}$.

Sample Problem 4.8

A pole is supported by three cables as shown in the figure to the left. The tension in the cable AB is measured by a strain gage to be 3 kN. If the moment at the base of the pole O due to the tensions in the three cables is zero, determine the tension in the other two cables.

Solution Set up unit vectors along each of the three cables:

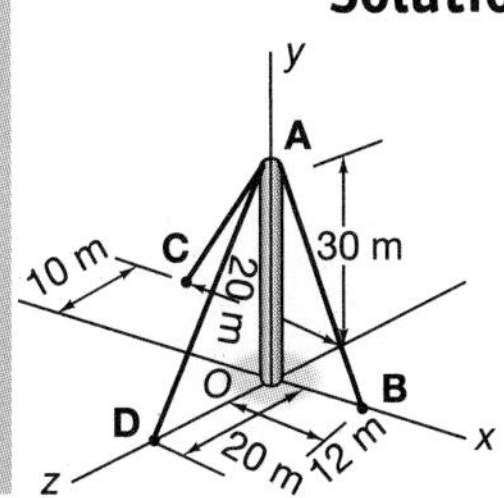

$$\mathbf{AB} = 12\hat{\mathbf{i}} - 30\hat{\mathbf{j}} \quad \hat{\mathbf{e}}_{AB} = \frac{\mathbf{AB}}{|\mathbf{AB}|} \quad \hat{\mathbf{e}}_{AB} = 0.371\hat{\mathbf{i}} - 0.928\hat{\mathbf{j}}$$

$$\mathbf{AC} = -20\hat{\mathbf{i}} - 30\hat{\mathbf{j}} - 10\hat{\mathbf{k}} \quad \hat{\mathbf{e}}_{AC} = \frac{\mathbf{AC}}{|\mathbf{AC}|} \quad \hat{\mathbf{e}}_{AC} = -0.535\hat{\mathbf{i}} - 0.802\hat{\mathbf{j}} - 0.267\hat{\mathbf{k}}$$

$$\mathbf{AD} = -30\hat{\mathbf{j}} + 20\hat{\mathbf{k}} \quad \hat{\mathbf{e}}_{AD} = \frac{\mathbf{AD}}{|\mathbf{AD}|} \quad \hat{\mathbf{e}}_{AD} = -0.832\hat{\mathbf{j}} + 0.555\hat{\mathbf{k}}$$

Establish a vector from the origin O to point A; $\mathbf{r}_{A/O} = 30\hat{\mathbf{j}}$. The moment about the origin of the tension in each cable can now be formed as:

$$\mathbf{M}_{\mathbf{AB}} = \mathbf{r}_{A/O} \times 3\hat{\mathbf{e}}_{\mathbf{AB}} = -33.42\hat{\mathbf{k}}$$

$$\mathbf{M}_{\mathbf{AC}} = \mathbf{r}_{A/O} \times T_{AC}\hat{\mathbf{e}}_{\mathbf{AC}} = T_{AC}(-8.018\hat{\mathbf{i}} + 16.64\hat{\mathbf{k}})$$

$$\mathbf{M}_{\mathbf{AD}} = \mathbf{r}_{A/O} \times T_{AD}\,\hat{}\mathbf{e}_{\mathbf{AD}} = T_{AD}16.64\hat{\mathbf{i}}$$

Setting the total moment at O to zero, yields:

$$T_{AC} = 2.08 \text{ kN}$$
$$T_{AD} = 1.00 \text{ kN}$$

This problem could also have been solved without writing the moment equations if you realize that if the sum of the tensions in the cable is in the y direction, the moment at the origin is zero.

$$3\hat{\mathbf{e}}_{\mathbf{AB}} + T_{AC}\hat{\mathbf{e}}_{\mathbf{AC}} + T_{AD}\hat{\mathbf{e}}_{\mathbf{AD}} = C\hat{\mathbf{j}}$$

where C is the compression in the pole.

$$3(0.371\hat{\mathbf{i}} - 0.928\hat{\mathbf{j}}) + T_{AC}(-0.535\hat{\mathbf{i}} - 0.802\hat{\mathbf{j}} - 0.267\hat{\mathbf{k}})$$
$$+ T_{AD}(-0.832\hat{\mathbf{j}} + 0.555\hat{\mathbf{k}}) = C\hat{\mathbf{j}}$$

Solving the vector equation, yields:

$$T_{AC} = 2.08 \text{ kN}$$
$$T_{AD} = 1.00 \text{ kN}$$

Sample Problem 4.9

It is known that the two cables produce a force perpendicular to the wall of 3.6 kN and a moment of magnitude of 10 kN m at point D. Determine the tension in the two cables.

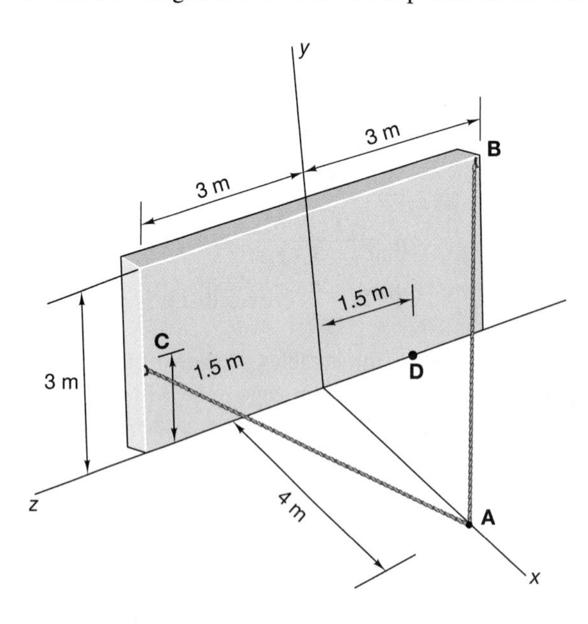

Solution Create relative position vectors from B to A and from C to A.

$$\mathbf{r}_{A/B} = 4\hat{\mathbf{i}} - 3\hat{\mathbf{j}} + 3\hat{\mathbf{k}}$$

$$\mathbf{r}_{A/C} = 4\hat{\mathbf{i}} - 1.5\hat{\mathbf{j}} - 3\hat{\mathbf{k}}$$

Form unit vectors along these two relative position vectors.

$$\hat{\mathbf{e}}_{A/B} = 0.686\hat{\mathbf{i}} - 0.514\hat{\mathbf{j}} + 0.514\hat{\mathbf{k}}$$
$$\hat{\mathbf{e}}_{A/C} = 0.766\hat{\mathbf{i}} - 0.287\hat{\mathbf{j}} - 0.575\hat{\mathbf{k}}$$

The tensions in the two cables can be written as:

$$\mathbf{T_B} = T_B\hat{\mathbf{e}}_{A/B}$$
$$\mathbf{T_C} = T_C\hat{\mathbf{e}}_{A/C}$$

The known fact that the cables produce a force perpendicular to the wall of 3.6 kN can be expressed mathematically as:

$$0.686T_B + 0.766T_C = 3.6$$

The second known fact that the magnitude of the moment at D is 15 kN m is more difficult to express mathematically. Form a position vector from D to A.

$$\mathbf{r}_{A/D} = 4\hat{\mathbf{i}} + 1.5\hat{\mathbf{k}}$$

We can now form the moment of each of the tensions about point D.

$$\mathbf{M_B} = \mathbf{r}_{A/D} \times T_B\hat{\mathbf{e}}_{A/B}$$
$$\mathbf{M_C} = \mathbf{r}_{A/D} \times T_C\hat{\mathbf{e}}_{A/C}$$

We know that the magnitude of the sum of these two vectors is 10 kN m, that is

$$|\mathbf{M_B} + \mathbf{M_C}| = 10$$

This is mathematically equivalent to writing

$$(\mathbf{M_B} \cdot \mathbf{M_B}) + 2(\mathbf{M_B} \cdot \mathbf{M_C}) + (\mathbf{M_C} \cdot \mathbf{M_C}) = 100$$

Expanding this equation yields

$$5.89T_B^2 - 1.7T_BT_C + 13.397T_C^2 - 100 = 0$$

This equation has the two unknown tensions as squares or a product and will have to be solved numerically with the first linear equation either by use of a calculator or computer. Write the two equations such that the tension T_C is a function of T_B

$$T_{C1}(T_B) = \frac{1}{0.766}(3.6 - 0.686T_B)$$
$$T_{C2}(T_B) = \frac{1}{2*13.397}\left[1.7T_B + \sqrt{1.7^2T_B^2 - 4*13.397*(5.89T_B^2 - 100)}\right]$$

We can obtain a solution graphically by plotting these two curves.

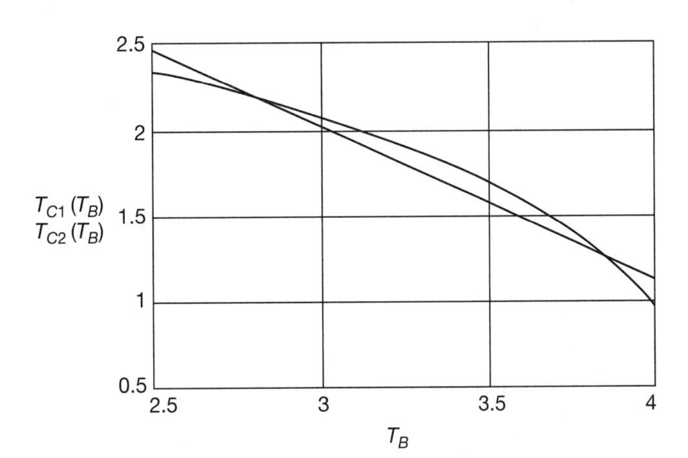

We can note that there are two solutions to the problem; one at each of the two intersections. The tensions in the cables are:

$$T_B = 2.8 \qquad 3.9 \text{ kN}$$

or

$$T_C = 2.2 \qquad 1.2 \text{ kN}$$

Problems

4.11 Calculate the moment for the system of Problem 4.5, using the vector definition.

4.12 Consider again the wrench and bolt of Problem 4.7, shown again in Figure P4.12. Calculate the angle θ such that the moment about point O caused by the 250-N force **F** has a maximum value, using the vector definition of a moment. Comment on the moment generated by the x component of **F**.

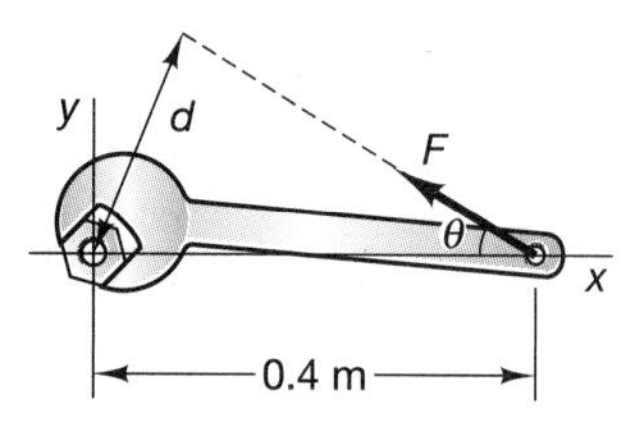

Figure P4.12

4.13 Repeat Problem 4.8 (see Figure P4.13) using vector notation.

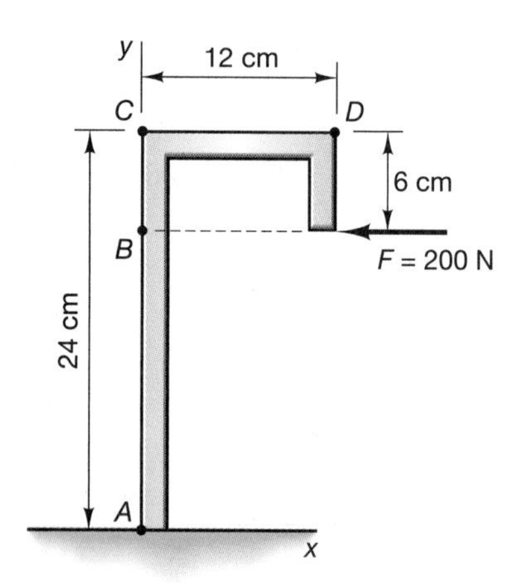

Figure P4.13

4.14 Consider the automobile towing bar illustrated in Figure P4.14. A 4000-N force is applied to the trailer hitch as illustrated and causes the force to change through the angle θ as the lead car pulls on the bar AB. Calculate the moments about the bumper connection point B as a function of the angle θ. Calculate the value of $\mathbf{M}_B$ at $\theta = 0$, 45°, and 90°.

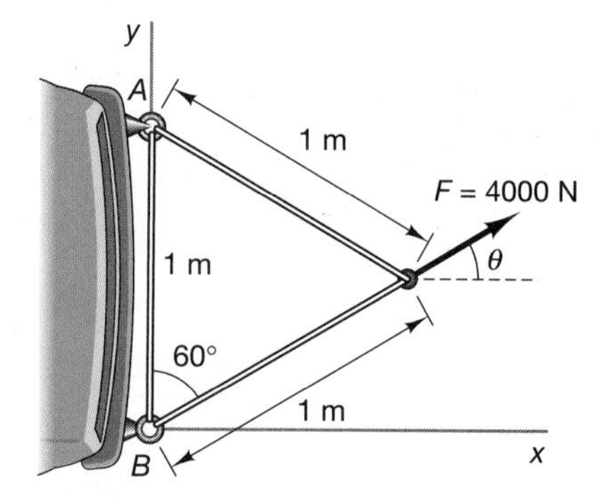

Figure P4.14

4.15 A position vector from point O to the force vector $\mathbf{F} = \hat{\mathbf{i}} - 2\hat{\mathbf{j}} + 5\hat{\mathbf{k}}$ kN is given by $\mathbf{r} = 2\hat{\mathbf{i}} + 3\hat{\mathbf{j}} - 4\hat{\mathbf{k}}$ m. Calculate the moment caused by **F** acting at **r** about point O.

4.16 A 300-N force acts on a rod, as shown in Figure P4.16. The coordinates are given in centimeters. Calculate the moment **F** exerts at the origin.

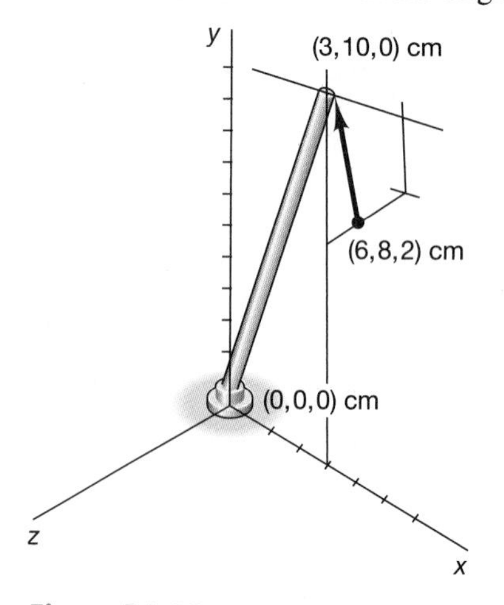

Figure P4.16

4.17 A rope hooked onto a bracket applies a force of 9000 N along its length to the bracket. (See Figure P4.17.) Calculate the moment produced at the point where the bracket connects to the tractor, and calculate the moment created at the elbow in the bracket.

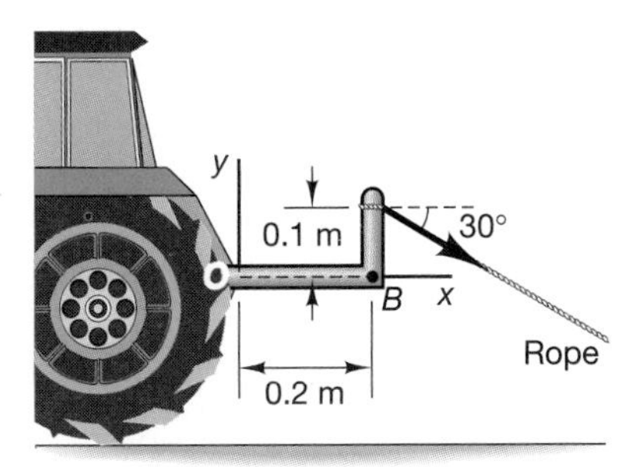

Figure P4.17

4.18 A crane lifts the end of a drainpipe, as shown in Figure P4.18. If the crane provides a force of 6000 N acting along the line CB, calculate the moment created about the point A, using the coordinate system indicated.

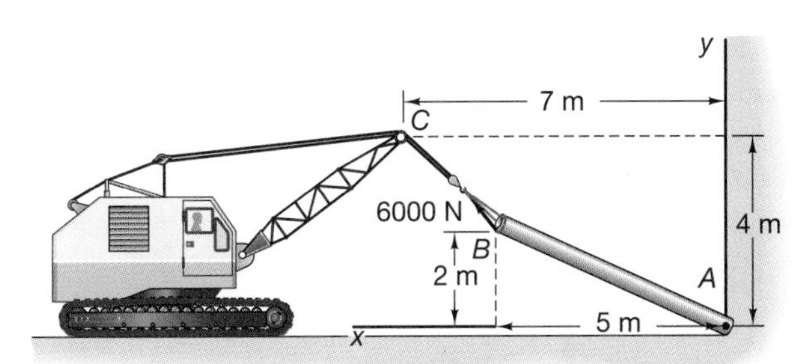

Figure P4.18

4.19 Suppose you are pushing on a lug wrench to tighten a lug nut. (a) With the configuration shown in Figure P4.19 and $\theta = 25°$, calculate the moment applied by the force—not about the lug nut, but rather about the center of the wheel at O. (b) Calculate the moment about O as the angle θ changes from 0 to 90° in 10° increments.

4.20 An 18-kN force is applied perpendicularly to the top of a concrete monorail support along the direction and at the point illustrated in Figure P4.20. For $\theta = 45°$, calculate the moment that this force creates about point A and again about point B.

4.21 A force $\mathbf{F} = 8\hat{\mathbf{i}} + 3\hat{\mathbf{j}} - \hat{\mathbf{k}}$ kN is applied to a part machined from a piece of aluminum 200 × 200 × 300 mm in size, at point A in Figure P4.21. Calculate the moment of $\mathbf{F}$ about the point O.

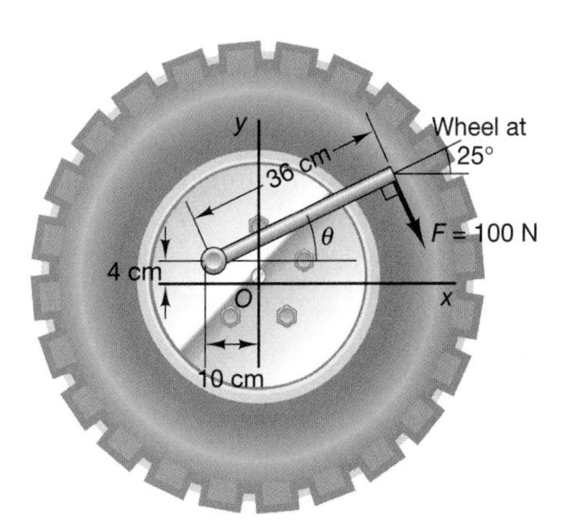

Figure P4.19

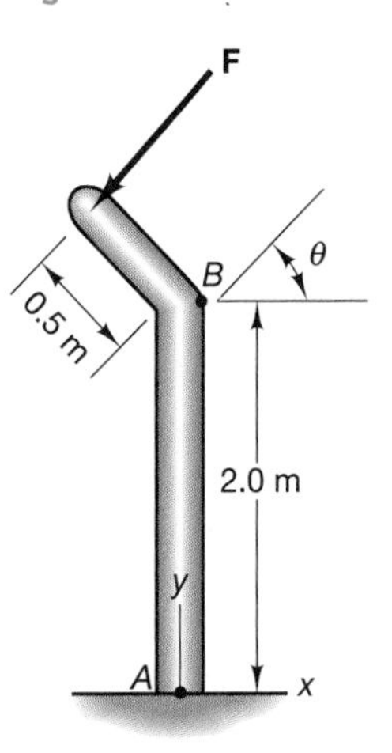

Figure P4.20

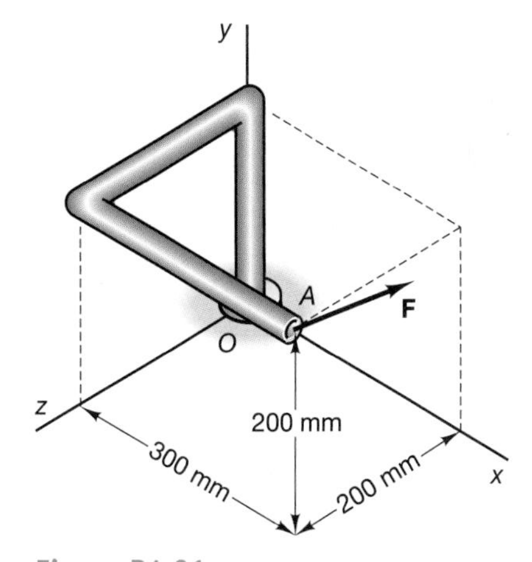

Figure P4.21

4.22 A force $\mathbf{F} = -\hat{\mathbf{i}} - \hat{\mathbf{j}} + 3\hat{\mathbf{k}}$ N acts through a moment arm from a point O to the line of action of $\mathbf{F}$ defined by $\mathbf{r} = \hat{\mathbf{i}} + \hat{\mathbf{j}} + 3\hat{\mathbf{k}}$ m. Calculate the resulting moment about point O.

4.23 A 50-N force lies along the line of action from point A to point B in Figure P4.23. Calculate the moment about point O in two ways: first using the vector from O to point B and then using the vector from O to point A. Are these moments the same? Explain why they should or shouldn't be the same by appealing to the definition of a moment.

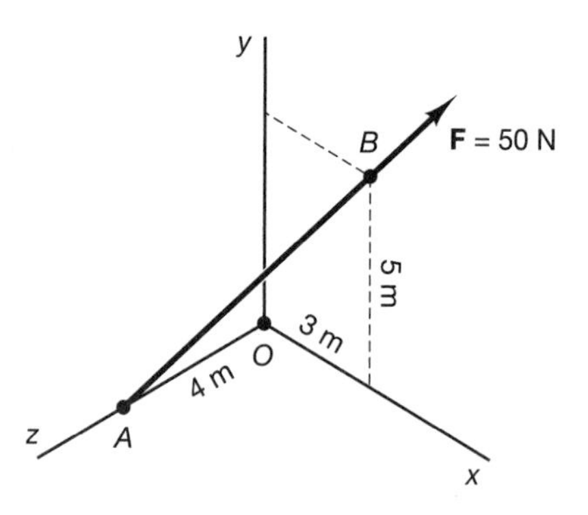

Figure P4.23

4.24 The edge of a box is lifted by pushing on a lever with a 600-N force along the direction indicated in Figure P4.24. Calculate the moment exerted by the force about point A on the box.

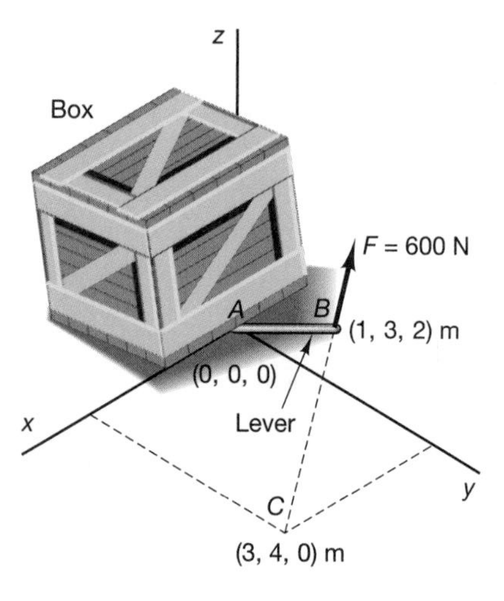

Figure P4.24

4.25 A force $\mathbf{F}_1 = 40\hat{\mathbf{i}} + 20\hat{\mathbf{j}} - 15\hat{\mathbf{k}}$ N is applied by a pilot to the bottom of an airplane steering wheel. (See Figure P4.25.) Calculate the moment of the force $\mathbf{F}_1$ about point O.

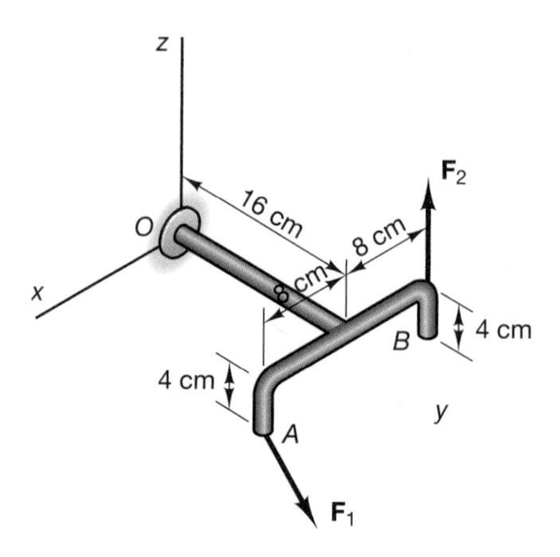

Figure P4.25

4.26 Suppose a pilot applies forces $\mathbf{F}_1 = 40\hat{\mathbf{i}} + 20\hat{\mathbf{j}} - 10\hat{\mathbf{k}}$ N to one end of the steering wheel of Figure P4.25 at point A and $\mathbf{F}_2 = 25\hat{\mathbf{k}}$ N to the other end at point B, in an attempt to turn the wheel. Calculate the moment about point O caused by these two forces.

4.27 Two forces of 10 kN each are applied to a pole, as illustrated in Figure P4.27, in two different directions. Calculate the total moment applied at point O by the two forces.

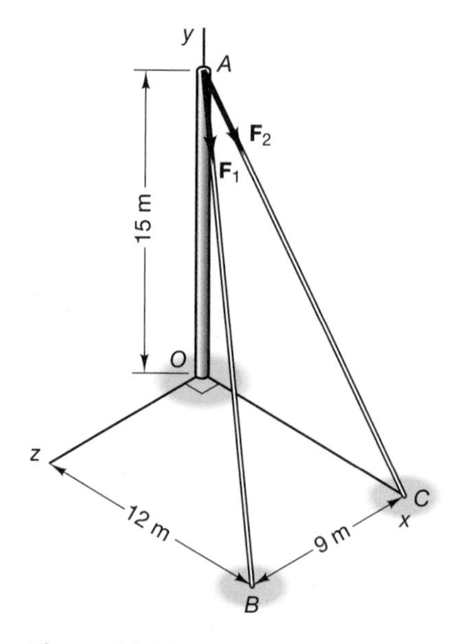

Figure P4.27

4.28 Two forces are applied to a crank mechanism as shown in Figure P4.28. The forces are $\mathbf{F}_1 = 50\hat{\mathbf{i}}$ N and $\mathbf{F}_2 = -50\hat{\mathbf{k}}$ N, expressed in the coordinate system indicated. (a) Which force produces the largest moment around point O in the $\hat{\mathbf{j}}$-direction? (b) Calculate the total moment around O of both forces. Note that only the $\hat{\mathbf{j}}$-component of moment is useful for turning the crank around the y-axis.

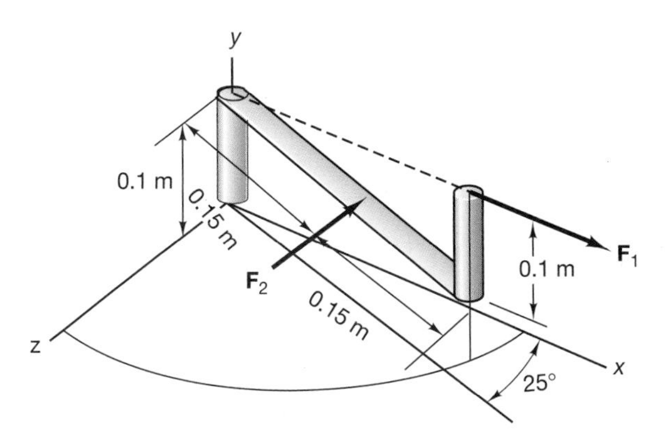

Figure P4.28

4.29 Repeat Problem 4.28 (b) using $\mathbf{F}_1 = \mathbf{F}_2 = -50\hat{\mathbf{k}}$ N.

4.30 Two wires are connected to a telephone pole to help support it. (See Figure P4.30.) Each wire exerts a force of 2.1 kN with a line of action along the wire. Compute the moment about O caused by the two forces.

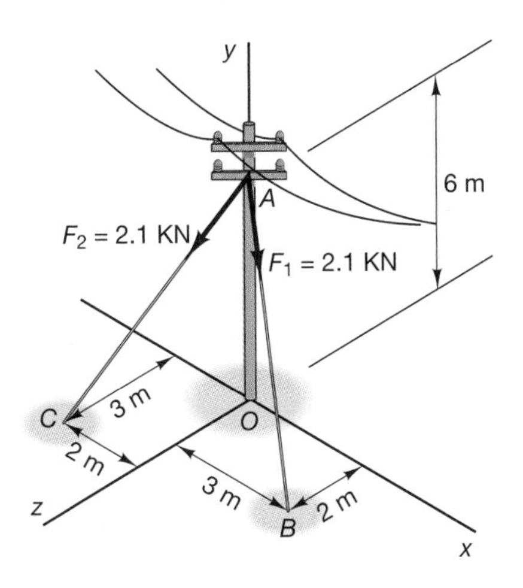

Figure P4.30

4.31 Suppose that you want to produce a 300-N · m torque about a point O in the $\hat{\mathbf{j}}$-direction, using a force of $\mathbf{F} = 100\hat{\mathbf{i}} + 40\hat{\mathbf{k}}$ N. (Note that $\mathbf{F} \cdot \mathbf{M} = \mathbf{0}$, as it must.) Calculate a value of the position vector $\mathbf{r}$ that, combined with the force $\mathbf{F}$, will produce the desired moment.

4.32 Given that $\mathbf{M}_O = 20\hat{\mathbf{i}} + 10\hat{\mathbf{j}} + 5\hat{\mathbf{k}}$ kN · m and $\mathbf{F} = 10\hat{\mathbf{i}} - 10\hat{\mathbf{j}} - 20\hat{\mathbf{k}}$ kN, where $\mathbf{M}_O \cdot \mathbf{F} = 0$, calculate the value of the shortest moment arm $\mathbf{r}$ that will produce the given moment by application of the given force.

4.33 Suppose that you need to find a force $\mathbf{F}$ that, when applied through a position defined by $\mathbf{r}_0 = 10\hat{\mathbf{i}} + 10\hat{\mathbf{j}} - 5\hat{\mathbf{k}}$ m, produces a moment $\mathbf{M}_O = 10\hat{\mathbf{i}} + 5\hat{\mathbf{j}} + 30\hat{\mathbf{k}}$ N · m. Check first to see whether a solution is possible; then calculate the smallest $\mathbf{F}$ if possible.

4.34 Repeat Problem 4.33 for a position vector $\mathbf{r}_O = 3\hat{\mathbf{i}} + 2\hat{\mathbf{k}}$ m and a moment $\mathbf{M}_O = 20\hat{\mathbf{i}} - 30\hat{\mathbf{k}}$ N · m.

4.35 Calculate the smallest force $\mathbf{F}$ that, when applied through a position defined by $\mathbf{r}_O = 3\hat{\mathbf{i}} + 5\hat{\mathbf{j}} + 2\hat{\mathbf{k}}$ m, produces a moment $\mathbf{M}_O = 20\hat{\mathbf{i}} - 30\hat{\mathbf{k}}$ N · m. Check first to see whether a solution is possible. How does this compare with Problem 4.34?

4.36 Calculate the force $\mathbf{F}$ that, when applied through a position defined by $\mathbf{r}_O = 3\hat{\mathbf{i}} + 5\hat{\mathbf{j}} + 2\hat{\mathbf{k}}$ m, produces a moment $\mathbf{M}_O = 20\hat{\mathbf{i}} - 35\hat{\mathbf{j}} + 6\hat{\mathbf{k}}$ N · m. Check first to see whether a solution is possible; if so, calculate $\mathbf{F}$.

4.37 Using the properties of the cross-product, determine a vector from the origin perpendicular to the line AB as shown in Figure P4.37.

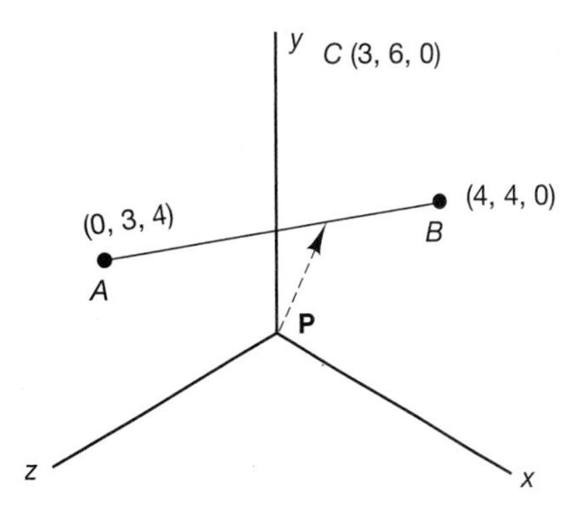

Figure P4.37

4.38 Repeat Problem 4.37 and determine a vector from the point C perpendicular to the line AB.

4.39 Given the tensions T_A, T_B, and T_C and the attachment coordinates a_x, a_y, b_x, b_y, and h, as shown in Figure P4.39, determine the attachment point of cable C such that the moment of the cable tensions about the base of the pole is zero. Calculate the compression force in the pole.

$T_A = 200$ N	$a_x = 1$ m	$a_y = -2$ m
$T_B = 100$ N	$b_x = -3$ m	$b_y = -1$ m
$T_C = 300$ N		
$h = 5$ m		

4.40 In Problem 4.39, by specifying tensions in C ranging from 100 to 1000 N, show that the attachment of C follows along a line $C_y = \lambda C_x$ and determine λ.

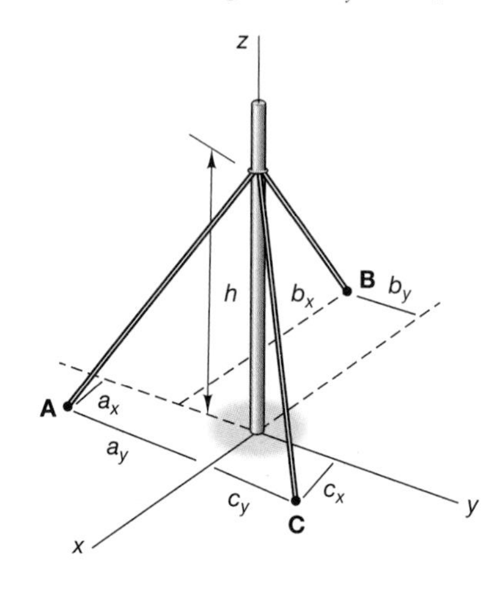

Figure P4.39

4.5 MOMENT OF A FORCE ABOUT AN AXIS

In Section 4.3, we discussed the moment of a force about a point in space and showed that the moment could be easily obtained by use of the cross product. In many practical applications, it is necessary to determine the component of that moment in a specific direction. If a line is drawn through the point in space, parallel to the desired component, this moment component is referred to as the ***moment of a force*** about that line in space. In Section 2.8, the dot, or scalar, product was used to determine the components of a vector parallel and perpendicular to a line in space. The same mathematical method will be used here to determine the moment of a force about a line in space.

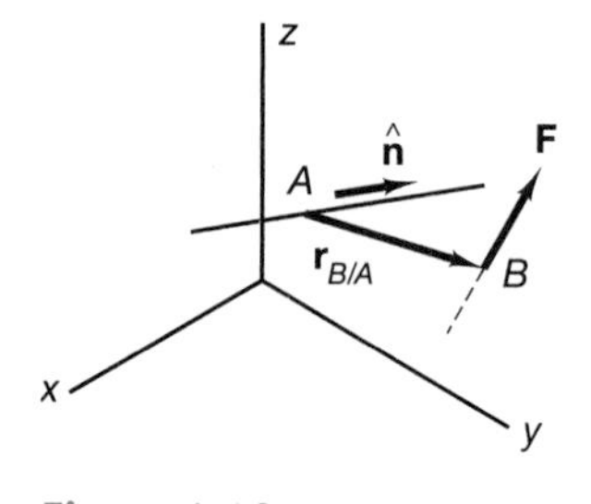

Figure 4.12

Consider a specific line in space and a force applied to a point not on the line, as shown in Figure 4.12. Let A be any point on the line in space, and let B be any point on the line of action of the force $\mathbf{F}$. The moment of the force about the point A is

$$\mathbf{M}_A = \mathbf{r}_{B/A} \times \mathbf{F} \tag{4.17}$$

where $\mathbf{r}_{B/A}$ is the vector from A to B, or the position vector of point B relative to point A. The direction of the line in space is given by the unit vector $\hat{\mathbf{n}}$, and the component of the moment parallel to the line may be found using the dot product. Now, by the commutative property of the dot product, the absolute value of the moment in the $\hat{\mathbf{n}}$ direction may be written as the projection of the moment in that direction:

$$|\mathbf{M}_n| = \mathbf{M}_A \cdot \hat{\mathbf{n}} = \hat{\mathbf{n}} \cdot (\mathbf{r}_{B/A} \times \mathbf{F}) \tag{4.18}$$

This is the magnitude of the moment of the force about the line. The moment is positive if it is in the direction of $\hat{\mathbf{n}}$. (The direction of the moment vector is determined by the right-hand rule.) The magnitude of this component is a scalar triple product, which, by means of Eq. (2.120), may be written as a determinant:

$$|\mathbf{M}_n| = \hat{\mathbf{n}} \cdot (\mathbf{r}_{B/A} \times \mathbf{F}) = \begin{vmatrix} n_x & n_y & n_z \\ r_{B/Ax} & r_{B/Ay} & r_{B/Az} \\ F_x & F_y & F_z \end{vmatrix} \tag{4.19}$$

Since the magnitude of the moment of a force about a line is a scalar triple product, the identities given in Eq. (2.121) are of use in some applications. Using the unit vector $\hat{\mathbf{n}}$, we may write Eq. (4.19) as

$$\mathbf{M}_n = |\mathbf{M}_n|\hat{\mathbf{n}} = |\hat{\mathbf{n}}\cdot(\mathbf{r}_{b/a} \times \mathbf{F})|\hat{\mathbf{n}} \tag{4.20}$$

The moment about a line is the component of the moment about point A that is parallel to the line. But there is also a component that is perpendicular to the line. This component is obtained by subtracting the parallel moment from the total moment, or by the use of Eq. (2.127):

$$\begin{aligned} \mathbf{M}_p &= \mathbf{M}_A - \mathbf{M}_n \\ \mathbf{M}_P &= (\mathbf{r}_{b/a} \times \mathbf{F}) - [\hat{\mathbf{n}}\cdot(\mathbf{r}_{B/A} \times \mathbf{F})]\hat{\mathbf{n}} \\ &= \hat{\mathbf{n}} \times [(\mathbf{r}_{B/A} \times \mathbf{F}) \times \hat{\mathbf{n}}] \end{aligned} \tag{4.21}$$

The development here is mathematically the same as that shown in Sections 2.6 and 2.9.

The concept of a moment about a line is very useful in determining the torque on a shaft and the turning effect of a screwdriver, as well as in other mechanical applications. The dot product projects the moment vector onto the line and is equal to

$$|\mathbf{M}_n| = |\mathbf{M}_A| \cos\theta \tag{4.22}$$

where θ is the angle between the moment about point A and the line. The maximum moment about the line occurs when θ equals zero, and there is no turning effect about the line when θ is 90°.

Sample Problem 4.10

A tire lug wrench is used to remove the lug nuts on a tire. Compute the useful torque being exerted on the lug nut when a 200-N force is applied. (See the figure below.)

Solution

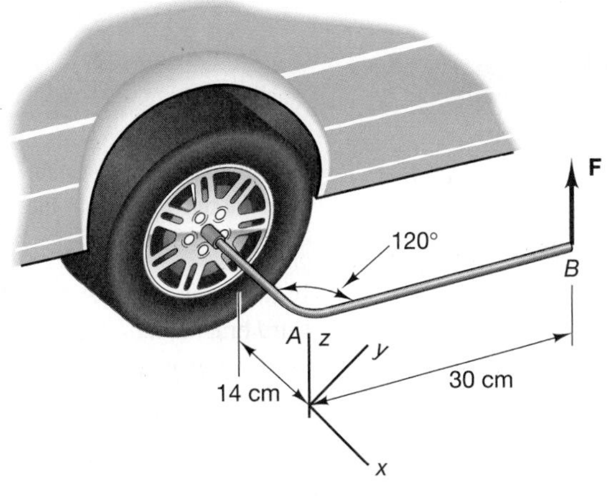

The only useful turning effect of the lug wrench is the moment of force $\mathbf{F}$ about a line in the x-direction through point A; the components of the moment at A about the y- and z-axes would not turn the lug nut. First, let us determine the moment of the force about point A. We construct a vector from A to B, in feet, as follows:

$$\begin{aligned} \mathbf{r}_{B/A} &= 0.3(\sin 30\hat{\mathbf{i}} + \cos 30\hat{\mathbf{j}})\ \text{m} \\ \mathbf{F}_B &= 200\hat{\mathbf{k}}\ \text{N} \\ \mathbf{M}_A &= \mathbf{r}_{B/A} \times \mathbf{F}_B = (0.15\hat{\mathbf{i}} + 0.25\hat{\mathbf{j}}) \times 200\hat{\mathbf{k}} \\ &= 52\hat{\mathbf{i}} - 30\hat{\mathbf{j}}\ \text{N}\cdot\text{m} \end{aligned}$$

The useful torque is the component parallel to the lug nut, or the x-direction, and is

$$\mathbf{M}_x = (\mathbf{M}_A \cdot \hat{\mathbf{i}})\,\hat{\mathbf{i}} = 52\hat{\mathbf{i}}\ \text{N}\cdot\text{m}$$

This solution may be obtained directly by use of computational software.

Sample Problem 4.11 Suppose that $\mathbf{F}_r$ and $\mathbf{F}_l$ represent the forces applied to the right and left ends, respectively, of the handlebar of a bicycle, as shown in the diagram at the left. If the forces on the handlebar are $\mathbf{F}_r = (-80\hat{\mathbf{i}} - 80\hat{\mathbf{j}})$ N and $\mathbf{F}_l = (-80\hat{\mathbf{i}} - 80\hat{\mathbf{j}} + 40\hat{\mathbf{k}})$ N, determine the moment exerted by the rider at the center of the wheel axis. Then determine the component of the moment about the y-axis.

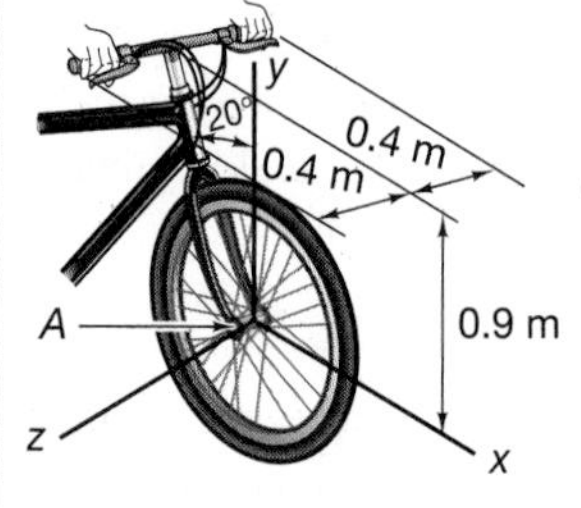

Solution A position vector from A to the right handlebar grip is

$$\mathbf{r}_{r/A} = -(0.90 \sin 20°)\hat{\mathbf{i}} + (0.90 \cos 20°)\hat{\mathbf{j}} + 0.40\hat{\mathbf{k}}$$

Similarly, a position vector from A to the left handlebar grip is

$$\mathbf{r}_{l/A} = -(0.90 \sin 20°)\hat{\mathbf{i}} + (0.90 \cos 20°)\hat{\mathbf{j}} - 0.40\hat{\mathbf{k}}$$

The moment at A is

$$\begin{aligned}\mathbf{M}_A &= \mathbf{r}_{r/A} \times \mathbf{F}_r + \mathbf{r}_{l/A} \times \mathbf{F}_l \\ &= [-(0.90 \sin 20°)\hat{\mathbf{i}} + (0.90 \cos 20°)\hat{\mathbf{j}} + 0.40\hat{\mathbf{k}}] \times (-80\hat{\mathbf{i}} - 80\hat{\mathbf{j}}) \\ &\quad + [-(0.90 \sin 20°)\hat{\mathbf{i}} + (0.90 \cos 20°)\hat{\mathbf{j}} - 0.40\hat{\mathbf{k}}] \times (-80\hat{\mathbf{i}} - 80\hat{\mathbf{j}} + 40\hat{\mathbf{k}}) \\ &= 33.8\hat{\mathbf{i}} + 12.3\hat{\mathbf{j}} + 184.6\hat{\mathbf{k}}\end{aligned}$$

The moment about the y-axis is 12.3 N · m.

Computational software may be used to perform these calculations.

Problems

4.41 A force $\mathbf{F} = 12\hat{\mathbf{i}} + 12\hat{\mathbf{j}} - 8\hat{\mathbf{k}}$ N is applied to a handrail. (See Figure P4.41.) Referring to the dimensions given (in centimeters), calculate the moment of the force F about the point A and about the AB-axis.

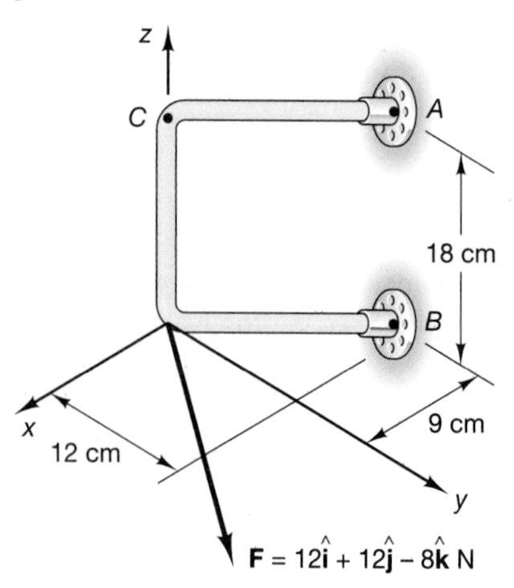

Figure P4.41

4.42 Consider the system of Figure P4.41, and compute the moment due to **F** about the line CA.

4.43 An electrical conduit is supported by two brackets at points A and B, as shown in Figure P4.43. A third bracket requires the conduit to be bent, and the bracket at C results in a force $\mathbf{F} = -60\hat{\mathbf{i}} + 20\hat{\mathbf{j}} - 30\hat{\mathbf{k}}$ N being

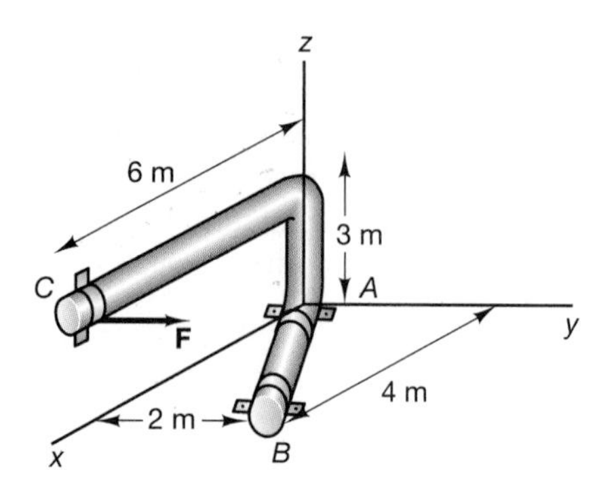

Figure P4.43

applied to point C. Calculate the moment about the line AB; that is, determine the rotational effect of the force at C about the conduit passing through the two lower brackets.

4.44 Calculate the moment exerted on the sprocket shaft of a bicycle as the rider's foot provides the force indicated on the pedal. (See Figure P4.44.)

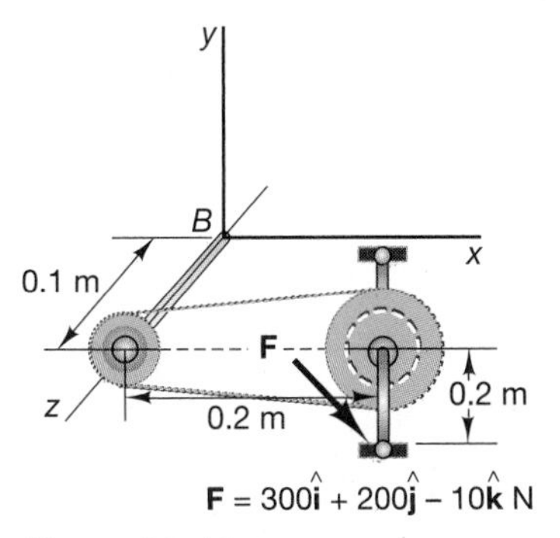

Figure P4.44

4.45 A 1.5-m. flagpole is stabilized by a cable of tension $T = 1$ kN, as shown in Figure P4.45. Calculate the moment of the force about the z-axis passing through the origin.

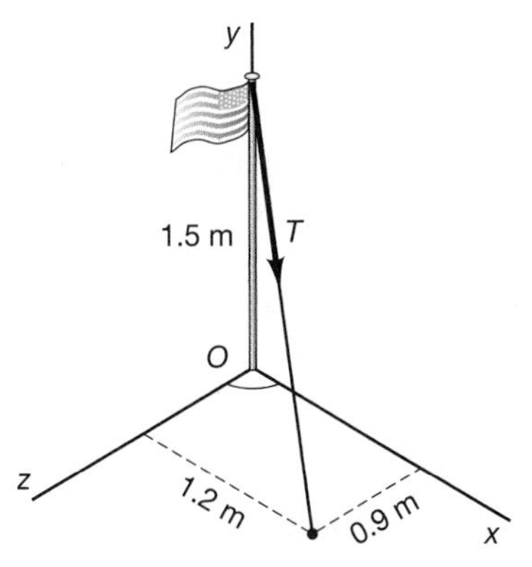

Figure P4.45

4.46 A 40-N force is applied at point B to the hand crank of a window-opening mechanism in the direction of the x-axis, as illustrated in Figure P4.46. (a) Calculate the value of the moment produced by the force about the z-axis. (b) Suppose that the 40-N force is moved and applied at point B in the direction 45° from the negative x-axis, but still in the x–y plane. Again calculate the moment about the z-axis.

4.47 A force $\mathbf{F} = -360\hat{\mathbf{i}} - 480\hat{\mathbf{j}} + 400\hat{\mathbf{k}}$ N is applied at a point A to a metal bracket machined out of a 500-mm cube with a mounting pin through one side. (See Figure P4.47) The centerline of the mounting pin lies parallel to the y-axis, 250 mm up from the x–y plane and 10 mm in from the face of the bracket. Caculate the moment about the center pp' through the mounting pin.

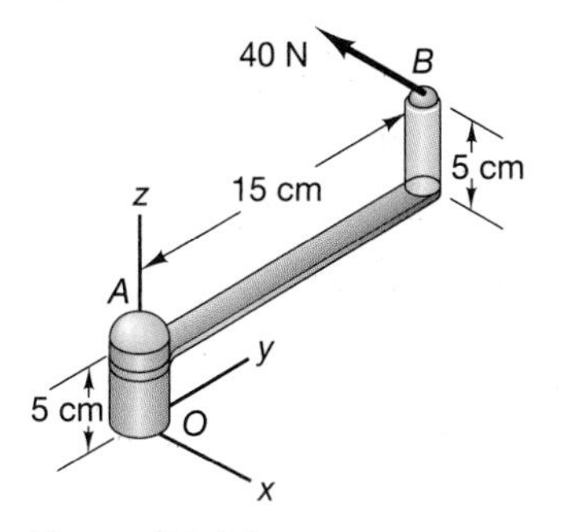

Figure P4.46

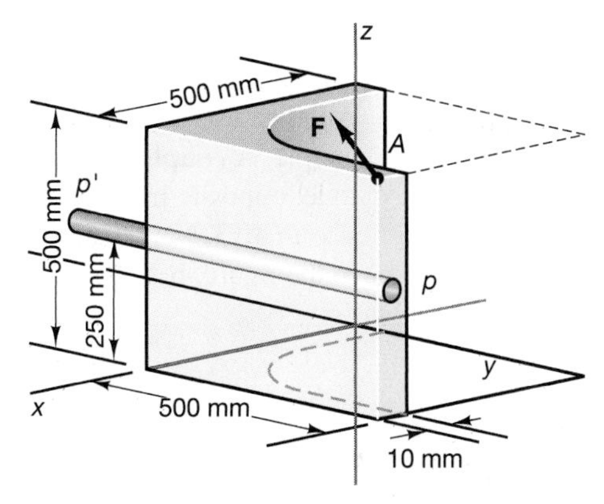

Figure P4.47

4.48 A 500-mm cube of metal has a mounting pin running from one corner of the cube to the center of one of the cube's surfaces, as shown in Figure P4.48. A force $\mathbf{F} = 100\hat{\mathbf{i}} - 100\hat{\mathbf{j}} - 30\hat{\mathbf{k}}$ N is applied at the center of the top surface of the cube. Calculate the moment generated by the force along the mounting pin PP'.

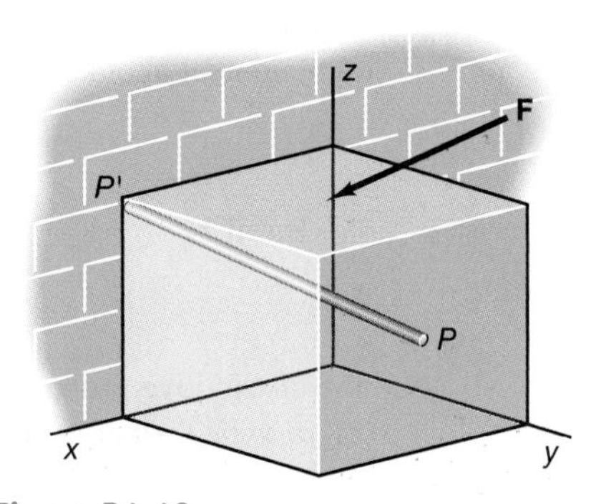

Figure P4.48

4.49 Find a force **F** acting at C on the crank mechanism illustrated in Figure P4.49. such that the moment produced around point O is $\mathbf{M}_O = 12\hat{\mathbf{i}} - 15\hat{\mathbf{j}} - 30\hat{\mathbf{k}}\ \text{N}\cdot\text{m}$, for the case when $\theta = 90°$.

4.50 Consider the crank system of Figure P4.49. Let $\mathbf{F} = -20\hat{\mathbf{j}} - 20\hat{\mathbf{k}}$ N, and examine the effect of θ (between 0 and 90°) on the moment. For what value of θ is the moment about point O the maximum?

4.51 Repeat Problem 4.50, but with all the force applied only along the z-direction. That is, examine the effect of varying θ between 0 and 90° with a force of the same magnitude, but directed along z.

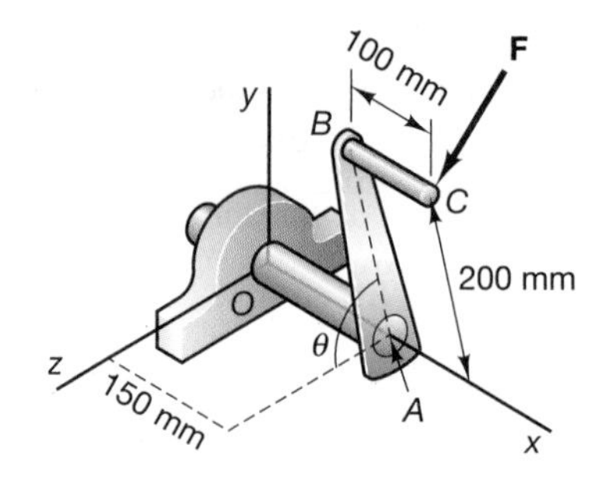

Figure P4.49

4.6 MOMENT OF A COUPLE

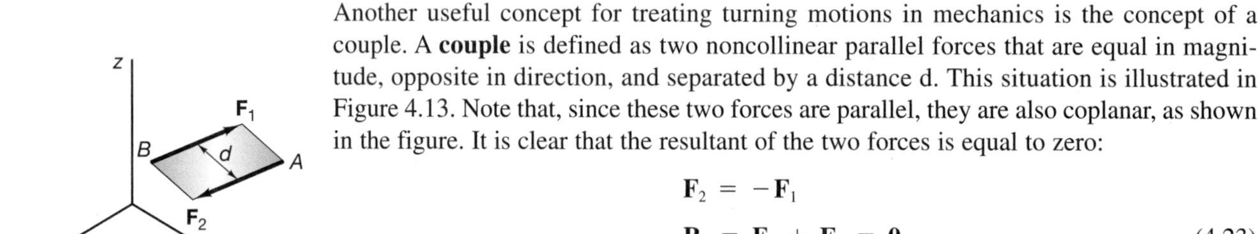

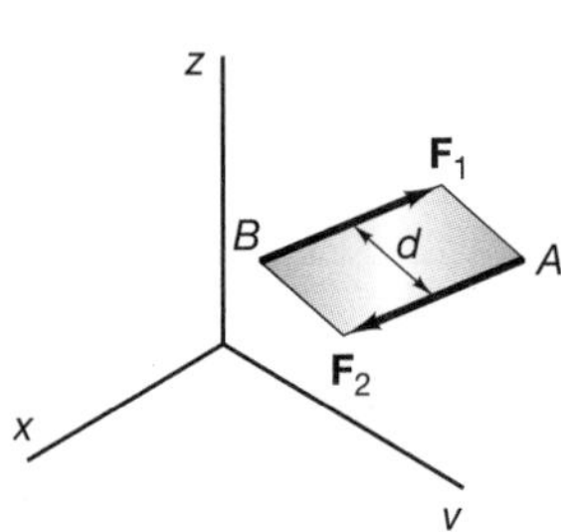

Figure 4.13

Another useful concept for treating turning motions in mechanics is the concept of a couple. A **couple** is defined as two noncollinear parallel forces that are equal in magnitude, opposite in direction, and separated by a distance d. This situation is illustrated in Figure 4.13. Note that, since these two forces are parallel, they are also coplanar, as shown in the figure. It is clear that the resultant of the two forces is equal to zero:

$$\mathbf{F}_2 = -\mathbf{F}_1$$
$$\mathbf{R} = \mathbf{F}_1 + \mathbf{F}_2 = \mathbf{0} \tag{4.23}$$

Since the resultant of a system of forces is associated with the translational effect the forces have on a body, a couple has no tendency to translate the body. Conceptually, the couple does have a rotational effect upon the body. The fact that two forces which are equal in magnitude, opposite in direction, and noncollinear produce only a rotational effect would, in itself, make couples an interesting concept to study. However, as will be seen later, other characteristics of couples make them an important "tool" for the study of statics. We will show that we can reduce a group of forces to an equivalent system consisting of a single resultant and a couple.

Consider the moment of two forces $\mathbf{F}_1$ and $\mathbf{F}_2$ about the origin, where the points A and B may be any points on the line of action of $\mathbf{F}_1$ and $\mathbf{F}_2$, respectively. If a coordinate system is chosen and position vectors to points A and B established as shown in Figure 4.14, the moment of the two forces can be obtained by the use of the cross product. The total moment about the origin is the sum of the moment of each force about the origin:

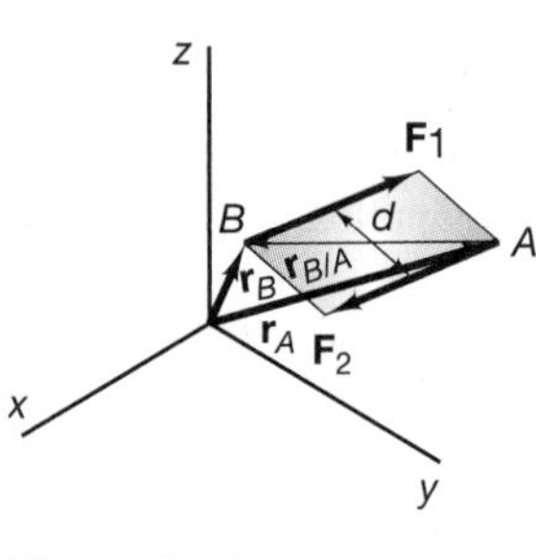

Figure 4.14

$$\mathbf{M}_O = \mathbf{r}_B \times \mathbf{F}_1 + \mathbf{r}_A \times \mathbf{F}_2 \tag{4.24}$$

But if $\mathbf{F}_2 = -\mathbf{F}_1$, Eq. (4.24) may be written as

$$\mathbf{M}_O = \mathbf{r}_B \times \mathbf{F}_1 - \mathbf{r}_A \times \mathbf{F}_1 \tag{4.25}$$

The distributive property of the cross product allows the moment of the couple about the origin to be further simplified to

$$\mathbf{M}_O = (\mathbf{r}_B - \mathbf{r}_A) \times \mathbf{F}_1 \tag{4.26}$$

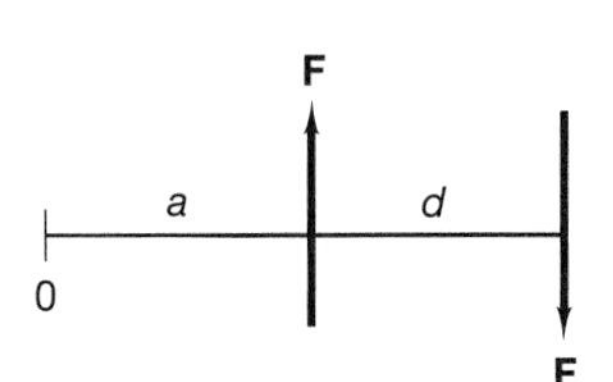

Figure 4.15

But ($\mathbf{r}_B - \mathbf{r}_A$) is the position of B relative to A, viz., $\mathbf{r}_{B/A}$. Therefore, the moment of the couple about the origin can be reduced to the simple expression

$$\mathbf{M}_O = \mathbf{r}_{B/A} \times \mathbf{F}_1 \tag{4.27}$$

Eq. (4.27) proves that the moment of the two forces forming the couple depends only upon the forces and the relative position vector between them, not upon their positions relative to the reference origin. A simple two-dimensional example will clearly illustrate this dependency, as shown in Figure 4.15.

From the figure, the moment about point O can be written as

$M_O = -Fa + F(a + d)$, where clockwise is taken to be positive. Thus,

$$M_O = Fd \tag{4.28}$$

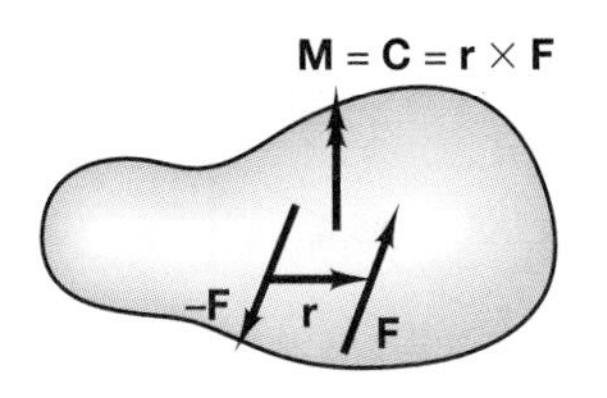

Figure 4.16 The moment of the couple **C**, formed by the two forces **F** separated by a distance **r**.

We see, then, that the moment of the couple does not depend upon the distance a and depends only upon the perpendicular distance d between the two forces forming the couple. The **moment of a couple** may be considered a **free vector**, as it is the same about any point in space. The two parallel forces forming the couple also define a plane in space, and the moment of the couple vector is perpendicular to this plane, as shown in Figure 4.16. Let

$$\mathbf{C} = \mathbf{r} \times \mathbf{F} \tag{4.29}$$

where **r** is any vector between the lines of action of the two forces. This special moment is referred to as the moment of the couple. It is some times called the couple moment and also called the couple.

Two couples are said to be equivalent if their moment vectors are equal. This situation is illustrated in Figure 4.17. The moment of the couple formed by $\mathbf{F}_1$ is equivalent to that formed by $\mathbf{F}_2$ if

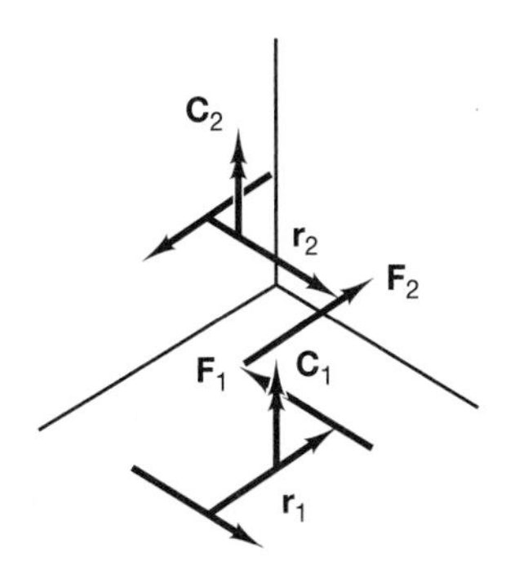

Figure 4.17 Two couples are equivalent if their moment vectors are equal.

$$\mathbf{C}_1 = \mathbf{C}_2$$

$$\mathbf{r}_1 \times \mathbf{F}_1 = \mathbf{r}_2 \times \mathbf{F}_2 \tag{4.30}$$

These identities require that the force pairs $\mathbf{F}_1$ and $\mathbf{F}_2$ lie in parallel planes and that the products of their magnitudes and perpendicular distances be equal; that is, for two couples to be equivalent, we must have

$$F_1 d_1 = F_2 d_2 \tag{4.31}$$

where d_1 and d_2 are the perpendicular distances between the paired vectors.

Since couples are free vectors (having no unique point of application on the body), they may be added by the rules of vector addition and may be considered to be acting anywhere on the rigid body. In Figure 4.18, the moments of the couples $\mathbf{C}_1$, $\mathbf{C}_2$, and $\mathbf{C}_3$ may be added, to give a resultant moment

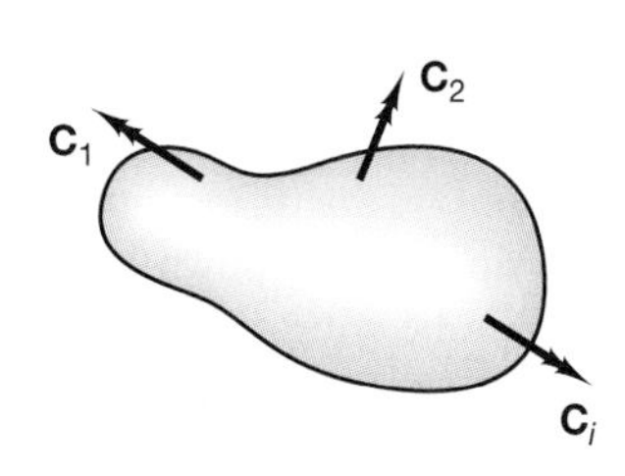

Figure 4.18

$$\mathbf{C} = \sum_i \mathbf{C}_i \tag{4.32}$$

The moment **C** is also a free vector and represents the combined turning effect on the body of all couples acting on the body.

Sample Problem 4.12

The foundation of a building constructed to sit in water is extended to spread the weight of the building over a large surface area. This configuration is called a "raft" or "mat" foundation. If the loading on the foundation is as shown in the following figure, calculate the moment of the couples acting on it.

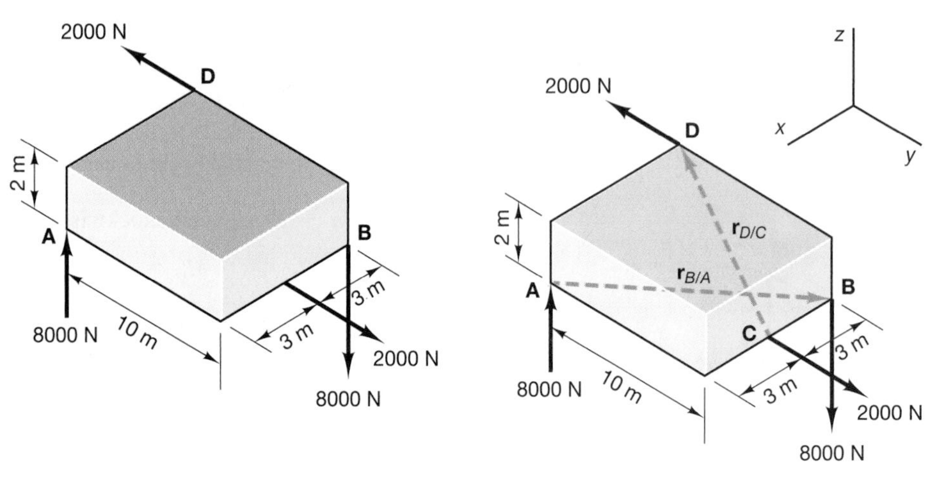

Solution We establish a coordinate reference system as shown in the figure. With this reference system, we can write the forces and position vectors of the couples as

$$\mathbf{r}_{B/A} = -6\hat{\mathbf{i}} + 10\hat{\mathbf{j}} \qquad \mathbf{F}_B = -8000\hat{\mathbf{k}}$$

$$\mathbf{r}_{D/C} = -3\hat{\mathbf{i}} - 10\hat{\mathbf{j}} + 2\hat{\mathbf{k}} \quad \mathbf{F}_D = -2000\hat{\mathbf{j}}$$

The moments of the couples and the total moment are then

$$\mathbf{M} = \mathbf{C}_{AB} + \mathbf{C}_{CD} = \mathbf{r}_{A/B} \times \mathbf{F}_B + \mathbf{r}_{D/C} \times \mathbf{F}_D$$

$$\mathbf{M} = (-48000\hat{\mathbf{j}} - 80000\hat{\mathbf{i}}) + (6000\hat{\mathbf{k}} + 4000\hat{\mathbf{i}})\ \text{N} \cdot \text{m}$$

$$\mathbf{M} = -76000\hat{\mathbf{i}} - 48000\hat{\mathbf{j}} + 6000\hat{\mathbf{k}}$$

The combined moment produces a rotational effect about all three axes. Computational software may be used to perform the vector calculations.

Problems

4.52 In Figure P4.52, compute the moment of the two equal and opposite forces about the point A and again about the point B. Convince yourself that the moment of the couple, which is the sum of the moments of two equal and opposite forces, is a free vector.

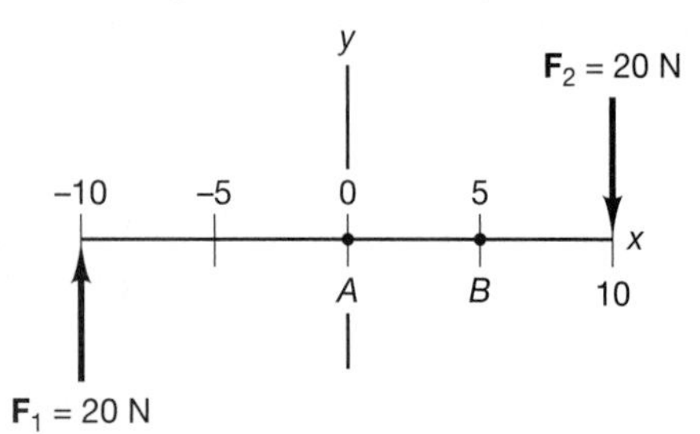

Figure P4.52

4.53 Calculate the moment of the couple due to the two equal and opposite forces applied to the steering wheel shown in Figure P4.53.

4.54 An airplane uses its ailerons to bank (a turning motion) by putting one aileron up and one down. (See Figure P4.54.) The air rushing past the wing and into the ailerons causes the two equal and opposite forces indicated in the figure. If these forces are each approximately 2400 N acting at 3 m from the center of the airplane, calculate the resulting moment of the couple.

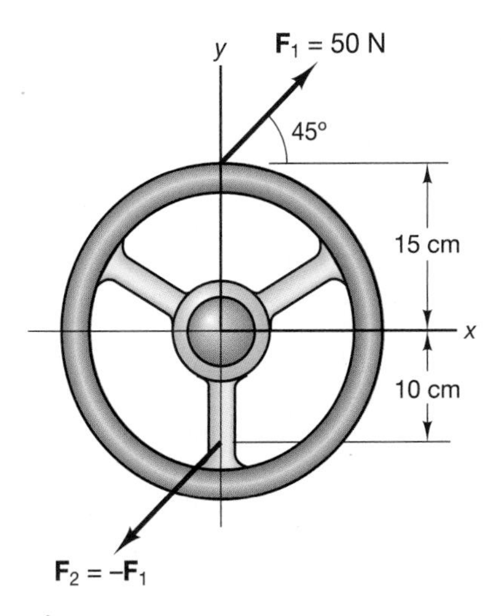

Figure P4.53

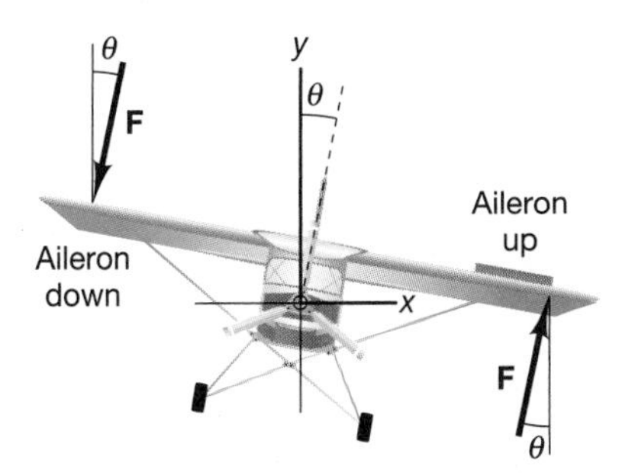

Figure P4.54

4.55 Consider the raft foundation of Sample Problem 4.10, repeated in Figure P4.55 with different values for the applied forces. Calculate the total moment generated by the two couples, and determine the resulting magnitude of these moments.

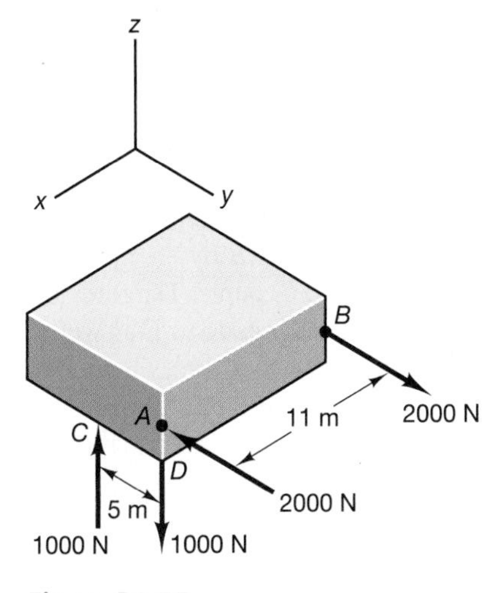

Figure P4.55

4.56 A T-bar is used to open a water valve by pushing down on one side and up on the other side with two 50-N forces aligned along the y-axis. (See Figure P4.56.) The point of application of the two equal and opposite forces is given by specifying the coordinates (x,y,z) of the ends of the handle, in centimeters. Calculate the moment of the couple and its magnitude and direction cosines.

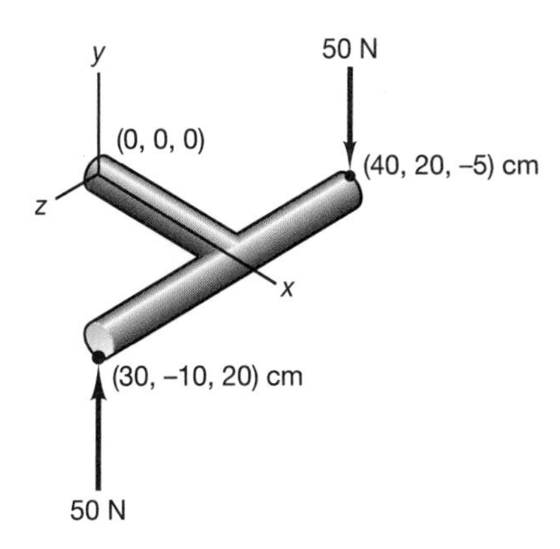

Figure P4.56

4.57 Figure P4.57 shows a mounting bracket loaded with two sets of forces. Determine the moment of each couple and then the resultant couple.

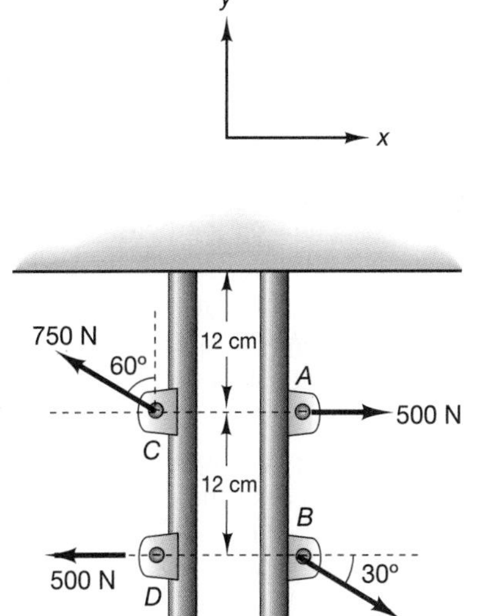

Figure P4.57

4.58 Two couples act on a piping system, as shown in Figure P4.58. Each force is aligned along either the x-axis or the z-axis. Calculate the resultant moment, as well as its magnitude and direction cosines. Then express the total moment of the couples as a magnitude times a unit vector along the resultant moment.

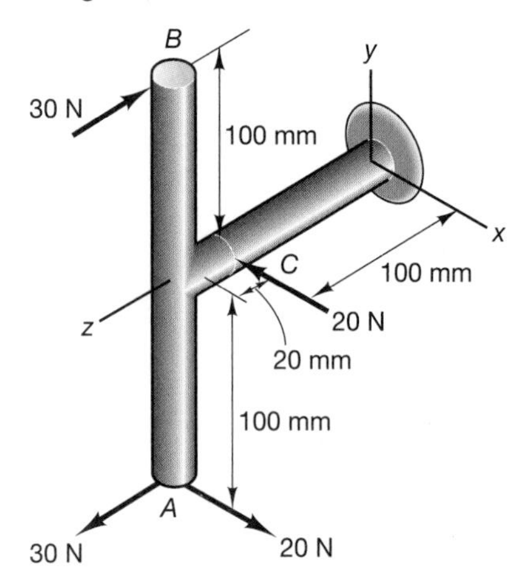

Figure P4.58

4.59 Two couples act on the keystone of an archway. (See Figure P4.59.) Determine the resultant moment, and express it as the product of the magnitude and a unit vector along the direction of the resultant.

4.60 Consider designing the keystone illustrated in figure P4.59 by choosing the thickness of stone β and the cut angle θ (i.e., γ) so that the total couple has the specific value of $\mathbf{C}_{\text{TOT}} = -100\hat{\mathbf{i}} + 100\hat{\mathbf{j}}$ Nm for the indicated forces.

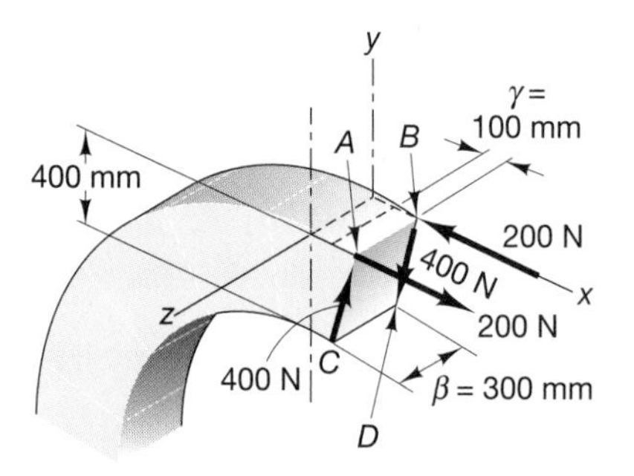

Figure P4.59

4.61 A couple consisting of two 300-N forces is applied to a crate. (See Figure P4.61.) The crate is a 500-mm cube, and the 300-N force at A has direction cosines of $\cos\theta_x = 0.408$, $\cos\theta_y = 0.408$, and $\cos\theta_z = 0.816$. Calculate the moment of the couple.

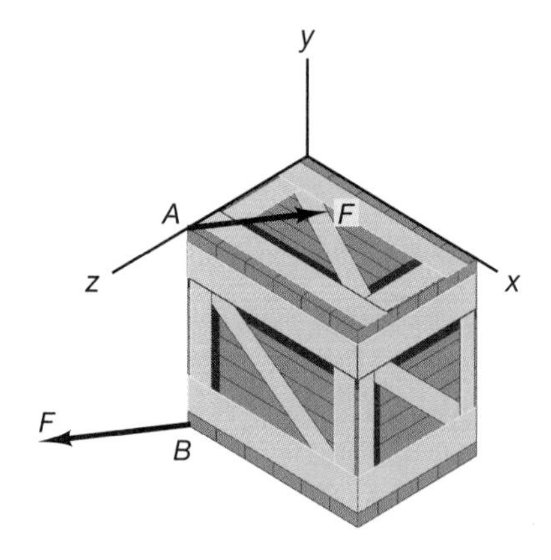

Figure P4.61

4.7 EQUIVALENT FORCE SYSTEMS

Forces may be moved to a different point on a rigid body by the use of couples, and the new system of forces and couples is called an ***equivalent*** force system. Two force systems are said to be equivalent if they produce the same resultant and the same moment about any point on the body. Consider the force $\mathbf{F}$ acting at point B on the body shown in Figure 4.19(a). Because the body is a rigid body and not a particle, the force has both a translational and a rotational effect about any other point A on the body. Point B and other points along the line of action of the force undergo only translational effects. To examine the translational and rotational effects at A, we can replace the single force at B by an equivalent force system consisting of the original force passing through A and a couple. The force $\mathbf{F}$ is "moved" to point A by adding two forces at A that are equal in magnitude to $\mathbf{F}$ as well as to each other, are opposite in direction, and are collinear, so that no net forces have been added to the body (Figure 4.19(b)). The resultant force and moment of the force about any point on the body have been unaltered by adding an equivalent zero force at A; that is,

$$\mathbf{F} + (-\mathbf{F}) = 0 \tag{4.33}$$

The force $\mathbf{F}$ at B and the force $-\mathbf{F}$ at A form a couple having the moment

$$\mathbf{C} = \mathbf{r}_{B/A} \times \mathbf{F} \tag{4.34}$$

where $\mathbf{r}_{B/A}$ is the position vector of B relative to A. (See Figure 4.19(c)).

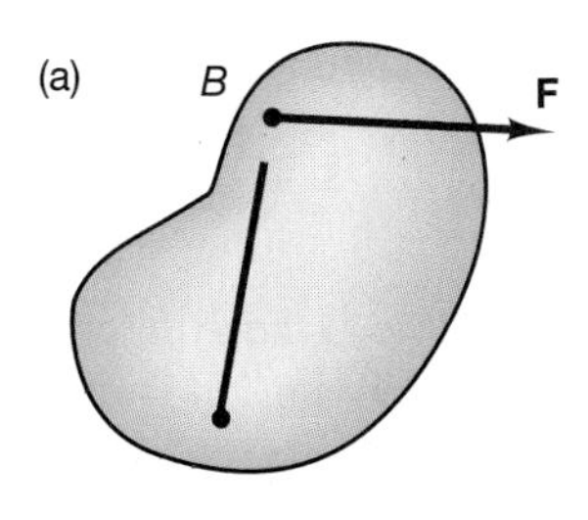

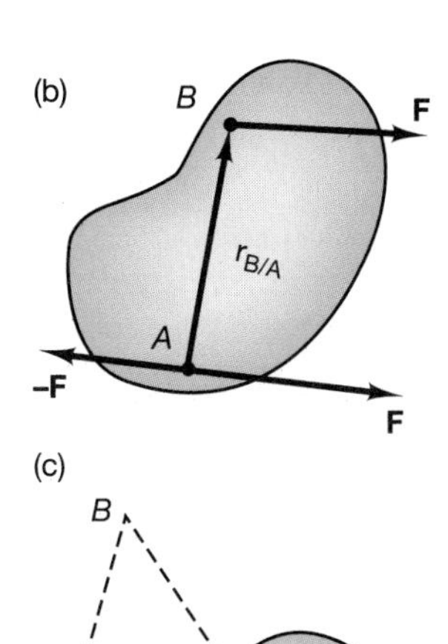

Figure 4.19 (a) A force **F** acting at point B on a rigid body. (b) An equivalent force system formed by adding equal and opposite collinear forces at point A. (c) An equivalent force system consisting of the force **F** acting at point A and a couple.

The three force systems shown in Figures 4.19 (a), (b), and (c) are equivalent force systems, because they produce the same translational and rotational effects on the body. Note that the local effects on the body *do* change from one system to the other, and the concept of equivalent force systems cannot be used if internal forces and deformations are of interest.

Thus, for a rigid body, an applied force at one point can be replaced by an equivalent force applied at a different point , together with the moment of a couple. The converse is also true. That is, any force and moment of a couple that are mutually perpendicular can be replaced by a single force. This will be discussed in detail in Section 4.8. Forming equivalent force systems is fundamental to the analysis of problems in dynamics in which the forces will be considered to be acting on the center of mass of the body.

We have seen that a single force may be moved to different points on a body by the use of a couple to form equivalent force systems. The same method may be used when many forces act on a single rigid body. Consider the body shown in Figure 4.20(a), which is subjected to the forces with different lines of action. An equivalent force system can be determined consisting of the resultant acting at point O and a moment. Note that the rotational and translational effects of the forces are still equivalent when the forces are resolved into a single resultant force and a moment. Each of the forces in Figure 4.20(a) may be moved to point O and the concurrent forces summed to give the resultant force. Moving each force produces a couple, and the vector sum of the moments of these couples gives a single resultant moment. Mathematically, the situation is expressed as

$$\mathbf{R} = \sum \mathbf{F}_i \tag{4.35}$$

$$\mathbf{C} = \sum \mathbf{c}_i = \sum \mathbf{r}_i \times \mathbf{F}_i \tag{4.36}$$

The force system in Figure 4.20(a) is reduced to the equivalent force system in Figure 4.20(b) consisting of a force and moment at point O. The choice of the point O is arbitrary, and O is usually chosen to be a point of some interest, as dictated by a particular problem or application.

If an equivalent force system is defined at point A, another equivalent force system may be determined for point B by moving the resultant at A to B and forming a couple. The former moment at point A is a free vector and may then be added to this new moment of the couple. Two such equivalent force systems are shown in Figure 4.21.

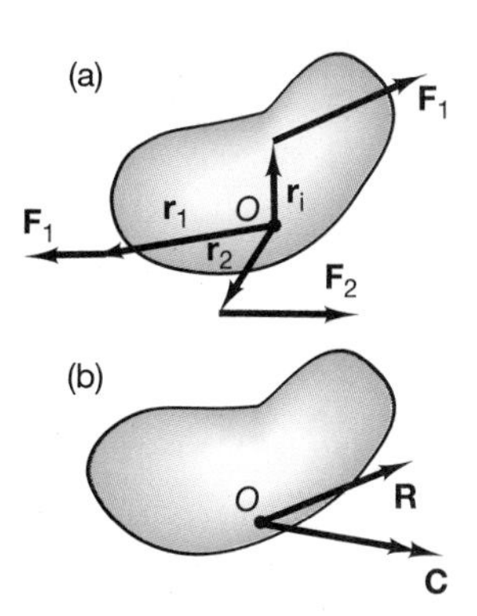

Figure 4.20 (a) A rigid body acted on by forces with different lines of action. (b) An equivalent force system consisting of a resultant and a couple.

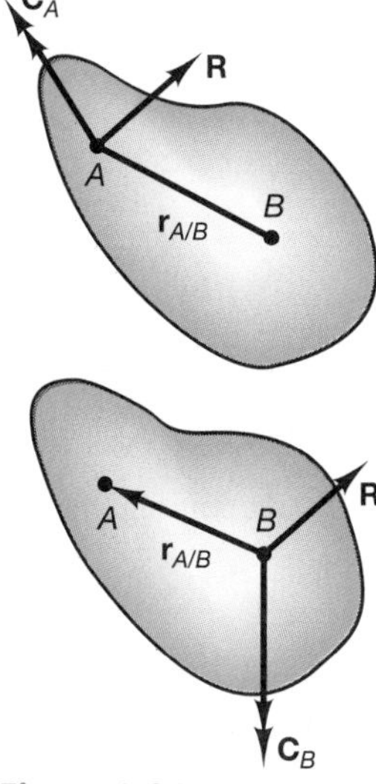

Figure 4.21

The resultant force is the same at both point A and point B. The moment at point B is related to the moment at A by

$$\mathbf{C}_B = \mathbf{C}_A + \mathbf{r}_{A/B} \times \mathbf{R} \tag{4.37}$$

The moment $\mathbf{C}_A$ is a free vector, which is the same at points A and B, and the difference between $\mathbf{C}_A$ and $\mathbf{C}_B$ is due to the moment of the couple formed by "moving" $\mathbf{R}$ from A to B.

Sample Problem 4.13

In the diagram at the left, determine the equivalent force systems consisting of a force and the moment of a couple if the 1000-N load applied at point A is moved to point B and then to points C and D.

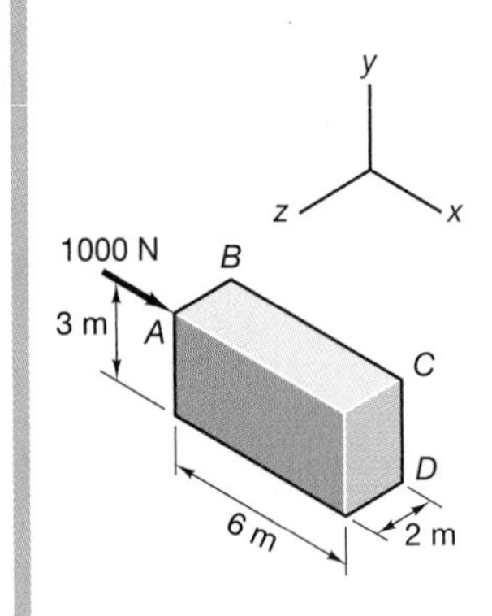

Solution The original force system at A is

$$\mathbf{F}_A = 1000\hat{\mathbf{i}}$$

A position vector from B to the line of action of the force is

$$\mathbf{r}_{A/B} = \mathbf{r}_A - \mathbf{r}_B = 2\hat{\mathbf{k}}$$

At point B, the equivalent force system becomes a force equal to the force at A and the moment of the couple, formed by the force at A and an equal negative force acting at B:

$$\mathbf{F}_B = 1000\hat{\mathbf{i}}$$

$$\mathbf{C}_B = \mathbf{r}_{A/B} \times \mathbf{F}_A = 2\hat{\mathbf{k}} \times 1000\hat{\mathbf{i}} = 2000\hat{\mathbf{j}}\ \text{N}\cdot\text{m}$$

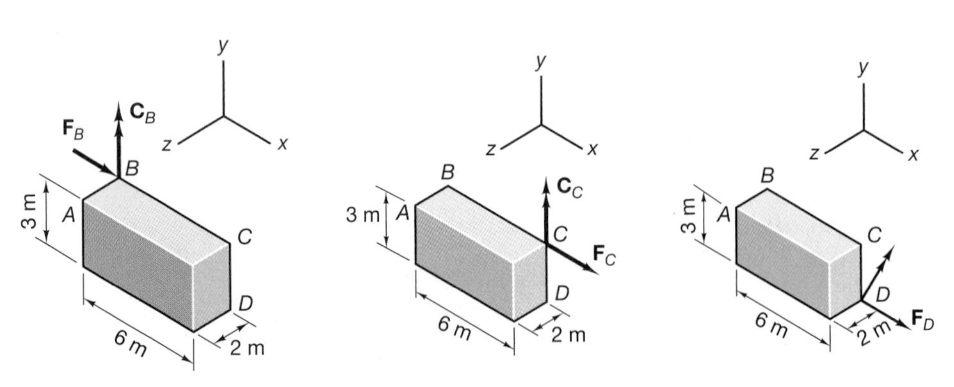

At point C, the equivalent force system becomes

$$\mathbf{F}_C = 1000\hat{\mathbf{i}}$$

$$\mathbf{C}_C = \mathbf{r}_{A/C} \times \mathbf{F}_A = (-6\hat{\mathbf{i}} + 2\hat{\mathbf{k}}) \times 1000\hat{\mathbf{i}} = 2000\hat{\mathbf{j}}\ \text{N}\cdot\text{m}$$

At point D, the equivalent force system becomes

$$\mathbf{F}_C = 1000\hat{\mathbf{i}}$$

$$\mathbf{C}_D = \mathbf{r}_{A/D} \times \mathbf{F}_A = (-6\hat{\mathbf{i}} + 3\hat{\mathbf{j}} + 2\hat{\mathbf{k}}) \times 1000\hat{\mathbf{i}} = (2000\hat{\mathbf{j}} - 3000\hat{\mathbf{k}})\ \text{N}\cdot\text{m}$$

The preceding figures show the direction and point of application of $\mathbf{F}_B$, $\mathbf{F}_C$, and $\mathbf{F}_D$.

Problems

4.62 A 25-N force is applied to a gearshift knob. (See Figure P4.62.) Replace this single force system with an equivalent force applied at point *B* and a couple.

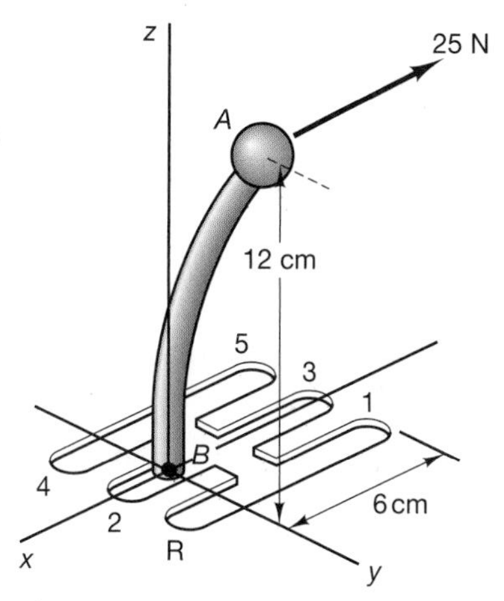

Figure P4.62

4.63 A 500-N force is applied to a 1-m square block at point *B*, as shown in Figure P4.63. Replace this single force at *B* with an equivalent force applied at *A* and a couple.

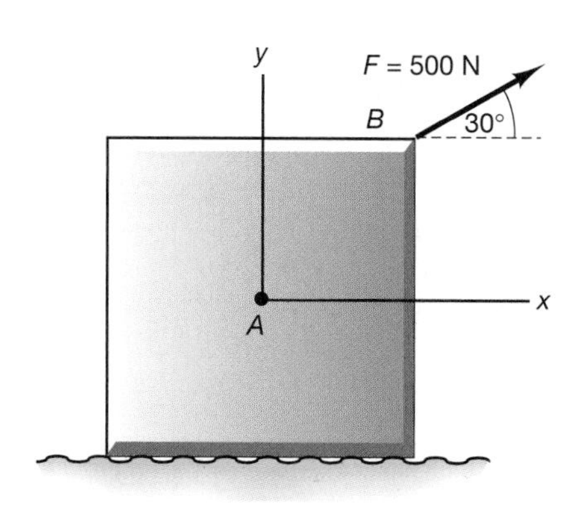

Figure P4.63

4.64 A 300-N force is applied to a 1-m square cube as indicated in Figure P4.64. Calculate the equivalent force system for (a) moving the force to point *C* and (b) moving the force to point *B*.

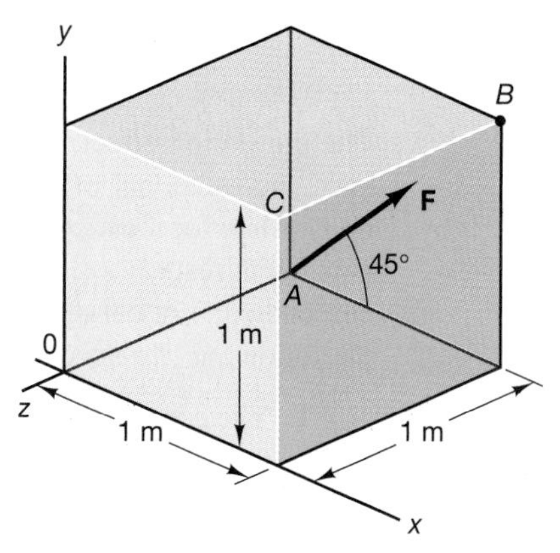

Figure P4.64

4.65 A force $\mathbf{F} = -20\hat{\mathbf{i}} - 10\hat{\mathbf{j}} - 10\hat{\mathbf{k}}$ N is applied to a sphere at point *A* in Figure P4.65. Replace this force with an equivalent force at *O*, and calculate the resulting moment from the couple created in the process.

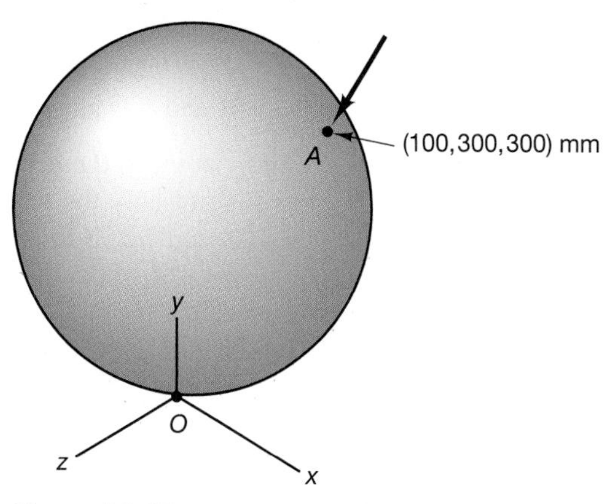

Figure P4.65

4.8 SPECIAL EQUIVALENT FORCE SYSTEMS

In general, any force system can be resolved into an equivalent force–moment system at any point *O* on a body representing the translation and rotational effects about the point *O*. For example, consider a body with *n* forces acting upon it and with position vectors $\mathbf{r}_i$ from the point *O* to the ith force. The body may also be subjected to *m* moments or moments of couples. (These moments represent free vectors and therefore do not have points of application.) The equivalent force–moment system is given by the resultant force and the sum of the moments of the couples formed by moving the forces to point O. Symbolically, this equivalence becomes

$$\mathbf{R} = \sum_{i=1}^{n} \mathbf{F}_i$$

$$\mathbf{C}_O = \sum_{i=1}^{n} \mathbf{r}_i \times \mathbf{F}_i + \sum_{i=1}^{m} \mathbf{C}_i \tag{4.38}$$

The translation effect, due to the resultant $\mathbf{R}$, is the same at all points on the body. However, the rotational effect, due to the moment $\mathbf{C}_O$, is different at different points on the body. Also, for most points on the body, the resultant $\mathbf{R}$ and the moment $\mathbf{C}$ will not be oriented with respect to each other in any special manner; that is, they will not necessarily be perpendicular or parallel to each other. This situaton is illustrated in Figure 4.22. However, the following two special cases do occur:

1. The equivalent system is a pure moment. In this case, $\mathbf{R} = 0$, and the moment $\mathbf{C}$ is the same for every point on the body. The forces acting on the body are part of couples and produce a purely rotational effect. This effect is the same for all points on the body.
2. A point may be found at which the equivalent force system has no rotational effect. In this case, there exists a point where $\mathbf{C} = 0$, and the equivalent force system for this point is only the resultant $\mathbf{R}$. If a single such point can be found, then there are an infinite number of these points lying on the line of action of $\mathbf{R}$ through this single point. If such a point or line exists, then the equivalent force–moment system at any point on the body that is not on the line will consist of the resultant $\mathbf{R}$ and a moment $\mathbf{C}$ that are perpendicular to each other. Such a system is shown in Figure 4.23, where A is a point at which the equivalent force system is only the resultant $\mathbf{R}$ and B is any other point on the body not on the line of action of $\mathbf{R}$ through A.

Any point on the line of action of $\mathbf{R}$ passing through A has an equivalent force system consisting of only the resultant force $\mathbf{R}$; that is, the moment $\mathbf{C}$ would be zero. Note that when the equivalent force system is found at point B, the moment at B is

$$\mathbf{C}_B = \mathbf{r}_{A/B} \times \mathbf{R} \tag{4.39}$$

Where $\mathbf{C}_B$ is perpendicular to both $\mathbf{R}$ and $\mathbf{r}_{A/B}$.

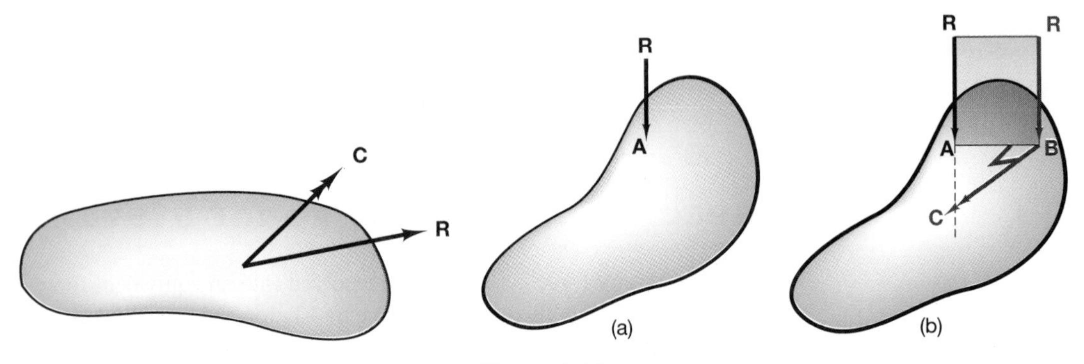

Figure 4.22

Figure 4.23

The reduction of a force system to an equivalent system consisting of only the resultant force with a unique line of action in space can always be accomplished if, at any point on the body, the equivalent force system consists of a moment and a force that are perpendicular. This realtionship will always occur in three special cases:

1. Concurrent force systems
2. Coplanar force systems
3. Parallel force systems

We consider each of these systems separately.

4.8.1 CONCURRENT FORCE SYSTEMS

A ***concurrent force system*** is a force system, all of whose lines of action of the forces intersect at a common point. In Chapter 2, we examined concurrent force systems and noted that such a system can be replaced by a single resultant force passing through the common point. The resultant is equal to the vector sum of all the concurrent forces. If the resultant force is moved from the concurrent point, the equivalent force system consists of the resultant force and a perpendicular moment vector.

4.8.2 COPLANAR FORCE SYSTEMS

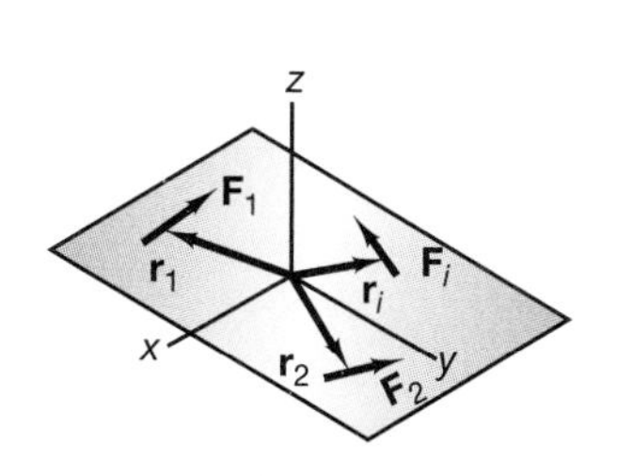

Figure 4.24 A coplanar force system.

As has been defined previously, a ***coplanar force system*** exists when all the forces in a given situation may be considered to act in a single plane in space. Since the coordinate system may be defined in any manner, no loss in generality results in choosing the plane of the forces to be the x–y plane, as shown in Figure 4.24. Each force has only x- and y-components, that is,

$$\mathbf{F}_i = F_{xi}\hat{\mathbf{i}} + F_{yi}\hat{\mathbf{j}} \tag{4.40}$$

Similarly, the position vector from the origin to each force also lies in the x–y plane:

$$\mathbf{r}_i = x_i\hat{\mathbf{i}} + y_i\hat{\mathbf{j}} \tag{4.41}$$

The moment of the forces about the origin then becomes

$$\mathbf{C}_O = \mathbf{M}_O = \sum_i \mathbf{r}_i \times \mathbf{F}_i = \sum_i (x_iF_{yi} - y_iF_{xi})\hat{\mathbf{k}} \tag{4.42}$$

and this moment is in the z-direction. The resultant force is

$$\mathbf{R} = \sum_i \mathbf{F}_i = \sum_i (F_{xi}\hat{\mathbf{i}} + F_{yi}\hat{\mathbf{j}}) \neq 0 \tag{4.43}$$

Therefore, the coplanar force system may be replaced by an equivalent force system consisting of a resultant **R** and a moment $\mathbf{C}_O$ at the origin. Note that the moment vector and the resultant force vector are perpendicular to each other. This can be verified by examining the scalar or dot product between the two vectors as well as by examining the two vectors themselves. The dot product will be zero:

$$\mathbf{R}\cdot\mathbf{C}_O = 0 \tag{4.44}$$

Since at one point in the body—the origin, in this case—there exists an equivalent force system consisting of the resultant force and a moment perpendicular to it, another point can be found in the plane where the equivalent force system is only the resultant force, as shown in Figure 4.25. Moving **R** away from point O to any other point in the x–y

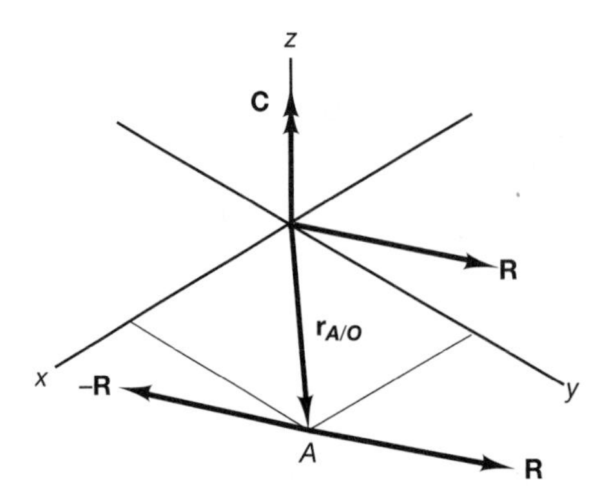

Figure 4.25

plane generates an additional moment of the couple that is perpendicular to **R** and in the z-direction. Conceptually, it can be seen that there exists a point A at which the additional moment of the couple is equal and opposite to the original moment **C**. Therefore, at this point A, the equivalent force system is

$$\mathbf{C}_A = \mathbf{C} + \mathbf{r}_{A/O} \times (-\mathbf{R}) = 0$$

$$\mathbf{R} = \mathbf{R} \tag{4.45}$$

The position of this special point A, at which the moment of the couple vanishes, can be obtained by examining Eq. (3.70). We have

$$\mathbf{C}_A = \mathbf{C} + \mathbf{r}_{A/O} \times (-\mathbf{R}) = 0$$

Therefore,

$$\mathbf{r}_{A/O} \times (\mathbf{R}) = \mathbf{C} \tag{4.46}$$

Since the resultant **R** and the moment **C** are known, the position vector from O to A may be obtained by expanding this vector equation into scalar form, yielding

$$x_{A/O}R_y - y_{A/O}R_x = C \tag{4.47}$$

Since **C** has only one nonzero component in this coordinate system along the z-axis, this last expression is in the form of an equation of a line in the x–y plane:

$$ax + by = c \tag{4.48}$$

This linear relationship should not be surprising, as A may be any point on the line of action of **R**. Therefore, any coplanar force system may be replaced by an equivalent force system consisting of the resultant acting along a unique line in the plane.

One particular point on that line may be determined by using a direct vector solution. This point would be the perpendicular distance from point A to the origin—that is, the vector $\mathbf{r}_{A/O} = \mathbf{p}_{A/O}$, which is perpendicular to the resultant force **R**. This vector would then be perpendicular to both **R** and **C**. Equation (4.46) may be solved directly for the point in question by using the vector triple product. We take the cross product of Eq. (4.46) with the resultant vector **R**:

$$\mathbf{R} \times (\mathbf{p}_{A/O} \times \mathbf{R}) = \mathbf{R} \times \mathbf{C}$$

Expanding the vector triple product (see Eq. (2.123)) yields

$$\mathbf{p}_{A/O}(\mathbf{R} \cdot \mathbf{R}) - \mathbf{R}(\mathbf{p}_{A/O} \cdot \mathbf{R}) = \mathbf{R} \times \mathbf{C}$$

But $(\mathbf{p}_{A/O} \cdot \mathbf{R}) = 0$, because $\mathbf{p}_{A/O}$ and $\mathbf{R}$ are perpendicular. Therefore,

$$\mathbf{p}_{A/O} = \frac{\mathbf{R} \times \mathbf{C}}{(\mathbf{R} \cdot \mathbf{R})} \tag{4.49}$$

4.8.3 PARALLEL FORCE SYSTEMS

A ***parallel force system*** is a force system in which the lines of action of all the forces acting on a body are parallel. In general, these forces do not lie in a plane, but are perpendicular to a common plane. A parallel force system is shown in Figure 4.26 in which all the forces are in the z-direction and, therefore, are perpendicular to the x–y plane. Each of the forces would have only a component in the z-direction and hence can be written as

$$\mathbf{F}_i = F_i\hat{\mathbf{k}} \tag{4.50}$$

Therefore, the resultant also has only a component in the z-direction:

$$\mathbf{R} = \sum_i F_i\hat{\mathbf{k}} \tag{4.51}$$

A position vector may be formed from the origin to the intercept of the line of action of each force with the x–y plane:

$$\mathbf{r}_{i/O} = x_i\hat{\mathbf{i}} + y_i\hat{\mathbf{j}} \tag{4.52}$$

The moment generated by moving the forces to the origin is thus

$$\begin{aligned}\mathbf{C} &= \sum_i \mathbf{r}_{i/O} \times \mathbf{F}_i = \sum_i y_iF_i\hat{\mathbf{i}} - \sum_i x_iF_i\hat{\mathbf{j}} \\ &= C_x\hat{\mathbf{i}} + C_y\hat{\mathbf{j}}\end{aligned} \tag{4.53}$$

This moment lies in the x–y plane, perpendicular to the resultant force vector. Note that although the position vectors of the forces were taken in the x–y plane, they could have been chosen to be any point on the line of action of the force, and the resulting moment would have been the same, since $\hat{\mathbf{k}} \times \hat{\mathbf{k}} = 0$. The equivalent force system at the origin consists of the resultant force $\mathbf{R}$ and the moment $\mathbf{C}$, which is perpendicular to $\mathbf{R}$.

Again, an equivalent force system exists consisting only of the resultant force $\mathbf{R}$, having a unique line of action in space with a unique intercept with the x–y plane, as shown in Figure 4.27. When $\mathbf{R}$ is moved to this unique intercept, point A, the equivalent force system is

$$\begin{aligned}\mathbf{R} &= \mathbf{R} \\ \mathbf{C}_A &= \mathbf{C} + \mathbf{r}_{A/O} \times (-\mathbf{R}) = \mathbf{0}\end{aligned} \tag{4.54}$$

The point A may be located by solving the vector equation

$$\mathbf{r}_{A/O} \times (\mathbf{R}) = \mathbf{C} \tag{4.55}$$

Expanding this vector equation into its scalar components yields

$$(x_{A/O}\hat{\mathbf{i}} + y_{A/O}\hat{\mathbf{j}}) \times R\hat{\mathbf{k}} = C_x\hat{\mathbf{i}} + C_y\hat{\mathbf{j}}$$

$$x_{A/O} = -\frac{C_y}{R} \qquad y_{A/O} = \frac{C_x}{R} \tag{4.56}$$

Alternatively, since the three vectors in Eq. (4.55) are mutually perpendicular, this equation again could be solved by a direct vector solution using the vector triple product:

$$\mathbf{R} \times (\mathbf{p}_{A/O} \times \mathbf{R}) = \mathbf{R} \times \mathbf{C}$$

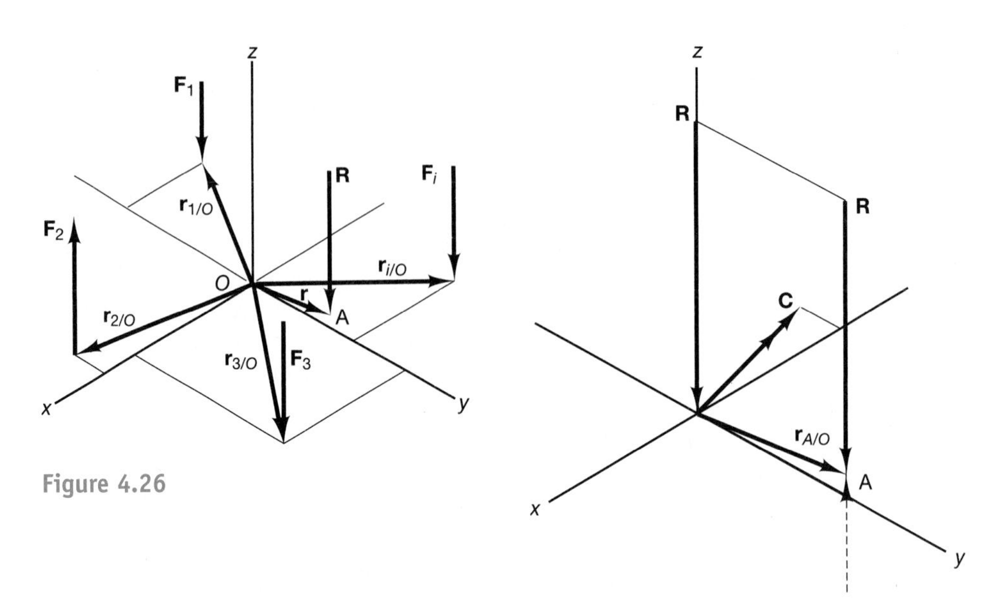

Figure 4.26

Figure 4.27

Expanding the vector triple product (see Eq. (2.123)) results in

$$\mathbf{p}_{A/O}(\mathbf{R}\cdot\mathbf{R}) - \mathbf{R}(\mathbf{p}_{A/O}\cdot\mathbf{R}) = \mathbf{R}\times\mathbf{C} \tag{4.57}$$

But $(\mathbf{p}_{A/O}\cdot\mathbf{R}) = 0$, because $\mathbf{p}_{A/O}$ and $\mathbf{R}$ are perpendicular. Therefore,

$$\mathbf{p}_{A/O} = \frac{\mathbf{R}\times\mathbf{C}}{(\mathbf{R}\cdot\mathbf{R})}$$

This is the position vector relative to the unique intercept of the resultant with the plane such that the equivalent force system consists of only the resultant force. Equation (4.56) is the same as Eq. (4.47) for the coplanar force system.

One of the most important applications of a parallel force system occurs when gravitational attraction alone acts upon a body. In this situation, the equivalent force system consisting of only the resultant is used to define the center of gravity of the body.

Sample Problem 4.14

A coplanar force system consists of the following forces with position vectors from the origin, as given:

$$\mathbf{F}_1 = 100\hat{\mathbf{i}} + 100\hat{\mathbf{j}}\ (\text{N}) \qquad \mathbf{r}_{1/O} = 2\hat{\mathbf{i}} + 3\hat{\mathbf{j}}\ (\text{m})$$

$$\mathbf{F}_2 = 50\hat{\mathbf{j}} \qquad \mathbf{r}_{2/O} = -10\hat{\mathbf{i}} + 2\hat{\mathbf{j}}$$

$$\mathbf{F}_3 = 300\hat{\mathbf{i}} - 450\hat{\mathbf{j}} \qquad \mathbf{r}_{3/O} = 4\hat{\mathbf{i}} - 4\hat{\mathbf{j}}$$

Determine the equivalent force system consisting of a single resultant, and determine the line of action of the resultant in space.

Solution Since all three forces lie in the x–y plane, this problem is an example of case 2, discussed on page 153 earlier. The resultant force is

$$\mathbf{R} = \sum\mathbf{F} = 400\hat{\mathbf{i}} - 300\hat{\mathbf{j}}$$

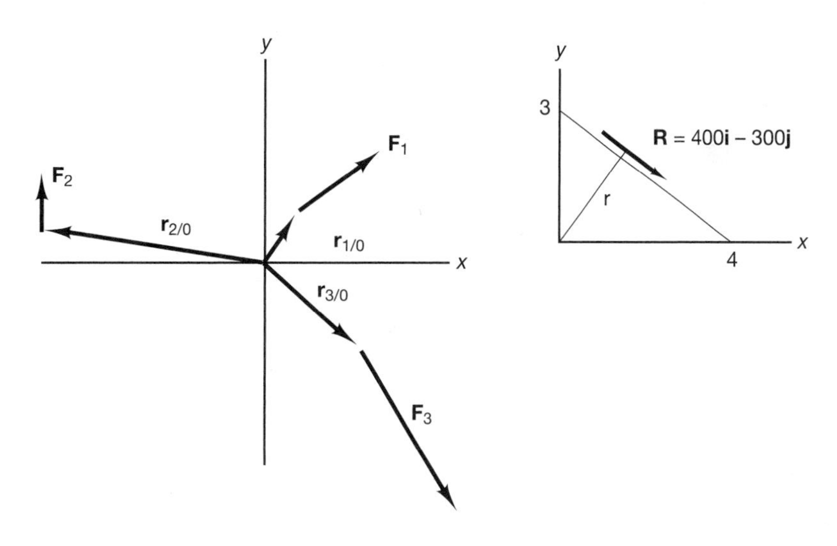

The couple at, or the moment of the forces about, the origin, is

$$\mathbf{C} = \sum \mathbf{r}_{i/O} \times \mathbf{F}_i = (2\hat{\mathbf{i}} + 3\hat{\mathbf{j}}) \times (100\hat{\mathbf{i}} + 100\hat{\mathbf{j}}) + (-10\hat{\mathbf{i}} + 2\hat{\mathbf{j}}) \times (50\hat{\mathbf{j}})$$

$$+ (4\hat{\mathbf{i}} - 4\hat{\mathbf{j}}) \times (300\hat{\mathbf{i}} - 450\hat{\mathbf{j}}) = -1200\hat{\mathbf{k}}$$

To find the unique line of action of the equivalent force system consisting of only the resultant **R**, we need to solve the equation

$$\mathbf{r}_{A/O} \times \mathbf{R} = \mathbf{C}$$

This reduces to the scalar equation

$$-300x_A - 400y_A = -1200 \quad \text{or} \quad 3x + 4y = 12$$

which is a line in the $x-y$ plane intercepting the x-axis 4 meters from the origin and intercepting the y-axis 3 meters from the origin. The vector **r** perpendicular to the resultant may be determined directly from Eq. (4.8) (see diagram at upper right):

$$\mathbf{r} = \frac{\mathbf{R} \times \mathbf{C}}{\mathbf{R} \cdot \mathbf{R}} = \frac{(400\hat{\mathbf{i}} - 300\hat{\mathbf{j}}) \times (-1200\hat{\mathbf{k}})}{(400\hat{\mathbf{i}} - 300\hat{\mathbf{j}}) \cdot (400\hat{\mathbf{i}} - 300\hat{\mathbf{j}})}$$

$$= \frac{360{,}000\hat{\mathbf{i}} + 480{,}000\hat{\mathbf{j}}}{250{,}000} = 1.44\hat{\mathbf{i}} + 1.92\hat{\mathbf{j}} \text{ m}$$

The vector algebra may be done using computational software, as shown in the Computational Supplement.

Sample Problem 4.15

Determine the center of gravity of the human body placed horizontally in a gravitational field, as shown in the figure on the next page. The sagittal plane of the body is a plane passing through the middle, separating the right side from the left side. The weight of the body can be considered to be symmetrically distributed about that plane.

The weight of the various body segments and their location from the origin are as follows:

$$F_{\text{legs}} = 0.12 \text{ body weight} \qquad \mathbf{r} = 50\hat{\mathbf{i}} + 50\hat{\mathbf{j}} \text{ (cm)}$$

$$F_{\text{thighs}} = 0.16 \text{ B.W.} \qquad \mathbf{r} = 30\hat{\mathbf{i}} + 30\hat{\mathbf{j}}$$

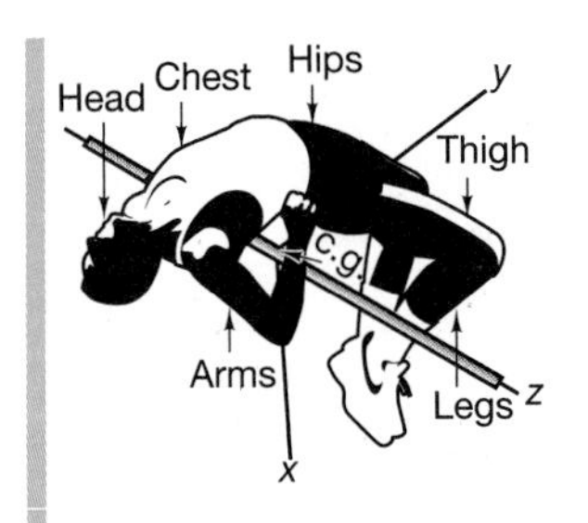

$$F_{\text{hips}} = 0.23 \text{ B.W.} \qquad \mathbf{r} = 12.5\hat{\mathbf{i}} + 12.5\hat{\mathbf{j}}$$

$$F_{\text{chest}} = 0.32 \text{ B.W.} \qquad \mathbf{r} = 16.25\hat{\mathbf{i}} - 27.5\hat{\mathbf{j}}$$

$$F_{\text{arms}} = 0.10 \text{ B.W.} \qquad \mathbf{r} = 50\hat{\mathbf{i}} - 75\hat{\mathbf{j}}$$

$$F_{\text{head}} = 0.07 \text{ B.W.} \qquad \mathbf{r} = 42.5\hat{\mathbf{i}} - 60\hat{\mathbf{j}}$$

$$\sum F_i = 1.00 \text{ B.W.}$$

Solution From Eq. (4.55), the center of gravity in the x–y plane is

$$\sum x_i F_i = 50(0.12) + 30(0.16) + 12.5(0.23) + 16.25(0.32) + 50(0.10) + 42.5(0.07)$$

$$\sum y_i F_i = 50(0.12) + 30(0.16) + 12.5(0.23) - 27.5(0.32) - 75(0.10) - 60(0.07)$$

$$x_{c.g.} = \frac{C_y}{R} = -\frac{\sum x_i F_i}{\sum F_i} = 27.95 \text{ cm}$$

$$y_{c.g.} = \frac{C_x}{R} = -\frac{\sum y_i F_i}{\sum F_i} = -6.825 \text{ cm}$$

Notice that the center of gravity is 28 cm below the body (in the positive x-direction). This is why the center of gravity of the body may actually pass underneath the bar in high jumping using the "Fosbury Flop." Dick Fosbury developed a new style of jumping in high school and won a gold medal in the 1968 Olympics in Mexico City with a jump of 2.24 m (7 ft 4 1/4 in.). He developed the difficult technique of rotating his body in such a manner that he went over the bar backwards.

4.9 GENERAL EQUIVALENT FORCE SYSTEMS

In Section 4.7, we defined equivalent force systems to be systems that have the same moment and resultant force at any point on a body. Equivalent force systems are formed by the use of couples, which allow movement of the resultant force. The moment of the couple formed in this manner is *always perpendicular to the resultant force.* Therefore, an equivalent force system always exists that has only a moment that is parallel to the resultant force. This system, called a **wrench**, is discussed later in the section. When a force system is such that the original system consists of a resultant force and a moment that are perpendicular to each other, there exists an equivalent force system that is only the resultant force with a specific line of action, as was shown in Section 4.8.

There are applications in which an equivalent system composed of the resultant force and a moment in a specific direction is convenient to describe the effects of the force system on a rigid body. This moment always has a component $\mathbf{C}_{\parallel}$ equal to the component of the original moment parallel to the resultant force. Consider the case where the original force system is composed of a moment and resultant force at point O. This system will be specified by the two vectors $\mathbf{R}$ and $\mathbf{M}_O$, where $\mathbf{R} \neq 0$ acts through point O and $\mathbf{M}_O$ is a free vector. Suppose an equivalent force system is desired that is composed of the resultant force acting at point A and a moment $\mathbf{T}$ of specified direction but unknown magnitude,

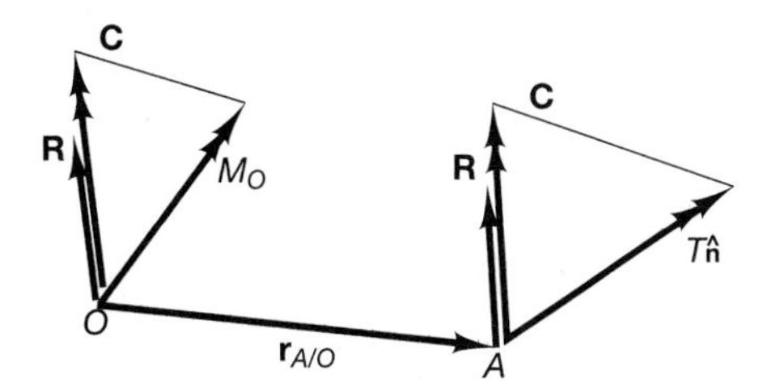

Figure 4.28

as shown in Figure 4.28. If the force system at A is equivalent to the force system at O, the problem may be formulated mathematically as

$$\mathbf{r}_{A/O} \times \mathbf{R} + T\hat{\mathbf{n}} = \mathbf{M}_O \tag{4.58}$$

where $\mathbf{r}_{A/O}$ is an unknown vector from O to point A on the line of action of the resultant in the equivalent force system and $\hat{\mathbf{n}}$ is a unit vector in the direction of the desired moment $\mathbf{T}$. Equation (4.58) may be rewritten as

$$\mathbf{r}_{A/O} \times \mathbf{R} = \mathbf{M}_O - T\hat{\mathbf{n}} \tag{4.59}$$

Equation (4.59) may be viewed as a vector equation for the unknown position vector $\mathbf{r}_{A/O}$ if the magnitude T is first determined. The vector $\left(\mathbf{M}_O - T\hat{\mathbf{n}}\right)$ on the right-hand side of this equation must be perpendicular to both $\mathbf{r}_{A/O}$ and $\mathbf{R}$ if a solution exists. The vector $\mathbf{T}$ must be such that it removes the component of $\mathbf{M}_O$ that is parallel to $\mathbf{R}$; that is,

$$\mathbf{R}\cdot\mathbf{M}_O - \mathbf{R}\cdot T\hat{\mathbf{n}} = 0$$

Therefore,

$$T = \frac{\mathbf{R}\cdot\mathbf{M}_O}{\mathbf{R}\cdot\hat{\mathbf{n}}} \tag{4.60}$$

The magnitude of the unknown moment T can always be determined if the denominator of Eq. (4.60) is not zero—that is, if the vectors $\mathbf{R}$ and $\hat{\mathbf{n}}$ are not perpendicular. The vector $\mathbf{r}_{A/O}$ is a vector from point O to any point A on the line of action of $\mathbf{R}$; therefore, an infinite number of vectors satisfy this requirement. However, if a solution exists, then a particular position vector $\mathbf{p}_O$ perpendicular to both $\mathbf{R}$ and $\left(\mathbf{M}_O - T\hat{\mathbf{n}}\right)$ may be found by a direct vector calculation. As shown in Eqs. (4.6)–(4.8), we take the vector product of both sides of Eq. (4.59) with the resultant force $\mathbf{R}$, yielding

$$\mathbf{R} \times (\mathbf{p}_O \times \mathbf{R}) = \mathbf{R} \times (\mathbf{M}_O - T\hat{\mathbf{n}}) \tag{4.61}$$

Using the vector identity, Eq. (2.123), for the vector triple product yields

$$\mathbf{p}_O(\mathbf{R}\cdot\mathbf{R}) - \mathbf{R}(\mathbf{R}\cdot\mathbf{p}_O) = \mathbf{R} \times (\mathbf{M}_O - T\hat{\mathbf{n}})$$

But

$$(\mathbf{R}\cdot\mathbf{p}_O) = 0$$

Therefore,

$$\mathbf{p}_O = \frac{\mathbf{R} \times (\mathbf{M}_O - T\hat{\mathbf{n}})}{(\mathbf{R}\cdot\mathbf{R})} \tag{4.62}$$

4.9.1 THE WRENCH

Equation (4.60) shows that a general equivalent force system may be created that consists of the resultant force and a moment in a specified direction if the scalar product of the unit vector in that specified direction and the resultant force is not zero. Every generalized force system can be resolved into an equivalent force system consisting of the resultant force and a moment vector *parallel* to the resultant. This system, which has a unique line of action in space and is called a ***wrench***, is illustrated in Figure 4.29. If the resultant **R** and the parallel moment **C** have the same sense, the wrench is said to be positive (Figure 4.29(a)). If the resultant force and the parallel moment are of opposite sense, the wrench is said to be negative (Figure 4.29(b)).

Consider a force system consisting of a resultant force **R** and a moment $\mathbf{M}_O$ at the origin, as shown in Figure 4.30. In general, the free vector $\mathbf{M}_O$ is not parallel to the resultant force vector. Rather, $\mathbf{M}_O$ has a component parallel to **R**, denoted by $\mathbf{C}_{\|}$, and a component perpendicular to **R**, denoted by $\mathbf{C}_{\perp}$. Thus, $\mathbf{M}_O$ can be uniquely resolved into components parallel and perpendicular to **R**.

Recall that if **R** is moved to any point in space, the additional moment of the couple generated by that move must be perpendicular to **R**. Therefore, a line in space, called the ***wrench axis***, can be found parallel to **R**, along which **R** may be considered to be acting, and such that the associated moment of the couple will be equal in magnitude, but opposite in direction, to the component $\mathbf{C}_{\perp}$ that is perpendicular to **R**. This is the same process that we used to resolve coplanar and parallel force systems into equivalent systems containing only the resultant **R**. However, in those cases, it was shown that the moment $\mathbf{M}_O$ was perpendicular to **R**. In the generalized case, no movement of **R** will produce a moment component that is parallel to **R**. Therefore, the most general equivalent force system is the resultant **R** and a *parallel* moment $\mathbf{C}_{\|}$. This combination, the wrench, has a unique line of action in space, the wrench axis, for the specific force system, as shown in Figure 4.31.

A similar approach can be taken to analyze the three-dimensional motion of a body and is sometimes referred to as "screw axis" and "helical axis." These terms are applied to the analysis of force systems, but actually, they apply only to kinematics, and the term "wrench" should be used for force systems.

To determine the wrench, it is first necessary to resolve the moment $\mathbf{M}_O$ into components parallel and perpendicular to **R**. The process of resolving a vector into parallel and

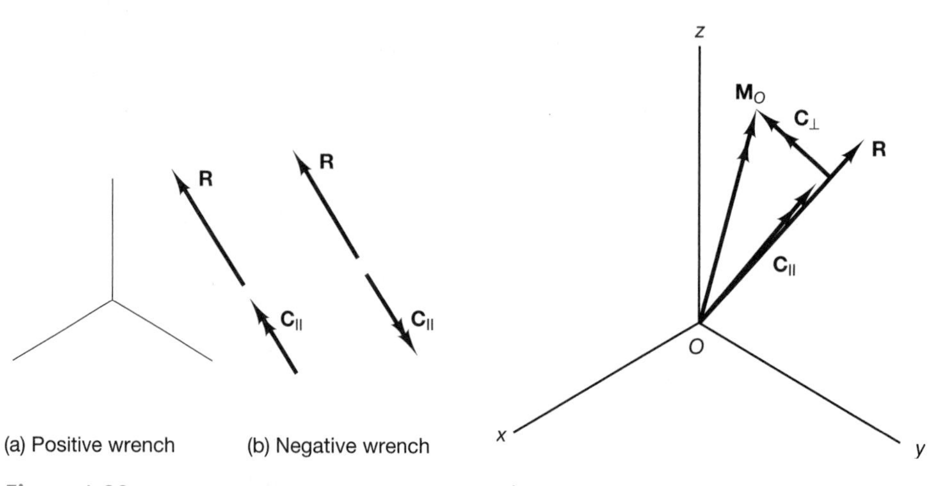

(a) Positive wrench

(b) Negative wrench

Figure 4.29

Figure 4.30

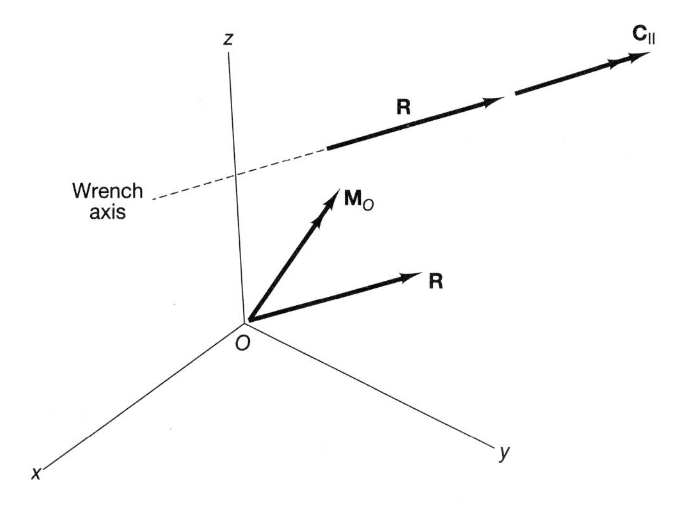

Figure 4.31

perpendicular components along a line in space was explained in Sections 2.8, 2.9 and 4.5. The direction of **R** will be designated by a unit vector defined as

$$\hat{\mathbf{e}}_R = \frac{\mathbf{R}}{|\mathbf{R}|} \tag{4.63}$$

The component of $\mathbf{M}_O$ parallel to **R** is the projection of $\mathbf{M}_O$ onto the line of action of **R** multiplied by the unit vector along **R**, or

$$\mathbf{C}_{\parallel} = (\hat{\mathbf{e}}_R \cdot \mathbf{M}_O)\hat{\mathbf{e}}_R \tag{4.64}$$

The component of $\mathbf{M}_O$ perpendicular to **R** is found by subtracting the parallel component from $\mathbf{M}_O$, yielding

$$\mathbf{C}_{\perp} = \mathbf{M}_O - \mathbf{C}_{\parallel} \tag{4.65}$$

Now, a line in space for which the perpendicular component of $\mathbf{M}_O$ is zero can be found by applying the methods used for parallel and coplanar force systems. Let **r** be a vector from the origin to an arbitrary point on the wrench axis. Then

$$\mathbf{r} \times \mathbf{R} = \mathbf{C}_{\perp} \tag{4.66}$$

This equation may be written in scalar form as the three scalar equations

$$\begin{aligned} yR_z - zR_y &= C_{\perp x} \\ -xR_z + zR_x &= C_{\perp y} \\ xR_y - yR_x &= C_{\perp z} \end{aligned} \tag{4.67}$$

Although Eqs. (4.67) appear to be three equations for the unknown components x, y, and z of the vector **r**, and easily solved, this is not the case, because the equations are not linearly independent. Their dependence may be seen if the equations are written in matrix notation as

$$\begin{bmatrix} 0 & R_z & -R_y \\ -R_z & 0 & +R_x \\ R_y & -R_x & 0 \end{bmatrix} \begin{bmatrix} x \\ y \\ z \end{bmatrix} = \begin{bmatrix} C_{\perp x} \\ C_{\perp y} \\ C_{\perp z} \end{bmatrix} \tag{4.68}$$

The determinant of the coefficient matrix is zero, and the matrix does not have an inverse. That is, the coefficient matrix is singular, indicating that the three equations not linearly independent. The value for x is

$$x = \frac{\left|\begin{bmatrix} C_{\perp x} & R_z & -R_y \\ C_{\perp y} & 0 & +R_x \\ C_{\perp z} & +R_x & 0 \end{bmatrix}\right|}{\left|\begin{bmatrix} 0 & R_z & -R_y \\ -R_z & 0 & +R_x \\ R_y & -R_x & 0 \end{bmatrix}\right|} \tag{4.69}$$

The denominator is the determinant of the coefficient matrix and is zero. This is not unexpected, because any point on the wrench axis would be a solution for Eq. (4.66). A specific point on the wrench axis must be selected in order to solve the equation. One way would be to choose intercepts of the wrench axis with the coordinate planes. For example, to determine the intercept of the wrench axis with the x–y plane, set $z = 0$ in Eq. (4.67), and the coordinates to the intercept point are

$$\begin{aligned} x &= \frac{C_{\perp y}}{R_z} \\ y &= \frac{C_{\perp x}}{R_z} \\ z &= 0 \end{aligned} \tag{4.70}$$

In a similar manner, the intercepts with the y–z plane and the x–z plane are, respectively,

$$\begin{aligned} x &= 0 & \qquad x &= \frac{C_{\perp z}}{R_y} \\ y &= \frac{C_{\perp z}}{R_z} & \qquad y &= 0 \\ z &= \frac{C_{\perp y}}{R_x} & \qquad z &= -\frac{C_{\perp x}}{R_y} \end{aligned} \tag{4.71}$$

Each of these intercepts is a point on the wrench axis and can be used to determine this axis in space.

A direct vector solution method can also be used to obtain a vector from the origin that is perpendicular to the wrench axis. This is a particular choice of the vector $\mathbf{r}$ in Eq. (4.66) that is mutually perpendicular to the other two vectors. Again, a vector triple product is formed, such as

$$\begin{aligned} \mathbf{R} \times (\mathbf{r} \times \mathbf{R}) &= \mathbf{R} \times \mathbf{C}_\perp \\ \mathbf{r}(\mathbf{R} \cdot \mathbf{R}) - \mathbf{R}(\mathbf{r} \cdot \mathbf{R}) &= \mathbf{R} \times \mathbf{C}_\perp \end{aligned}$$

Since $\mathbf{r}$ and $\mathbf{R}$ are perpendicular,

$$\begin{aligned} \mathbf{r}(\mathbf{R} \cdot \mathbf{R}) &= \mathbf{R} \times \mathbf{C}_\perp \\ \mathbf{r} &= \frac{\mathbf{R} \times \mathbf{C}_\perp}{\mathbf{R} \cdot \mathbf{R}} \end{aligned} \tag{4.72}$$

The latter equation locates a particular point on the wrench axis, and the entire axis may be constructed by observing that the resultant **R** acts through this point.

Sample Problem 4.16

In biomechanical studies of human motion, a dynamometer in the form of a force plate is placed in the floor to measure the forces between the foot and the ground when an individual walks. The force plate uses strain gages to measure the three components of force and the three components of the moment about the center of the instrument, which is 40 mm below the surface of the force plate, as shown in the following diagram:

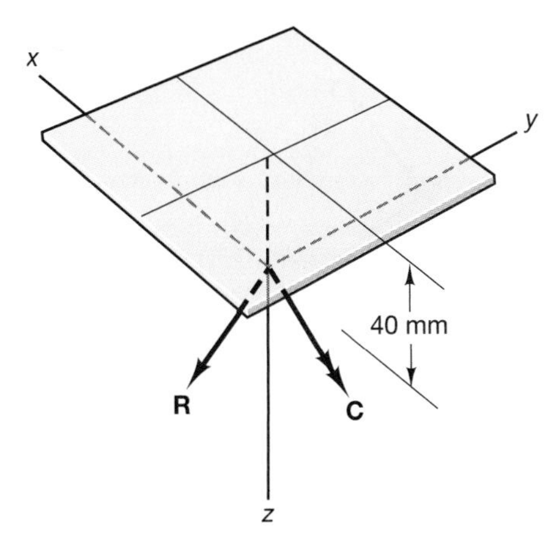

Suppose the force plate measures a resultant force and a moment of

$$\mathbf{R} = 50\hat{\mathbf{i}} + 150\hat{\mathbf{j}} + 800\hat{\mathbf{k}} \text{ (N)}$$

$$\mathbf{M}_O = 80\hat{\mathbf{i}} + 10\hat{\mathbf{j}} + 10\hat{\mathbf{k}} \text{ (N} \cdot \text{m)}$$

First, determine the wrench force and moment (called the ground reaction torque in this application). Then determine the intercept of the wrench axis with the surface of the force plate. (This intercept point, called the "center of pressure" or "center of force," is the point at which the force is concentrated between the foot and the plate.)

Solution

We first determine a unit vector in the **R** direction:

$$\hat{\mathbf{e}}_\mathbf{R} = \frac{\mathbf{R}}{|\mathbf{R}|} = \frac{50\hat{\mathbf{i}} + 150\hat{\mathbf{j}} + 800\hat{\mathbf{k}}}{815.5}$$

$$\hat{\mathbf{e}}_\mathbf{R} = 0.061\hat{\mathbf{i}} + 0.184\hat{\mathbf{j}} + 0.981\hat{\mathbf{k}}$$

The component of the couple parallel to the reaction (the ground reaction torque) is

$$\mathbf{T} = \mathbf{C}_\perp = (\mathbf{M}_O \cdot \hat{\mathbf{e}}_R)\hat{\mathbf{e}}_R = 16.53\hat{\mathbf{e}}_R = 1.01\hat{\mathbf{i}} + 3.04\hat{\mathbf{j}} + 16.33\hat{\mathbf{k}} \text{ (N} \cdot \text{m)}$$

The perpendicular component of the moment $\mathbf{M}_O$ is

$$\mathbf{C}_\parallel = \mathbf{M}_O - \mathbf{T} = 78.99\hat{\mathbf{i}} + 6.96\hat{\mathbf{j}} - 6.22\hat{\mathbf{k}}$$

The position vector from the origin to the wrench axis intercept with the surface of the force plate (center of pressure) is

$$\mathbf{r} = x\hat{\mathbf{i}} + y\hat{\mathbf{j}} - 0.04\hat{\mathbf{k}}$$

and is determined by

$$\mathbf{r} \times \mathbf{R} = \mathbf{C}_{\perp}$$

$$(x\hat{\mathbf{i}} + y\hat{\mathbf{j}} - 0.04\hat{\mathbf{k}}) \times (50\hat{\mathbf{i}} + 150\hat{\mathbf{j}} + 800\hat{\mathbf{k}}) = (78.99\hat{\mathbf{i}} + 6.96\hat{\mathbf{j}} - 6.22\hat{\mathbf{k}})$$

$$x = -0.011\text{m} \qquad y = 0.091 \text{ m}$$

The equivalent force system in the form of a wrench at the surface of the force plate is as follows:

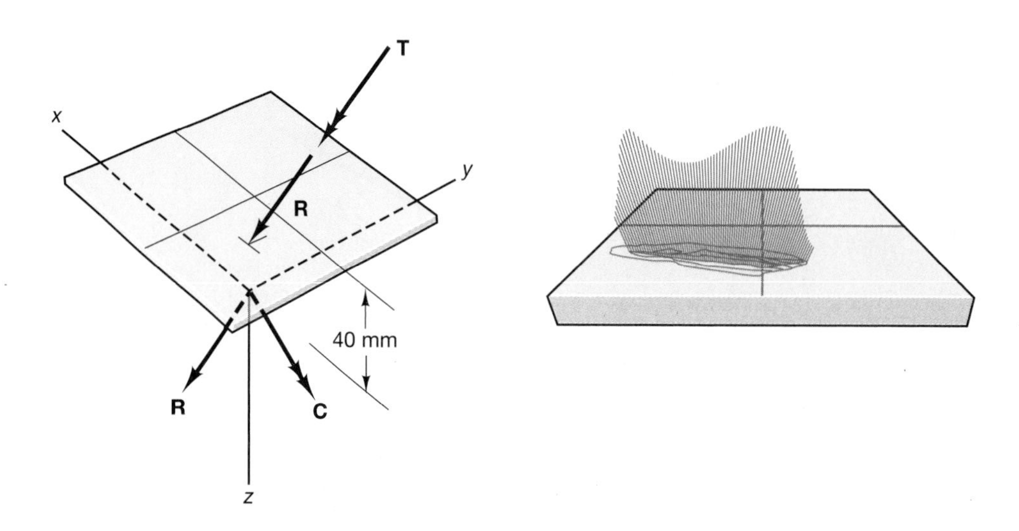

This information is interpreted to be the point of application of the ground reaction force on the plantar surface (the bottom) of the foot and may be used to calculate the loading at the ankle, knee, and hip joints. The wrench calculation involves many vector calculations, which are shown in the Computational Supplement.

In a gait analysis—that is, an analysis of the forces acting on the body when a person is walking—the wrench intercept is computed every 10 ms during contact of the foot with the plate. An actual image pattern of a center-of-pressure plot from gait analysis is shown in the above diagram on the right.

Sample Problem 4.17

It is known that only a moment in the z-direction can be generated on the surface of a force plate in a gait analysis laboratory. The force and moment at the instrument's center are, respectively,

$$\mathbf{R} = 50\hat{\mathbf{i}} - 150\hat{\mathbf{j}} + 450\hat{\mathbf{k}} \text{ N}$$

$$\mathbf{M}_O = 13.5\hat{\mathbf{i}} + 9\hat{\mathbf{j}} - 6\hat{\mathbf{k}} \text{ N} \cdot \text{m}$$

Determine an equivalent force system composed of the resultant force and a moment in the z-direction. Also, determine a perpendicular vector from the center of the instrument to the line of action of the resultant in the equivalent system.

Solution The magnitude of the moment in the z-direction is found from Eq. (4.60), where

$$\hat{\mathbf{n}} = \hat{\mathbf{k}}$$

Therefore,

$$T = \frac{\mathbf{R} \cdot \mathbf{M}_O}{\mathbf{R} \cdot \hat{\mathbf{n}}}$$

$$T = (50 \times 13.5 - 150 \times 9 - 450 \times 6)/450 = -7.5 \text{ N} \cdot \text{m}$$

From Eq. (4.62), the perpendicular from O to the line of action of $\mathbf{R}$ is

$$\mathbf{P}_O = \frac{\mathbf{R} \times (\mathbf{M}_O - T\hat{\mathbf{n}})}{\mathbf{R} \cdot \mathbf{R}}$$

$$\mathbf{p}_O = -0.019\hat{\mathbf{i}} + 0.026\hat{\mathbf{j}} + 0.011\hat{\mathbf{k}} \text{ m}$$

The solution using software is shown in the Computational Supplement.

Problems

4.66 Two forces are applied as shown in Figure P4.66 to the top and base of a laptop computer. (a) Calculate the resultant force system at the hinge, denoted as point A, for $\theta = 30°$. (b) Recompute the couple for other angles: $\theta = 10°, 20°, 40°, 50°, \ldots, 90°$.

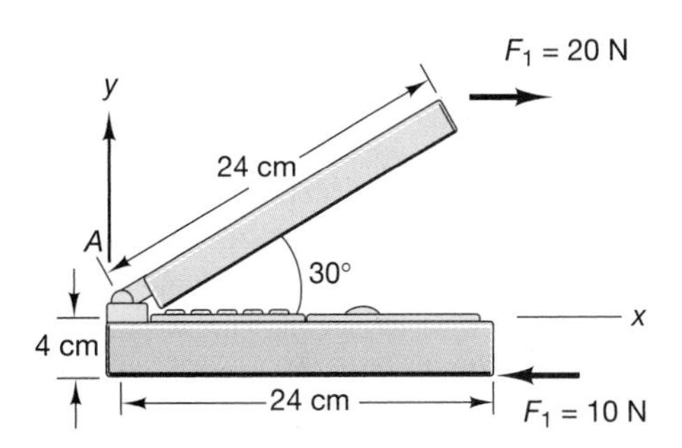

Figure P4.66

4.67 Four forces act on a cross-sectional piece of a truck frame. (See Figure P4.67.) Calculate the resultant force and moment at point A, and find a point B somewhere along the x-axis where the resultant alone will represent the effect of all four forces. Here, let $F_1 = 10$ kN, $F_2 = 10$ kN, $F_3 = 3$ kN, and $F_4 = 4$ kN.

4.68 Consider the system of Problem 4.67, and calculate the resultant and moment about point O. Find a point B somewhere along the x-axis where the resultant force alone represents the result of all four forces.

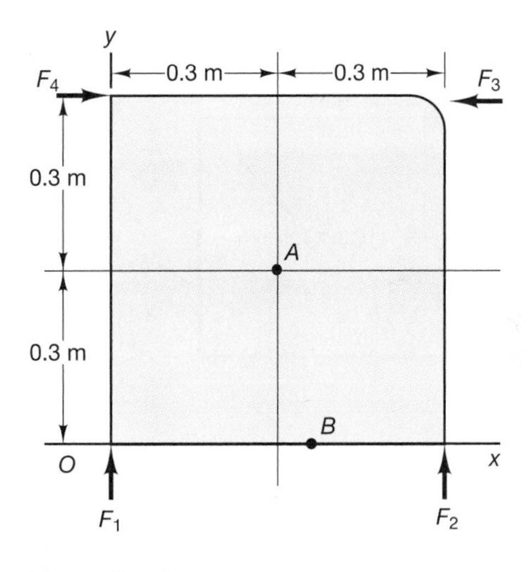

Figure P4.67

4.69 Consider the system of Problem 4.67, and compute the position where the resultant alone represents the system and the position vector from the origin to the tail of $\mathbf{R}$ is perpendicular to $\mathbf{R}$. (That is, use the direct vector solution.)

4.70 Calculate the center of gravity (which acts down along z) of the machine part shown in Figure P4.70 (on the next page) by using the direct vector approach. The position of the weight distribution of each part is given relative to the origin in the x–y plane.

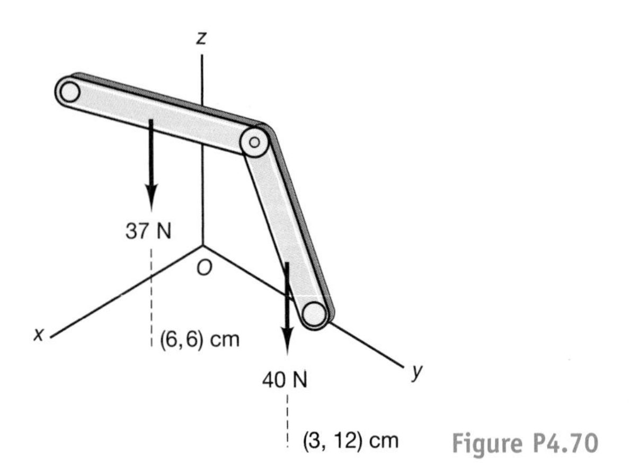

Figure P4.70

4.71 A shipping crate has four forces and a couple applied to it. (See Figure P4.71.) Calculate the resultant force and moment about the origin; then calculate a point at which the resultant force alone will produce the equivalent effect on the rigid body.

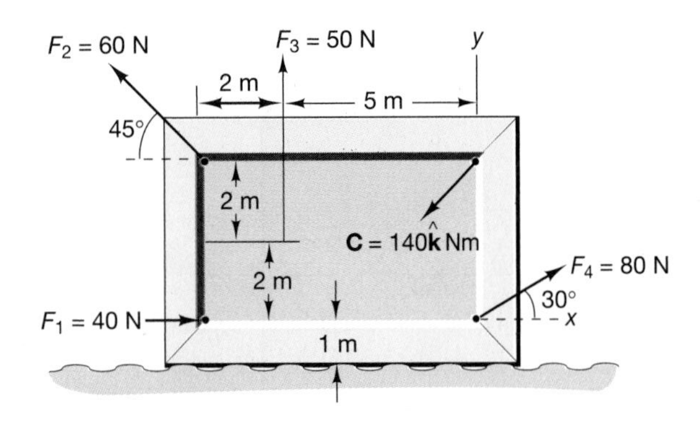

Figure P4.71

4.72 Two workers are pushing and pulling a box, as shown in Figure P4.72. The force $\mathbf{F}_1$ is 1000 N at 45° from x, and the force $\mathbf{F}_2$ is 500 N, 30° up from the x-axis. (a) Calculate the resultant at point O. (b) Compute a point A at which the resultant force alone produces an equivalent effect.

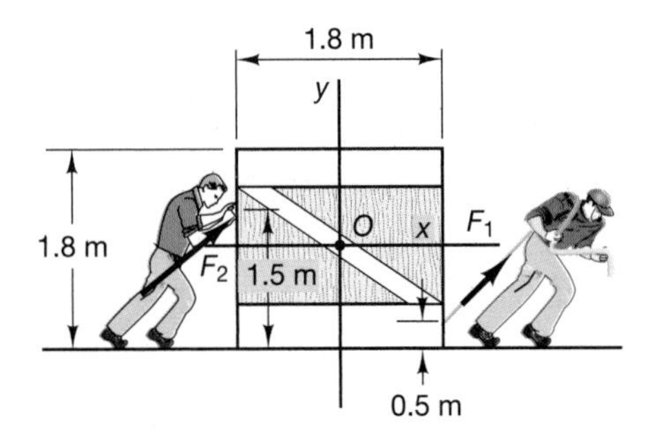

Figure P4.72

4.73 Three loads are applied to a bridge truss section. (See Figure P4.73.) (a) Calculate the equivalent force system (i.e., calculate the resultant and a point at which it should act). Here, $\mathbf{F}_1 = -25\hat{\mathbf{i}}$ kN, $\mathbf{F}_2 = 10\hat{\mathbf{i}} - 17.32\hat{\mathbf{j}}$ kN, and $\mathbf{F}_3 = -30\hat{\mathbf{j}}$ kN. (b) Calculate the equation of the line determined by $\mathbf{R}$ and the point of application calculated in (a), and use this equation to find a point on the structure where the resultant can be applied to produce an equivalent system.

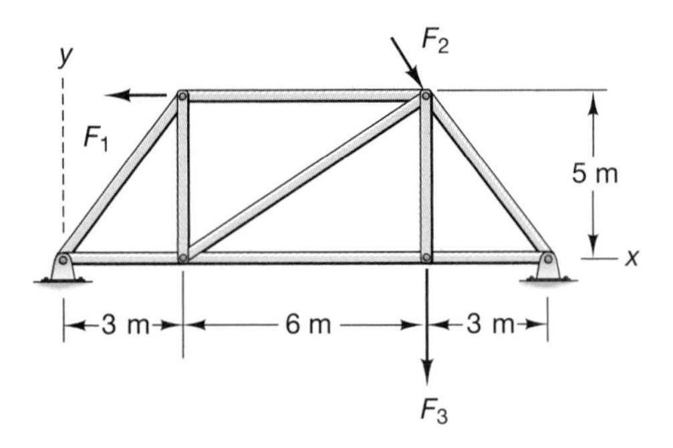

Figure P4.73

4.74 Given vectors $\mathbf{R} = 500\hat{\mathbf{i}} + 250\hat{\mathbf{j}} - 100\hat{\mathbf{k}}$ N and $\mathbf{C} = 10\hat{\mathbf{i}} + 40\hat{\mathbf{j}} + 30\hat{\mathbf{k}}$ Nm, calculate the components of vector $\mathbf{C}$ parallel to and perpendicular to $\mathbf{R}$.

4.75 Given vectors $\mathbf{R} = 1000\hat{\mathbf{i}} - 270\hat{\mathbf{j}} + 500\hat{\mathbf{k}}$ N and $\mathbf{C} = 50\hat{\mathbf{i}} + 50\hat{\mathbf{j}} - 50\hat{\mathbf{k}}$ N · m, compute the components of vector $\mathbf{C}$ parallel to and perpendicular to $\mathbf{R}$.

4.76 Three forces are applied to a rigid body, as indicated in Figure P4.76. Calculate the resultant force and moment about the origin, and resolve the moment into components parallel and perpendicular to the resultant. Use this information to calculate the wrench. Let $\mathbf{F}_1 = \mathbf{F}_2 = -10\hat{\mathbf{i}} + 20\hat{\mathbf{j}}$ N and $\mathbf{F}_3 = 30\hat{\mathbf{i}} + 30\hat{\mathbf{j}} + 30\hat{\mathbf{k}}$ N.

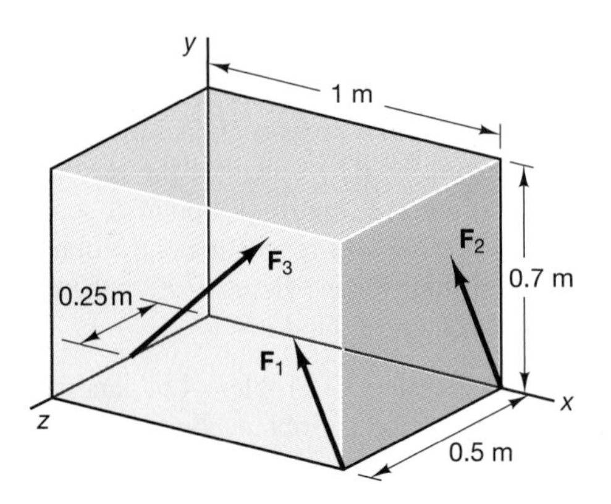

Figure P4.76

4.77 In Figure P4.77, a person standing on the deck and leaning on the railing exerts the forces $\mathbf{F}_1 = -750\hat{\mathbf{j}}$ N, $\mathbf{F}_2 = 40\hat{\mathbf{i}} - 40\hat{\mathbf{j}} + 40\hat{\mathbf{k}}$ N, and $\mathbf{F}_3 = -20\hat{\mathbf{i}} - 35\hat{\mathbf{j}} + 60\hat{\mathbf{k}}$ N. Calculate the resultant wrench, and find the vector from the origin to the line of action of the wrench.

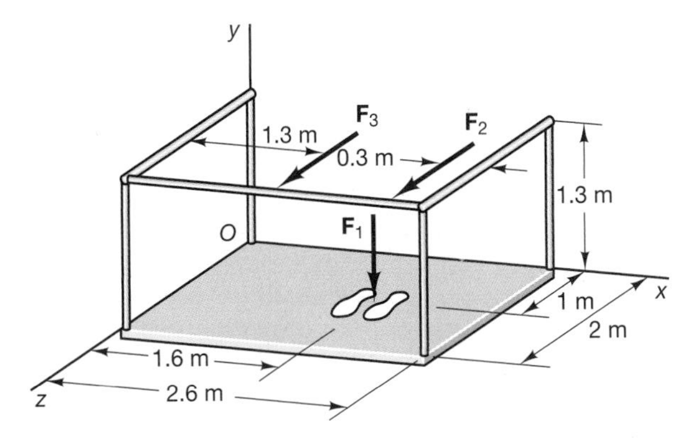

Figure P4.77

4.78 Three forces act on a mounting bracket, as shown in Figure P4.78. For $\mathbf{F}_1 = -10\hat{\mathbf{i}} - 3\hat{\mathbf{j}} - 4\hat{\mathbf{k}}$ N, $\mathbf{F}_2 = 10\hat{\mathbf{i}} + 10\hat{\mathbf{j}} + 10\hat{\mathbf{k}}$ N, and $\mathbf{F}_3 = 10\hat{\mathbf{i}} + 10\hat{\mathbf{j}}$ N, calculate the wrench and the vector $\mathbf{r}$ from the origin to the wrench that is perpendicular to the wrench axis.

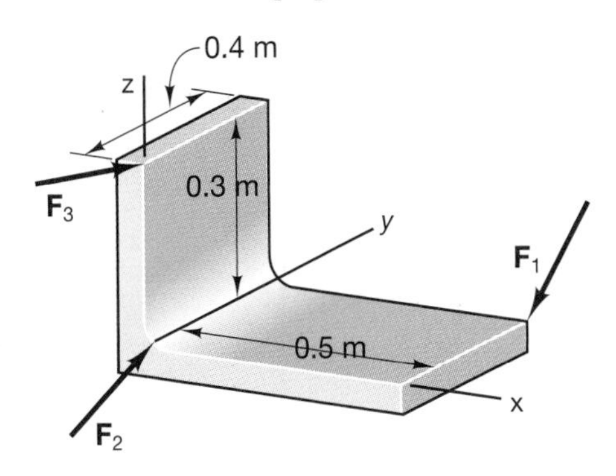

Figure P4.78

4.79 Consider a resultant $\mathbf{R} = 3\hat{\mathbf{i}} + 6\hat{\mathbf{j}} + 2\hat{\mathbf{k}}$ N and couple $\mathbf{M} = 12\hat{\mathbf{i}} + 4\hat{\mathbf{j}} + 6\hat{\mathbf{k}}$ N · m. Determine the equivalent wrench and the vector $\mathbf{r}$ that extends from the origin to the wrench axis and is perpendicular to the wrench axis.

4.80 Consider a resultant force $\mathbf{F} = 2587\hat{\mathbf{i}} - 1232\hat{\mathbf{j}} - 500\hat{\mathbf{k}}$ N acting at the origin and a resultant moment given by $\mathbf{C} = 2700\hat{\mathbf{i}} + 300\hat{\mathbf{j}} + 100\hat{\mathbf{k}}$ N · m. Determine the equivalent wrench and the vector $\mathbf{r}$ that extends from the origin to the wrench axis that is perpendicular to the wrench.

4.81 Calculate the position vector $\mathbf{p}_O$ and the moment $\mathbf{T}$ that locates the resultant vector $\mathbf{R} = 25\hat{\mathbf{i}} - 10\hat{\mathbf{j}} + 3\hat{\mathbf{k}}$ N such that the system described by $\mathbf{R}$ and the moment $\mathbf{M}_O = 5\hat{\mathbf{i}} + 5\hat{\mathbf{j}} + 5\hat{\mathbf{k}}$ N · m is represented by $\mathbf{R}$ at $\mathbf{r}$ with moment totally in the $\hat{\mathbf{i}}$ direction.

4.82 A lever mechanism has a resultant force $\mathbf{R} = 25\hat{\mathbf{i}} + 5\hat{\mathbf{j}} + 3\hat{\mathbf{k}}$ N and moment $\mathbf{M}_O = -10\hat{\mathbf{i}} + 27\hat{\mathbf{k}}$ N · m. Determine the position of $\mathbf{R}$ and a value of the moment $\mathbf{T}$ such that $\mathbf{R}$ placed at the position described by $\mathbf{p}_O$ produces a net moment in the z-direction. (See Figure P4.82.)

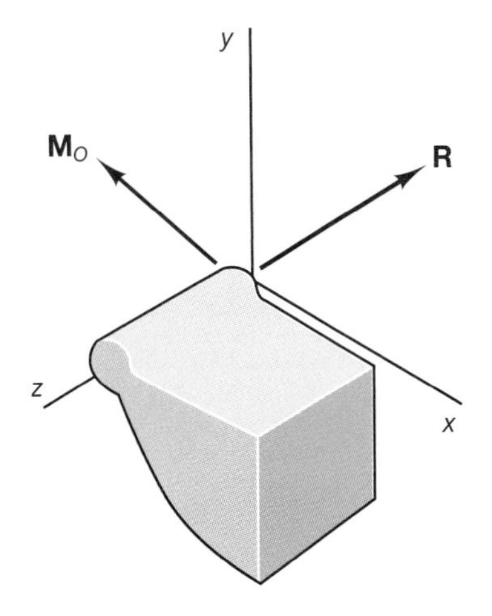

Figure P4.82

4.83 The mechanism in Figure 4.82 has a resultant force of $\mathbf{R} = 10\hat{\mathbf{i}} + 10\hat{\mathbf{j}} + 10\hat{\mathbf{k}}$ N and a moment of $\mathbf{M}_O = -5\hat{\mathbf{i}} - 5\hat{\mathbf{j}} - \mathbf{k}$ N · m applied to it. Compute the position of $\mathbf{R}$ and a value of the moment $\mathbf{T}$ such that $\mathbf{R}$ placed at the position described by the vector $\mathbf{p}_O$ produces a net moment at a 45° angle in the x–y plane.

4.84 Consider the resultant force $\mathbf{R} = 10\hat{\mathbf{i}}$ N and moment $\mathbf{M}_O = 5\hat{\mathbf{i}}$ N · m, shown in Figure P4.84. Calculate the position of $\mathbf{R}$ and the moment $\mathbf{T}$ such that $\mathbf{T}$ is applied along the line a in the figure.

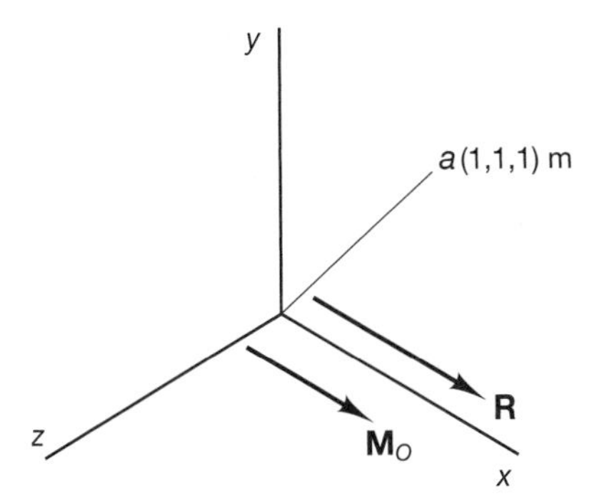

Figure P4.84

Chapter Summary

This chapter discussed the forces and moments that act on rigid bodies. The point of application of the force acting on a rigid body will change the moment (turning effect) of that force about different points on the body. The moment was defined by the cross product of a position vector from the point of interest to the point of application of the force with the force vector ($\mathbf{M} = \mathbf{r} \times \mathbf{F}$). The moment of a force about an axis in space was defined vectorially. A couple was defined as two noncollinear parallel forces that are equal in magnitude, opposite in direction, and separated by a distance d.

Using the concept of the couple, equivalent force systems were defined. The most general equivalent force system is the wrench and applications of the wrench were shown for gait analysis force plates.

Chapter 5

DISTRIBUTED FORCES: CENTROIDS AND CENTER OF GRAVITY

The air bag distributes the forces acting on a vehicle's occupants during a traffic accident. (Photo courtesy of Richard Olivier/Corbis Canada)

5.1 INTRODUCTION

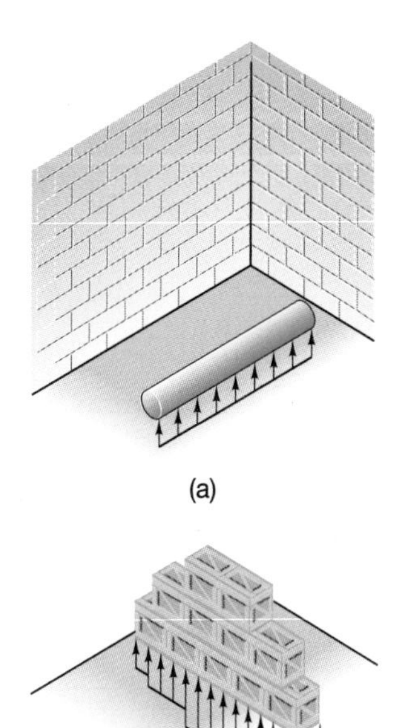

Figure 5.1

In the previous chapters, we treated forces as ***concentrated forces*** applied to a particular point on a body and having a specific line of action. These forces could then be represented by a single vector with magnitude, direction, sense, and a specified line of action or point of application. However, a concentrated force is really only a mathematical simplification, similar to modeling an object as a particle of zero size. A concentrated force would have to be distributed across an area of zero in the form of some distributed pressure. The product of that pressure with the area of zero would be equal to the magnitude of the concentrated force. The only pressure that would satisfy this condition would be of infinite value. Therefore, a concentrated force is, ideally, an infinite pressure acting on a zero area. We will investigate mathematically such a concept in Chapter 8 when the Dirac delta function is introduced.

In many applications, we cannot assume the force to be concentrated, but must consider it as distributed along a line or over a surface. Examples of ***distributed forces*** are shown in Figure 5.1. The force in Figure 5.1(a) is distributed uniformly along the line of contact between the cylinder and the flat surface (such as a water pipe lying on a roof). In Figure 5.1(b), the crates on the loading dock distribute the force nonuniformly across that surface in an unequal manner.

In any particular problem, a force may be modeled as distributed or concentrated, depending upon the application and the required analysis. This choice is usually based upon a comparison of the size of the area over which the force is distributed with the dimensions of the body. For example, in Figure 5.2, the forces between the table legs and the floor are distributed across the ends of each leg. However, in analyzing the entire table, we may consider the forces as concentrated without loss of accuracy.

A distributed force is specified by its magnitude per unit length or per unit area, together with its direction, similar to the manner of specifying a concentrated force. The distributed force may be normal (perpendicular) to the surface and, if caused by gases or fluids, is then called the pressure. Distributed forces may also act tangent to a surface, as in the case of friction or shear. The distributed forces internal to a body are called stresses and are studied in courses in mechanics of materials.

Distributed forces along a line have units of force per unit length (N/m or lb/ft), and distributed forces over an area have units of force per unit area (N/m^2 or lb/in^2).

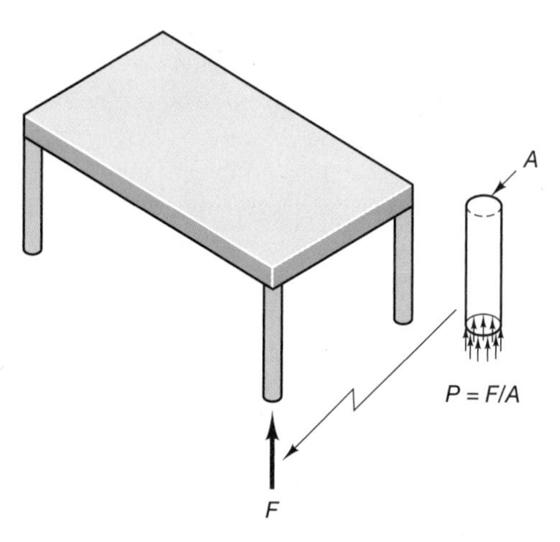

Figure 5.2

Figure 5.3

The SI units N/m^2 are called ***pascals*** (abbreviated as Pa). The U.S. customary units lb/in^2 are frequently written as psi. Pressures are also specified in millimeters of mercury or atmospheres:

$$1 \text{ mm Hg} = 0.019 \text{ psi} = 133 \text{ N/m}^2\text{(Pa)}$$

$$1 \text{ atm} = 14.7 \text{ psi} = 760 \text{ mm Hg} = 1.01 \times 10^5 \text{ Pa}$$

One atmosphere is the pressure exerted on the surface of the Earth at sea level by the air or the atmosphere.

A common example of a distributed force is the force between the tire of a car and the ground, as shown in Figure 5.3. The portion of the car's weight carried by that particular tire is equal to the tire pressure times the area of contact of the tire with the road surface:

$$W = pA$$

The area of contact can vary, depending upon the type of tire and the particular tread, as well as on the pressure to which the tire is inflated. Using the preceding expression, we can determine the weight of the car by measuring the surface area A and the tire pressure p. In the past, this technique was sometimes used by Department of Transportation officers to determine the weight of trucks using back roads where there were no weigh stations.

Other forces distributed over the volume of a body are called ***body forces***. The most common body force is the force of gravitational attraction, but body forces may also arise from magnetic or electrical effects. The gravitational attraction of a body on the surface of the Earth is determined by Newton's law of gravitational attraction. A small mass on the surface of the Earth is shown in Figure 5.4. The gravitational force is

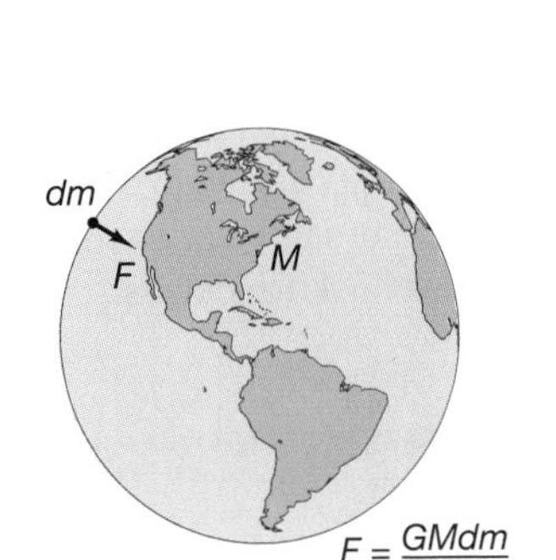

Figure 5.4

$$F = \frac{GMdm}{R^2} = gdm$$

where G is the universal gravitational constant, M is the mass of the Earth, R is the radius of the Earth, dm is a differential element of mass, and g, called the gravitational acceleration, is equal to

$$g = \frac{GM}{R^2} = 9.807 \text{ N/kg} = 32.17 \text{ lb/slug}$$

which is the force per unit mass caused by the gravitational attraction of the Earth. The constant g, has the same units as the units of acceleration—that is, m/s^2 or ft/s^2. It was first introduced in Chapter 1, and will be used here to define the center of gravity of a body.

In applications that use a rigid-body model, we can replace the distributed body force of gravitational attraction with a concentrated force acting through the center of gravity. Other distributed forces, such as the contact pressure between the cylinder and the roof and between the crates and the loading dock in Figure 5.1, can be replaced by a concentrated force. The location and magnitude of the concentrated force is determined by creating an equivalent force system that has the same resultant and moment as the distributed force system.

In Chapter 4, we discussed the concept of the moment of a force as a measure of the turning effect of a force about a point or line. The moment about a point or line was defined to be the vector product of a position vector from the point or line to the force's line of action and the force itself. A similar position vector is mathematically useful in describing the position of the geometric center of a line, area, or volume in space. There are an infinite number of points on a line or in an area or volume. We will determine the geometric center of a line, area, or volume, called the ***centroid***, using the concept of a moment of a line, area, or volume. These moments are frequently called the *first moment* and are linearly dependent upon the distance from a point or line of reference to an element of the line, area or volume. If the mass density is constant in a body modeled as a line, a two-dimensional area, or a volume, the center of gravity coincides with the centroid. Later in the text, second moments of areas will be examined that depend upon the aforementioned distance squared. This analogy with the moment of a force, although not involving cross products, is still very useful in determining the center of mass of a body or the centroid of a line, area, or volume.

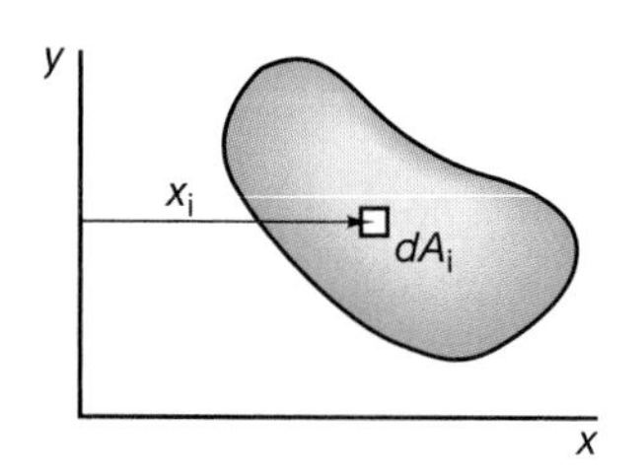

Figure 5.5

For example, the first moment of an area about the y-axis is defined as the sum of the moments of each differential element of the area about the y-axis, as shown in Figure 5.5. Using the standard notation for a differential area from calculus, we see that the moment of the ith element is

$$dM_i = x_i\, dA_i \tag{5.1}$$

The first moment of the area about the y-axis is written as an integral over an area:

$$M_y = \int_A x dA \tag{5.2}$$

The first moment of a line and of a volume will be defined in a similar manner. The first moment of the mass distribution will also be used to determine a particular point in the body called the *center of mass,* as discussed in the next section.

The concepts of centroid, center of mass, and center of gravity discussed in the sections that follow depend heavily on integral calculus. In particular, recall that integration is defined in terms of a limit of a summation as the size of the elements being summed approaches zero. It is important to remember the methods of integral calculus and, in particular, the use of differential or infinitesimal elements of length (ds), area ($dx\, dy$), and volume ($dx\, dy\, dz$). Centroids, the center of mass, and the center of gravity are usually discussed as examples of multiple integration; you may wish to consult a calculus book for details.

5.2 CENTER OF MASS AND CENTER OF GRAVITY

5.2.1 CENTER OF MASS

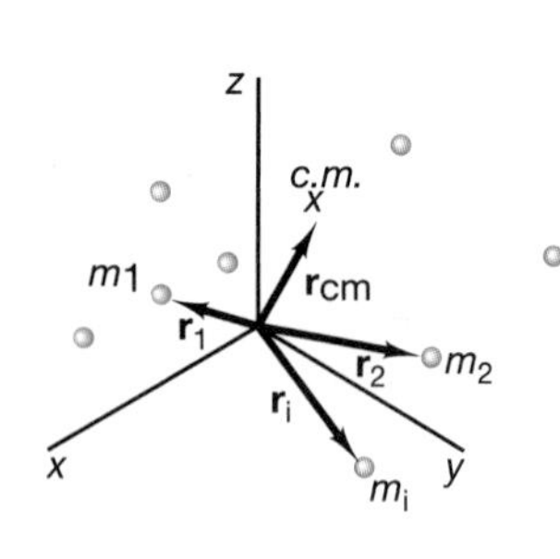

Figure 5.6

Consider a system of mass particles distributed in space, as shown in Figure 5.6. Such a system can often be treated as equivalent to a lumped mass concentrated at a specific point for some dynamic analyses. The ***center of mass*** is defined to be the point in space where the mass of all the particles can be considered to be concentrated. Let

$$M = \sum_i m_i \tag{5.3}$$

be the total mass of all the particles. The position vector from the origin to the center of mass is defined as

$$\mathbf{r}_{c.m.} = \frac{\sum_i m_i \mathbf{r}_i}{M} \tag{5.4}$$

The position vector to the center of mass can be considered to be the weighted average of the individual position vectors $\mathbf{r}_i$. The vector equation (5.4) can be written in terms of its scalar components as

$$x_{c.m.} = \frac{\sum_i m_i x_i}{M} \qquad y_{c.m.} = \frac{\sum_i m_i y_i}{M} \qquad z_{c.m.} = \frac{\sum_i m_i z_i}{M} \tag{5.5}$$

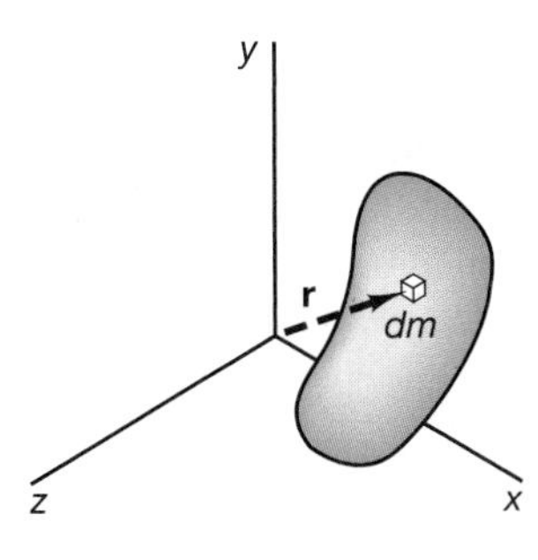

Figure 5.7

If the mass is continuously distributed in a rigid body, instead of being distributed as a system of mass particles, the summations become integrals. Then the total mass and center of mass for the body, shown in Figure 5.7, become

$$\begin{aligned} M &= \int_m dm \\ \mathbf{r}_{c.m.} &= \frac{1}{M}\int_m \mathbf{r}\, dm \end{aligned} \tag{5.6}$$

where dm denotes a differential element of distributed mass. If the body has a uniform mass density ρ, the differential mass element dm is the mass contained in the differential element of volume dV, so that

$$dm = \rho\, dV$$

The vector $\mathbf{r}$ is the vector from the origin to the element of mass dm. The corresponding scalar equations for the position of the center of mass are

$$x_{c.m.} = \frac{1}{M}\int_m xdm \quad y_{c.m.} = \frac{1}{M}\int_m ydm \quad z_{c.m.} = \frac{1}{M}\int_m zdm \tag{5.7}$$

The element of mass dm may be written in terms of the mass density r and a differential volume element dV as

$$\begin{aligned} dm &= \rho dV \\ M &= \int_V \rho dV \\ \mathbf{r}_{c.m.} &= \frac{1}{M}\int_V \rho \mathbf{r}\, dV \end{aligned} \tag{5.8}$$

If the mass density does not vary over the volume, it is a constant and may be taken outside the integral. Thus, for a constant mass density, the total mass and the center of mass become, respectively,

$$\begin{aligned} M &= \rho\int_V dV = \rho V \\ \mathbf{r}_{c.m.} &= \frac{\rho}{M}\int_V \mathbf{r}\, dV = \frac{\rho}{\rho V}\int_V \mathbf{r}\, dV = \frac{1}{V}\int_V \mathbf{r}\, dV \end{aligned} \tag{5.9}$$

It will be shown that the last integral defines the *centroid* of the volume. Therefore, if the density is constant in a rigid body, the center of mass coincides with the centroid of the volume.

5.2.2 CENTER OF GRAVITY

The weight of a body resting on the Earth's surface is due to the gravitational attraction between the Earth and the mass of each part of the body. This attraction, therefore, produces body forces distributed throughout the body. If the body is small in comparison to the Earth's radius, the distance from each element of mass to the center of the Earth is approximately the same value. Therefore, the magnitude of the body force on each differential element of mass dm is

$$dF = gdm \tag{5.10}$$

where g is the gravitational acceleration, as defined in Chapter 1.

All these attraction forces acting on the body are concurrent at the center of the Earth, but since the radius of the Earth is large compared to the dimensions of the body, the forces can be considered to be parallel. Hence, the gravitational body forces form a parallel force system and may be replaced by an equivalent force system consisting of a single resultant force with a unique line of action. This equivalent resultant force, **R**, is equal to the sum of all the parallel forces, as was shown in Chapter 4. Since the body forces act on each element of mass *dm*, the sum of all the forces may be represented as an integral, viz.,

$$\mathbf{R} = \int_m g\, dm\, \hat{\mathbf{e}}_R \tag{5.11}$$

where $\hat{\mathbf{e}}_R$ is a unit vector directed toward the center of the Earth. This resultant force is the *weight of the body*. Although the body has mass, it has weight only because of gravitational attraction.

The ***center of gravity*** is defined as the point in the body at which the total weight is considered to be concentrated. If the body forces due to gravitational attraction are considered as a parallel force system, the methods shown in Chapter 4 (Eqs. 4.50–56) may be used as a guide to determine the center of gravity. Since the mass is continuously distributed, the summations in those equations are replaced by integrals. The moment about the origin of the weight of the body is equal to the integral of moments of all of the gravitational forces due to each differential element of mass. *The position vector to the center of gravity can be expressed as*:

$$\mathbf{r}_{c.g.} \times \mathbf{R} = \int_m \mathbf{r} \times (gdm)\hat{\mathbf{e}}_R$$

If $\hat{\mathbf{e}}_R$ is considered to be in the $-z$ direction, then

$$\hat{\mathbf{e}}_R = -\hat{\mathbf{k}} \tag{5.12}$$

and

$$x_{c.g.}\int_m gdm = \int_m xgdm$$

$$y_{c.g.}\int_m gdm = \int_m ygdm$$

Therefore, since g is considered constant across a small body on the surface of the Earth,

$$x_{c.g.} = \frac{\int xdm}{\int dm} = \frac{1}{M}\int xdm$$

$$y_{c.g.} = \frac{\int ydm}{\int dm} = \frac{1}{M}\int ydm \tag{5.13}$$

The choice of coordinate directions is arbitrary, and it is easily seen that

$$z_{c.g.} = \frac{1}{M}\int_m zdm \tag{5.14}$$

The center of gravity of the body is located at the same point as the center of mass in this case. Therefore, the point may be referred to as either the center of mass or the center of gravity.

Table 5.1

	Mass	Radius
Earth	597.6×10^{22} kg	6.371×10^6 m
Moon	7.35×10^{22}	1.738×10^6 m

Although the mass of the object is constant, its weight will depend upon which planet or moon the object is on and its height above the center of the planet or moon. For example, the mass and the radius of the moon are different from those of the Earth, as shown in Table 5.1. Using the values given in the table and the value of G, we find that the gravitational acceleration constant g on the Earth is

$$g_{\text{earth}} = 9.81 \text{ m/s}^2$$

and on the moon is

$$g_{\text{moon}} = 1.62 \text{ m/s}^2$$

Therefore, the weight of an object on the surface of the moon would be only 16.5% of its weight on the surface of the Earth. Thus, a 200-lb (91 kgf) man taking a "moon walk" would weigh only 33 lb (15 kgf), which explains his ability to make large leaps across the surface of the moon. His muscles have developed here on Earth to allow him to move against a much higher gravitational attraction. In this regard, there is some concern that, for extended periods in space, astronauts' muscles may weaken and their joints loosen. In fact, astronauts gain height owing to a decrease in pressure on the disks of the spine. The condition of low gravity—microgravity—also may causes heart weakness and loss of balance. A person's balance is maintained, in part, by the inner ear, or the vestibular system sensing head accelerations. In 1991, NASA flew a shuttle mission called Microgravity Vestibular Investigations (MVI) to study these effects. (See Figure 5.8.)

Figure 5.8 NASA logo for the MVI mission. (Courtesy of NASA)

5.3 AVERAGE POSITION: CENTROIDS OF AREAS, VOLUMES, AND LINES; THE FIRST MOMENT

5.3.1 CENTROID OF AN AREA

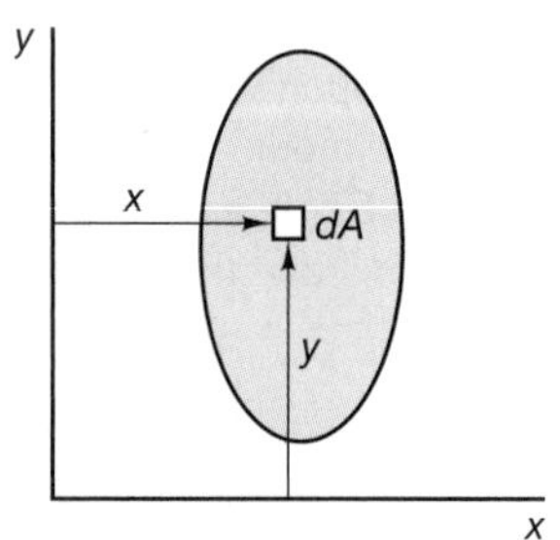

Figure 5.9

The centroid of an area A is the geometric center of the area. Its position can be determined by averaging the first moment of the area over the entire area, as shown in Figure 5.9.

The centroid of the area is located relative to the x and y reference axes by the equations

$$x_c = \frac{1}{A}\int_A x dA$$
$$y_c = \frac{1}{A}\int_A y dA \tag{5.15}$$

where

$$A = \int_A dA$$

is the area.

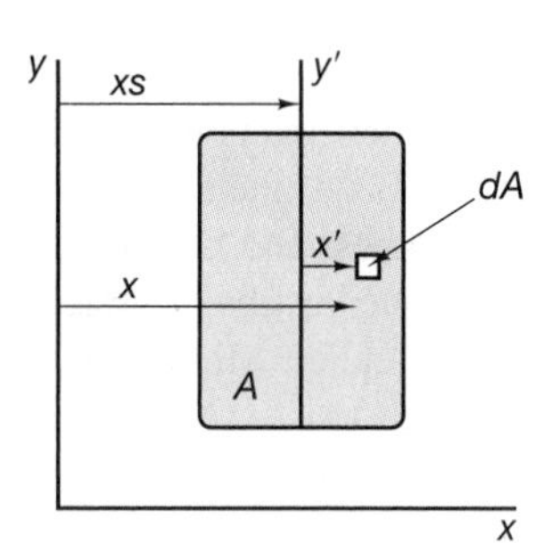

Figure 5.10

Axes of symmetry are always centroidal axes, as can be seen by examining the area in Figure 5.10. Let x and y be any reference axes in space, and let y' be an axis of symmetry for the area A. An axis of symmetry is a line that divides the area into two parts that have exactly the same geometry. Let xs be the distance from the reference axis to the axis of symmetry. The x-coordinate of an arbitrary differential element of area dA is related to xs and x' by the equation

$$x = xs + x'$$

If the variable x is replaced by the sum of the constant xs and the variable x', the distance to the centroid is

$$\begin{aligned} x_c &= \frac{1}{A}\int x dA \\ &= \frac{1}{A}\int_A (x' + xs)\, dA = \frac{1}{A}\int_A x' dA + \frac{xsA}{A} \\ &= \frac{1}{A}\int_A x' dA + xs \end{aligned}$$

Note, however, that the integral

$$\int_A x' dA$$

is the integral of an odd function between symmetrical limits and is always zero. (See Math Window 5.1.) Therefore,

$$x_c = xs$$

and the axis of symmetry is a *centroidal axis*—that is, an axis that passes through the centroid. Note that there are an infinite number of centroidal axes, but if an area has two or more axes of symmetry, the intersection of these axes is the centroid of the area.

MATHEMATICS WINDOW 5.1:

Properties of Even and Odd Functions An even function of the variable x satisfies the equation

$$f_e(-x) = f_e(x)$$

and an odd function satisfies the condition

$$f_o(-x) = -f_o(x)$$

Examples of even functions are

$$x^2, \cos x, x \sin x, x^{2n}$$

Examples of odd functions are

$$x, \sin x, x \cos x, x^{\text{odd power}}$$

The integral of an even function results in an odd function, and the integral of an odd function results in an even function. In a similar manner, the differential of an even function is odd, and that of an odd function is even.

The integral of an even function across symmetrical limits is

$$\int_{-a}^{+a} f_e(x)dx = 2\int_0^a f_e(x)dx$$

The integral of an odd function across symmetrical limits is

$$\int_{-a}^{+a} f_o(x)dx = 0$$

Not all functions are even or odd; for example, $(x + x^2)$ is neither even nor odd. However, any function may be written as the sum of an even function and an odd function.

Let $g(x)$ be a function that is neither even nor odd. Then g may be written as

$$g(x) = \underbrace{\frac{g(x) + g(-x)}{2}}_{\text{even}} + \underbrace{\frac{g(x) - g(-x)}{2}}_{\text{odd}}$$

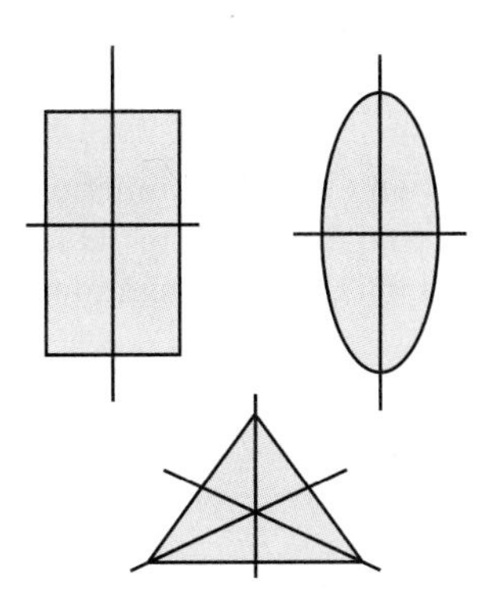

Figure 5.11

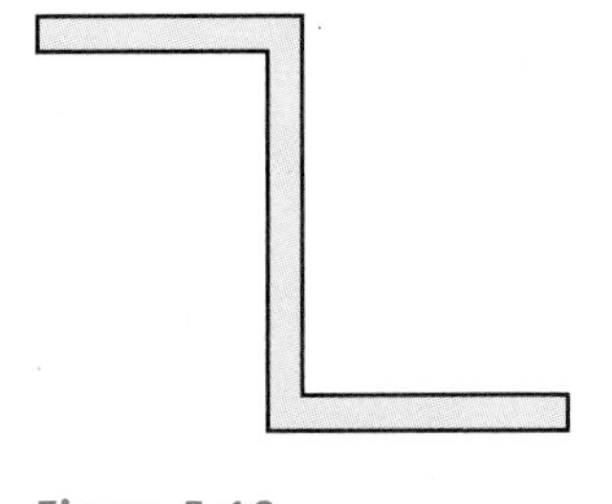

Figure 5.12

Frequently, the centroid of an area may be obtained by inspection and recognition of axes of symmetry. Examples are shown in Figure 5.11. Symmetry can sometimes be deceiving, however, as in the case of the area shown in Figure 5.12, which has no axis of symmetry.

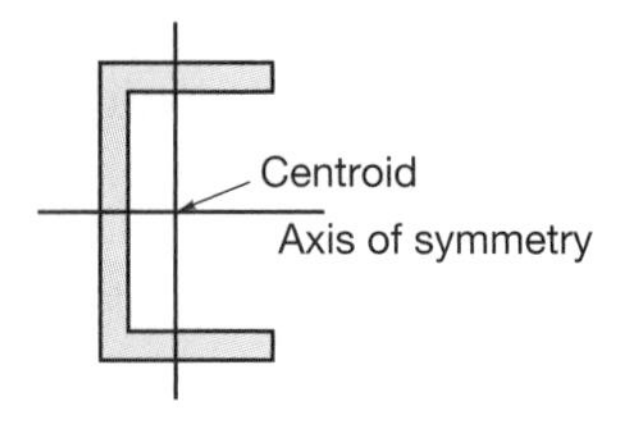

Figure 5.13

The centroid does not always lie within the area itself, as can be seen in the case of a channel area, as shown in Figure 5.13.

The centroid of an area is defined by Eq. (5.15) as an integral over the area. Integration of areas can be accomplished by the use of either single integrals or double integrals. Consider the triangular area shown in Figure 5.14. In Figure 5.14(a), a line element of integration has been chosen. By similar triangles, we see that

$$\frac{\Delta x}{h - y} = \frac{b}{h}$$

Since the area of the triangle is $bh/2$, the distance y_c from the x-axis to the centroid is

$$\left(\frac{1}{2}hb\right)y_c = \int_0^h y(xdy) = \int_0^h y\frac{b(h-y)}{h}dy$$

$$y_c = \frac{2}{hb}\left[\frac{b}{h}\left(\frac{hy^2}{2} - \frac{y^3}{3}\right)\right]_0^h = \frac{h}{3}$$

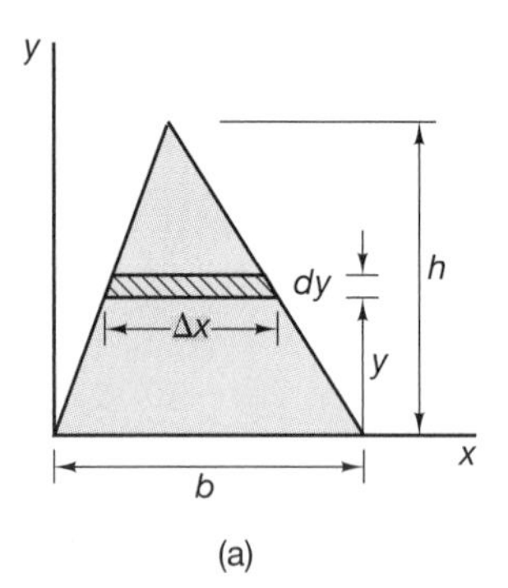

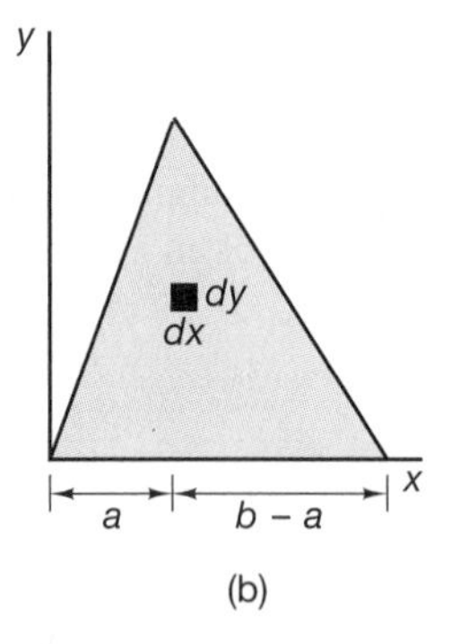

Figure 5.14 (a) First-order element of integration. (b) Second-order element of integration.

The centroid may also be obtained by double integration using Figure 5.14(b):

$$y_c = \frac{2}{bh}\int_0^h\left(\int_{\frac{ay}{h}}^{b-\frac{(b-a)y}{h}} dx\right)ydy = \frac{2}{bh}\int_0^h \frac{b(h-y)}{h}ydy$$

$$y_c = \frac{h}{3}$$

Most computational software packages will allow integration either by numerical methods or by analytical techniques, replacing the need to find integrals in tables of integration. The use of computational software is shown in the Computational Supplement.

Sample Problem 5.1

Find the centroid of the semicircular area shown at the left.

Solution

We will choose polar coordinates and use double integration to obtain the centroid. The y-axis is an axis of symmetry and is therefore a centroidal axis. The element of area in polar coordinates is $r\,dr\,d\theta$, and the area of the semicircle is

$$\int_0^\pi\int_0^R rdrd\theta = \pi R^2/2$$

The distance to the element of area is $y = r\sin\theta$, and the distance from the origin to the centroid on that axis is

$$y_c = \frac{1}{A}\int_A ydA = \frac{2}{\pi R^2}\int_0^\pi\int_0^R r\sin\theta(rdr\,d\theta) = \frac{2R}{3\pi}\int_0^\pi \sin\theta d\theta = \frac{4R}{3\pi}$$

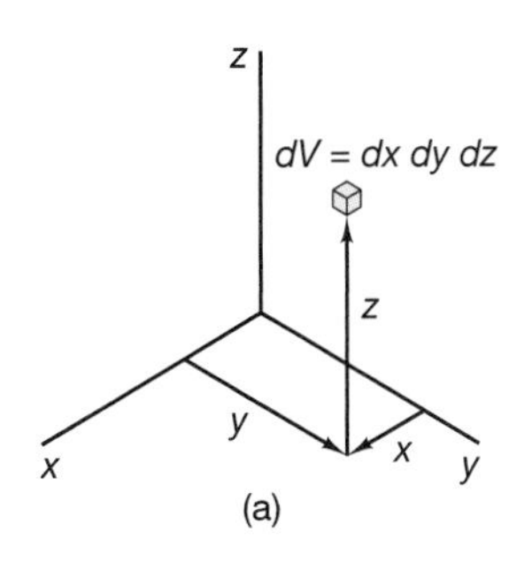

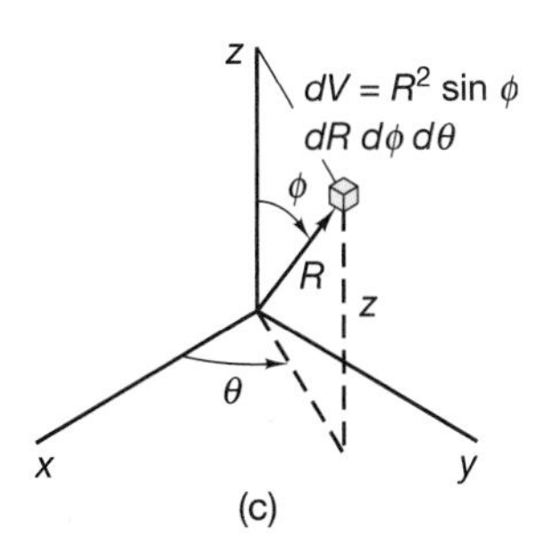

Figure 5.15 (a) Rectangular coordinates. (b) Cylindrical coordinates. (c) Spherical coordinates.

5.3.2 CENTROID OF A VOLUME

The centroid of a volume is defined as the geometric center of the volume. It can be obtained by considering the ***first moment*** of the volume, averaged over the entire volume. Mathematically,

$$x_c = \frac{1}{V}\int_V x dV \quad y_c = \frac{1}{V}\int_V y dV \quad z_c = \frac{1}{V}\int_V z dV \tag{5.16}$$

where the first moments of the volume are

$$Q_x = \int_V x dV \quad Q_y = \int_V y dV \quad Q_z = \int_V z dV \tag{5.17}$$

Equation (5.16) can be written in vector notation as

$$\mathbf{r}_c = \frac{1}{V}\int_V \mathbf{r} dV \tag{5.18}$$

It was shown in Section 5.2 that if a physical body has a constant mass density, its center of mass is at the same point as the centroid of its volume. If the mass density or specific weight of the body is variable across its volume, the center of mass (or the center of gravity) will not coincide with the centroid of the volume.

Eqs. (5.16–5.18) can be solved for any volume by single or triple integration. If single integration is used, an element of volume is chosen that will cover the entire volume when integrated over one variable. If triple integration is used, a differential of volume is given for a particular choice of a coordinate system. For example, if rectangular coordinates (x, y, z) are used, the element of volume is $dV = dx\, dy\, dz$. If cylindrical coordinates (r, θ, z) are used, the element of volume is $dV = r\, dr\, d\theta\, dz$, and for spherical coordinates (R, ϕ, θ), the element of volume is $dV = R^2 \sin\phi\, dR\, d\phi\, d\theta$. These coordinate systems are shown in Figure 5.15. Note that the notation for the angles in spherical coordinates varies in the literature; that is, many textbooks will designate the angle from the z-axis to the R position vector as θ and the angle in the x–y plane as ϕ.

It is sometimes more difficult to determine a single-volume-element for single integration method than to use a differential volume element for the coordinate system together with triple integration techniques.

Sample Problem 5.2

Determine the centroid of a hemisphere of radius R.

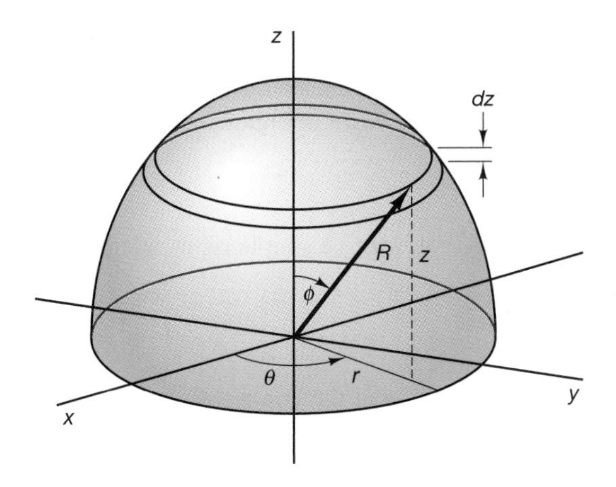

Solution The centroid will lie on the axis of symmetry—that is, the z-axis—so only the position on that axis needs to be determined. We can establish a volume element of $\pi\, r^2\, dz$ and use single integration:

$$r^2 + z^2 = R^2$$
$$\pi r^2 dz = \pi(R^2 - z^2)dz$$

This volume element may be integrated along the z-axis from 0 to R to determine the volume and the centroid:

$$V = \int_0^R \pi(R^2 - z^2)\, dz = \frac{2}{3}\pi R^3$$
$$z_c = \frac{3}{2\pi R^3}\int_0^R \pi(R^2 - z^2)\, z\, dz = \frac{3R}{8}$$

The volume and the location of the centroid may also be obtained by triple integration by using the element of volume in spherical coordinates:

$$V = \int_0^R \int_0^{2\pi} \int_0^{\frac{\pi}{2}} (r^2 \sin\phi)\, d\phi d\theta dr = \frac{2}{3}\pi R^3$$
$$z = r\cos\phi$$
$$z_c = \frac{3}{2\pi R^3}\int_0^R \int_0^{2\pi} \int_0^{\frac{\pi}{2}} (r^3 \sin\phi \cos\phi)\, d\phi d\theta dr = \frac{3R}{8}$$

5.3.3 CENTROID OF A LINE

The centroid of a line with differential element of length ds is also defined by the first moment of a line, divided by the length of the line:

$$x_c = \frac{1}{L}\int_L x ds \quad y_c = \frac{1}{L}\int_L y ds \quad z_c = \frac{1}{L}\int_L z ds \tag{5.19}$$

Here, $L = \int_L ds$ is the length of the line.

The centroid of a straight line is the midpoint of the line and can be found by inspection. A simple example is shown in Figure 5.16. The line is defined by the equation

$$y = 1 - x \quad 0 \le x \le 1$$

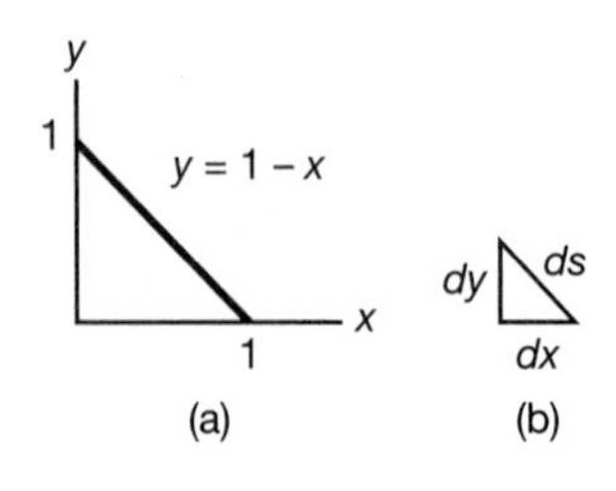

Figure 5.16 (a) A line in a plane. (b) The differential length ds.

The differential length ds forms a simple right triangle with the differential increments dx and dy, so that

$$ds = \sqrt{dx^2 + dy^2}$$

Differentiating the equation of the line yields

$$dy = -dx$$

Substituting for dy, the differential length ds is written in terms of the variable x, and the length and centroid of the line are

$$ds = \sqrt{2}dx$$

$$L = \int_L ds = \sqrt{2}\int_0^1 dx = \sqrt{2}$$

$$x_c = \frac{1}{L}\int_0^1 xds = \frac{1}{\sqrt{2}}\int_0^1 \sqrt{2}xdx = \frac{x^2}{2}\bigg|_0^1 = \frac{1}{2}$$

$$y_c = \frac{1}{L}\int_0^1 yds = \frac{1}{\sqrt{2}}\int_0^1 \sqrt{2}(1 - x)dx = \left(x - \frac{x^2}{2}\right)\bigg|_0^1 = \frac{1}{2}$$

The curve is initially defined in terms of a differential parameter ds along the line. Although we eliminated this parameter by defining the line as a function of x, we will see that a three-dimensional curve is defined in terms of a parameter along its length.

Sample Problem 5.3

Determine the centroid of the family of curves $y = x^n$.

Solution

We will define the differential element of the line ds in terms of dx and dy by differentiating the equation of the curve

$$dy = nx^{n-1}$$

The differential length along the curve is

$$ds = \sqrt{n^2x^{2(n-1)} + 1}dx$$

We consider curves defined over the region $0 \le x \le 1$, and the length of the curve is

$$L = \int_0^1 \sqrt{n^2x^{2(n-1)} + 1}dx$$

The centroid is

$$x_c = \frac{1}{L}\int_0^1 x\sqrt{n^2x^{2(n-1)} + 1}dx$$

$$y_c = \frac{1}{L}\int_0^1 x^n\sqrt{n^2x^{2(n-1)} + 1}dx$$

Computational software may be used to evaluate the integrals for different values of n. (See the Computational Supplement.)

5.3.4 CENTROID OF A CURVE IN SPACE

A curve in space has a definite geometric center: the centroid. To determine the centroid and the length of a general curve in space, a differential length ds along the curve is again introduced. We will first consider curves that lie in a plane. A portion of a planar curve is shown in Figure 5.17. The unit vector $\hat{\mathbf{t}}$ is tangent to the curve at any point and is a function of the angle θ. The parameter s is a measure of the length of the curve at any point, and the angle θ is a function of s. The tangent vector can be written in terms of the differential lengths as

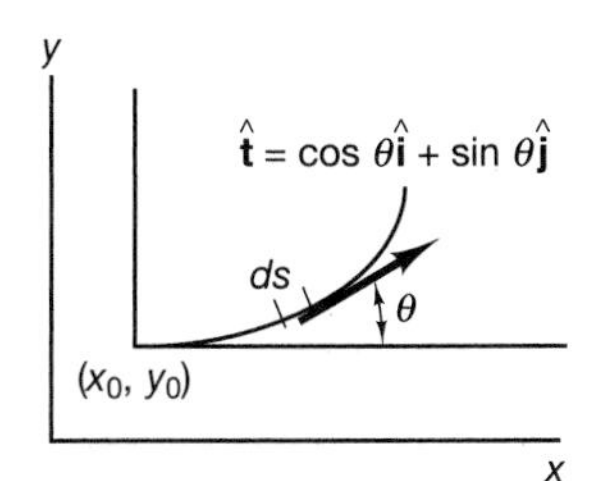

Figure 5.17

$$\hat{\mathbf{t}} = \frac{dx}{ds}\hat{\mathbf{i}} + \frac{dy}{ds}\hat{\mathbf{j}}$$

The curve is defined by the functional dependency of θ upon s. For example, a circle with radius R and center at $(0,R)$ is defined by the parametric relationship

$$\theta(s) = s/R$$

The curve shown in Figure 5.17 starts at (x_0,y_0), and the coordinates of a point s on the curve, defined by $\theta(s)$, are

$$x(s) = x_0 + \int_0^s \cos\theta(\eta)d\eta$$
$$y(s) = y_0 + \int_0^s \sin\theta(\eta)d\eta \tag{5.20}$$

A dummy variable of integration, η, is introduced in Eq. (5.20) for clarity. The length of the curve is given by the maximum value of the parameter s. Note that the units of θ are radians, so that the function $\theta(s)$ must be a function of s normalized by a length. In the case of a circle, the function was s/R. An arc length of the curve ds for a change of $d\theta$ is shown in Figure 5.18. As the unit tangent vector changes from $\hat{\mathbf{t}}$ to $\hat{\mathbf{t}}'$ moving along a distance ds on the curve, the radius of curvature of the curve at that point is

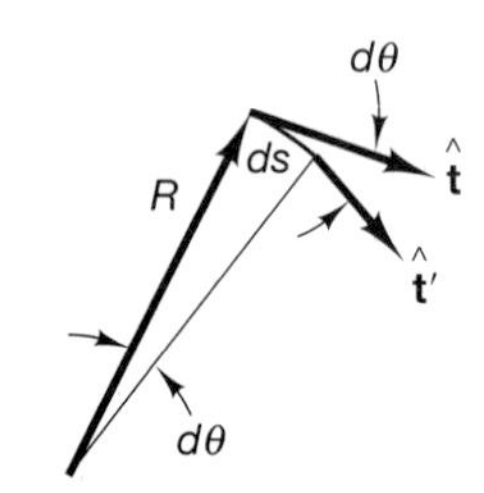

Figure 5.18

$$ds = Rd\theta$$

The radius of curvature R at any point on the curve is related to the function $\theta(s)$ by

$$\frac{1}{R} = \frac{d\theta(s)}{ds} \tag{5.21}$$

The centroid of the planar curve of length L is defined by Eq. (5.19) and, using Eq. (5.20), may be written as

$$x_c = x_0 + \frac{1}{L}\int_0^L x(s)ds \qquad y_c = y_0 + \frac{1}{L}\int_0^L y(s)ds \tag{5.22}$$

Integration of Eqs. (5.20–22) is difficult in most cases, and indefinite integrals usually are not found in tables. Therefore, most centroids will be determined by numerical integration.

The length and the centroid of a general curve in three-dimensional space are determined in a similar manner. A unit tangent vector $\hat{\mathbf{t}}$ to the curve at any point s on the curve is shown in Figure 5.19. The direction of the tangent vector is given by the angle $\beta(s)$ between $\hat{\mathbf{t}}$ and its projection on the x–y plane and the angle $\theta(s)$ between the x–y projection and the x-axis. Both of these angles are functions of the position s along the curve. The unit tangent vector, expressed as a function of the curve length s, is

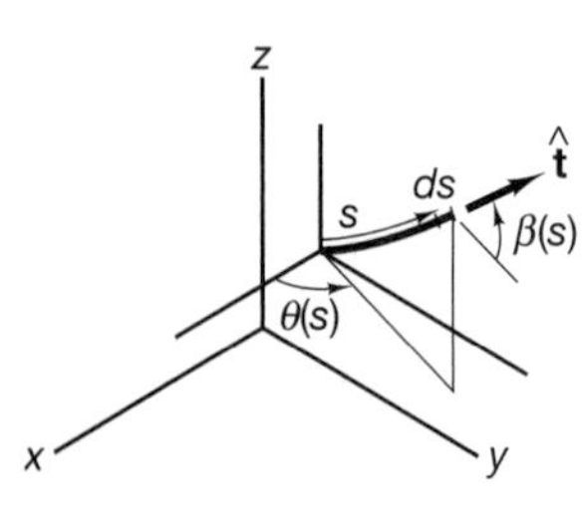

Figure 5.19

$$\hat{\mathbf{t}}(s) = \cos\theta(s)\cos\beta(s)\hat{\mathbf{i}} + \sin\theta(s)\cos\beta(s)\hat{\mathbf{j}} + \sin\beta(s)\hat{\mathbf{k}} \tag{5.23}$$

The functional dependency of θ and β on s specifies a given curve in space. The x-, y-, and z-components of $\hat{\mathbf{t}}$ are the slopes dx/ds, dy/ds, and dz/ds, respectively. Therefore, the coordinates of a point s on the curve are

$$x(s) = x_0 + \int_0^s \cos\theta(s)\cos\beta(s)ds$$
$$y(s) = y_0 + \int_0^s \sin\theta(s)\cos\beta(s)ds \quad 0 \le s \le L \text{ (length of curve)} \tag{5.24}$$
$$z(s) = z_0 + \int_0^s \sin\beta(s)ds$$

For example, a helical spiral of radius ($R \cos \beta$) with a constant pitch (β = constant) is described by the parametric equations

$$\theta = s/R \qquad \beta = \text{constant}$$

The pitch of the helical spiral is the constant slope in the *z*-direction. The helical spiral is the basis of the design and manufacture of coil springs, which are found in many machines. The helical curve is shown in the Computational Supplement.

Eq. (5.19) gives the length of the curve and its centroid as

$$
\begin{aligned}
L &= \int_0^L ds \\
x_c &= x_0 + \frac{1}{L}\left(\int_0^L\left(\int_0^s \cos\theta(s)\cos\beta(s)ds\right)ds\right) \\
y_c &= y_0 + \frac{1}{L}\left(\int_0^L\left(\int_0^s \sin\theta(s)\cos\beta(s)ds\right)ds\right) \qquad (5.25) \\
y_c &= y_0 + \frac{1}{L}\left(\int_0^L\left(\int_0^s \sin\beta(s)\,ds\right)ds\right)
\end{aligned}
$$

The definitions and formulas for centroids of lines, areas, and volumes are very similar and differ only in the differential element that appears in the integrand. For a line, the differential element is that of the length of the line, *ds*. For areas, the differential element is that of the area *dA*, and for volumes, it is *dV*. These differential elements depend upon the coordinate system, and the choice of coordinate system will vary with the object's shape. The resulting integrals can be evaluated using calculus skills, tables of integrals, numerical integration by means of computational software, or analytical techniques via a symbolic manipulator computer code.

The computations of centroids, center of mass, and center of gravity reduce to integration; and therefore, these computations often form a topic of application in most courses in integral calculus. In general, the integrals are multiple integrals requiring integration over two or three coordinate directions. Examples may be found in any calculus text.

Problems

Centroids of Lines

5.1 Determine the centroid of the line $y = x$ stretching from the origin to the point (1,1) m. (See Figure P5.1.)

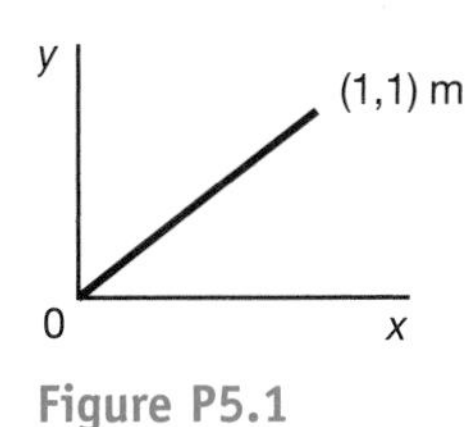

Figure P5.1

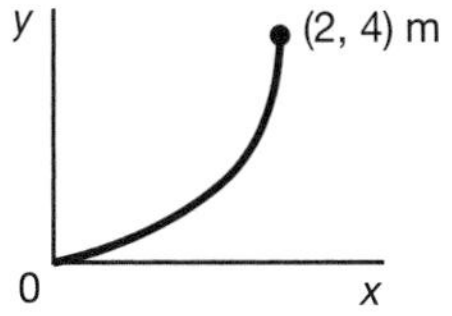

Figure P5.2

5.2 Determine the centroid of the line $y = x^2$ stretching from the origin to the point (2,4) m. (See Figure P5.2.)

5.3 Repeat Problem 5.2 for the case where the line extends to the point $x = 10$, $y = 100$ m.

5.4 Determine the centroid of the line $y = x^3$ stretching from the origin to the point (2,8) m.

5.5 Determine the centroid of the line $y = \sin x$ stretching from the origin to the point ($\pi/2$,1) mm. (See Figure P5.5.) Be sure to label the units.

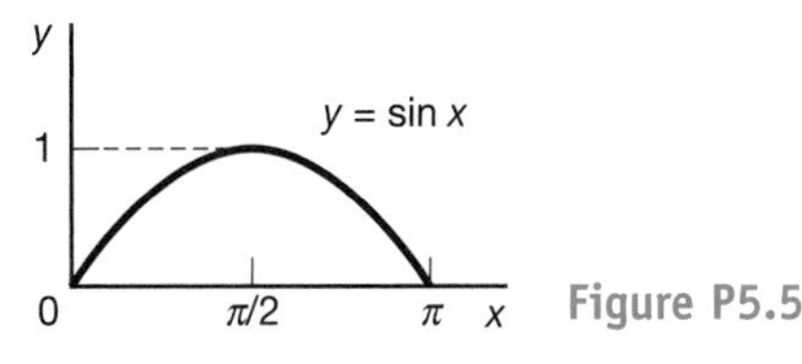

Figure P5.5

5.6 Repeat Problem 5.5, but change the length of the line to extend to (a) $x = \pi$ rad and (b) $x = 2\pi$ rad. (c) What is the length of one cycle of a sine curve of maximum amplitude 1 mm?

5.7 Calculate the centroid of the semicircular "line" used as a handle on a vehicle. (See Figure P5.7.) The choice here is to write the equation of the circle and use rectangular coordinates or polar coordinates for the integration. Set the radius at $R = 10$ cm.

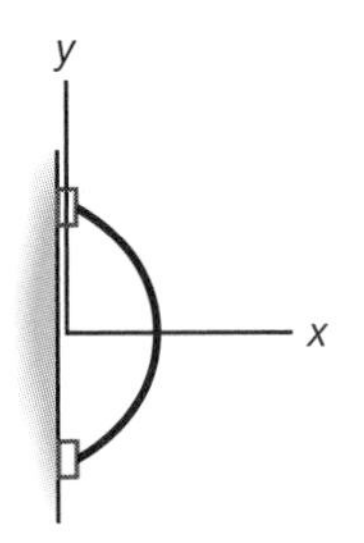

Figure P5.7

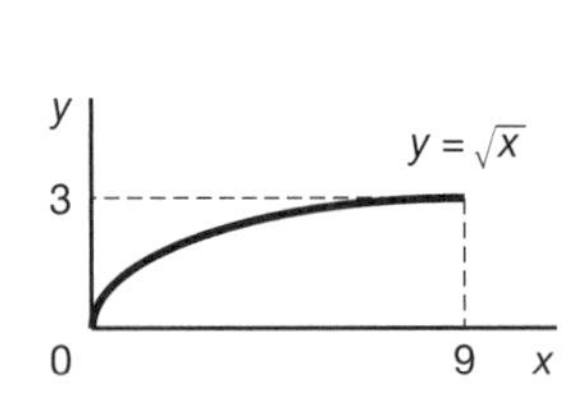

Figure P5.8

5.8 Calculate the centroid of the line $y = \sqrt{x}$ stretching from the origin to the point $x = 9$ m. (See Figure P5.8.)

Centroids of Areas

5.9 Calculate the centroid of a rectangular area by direct integration, and verify the statement in the text that a centroid always lies on the axis of symmetry. (See Figure P5.9.)

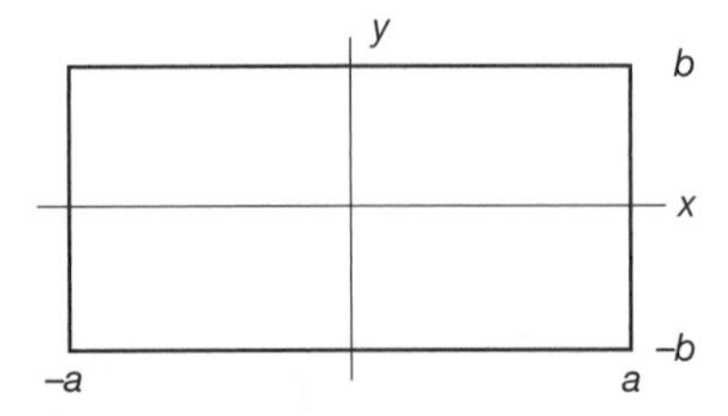

Figure P5.9

5.10 Calculate, by direct integration, the centroid of the rectangular area relative to the coordinate system illustrated in Figure P5.10.

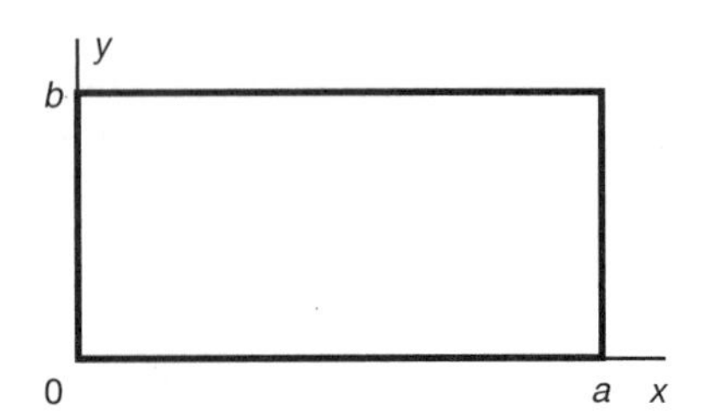

Figure P5.10

5.11 Compute the centroid of the area of a triangle of height h and base b, using the coordinate system illustrated in Figure P5.11.

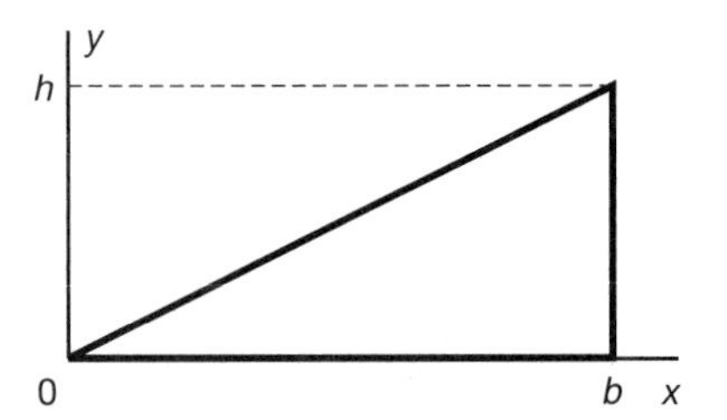

Figure P5.11

5.12 Calculate the centroid of the triangular area shown in Figure P5.12.

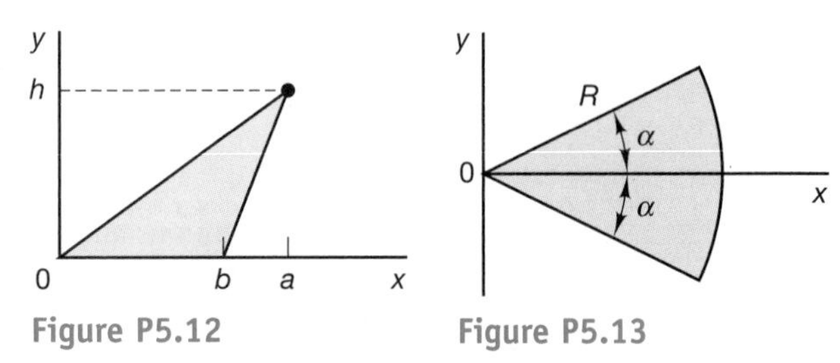

Figure P5.12

Figure P5.13

5.13 In Figure P5.13, calculate the centroid of the circular arc of radius R centered on the x-axis.

5.14 Calculate the centroid of the quadrant of the circle of radius R shown in Figure P5.14.

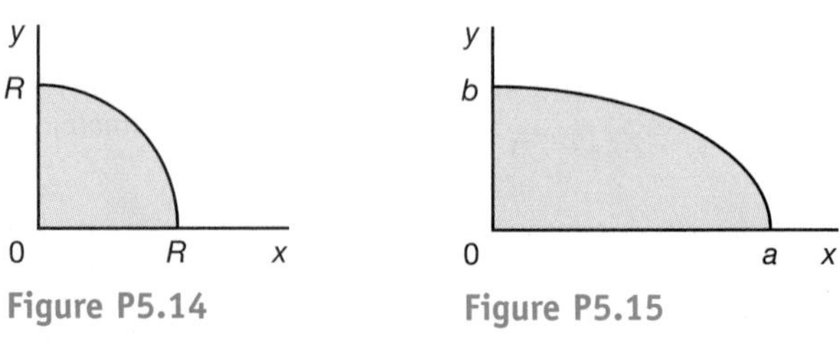

Figure P5.14

Figure P5.15

5.15 Calculate the centroid of the area formed by the quadrant of an ellipse. (See Figure P5.15.) The equation of an ellipse is $\left(\frac{x}{a}\right)^2 + \left(\frac{y}{b}\right)^2 = 1$.

5.16 Calculate the centroid of a parabolic spandrel—that is, the area formed by the parabola $y = x^2$ and the x-axis. (See Figure P5.16.)

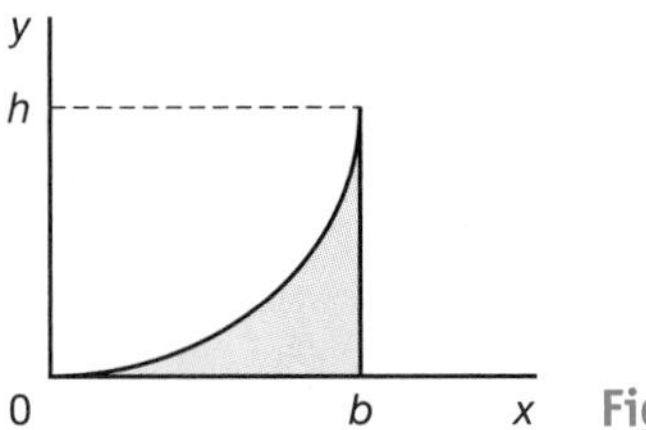

Figure P5.16

5.17 Calculate the centroid of the area between the curve $y = h\sqrt{x/b}$ and the x-axis. (See Figure P5.17.)

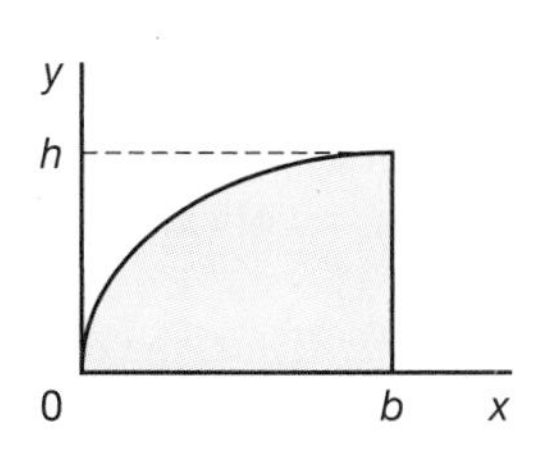

Figure P5.17

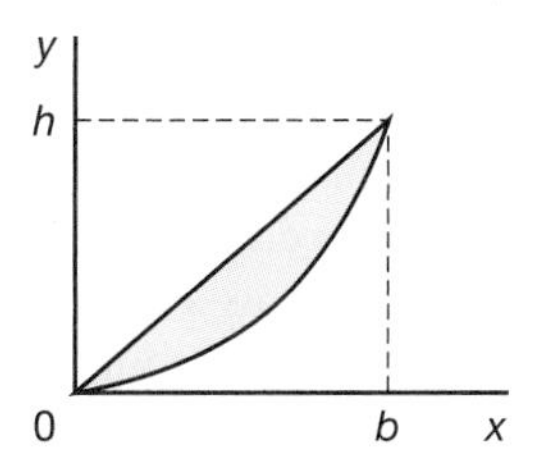

Figure P5.18

5.18 Calculate the area and the centroid of the area between the two curves illustrated in Figure P5.18. The top curve is $y = \frac{h}{b}x$ and the bottom curve is $y = \frac{h}{b^2}x^2$, between the origin and the point (b, h).

5.19 Calculate the area and the centroid of the area enclosed between the line $x = a$, the origin, the curve $x = ky^3$, and the x-axis. (See Figure P5.19.)

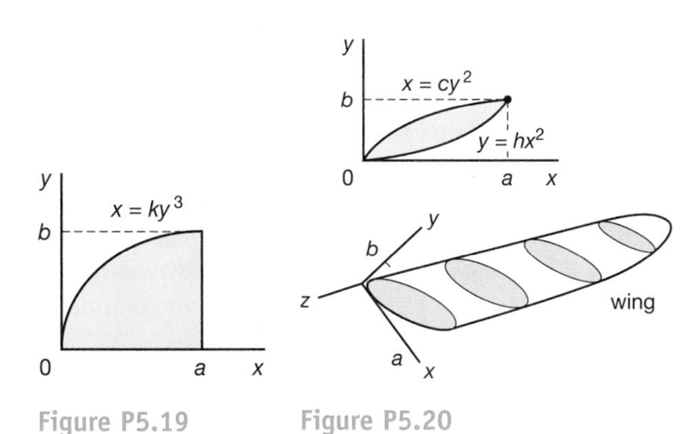

Figure P5.19 Figure P5.20

5.20 A flat plate is machined out of aluminum to form a frame for an aircraft wing. The plate has the shape shown between the two curves in Figure P5.20. The area of the piece is needed to calculate its contribution to the mass of the airplane. The part needs to be reproduced for a number of different pairs of values of a and b as the wing tapers along its length. In addition, the location of the centroid needs to be known so that the machinist can clamp the part at that location. (a) Calculate the area of the part and the location of the centroid. Compute your answer in terms of a and b. (b) What are the area and the location of the centroid for the specific values $a = 3$ m and $b = 1$ m?

Volumes

5.21 Compute the centroid and the volume of the solid circular cylinder of radius R showin in Figure P5.21.

5.22 Calculate the volume and the centroid of the rectangular solid illustrated in Figure P5.22.

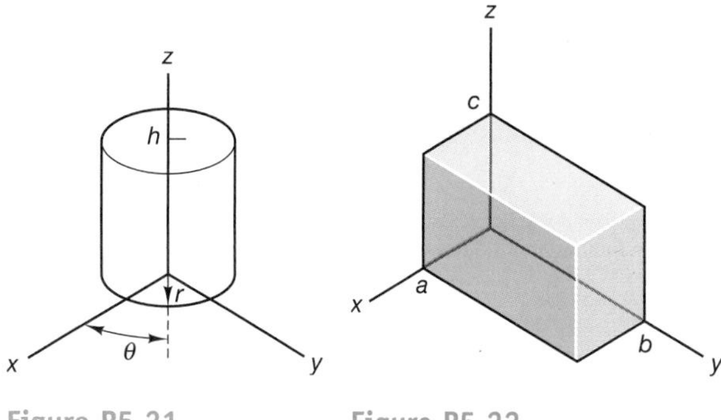

Figure P5.21 Figure P5.22

5.23 Compute the volume and the centroid of the cone shown in Figure P5.23.

5.24 Calculate the volume and the centroid of the parabolic "cone" illustrated in Figure P5.24. The equation of the line in the x–y plane is $y^2 = \frac{R^2}{h}x$.

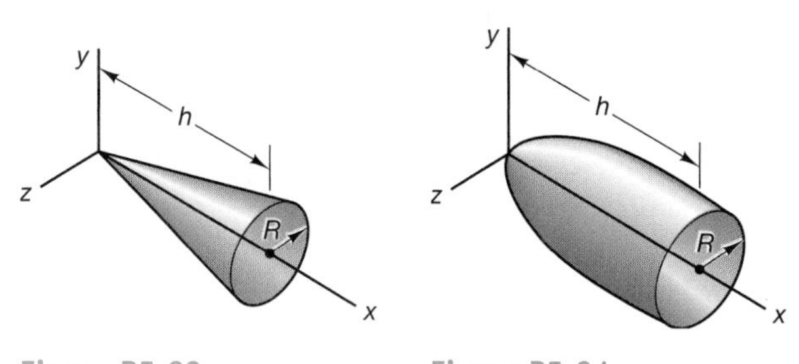

Figure P5.23 Figure P5.24

5.25 Calculate the volume and the centroid for the half-sphere of radius R shown in Figure P5.25. Note that the equation of the side of the element lying in the x–y plane is that of a circle.

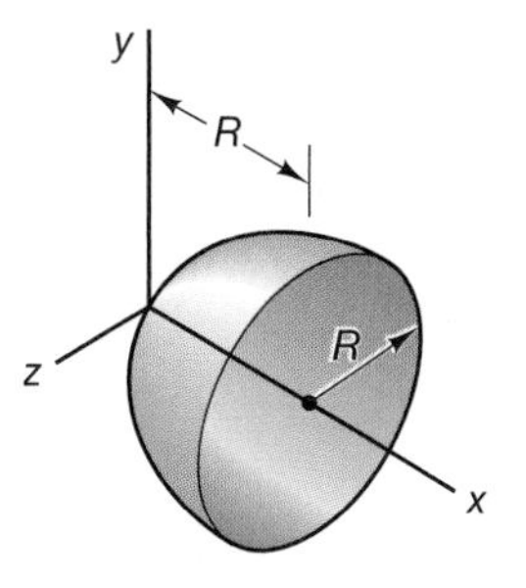

Figure P5.25

5.4 THEOREMS OF PAPPUS AND GULDINUS

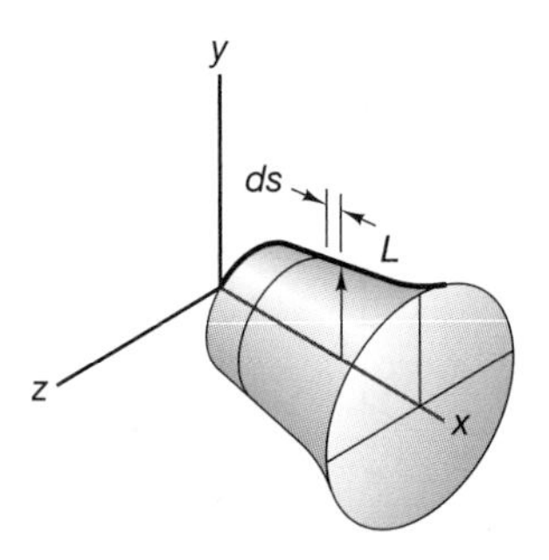

Figure 5.20a

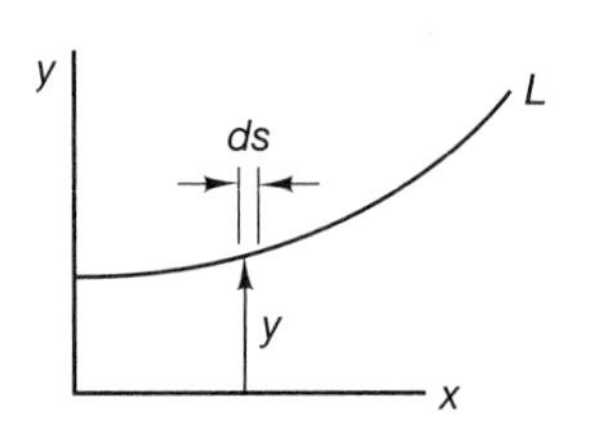

Figure 5.20b

Two theorems involving surfaces and volumes of revolution were developed by the Greek geometer Pappus Alexandria (300 A.D.), long before the creation of calculus in the seventeenth century. Later, authorship was claimed by the Swiss mathematician Paul Guldin (1577–1643), although Pappus's work was well known at that time.

Consider a curve L revolved about a nonintersecting axis, as shown in Figure 5.20. This axis can be called the x-axis without loss of generality. As the line L is revolved about the x-axis, each differential line element ds forms a ring about that axis. The surface area of the ring is the circumference of the ring times the line element length ds. Thus, the differential surface area is

$$dA = 2\pi\, y\, ds$$

A surface area A is formed by revolving this plane curve about the nonintersecting x-axis. The area of this surface of revolution is the integral of dA over the length L, or

$$A = 2\pi \int_L y(s)ds$$

The centroid of the line is defined by Eq. (5.19) as

$$y_c L = \int_L y(s)ds$$

The surface area can then be written as

$$A = 2\pi y_c L \tag{5.26}$$

This calculation is the basis of the following theorem:

Pappus and Guldinus Theorem 1: *The area* A *of a surface of revolution generated by revolving a plane curve of length* L *about any nonintersecting axis in its plane is equal to the product of the length of the curve and the length of the path traveled by the centroid of the curve (the length of the generating curve times the distance traveled by the centroid of the generating curve).*

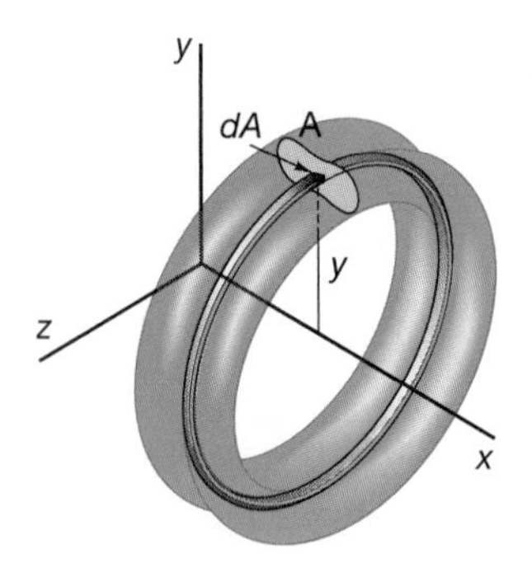

Figure 5.21

The previous argument can be extended to revolving areas as well. If an area is revolved about a nonintersecting line in its plane, a volume will be generated, as shown in Figure 5.21.

If an element of area dA is revolved about the axis, a ring of volume

$$dV = 2\pi\, y\, dA$$

will be formed. Then the total volume generated by revolving the area about the x-axis is

$$V = 2\pi \int_A y dA$$

Using Eq. (5.15), we find that the centroid of the generating area is

$$y_c A = \int_A y dA$$

The total volume generated by revolving the area is

$$V = 2\pi y_c A \tag{5.27}$$

This establishes the following theorem:

Pappus and Guldinus Theorem 2. *The volume* V *of the solid of revolution generated by revolving a plane area* A *about any nonintersecting axis in its plane is equal to the product of the area and the length of the path traveled by the centroid of the area (the product of the generating area times the distance traveled by the centroid of the generating area).*

Sample Problem 5.4

Determine the surface area of the surface of revolution of a half sine curve of amplitude 0.5 m beginning at (0, 2)m. An equation for this curve is

$$y(x) = 2 + 0.5\sin(\pi \times x) \text{ from } x = 0 \text{ to } x = 1$$

Computational Solution:

$$x := 0, 0.05 \ldots 1$$

$$y(x) := 2 + 0.5 \cdot \sin(\pi \cdot x)$$

$$ds(x) := \sqrt{1 + (0.5 \cdot \pi \cdot \cos(\pi \cdot x))^2}$$

$$L = \int_0^1 ds(x)\,dx$$

$$L = 1.464$$

$$y_c = \frac{1}{L} \cdot \int_0^1 y(x) \cdot ds(x)\,dx$$

$$y_c = 2.288$$

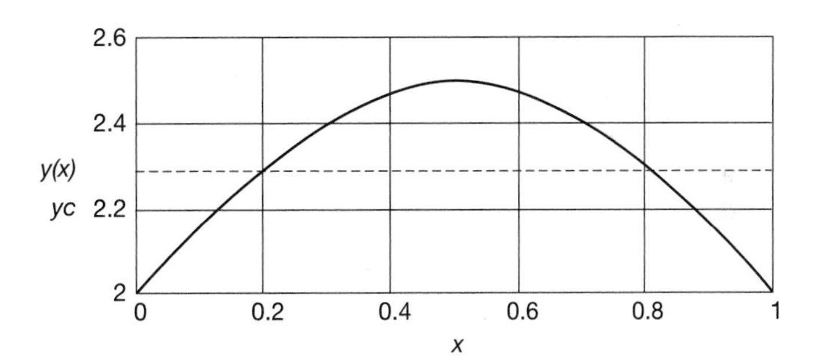

Sample Problem 5.5

Determine the surface area and volume of the solid that is generated by revolving a circle with radius r and center $(0,R)$ about the x-axis. (See figure at left.) This solid is called a torus.

Solution

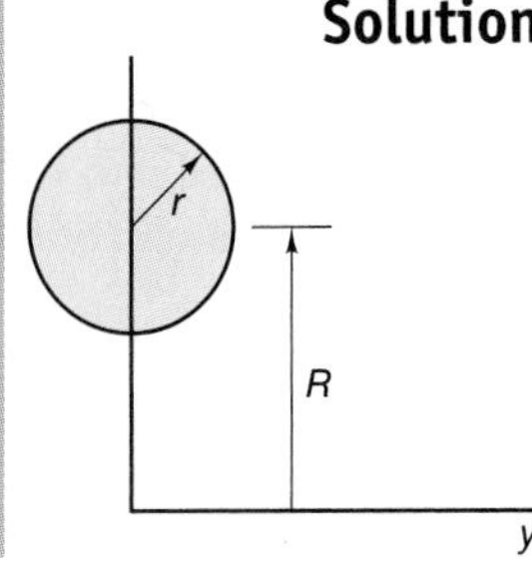

The centroid of the line enclosing the circle and the centroid of the area of the circle is

$$y_c = R$$

The length of the line generating the surface area is the circumference of the circle:

$$L = 2\pi r$$

The area of the circle is

$$A = \pi r^2$$

Therefore, if we use the second theorem of Pappus and Guldinus, we find that the surface area of the torus and the volume of the torus are, respectively,

$$\text{Area} = 2\pi R(2\pi r) = 4\pi^2 Rr$$

$$\text{Volume} = 2\pi R(\pi r^2) = 2\pi^2 Rr^2$$

Problems

Areas

5.26 In Figure P5.26, compute the surface area of the object formed by rotating the line about the x-axis from $x = 0$ to $x = 5$.

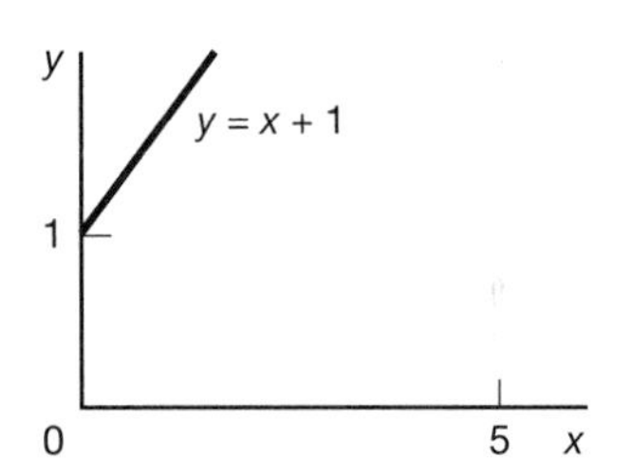

Figure P5.26

5.27 Use the first theorem of Pappus and Guldinus to compute the surface area of a sphere by rotating a semicircular arc about the x-axis. (See Table 5.2 for the centroid and arc length of a semicircular curve.)

5.28 Calculate the surface area of the line $y = \sin x$ rotated about the x-axis, using the first theorem of Pappus and Guldinus. (See Figure P5.28.)

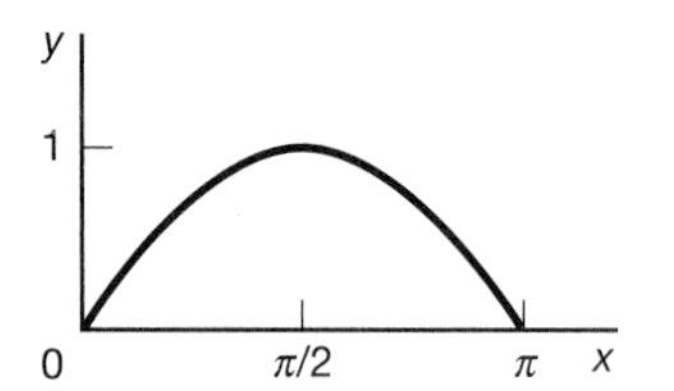

Figure P5.28

Volumes

5.29 Compute the volume formed by rotating the area of the rectangle in Figure P5.10 about the x-axis. Does your solution make sense?

5.30 Referring to the area and centroid of a triangle (see Problem 5.11 and Table 5.2), compute the volume of a cone.

5.31 Compute the volume of the solid formed by rotating the ellipse of Figure P5.15 about the x-axis.

5.32 Compute the volume formed by revolving the oblique triangle of Figure P5.12 about the x-axis. The volume is a cone with a second smaller cone cut out of one end. Does your answer make sense?

5.5 CENTROIDS OF COMPOSITE BODIES

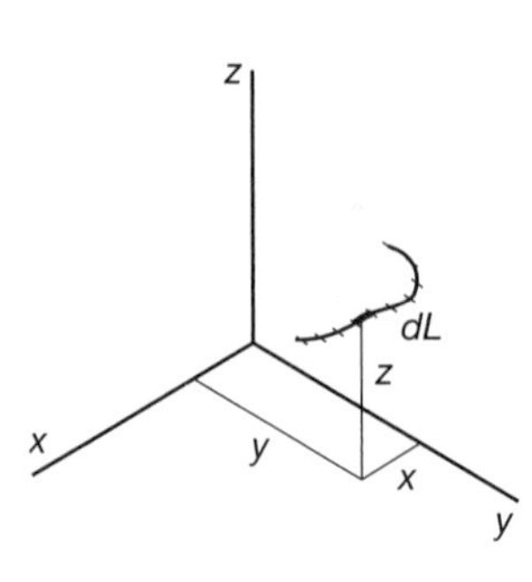

Figure 5.22

This section discusses the determination of centroids of lines, areas, and volumes of composite bodies. A composite body is an object made up of several other objects. If the components of the composite body are simple geometric shapes whose individual centroids are known or given in tables, the centroid of the composite can be found without integration. Centroids of volumes, areas, and lines are defined in terms of integrals. Numerical analysis methods approach the solution of these integrals by dividing the volume, area, or line into a finite number of subunits. For example, a line in space may be subdivided into subunits as shown in Figure 5.22. The length of the line may then be written as the sum of the lengths of the subunits:

$$L = \sum_i dL_i \tag{5.28}$$

The centroid of the line element dL_i has coordinates x_i, y_i, and z_i. Therefore, the centroid of the entire line becomes

$$x_c = \frac{1}{L}\sum_i x_i dL_i \quad y_c = \frac{1}{L}\sum_i y_i dL_i \quad z_c = \frac{1}{L}\sum_i z_i dL_i \tag{5.29}$$

Note that the individual line elements do not need to have the same length. As the lengths of the line elements are taken smaller, the summation approaches the integral definition given earlier. If the exact centroid is specified for each line element, the centroid of the entire line computed using the composite sum will also be exact. These composite summation formulas are useful for computing the centroid of a line if the centroids of the separate lengths of the line are known. This situation is illustrated in Sample Problem 5.6.

In a similar manner, the centroids of areas can be determined by considering them to be composites of areas A_i with known centroids. The total area and its centroid are defined by the summations of the various composite pieces

$$A = \sum_i A_i$$

$$\begin{aligned} x_c &= \frac{1}{A}\sum_i x_i A_i \\ y_c &= \frac{1}{A}\sum_i y_i A_i \\ z_c &= \frac{1}{A}\sum_i z_i A_i \end{aligned} \tag{5.30}$$

If there is a hole or opening of known shape in an area, the area without the hole may be treated as one element, and the hole itself may be treated as a second element of "negative" area by assigning a minus sign to that particular value of A_i. This approach of dividing an area into a composite of known areas avoids complex integration and greatly simplifies the determination of centroids.

The same method may be applied to determine the centroids of volumes, treated as composite volumes. The relevant formulation is

$$V = \sum_i V_i$$

$$\begin{aligned} x_c &= \frac{1}{V}\sum_i x_i V_i \\ y_c &= \frac{1}{V}\sum_i y_i V_i \\ z_c &= \frac{1}{V}\sum_i z_i V_i \end{aligned} \tag{5.31}$$

To facilitate dividing areas and volumes into composites, the centroids of simple areas and volumes are given in Table 5.2.

Table 5.2 *Centroids of Common Shapes*

Shape	$\bar{x}$	$\bar{y}$	Area
Triangular area	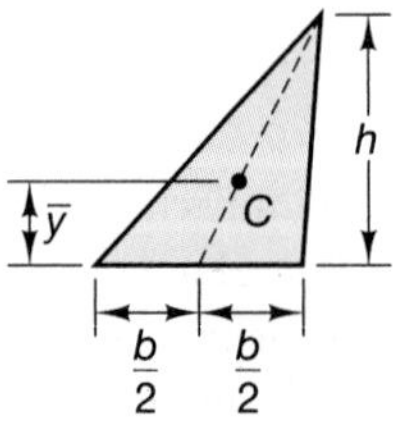	$\frac{h}{3}$	$\frac{bh}{2}$
Quarter-circular area	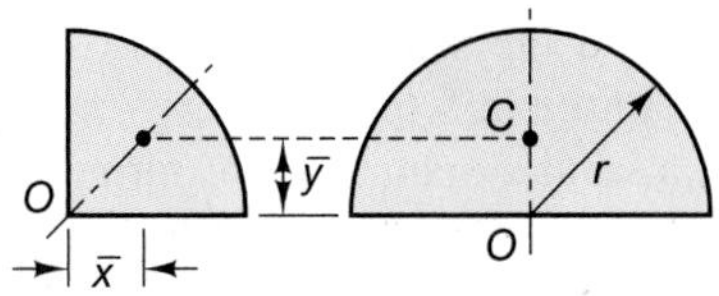 $\frac{4r}{3\pi}$	$\frac{4r}{3\pi}$	$\frac{\pi r^2}{4}$
Semicircular area	0	$\frac{4r}{3\pi}$	$\frac{\pi r^2}{2}$
Semiparabolic area	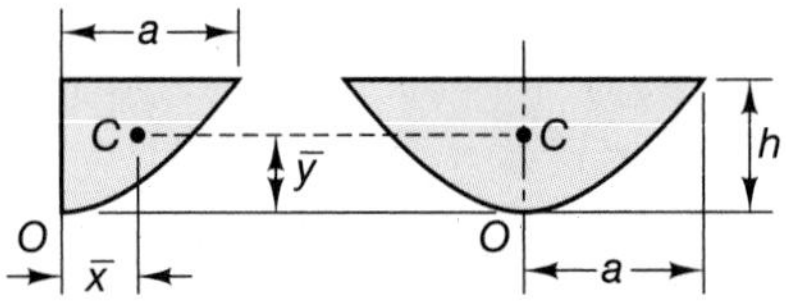 $\frac{3a}{8}$	$\frac{3h}{5}$	$\frac{2ah}{3}$
Parabolic area	0	$\frac{3h}{5}$	$\frac{4ah}{3}$
Parabolic spandrel	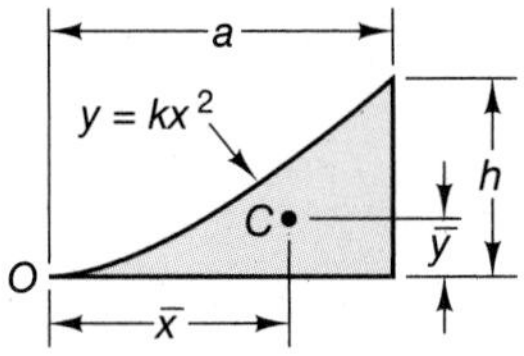 $\frac{3a}{4}$	$\frac{3h}{10}$	$\frac{ah}{3}$
Circular sector	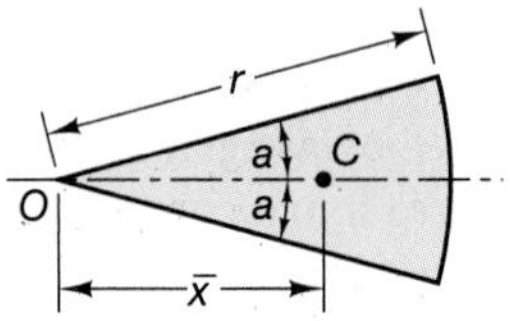 $\frac{2r \sin \alpha}{3\alpha}$	0	αr^2
Quarter-circular arc	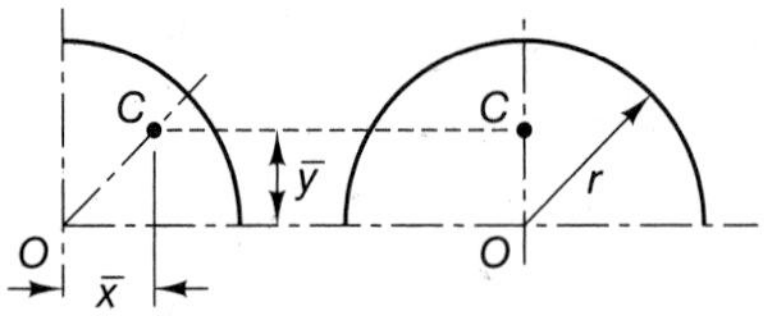 $\frac{2r}{\pi}$	$\frac{2r}{\pi}$	$\frac{\pi r}{2}$
Semicircular arc	0	$\frac{2r}{9}$	πr
Arc of circle	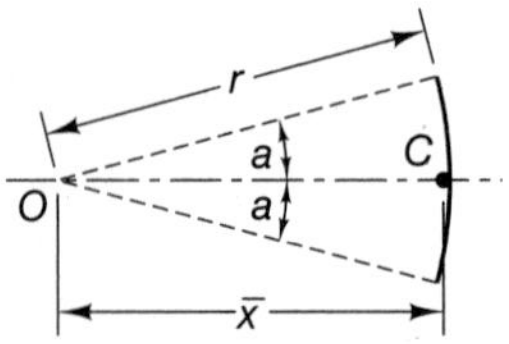 $\frac{r \sin \alpha}{\alpha}$	0	$2\alpha r$

Shape	$\bar{x}$	Volume
Hemisphere	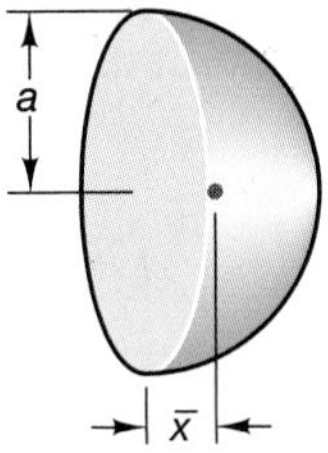 $\frac{3a}{8}$	$\frac{2}{3}\pi a^3$
Semiellipsoid of revolution	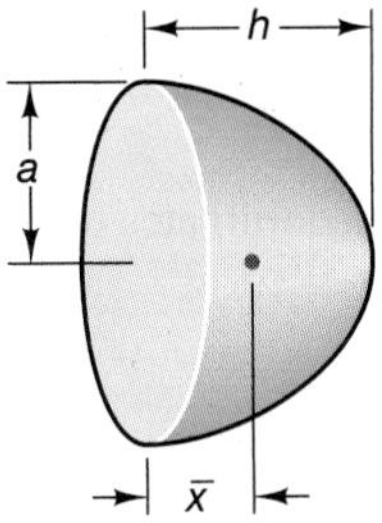 $\frac{3h}{8}$	$\frac{2}{3}\pi a^2 h$
Paraboloid of revolution	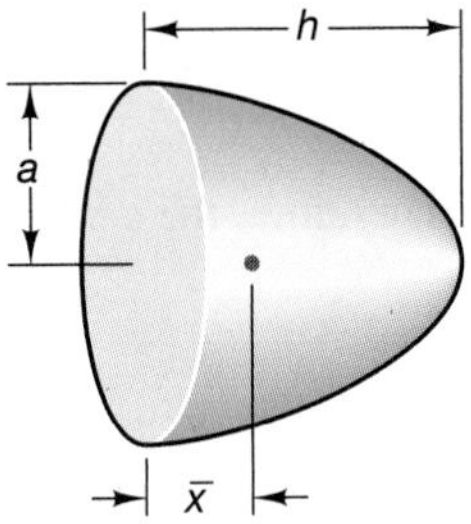 $\frac{h}{3}$	$\frac{1}{2}\pi a^2 h$
Cone	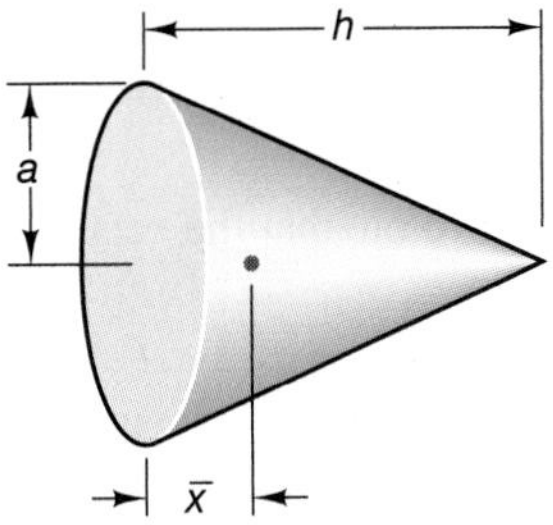 $\frac{h}{4}$	$\frac{1}{3}\pi a^2 h$
Pyramid	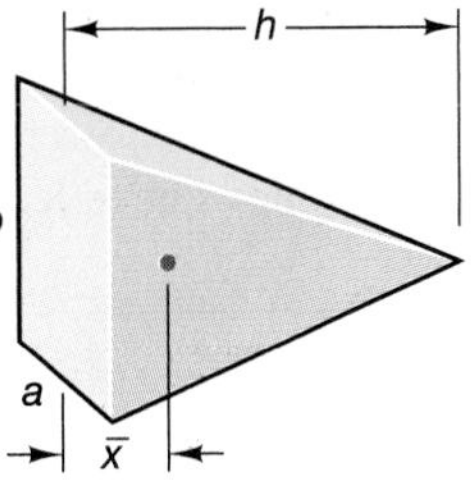 $\frac{h}{4}$	$\frac{1}{3}abh$

Sample Problem 5.6

Find the centroid of the line segment shown at left. (All units are in meters.)

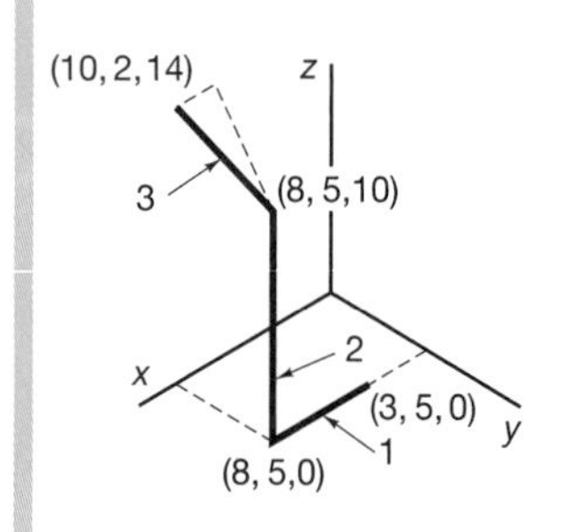

Solution

The lengths of the line segments are

$$L_1 = \sqrt{(8 - 3)^2 + (5 - 5)^2 + (0 - 0)^2} = 5 \text{ m}$$
$$L_2 = \sqrt{(8 - 8)^2 + (5 - 5)^2 + (10 - 0)^2} = 10 \text{ m}$$
$$L_3 = \sqrt{(10 - 8)^2 + (2 - 5)^2 + (14 - 10)^2} = 5.39 \text{ m}$$

The length of the line is

$$L = \sum_i L_i = 20.39 \text{ m}$$

The coordinates to the midpoint or centroid of each line segment are as follows:

Line Segment	L_i, m	X_i, m	Y_i, m	Z_i, m
1	5	5.5	5	0
2	10	8	5	5
3	5.39	9	3.5	12

Therefore, the centroid of the entire line is

$$x_c = \frac{1}{L}\sum_i x_i L_i = \frac{1}{20.39}(5.5 \cdot 5 + 8 \cdot 10 + 9 \cdot 5.39) = 7.65 \text{ m}$$

$$y_c = \frac{1}{L}\sum_i y_i L_i = \frac{1}{20.39}(5 \cdot 5 + 5 \cdot 10 + 3.5 \cdot 5.39) = 4.60 \text{ m}$$

$$z_c = \frac{1}{L}\sum_i z_i L_i = \frac{1}{20.39}(0 \cdot 5 + 5 \cdot 10 + 12 \cdot 5.39) = 5.62 \text{ m}$$

Note that the centroid is not on any portion of the line. (See figure at left.)

Sample Problem 5.7

Determine the centroid for the area shown at the left. (All dimensions are in mm.)

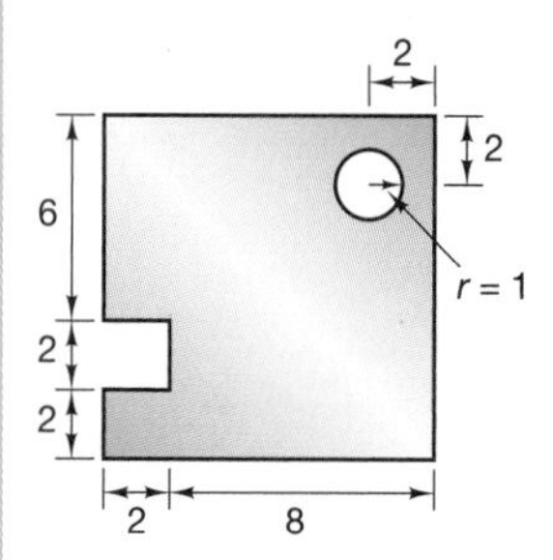

Solution

We divide the area into three parts, two of which are negative to account for the square cutout on the lower left and the round hole on the upper right. The reference axes are placed at the lower left of the area, as shown. The centroid may be found by completing the following table:

Part Number	Area A_1, mm²	X_{ic}, mm	Y_{ic}, mm	$X_{ic}A_i$, mm³	$Y_{ic}A_i$, mm³
1	100	5	5	500	500
2	−4	1	3	−4	−12
3	−3.14	8	8	−25.13	−25.13
Σ	92.86			470.87	462.87

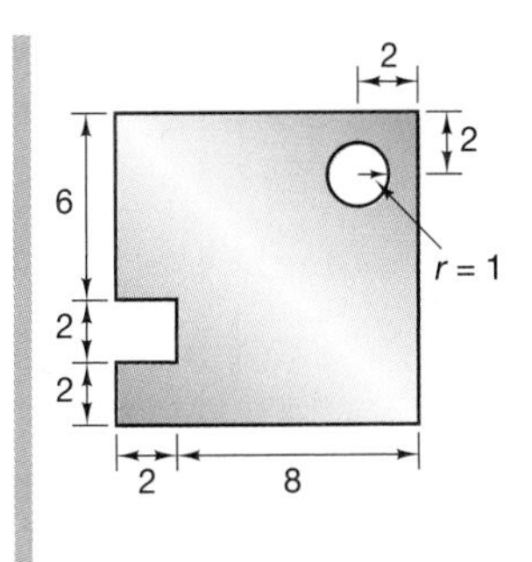

The centroid of the composite area is

$$x_c = \frac{\sum x_{ic} dA_i}{\sum dA_i} = \frac{470.87}{92.86} = 5.07 \text{ mm}$$

$$y_c = \frac{\sum y_{ic} dA_i}{\sum dA_i} = \frac{462.87}{92.86} = 4.98 \text{ mm}$$

Here, the centroid of each part is computed by using the shape from the table of common shapes, such as a square, and then relating the position of this centroid to a common origin.

Centroids are a critical part of machine design and force analysis. If a part is held in place by a number of fasteners, the load should pass through the centroid of the fasteners in order to prevent a rotational effect resulting in uneven loading of the fasteners and possible failure.

Problems

5.33 Calculate the centroid of the three line segments illustrated in Figure P5.33. These line segments represent a tripod. The dimensions are in meters, and the thickness is ignored.

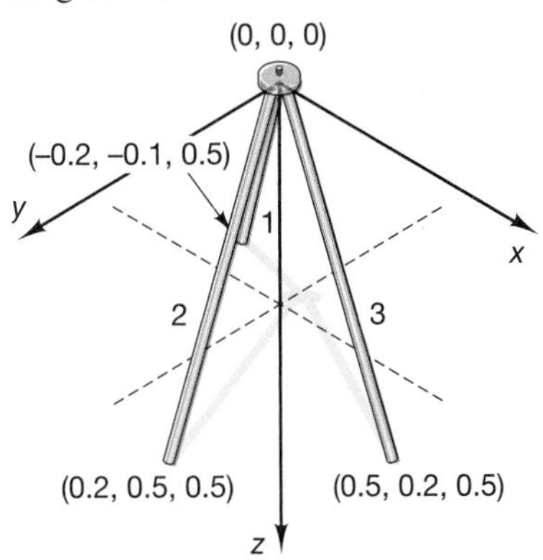

Figure P5.33

5.34 A spring clip used to hold a watch cover in place is shown in Figure P5.34. Calculate the object's centroid, assuming it to have no thickness (i.e., assume that it is a line). The dimensions are in centimeters.

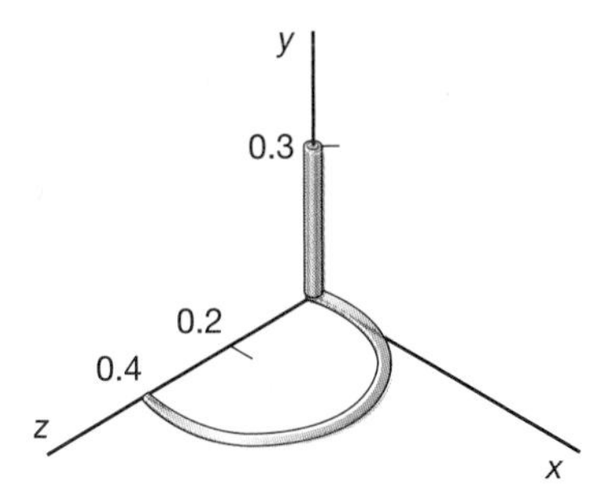

Figure P5.34

5.35 Calculate the centroid of the line segment shown in Figure P5.35. The dimensions are in meters.

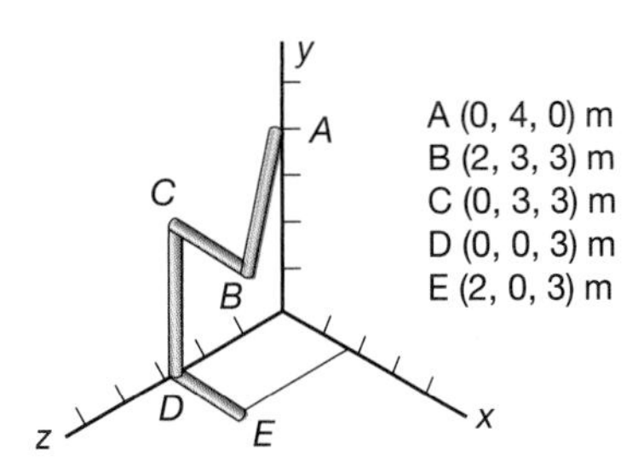

Figure P5.35

5.36 Calculate the centroid of the security fence illustrated in Figure P5.36. Compare this to the centroid of the fence without the "V" connection on the top.

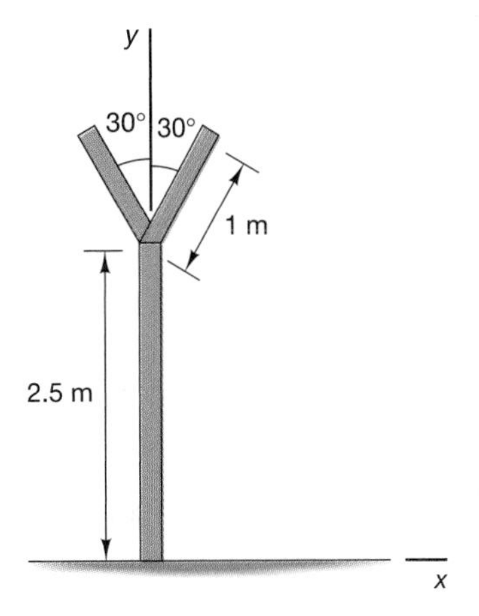

Figure P5.36

5.37 Calculate the centroid of the wire-frame triangle illustrated in Figure P5.37. How does it compare to the centroid of a solid triangle? Dimensions are in meters.

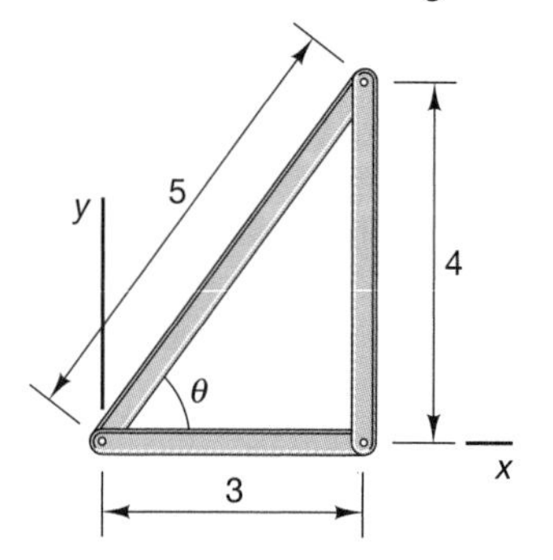

Figure P5.37

Area Centroids

5.38 A key is designed from thin stock with the dimensions shown in Figure P5.38, in centimeters. Use the formulation for composite systems to compute the centroid of the key.

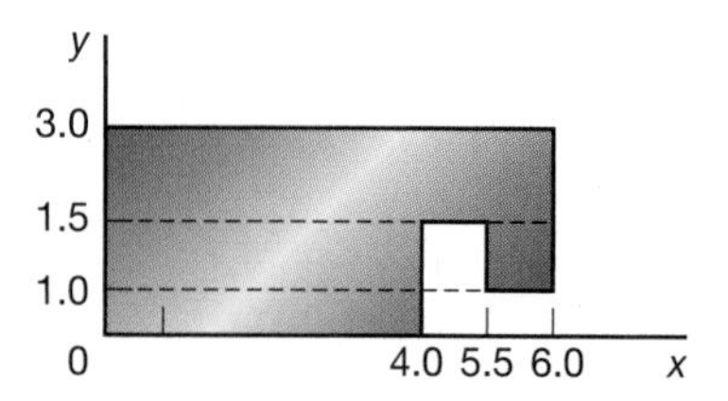

Figure P5.38

5.39 A mounting bracket is machined out of a flat piece of metal. (See Figure P5.39.) This changes the centroid and, hence, the center of mass of the piece. Compute the new centroid, and compare it to that of the same piece without the holes. The dimensions in the figure are all in meters. The holes are all of radius 0.05 m.

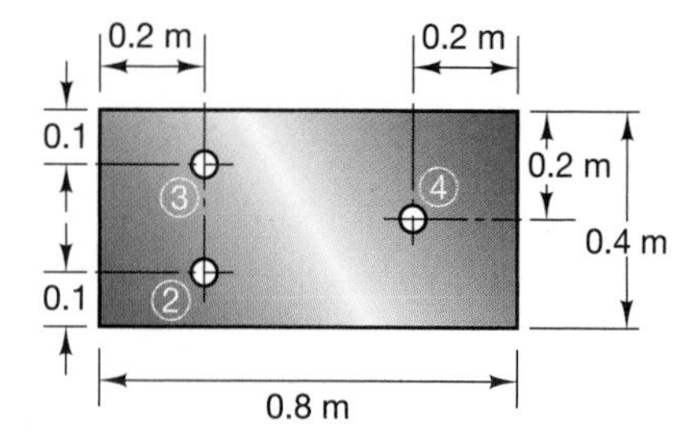

Figure P5.39

5.40 Repeat Problem 5.39, but (a) make the hole at 4 have a radius of 0.175 m, and (b) omit the hole at 4, and compare your answer to that of the metal sheet before the holes are drilled.

5.41 Compute the centroid of the concrete section of the monorail support member shown in Figure P5.41. Is the centroid located on the structure?

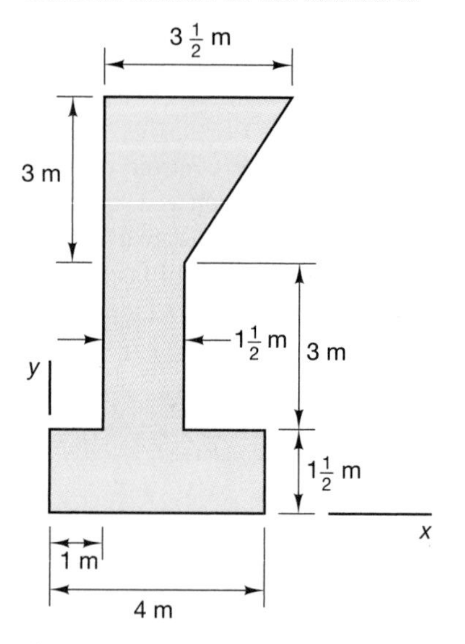

Figure P5.41

5.42 Calculate the centroid of the cross section of the "I" beam illustrated in Figure P5.42.

5.43 A mounting bracket is manufactured out of 5 mm-wide steel and formed into the shape illustrated in Figure P5.43. Compute the centroid. Does the centroid lie on the object?

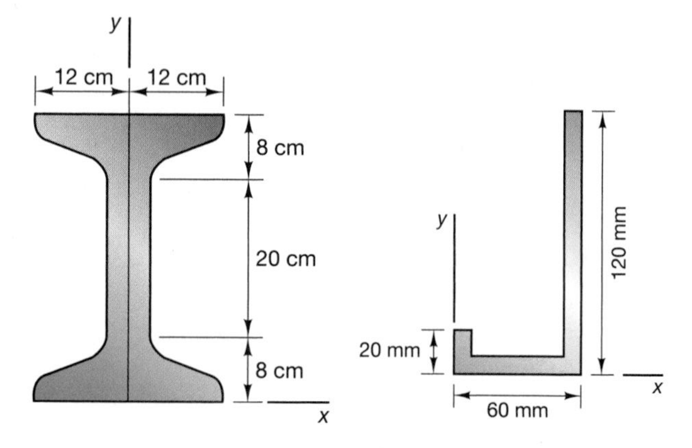

Figure P5.42 Figure P5.43

5.44 A concrete retaining wall is constructed to the cross-sectional dimensions illustrated in Figure P5.44. Before the wall is "backfilled," the engineer would like to know the centroid. (a) Calculate the centroid. (b) If the resulting force analysis requires the center of mass to be changed from the value you calculate, suggest three ways to redesign the wall to move its center of mass. Use the coordinate system suggested.

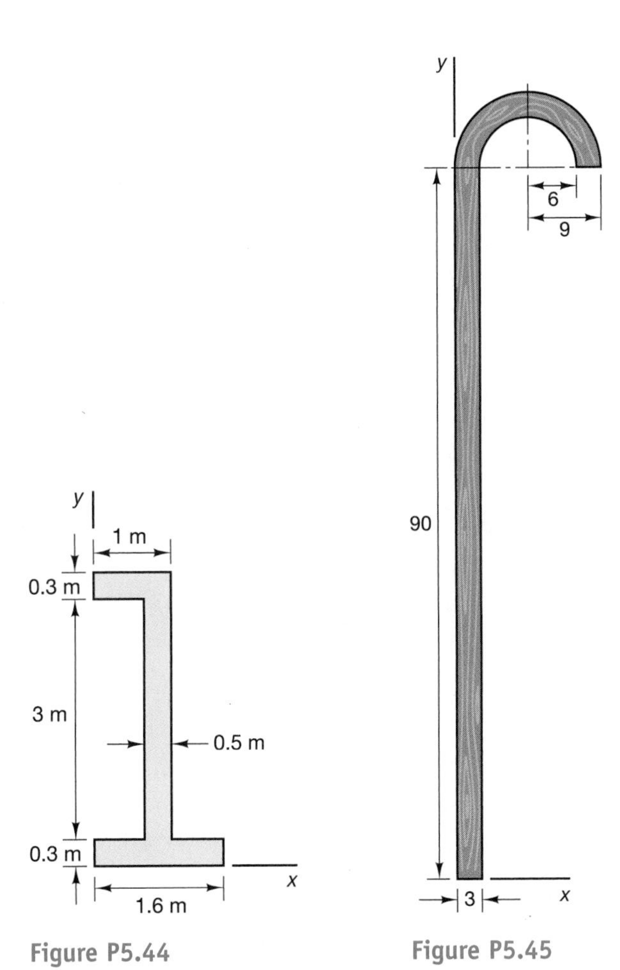

Figure P5.44

Figure P5.45

5.45 Compute the centroid of the cross section of a walking cane. (See Figure P5.45.) The dimensions are in centimeters. Is the centroid on the cane?

5.46 Compute the area centroid of the wooden shelf illustrated in Figure P5.46. The boards are cut from wood that is 1 cm thick.

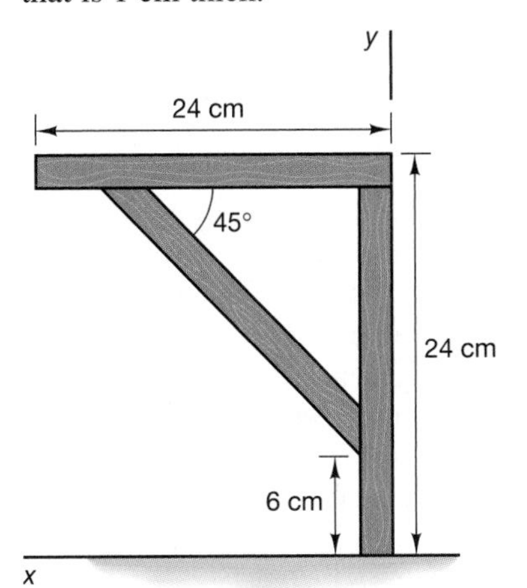

Figure P5.46

Volumes

5.47 Two concrete columns are connected to form a support structure for a dock. (See Figure P5.47.) Compute the centroid of the composite body. The dimensions are in meters.

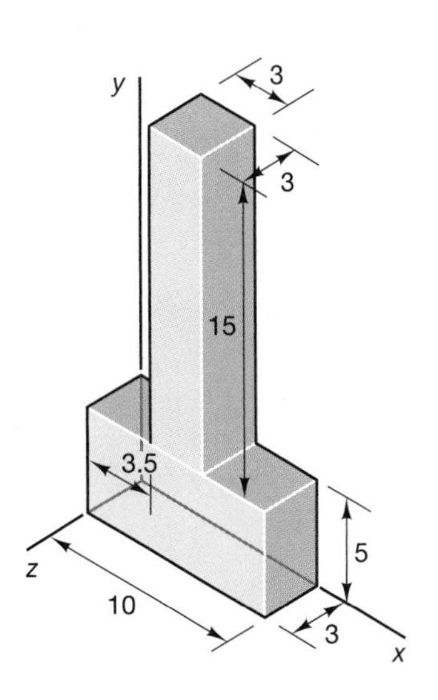

Figure P5.47

5.48 Compute the centroid of the machine part illustrated in Figure P5.48 after two holes are cut into it. Hole 2 is 1 cm in diameter and runs all the way through the block. The other hole, labeled 3, is a 1-cm-diameter hole drilled to a depth of 2 cm Do the holes alter the centroid of the block very much?

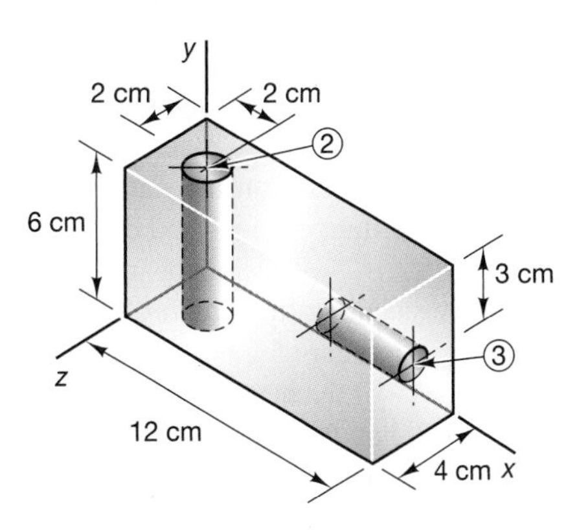

Figure P5.48

5.49 A paraboloid is cut out of a rectangular block of metal. (See Figure P5.49.) The paraboloid is 5 cm deep, and the rectangular parallelpiped is 7 cm high. The radius of the paraboloid is 2 cm. Calculate the centroid and volume of the machined part.

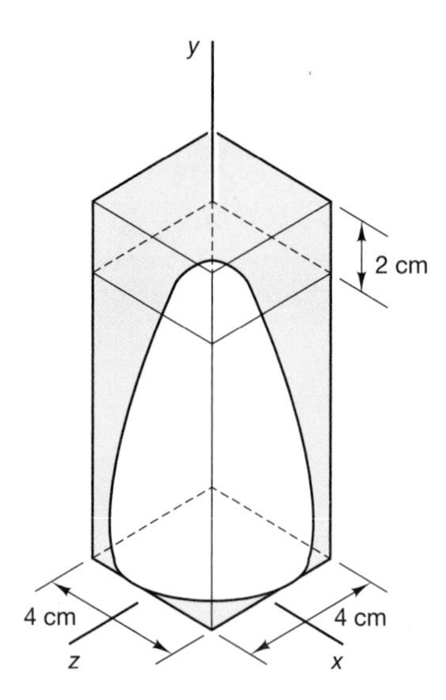

Figure P5.49

5.50 Three rectangular pieces of 6-mm-stock steel are welded together using materials of roughly the same density. (See Figure P5.50.) Approximate the weld by a solid rectangular tetrahedron at the intersection of the three plates, and compute the centroid and total volume of the entire piece.

5.51 The jetway leading to an aircraft is a telescopic device formed by a series of interconnected rectangular parallelpipeds. The designer of this device needs to know where the center of gravity is at each position as the jetway telescopes out along the x-axis. The length of each rectangular parallelpiped is 8 m, and the walls are 0.08 m thick. The first section is 8 m $\times$ 4 m $\times$ 4 m. Using the centerline coordinate system illustrated in Figure P5.51, compute the position of the centroid as the section extends from zero to 6 m past the edge of the first section.

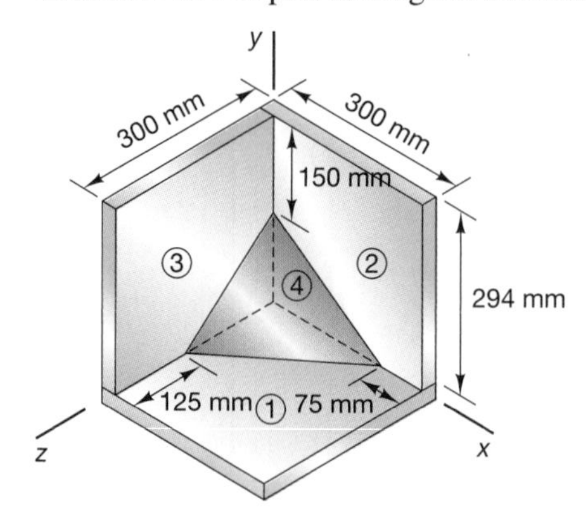

Figure P5.50

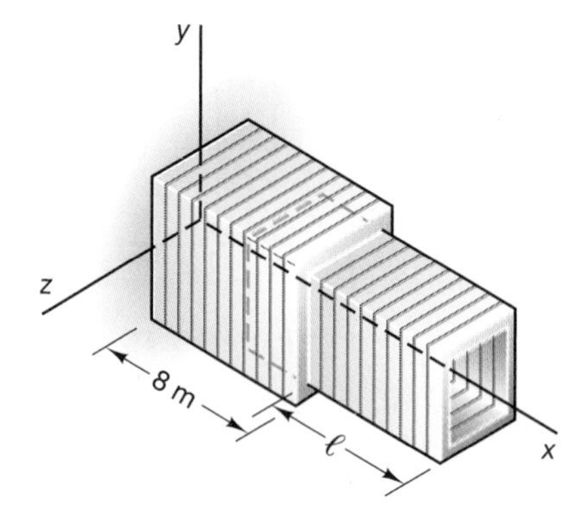

Figure P5.51

5.6 DISTRIBUTED LOADS ON BEAMS

A structural member common in many engineering applications is the beam. A ***beam*** is defined as a member whose length is much greater than its depth or width. Beams are usually straight members and are used to support loads that are perpendicular to their long axis. These members support bending loads and therefore are treated as a special case in courses on mechanics of materials or structural analysis. A beam that is initially curved must be analyzed differently than a straight beam.

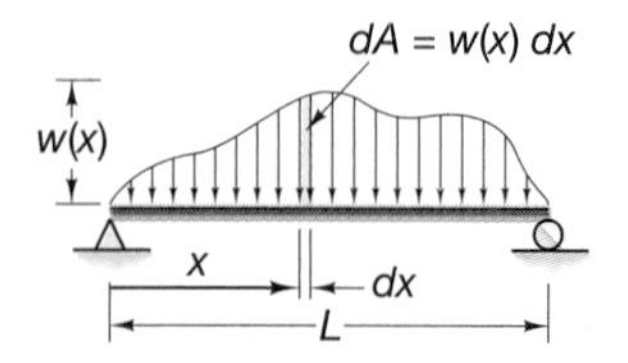

Figure 5.23

A straight beam is modeled as a two-dimensional member with the loading distributed along the length of the beam in the direction perpendicular to its long axis, as shown in Figure 5.23. The distributed load is shown as a function of x, the position along the length of the beam, and is written as $w(x)$, given in N/m. Note that this distribution can be treated as the area A under the loading curve with an element of area $dA = w(x)\,dx$. The total equivalent load on a beam of length L is

$$W = \int_0^L w(x)dx = A \tag{5.32}$$

The distributed load on the beam can be considered as a continuous set of parallel forces. Thus, an equivalent force system can be found consisting of a single force W acting at a specific point on the beam. The development of this type of equivalent force system was presented in Section 4.9. The single force must produce the same moment about the origin as the distributed force system; therefore,

$$x_c W = \int_0^L w(x)x dx$$
$$x_c = \frac{1}{W}\int_0^L w(x)x dx \qquad (5.33)$$

Since W is equal to the area A under the loading curve, and since wdx is an element of this area dA, Eq. (5.33) is equivalent to determining the centroid of the area under the loading curve.

Therefore, if only the external support reactions are required, a distributed load on a beam can be replaced by an equivalent force system consisting of a resultant equal to the area under the loading curve acting through the centroid of that area.

If the internal forces in the beam are to be calculated, a similar approach is used when the beam is divided into sections. (See Chapter 8.)

To analyze the internal stresses (the internal distributed loading per unit area) in beams, the centroid of the cross-sectional area of the beam must be determined. It is important to realize that this centroid has nothing to do with the centroid of the loading area. The replacement of the distributed load with an equivalent single force is useful in computing the reactions acting on the beam, but this equivalency cannot be used in determining the internal forces acting on the beam or the deformation of the beam.

Sample Problem 5.8

Determine the equivalent force system consisting of a single concentrated load for the beam loaded as shown in the figure at left.

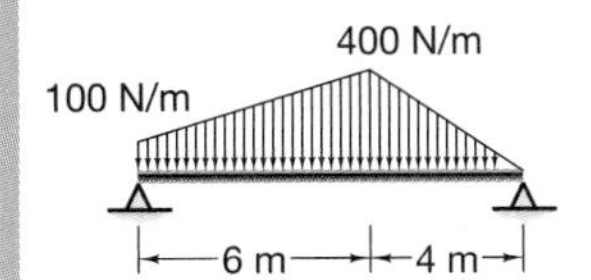

Solution Divide the beam loading into three regions, as follows:

The equivalent force system can be found from the accompanying table, similarly to the case of a centroid of an area. We have

$$A_1 = 100(6) = 600 \text{ N} \qquad x_1 = 3 \text{ m}$$
$$A_2 = 1/2\ 300(6) = 900 \text{ N} \qquad x_2 = 2/3(6) = 4 \text{ m}$$
$$A_3 = 1/2\ 400(4) = 800 \text{ N} \qquad x_3 = 6 + 1/3(4) = 7.33 \text{ m}$$

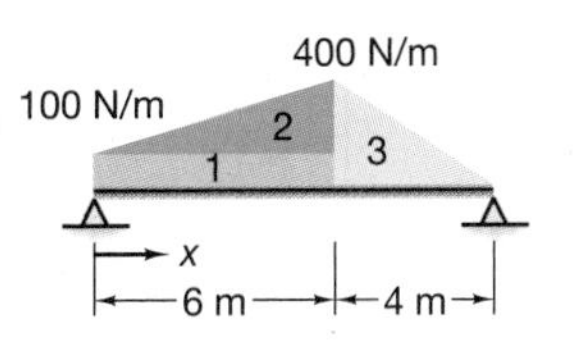

Load Area	*Area dA_i*	*Distance to Centroid X_i*	*$X_i dA_i$*
1	600 N	3 m	1800 Nm
2	900 N	4 m	3600 Nm
3	800 N	7.33 m	5864 Nm
Σ	2300 N		11,264 Nm

The equivalent load of 2300 N acts at a distance $x_c = 11{,}264/2300 = 4.9$ m from the left end of the beam. The external support reactions can now be computed using this equivalent load.

Problems

5.52 Calculate the equivalent force system for the load shown in Figure P5.52.

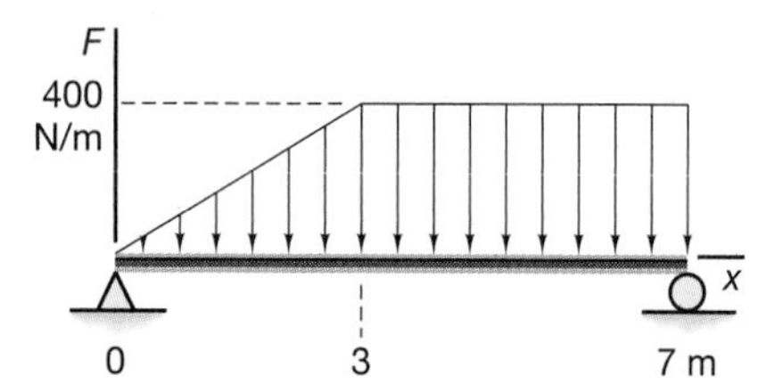

Figure P5.52

5.53 In Figure P5.53, compute the equivalent load and its point of application.

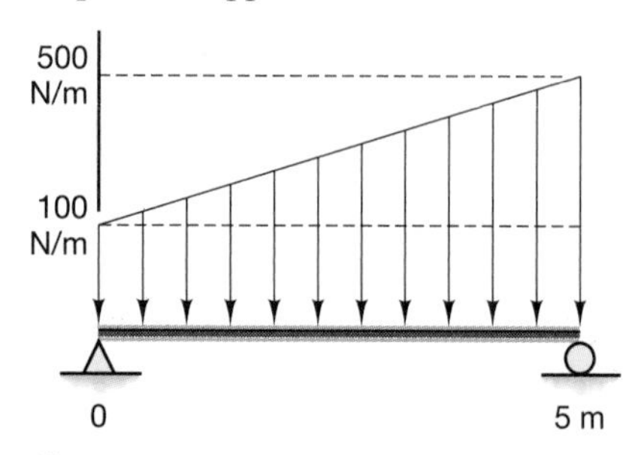

Figure P5.53

5.54 A load of sand is distributed along a truck bed. (See Figure P5.54.) Compute the equivalent load and its point of application.

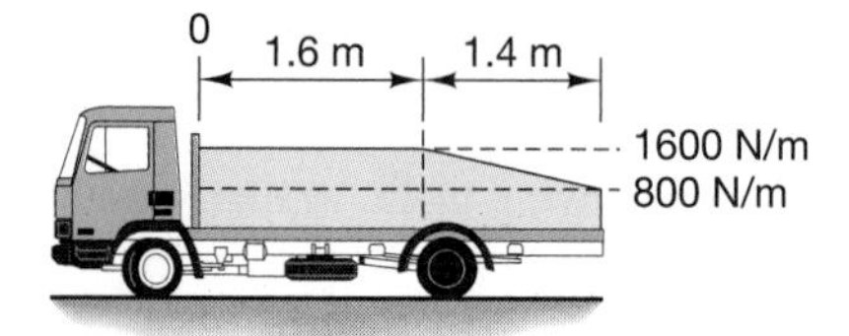

Figure P5.54

5.55 Replace the load distributed along the beam shown in Figure P5.55 with a single force acting at a particular point.

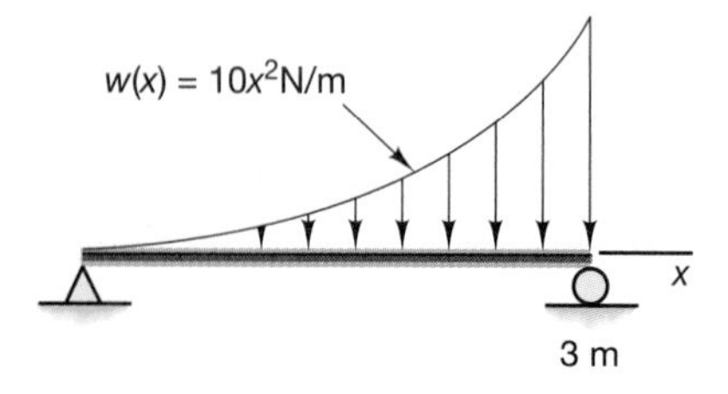

Figure P5.55

5.56 A wind load is applied to an eight-story building. (See Figure P5.56.) It is known that the wind load may be modeled by $w(x) = 10(x^3 + 10x^2)$ N/m. Calculate the equivalent point load and its position of application.

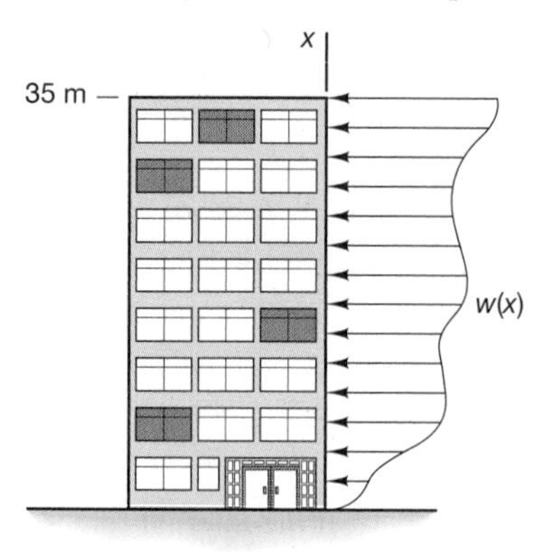

Figure P5.56

5.57 Repeat Problem 5.56, but now assume that the wind load is given by $w(x) = 10\left(1 + \sin\left(\frac{\pi x}{50}\right)\right)$ N/m.

5.58 Repeat Problem 5.56 with a windload of $w(x) = 10(1 + \sin x)$ N/m.

5.59 Calculate an equivalent force and point of application for the lift pressure applied to an airplane wing for the case $p(x) = 1000\, x \sin(x)$ N/m. (See Figure P5.59.)

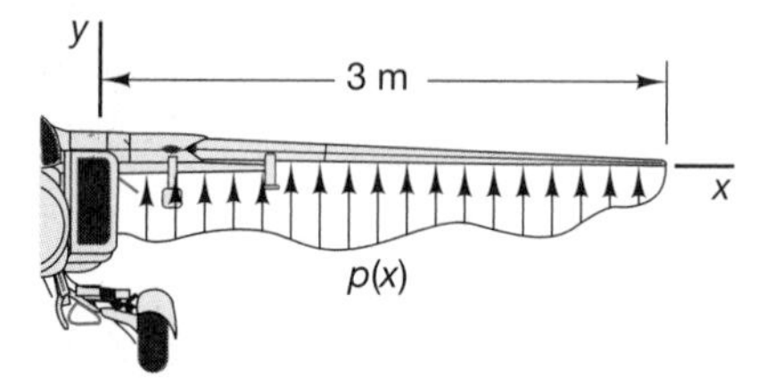

Figure P5.59

5.60 A side-impact air bag for an automobile may be modeled as a distributed load by $w(x) = 100(1 + \cos x)$ N/m. Compute the equivalent force and point of action for this system. Use Figure P5.60.

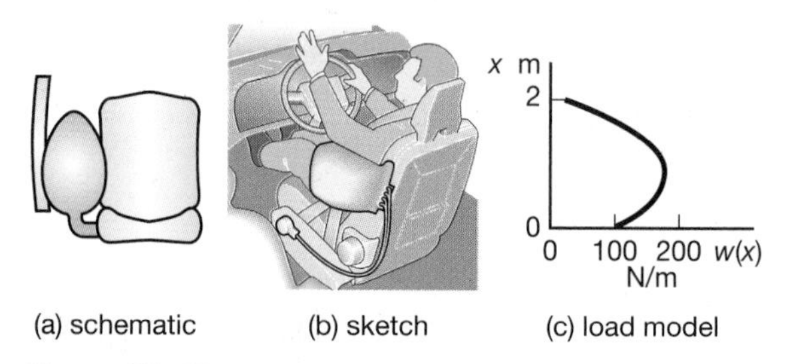

Figure P5.60

5.61 Snow and ice have accumulated on a 8-m-long flat roof. (See Figure P5.61.) The resulting distributed load is $w(x) = 1000\left(x \cos \frac{\pi x}{40}\right)^2$ N/m. Compute the equivalent point-force system.

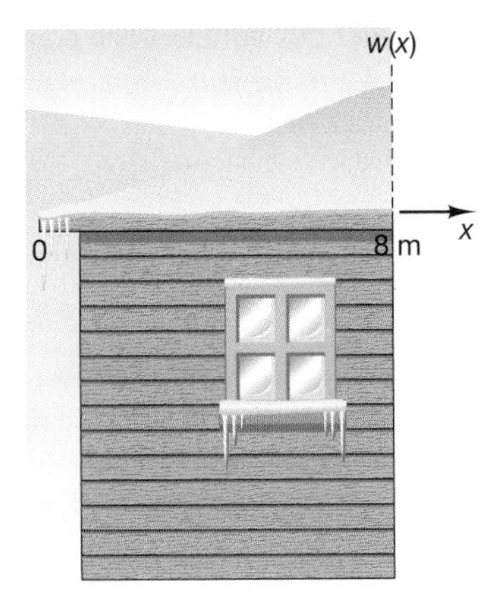

Figure P5.61

5.62 Calculate the equivalent load on the glass skylight shown in Figure P5.62 due to the distributed snow load, assuming that $w_1(x) = 10 \sin(x)$ N/m and $w_2(x) = 2 \sin(x)$ N/m representing the "missing" load due to the curvature of the glass.

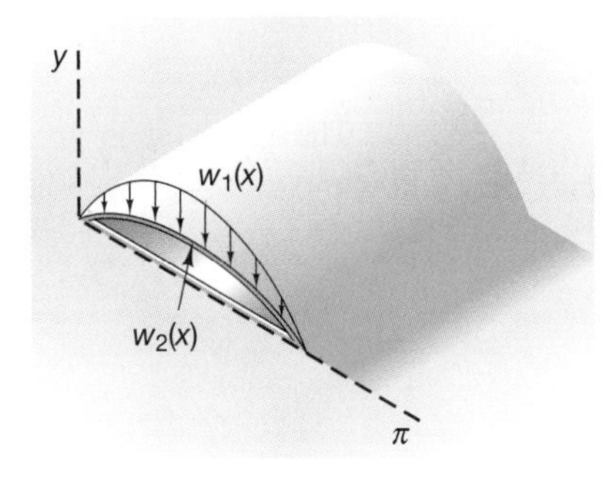

Figure P5.62

5.7 FORCES DUE TO FLUID PRESSURE ACTING ON A SUBMERGED SURFACE

A fluid at rest exerts a *hydrostatic pressure* at any point in it. According to Pascal's law, this pressure is equal in all directions. The magnitude of the pressure at any point is equal to the surface pressure plus the weight of the fluid above the point. Therefore, the absolute pressure on a point A at a depth d below the surface of a fluid is

$$p_A = p_s + \gamma d \tag{5.34}$$

where p_s is the atmospheric pressure on the surface, γ is the specific weight of the fluid, and d is the depth of point A below the surface. Equation (5.34) can be written in terms of the density of the fluid, ρ, as

$$p_A = p_s + \rho g d \tag{5.35}$$

where g is the gravitational acceleration constant. The specific weight of fresh water is 9800 N/m^3, and the density of water is 1000 kg/m^3 in SI units. Although Pascal's law is true for either a gas or a liquid, Eqs. (5.34) and (5.35) hold only for incompressible fluids, since the density of a gas varies with both altitude and temperature.

Consider the force (the resultant of the hydrostatic pressure) on the submerged rectangular plate whose edge is shown in Figure 5.24. The pressure p at any point x along the plate is

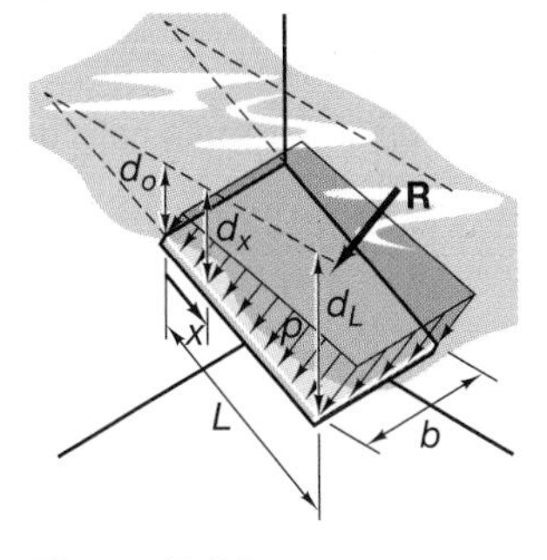

Figure 5.24

$$p = \rho g d_x$$

where d_x is the distance to the free surface of the liquid at point x and ρ and g are as defined before. The force on a differential element dx is

$$dF = \rho g d_x b dx \tag{5.36}$$

where b is the uniform width of the plate. The loading volume on the surface of the plate is trapezoidal, and the resultant force passes through the centroid of this volume. The point of application of the resultant force on the plate is called the *center of pressure.* If d_0 is the depth of the plate at $x = 0$ and d_L is the depth of the plate at $x = L$, then

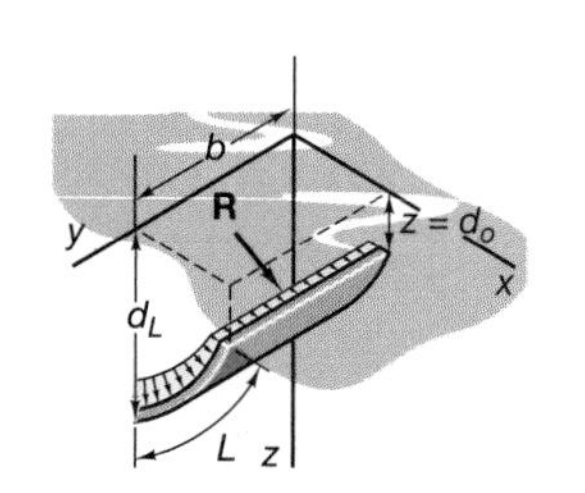

Figure 5.25

$$R = 1/2(d_0 + d_L)\rho gLb \tag{5.37}$$

The resultant is normal to the plate, acting through the centroid of the volume, given by

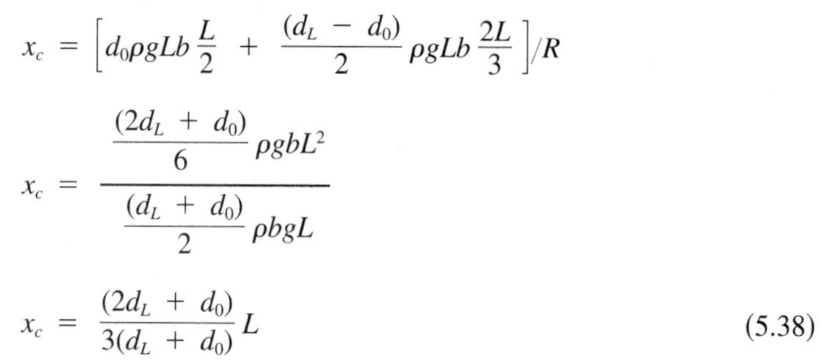

$$x_c = \left[d_0\rho gLb\frac{L}{2} + \frac{(d_L - d_0)}{2}\rho gLb\frac{2L}{3}\right]/R$$

$$x_c = \frac{\dfrac{(2d_L + d_0)}{6}\rho gbL^2}{\dfrac{(d_L + d_0)}{2}\rho bgL}$$

$$x_c = \frac{(2d_L + d_0)}{3(d_L + d_0)}L \tag{5.38}$$

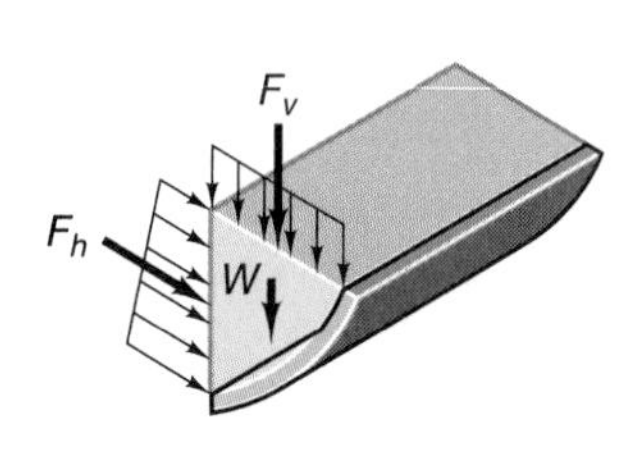

Figure 5.26

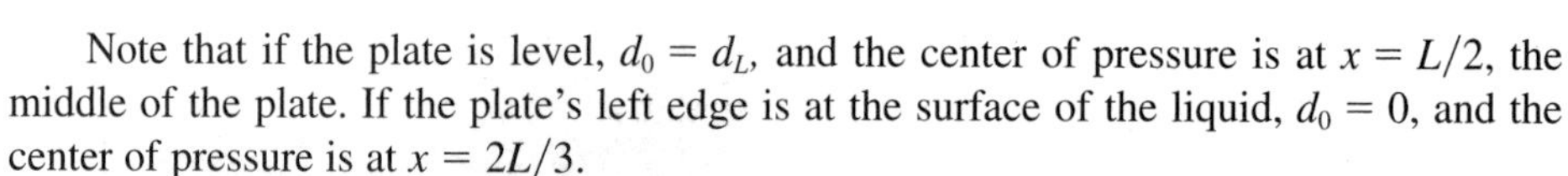

Note that if the plate is level, $d_0 = d_L$, and the center of pressure is at $x = L/2$, the middle of the plate. If the plate's left edge is at the surface of the liquid, $d_0 = 0$, and the center of pressure is at $x = 2L/3$.

If a curved plate of constant width b is submerged in a liquid, the pressure acting normal to the surface of the plate changes direction, such that it remains normal to the surface, as shown in Figure 5.25. The calculation of the resultant force $\boldsymbol{R}$ and its line of action is more difficult: Integration must be used to determine the volume under the pressure loading and the centroid of this volume.

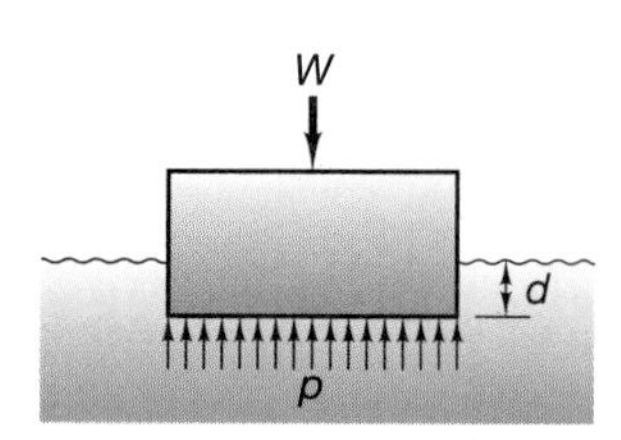

Figure 5.27

Another method to determine these values is to construct a free-body diagram of the plate and the volume of liquid bounded by it, as shown in Figure 5.26. The force $\boldsymbol{F}_v$ is the vertical force of the fluid above the volume. The force $\boldsymbol{F}_h$ is the horizontal force caused by the pressure acting horizontal to the volume. Both of these forces can be calculated, since they act on plane surfaces. $\boldsymbol{W}$ is the weight of the fluid contained in the volume and may be computed if the curve of the plate is known. These three forces have known lines of action and may be considered as coplanar forces. Thus, an equivalent resultant $\boldsymbol{R}$, with a specific line of action, may be determined by using the methods outlined in Section 4.9, yielding

$$R = F_v + F_h + W \tag{5.39}$$

If the curve of the plate can be expressed as a simple function, the resultant force and the center of pressure can be easily determined.

5.7.1 BUOYANCY

Archimedes discovered the principle of buoyancy and understood why an object could float in any fluid, gas or liquid. Consider a box floating in a liquid, as shown in Figure 5.27. The pressure on the bottom of the box is

$$p = \rho gd \tag{5.40}$$

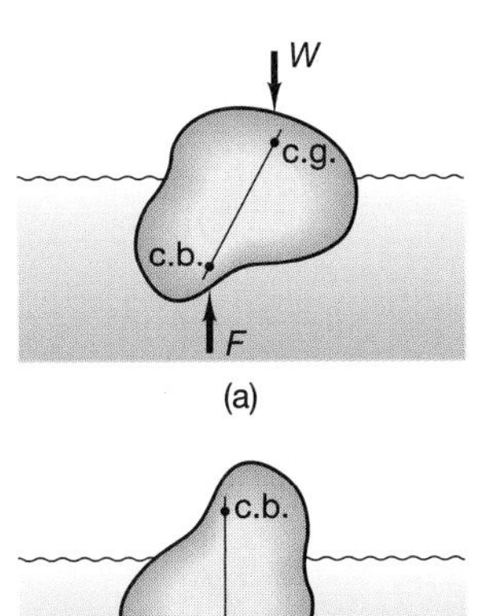

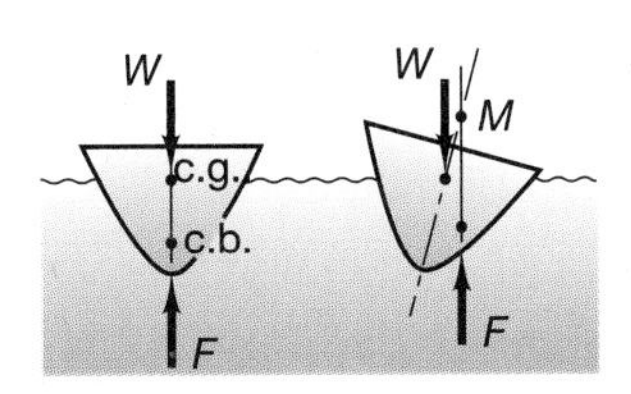

Figure 5.28 If an object's center of gravity is not on a vertical line with the center of buoyancy (a), the object rotates until the two centers line up vertically (b).

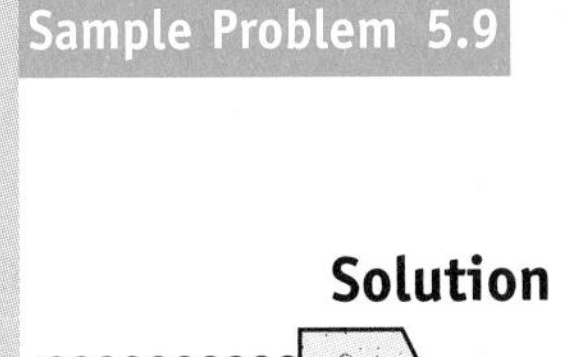

Figure 5.29

The total resultant force acting across the bottom of the box is equal to

$$R = pA = \rho\, gdA \tag{5.41}$$

where A is the area of the bottom. The box floats and is in equilibrium when the resultant, called the force of *buoyancy*, equals the weight of the box. However, dA is the volume of water displaced by the box. Therefore, the force of buoyancy is equal to the weight of the fluid that was displaced, and its line of action passes through the center of gravity of the displaced water. If the weight of the box is increased, the box will settle farther beneath the surface, until the weight of the displaced fluid is equal to the weight of the box. If the weight of the box is greater than the weight of an equal volume of water, the box will sink.

The line of action of the force of buoyancy passes through the center of gravity of the displaced water, called the center of buoyancy. If the center of gravity of the object is not on a vertical line with the center of buoyancy, the object will roll until the two centers are on a vertical line, as shown in Figure 5.28(b). Note that, in Figure 5.28(a), the weight and the force of buoyancy form a couple. The object rolls clockwise until an equilibrium orientation is reached that coincides with the center of buoyancy located vertically above the center of gravity.

The locations of these two centers are of great importance in analyzing the stability of a floating object, particularly ships, boats, and so on. The center of gravity of a ship remains at the same point as the ship rolls, but the center of buoyancy moves as the shape of the displaced volume changes. (Note that the displaced volume is constant, but as the shape of the volume changes, the center of buoyancy changes.) Ships are designed such that they are stable for lists up to about 20°. This angle of list is illustrated for the cross section of the hull shown in Figure 5.29.

The point of intersection of the vertical line through the center of buoyancy with the centerline of the cross section of the hull is called the ***metacenter*** M, and the distance h of the metacenter above the center of gravity is called the metacentric height. For most ship designs, this height remains almost constant for lists up to 20°. When M is above the center of gravity, the resulting couple tends to right the ship, and the hull design is stable. If the metacenter is below the center of gravity when the ship lists, the moment of the couple is in the direction of the list and causes a greater list—clearly an unstable situation.

Sample Problem 5.9

The water behind a dam is 50 meters deep, as shown in the figure to the left. Determine the magnitude of the resultant force **R** exerted on a 10-m length of the dam by the water pressure. Determine the distance from the surface of the water to the center of pressure.

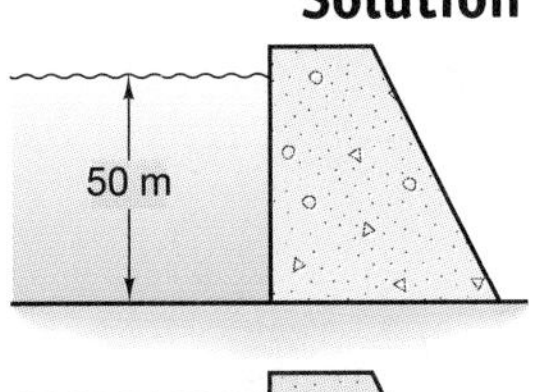

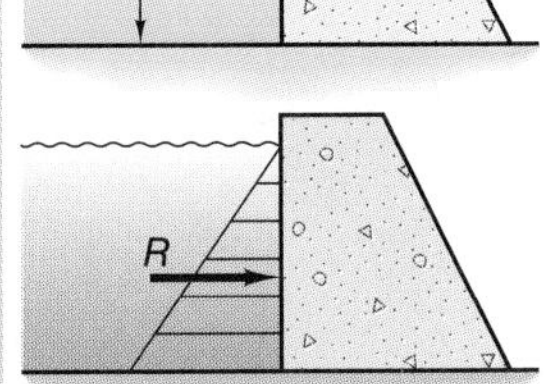

Solution

Neglecting atmospheric pressure, we find that the pressure at the base of the dam is

$$p = \rho\, gd = (1000)(9.81)(50) = 490.5 \text{ kN/m}^2$$

Therefore, the resultant **R** on a 10-m length of the dam is

$$R = 1/2(490{,}500)(50)(10) = 122.6 \text{ MN}$$

The resultant acts at two-thirds the depth of the dam:

$$d_R = 2/3\,(50) = 33.3 \text{ m}$$

Problems

5.63 A seawall is constructed along a freshwater channel from a flat, vertical surface. (See Figure P5.63.) Compute the magnitude of the resultant force $\boldsymbol{R}$ along a 25-m length of the channel and the distance from the surface of the water to the center of pressure.

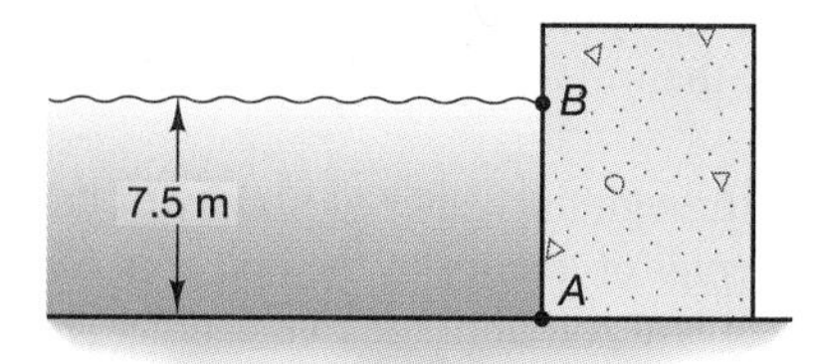

Figure P5.63

5.64 A seawall consisting of a slanted wall at 45° is constructed along a freshwater channel. (See Figure P5.64.) Compute the magnitude of the resultant force $\boldsymbol{R}$ along a 25-m length of the channel.

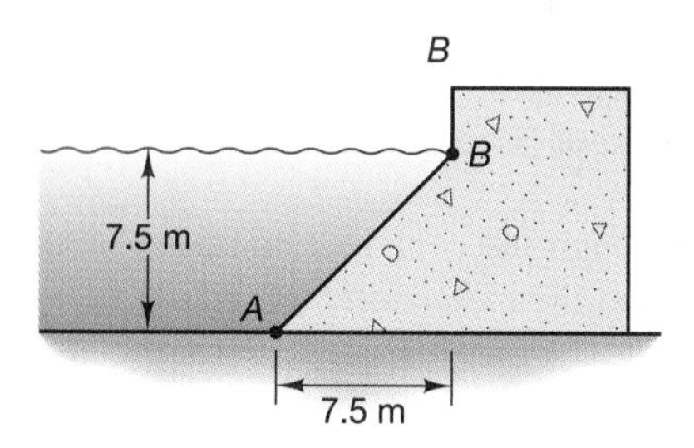

Figure P5.64

5.65 A seawall consisting of a circular wall of radius 7.5 m is constructed along a freshwater channel. (See Figure P5.65.) Compute the magnitude of the resultant force $\boldsymbol{R}$ along a 25-m length of the channel.

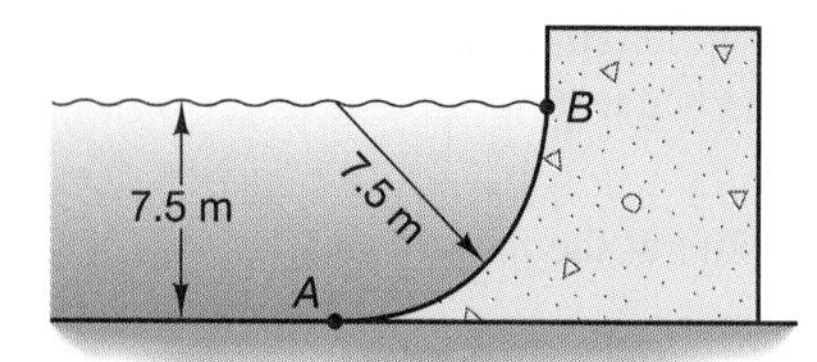

Figure P5.65

5.66 A seawall is constructed along a freshwater channel ($\gamma = 9800\ \text{N/m}^3$) in the cross-sectional shape of a parabola with vertex at point A, as shown in Figure P5.66. Compute the magnitude of the resultant force $\boldsymbol{R}$ along a 2.5-m length of channel and a point through which this force must act.

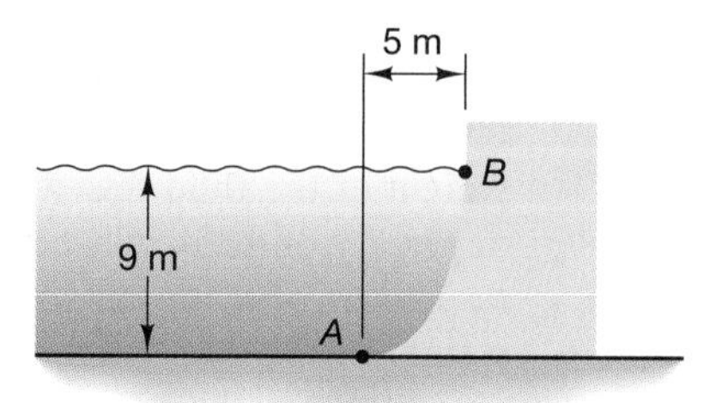

Figure P5.66

5.67 A glass box is filled with colored water in order to be used as a bookend. (See Figure P5.67.) Compute the location and magnitude of the resultant force acting on each side and on the bottom of the box.

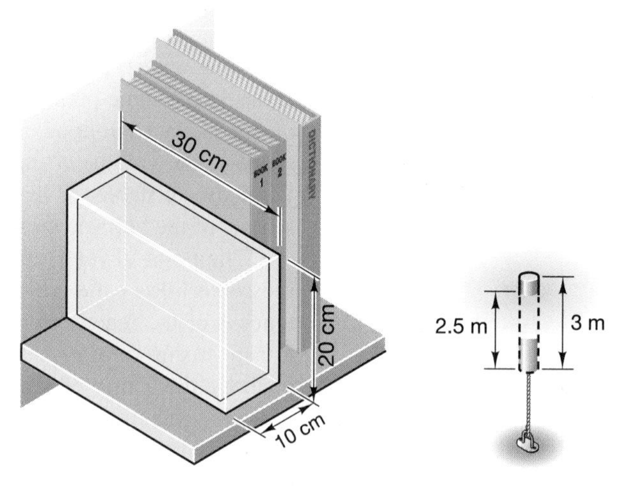

Figure P5.67

Figure P5.68

5.68 A marker buoy is held in a freshwater lake by a rope. Compute the tension in the cable that holds the buoy in the position shown in Figure P5.68. Consider the buoy to be a hollow cylinder of mass 60 kg and radius 0.1 m.

5.69 How much weight placed at the center can the rectangular "raft" shown in Figure P5.69 hold before

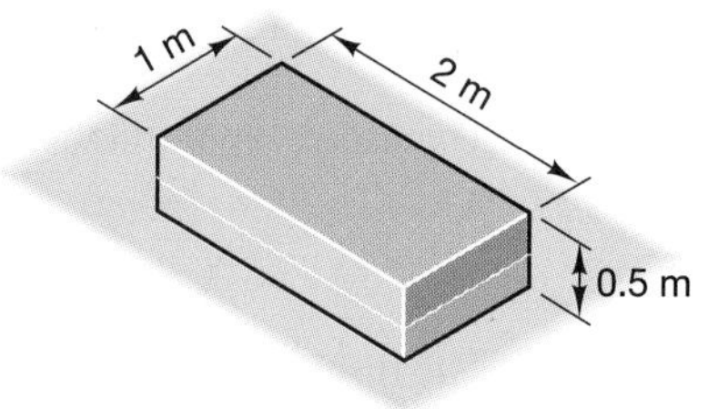

Figure P5.69

submerging? The raft is a hollow $1 \times 2 \times 0.5$-m box and has a mass of 40 kg.

5.70 A hollow, plastic cylindrical drum of negligible weight is to be designed to hold up a floating dock. (See Figure P5.70.) Each drum must support 2000 N and be no longer than 1 m. Calculate the radius of the cylinder that must be used to float the dock.

Figure P5.70

Chapter Summary

Centroid The centroid is the geometric center of a line, area, or volume. The centroid is determined by using the first moment of the line, area, or volume, defined as

$$\mathbf{M} = \int_L \mathbf{r}\, dL \quad \text{(line)}$$

$$\mathbf{M} = \int_A \mathbf{r}\, dA \quad \text{(area)}$$

$$\mathbf{M} = \int_V \mathbf{r}\, dV \quad \text{(volume)}$$

The centroid is defined as

$$\mathbf{r}_c = \frac{1}{L}\int_L \mathbf{r}\, dL \quad \text{(line)}$$

$$\mathbf{r}_c = \frac{1}{A}\int_A \mathbf{r}\, dA \quad \text{(area)}$$

$$\mathbf{r}_c = \frac{1}{V}\int_V \mathbf{r}\, dV \quad \text{(volume)}$$

Center of Mass The center of mass of a rigid body is the point at which all the mass is considered to be concentrated and is defined as

$$\mathbf{r}_{cm} = \frac{1}{M}\int \mathbf{r}\, dm$$

where M is the total mass.

Theorems of Pappus and Guldinus

Pappus and Guldinus Theorem 1:

The area A *of a surface of revolution generated by revolving a plane curve of length* L *about any nonintersecting axis in its plane is equal to the product of the length of the curve and the length of the path traveled by the centroid of the curve (the length of the generating curve times the distance traveled by the centroid of the generating curve).*

Pappus and Guldinus Theorem 2:

The volume V *of the solid of revolution generated by revolving a plane area* A *about any nonintersecting axis in its plane is equal to the product of the area and the length of the path traveled by the centroid of the area (the product of the generating area times the distance traveled by the centroid of the generating area).*

Centroids of Composite Bodies If the centroids of simple geometric shapes are known and a line, area, or volume can be considered as a composite of these simple shapes, the centroid may be defined as a finite sum rather than an integral:

$$\mathbf{r}_c = \frac{1}{L}\sum_i \mathbf{r}_i dL_i \quad \text{(line)}$$

$$\mathbf{r}_c = \frac{1}{A}\sum_i \mathbf{r}_i dA_i \quad \text{(area)}$$

$$\mathbf{r}_c = \frac{1}{V}\sum_i \mathbf{r}_i dV_i \quad \text{(volume)}$$

Distributed Loads on a Beam The distributed load on a beam is given as a function $w(x)$ N/m (lb•ft) of the position x along the beam. The total equivalent load is

$$W = \int_0^L w(x)dx = A$$

or the area under the load distribution curve. The equivalent load can be considered to act at a point on the beam such that the moment of the equivalent load is equal to the moment of the distributed load. This point is coincident with the centroid of the area under the load distribution curve, so that

$$x_c W = \int_0^L w(x)xdx$$

$$x_c = \frac{1}{W}\int_0^L w(x)xdx$$

The equivalent load may be used only to determine the reactions at the beam supports. (See Chapter 7 for the determination of the internal forces and moments in beams.)

Forces on Submerged Surfaces A fluid at rest exerts a hydrostatic pressure on each point on a submerged surface. This pressure is equal to the surface pressure plus the weight of the fluid above the point on the submerged surface. The volume under the pressure distribution plot is equal to the total force acting on the submerged surface. This equivalent total force will pass through the center of pressure, a point that is coincident with the centroid of the volume under the pressure distribution plot.

Buoyancy When an object floats in a liquid, the object will be held in equilibrium by a vertical force of buoyancy equal in magnitude to the weight of the liquid displaced by the object. The object will float if the force of buoyancy is equal to the weight of the object. The maximum force of buoyancy is equal to the weight of a volume of liquid equal to the volume of the object. If the weight of the object is greater than the maximum force of buoyancy, the object will sink in the liquid. The line of action of the force of buoyancy passes through the centroid of the volume of the displaced liquid, which is the center of buoyancy. When an object floats, the vertical line of action of the force of buoyancy must pass through the center of buoyancy and the center of gravity of the object.

The centroid of an area and the center of gravity of a body are defined mathematically in this chapter. The centroid of an area is very important in the study of distributed loads and will be used in courses in Mechanics or Strength of Materials. It is presented in Statics as preparation for advanced courses. In a similar manner, the center of gravity is defined for the purpose of locating the center of mass in dynamics problems and to designate the point of application of the weight vector when analyzing equilibrium in rigid bodies.

This chapter is dependent upon integral calculus and many of the concepts are introduced in calculus courses. Applications to distributed forces on beams and fluid pressure and buoyancy are presented.

Chapter 6

EQUILIBRIUM OF RIGID BODIES

The cargo container is held in equilibrium by the crane. (Photo cortesy of Tomasz Gulla/ Shutterstock)

6.1 INTRODUCTION

The concept of equilibrium of a particle was introduced in Chapter 3. Newton's second law states that if there is a net force acting upon a particle, there will be a change in the motion of that particle. If it is observed that an object (a rigid body or a particle) is at rest or has no change in its state of motion, then the object is said to be in equilibrium. Newton's second law then implies that there is no net force acting upon the body. For a single particle, we wrote three scalar equations of *equilibrium.* This system of three equations was solved for three unknown reactions. In Chapter 4, we examined equivalent force systems, and the turning effect of a force was defined as the moment acting on a body about a point on the body. When an object is modeled as a particle, all forces acting on the particle are concurrent, and the moment of these forces about the concurrent point is zero, as all forces pass through this point. The particle is then in equilibrium if the resultant of the concurrent force system is zero.

The most general force system acting upon a rigid body may be expressed as a resultant force **R** with a specific line of action and the moment of a couple **C**. The value of the moment of the couple changes if a different equivalent force system is developed such that **R** has a different point of application and a different line of action. For a rigid body in a state of equilibrium, the resultant force and the moment of the couple of any equivalent force system acting upon that body is zero. This yields the two vector equations of equilibrium:

$$\mathbf{R} = \sum F_x\hat{\mathbf{i}} + \sum F_y\hat{\mathbf{j}} + \sum F_z\hat{\mathbf{k}} = 0$$

$$\mathbf{M_O} = \sum \mathbf{r} \times \mathbf{F} = \sum M_x\hat{\mathbf{i}} + \sum M_y\hat{\mathbf{j}} + \sum M_z\hat{\mathbf{k}} = 0 \qquad (6.1)$$

Note that a vector is zero if and only if each of its components is zero. Hence, these two vector equations of equilibrium may be written as the six scalar equations

$$\sum F_x = 0 \quad \sum M_x = 0$$

$$\sum F_y = 0 \quad \sum M_y = 0$$

$$\sum F_z = 0 \quad \sum M_z = 0 \qquad (6.2)$$

Equations (6.1) and (6.2) state that if a rigid body is in equilibrium, the sum of the forces in each direction is zero, and the sum of the moments about the x-, y-, and z-axes through any point on the body is zero. In general, the equilibrium of a single rigid body is expressed by the system of six equations shown in Eq. (6.2). Matrix notation was introduced in Chapter 2 to facilitate the solution of systems of linear equations.

If a force system is represented by the resultant force **R** at a point A and a moment $\mathbf{C}_A$, an equivalent force system at point B has a moment

$$\mathbf{C_B} = \mathbf{C_A} + \mathbf{r}_{A/B} \times \mathbf{R} \qquad (6.3)$$

If the moment $\mathbf{C}_A$ and the resultant force **R** are both zero (the equilibrium condition), then the moment $\mathbf{C}_B$ is zero for all values of the relative position vector $\mathbf{r}_{A/B}$. Therefore, for a rigid body to be in equilibrium, the moment about any point on the body must be zero. Although equations may be written for the moment about many points in the body, there are only *six independent equations of static equilibrium* for a single rigid body. Thus, taking the moment about two different points does not necessarily generate new independent equations.

Forces acting upon a body are either body forces, such as gravitational forces, or surface forces. The surface forces occur at points of contact with other bodies and at supports. The body forces and the surface forces at points of contact with other bodies are generally known, and the forces and moments at points of support will, in general, be unknown. Since there are six scalar equations in Eq. (6.2), six unknown components of support forces or moments may be determined by the equations of equilibrium. Still, conditions will be encountered in which there are more supports than are necessary to maintain equilibrium, and problems involving such conditions will be classified as *statically indeterminate* problems. (We will discuss these problems in more detail in Section 6.7.)

Before we attempt to solve equilibrium equations, an object must be modeled as either a particle or a rigid body, and then a *free-body diagram* must be created showing all the applied forces and moments and the support reactions. Modeling the body in this manner is an art based upon experience. The free-body diagram is a worksheet that will be used to write the equations of equilibrium. Therefore, the **free-body diagram** should show all assumptions that have been made, all important geometrical dimensions, and the locations and directions of all forces acting on the body. This is a very important aspect of the analysis, for if a correct free-body diagram is made, the solution of any problem of static equilibrium can be reduced to the solution of simultaneous linear equations. Thus, once the correct free-body diagram is completed, the solution becomes systematic. Methods of solving a system of linear equations using matrix notation were discussed in Section 2.7.

We will first examine equilibrium problems in which the body can be modeled as a two-dimensional object, and then the analysis will be extended to three dimensions. In a two-dimensional problem, the forces will be coplanar, and the moment of these forces taken about any point will be perpendicular to the coplanar forces and, therefore, perpendicular to the plane of the forces. Therefore, the problem is three dimensional in a vector sense. However, in the two-dimensional problem, the moment has only one component. Hence, only three of the six equations of static equilibrium are involved in the special case of a two-dimensional problem. If a coordinate system is chosen such that the x–y plane contains all the forces, the only component of the moment is in the z-direction. The three nontrivial scalar equilibrium equations are

$$\sum F_x = 0$$

$$\sum F_y = 0 \qquad (6.4)$$

$$\sum M_z = 0$$

6.2 SUPPORTS FOR A TWO-DIMENSIONAL MODEL

If we model a rigid body as a two-dimensional object, all the forces acting on it lie in a common plane, and all moments acting on the body must be perpendicular to that plane. The ***supports*** also must be treated as two dimensional. Supports on a body restrict movement of the body at the point where they act. In the two-dimensional model, the supports may restrict translation in one or two directions, or they may restrict rotation about an axis perpendicular to the plane of the body. It is extremely important to become familiar with the various types of supports discussed and to learn the types of constraining forces or moments that occur at a support. We will first consider supports that restrict motion in one direction only.

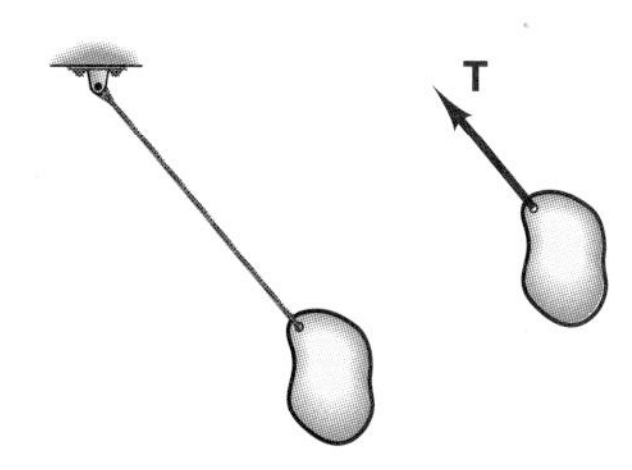

Figure 6.1

The cable shown in Figure 6.1 restricts the motion of the body such that the length of the cable remains constant. The supporting force is the tension in the cable. Many devices, such as suspension bridges, pulley systems, and so on, are supported by ropes or cables. The support of a flexible cord, cable, or rope shown in Figure 6.1 can be represented by a tensile force acting along the cord, running from its point of attachment on the object to

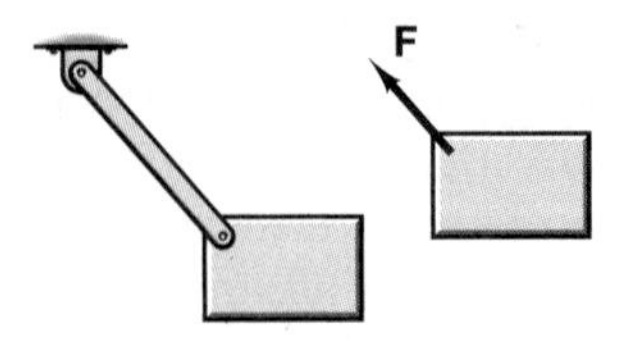

Figure 6.2

the anchor point. Note that this type of support cannot transmit compression; therefore, both the direction and the sense of the support force are known. If the solution for the support force yields a negative value, indicating that the cable must be in compression, then either the model is incorrect or there is an error in the calculations, as this type of support cannot support compression. Such a support was considered in detail in Chapter 3, and the direction of the force is represented by a unit vector acting along the cable from the body to the support connection. This unit vector is obtained from the geometry of the system.

A support in the form of a rigid link can transmit either tension or compression along the axis of the link, as shown in Figure 6.2. The line of action of the force, but not the sense of the vector, is known. The member may be either in tension (as shown) or in compression, and the solution for the force will yield a positive value if the assumed sense is correct or a negative value if the force vector is acting in the opposite sense. There is only one unknown reaction acting at this support.

The ground reaction force between a roller, rocker, or contact with a smooth surface (no friction), as shown in Figure 6.3, can transmit only a ***normal force*** of compression between the object and the surface. The force is called a normal force because it acts perpendicular, or normal, to the contacting surface.

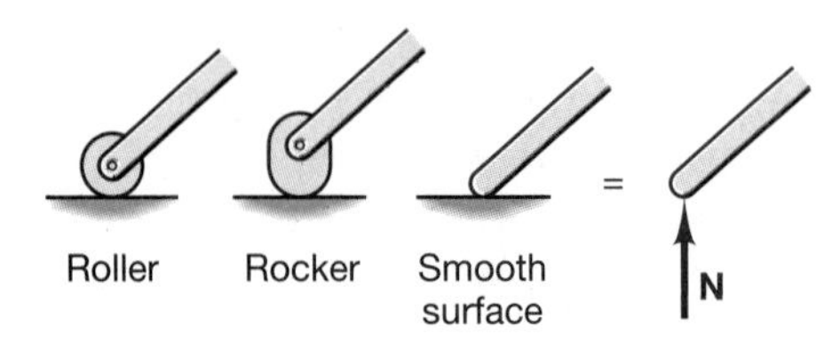

Figure 6.3

Therefore, there is only one unknown reaction at this support, and it can act only in a normal direction, to oppose motion. When the body is supported by contact with inclined surfaces, the direction of the normal force can be obtained from the geometry of the body or by the use of a cross product. An example of the use of trigonometry to obtain a normal unit vector for a two-dimensional problem is shown in Figure 6.4. If the surface is horizontal (in the x, z plane), the normal is in the positive y-direction. Consequently, if the surface makes an angle of θ with the horizontal x-axis, the unit normal vector makes an angle θ with the vertical y-axis. The unit normal vector is

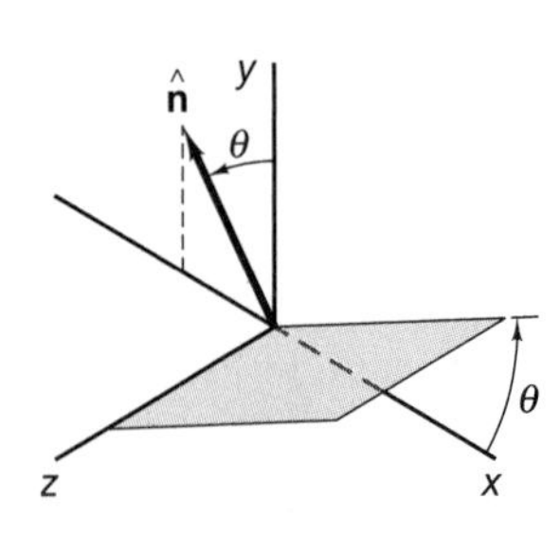

Figure 6.4

$$\hat{\mathbf{n}} = -\sin(\theta)\hat{\mathbf{i}} + \cos(\theta)\hat{\mathbf{j}} \tag{6.5}$$

In this case, the normal was found by examining the drawing.

Another approach to determining the unit vector normal to the surface, which can easily be extended to three-dimensional problems, uses the fact that the cross product is perpendicular to the plane formed by the two vectors in vector multiplication. Consider the surface shown in Figure 6.5, which makes an angle of θ degrees with the horizontal x–z plane. Note that this is the same surface shown in Figure 6.4. The unit normal vector is

$$\hat{\mathbf{n}} = \frac{\hat{\mathbf{T}} \times \hat{\mathbf{t}}}{|\hat{\mathbf{T}} \times \hat{\mathbf{t}}|} \tag{6.6}$$

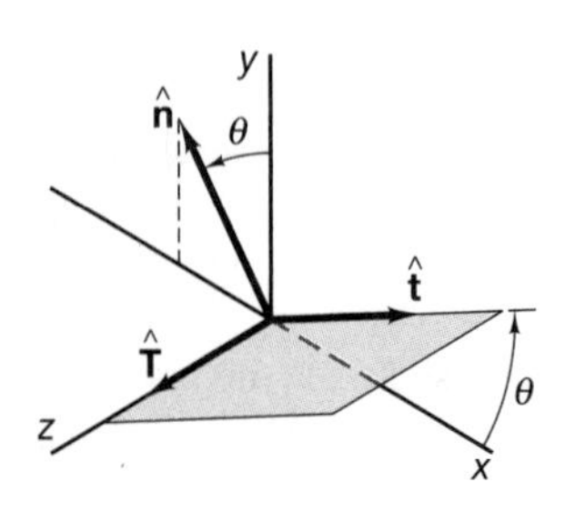

Figure 6.5

The unit vectors tangent to the plane are

$$\begin{aligned} \hat{\mathbf{t}} &= \cos(\theta)\hat{\mathbf{i}} + \sin(\theta)\hat{\mathbf{j}} \\ \hat{\mathbf{T}} &= \hat{\mathbf{k}} \end{aligned} \tag{6.7}$$

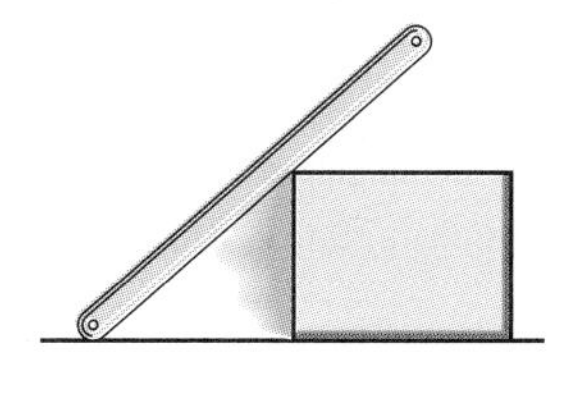

(a)

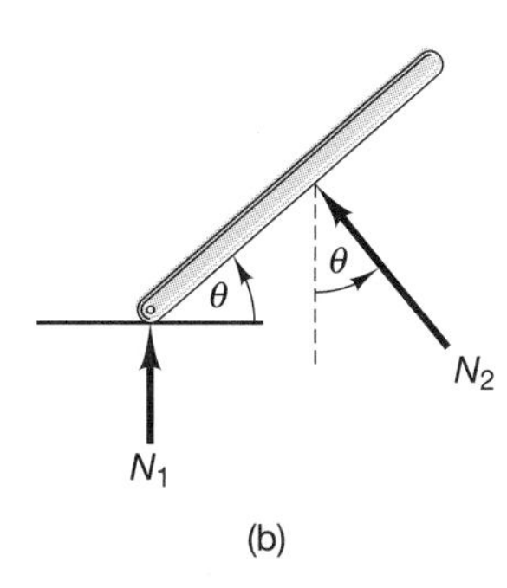

(b)

Figure 6.6

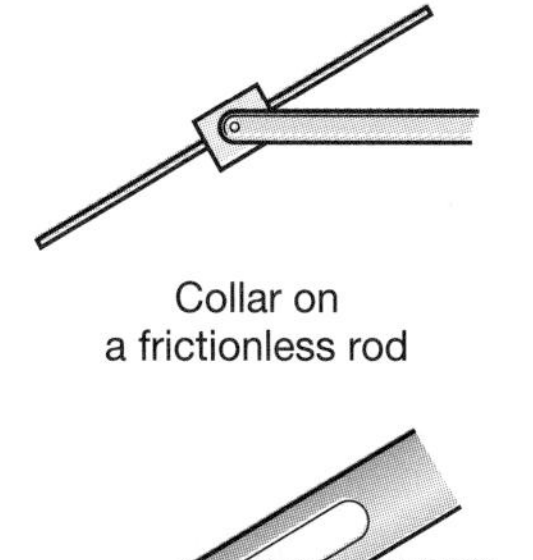

Collar on
a frictionless rod

Frictionless pin

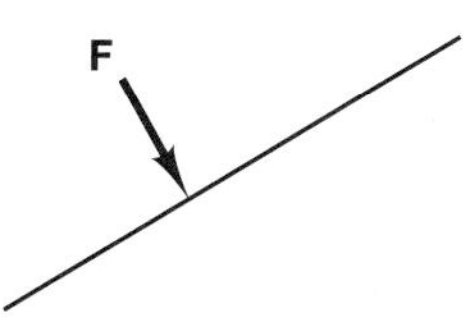

Force is normal
to the contacting surface

Figure 6.7

In this case, the two vectors tangent to the plane are orthogonal, and there is no need to divide by the magnitude of the cross product in Eq. (6.6). The unit normal vector is

$$\hat{\mathbf{n}} = \hat{\mathbf{T}} \times \hat{\mathbf{t}} = \hat{\mathbf{k}} \times [\cos(\theta)\hat{\mathbf{i}} + \sin(\theta)\hat{\mathbf{j}}$$
$$= -\sin(\theta)\hat{\mathbf{i}} + \cos(\theta)\hat{\mathbf{j}}$$

The latter equation is the result obtained in Eq. (6.5) and is an alternative approach to trigonometry.

It is important to recognize which is the normal surface between contacting bodies. Consider a bar lying against the corner of a box, as shown in Figure 6.6(a). Two normal forces act on the bar, as indicated in Figure 6.6(b). The normal force at the bottom of the bar is perpendicular to the floor, as movement of the box would be tangent to the floor. The normal force at the contact with the box is perpendicular to the bar, as the movement at this point would be tangent to the surface of the bar.

Normal forces also occur in collar or pin-slot supports, as shown in Figure 6.7. A normal force exists, for these cases, between the collar and the rod or between the pin and the slot. The force is normal to the contacting surface (the surface of the rod or the surface of the slot) and perpendicular to the direction of movement. However, the contact may occur on either side of the rod or slot, and the sense of the normal force is not known. If the slot or rod is curved, the normal force will be perpendicular to the tangent of the curve.

We will now consider supports that can constrain motion in two directions. Both a normal force and a friction force occur at the point of contact between a rigid body and a rough surface, as shown in Figure 6.8. Friction forces are usually introduced in basic physics courses and will be discussed in detail in Chapter 9. There are two unknown reactions at the point of contact with a rough surface. In some cases, it is advantageous to treat the normal force and the friction force as a single unknown reaction force acting at an unknown angle α.

Motion is constrained in both the horizontal and vertical directions at a hinge joint when friction can be neglected, as shown in Figure 6.9. If friction is neglected, the hinge is free to rotate. The two components of the reaction force **R** can be treated as two support forces with known directions and unknown magnitudes, or the reaction **R** may be treated as a single unknown reaction whose magnitude and direction are unknown. If friction must be considered, the rotation is restricted by a moment about the pin. The horizontal and vertical components of the reaction force may be assumed in either direction, and the sign will confirm or contradict this initial assumption.

The fixed or built-in support shown in Figure 6.10 restricts horizontal motion, vertical motion, and rotation of the body at the point of support.There will be an unknown reaction force and moment acting on the body at this point. The sense of the components of this force and of the moment is not known and may be initially assumed and then confirmed by the signs of the results.

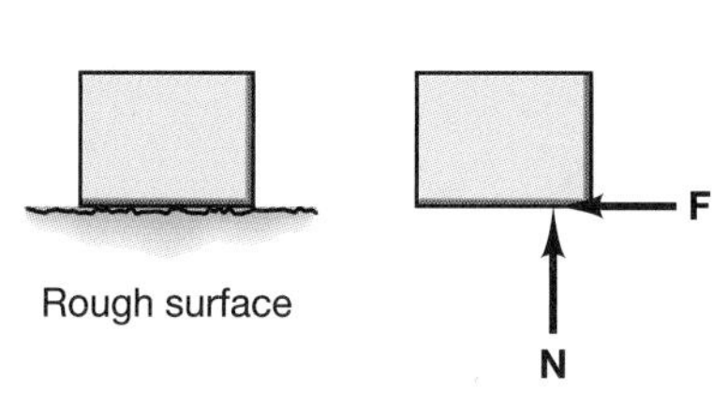

Figure 6.8

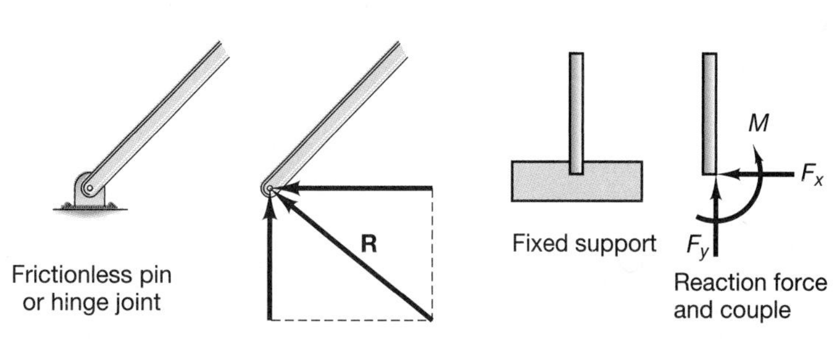

Figure 6.9

Figure 6.10

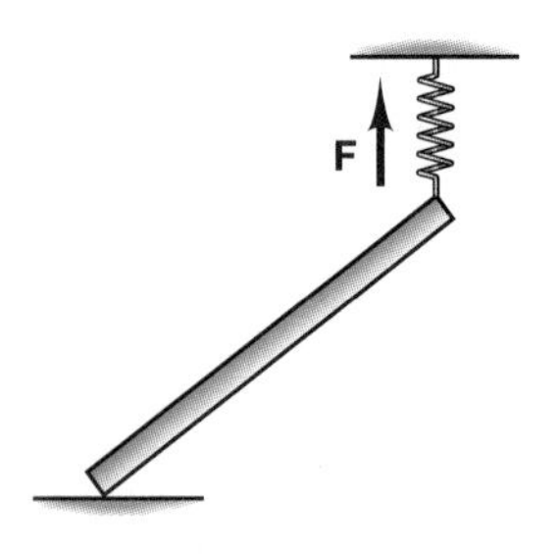

Figure 6.11

Some supports should be considered as deformable cables or ropes and can be modeled as a spring, as shown in Figure 6.11. The spring supporting the bar will exert a force along the axis of the spring, and the magnitude of this force will depend upon how much the spring has been stretched. Assuming a linear spring with a spring constant k, the spring force will be

$$F = kd$$

where d is the amount that the spring has been stretched.

A rigid body may have been supported by any combination of these supports to restrict its motion. Once the body has been isolated from its surroundings, the supports are replaced by the supporting forces and moments. A common source of error is incorrectly representing these supporting forces or moments.

6.3 SUPPORTS FOR A THREE-DIMENSIONAL MODEL

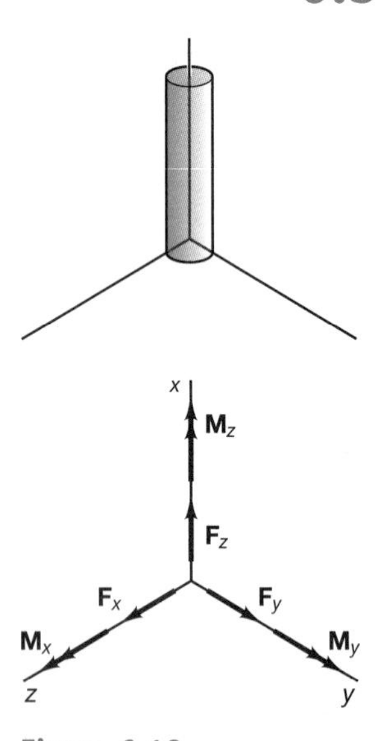

Figure 6.12

In Section 6.2, supports of structures that could be modeled as two-dimensional bodies were discussed. These supports could exert two components of force and one component of a moment on the body and provide constraints to the two possible translations and the one rotation of the body. Therefore, the three degrees of freedom of the body are constrained. All bodies are essentially three dimensional, and only a limited number can be modeled in a two-dimensional manner. A three-dimensional body has six ***degrees of freedom***—three translations and three rotations—and the supports must constrain these movements. As in the case of two-dimensional models, different types of supports offer different types of constraint. The most restrictive type of support is a fixed (embedded or welded) type of support restricting all six degrees of freedom by exerting three components of force and three components of moment on the body, as shown in Figure 6.12. The reactions at the fixed support are a reaction force that can be oriented in any direction to ensure equilibrium and a moment whose vector representation can be oriented in any direction. It is usually best to represent these two vectors by their coordinate components, as shown in the figure.

The least restrictive support is a cable or rope. We saw in the two-dimensional model that the only force in this type of support is tension along the cable, and the cable restricts only one degree of freedom—that is, translation in the direction of extension of the cable. The body is free to translate in directions perpendicular to the cable, even when the cable is in tension. We see that all constraints on, or supports for, a rigid body must be examined as a system to see whether the body is restrained in all six degrees of freedom.

A type of support similar to a roller or rocker is a ball support or a frictionless surface, either of which can exert only a normal force, as shown in Figure 6.13. If the contact is on an inclined surface, Eq. (6.6) can be used to determine a unit normal vector to the surface, and the normal force can be expressed in vector form as $\mathbf{N} = N\hat{\mathbf{n}}$. Consider the inclined surface shown in Figure 6.14. The unit tangent vectors $\hat{\mathbf{T}}$ and $\hat{\mathbf{t}}$ lie in the y–z and x–z planes, respectively, and are

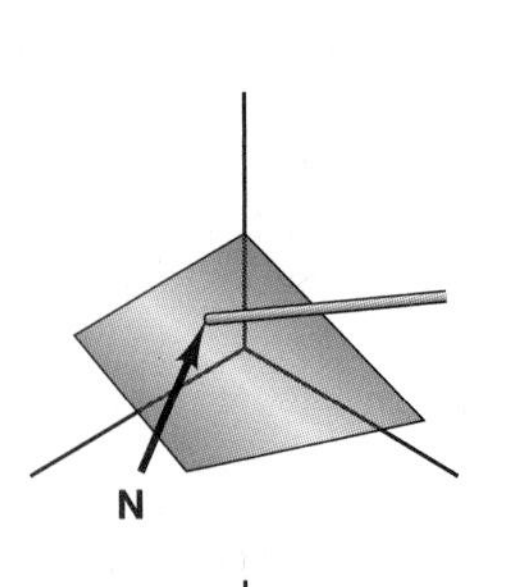

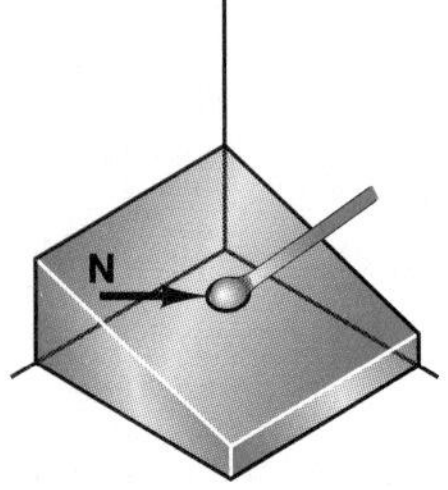

Figure 6.13

$$\hat{\mathbf{t}} = \cos(\theta)\hat{\mathbf{i}} + \sin(\theta)\hat{\mathbf{k}}$$

$$\hat{\mathbf{T}} = \cos(\beta)\hat{\mathbf{j}} + \sin(\beta)\hat{\mathbf{k}} \tag{6.8}$$

The scalar product between the two unit tangent vectors is not zero, and therefore, the two vectors are not orthogonal. The unit normal vector is

$$\hat{\mathbf{n}} = \frac{\hat{\mathbf{t}} \times \hat{\mathbf{T}}}{|\hat{\mathbf{t}} \times \hat{\mathbf{T}}|} = \frac{-\sin(\theta)\cos(\beta)\hat{\mathbf{i}} - \cos(\theta)\sin(\beta)\hat{\mathbf{j}} + \cos(\theta)\cos(\beta)\hat{\mathbf{k}}}{\sqrt{\sin^2\theta\cos^2\beta + \cos^2\theta}} \tag{6.9}$$

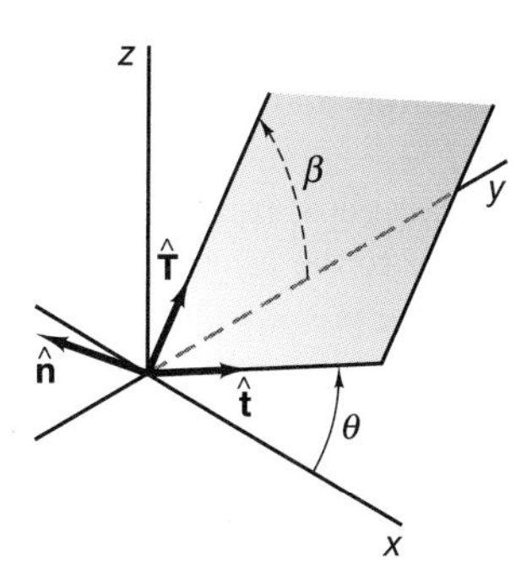

Figure 6.14

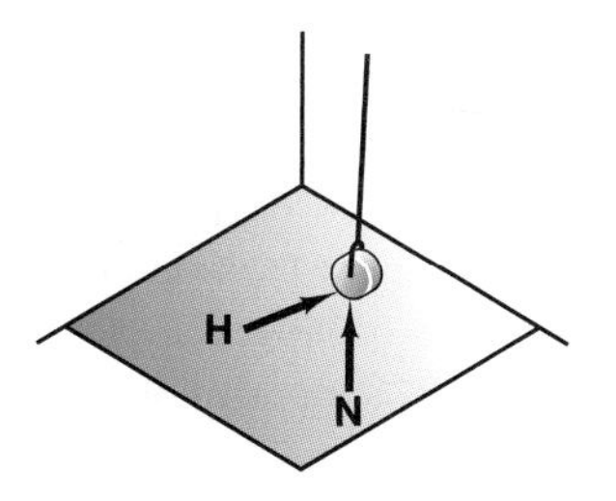

Figure 6.15

Note that this type of support partially constrains only one degree of freedom: the direction normal and into the surface of contact. The body is free to separate from the surface, as is the case of the normal force between your foot and the ground.

The roller support shown in Figure 6.15 can exert two components of force, one vertical and one horizontal. The horizontal force is due to resistance by the roller to sliding in a direction perpendicular to it. Therefore, the directions of both the normal and the horizontal forces are known.

When a body is in contact with a rough surface, the surface can exert three components of force restricting all translation of the body, as long as the normal force is directed toward the body. Again, the body can be separated from the surface, and in that case there would be no contacting forces. A type of support similar to a roller is a ball and half-socket or a ball-and-cup support, as shown in Figure 6.16. The ball can be withdrawn from the cup if it is pulled along the axis of the rod. Therefore, the normal force must act in compression along the axis of the rod. The cup constrains translations in the two directions perpendicular to the axis of the rod. Therefore, the ball-and-cup joint constrains two translations completely and the third translation partially, but the three degrees of rotational freedom are unconstrained. The human hip joint shown in Figure 6.17 is a ball-and-cup joint and requires ligaments and muscles to keep the ball (the head of the femur) in the cup (the acetabular cup of the pelvis). The normal force **N** pushes along the axis of the neck of the femur, and a vertical force **V** and an anterior–posterior force **AP** act perpendicular to this axis.

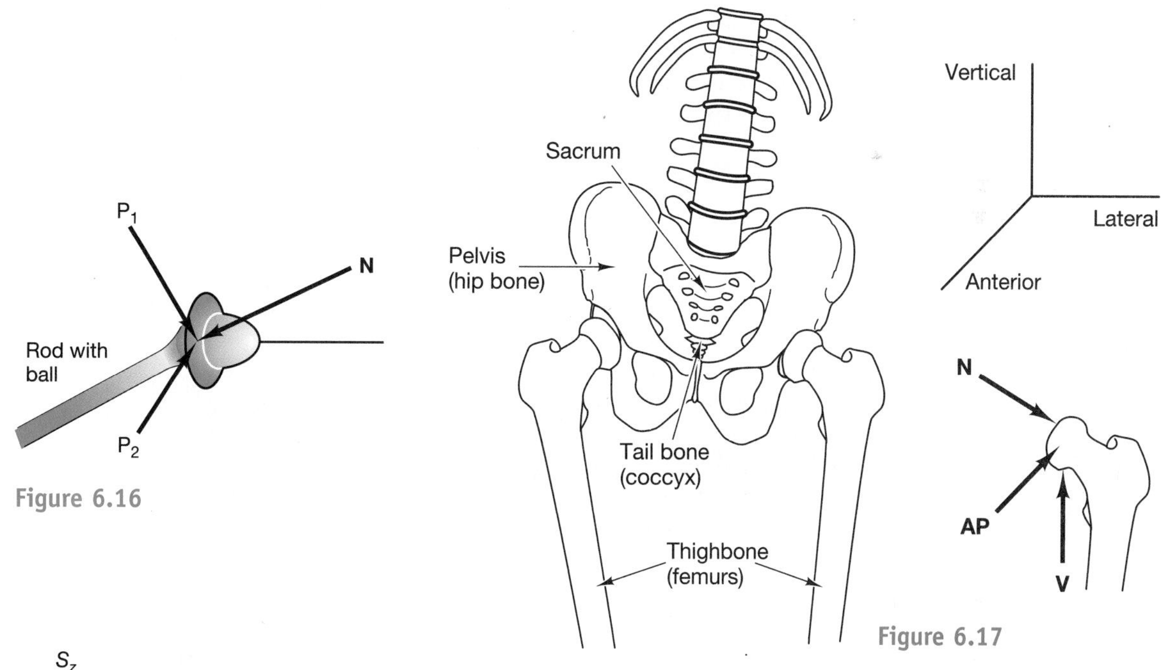

Figure 6.16

Figure 6.17

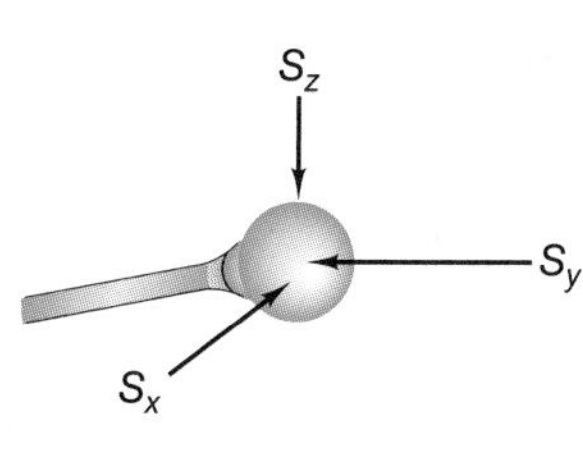

Figure 6.18

A ball-and-socket joint fully constrains all translations, but the three rotational degrees of freedom are unconstrained, as shown in Figure 6.18. In a ball-and-socket support, when the ball cannot be pulled from the socket, the components of the constraining force can be aligned with the coordinate axes.

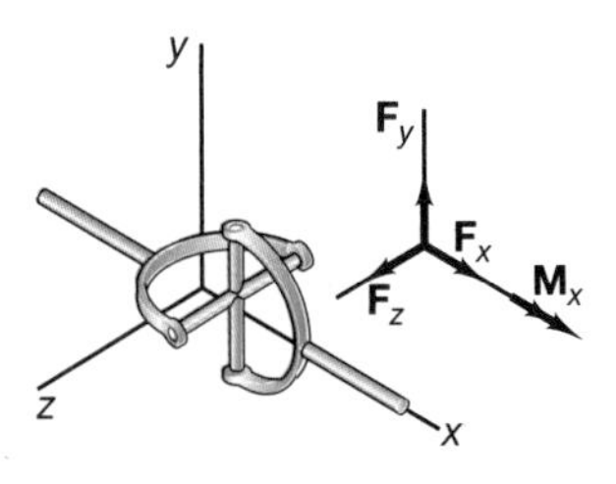

Figure 6.19

A universal joint constrains all translations and one rotation, as shown in Figure 6.19. The universal joint constrains translations in the x-, y-, and z-directions and constrains or transmits a moment about the x-axis.

Hinge and bearing supports are more complex, and although they allow rotation about the hinge or bearing axis, rotations about the two axes perpendicular to the hinge or bearing axis are constrained. A hinge or bearing with no thrust resistance is shown in Figure 6.20(a). In supports of this type, there is no constraint on translation along the hinge or shaft axis, nor is there any constraint on rotation about this axis. In analysis, hinges and bearings are usually simplified to constrain only translations in the two directions perpendicular to the hinge or shaft axis, and the moments are neglected. When hinges or bearings are modeled in this manner, they are called simple supports, and one translational and all three rotational degrees of freedom are unconstrained. Although this may appear to be an unrealistic assumption, examination of the ball bearing shown in Figure 6.20(b) shows that rotations about the y- and z-axes are not totally constrained. The moments about these axes are usually due to misalignment of the bearings or excessive bending in the shaft.

A thrust bearing or a hinge that constrains translation along its axis is shown in Figure 6.21. This type of support constrains five degrees of freedom, and only rotation about the hinge or shaft axis is unconstrained. Note the addition of F_x is in the free-body diagram. As in the case of ordinary hinges or bearings, thrust bearings or hinges are usually modeled such that they constrain only translations, and therefore, support moments are neglected.

As in the two-dimensional model, supports of a three-dimensional body may underconstrain or overconstrain the body. If the body is underconstrained, it will have at least one degree of freedom unconstrained. While this may be the desired design, the equilibrium equation associated with the unconstrained degrees of freedom will not be applicable. The body will be unstable with respect to those degrees of freedom, and the consequences of this should be considered in the design.

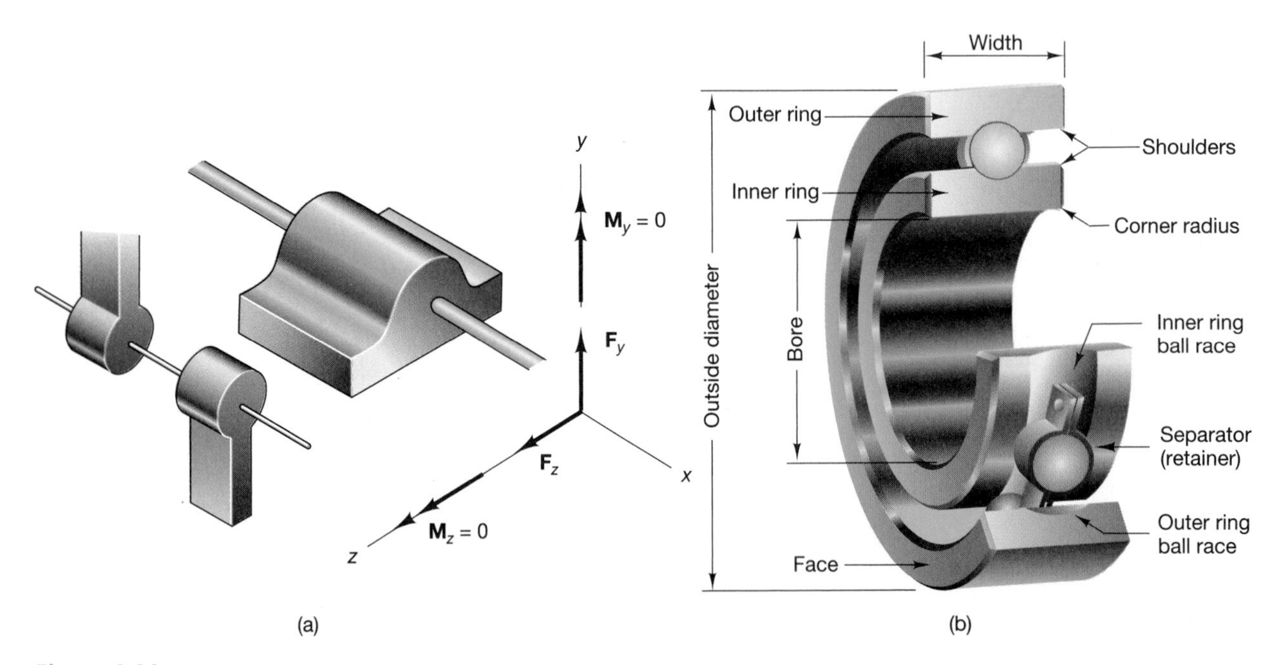

Figure 6.20

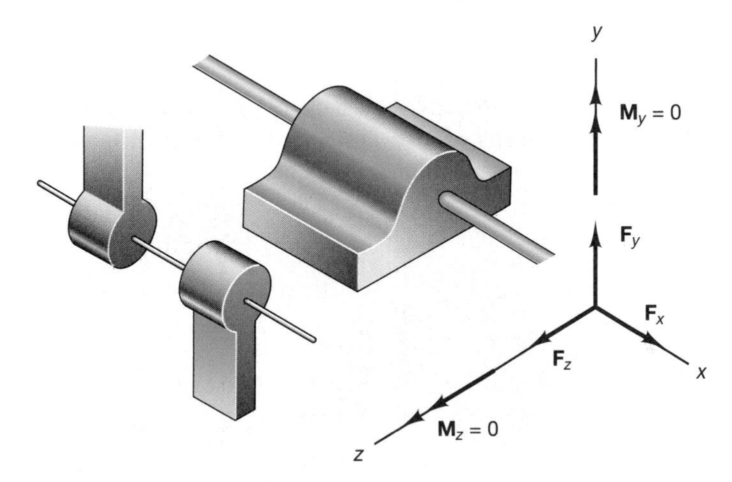

Figure 6.21

If a structure is overconstrained, the problem will be called a *statically indeterminate problem.* Only six equilibrium equations can be written, and if the number of unknown support forces and moments exceeds six, some of the support forces and moments cannot be determined by equilibrium considerations alone. This class of problems is usually solved by considering the deformation of the body and the deformation of the supports. Statically indeterminate systems will be discussed in Section 6.7

6.4 FREE-BODY DIAGRAM

We first introduced the concept of the free-body diagram in Section 3.9 and applied it to particles. A rigid body is modeled as a two-dimensional or a three-dimensional body, and the development of a free-body diagram becomes more difficult.

The process of developing a free-body diagram is as follows:

1. Clearly identify the object that is to be isolated in the free-body diagram.
2. Decide whether the object is to be modeled as a particle or a rigid body in two or in three dimensions.
3. Accurately represent the dimensions of the object, obtained from a space diagram, which may be a photograph, a sketch of the object, or a blueprint. Sketch the shape of the body.
4. Isolate the object from its supports and other bodies, and show all external forces or moments acting on the body at their points of application. These forces may be at the supports or may be due to gravitational attraction or contact with other bodies. If a force is assumed negligible, clearly state that assumption on the free-body diagram.
5. Count the number of unknown forces and moments that appear on the free-body diagram, and determine whether sufficient equations of equilibrium are available to solve the problem. If there are more unknowns than equations, the free-body diagram may be erroroneous, or the problem may be statically indeterminate. The unknown forces usually consist of the reactions or constraining forces or moments acting on the body at the supports, or, in some cases, they may be the forces due to contact with other bodies.

Sample Problem 6.1

Construct a free-body diagram for each of the objects shown.

a. A car is shown below. The normal force on the front tires and the back tires is depicted in the free-body diagram underneath, with the horizontal distances between the tires and the center of mass designated by the letters a and b. In this case, the car has been modeled as a two-dimensional object.

b. The lower end of the stick shown below rests on a rough floor. The stick leans against a smooth corner of a box. The free-body diagram of the stick also is drawn below.

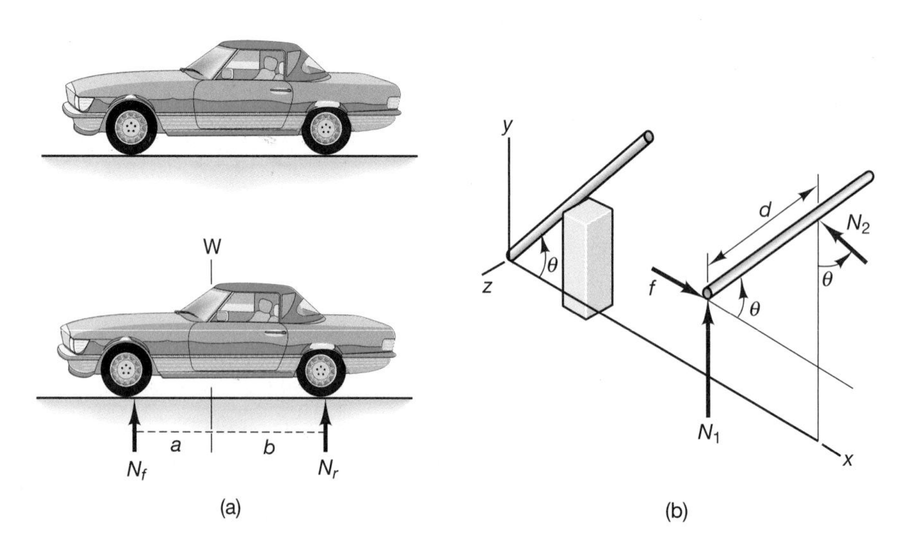

(a) (b)

c. A 50-kg sign is supported on the end of a rigid bar with a ball-and-socket joint at point A and three cables, as shown below (left). The free-body diagram is also below (right).

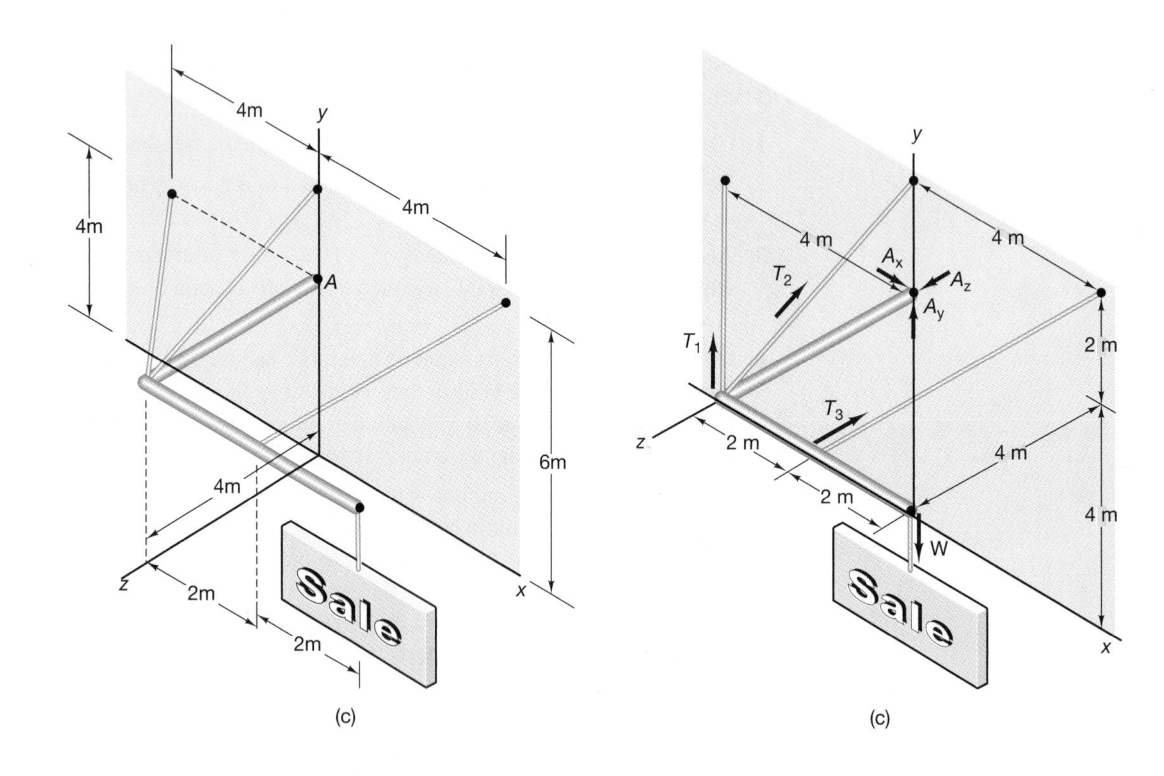

(c) (c)

d. A rod of length $3R$ is placed in a bowl of radius R, making an angle α with the horizontal. (See below (left).) The free-body diagram for the rod is shown below (right).

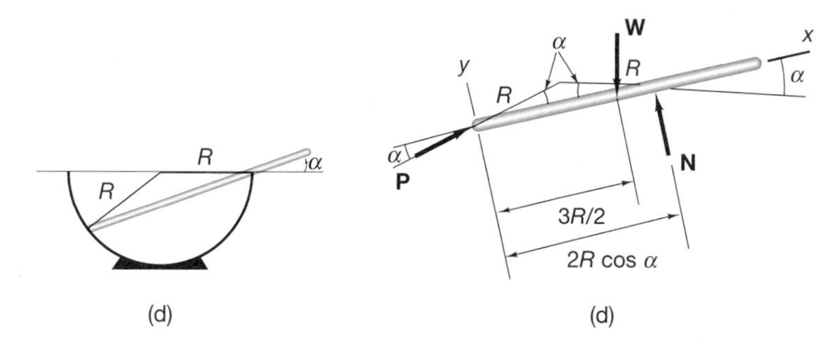

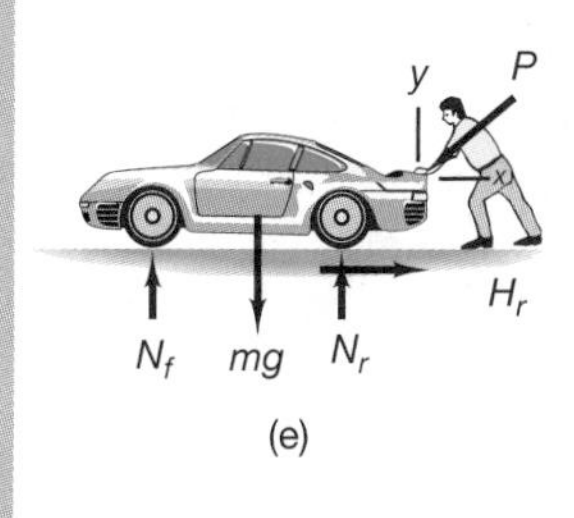

e. The man in the photo at the right is pushing on the rear of the car, and a horizontal force exists between the rear tires and the ground if the front tires are assumed to be free to rotate. Note that you need more information to complete the free-body diagram. The required geometric information includes:.

The wheelbase of the car (the distance between N_f and N_r).
The location of the center of mass of the car.
The point of application of the force beneath the man's hands.

You also would need to know the weight of the car and the amount of the force applied by the man. The unknown reaction forces at the tires could then be determined.

Problems

6.1 Consider the sketch of a movable crane shown in Figure P6.1. The wheels are 4 m apart, and the center of gravity is halfway between the wheels. Assume that the ballast is an external force acting 1 m behind the left axle and that the weight hanging off the crane at point B is also treated as an external force. Draw the free-body diagram of the crane, and label all the forces and dimensions. Be sure to include a coordinate axis.

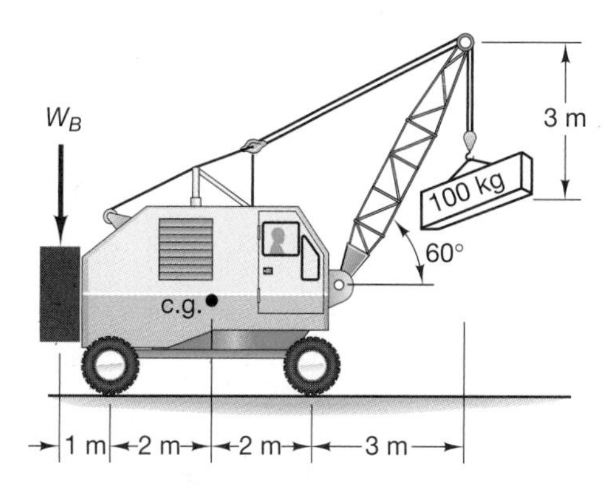

Figure P6.1

6.2 A small bridge on a walking path has a pin at one end and a roller at the other. (See Figure P6.2.) Separate the bridge from its supports, and make a free-body diagram. The center of gravity of the bridge is at its geometric center, midway between the supports.

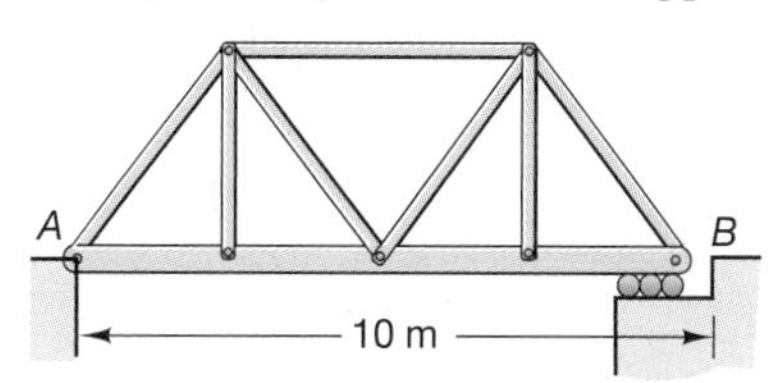

Figure P6.2

6.3 Construct free-body diagrams for the three human activities shown in Figure P6.3 Show all loads external to the skeleton and approximate dimensions.

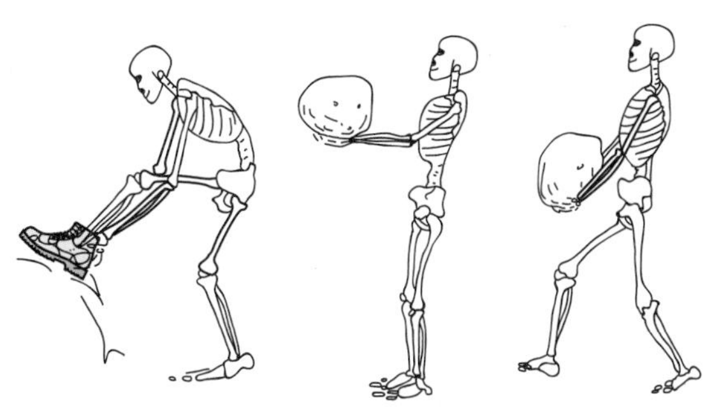

Figure P6.3

6.4 Draw a free-body diagram of the automobile side mirror shown in Figure P6.4, by isolating the mirror from the car body. Include the mirror's weight. The mirror forms a fixed connection with the car body.

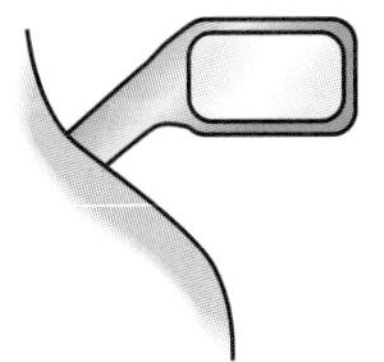

Figure P6.4

6.5 Make a free-body diagram of the beam used to hold up the sign illustrated in Figure P6.5. The sign has a mass of 10 kg, and the uniform beam's mass is 5 kg.

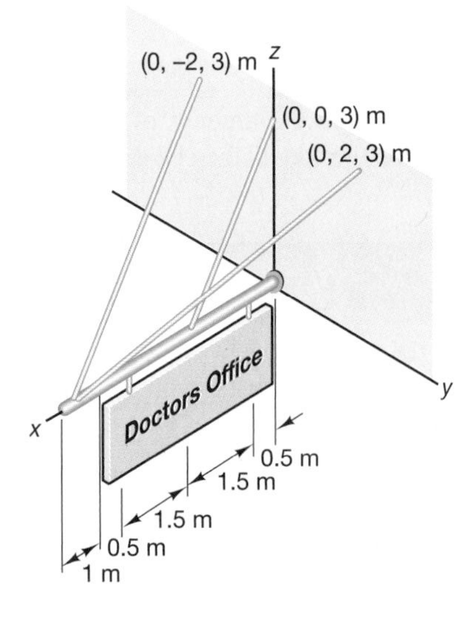

Figure P6.5

6.6 Figure P6.6 shows a uniform 4-meter-long slab supported by a cable at point A and two rigid links at points B and C. The slab has a mass of 5.1 kg. Draw a free-body diagram of the slab.

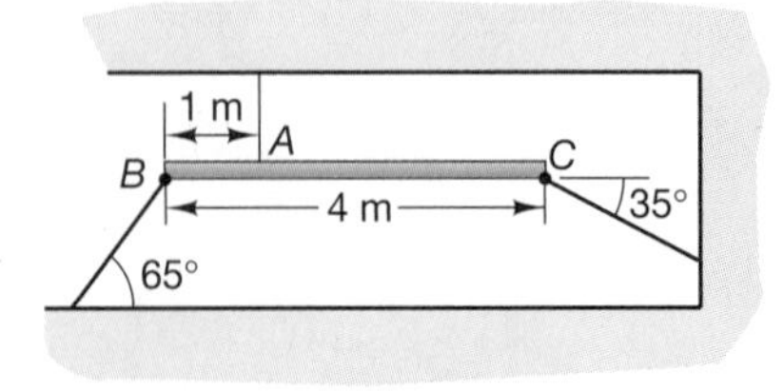

Figure P6.6

6.7 To understand spinal loading, it is necessary to construct a free-body diagram of the upper body. In Figure P6.7 two external loads, **B** upper body weight and **W** weight being lifted are shown. The internal loads are **M**, the back muscle force, **P**, the abdominal pressure, **C**, spinal vertebra compression, and **S**, shear across the disc. Estimate dimensions and discuss the changes in spinal loading as the angle θ increases.

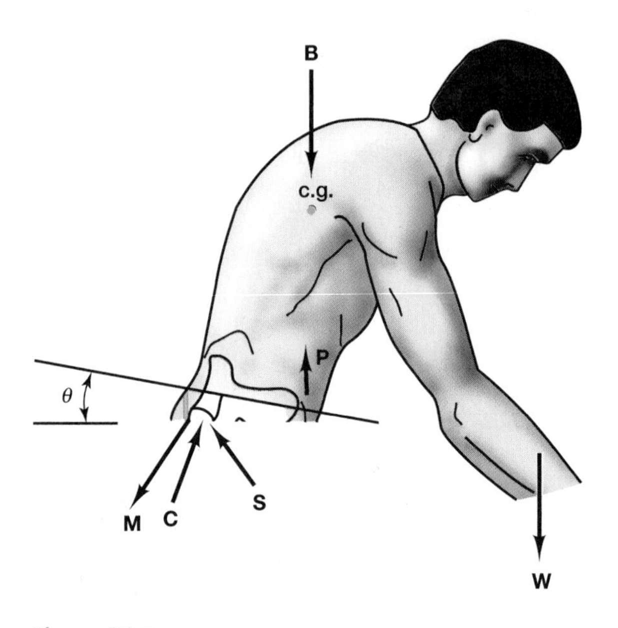

Figure P6.7

6.8 An old icebox is placed on a ramp having a slope of 15°, as shown in Figure P6.8. The box is 0.4 m by 1 m in cross section, with its center of gravity at its geometric center. The box is held in place by a small rib that can be modeled as applying a force along the ramp. Draw a free-body diagram of the box, which has a mass of 50 kg.

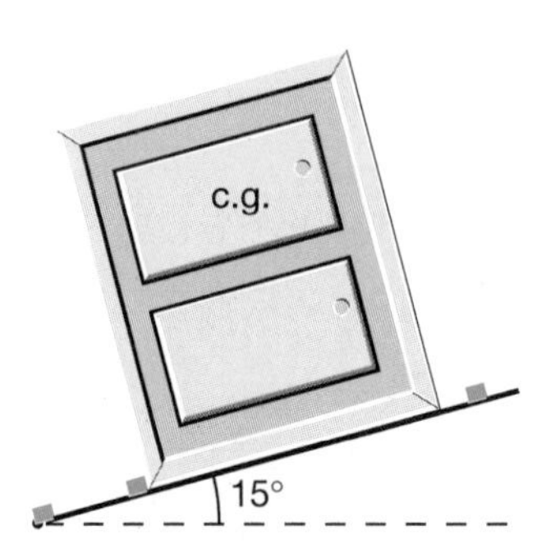

Figure P6.8

6.9 A 50-N bar is pinned at one end, *A*, and held up by a rope at *B*, as illustrated in Figure P6.9. A 100-N sign hangs by ropes. (a) Draw a free-body diagram of the sign, and (b) draw a free-body diagram of the bar.

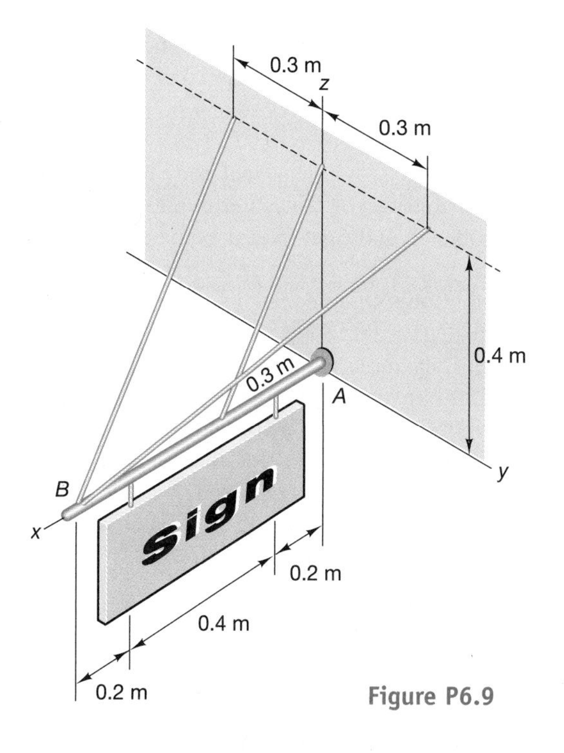

Figure P6.9

6.10 Draw a free-body diagram of the lifting mechanism illustrated in Figure P6.10. The mechanism has a mass of 10 kg.

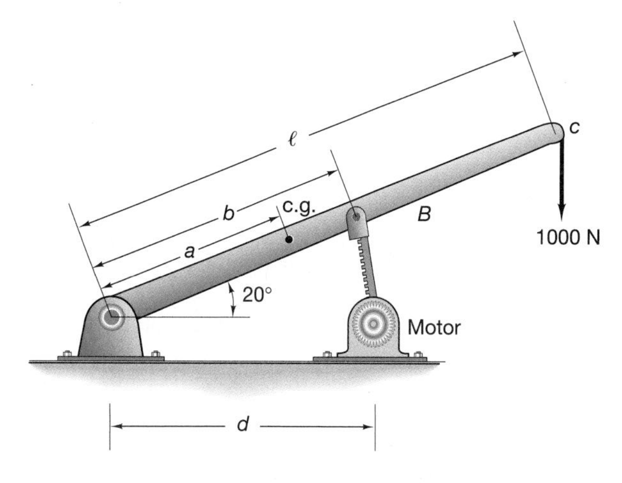

Figure P6.10

6.5 EQUILIBRIUM OF A RIGID BODY IN TWO DIMENSIONS

The equations of equilibrium, Eqs. (6.1), are simplified when a body can be modeled in two dimensions. If the x–y plane is considered to be the plane of the structure, the two-dimensional model requires that there be no forces acting in the z-direction and no couples with components in the x- or y-directions. In this case, there are only three nontrivial scalar equilibrium equations:

$$\sum F_x = 0$$
$$\sum F_y = 0 \qquad (6.10)$$
$$\sum M_z = 0$$

Before we discuss the solution of this reduced set of equations of equilibrium, it is important to consider when a rigid body can be modeled two dimensionally. The first case would be when the depth or thickness of the body is small compared to the other dimensions and the body appears to occupy a plane in space. Also, all forces acting on the body must be coplanar and lie in the plane of the body. Thin plates or beams satisfy this condition in certain states of loading. Even if a plate is thin, however, it cannot be modeled as a two-dimensional body if any of the forces are normal to its surface. An example of a two-dimensional model in which all the forces are coplanar is shown in Figure 6.22. Notice that

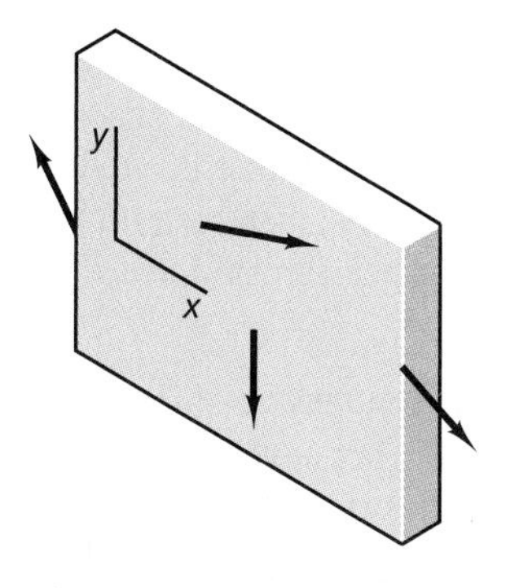

Figure 6.22

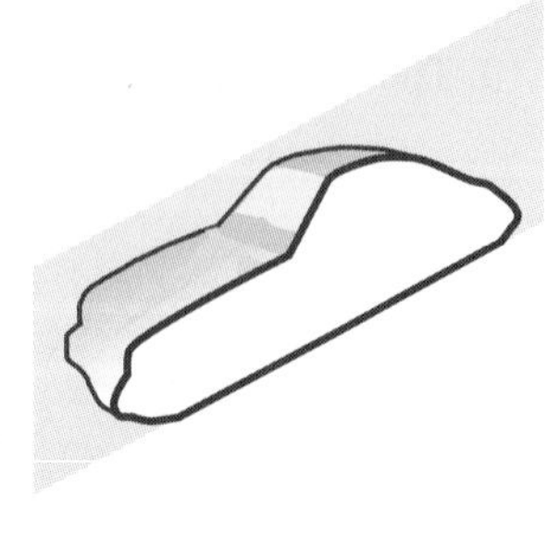

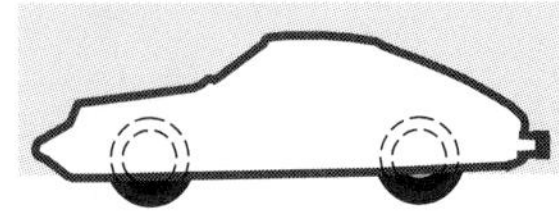

Figure 6.23

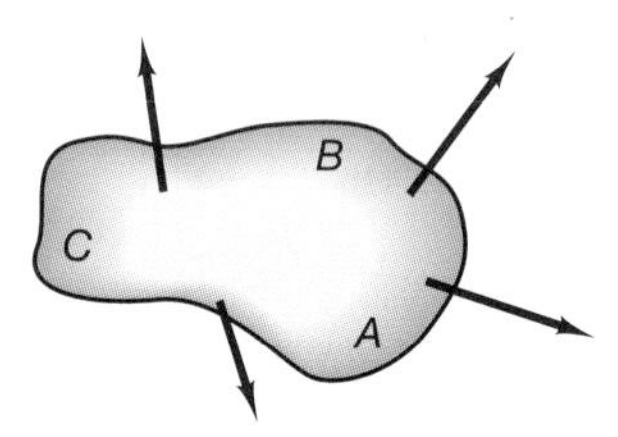

Figure 6.24

the body must be constrained by supports to resist translation within the plane and rotation about an axis perpendicular to the plane. Although supports may exist that prevent the body from translating perpendicularly to this plane and rotating out of the plane, reactions due to those supports are not shown on the free-body diagram, as they are assumed to be zero.

The second case when an object can be modeled as a two-dimensional body occurs when there is a plane of both geometric and loading symmetry in the body. The forces acting perpendicular to this plane of symmetry would cancel, and forces parallel to the plane of symmetry could be resolved into an equivalent force system coplanar with the plane of symmetry, as shown in Figure 6.23.

In some cases, an automobile can be treated as a two-dimensional body by using the midplane of the car as a plane of symmetry and considering the center of mass of the car to be in this plane, the external tire loading on the left and right front tires to be equal, and the left and right rear tires to be equally loaded.

Equilibrium in two dimensions requires that only three equations of equilibrium be satisfied (the other three are automatically satisfied, as that force component and the two moment components are zero), and therefore, only three unknowns can be determined from these three equations. The equations can be obtained by writing the vector equation, setting the resultant of the forces acting on the body equal to zero, and using the cross product to take the moment about any point in the body and setting the moment vector equal to zero. The scalar equilibrium equations also can be obtained, by setting the moment about more than one point equal to zero. Therefore, for the thin plate shown in Figure 6.24, the scalar equations of equilibrium may take the following forms:

$$\left|\begin{array}{l}\sum F_x = 0\\ \sum F_y = 0\\ \sum M_A = 0\end{array}\right| \quad \text{or} \quad \left|\begin{array}{l}\sum F_x = 0\\ \sum M_A = 0\\ \sum M_B = 0\end{array}\right| \quad \text{or} \quad \left|\begin{array}{l}\sum M_A = 0\\ \sum M_B = 0\\ \sum M_c = 0\end{array}\right| \tag{6.11}$$

These three sets of equations are linearly independent and may be used in any combination, but there are only three linearly independent equations, and therefore, only three unknowns may be determined. There is no advantage to taking moments about more than one point on the body. The only reason to attempt to use any combination of equations other than Eq. (6.10) is to reduce the effort needed to solve the system of three linear equations. For example, if the moments are taken about a support point, the unknown support forces will not enter into that equation. If modern computational tools are used to solve the resulting three equations for the three unknowns, the most straightforward approach is to write the two scalar force equations and the moment equation about any point and then solve the 3×3 system of linear equations.

6.5.1 SOLUTION STRATEGY

Equilibrium problems are solved in a systematic and organized manner following the creation of a free-body diagram. In the remaining sections of this chapter and in the chapters that follow, we will concentrate on the modeling aspects of the problems; that is, we will create accurate free-body diagrams and properly represent supports that restrain the motion of the rigid body or bodies. Once the free-body diagram has been created, we solve the problem according to the following steps:

1. Select a coordinate system, including an origin. This selection may, in some cases, lessen the difficulty of solving the equations of equilibrium, but any selection will lead to the same results. When the dynamics of rigid bodies is studied, the selection of the origin and the orientation of the coordinate axes is more restricted, as

the origin is usually placed at the center of mass of the body, and the coordinate axes are oriented to coincide with principal axes.

2. Write the force and couple vectors that act on the body in vector notation in the selected coordinate system. Write position vectors from the origin to the line of action of each force vector. Write all vectors in explicit vector notation.
3. Write the two vector equations of equilibrium: The sum of the forces equals zero, and the sum of the moments about the origin equals zero. These two vector equations yield three scalar equations in plane statics problems and six scalar equations in general equilibrium problems.
4. Solve the system of equations.
5. Examine the resulting solution to ensure consistency with the physical characteristics of the model.

6.5.2 A TWO-FORCE MEMBER

A special case of equilibrium occurs when just forces act at only two points on a rigid body. In most cases, the weight of the body is ignored. The rigid body is called a *two-force body,* and if it is in equilibrium, *the two forces must be equal in magnitude, be collinear, and have opposite sense,* as shown in Figure 6.25. Although more than one force may act at points A and B (for example, the components), they may be summed to produce a single equivalent force at each point, and these equivalent forces must satisfy the two-force member condition

$$\mathbf{F_A} = -\mathbf{F_B} \tag{6.12}$$

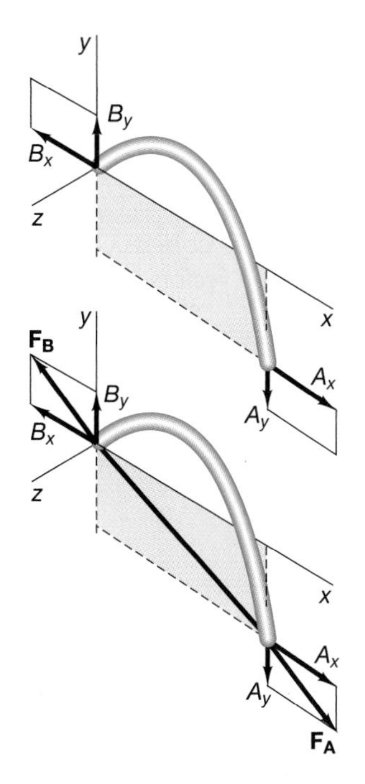

Figure 6.25

If the three scalar equilibrium equations are written for the two-force member, the result will show that the forces are equal, opposite, and collinear. Therefore, although not necessary for the analysis, the force relationship for a two-force member can be used to reduce the number of unknowns. This will be particularly true in the analysis of structures, presented in Chapter 7.

In many applications, the two-force member is a straight, slender member, and the line of action of the two forces is coincident with the axis of the member, as shown in Figure 6.26. The straight member in Figure 6.26(a) is in compression, and the straight member in Figure 6.26(b) is in tension. The trusses shown in Chapter 7 will be modeled as straight two-force members that are in either tension or compression.

6.5.3 A THREE-FORCE MEMBER

If a body is acted upon by only three forces, the body is called a *three-force body.* If a three-force member is in equilibrium, *the forces must be coplanar and be either concurrent or parallel.* It is easily seen that the forces acting on a three-force member must be coplanar. Any two of the three forces define a plane in space, and the third force must be coplanar; if it weren't, it would have a component perpendicular to that plane, and therefore, the body would not be in equilibrium. Examples of these two cases are shown in Figure 6.27. Since the forces on a three-force member are coplanar, the member may always be treated as a two-dimensional body. These observations concerning three-force members can be used to reduce the effort expended in solving the equations and are also the basis of some graphical or trigonometric solutions. As in the case of two-force members, three-force members can be analyzed as any other rigid body in equilibrium.

One of the most important, and certainly one of the simplest, examples of a three-force member is a lever. A first-class lever, which is shown in Figure 6.28, is an excellent example of the turning effects of forces—that is, the moments. The origin of the coordinate

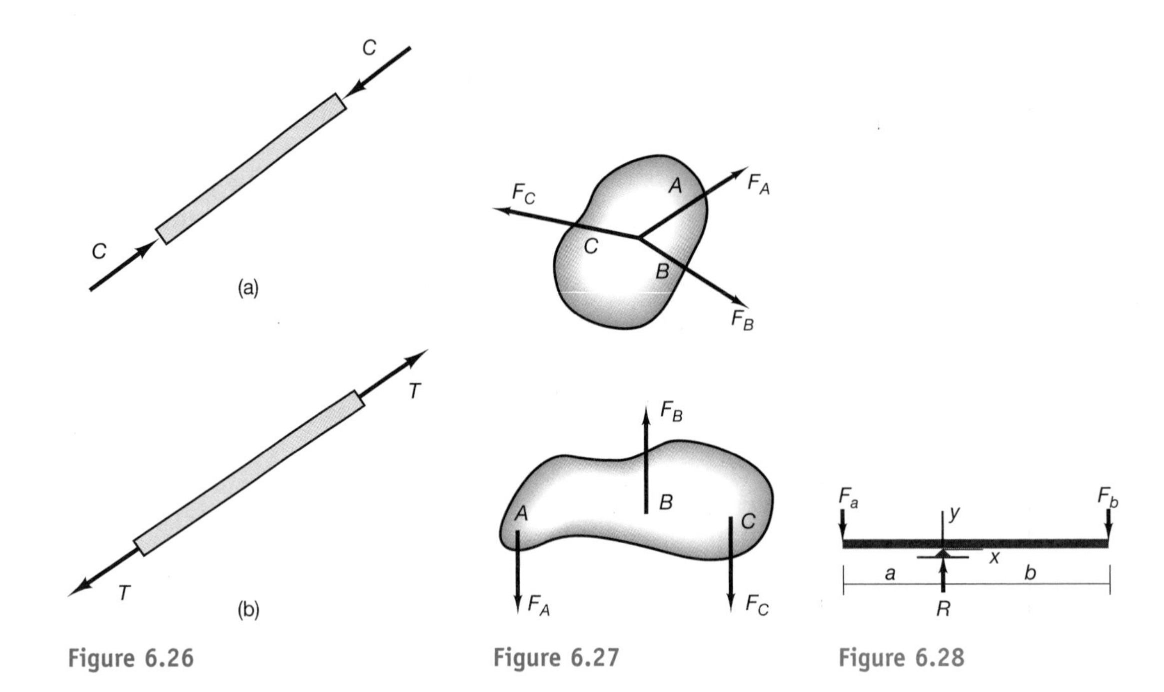

Figure 6.26

Figure 6.27

Figure 6.28

system is at the fulcrum of the lever, and the x-axis is parallel with the lever in its equilibrium position. The system is a parallel force system, and all the force vectors have only components in the y-direction. The force and position vectors are

$$\begin{aligned} \mathbf{F}_a &= -F_a\hat{\mathbf{j}} & \mathbf{r}_{a/O} &= -a\hat{\mathbf{i}} \\ \mathbf{R} &= R\hat{\mathbf{j}} & \mathbf{r}_{R/O} &= 0 \\ \mathbf{F}_b &= -F_b\hat{\mathbf{j}} & \mathbf{r}_{b/O} &= b\hat{\mathbf{i}} \end{aligned}$$

The scalar equilibrium equations are obtained by:
summing forces in the y-direction yielding

$$-F_a + R - F_b = 0$$

taking moments about the origin results in

$$b\,F_b - a\,F_a = 0$$

The reaction R at the fulcrum of the lever is equal to the sum of the forces F_a and F_b if the lever is in equilibrium. Summing moments about the fulcrum shows that the moment caused by F_a must be equal and opposite to the moment caused by F_b for equilibrium. Therefore, the required relationship between the forces is

$$F_a = (b/a)F_b \tag{6.13}$$

This relationship is observed by any child who has played on a seesaw. A common expression, "getting the short end of the stick," refers to having received a bad deal or a poor bargain. Notice that the force at point a "got the short end of the stick" and therefore has to exert a greater effort to balance the turning effect of the force at point b.

A different class of lever is illustrated in Figure 6.29, in which the fulcrum is at the left end of the lever.

From equilibrium in the y-direction, the reaction at the fulcrum is

$$\mathbf{R} = \mathbf{F_a} - \mathbf{F_b}$$

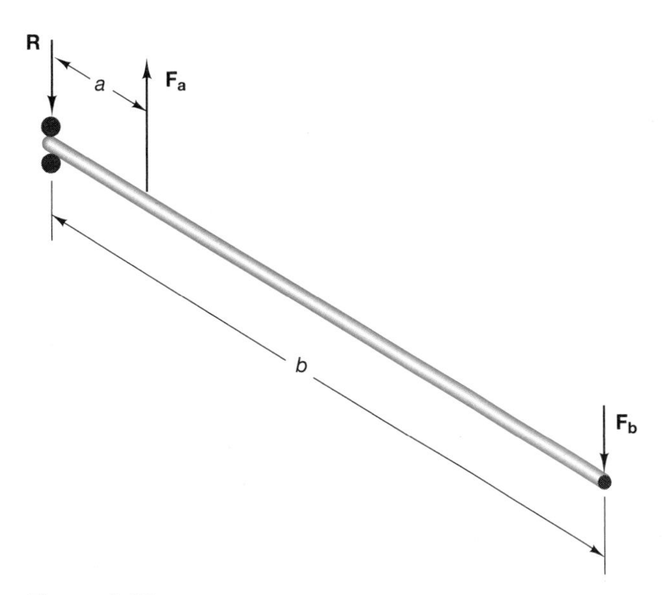

Figure 6.29

A summation of moments about the fulcrum leads to the same relationship given by Eq. (6.13), and $\mathbf{F}_a$ is much larger than $\mathbf{F}_b$.

The muscles in the human body "got the short end of the stick," as may be seen in Figure 6.30. The ratio b/a in the forearm is $8/1$ in the average human. Therefore, if a 20-pound object is held in the hand, the biceps muscle must exert a force of 160 pounds, and the compression at the elbow joint is 140 pounds.

Creating free-body diagrams of other body segments shows that most joints in the human body transmit forces far in excess of the weight of the individual.

Consider an individual seated and holding an object in the central or sagittal plane of the body as illustrated in 6.31a. We can treat this as a simple lever problem as shown in the free-body diagram in Figure 6.31b. The weight in the hand must be balanced by the moment produced about the spinal column by the posterior back muscles. Since in this case the ratio b/a can be 20 to 50, the muscle force can be 20 to 50 times greater than the weight and the spinal loading can be very large.

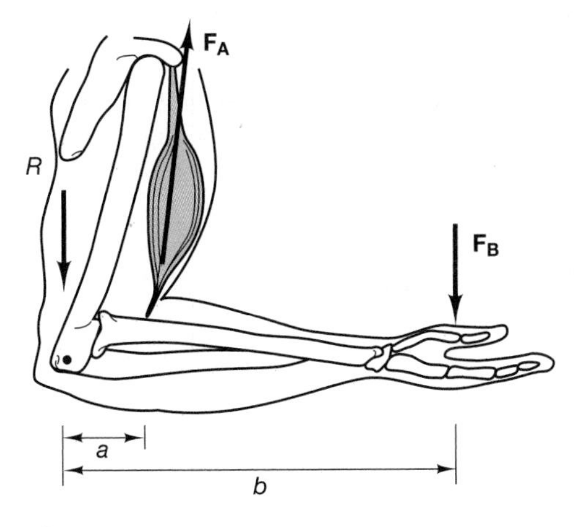

Figure 6.30

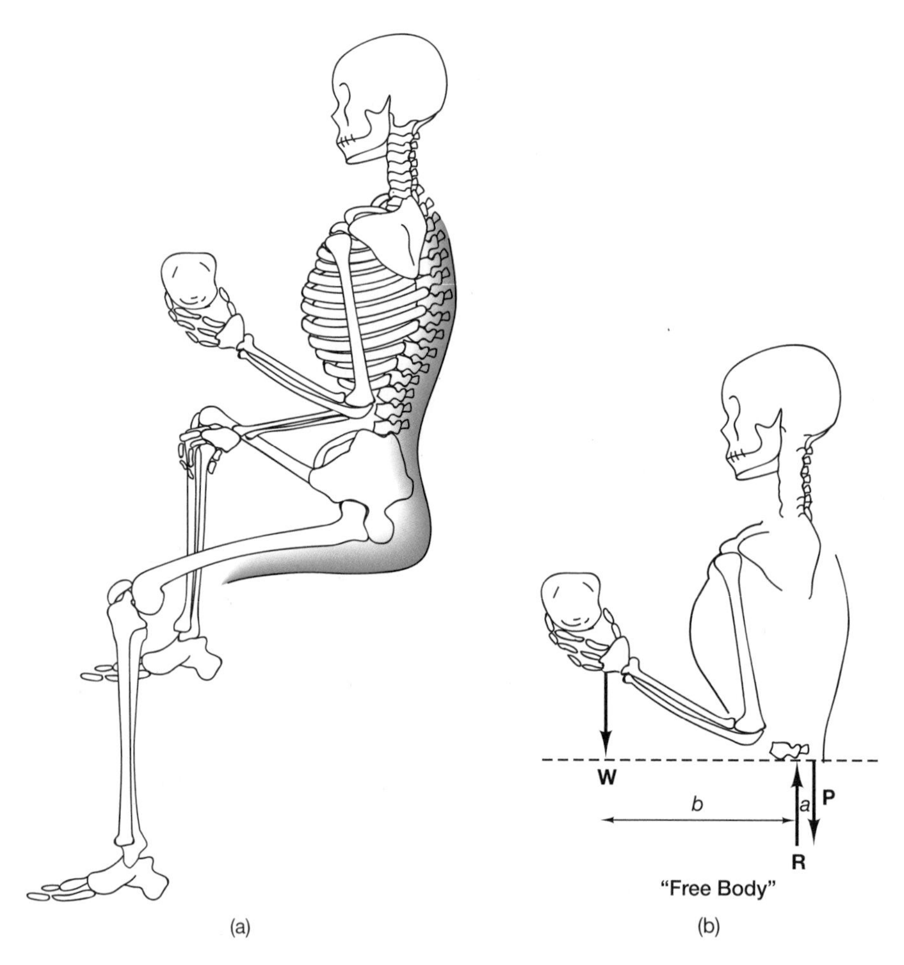

Figure 6.31

Sample Problem 6.2

A fold-down shelf is loaded with a uniform row of books, as shown in the diagram on the next page. Determine the reactions at the shelf supports. The books weigh 180 N altogether, and the center of mass of the books is 12 centimeters from the wall. Ignore the weight of the shelf.

Solution We choose a coordinate system with origin at A and draw a free-body diagram. First, we write all the forces in vector notation and determine the position vector from the origin to a point on the line of action of each force:

$$\mathbf{A} = A_x\hat{\mathbf{i}} + A_y\hat{\mathbf{j}} \qquad \mathbf{r}_{A/A} = 0$$

$$\mathbf{B} = -B_x\hat{\mathbf{i}} \qquad \mathbf{r}_{B/A} = -4\hat{\mathbf{j}}$$

$$\mathbf{W} = -40\hat{\mathbf{j}} \qquad \mathbf{r}_{W/A} = -6\hat{\mathbf{i}}$$

The resultant of the force system $\mathbf{R} = \Sigma\mathbf{F} = 0$, and therefore, each component of the resultant is zero:

$$A_x - B_x = 0$$

$$A_y - 40 = 0$$

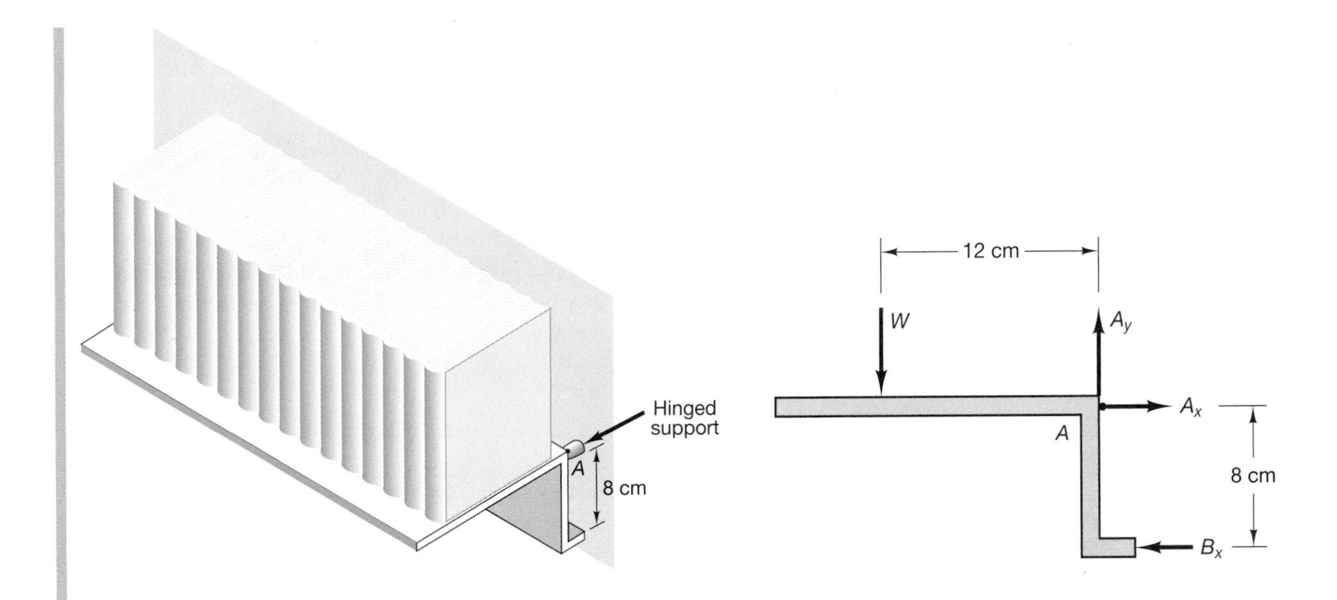

The moment may be computed about any point, and we will take it about the origin A to facilitate the solution. (Only one unknown, B_x, will appear in the equation.) We have

$$\mathbf{r}_{B/A} \times \mathbf{B} + \mathbf{r}_{W/A} \times \mathbf{W} = [12(180) - 8B_x]\hat{\mathbf{k}} = 0$$

$$B_x = 270 \text{ N}$$

This result can be substituted into the first equation, yielding

$$A_x = 270 \text{ N}$$

From the second equation, we obtain

$$A_y = 180 \text{ N}$$

Placing the origin at the point A and summing moments about that point made the solution of the 3×3 system of equations easier, but the origin could have been placed anywhere.

Alternate Solution

The shelf may be considered a three-force and the three forces are concurrent.

$$A(\cos 33.7\,\hat{\mathbf{i}} + \sin 33.7\,\hat{\mathbf{j}}) - 180\hat{\mathbf{j}} - B\hat{\mathbf{i}} = 0$$

$$\begin{aligned} A \sin 33.7 &= 180 \\ A &= 238 \text{ N} \\ A \cos 33.7 &= B \\ B &= 270 \text{ N} \end{aligned}$$

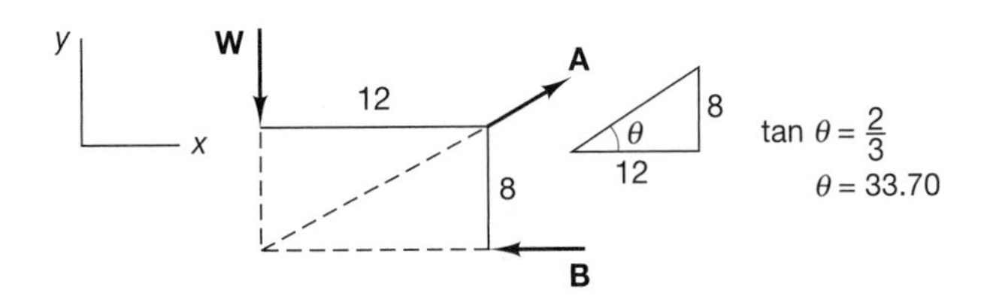

Sample Problem 6.3

If the center of mass of the front-wheel-drive car shown to the left (top) is 30 percent of the wheel base behind the front wheels, determine the load on the front and rear tires.

Solution

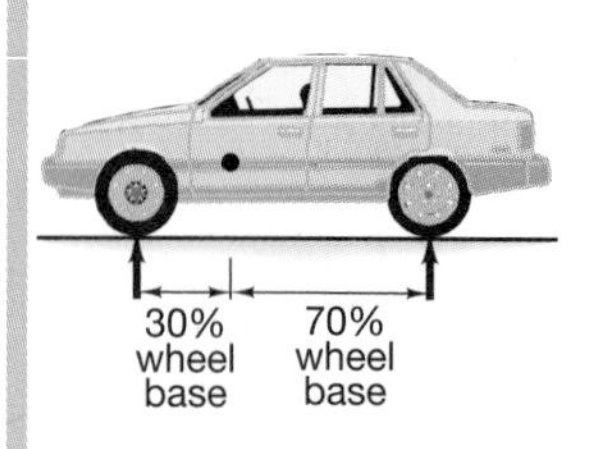

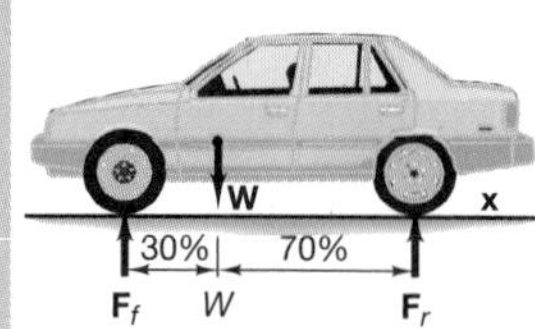

Assuming that the center of mass lies on the plane of geometric symmetry of the car, we can treat this problem as a case of parallel and coplanar forces. We will select the origin of the coordinate system at the point of contact between the front wheels and the ground. Once the free-body diagram has been drawn (see figure at the left (bottom)), the force and position vectors can be written in vector notation. The force vectors on the tires represent two front tires or two rear tires, and we have

$$\mathbf{F}_r = 2F_r\hat{\mathbf{j}} \qquad \mathbf{r}_{r/f} = 1.00\hat{\mathbf{i}}$$
$$\mathbf{F}_f = 2F_f\hat{\mathbf{j}} \qquad \mathbf{r}_{f/f} = 0$$
$$\mathbf{W} = -W\hat{\mathbf{j}} \qquad \mathbf{r}_{W/f} = 0.30\hat{\mathbf{i}}$$

The two nontrivial equilibrium equations are

$$2\,F_f + 2\,F_r - W = 0$$
$$2.00\,F_r - 0.30\,W = 0$$

Solving this linear system, we get

$$F_r = 0.15\,W$$
$$F_f = 0.35\,W$$

Therefore, each front tire carries 35 percent of the weight of the car, and each rear tire carries 15 percent of the weight of the car. Front-wheel-drive vehicles can generate more traction because of the greater percentage of weight on the front tires.

Sample Problem 6.4

If a rod of length $3R$ is placed in a hemispherical bowl of radius R, determine the angle α the rod will make with the horizontal for the rod to be in equilibrium. Neglect friction between the rod and the bowl, and assume that the bowl does not rock. (See figure at left (top).)

Solution

The reactions must be normal to the contacting surfaces. The free-body diagram is shown at the left (bottom). We have chosen the origin of the coordinate system at the left end of the rod, which is in contact with the inside of the bowl, and the x-axis is parallel to the rod and the y-axis is perpendicular to the rod. Note that trigonometry has been used to find the distance from the origin to the normal force **N**. Frequently, the greatest difficulty encountered is finding required geometric parameters for the free-body diagram. The force vectors and position vectors from the origin to the line of action of each force are

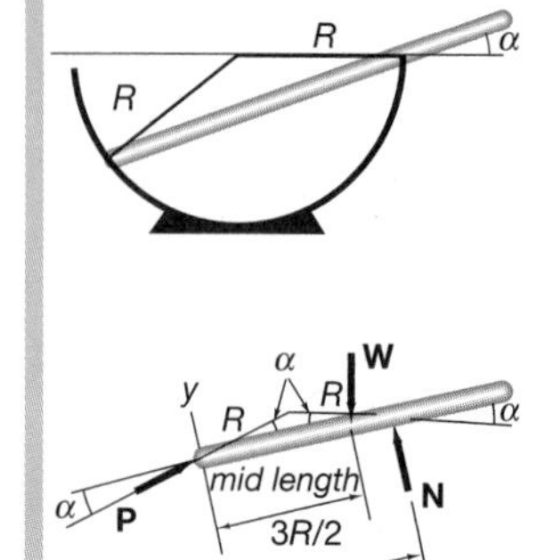

$$\mathbf{P} = P(\cos\alpha\hat{\mathbf{i}} + \sin\alpha\hat{\mathbf{j}}) \qquad \mathbf{r}_{P/O} = 0$$
$$\mathbf{W} = W(-\sin\alpha\hat{\mathbf{i}} - \cos\alpha\hat{\mathbf{j}}) \qquad \mathbf{r}_{W/O} = 3R/2\hat{\mathbf{i}}$$
$$\mathbf{N} = N\hat{\mathbf{j}} \qquad \mathbf{r}_{R/O} = (2\,\mathrm{R}\cos\alpha)\hat{\mathbf{i}}$$

The equilibrium equations are

$$\sum F_x: P\cos\alpha - W\sin\alpha = 0 \quad \text{(SP6.4.1)}$$

$$\sum F_y: P\sin\alpha - W\cos\alpha + N = 0 \quad \text{(SP6.4.2)}$$

$$\sum M_O: -(3R/2)W\cos\alpha + (2R\cos\alpha)N = 0 \quad \text{(SP6.4.3)}$$

Equations (SP6.4.1) thorugh (SP6.4.3) constitute a system of nonlinear equations for P, N, and α in terms of W. The moment equation, Eq. (SP6.4.3), can be solved directly for the normal force at the point of contact of the rod with the lip of the bowl:

$$N = 3/4W \quad \text{(SP6.4.4)}$$

Eqs. (SP6.4.1) and (SP6.4.2) give two solutions relating P to W and α:

$$P = W\tan\alpha$$

$$P = \frac{(\cos\alpha - 0.75)W}{\sin\alpha} \quad \text{(SP6.4.5)}$$

Equating these two values of P gives a transcendental equation for α:

$$\frac{\sin\alpha}{\cos\alpha} = \frac{\cos\alpha - 0.75}{\sin\alpha}$$

Thus,

$$\sin^2\alpha - \cos^2\alpha + 0.75\cos\alpha = 0 \quad \text{(SP6.4.6)}$$

This equation can be solved by using trigonometric identities, thereby eliminating the $\sin^2\alpha$ terms, or by utilizing computational software to find the roots of the equation. From trigonometry, we obtain a quadratic equation for $\cos\alpha$:

$$1 - 2\cos^2\alpha + 0.75\cos\alpha = 0$$

This equation is in the form $ax^2 + bx + c = 0$, with solutions given by

$$x = \frac{-b \pm \sqrt{b^2 - 4ac}}{2a}$$

The solutions of the equation are

$$\cos\alpha = \begin{pmatrix} 0.919 \\ -0.544 \end{pmatrix}$$

or

$$\alpha = \begin{pmatrix} 23.2° \\ 123° \end{pmatrix}$$

The value of 123° is physically unrealistic, so the correct equilibrium position is 23.2°. The normal contact force P at the end of the rod may now be obtained:

$$P = W\tan\alpha = 0.429\,W$$

The computational solution is shown in the Computational Supplement.

Sample Problem 6.5

A uniform 2-m, 50-kg bar is supported by a spring with a spring constant of 500 N/m. The spring is unstretched in the vertical position with a length of 2 m. Determine the equilibrium angle, α, of the bar.

Solution A free-body diagram of the bar in the equilibrium position is shown in Figure SP6.5.2 The tension in the spring may be written as:

$$\mathbf{T}(\alpha) = T(\alpha)\left[-\frac{\sin\alpha}{\sqrt{5 - 4\cos\alpha}}\hat{\mathbf{i}} + \frac{2 - \cos\alpha}{\sqrt{5 - 4\cos\alpha}}\hat{\mathbf{j}}\right]$$

where $T(\alpha) = kl\left[\sqrt{5 - 4\cos\alpha} - 1\right]$

The weight vector may be written as:

$$W = 50g$$
$$\mathbf{W} = -W\hat{\mathbf{j}}$$

The position vectors from point C to $\mathbf{T}$ and $\mathbf{W}$ are:

$$\mathbf{r}_{T/C} = l(\sin\alpha\hat{\mathbf{i}} + \cos\alpha\hat{\mathbf{j}})$$
$$\mathbf{r}_{W/C} = \frac{l}{2}(\sin\alpha\hat{\mathbf{i}} + \cos\alpha\hat{\mathbf{j}})$$

The moment about point C is

$$\mathbf{M}_C = \mathbf{r}_{T/C} \times \mathbf{T} + \mathbf{r}_{W/C} \times \mathbf{W} = \mathbf{0}$$

$$\mathbf{M}_C = \left\{T(\alpha)l\left[\frac{2\sin\alpha - \sin\alpha\cos\alpha + \sin\alpha\cos\alpha}{\sqrt{5 - 4\cos\alpha}}\right] - \frac{Wl}{2}\sin\alpha\right\}\hat{\mathbf{k}} = \mathbf{0}$$

Figure SP6.5.1

Factoring out ($l\sin\alpha$) the solution may be written in the form of the root of a function of α:

$$f(\alpha) = T(\alpha)\frac{2}{\sqrt{5 - 4\cos\alpha}} - \frac{W}{2} = 0$$

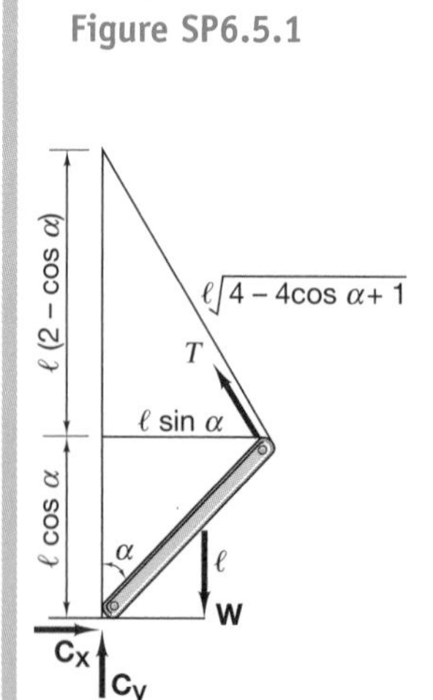

Figure SP6.5.2

This function can be plotted against values of the angle in radians as shown in Figure SP6.5.3.

$$\alpha := 0, \frac{\pi}{100} \,..\, \frac{\pi}{5}$$

radians

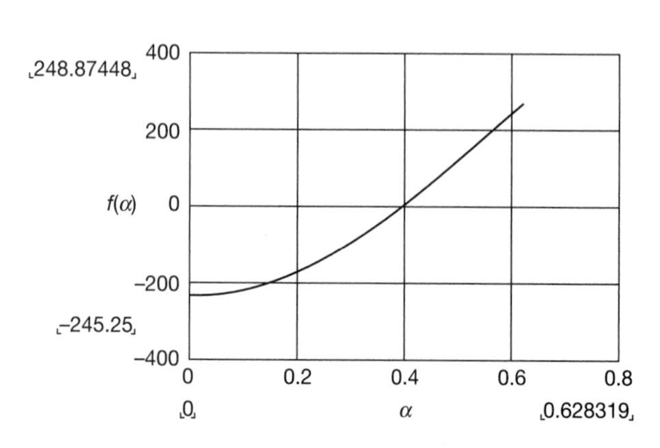

Figure SP6.5.3

The function is zero at 0.389 radians or 22.3°. Therefore the equilibrium angle is

$$\alpha = 22.3°$$

Problems

6.11 Two children are sitting on swings hanging from a 4-m-long jungle gym of mass 25 kg. (See Figure P6.11.) The child at *C* has a mass of 35 kg, and the child at *D* has a mass of 30 kg. Calculate the reactions of the pin at *A* and the roller at *B*.

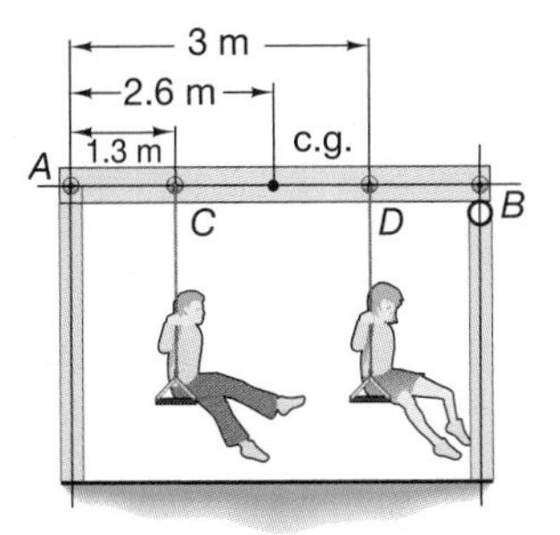

Figure P6.11

6.12 In Figure P6.12, a small pedestrian bridge is supported by a rocker at one end (*B*) and a pin at the other (*A*). A 100-kg person is standing 1 m out from point *A*. The bridge has a mass of 250 kg. Assuming that the center of gravity is in the middle of the bridge, calculate the reaction at points *A* and *B*.

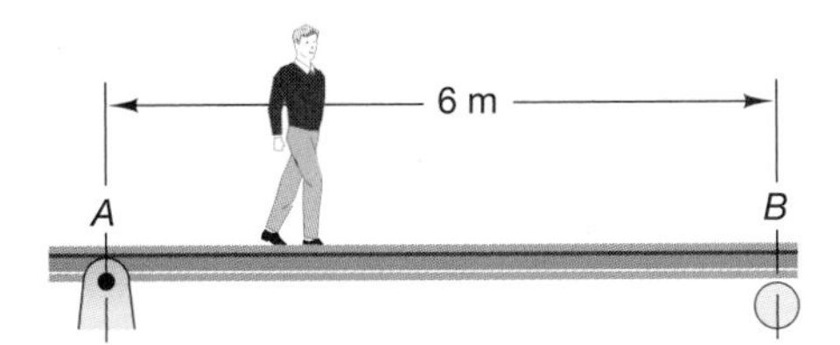

Figure P6.12

6.13 Consider a deck beam of mass 45 kg with a 90-kg man standing near the edge as shown in Figure P6.13. Calculate the reaction forces at the pin at point *A* and the roller at point *B*.

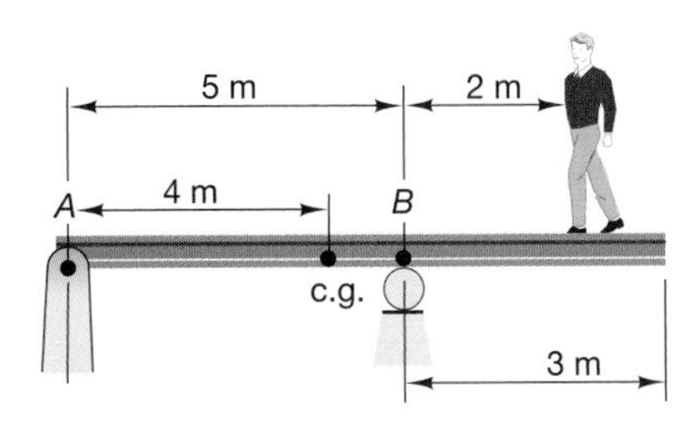

Figure P6.13

6.14 An air-conditioning unit having a mass of 200 kg sits on a shelf supported by a pin-and-cable arrangement, as illustrated in Figure P6.14. Assume that the geometric center and center of gravity of the shelf coincide and that the platform has a mass of 25 kg. Calculate the reactions at *A* and *B*.

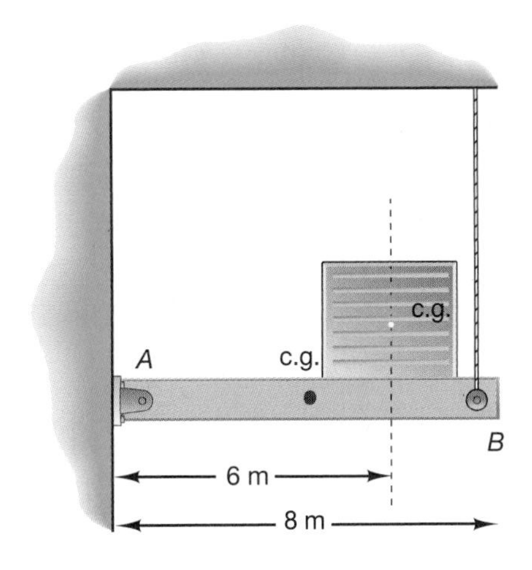

Figure P6.14

6.15 Consider the generic "two-dimensional" automobile. In road-test information provided in magazines and newspapers, the center of gravity is usually indicated as a percentage of the weight distribution, from front to rear. With the wheel base, the car's weight, and the weight distribution given in the following table, calculate the distance to the front and rear tires from the center of mass for each car listed.

CAR MAKE	CURB MASS	WHEEL BASE	WEIGHT DISTRIBUTION (F/R)
Lincoln Continental	1805 kg	2.77 m	62.4/37.6%
BMW 318 ti	1260 kg	2.70 m	51.4/48.6%
Porsche 911	1440 kg	2.27 m	50/50%
Chevrolet Blazer	1912 kg	2.72 m	56/44%

6.16 A sign fixture consists of a 40-kg beam pinned at point D and fixed by a cable at point A. (See Figure P6.16.) The uniform beam is 1 m long with center of mass at its geometric center. The sign is centered, is 0.8 meter long, has a mass of 10 kg with center of mass at its geometric center, and is held to the support beam by two cables at points B and C. Calculate the reaction at D and the tension in the cable at point A.

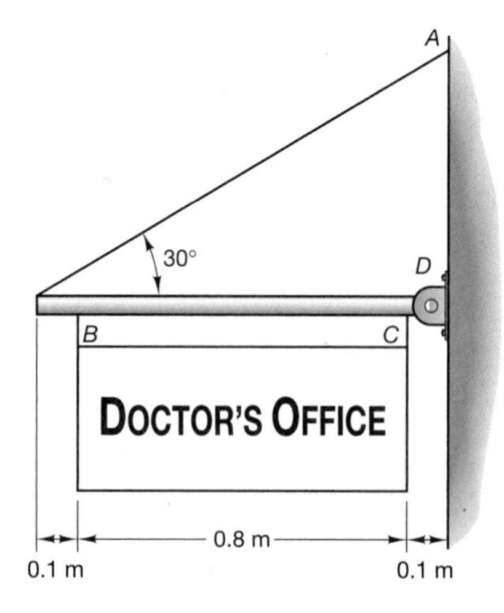

Figure P6.16

6.17 Calculate the reaction forces for the 100-kg cart illustrated in Figure P6.17. The points A, B, and C are all frictionless points of contact.

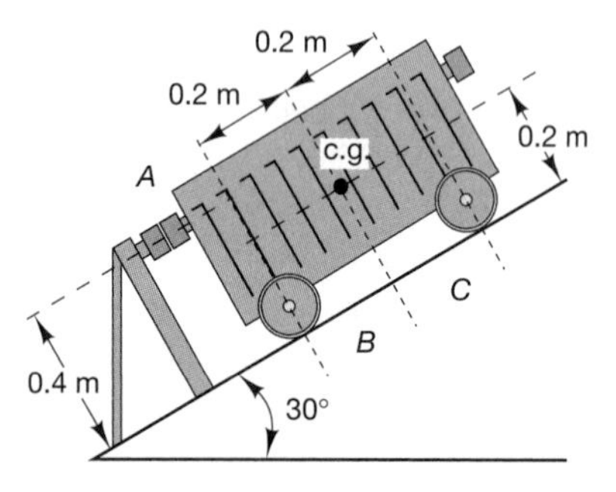

Figure P6.17

6.18 The bridge support structure shown in Figure P6.18 has a mass of 101.94 kg with center of gravity located midway between A and B. Calculate the reaction of the bridge supports if a 3-kN load is applied at the point indicated in the figure.

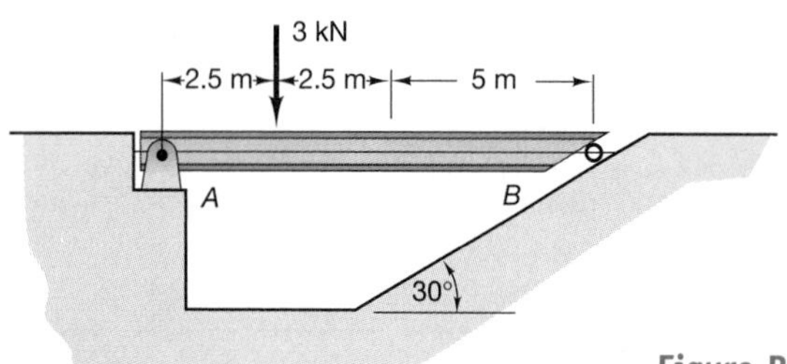

Figure P6.18

6.19 A storage door is held open by a cable. (See Figure P6.19.) The manufacturer of the door needs to specify to the customer what type of cable must be used to hold the door open. Calculate the tension in the cable and the reactions at the hinge A for the case $\theta = 30°$. Assume that the cable is vertical at this angle and that the mass of the door is 200 kg.

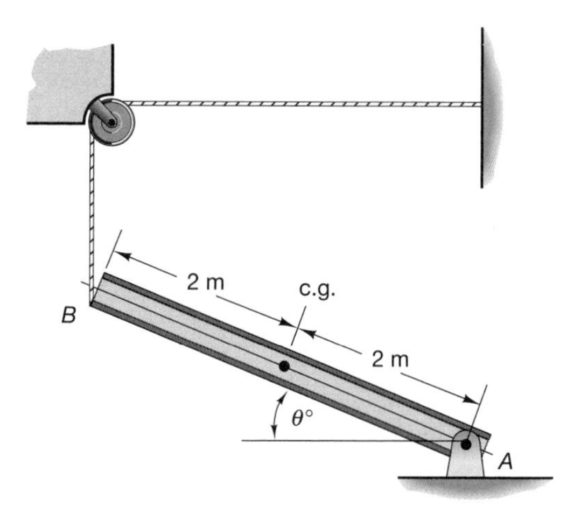

Figure P6.19

6.20 The manufacturer of a storage bin uses a winding mechanism to open and close the door of the bin. (See Figure P6.20.) If the uniform door has a mass of 50 kg, calculate the tension in the cable and the reaction force on the hinge for angles θ ranging from 5 to 40° in increments of 5°. Assume that the center of mass of the door is its geometric center. This information is needed to properly design both the hinge and the cable.

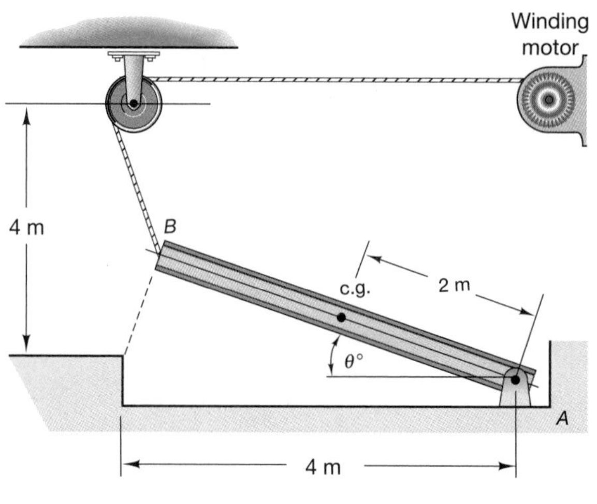

Figure P6.20

6.21 A truss structure is used to hold power lines, as illustrated in Figure P6.21. Calculate the reaction forces at the base of the tower. Note that if the applied forces result in a net counterclockwise moment, the tower will tip.

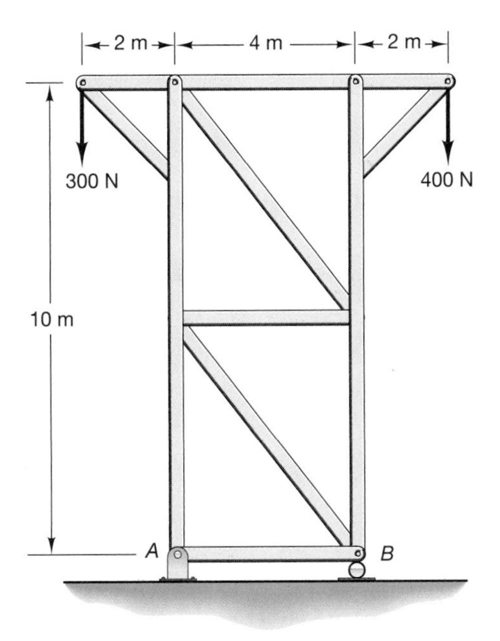

Figure P6.21

6.22 A truss structure is used to hold power lines, as illustrated in Figure P6.22. Unlike the situation in the previous problem, the power lines provide forces skewed to the tower. Calculate the reaction forces at the base of the tower.

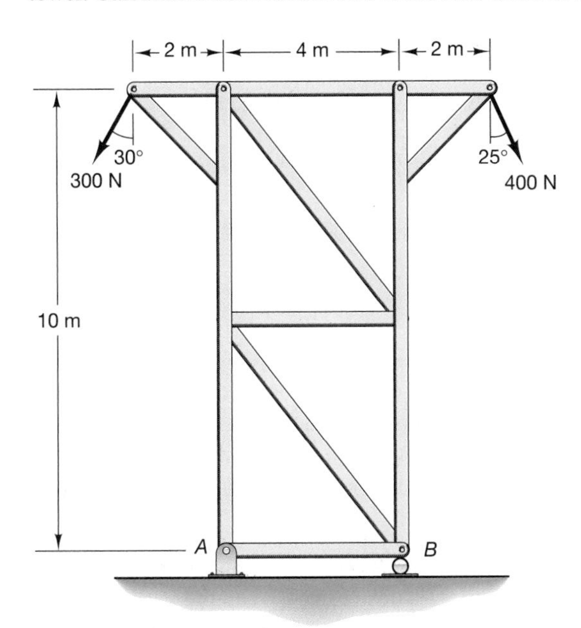

Figure P6.22

6.23 A 1-m-long bar of mass 10 kg is held up by three cables (F, T_1, and T_2), as shown in Figure P6.23. Determine the force in the three cables for $\alpha = 45°$ and $\beta = 30°$.

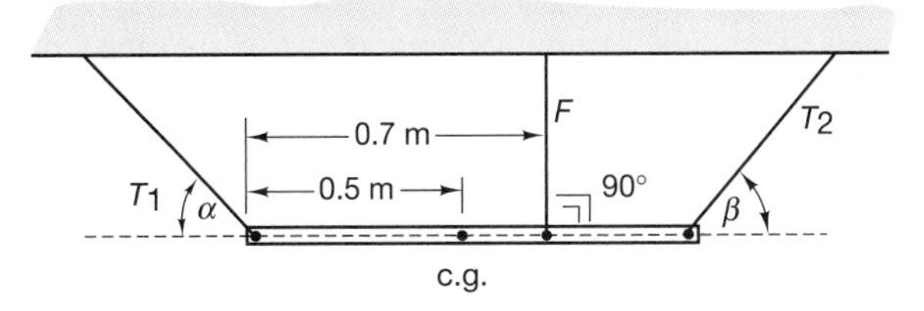

Figure P6.23

6.24 A uniform beam is pinned at one end and supported by a cable at the other end. (See Figure P6.24.) The beam has a mass of 20 kg and is 4 m long. Compute the reactions at the pinned end and the tension in the cable for the case $\theta = 30°$ Treat the beam as a three-force member.

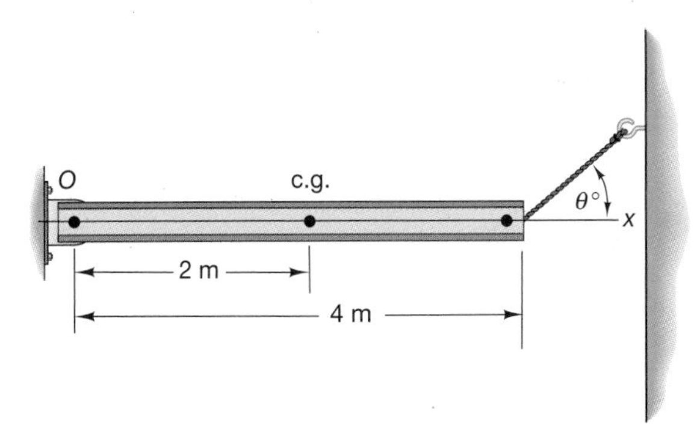

Figure P6.24

6.25 Consider the system of Figure P6.24. Let the angle θ take on several values between 10° and 90°, in increments of, say, 10°. Compute the reaction forces and the tension. Comment on what happens at $\theta = 0$ and $\theta = 90°$.

6.26 An overhanging roof for an outdoor cafe consists of a beam 5 m long of mass 100 kg. (See Figure P6.26.) Consider designing the support system. That is, choose where to attach the cable (pick ℓ) so that the tension in the cable and the reaction force at the pin are the smallest. Compute values of the reaction forces for $\ell = 0.5, 1, 1.5, \ldots, 5$.

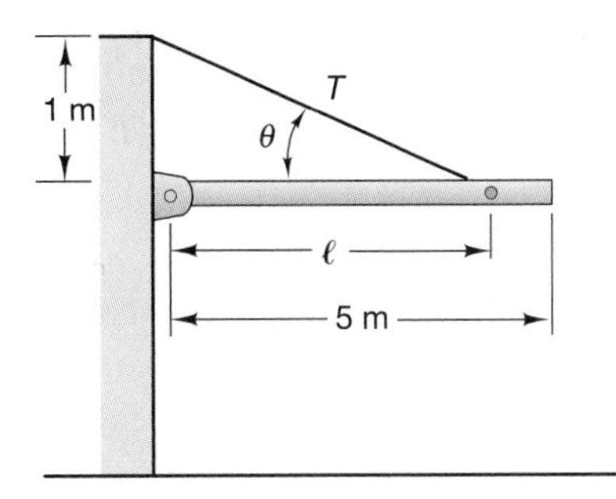

Figure P6.26

6.27 A support is held in place by a collar on a frictionless rod and a pin as shown in Figure P6.27. The center of mass of the 1000-kg support is at its geometric center. (a) Determine the reactions at A and B if a 4000-kg mass is suspended from the support at C. (b) Point C is attached to a moving track that runs between points E and F. Compute the reactions at A and B for each position of the hanging mass, starting at F and placing it at 0.5-m increments until point E is reached.

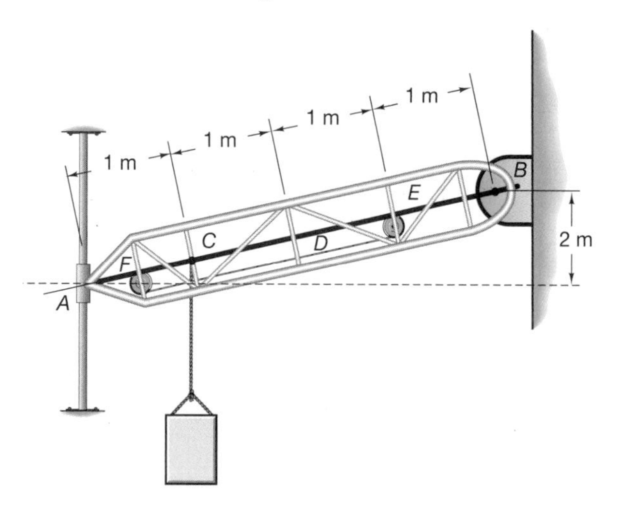

Figure P6.27

6.28 The lift bar AC of a tow truck is modeled as a pinned beam. (See Figure P6.28. The hoist cable CD is assumed to undergo no friction as it goes over the top of the lift bar. Calculate the forces in the supporting cable (at B) and at the hinge of the lift bar (point A). Assume the lift bar to be 2 m in length, ignore its mass, and assume that the support cable is attached two-thirds of the way up the bar.

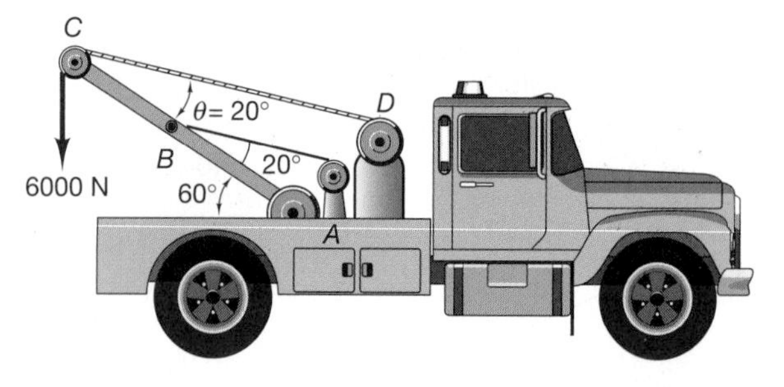

Figure P6.28

6.29 A lifting mechanism consists of a uniform beam pinned at point A and positioned by a roller at point B. (See Figure P6.29.) As hydraulic motion moves the roller up, the point of control ℓ changes. (a) Calculate the symbolic relationship for the reaction forces at A and B in terms of the length ℓ, the weight W, and the angle β. The center of mass is at the geometric center of the beam. (b) Evaluate these forces for the case $\beta = 30°$, $\ell = 1.5$ m, and $W = 400$ N.

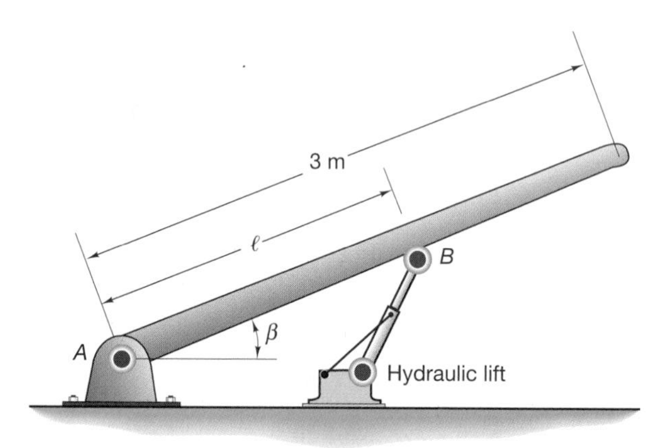

Figure P6.29

6.30 A uniform bar of weight W and length ℓ is supported by a spring with spring constant k and unstretched length ℓ when the bar in an upright position. (See Figure P6.30). Consider the pin at the base of the bar to be frictionless so the bar in the upright position is unstable and the bar will fall to the right to an angle α. Determine the angle α in terms of W, k, and ℓ.

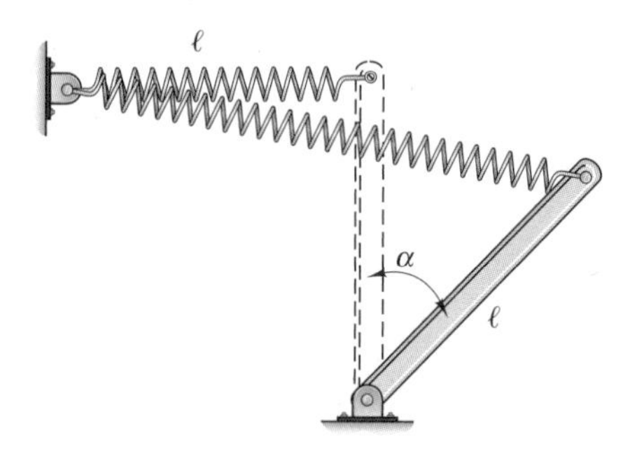

Figure P6.30

6.31 A uniform bar of mass m and length ℓ is supported by two springs as shown in Figure P6.31. The springs are unstretched when the bar is horizontal. Determine the equilibrium angle α in terms of the parameters given.

6.32 A uniform bar of mass m and length ℓ is supported by a spring with spring constant k as shown in Figure P6.32. The spring is unstretched when the bar is vertical. Determine the equilibrium angle θ.

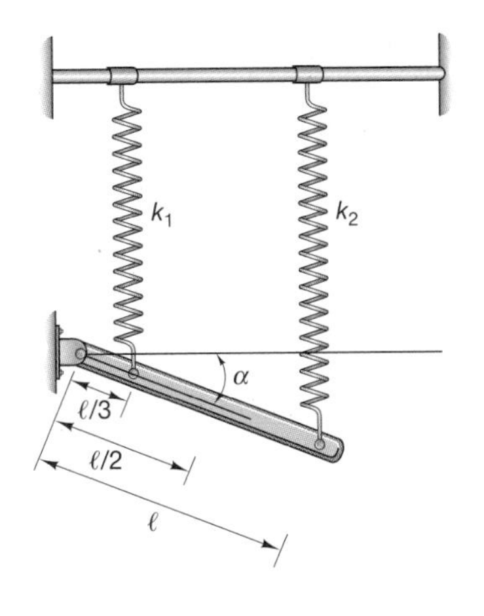

Figure P6.31

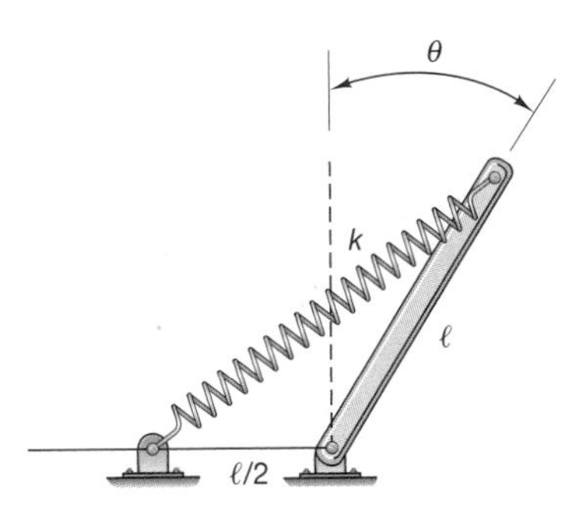

Figure P6.32

6.33 When a weight lifter lifts weight, the force between each foot and the ground is equal to $F = \frac{1}{2}(BW + W)$ where BW is the body weight and W is the weight lifted. A detail of the knee joint is shown in Figure P6.33. The moment of the force on the foot is balanced by the moment in the tension in the patella tendon T. Examination of anatomical dimensions gives the order of magnitude of D and d:D = 25 to 30 cm and d = 5 to 6 cm. Calculate the tension in the patella tendon and choose realistic numbers for the body weight and weight lifted. When the weight is accelerated up, the force between the foot and the ground is greater. Sometimes this results in the rupture of the patella tendon.

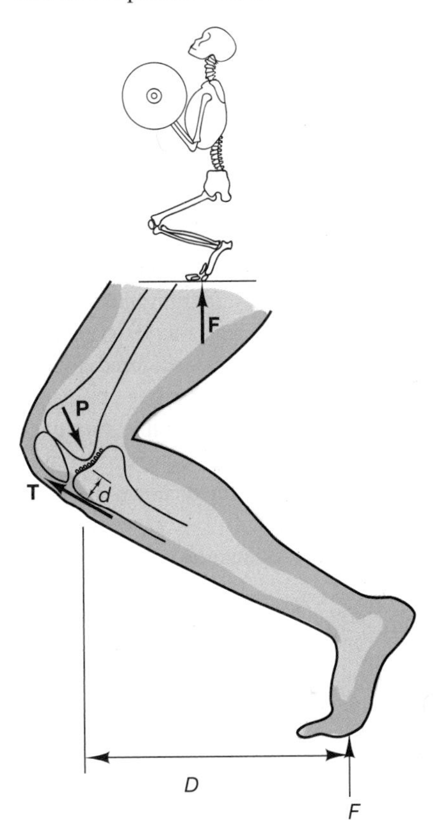

Figure P6.33

6.6 EQUILIBRIUM OF A RIGID BODY IN THREE DIMENSIONS

In Section 6.1, we showed that the resultant force and the moment of all applied forces and reactions must be zero to maintain static equilibrium of a rigid body in three dimensions. This is written as two vector equations

$$\sum \mathbf{F} = 0$$
$$\sum \mathbf{M}_O = 0 \tag{6.14}$$

Equation (6.14) may also be written as

$$\sum \mathbf{F} = 0$$
$$\sum (\mathbf{r} \times \mathbf{F}) + \mathbf{C} = 0 \tag{6.15}$$

The couple **C** is the sum of all applied couples on the rigid body. The origin or point about which the moments of the forces are taken can be any point on the rigid body or a fixed point in space. Equation (6.14) may be written as the six scalar equations.

$$\sum F_x = 0 \qquad \sum M_x = 0$$
$$\sum F_y = 0 \qquad \sum M_y = 0 \qquad (6.16)$$
$$\sum F_z = 0 \qquad \sum M_z = 0$$

These six equations are solved for six unknown reactions or applied forces to maintain equilibrium of the rigid body. If there are more than six unknown forces or couples, the problem is statically indeterminate.

The strategy for solving equilibrium problems for a single rigid body in three dimensions is the same as that for solving coplanar equilibrium problems. First, a proper free-body diagram must be constructed before analysis is begun. All assumptions made in constructing this diagram must be stated; for example, friction is ignored, the weight of the body is considered small compared to the other applied loads, etc. If a proper free-body diagram has been constructed, an origin and a coordinate system can be established. Each force and couple should be written as a vector with components in the coordinate system. Position vectors from the origin to the line of action of each force should be expressed in component notation. Then the two vector equations of static equilibrium can be written. Usually, the problem has by now been reduced to the solution of six scalar equations for six unknowns.

Before the advent of modern computational methods to solve systems of equations, many tricks were developed to reduce the computational difficulties. Moments were taken about points where some of the unknown forces acted, so these unknowns did not explicitly appear in the equations, and the numerical solution of the equations was made simpler. However, if computational tools are used, it is clear that the moment equation may be taken about any point on the rigid body, with no increase in difficulty in the solution.

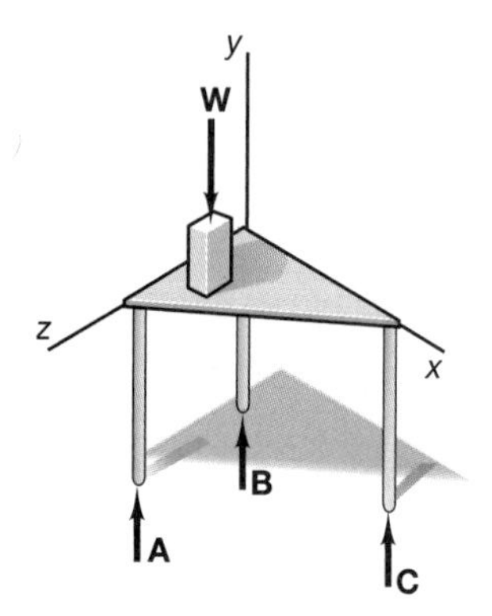

Figure 6.32

As in the case of coplanar force systems on a body modeled in two dimensions, several special force systems are worth noting in the case of three-dimensional problems. If all the forces applied to the rigid body in three dimensions are parallel, only three nontrivial scalar equilibrium equations remain. This parallel force system was discussed in Chapter 4. For example, if all the forces are parallel to the z-axis, summing the forces in the x- and y-directions yields trivial equations of zero equals zero. This situation is shown in Figure 6.32, of a platform supported by three posts and loaded with a box. Forces with only z-components cannot produce moments about that axis, so the summation of moments about the z-axis is also a trivial equation. Therefore, even when the forces and moments are formulated in vector notation, only three scalar equilibrium equations need to be considered.

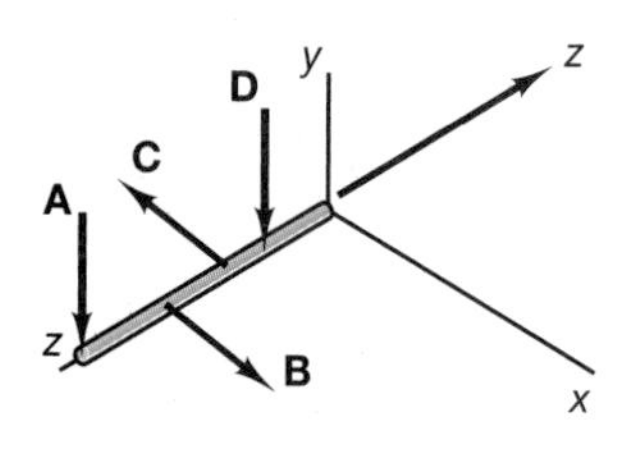

Figure 6.33

If all the forces intersect a common axis in space, the component of the moment of these forces along that axis is zero, and rotation about that axis is not constrained. For example, in Figure 6.33, all the forces intersect the axis of a shaft. If the axis of the shaft is the z-coordinate axis, none of the forces can produce a moment about that axis. Again, this will be apparent if the problem is formulated using vector notation, and only five scalar equilibrium equations will need to be considered.

A third case of a concurrent force system was considered in detail in Chapter 3, where the equilibrium of a particle was discussed. None of the forces would cause a moment about this concurrent point, and only three equilibrium equations need to be considered. However, the three rotational degrees of freedom are unconstrained.

6.6.1 CONSTRAINTS

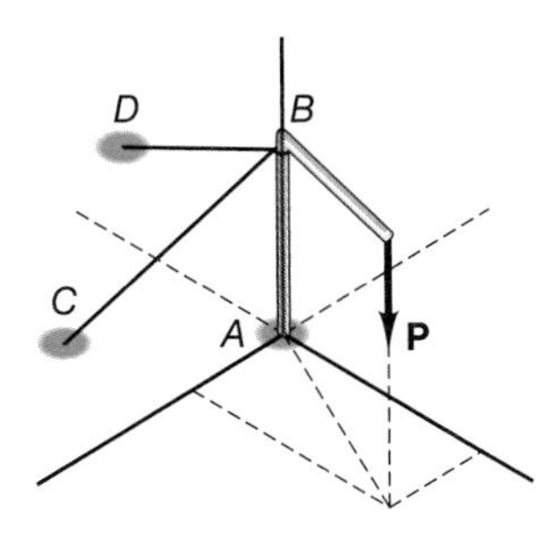

Figure 6.34

The supports on a rigid body act as constraints on the six degrees of freedom of that body—that is, the three translations and the three rotations. If there are fewer than six constraints, the body is underconstrained and is free to move in some manner. If there are more than six constraints on the single rigid body, the body is overconstrained and is statically indeterminate. In design and when one models constraints or supports, it is important to recognize underconstrained and overconstrained rigid bodies.

Consider the pole fixed at the base, as shown in Figure 6.34. The built-in support at *A* constrains all six degrees of freedom, and there are the three components of force and the three components of the moment at this support. If a cable *CB* or *DB* is added, the system is overconstrained and statically indeterminate. Now consider the case when the support at *A* is replaced by a ball-and-socket joint capable of constraining only translations with three components of the reaction force. Now, if only the cables *CB* and *DB* are added to the pole, the pole is underconstrained and free to rotate about its axis, as well as toward *C* and *D* about an axis parallel to a line from *C* to *D*.

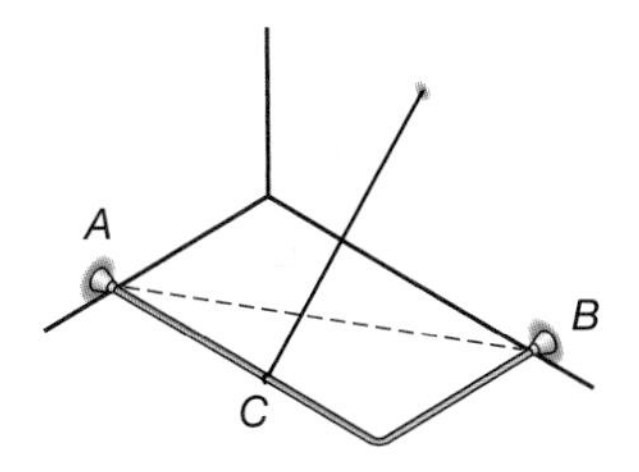

Figure 6.35

Consider the angle iron with ball-and-socket joints at *A* and *B* and supported by the cable from *C* to *D*, as shown in Figure 6.35. A ball-and-socket joint usually resists three translations with three corresponding forces (in the *x*-, *y*-, and *z*-directions). In this case, if we draw a straight line between the two ball and socket joints, we can note that a force along this line could be supported by either joint. Therefore, the bar could be considered improperly constrained in this direction and the force along this line can be constrained uniquely at only one of the two joints. A reasonable assumption would be the joint that was placed in compression would be the active one. That is, the joint where the force is pushing the ball into the socket. Modeled in this manner, there are only five constraints. The cable then acts as the sixth constraint and also restricts rotation about the line from *A* to *B* but only in a clockwise direction viewed from the right. If the cable were replaced by a link, rotation about this line would be totally constrained. The bar will be properly constrained with this change and the six unknown forces can be determined. One can see how the free-body diagram can influence design of a constraint system.

Sample Problem 6.6

The bar shown at the bottom of this page is supported by a ball-and-socket joint at *A*, by two cables *CG* and *BE*, and by a slender rod *BF* that is attached by means of ball-and-socket joints at both ends. Determine the reactions at the ball-and-socket joint at *A*, the tension in the two cables, and the force in the rod *BF* when the system is subjected to a force **P** of magnitude 1000 N. Neglect the weight of the bar.

Solution First we will construct the free-body diagram of the bar *ABCD*. The rod *BF* is a two-force member and transmits either tension or compression along its axis. The two cables can

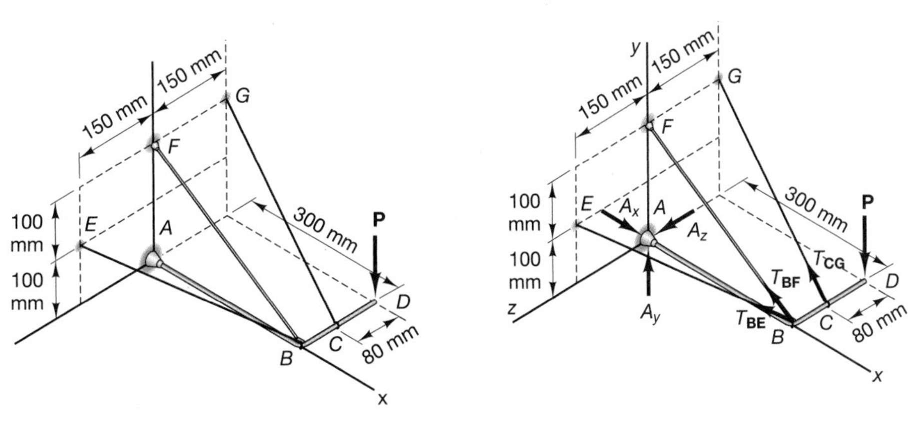

transmit only tension. We will assume a direction for all of the forces, and a negative value in the solution will indicate that the particular force acts in the opposite direction. The free-body diagram is shown on previous page. We first must write each force in vector notation and determine a position vector from the origin at A to a point on the line of action of each force. The forces in the cable and the rod will be written with an unknown magnitude, multiplied by a unit vector along the cable or rod. We obtain these unit vectors by first finding a position vector from the attachment at B or C to the attachment on the wall. The relative position vectors are

$$\begin{aligned}
\mathbf{r}_{E/B} &= -300\hat{\mathbf{i}} + 100\hat{\mathbf{j}} + 150\hat{\mathbf{k}} \\
\mathbf{r}_{F/B} &= -300\hat{\mathbf{i}} + 200\hat{\mathbf{j}} \\
\mathbf{r}_{G/C} &= -300\hat{\mathbf{i}} + 200\hat{\mathbf{j}} - 80\hat{\mathbf{k}} \\
\hat{\mathbf{e}}_{E/B} &= \frac{\mathbf{r}_{E/B}}{|\mathbf{r}_{E/B}|} \\
\hat{\mathbf{e}}_{F/B} &= \frac{\mathbf{r}_{F/B}}{|\mathbf{r}_{F/B}|} \\
\hat{\mathbf{e}}_{G/C} &= \frac{\mathbf{r}_{G/C}}{|\mathbf{r}_{G/C}|}
\end{aligned}$$

Using these unit vectors, we may write the forces as

$$\begin{aligned}
\mathbf{A} &= A_x\hat{\mathbf{i}} + A_y\hat{\mathbf{j}} + A_z\hat{\mathbf{k}} \\
\mathbf{P} &= -1000\hat{\mathbf{j}} \\
\mathbf{T}_{BE} &= T_{BE}(-0.857\hat{\mathbf{i}} + 0.286\hat{\mathbf{j}} + 0.429\hat{\mathbf{k}}) \\
\mathbf{T}_{BF} &= T_{BF}(-0.832\hat{\mathbf{i}} + 0.555\hat{\mathbf{j}}) \\
\mathbf{T}_{CG} &= T_{CG}(-0.812\hat{\mathbf{i}} + 0.542\hat{\mathbf{j}} - 0.217\hat{\mathbf{k}}
\end{aligned}$$

The position vectors from the origin to the line of action of the forces are

$$\mathbf{r}_B = 0.300\hat{\mathbf{i}} \qquad \mathbf{r}_C = 0.300\hat{\mathbf{i}} - 0.070\hat{\mathbf{k}} \qquad \mathbf{r}_D = 0.300\hat{\mathbf{i}} - 0.150\hat{\mathbf{k}}$$

The vector equations of equilibrium are

$$\mathbf{P} + \mathbf{A} + \mathbf{T}_{BE} + \mathbf{T}_{BF} + \mathbf{T}_{CG} = \mathbf{0}$$

$$\Sigma\mathbf{M}_A = \mathbf{r}_{B/A} \times (\mathbf{T}_{BE} + \mathbf{T}_{BF}) + \mathbf{r}_{C/A} \times \mathbf{T}_{CG} + \mathbf{r}_{D/A} \times \mathbf{P} = \mathbf{0}$$

The six scalar equations of equilibrium are

$$\begin{aligned}
&A_x - 0.857T_{BE} - 0.832T_{BF} - 0.812T_{CG} = 0 \\
&A_y + 0.286T_{BE} + 0.555T_{BF} + 0.542T_{CG} - 1000 = 0 \\
&A_z + 0.429T_{BE} - 0.217T_{CG} = 0 \\
&0.0379T_{CG} - 0.150(1000) = 0 \\
&-0.1286T_{BE} + 0.1218T_{CG} = 0 \\
&0.0857T_{BE} + 0.1664T_{BF} + 0.1625T_{CG} - 0.300(1000) = 0
\end{aligned}$$

Solving this system of simultaneous equations yields

$$A_x = 3106\text{ N} \qquad A_y = -2.13\text{ N} \qquad A_z = -750\text{ N}$$

$$T_{BE} = 3750\text{ N} \qquad T_{BF} = -3991\text{ N (bar is in compression)} \qquad T_{CG} = 3957\text{ N}$$

The solution of this problem using computational aids is presented in the Computational Supplement.

Sample Problem 6.7

The mechanical system shown in the first figure below on the left is used to lift a 100-kg mass. A shaft with a radius of 8 cm is supported by a thrust bearing at A and a nonthrust bearing at B. Determine the bearing forces as a function of the angle θ of the crank handle.

Solution

As has been discussed, the bearings can constrain rotations about the two axes perpendicular to the shaft. Therefore, if the middle bearing is a thrust bearing and the end bearing is a nonthrust bearing, there are five unknown bearing forces and four unknown bearing moments, in addition to the unknown force that must be exerted on the crank handle. Thus, the problem is clearly statically indeterminate. For a first approximation, the bearings will be assumed to be "simple supports," that is, the moments at these supports will be neglected and the problem solved by static equilibrium.The equilibrium equations are

$$\mathbf{P} + \mathbf{A} + \mathbf{W} + \mathbf{B} = 0$$

$$\Sigma \mathbf{M}_A = \mathbf{r}_{p/a} \times \mathbf{P} + \mathbf{r}_{w/a} \times \mathbf{W} + \mathbf{r}_{b/a} \times \mathbf{B} = 0$$

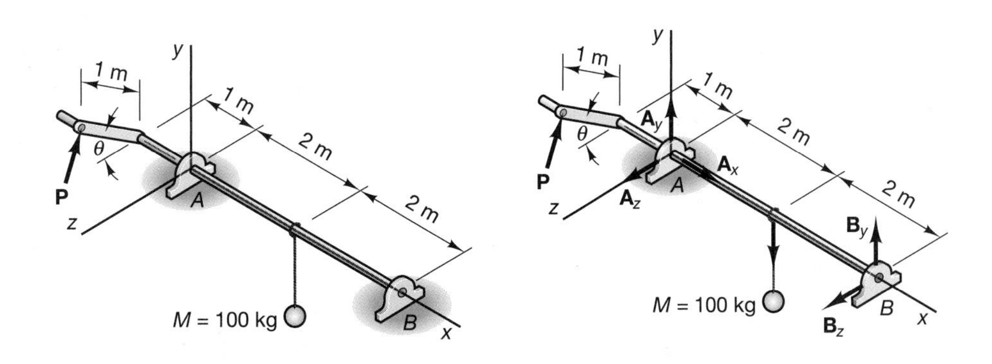

The force and position vectors are

$$\mathbf{P} = P(\cos\theta\hat{\mathbf{j}} - \sin\theta\hat{\mathbf{k}})$$
$$\mathbf{A} = A_x\hat{\mathbf{i}} + A_y\hat{\mathbf{j}} + A_z\hat{\mathbf{k}}$$
$$\mathbf{W} = -100(9.81)\hat{\mathbf{j}}$$
$$\mathbf{B} = B_y\hat{\mathbf{j}} + B_z\hat{\mathbf{k}}$$
$$\mathbf{r}_{p/a} = -\hat{\mathbf{i}} + \sin\theta\hat{\mathbf{j}} + \cos\theta\hat{\mathbf{k}}$$
$$\mathbf{r}_{m/a} = 2\hat{\mathbf{i}} + 0.08\hat{\mathbf{k}}$$
$$\mathbf{r}_{b/a} = 4\hat{\mathbf{i}}$$

Substitution of these vectors into the equilibrium equations yields

$$\sum \mathbf{F} = A_x\hat{\mathbf{i}} + (P\cos\theta + A_y - 981 + B_y)\hat{\mathbf{j}} + (-P\sin\theta + A_z + B_z)\hat{\mathbf{k}} = 0$$

$$\sum \mathbf{M}_A = (-P\sin^2\theta - P\cos^2\theta + 78.5)\hat{\mathbf{i}} + (-4B_z - P\sin\theta)\hat{\mathbf{j}}$$
$$+ (4B_y - P\cos\theta - 1962)\hat{\mathbf{k}} = 0$$

Setting the scalar components of the two vector equations equal to zero results in

$$\begin{aligned} P &= 78.5\text{ N} \\ A_x &= 0 & B_y &= 490.5 + 19.63\cos\theta \\ A_y &= 490.5 - 98.13\cos\theta & B_z &= -19.63\sin\theta \\ A_z &= 98.13\sin\theta \end{aligned}$$

The bearing forces can be graphed for a complete shaft revolution to determine when they will be a maximum.

Sample Problem 6.8

A cam shaft acts against a spring-loaded valve as shown in the top diagram to the left. Determine the bearing reactions at A and B and the torsion resisted by the motor for one revolution of the shaft. The bearing at A is a thrust bearing, and the bearing at B does not resist thrust. Model both bearings as simple supports (neglecting bearing moments). The cam is shown in detail just below the shaft. The spring constant $k = 200$ N/m, and the uncompressed length of the spring is to the shaft centerline.

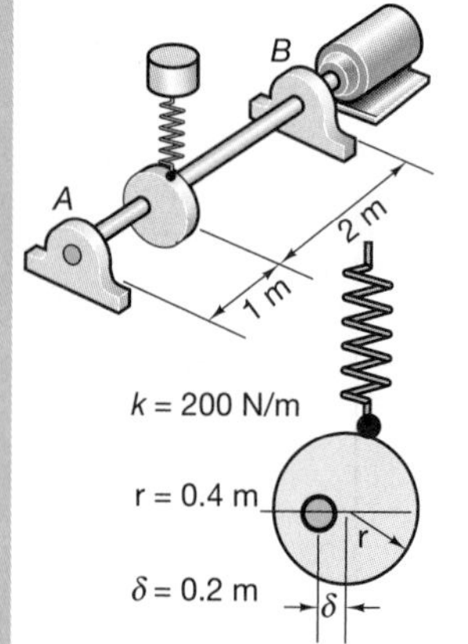

Solution

The origin of the coordinate system will be taken at the shaft center on the cam, as shown in the bottom figure to the left. A position vector to the point of contact of the cam with the plate, for any angle of rotation, θ, is

$$x_s = \delta\cos\theta \quad y_s = r + \delta\sin\theta \quad z_s = 0$$

The magnitude of the spring force is $F = k(y_s)$.

The force and position vectors are

$$\mathbf{F}_s = \begin{pmatrix} 0 \\ -ky_s \\ 0 \end{pmatrix} \quad \mathbf{r}_s = \begin{pmatrix} x_s \\ y_s \\ 0 \end{pmatrix} \quad \mathbf{A} = \begin{pmatrix} A_x \\ A_y \\ A_z \end{pmatrix} \quad \mathbf{r}_\mathbf{A} = \begin{pmatrix} 0 \\ 0 \\ 1 \end{pmatrix}$$

$$\mathbf{B} = \begin{pmatrix} B_x \\ B_y \\ 0 \end{pmatrix} \quad \mathbf{r}_\mathbf{b} = \begin{pmatrix} 0 \\ 0 \\ -2 \end{pmatrix} \quad \mathbf{T}_{\text{motor}} = \begin{pmatrix} 0 \\ 0 \\ T \end{pmatrix}$$

The equilibrium equations are

$$\begin{aligned} &A_x + B_x = 0 \\ &-k\,y_s + A_y + B_y = 0 \\ &A_z = 0 \\ &-A_y + 2\,B_y = 0 \\ &A_x - 2\,B_x = 0 \\ &-k\,x_s\,y_s + T = 0 \end{aligned}$$

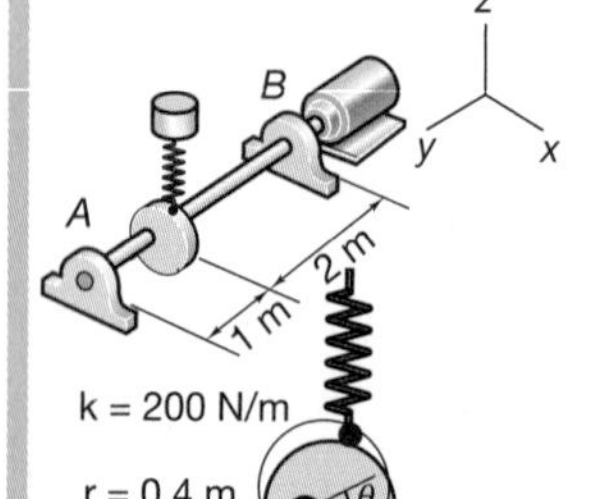

The solution of the system of six equations for the six unknowns is

$$A_x = 0 \quad B_x = 0 \quad A_y = 2/3(k\,y_s) \quad B_y + 1/3\,(k\,y_s) \quad A_z = 0 \quad T = k\,x_s\,y_s$$

Although these equations are easily found and solved by hand, they may be generated and solved analytically if desired. This approach is shown in the Computational Supplement.

Sample Problem 6.9

A gate is hinged on the left side as shown below. Construct a free-body diagram of the gate and determine the forces at the hinges in terms of the weight of the gate.

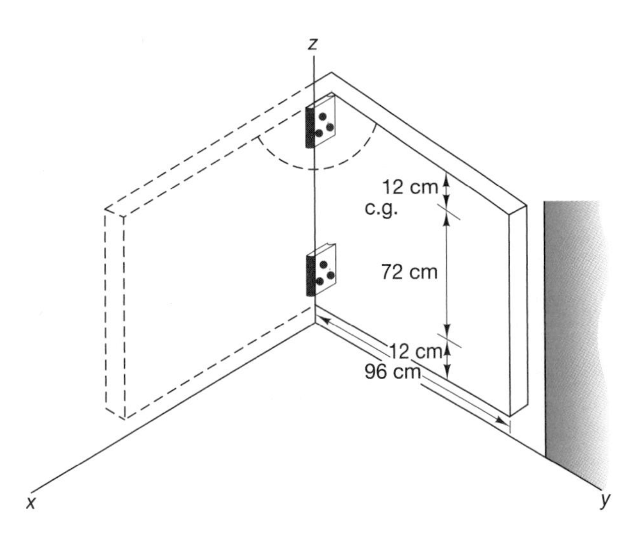

Solution The free-body diagram is shown to the left. Note that vertical forces are shown at both hinges as the vertical load could be carried at either point. There is no way to determine the distribution between the two points. Summing forces in the y direction yields

$$A_y - B_y = 0$$

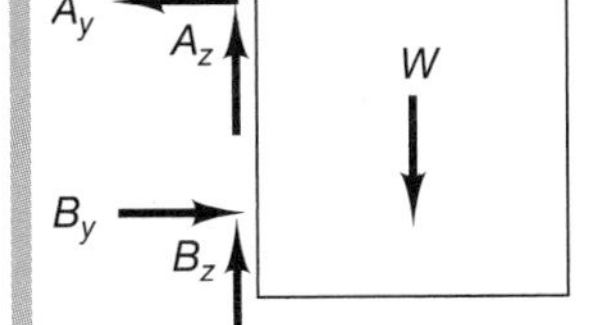

Summing moments around point **B** yields

$$72A_y - 48W = 0$$
$$A_y = B_y = \frac{2}{3}W$$

Summing forces in the z direction gives the only information on the vertical loads that can be determined.

$$A_z + B_z = W$$

The exact distribution of the weight between the two hinges cannot be determined but for design purposes, each hinge should be designed to carry the full weight.

6.7 STATICALLY INDETERMINATE REACTIONS AND IMPROPER CONSTRAINTS

In Section 6.6, we discussed some special cases where a rigid body was underconstrained or overconstrained. It is important to understand how constraints restrict the movement of the body before we attempt to design any structure or machine. If a body is improperly constrained or overconstrained, an equilibrium analysis to determine the reactions will not be possible. Improper constraints may result in failure of the structure and therefore must be recognized early in the design process. It is easiest to consider the role of constraints in bodies that may be modeled as two dimensional, before considering three-dimensional models. A two-dimensional body is considered to act in a single plane and, if unconstrained, may translate along two coordinate directions in that plane and rotate about an axis perpendicular to the plane. The two-dimensional body is then said to have three degrees of freedom. The constraints place the body in equilibrium and reduce its degrees of freedom. Different types of supports restrict different types of movements, as was shown in Section 6.3. For example, the reactions at a hinge joint can restrict horizontal and vertical movement of a body, but cannot restrict rotation about the pin at the hinge joint. A fixed support can restrict all three degrees of freedom in the two-dimensional model of a flagpole.

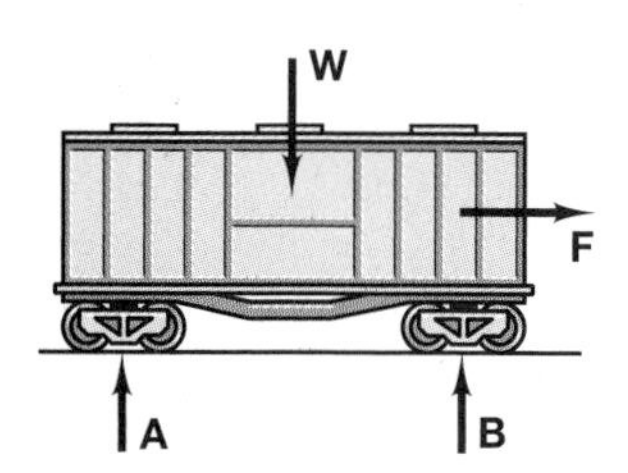

Figure 6.36

It was shown in Section 6.5 that, for two-dimensional models, only three equilibrium equations are applicable to determine the reactions. These three equations can be considered to coincide with the three degrees of freedom; that is, equilibrium in the horizontal direction constrains translation in the horizontal direction, equilibrium in the vertical direction constrains vertical translation, and rotational equilibrium about an axis perpendicular to the plane constrains rotation about that axis. Therefore, if a support system cannot produce a reaction in the horizontal direction, it cannot resist motion in that direction. Such a case is shown in Figure 6.36. The wheels underneath the railcar constrain the downward vertical movement and rotation of the car, but do not constrain its horizontal movement. If the brakes are set, then horizontal forces will act at both supports, and horizontal movement is constrained. When the brakes are released, no restrictions to horizontal movement are desired, as the car is designed to be pulled with minimum effort by the engine.

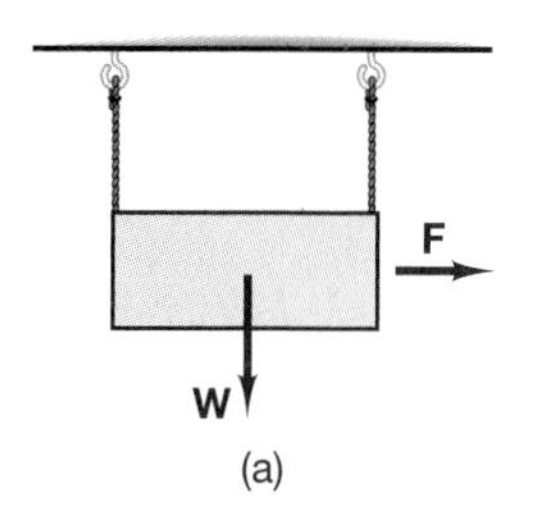

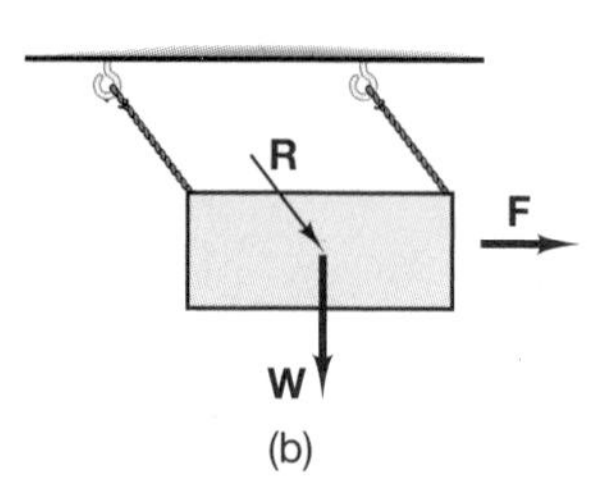

Figure 6.37

If a horizontal force is placed upon the body shown in Figure 6.37 supported by two cables, equilibrium cannot be maintained, and the body will move in a horizontal direction to another position of equilibrium. The equilibrium position shown in Figure 6.37(b) is such that the two cables are parallel to the resultant $\mathbf{R} = \mathbf{F} + \mathbf{W}$. Although equilibrium is possible in this case, it cannot be attained when the body is hanging in a vertical position. If the system is supported by a hinge at the left corner and the cable at the right, equilibrium is attained as shown in Figure 6.38(a). A free-body diagram of the body is shown in Figure 6.38(b), and the three reactions restrict all three degrees of freedom. However, note that if the horizontal force **F** is increased sufficiently that the counterclockwise moment about point A due to **F** is greater than the clockwise moment produced by **W**, the cable B, which can only carry tension, cannot maintain equilibrium, and the body rotates. In this case, the hinge support and the cable properly constrain the body under a restricted type of loading.

However, if the cable is attached at B in a horizontal direction, and only the weight is supported, the system would be improperly constrained, as shown in Figure 6.39.

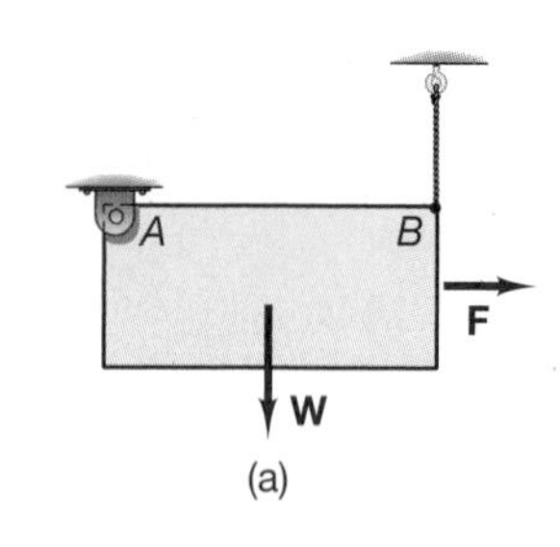

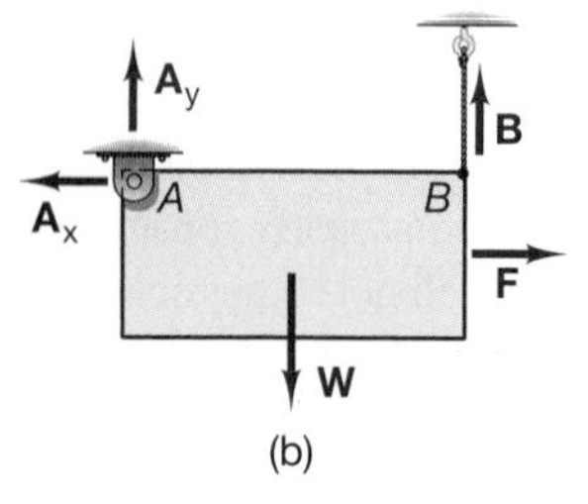

Figure 6.38

Figure 6.37(b) is the free-body diagram of the body, assuming that the cable at B is held in a horizontal direction. It is apparent that the weight causes a moment around point A in a clockwise direction, and the force in cable B passes through point A and therefore cannot resist this motion. Since the moment arm of force **B** about point A is zero, force **B** would have to be infinite to produce a resistance to rotation. In an actual case using this design, the cable would stretch, and the right side of the body would drop such that the line of action of force **B** would pass beneath point A. The moment arm of this force would still be small, and the reaction at B would have to be many times the weight of the object. This results in a very poor design and could lead to failure of the cable. A body supported in an improper manner is sometimes referred to as being ***unstable***. Since stability is used in many different analyses, it is better to refer to this condition as ***geometric instability***, as the instability comes from the geometry of the situation and results in movement.

A general method to determine whether a rigid body modeled two dimensionally is properly constrained is to examine the lines of action of the reaction forces. *If the reaction forces form either a concurrent or parallel force system, the body is improperly constrained.*

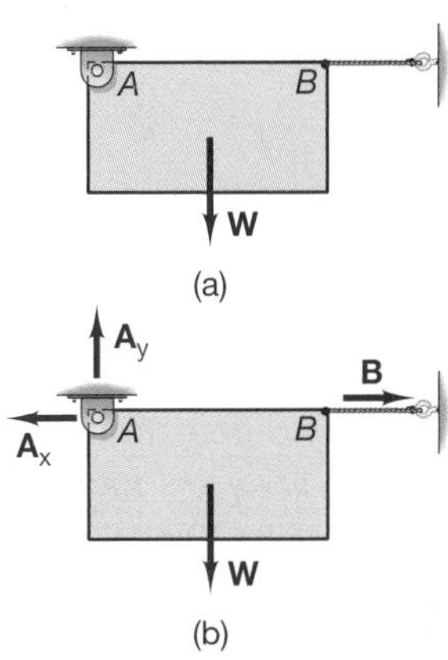

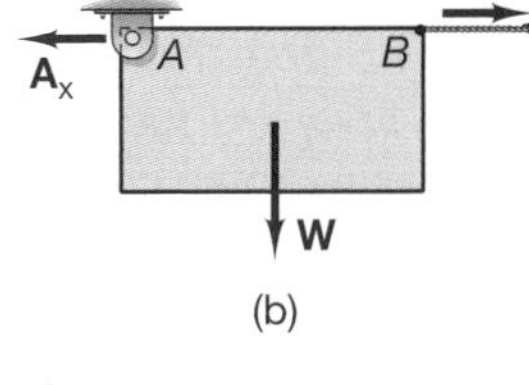

Figure 6.39

Figure 6.40

Many structures are overconstrained, and the reactions at the supports cannot be determined by the equations of equilibrium. These structures are called statically indeterminate. An example of a statically indeterminate body is shown in Figure 6.40. Again, Figure 6.40(b) is a free-body diagram of Figure 6.40(a). There are four unknown support forces and only three scalar equations of static equilibrium, so the system has one degree of indeterminacy. When a system is statically indeterminate, the reactions cannot be determined by static equilibrium alone, and the deformations of the support structures must be examined.

Statically indeterminate systems are quite common and are not necessarily the result of "bad" design. However, overconstraining a system may not always achieve the desired result of a better design. This is illustrated in the examples that follow.

The normal assumption in the case of redundant supports is that "more is always better," but this may not be so. Consider the beam supported by three equal columns in Sample Problem 6.10.

Sample Problem 6.10

Determine the reactions at the supports A, B, and C for the rigid beam loaded as shown in the top unnumbered figure.

Solution

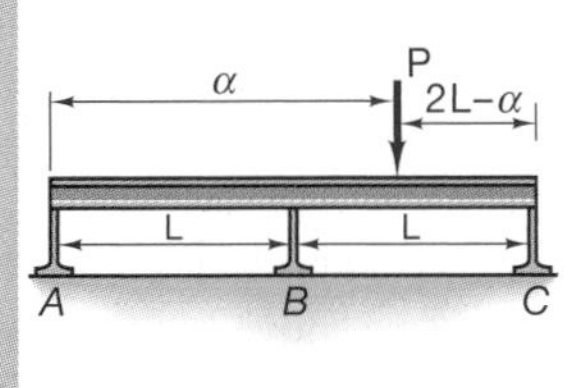

There are three vertical reactions, and only two equations of equilibrium may be written (summation of forces in the vertical direction and summation of moments). If the beam is modeled as a rigid body supported by three springs with equal stiffness—that is, with equal spring constants—the spring forces will be related to the spring deformations by

$$F = kd \quad \text{or} \quad d = F/k \tag{SP6.10.1}$$

Now consider the beam in a deformed geometry as the springs deflect (see bottom unnumbered figure). Two similar triangles are formed by the deformations at B and C, and an equation of geometric compatibility relating the deformations is

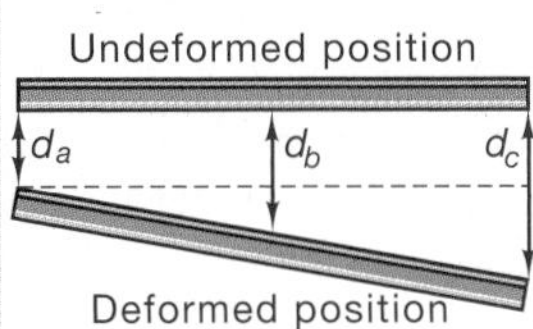

$$(d_b - d_a)/L = (d_c - d_a)/2\,\mathrm{L} \tag{SP6.10.2}$$

This may be simplified to

$$-d_a + 2\,d_b - d_c = 0 \tag{SP6.10.3}$$

If the spring constants are equal, Eq. (SP6.10.3) can be written in terms of the unknown vertical forces as

$$-F_a + 2F_b - F_c = 0 \tag{SP6.10.4}$$

If the origin is chosen at A, the two equilibrium equations are

$$\begin{aligned} F_a + F_b + F_c - P &= 0 \\ LF_b + 2LF_c - \alpha P &= 0 \end{aligned} \tag{SP6.10.5}$$

For any specified load **P** placed a distance α from the left end of the beam, Eqs. (SP6.10.4) and (SP6.10.5) form a system of three equations for the three unknown support forces. These equations are solved analytically using computational software. The result is

$$F_a = -\frac{P(-5L + 3\alpha)}{6L}$$

$$F_b = \frac{P}{3}$$

$$F_c = \frac{P(-L + 3\alpha)}{6L}$$

Note that the support at B always carries one-third of the force **P**, and if the load was placed above support B, each support carries one-third of the load. If the load was placed halfway between B and C, the support forces are

$$F_a = 0.084\text{ P} \qquad F_b = 0.333\text{ P} \qquad F_c = 0.583\text{ P}$$

Therefore, if the purpose of the structure was to support the load P applied at this point, the design would be better with only the two supports at B and C. The support forces would then each equal $0.5P$. The addition of the third support increases the maximum compressive force from $0.5P$ to $0.583P$ or a 17% increase with no apparent offsetting advantages.

The geometric compatibility equation can be obtained without using similar triangles by observing that the equation of a line is $y = ax + d$. Therefore, the deformed position of the beam can be written as

$$d = \xi x + d_a \tag{SP6.10.6}$$

where ξ is the slope of the beam and d is the deflection at any point. Now the deformation at b and c is

$$\begin{aligned} d_b &= \xi L + d_a \\ d_c &= \xi 2L + d_a \end{aligned}$$

Replacing the deflections at each point by F/k and eliminating the slope ξ from the two equations yields Eq. (SP6.10.4), and the solution of the linear system of equations can be obtained.

An equivalent example is shown in Figure 6.41, where the board is nailed at points a, b, and c, and the load is placed between b and c. If the load were to be moved to the left during use, the third nail at point a would be appropriate. These are very important considerations in the design of bolted or riveted attachments. "Don't just pound another nail in the board for no reason." Supports must be designed to serve a purpose. This is not to say that many statically indeterminate structures are not designed in this manner to resist other external forces. Statically indeterminate structures are considered in detail in the study of the mechanics of deformable bodies.

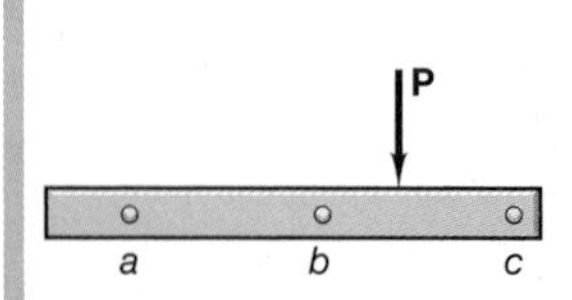

Figure 6.41

Sample Problem 6.11

A table has a 30-kg package placed on it, as shown in the top figure at the left. If the table has a mass of 50 kg, determine the force in each table leg.

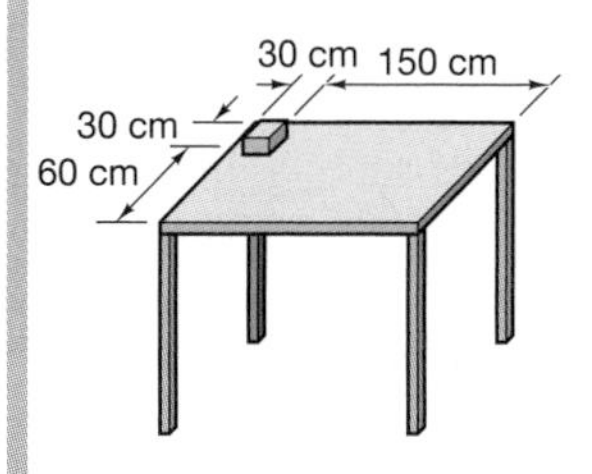

Solution The free-body diagram of the tabletop is shown in the bottom figure at the left. We have taken the origin of the coordinate system to be at point A, and the following force and position vectors can now be written for the system:

$$\begin{aligned} \mathbf{A} &= A\hat{\mathbf{k}} & \mathbf{r}_a &= 0 \\ \mathbf{B} &= B\hat{\mathbf{k}} & \mathbf{r}_b &= 180\hat{\mathbf{j}} \\ \mathbf{C} &= C\hat{\mathbf{k}} & \mathbf{r}_c &= 90\hat{\mathbf{i}} + 180\hat{\mathbf{j}} \\ \mathbf{D} &= D\hat{\mathbf{k}} & \mathbf{r}_d &= 90\hat{\mathbf{i}} \\ \mathbf{P} &= -30(9.81)\hat{\mathbf{k}} & \mathbf{r}_p &= 30\hat{\mathbf{i}} + 30\hat{\mathbf{j}} \\ \mathbf{W} &= -50(9.81)\hat{\mathbf{k}} & \mathbf{r}_w &= 45\hat{\mathbf{i}} + 90\hat{\mathbf{j}} \end{aligned}$$

Since this is a parallel force system, only three equilibrium equations can be written:

$$\sum F_x = A + B + C + D - 80(9.81) = 0$$

$$\sum M_x = 180B + 180C - 30(30)(9.81) - 90(50)(9.81) = 0$$

$$\sum M_y = -90C - 90D + 30(30)(9.81) + 45(50)(9.81) = 0$$

The three equations involve four unknowns, and the problem is statically indeterminate. If the tabletop is considered a rigid body, it forms a plane in space, and the equation for a plane is

$$z = \alpha x + \beta y + z_0$$

If the legs are treated as equal springs, the deflection of each leg is

$$\begin{aligned} \delta_b &= 0\alpha + 180\beta + \delta_a \\ \delta_c &= 90\alpha + 180\beta + \delta_a \\ \delta_d &= 90\alpha + 0\beta + \delta_a \end{aligned}$$

Substituting F/k for each deflection yields

$$\begin{aligned} B &= 180\beta + A \\ C &= 90\alpha + 180\beta + A \\ D &= 90\alpha + A \end{aligned}$$

We now have six equations one each for the four forces A, B, C, and D and the two slopes α and β. This system may be solved by matrix inversion or by hand. The result is

$$A = 270\text{ N} \quad B = 172\text{ N} \quad C = 123\text{ N} \quad D = 221\text{ N}$$
$$\alpha = -0.545 \quad \beta = -0.545$$

Note that if the table had six legs, the procedure to solve the problem would have been exactly the same.

Sample Problem 6.12

Consider a bar supported by ball-and-socket joints at A and B, as shown in the top figure to the left. The bar is loaded by a mass of 100 kg. Determine the support forces at A and B and the tension in the cable CD.

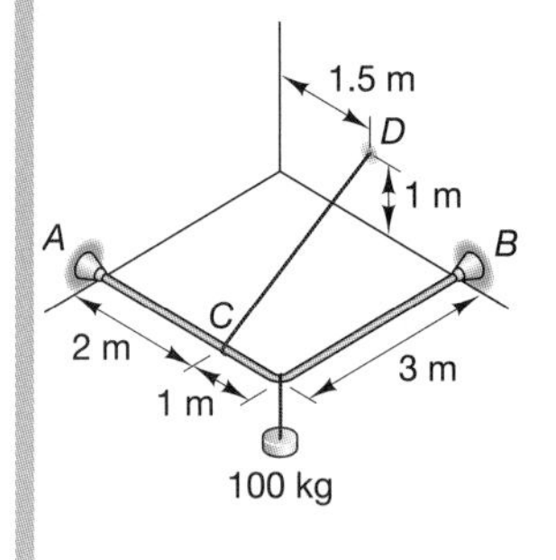

Solution

The free-body diagram for the bar is shown in the middle figure to the left. We will proceed to solve the equilibrium problem by expressing all the forces and position vectors in component notation, viz.,

$$\mathbf{A} = A_x\hat{\mathbf{i}} + A_y\hat{\mathbf{j}} + A_z\hat{\mathbf{k}} \qquad \mathbf{r}_A = 3\hat{\mathbf{k}}$$

$$\mathbf{B} = B_x\hat{\mathbf{i}} + B_y\hat{\mathbf{j}} + B_z\hat{\mathbf{k}} \qquad \mathbf{r}_B = 3\hat{\mathbf{i}}$$

$$\mathbf{T} = T\hat{\mathbf{e}}_T \qquad \mathbf{r}_T = 2\hat{\mathbf{i}} + 3\hat{\mathbf{k}}$$

$$\text{where } \mathbf{e}_T = \frac{-0.5\hat{\mathbf{i}} + \hat{\mathbf{j}} - 3\hat{\mathbf{k}}}{\sqrt{0.5^2 + 1 + 3^2}} = -0.156\hat{\mathbf{i}} + 0.312\hat{\mathbf{j}} - 0.937\hat{\mathbf{k}}$$

$$\mathbf{W} = -100(9.81)\hat{\mathbf{j}} \qquad \mathbf{r}_w = 3\hat{\mathbf{i}} + 3\hat{\mathbf{k}}$$

Summation of forces and moments yields

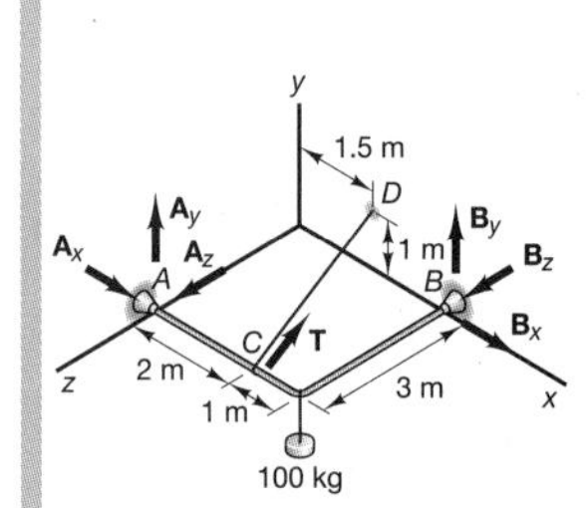

$$A_x + B_x - 0.156T = 0$$

$$A_y + B_y + 0.312T - 981 = 0$$

$$A_z + B_z = 0.937T = 0$$

$$-3A_y - 0.936T + 2943 = 0$$

$$3A_x - B_z + 1.406T = 0$$

$$3B_y + 0.624T - 2943 = 0$$

There are six equations for seven unknowns, and the system of equations cannot be solved. If we examine the free-body diagram, we can see that the indeterminacy is due to the ball-and-socket joints at A and B, which allow undetermined forces to act along a line between A and B. Either of these collinear forces could constrain movement of the system in the AB-direction. The indeterminacy can be removed by specifying that the supports at A and B are ball-and-cup joints. The revised free-body diagram is shown at the bottom to the left. A force A_n acts along the line AB at support A. This is the normal force at the ball-and-cup joint and must act away from the joint. If this force is found to have a negative value in the solution, the normal force then acts at ball-and-cup joint B. The two forces acting perpendicular to the line AB at each joint are A_y, A_p and B_y, B_p. The reaction force at A can now be written in terms of unit vectors along the line AB and unit vectors perpendicular to this line. Along the line, we have

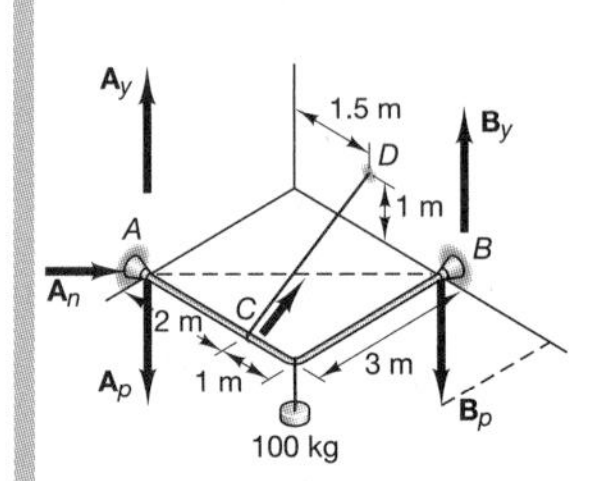

$$AB = 3\hat{\mathbf{i}} - 3\hat{\mathbf{k}}$$

$$\mathbf{AB} = \begin{pmatrix} 3 \\ 0 \\ -3 \end{pmatrix}$$

So the vector unit is

$$\hat{\mathbf{e}}_{AB} = \frac{\mathbf{AB}}{|\mathbf{AB}|}$$

$$= \begin{pmatrix} 0.707 \\ 0 \\ -0.707 \end{pmatrix}$$

The unit vector perpendicular to the line is

$$\hat{\mathbf{e}}_p = \hat{\mathbf{e}}_{AB} \times \hat{\mathbf{j}}$$

$$= \hat{\mathbf{e}}_{\mathbf{AB}} \times \begin{pmatrix} 0 \\ 1 \\ 0 \end{pmatrix}$$

$$= \begin{pmatrix} 0.707 \\ 0 \\ -0.707 \end{pmatrix}$$

The reactions at A and B are

$$\mathbf{A} = A_n\hat{\mathbf{e}}_{AB} + A_y\hat{\mathbf{j}} + A_p\hat{\mathbf{e}}_p$$

$$\mathbf{B} = B_y\hat{\mathbf{j}} + B_p\hat{\mathbf{e}}_p$$

The equilibrium equations are

$$0.707A_n + 0.707A_p + 0.707B_p - 0.156T = 0$$

$$A_y + B_y + 0.312T - 981 = 0$$

$$-0.707A_n + 0.707A_p + 0.707B_p - 0.937T = 0$$

$$-3A_y + 0.936T + 2943 = 0$$

$$2.121A_n + 2.121A_p - 2.121B_p - 1.406T = 0$$

$$3B_y - 0.624T - 2943 = 0$$

The linear system of equations is solved by hand or by writing the six equations in matrix notation and solving them using computational software. The solution is

$$A_n = -2.605 \times 10^3\text{N} \quad A_y = -490.5\text{ N} \quad A_p = 1.5621 \times 10^3\text{ N}$$

$$B_y = 0 \quad B_p = 2.084 \times 10^3\text{ N} \quad T = 4.716 \times 10^3\text{ N}$$

The normal force at A is negative, meaning that the bar is pushing against B, not A. Therefore,

$$\mathbf{A}_n = 0 \quad \mathbf{B}_n = -2605\,\hat{\mathbf{e}}_{\mathbf{AB}}\text{N}$$

The tension in the cable could have been determined in both the statically determinate and the statically indeterminate case by setting the component of the moment about the line AB equal to zero. The reactions at A and B do not produce moments about the line from A to B. This solution is used sometimes, but is based upon a special observation and is not the general solution. The moment about the line AB must be zero for equilibrium. This solution for the tension is shown in the Computational Supplement. Note that, although such an approach is used in many texts, it should not be considered as a method of solving statically indeterminate problems.

Problems

6.34 A uniform 10-kg sign is supported by a pole that is fixed at its base. (See Figure P6.34.) The sign is rigidly connected to the pole and is subjected to a wind horizontal wind load of 200 N normal to the center of the sign. Neglect the weight of the pole and determine the reactions at the base of the pole.

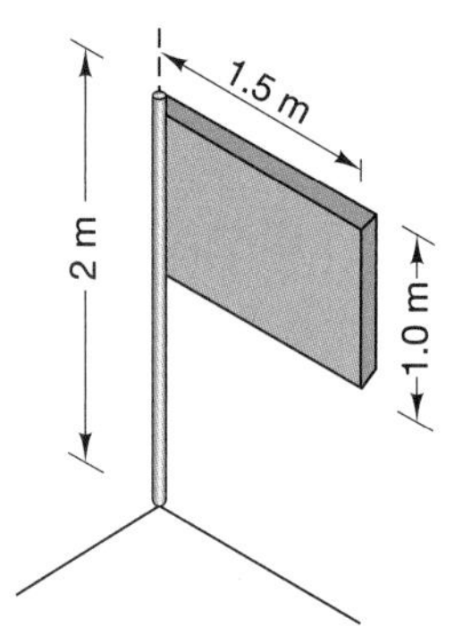

Figure P6.34

6.35 The pipe is supported by brackets at *A*, *B*, and *C* that do not support moments or forces along the axis of the pipe. (See Figure P6.35.) Determine the reactions at the brackets.

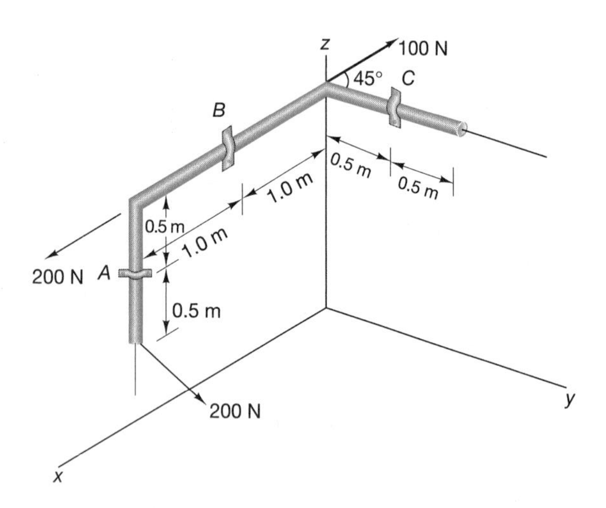

Figure P6.35

6.36 A uniform bar has a mass of 4-kg/m of length and is supported in a ball and socket joint at its base and leans against the corner of a wall as shown in Figure P6.36. Determine the reactions at the base and at the wall.

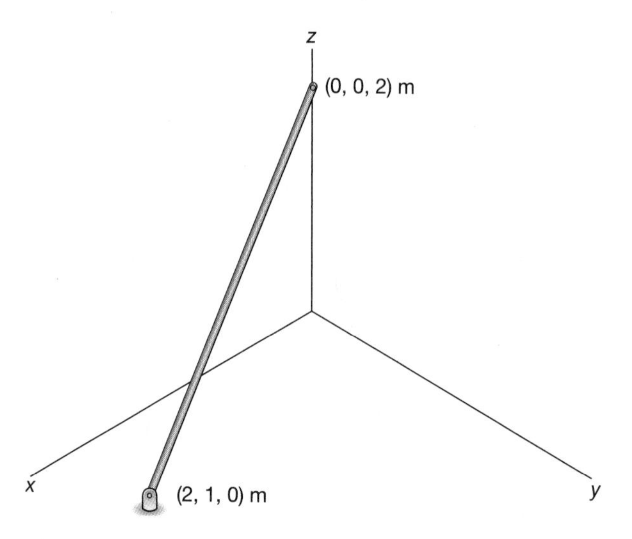

Figure P6.36

6.37 The lid on a baby-grand piano has a mass of 20 kg. Only the hinge at *A* resists force in the *x*-direction and neither hinge resists moments. Determine the force in the support bar *CD* and the reactions at the hinges. (See Figure P6.37.)

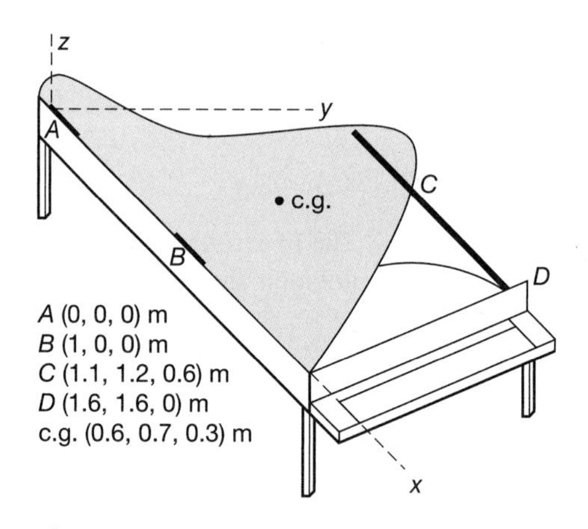

Figure P6.37

6.38 A uniform 13-m pole weights 200 N and is supported by a ball and socket joint at its base and two cables *AB* and *CD* as shown in Figure P6.38. Determine the reactions at the ball and socket joint and the tensions in the two cables.

6.39 A uniform 300-kg bar is supported by a ball and socket joint at *A* and cables *BC* and *BD* as shown in Figure P6.39. Determine the reactions at the ball and socket joint and the tensions in the two cables.

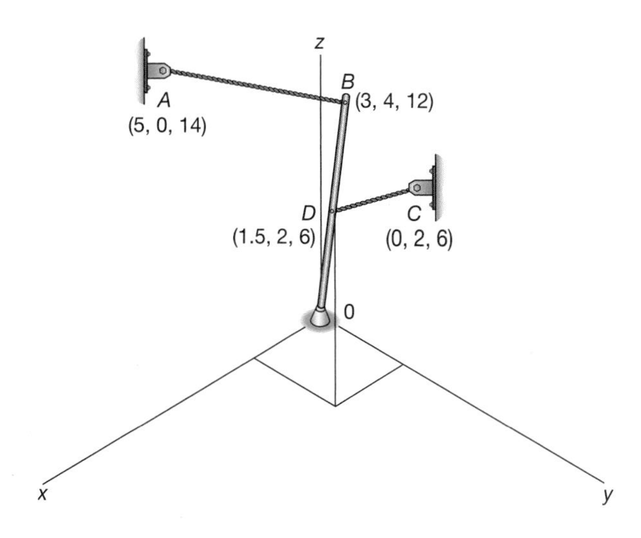

Figure P6.38

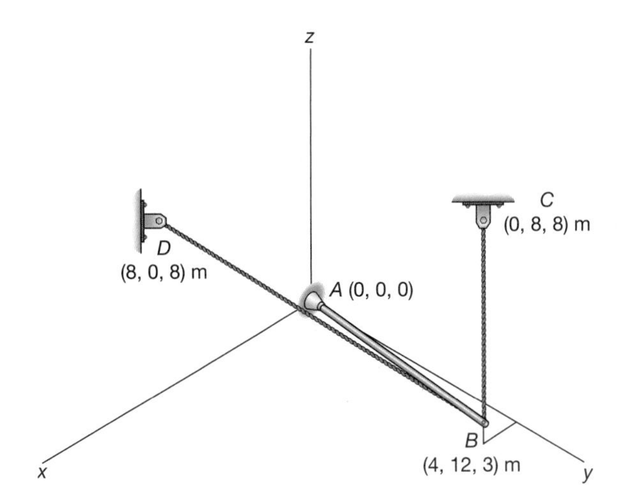

Figure P6.39

6.40 A uniform 2 × 2 m door weighs 200 N and is raised by a cable system by a motor on the wall located as shown in Figure P6.40. Only the hinge at A carries a load along the hinge axis and neither hinge will resists moments. Determine whether the tension in the lifting cable increases or decreases as the door is raised by examining the tension when the door is at 30° and 60°.

6.41 If the cable system in Problem and Figure P6.40 is replaced by a spring of initial length of 2 m and a spring constant of 200 N/m, determine the angle θ for equilibrium. (*Hint:* Consider only the moment about the y-axis.)

6.42 Determine the equilibrium angle for the system described in Problem and Figure P6.41 if the spring constant is 300 N/m.

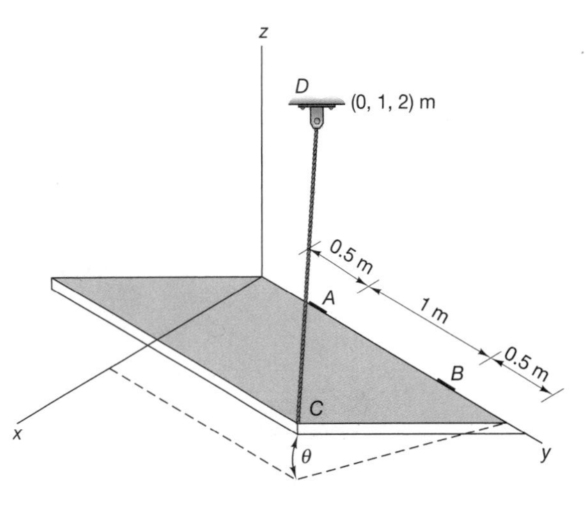

Figure P6.40

6.43 After determining the equilibrium angle in Problem 6.41, compute the spring force and the forces in the hinges if only hinge A carries the load along the hinge axis.

6.44 After determining the equilibrium angle in Problem 6.42, compute the spring force and the forces in the hinges if only hinge A carries the load the hinge axis.

6.1 Biomechanics Problems

6.45 A 660-N man is lifting a 500-N weight. The body is flexed 45° at the waist and rotated to the right. The upper body weighs 2/3 the weight of the body. For analytical purposes, the origin is taken at the L-3:L-4 junction and the coordinate system is aligned with the vertebral body with the x-axis anterior, the y-axis to the left and the z-axis superior as shown in Figure P6.45. The upper body weight vector, $\mathbf{B_W}$, the weight vector, $\mathbf{W}$, the muscle force of the deep back muscle, $\mathbf{M}$, the disc force, $\mathbf{S}$, and the disc torque, $\mathbf{T}$, are given as:

$$\begin{aligned}
\mathbf{B_W} &= 311\hat{\mathbf{i}} - 311\hat{\mathbf{k}}\ \text{N}\\
\mathbf{W} &= 353\hat{\mathbf{i}} - 353\hat{\mathbf{k}}\ \text{N}\\
\mathbf{M} &= M\hat{\mathbf{k}}\ \text{N}\\
\mathbf{S} &= S_x\hat{\mathbf{i}} + S_y\hat{\mathbf{j}} + S_z\hat{\mathbf{k}}\ \text{N}\\
\mathbf{T} &= T_x\hat{\mathbf{i}} + T_z\hat{\mathbf{k}}\ \text{Nm}
\end{aligned}$$

The position vectors to these forces are:

$$\begin{aligned}
\mathbf{r}_{\mathbf{BW}} &= 0.09\hat{\mathbf{i}} - 0.216\hat{\mathbf{j}} + 0.09\hat{\mathbf{k}}\ \text{m}\\
\mathbf{r}_{\mathbf{W}} &= 0.2\hat{\mathbf{i}} - 0.283\hat{\mathbf{j}} + 0.2\hat{\mathbf{k}}\ \text{m}\\
\mathbf{r}_{\mathbf{M}} &= -0.05\hat{\mathbf{i}}\ \text{m}\\
\mathbf{r}_{\mathbf{S}} &= 0
\end{aligned}$$

Determine the muscle and disk force and the disk torque.

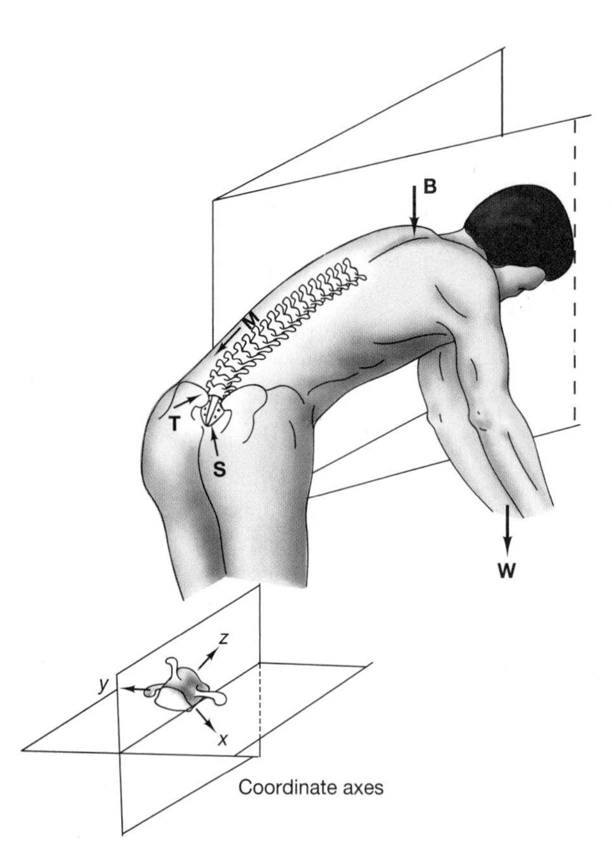

Figure P6.45

6.46 A runner comes into emergency with a torn (ruptured) Achilles tendon (heel cord H.C.), which occurred when he was going from the street up on a curb. A biomedical engineer is asked to explain to him what happened. The engineer explains that when he was moving from the street to the curb, his muscular system expected him to land on his heel with a force $\mathbf{F}_1$ as shown in Figure P6.46. The anterior tibialis muscle (T.A.) would fire at impact to counteract the plantar flexion moment around the ankle. Instead, the runner impacted with the ball of the foot with a force $\mathbf{F}_2$ as shown. The tibialis muscle would still have fired and the Achilles tendon would have to balance the dorsiflexion moment of both the impact force and the muscle force. If the impact force is three times body weight, compute first the expected muscle force and then compute the force on the tendon.

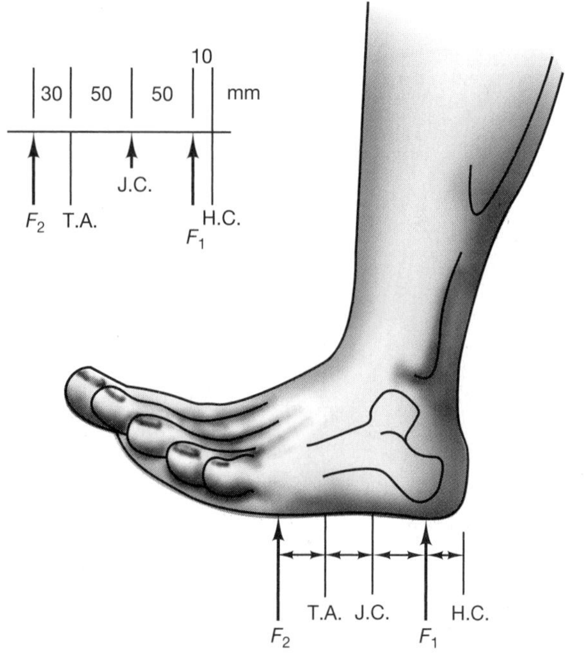

Figure P6.46

6.47 A basketball player comes down after a lay-up and lands with his right foot turned in as shown in Figure P6.47 with a force of 8 times body weight. The coordinates of the center of the ankle joint are (0.2, 0.2, 0.15) m measured from a coordinate system at the point of impact. Compute the ankle joint moments if the player has a mass of 100 kg. Which component of the moment produces an ankle sprain?

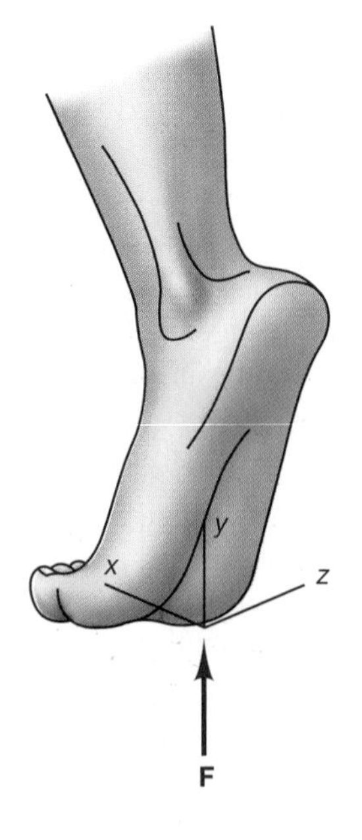

Figure P6.47

6.48 A man is shoveling sand that has a weight of 130 N. A coordinate system is placed on the third lumbar vertebrae with x anterior, y along the spine, and z toward the left of the body. (See Figure P6.48.) The coordinates of the load center are (2, 1, 0) m; compute the moment at the lumbar vertebrae. (*Note: You may treat the upper body and the shovel as a single body.*) We have not considered the weight of the torso, head or arms in this analysis. Would consideration of these weights increase or decrease the spinal loading?

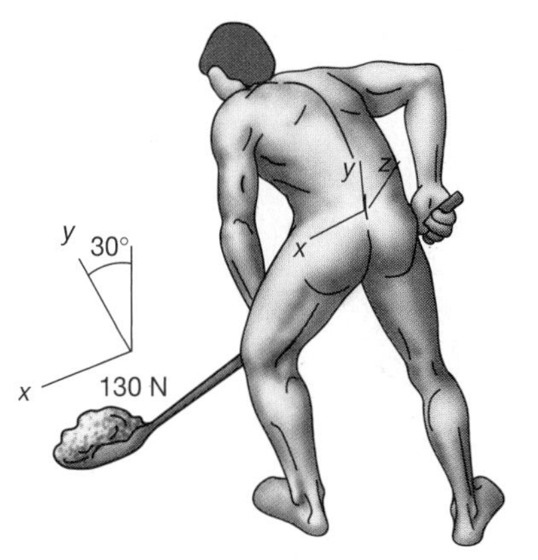

Figure P6.48

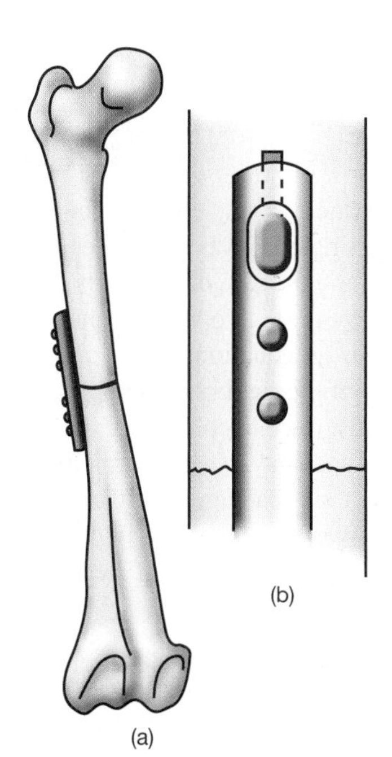

Figure P6.49

6.49 A fractured femur is set using a bone plate as shown in Figure P6.49. The load along the axis of the bone is W and is supported by two "springs" in parallel; that is, the bone and the plate. Assuming the spring constant of the plate is 3 times greater than that of the bone and that the fractured end of the bone are in contact before the load is applied, determine the portion of the load taken by the plate and that taken by the bone. (*Note:* Bone heals and remodels better under load; Wolfe's Law. This problem is known in orthopaedics as "stress shielding.")

6.50 The bone plate in Figure P6.49 (a) has a spring constant $k_p = 800$ N/mm and the bone has a spring constant of $k_b = 100$ N/mm. When the bone is set, there is a one millimeter gap at the fracture site, determine the force in the plate and in the bone when the bone is subjected to a 400 N force along the bone axis.

6.51 A bioengineer proposes a "compression plate" to address the problem of stress shielding as shown in Figure P6.49 (b). The lower screws are anchored to the bone below the fracture site and the top screw is inserted in the elongated hole at the top of the slot and a set screw is used to put compression across the fracture site. The bone is then in initial compression while the plane is in tension. When the 400 N load is placed on the bone, determine the initial compression of the bone if the bone and the plate each carry equal loads. Use the spring constants from Problem 6.49 and determine the initial compression in the bone Δ. (*Hint:* The bone is compressed an amount Δ and the plate is stretched an amount Δ.)

6.52 Two lumbar vertebrae are shown in Figure P6.52. Examine the ligament and disk forces if pure flexion occurs under the moment shown. Find the axis of rotation, the ligament and disk forces, and the angle of rotation if the force-deformation relations for the ligament and disk are:

$$F_D = K_D\delta_D \qquad k_D = 500 \text{ N/mm}$$
$$F_l = C\delta_l^2 \qquad C = 900 \text{ N/mm}^2$$

The vertebrae may be considered to be rigid bodies in the analysis.

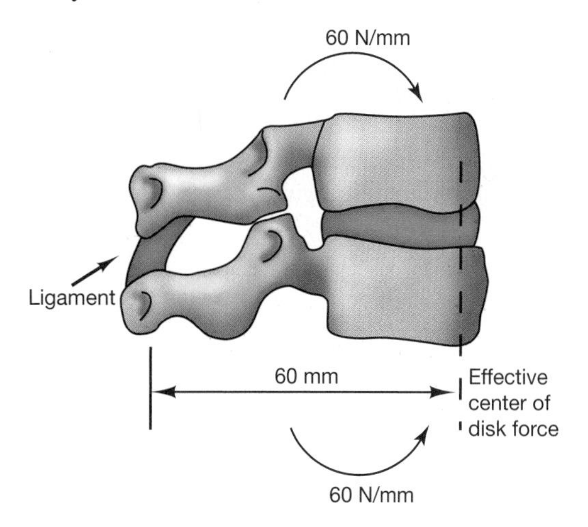

Figure P6.52

6.53 A 2 m-square rigid platform weighs 1000 N and is supported by four cables at its corners as shown in Figure P6.53. A crate weighing 400 N rests 0.5 m in from the edges of a corner. If the spring constants of the corner cables are 500 N/mm, determine the forces in the cables.

6.54 The platform in Figure P6.53 will be tilted due to the unequal tension in the cables, determine the "friction force" necessary to keep the crate from slipping.

6.55 For the platform shown in Figure P6.53, show that if the crate is removed, each cable will be in 250 N tension.

6.56 The maximum safe load for the cables in Problem and Figure P6.53 is 450 N. In an attempt to reduce the maximum cable tension, a fifth identical cable is placed on the edge halfway between cables *B* and *C*. Determine the tensions in the five cables and determine if it reduced the maximum cable tension.

6.57 Determine the tensions in the five cables in Problem P6.56 if the crate is removed.

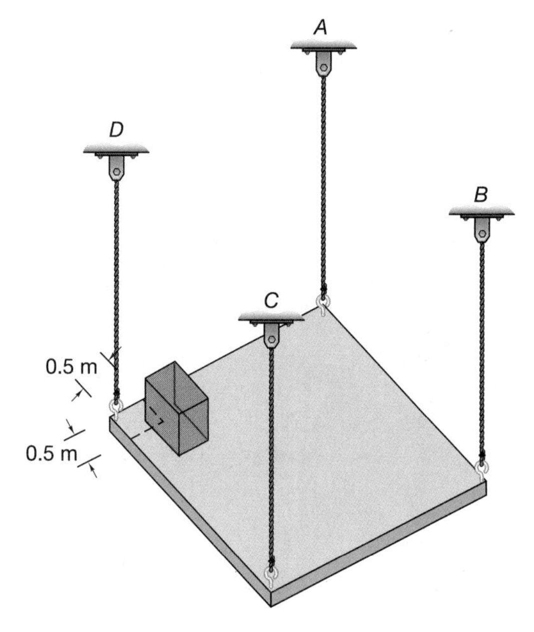

Figure P6.53

Chapter Summary

This chapter covers the concept of equilibrium of a rigid body. For the equilibrium of a particle, a single vector equation was satisfied. The sum of the forces acting concurrently on the particle were zero. For a single rigid body, two equilibrium equations must be satisfied; the sum of all the forces acting on the body must be zero and the sum of moments about any point on the body must be zero. In general, there are six scalar equations generated by these two vector equations. Therefore, there can be six unknown reactions that can be determined by solution of these six equations.

Modeling a rigid body presents the difficulty of determining the reactions at supports of the rigid body. Supports for two dimensions can be in the form of pins, roller, rocker, the normal force from a smooth surface, a collar, or fixed support. As might be expected, supports for a three-dimensional model are more complex and may range from ball and socket bearings to built in support that can have a three-dimensional force and moment or six reactions. Only six reactions are required for equilibrium of a three-dimensional rigid body. A rigid body can be over-supported and in this case will be statically indeterminate. The construction of the model or rigid body diagram is the most important step in the analysis of equilibrium of a rigid body.

Chapter 7

ANALYSIS OF STRUCTURES

The Golden Gate Bridge in San Francisco, California is an engineering structure that blends beauty and function. (Photo courtesy of Christophe Testi/Shutterstock)

7.1 INTRODUCTION

Structures are found everywhere in nature, from the spider web to the musculoskeletal system in mammals. Humans have built structures for housing, monuments, transportation, military installations, and many other applications. Some examples of early structures are the Roman aqueducts in Italy and France, the temples in Greece, and the tools and machines from archeological digs throughout the world. Aristotle (384–322 B.C.) and Archimedes (287–212 B.C.) established the early principles for the analysis of structures, forming the basis of statics. Leonardo da Vinci (1452–1519) formulated the early theories of structural engineering, and Galileo Galilei; (1564–1642) first published methods to analyze deformable materials in *Two New Sciences.* The equilibrium of single rigid bodies was examined in Chapter 6, observing that the resultant of all the external forces and couples acting on the body must be zero if the body is to remain at rest. Using the two vector equilibrium equations that were derived, the values of the external reactions that are necessary to maintain equilibrium can be determined. ***Structures*** are modeled as a number of rigid members or parts, and the forces transmitted from one part to another are determined. Although the structures examined here will be modeled as a connected group of rigid bodies, in general, structures are composed of deformable bodies. Structures may range in complexity and number of parts from the bridge to the pair of pliers shown in Figure 7.1. The bridge is made up of many parts, while the pair of pliers has only three parts. In order to properly design such structures, the forces acting on each part must be determined. These forces may be from external sources or they may be internal, arising from the action of one part upon another. The external forces are due to loads acting on the structure from other bodies and from the supports or constraints on the structure.

The focus of this chapter is on methods of determining the internal forces due to the action of one part of the structure on the another part of the structure. These internal forces occur in equal, opposite, and collinear pairs, acting at internal connections between the parts of the structure. To determine the internal forces, the structure will be taken apart and a free-body diagram of each part will be constructed. Every part will then be analyzed using the equations of equilibrium of a single rigid body. Since there are six such equations in three dimensions, and since the structure is composed of many rigid bodies, large systems of equations are encountered in the analysis of structures. The determination of the internal forces acting at connections in the structure is fundamental to the design of any structure.

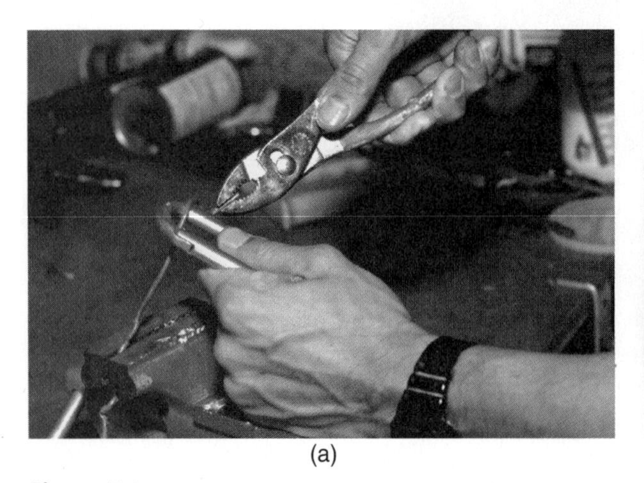
(a)

(b)

Figure 7.1 (Photo (a) courtesy of Dana White/Photo Edit. Photo (b) courtesy of Frederic Stevens/epa/Corbis Canada)

To simplify the presentation of methods of determining these forces, structures will be classified into three categories.

1. ***Trusses.*** These structures are designed to support loads and to prevent motion. The parts or members of the ***truss*** are composed of slender, straight pieces that are connected at their ends, forming joints. If these joints are idealized as connections with frictionless pins, and if the weight of the member is negligible compared to the forces transmitted at the joints, the members may be considered to be *two-force members*. This means that the only forces acting on the members are the forces at their joint connections, and these forces are transmitted along the axes of the members. The straight members are subjected to either tension or compression and are not subjected to bending or twisting. It is this fact that allows the design of very lightweight structures composed of long, slender members. If the member is pulled by the joint forces, the member is in tension, and if the joint forces push on the member, the member is in compression. A member in tension pulls away from the pin at the joint, and a member in compression pushes against the pin. Trusses may be planar, with all their members lying in a single plane, or they may be space trusses, having members not in a single plane and capable of supporting loads in any direction. A plane truss is shown in Figure 7.2(a), and a space truss is shown in Figure 7.2(b).

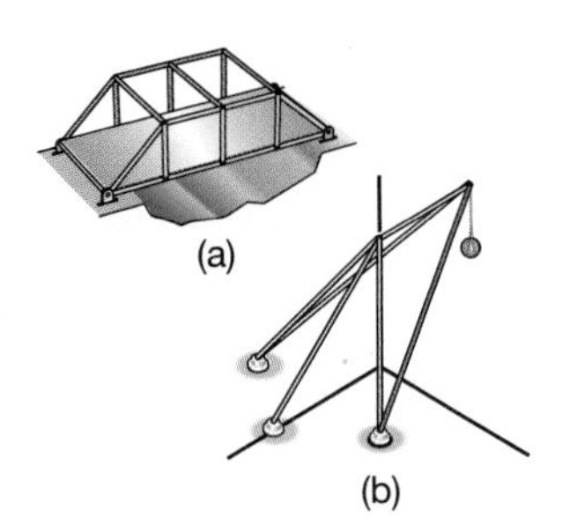

Figure 7.2

2. ***Frames.*** A ***frame*** is designed to support loads and to prevent motion, but, unlike a truss, it has at least one member with more than two forces acting upon it. This means that some parts of the frame cannot be modeled as simple tension or compression members, and the designer will have to consider the effects of bending and torsion. These multiforce members may arise because a part is connected to other parts at points other than endpoints or because the weight of a member, acting at its center of gravity, is large enough to be considered. A simple frame is shown in Figure 7.3.

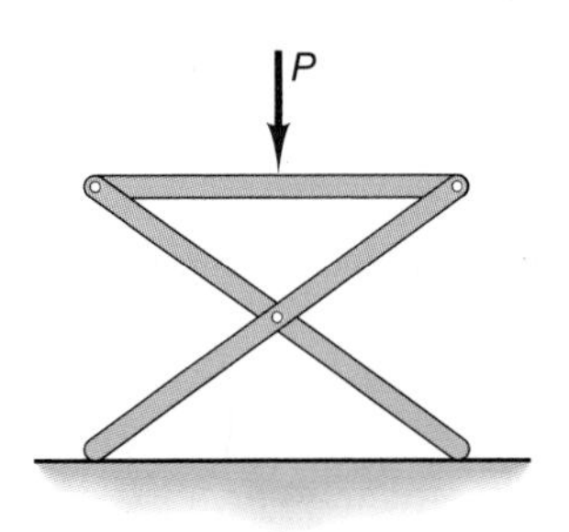

Figure 7.3

3. ***Machines or mechanisms.*** Webster's dictionary defines a ***machine*** as "an assemblage of parts that transmit forces, motion and energy one to another in a predetermined manner." Machines, by definition, contain moving parts and will always have at least one multiforce member. Thus, a pair of pliers is classified as a machine. Although the simple levers analyzed in Chapter 6 transmit forces and motion, they are not usually classified as machines, as they are single members.

7.2 PLANAR TRUSSES

The analysis of trusses is based upon a model which assumes that all members or parts of a truss are two-force members. A member is a straight, rigid element that is pinned to one or more other elements at connections called joints. A member is called a binary link in machine design and a member in structural design. Two-force members were introduced in Section 6.5, and straight two-force members are shown in Figure 7.4. Note that if the member *ab* is in equilibrium, the forces are equal in magnitude, opposite in direction, and collinear. Thus,

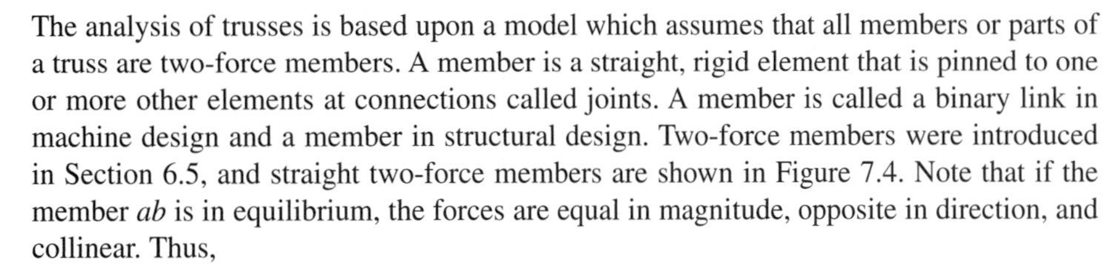

Figure 7.4

$$\mathbf{F}_a + \mathbf{F}_b = 0 \tag{7.1}$$

A straight two-force member can transmit either tension or compression, and lines of action of the forces are along the long axis of the member.

A two-force member that is not straight is shown in Figure 7.5. Again, if the member is in equilibrium, the forces at *a* and *b* are equal in magnitude, opposite in direction,

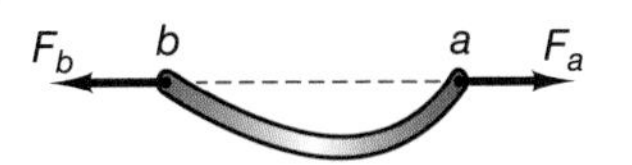

Figure 7.5

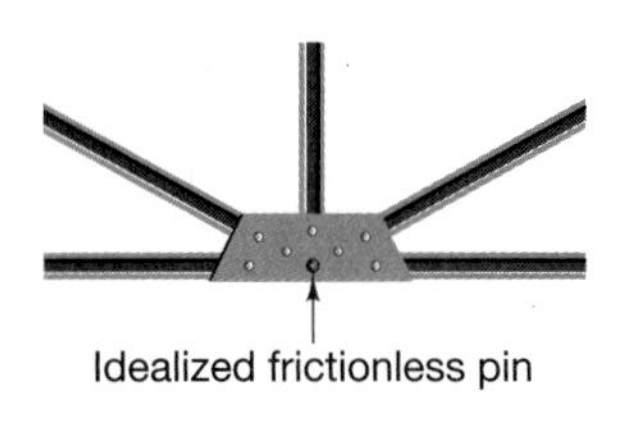

Figure 7.6

and collinear. However, in this case, it is clear that the bar will be subjected to bending. The internal forces are more complex than for a straight member, but they can be treated in the same manner as they are for a straight member in static analysis.

When the truss is composed of only two-force members, the connections between the members at the joints must be modeled as frictionless pin joints. Otherwise, moments at the joints would restrict rotation, and the truss members would not be two-force members. In real trusses, the members are bolted, welded, or connected by riveted gusset plates, as shown in Figure 7.6. Gusset plates can transmit bending moments between the members, which would invalidate the assumption of two-force members. However, for a first approximation of the forces in the members, this joint will be modeled as a frictionless pin joint.

Most engineering analyses start with the simplest model and progress to more complex models as required. Although this may seem to be the wrong approach, the simple models give *first-order effects*, and higher order effects, such as the moments at the joints, can be examined separately. Therefore, these simple models are powerful design tools. Before the use of modern computational tools, simple models provided the basis of most designs. Most design methods now employ computers to solve more complex models, so the methods presented here serve as only a starting point.

If the truss is modeled as pinned two-force members, all external loads and supports must act directly on the joints. Frequently, the weight of the individual members is neglected, but if the weight of the members is considered, the weights cannot be modeled as acting at the center of mass of the members. If the weight is considered to act at the center of mass, the member would not be a two-force member, and the structure would be analyzed in the manner presented in Section 7.9. Neglecting the weight of the members may be acceptable for small lightweight trusses, but it clearly is a poor model for a large steel bridge. The weight of the members can be included in the model as an approximation by considering that half of the weight of each individual member acts at its ends and treating this weight as additional external forces acting at the joints. Such as model is shown in Figure 7.7, where W_{AB}, for example, denotes the weight of the member AB.

Note that the weight of each member is considered to be acting at the joints; therefore, all members may be treated as two-force members and the entire structure modeled as a planar truss. Although this is an approximation, it is useful for examining the effects of the weights of the members.

Planar trusses are often used in pairs in bridges and other structures, and the loads are transmitted to the joints of the trusses by beams and stringers, as shown in Figure 7.8.

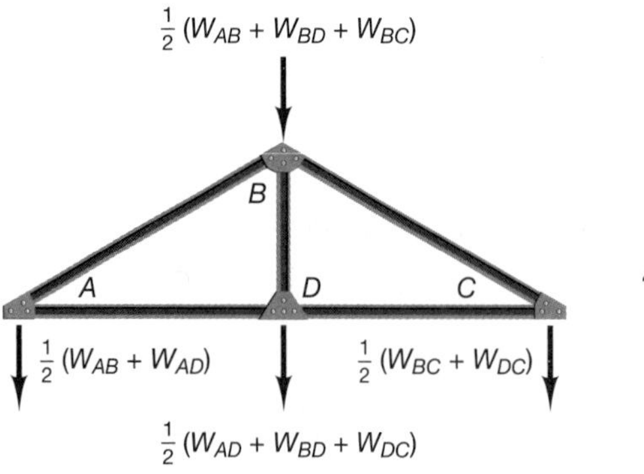

Figure 7.7

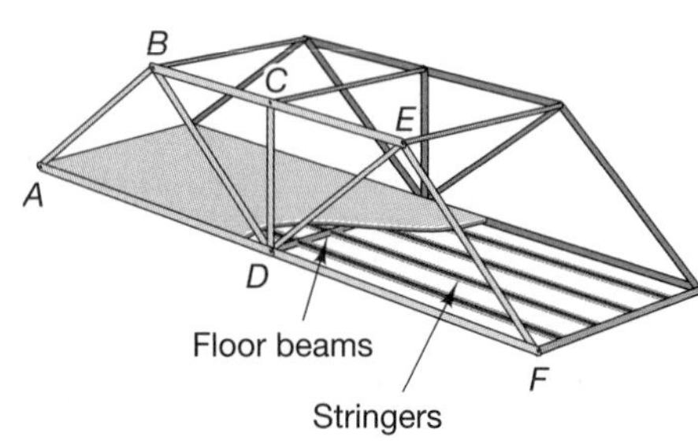

Figure 7.8

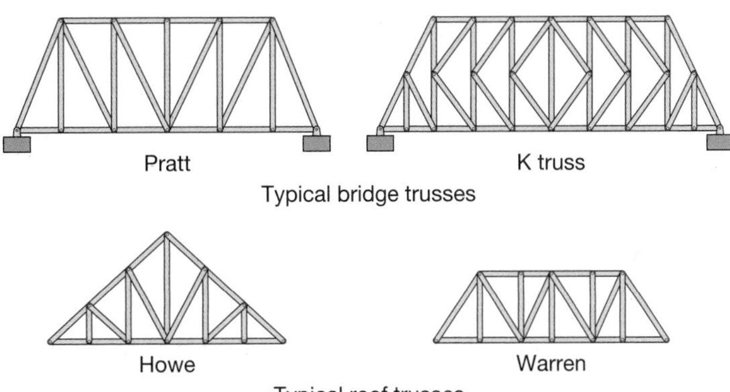

Figure 7.9

The sides of the bridge shown in the figure are composed of two identical plane trusses with joints at A, B, C, D, E, and F.

Some examples of commonly used trusses are illustrated in Figure 7.9.

7.3 SIMPLE TRUSSES

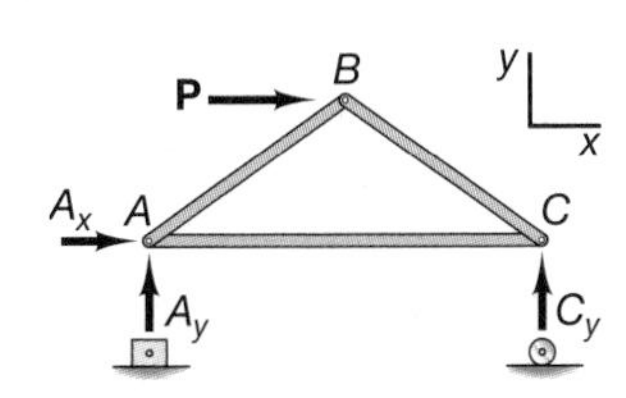

Figure 7.10

The basic configuration of members of a truss is a triangle. The triangle will retain its shape under an external load, even when removed from its supports, as shown in Figure 7.10. The lengths of the sides of the triangle are fixed, and this truss will not collapse even if the pin connections are loose.

The external reactions at A and C prevent movement of the truss in space and provide static equilibrium for the truss as a single rigid body. The truss ABC is said to be rigid and will deform only through the tension and compression in its three members. Note that, in the design of trusses, the compression members must receive special attention, as long, slender members under compressive loads may buckle. The buckling of members will be examined when the members are treated as deformable materials. The effect of buckling can easily be seen by pushing on the ends of a yardstick and watching it bow out. The bowed yardstick will be under both compression and bending.

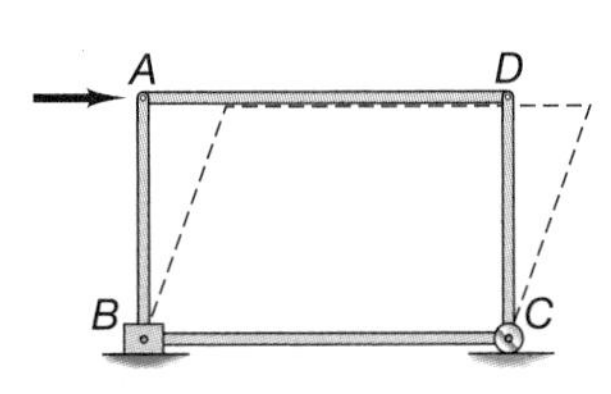

Figure 7.11

The simple truss shown in Figure 7.10 will be modeled as a two-dimensional structure, and only three support reactions restrain the structure. These three reactions may be determined by the use of the three scalar equations of static equilibrium, treating the truss as a single rigid body. If the support at C had not been a roller, but was pinned the same as the support at A, there would have been four support reactions, and the problem would have been statically indeterminate. In actual construction, one end of the truss is allowed to move or float on a rocker, so that expansion or contraction due to temperature changes can occur.

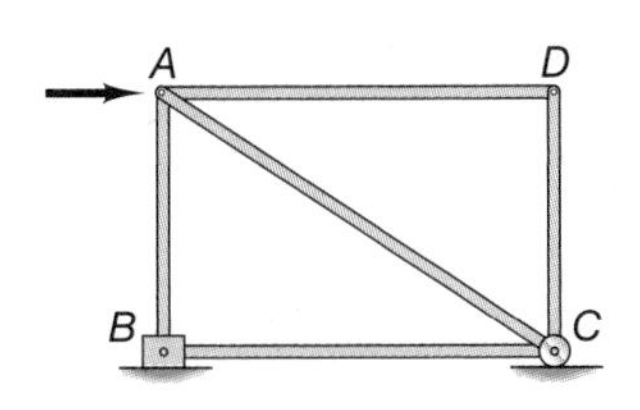

Figure 7.12

If the truss is composed of a four-member rectangle, it is not rigid; that is, it collapses under a load, as shown in Figure 7.11. When an additional member is added, forming two triangles, the truss becomes rigid, as shown in Figure 7.12. Notice that when member AC is added to the structures, the truss is now composed of two triangles and is rigid. The same result could have been accomplished by adding a member from B to D instead of the member AC. If, however, both members AC and BD are added, the truss is *overrigid*, and one of the additional members is said to be *redundant*. The internal forces in redundant members cannot be determined by static analysis, and the truss is now statically indeterminate.

The truss shown in Figure 7.12 may be developed from a triangular truss *ABC* by adding the joint *D* and two members, *AD* and *CD*. The development of more complex trusses could continue in this manner through the addition of a joint and two members, forming further triangles. A truss that can be constructed in this manner is called a ***simple truss*** and is made up of triangles developed such that, for each new joint, two members are added. The relationship between the number of members, m, and the number of joints, j, is

$$m = 2j - 3 \tag{7.2}$$

In Section 7.4, a simple method of truss analysis is presented. This method, used to determine the internal forces, is called the ***method of joints***. Two scalar equations of equilibrium of the joint pins are written for each joint of a simple plane truss. Therefore, Eq. (7.2) shows that there are sufficient equilibrium equations to solve for all m internal forces and the three external reactions. This equation provides a simple check on the rigidity and solvability of simple trusses. However, it is only a necessary and not a sufficient, condition for a truss to be stable: One or more members may be configured in such a manner as not to contribute to the stability of the truss. If $m + 3 > 2j$, there are more members than required, and the truss is statically indeterminate. If $m + 3 < 2j$, there are insufficient members for stability, and the truss or part of the truss will collapse under certain loads.

7.4 METHOD OF JOINTS

Early demands for methods to analyze trusses and the limitations on computational capabilities led to the development of special methods for analyzing simple trusses modeled as pinned members. These techniques are called *classical methods* and should be mastered before one studies the *modern matrix methods* that are used by structural engineers today. The *method of joints* is one classical method that can be adapted to use matrices, as will be shown. The modern ***matrix methods*** consider the deformation of the members and cannot be developed from the static equilibrium equations alone. However, if a truss is modeled as rigid two-force members connected with frictionless pins at joints, it can be analyzed considering only equilibrium. A free-body diagram of all the members and pins for the simple truss in Figure 7.10 is shown in Figure 7.13. Note that there are three members and three joints, so the condition for stability and solvability is satisfied.

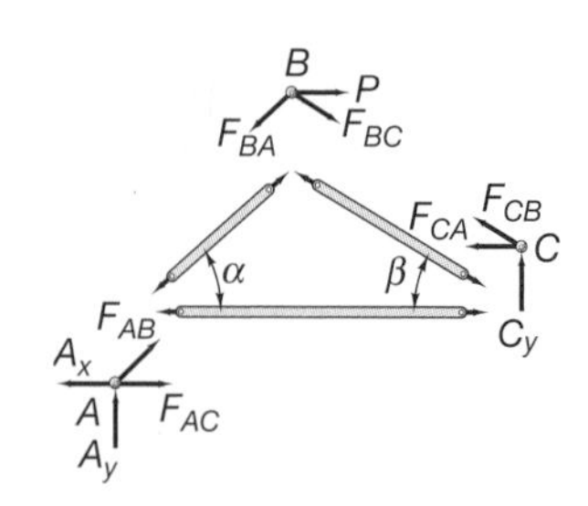

Figure 7.13

Assuming that the applied load P is known, there are six unknown forces: the internal forces F_{AB}, F_{AC}, and F_{BC}, and the three support reactions A_x, A_y, and C_y. Note that all the internal members are assumed to be in tension, and the equilibrium equations will be written consistent with that assumption. If, in the resulting solution of those equations, any of these force values are negative, it would indicate that the members are in compression rather than tension, as was assumed.

The method of joints is based upon examining the free-body diagram of each pin and noting that a concurrent force system acts on that pin. Therefore, for a planar truss, the following two scalar equations of equilibrium can be written for each pin:

$$\sum F_x = 0$$

$$\sum F_y = 0 \tag{7.3}$$

The orientations of each of the internal forces in the members are known, and only the magnitudes of these forces are unknown. Two unknown magnitudes may be determined at each joint. In Figure 7.13, the forces acting on pin A are

$$\begin{aligned} \mathbf{F}_{AB} &= F_{AB}(\cos\alpha\hat{\mathbf{i}} + \sin\alpha\hat{\mathbf{j}}) \\ \mathbf{F}_{AC} &= F_{AC} \\ \mathbf{R}_A &= A_x\hat{\mathbf{i}} + A_y\hat{\mathbf{j}} \end{aligned} \tag{7.4}$$

There are four unknowns acting at this joint, but only two equilibrium equations can be written for the joint. Thus, if we attempt a joint-by-joint solution, the analysis cannot start at pin A. If the entire truss is treated as a rigid body, the three scalar, two-dimensional equilibrium equations can be written for the entire truss, and the three reactions can be determined. The analysis could then start at pin A and proceed to pin C to complete the solution. In any approach, there will be only six independent equilibrium equations for this truss, corresponding to two scalar equations of equilibrium for each of the three pins.

As an alternative approach, the forces acting at joint B are

$$\begin{aligned} \mathbf{F}_{BA} &= -\mathbf{F}_{AB} = F_{AB}(-\cos\alpha\hat{\mathbf{i}} - \sin\alpha\hat{\mathbf{j}}) \\ \mathbf{F}_{BC} &= F_{BC}(\cos\beta\hat{\mathbf{i}} - \sin\beta\hat{\mathbf{j}}) \\ \mathbf{P} &= P\hat{\mathbf{i}} \end{aligned} \tag{7.5}$$

The applied force P is known, and equilibrium requires that

$$\mathbf{F}_{BA} + \mathbf{F}_{BC} + \mathbf{P} = 0 \tag{7.6}$$

This vector equation may be written as the two scalar equations

$$\begin{aligned} -F_{AB}\cos\alpha + F_{BC}\cos\beta + P &= 0 \\ -F_{AB}\sin\alpha - F_{BC}\sin\beta &= 0 \end{aligned} \tag{7.7}$$

These two equations may be solved for the unknown internal forces F_{AB} and F_{BC}, yielding

$$\begin{aligned} F_{BC} &= \frac{-P\sin\alpha}{\cos\beta\sin\alpha + \sin\beta\cos\alpha} \\ F_{AB} &= \frac{P\sin\beta}{\cos\beta\sin\alpha + \sin\beta\cos\alpha} \end{aligned} \tag{7.8}$$

The member BC is in compression (note the minus sign), and the member AB is in tension. Examining the free-body diagram of joint C, we can determine the force in AC and the single reaction at C. The force equilibrium yields

$$\begin{aligned} -F_{AC} - F_{BC}\cos\beta &= 0 \\ F_{BC}\sin\beta + C_y &= 0 \end{aligned}$$

Therefore,

$$\begin{aligned} F_{AC} &= \frac{P\sin\alpha\cos\beta}{\cos\beta\sin\alpha + \sin\beta\cos\alpha} \\ C_y &= \frac{P\sin\alpha\sin\beta}{\cos\beta\sin\alpha + \sin\beta\cos\alpha} \end{aligned} \tag{7.9}$$

The reactions at A are the only remaining unknowns, and they may be determined by examining equilibrium of the pin at A. We have

$$F_{AB}\cos\alpha + F_{AC} + A_x = 0$$
$$F_{AB}\sin\alpha + A_y = 0$$
$$A_x = -P \tag{7.10}$$
$$A_y = \frac{-P\sin\alpha\sin\beta}{\cos\beta\sin\alpha + \sin\beta\cos\alpha}$$

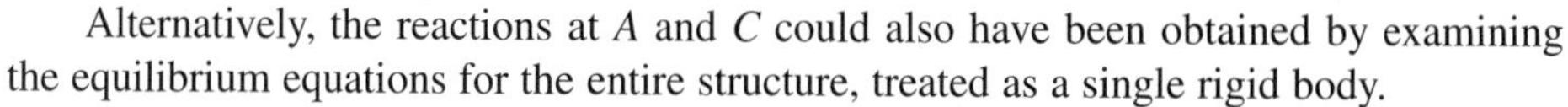

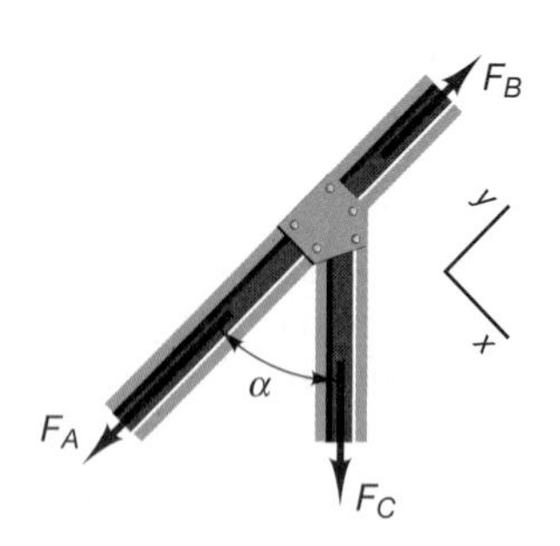

Figure 7.14

Alternatively, the reactions at A and C could also have been obtained by examining the equilibrium equations for the entire structure, treated as a single rigid body.

Some special joint connections can be analyzed and the forces in the members determined easily. If a joint involves only three members, and the axes of two of the members are collinear, the force in the third member can be determined by inspection if there are no external loads or reactions applied at the joint. Consider such a joint, shown in Figure 7.14. Equilibrium of the joint pin yields

$$-F_A\hat{\mathbf{j}} + F_B\hat{\mathbf{j}} + F_C(\sin\alpha\hat{\mathbf{i}} - \cos\alpha\hat{\mathbf{j}}) = 0$$

Since there are no external forces acting at the joint, inspection of the forces acting in the x-direction shows that

$$\mathbf{F}_c = 0 \tag{7.11}$$

and the forces $\mathbf{F}_A$ and $\mathbf{F}_B$ must be equal in magnitude and must be both tensile forces or both compressive forces. Hence,

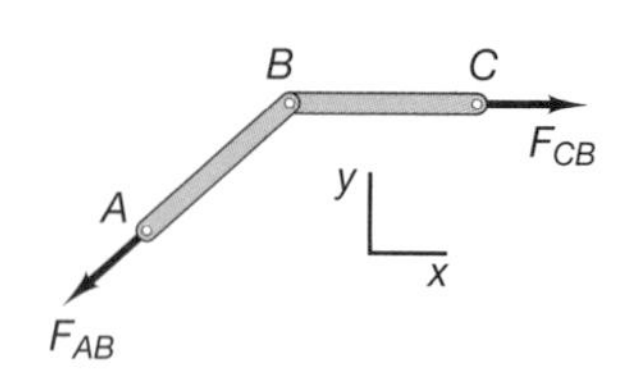

Figure 7.15

$$|\mathbf{F}_A| = |\mathbf{F}_B| \tag{7.12}$$

The member C appears to be unnecessary in this case, but it may be required to support an external load in a different loading environment.

If a joint involves only two members, and the joint is not loaded by external forces or reactions, the internal forces in the two members are zero, unless their axes are collinear. An example of this type of joint is shown in Figure 7.15. Summation of the forces in the y-direction shows that

$$\mathbf{F}_{AB} = 0 \tag{7.13}$$

and therefore, the force in BC must also be zero:

$$\mathbf{F}_{CB} = 0 \tag{7.14}$$

Again, this is true *if and only if there are no external forces or reactions acting at this joint.*

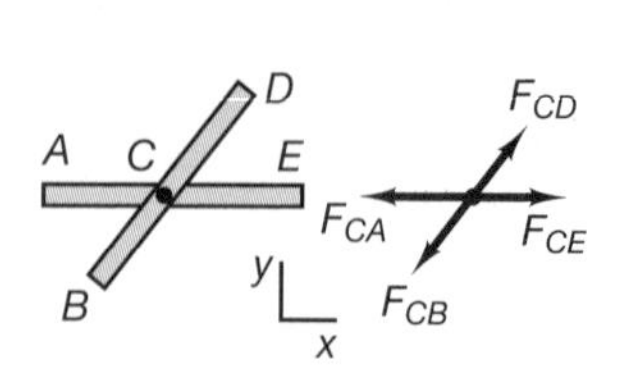

Figure 7.16

A final special case is a joint composed of four members whose axes are collinear with one of two intersecting lines, as shown in Figure 7.16. Summation of forces in the y-direction shows that the magnitudes of $\mathbf{F}_{CB}$ and $\mathbf{F}_{CD}$ are equal, and therefore, the magnitudes of $\mathbf{F}_{CA}$ and $\mathbf{F}_{CE}$ are also equal. Note that this is true if and only if there are no external loads or reactions at joint C.

Special situations such as these may be analyzed by inspection, but great care should be taken, as errors may be made by improperly applying the restrictions placed on these cases. Consider, for example, the truss shown in Figure 7.17. Inspection of joint B would indicate that the force in BC is zero and the magnitudes of the forces in AB and in BD are equal, with both forces being tensile or both being compressive. In comparison, the symmetric joint I must be analyzed, as there is an external force acting at that point.

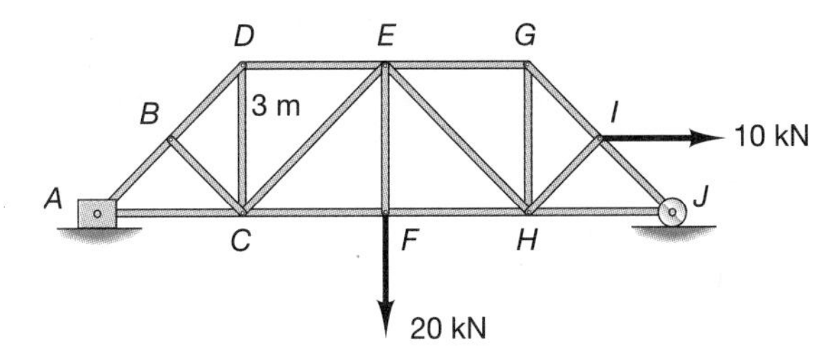

Figure 7.17

At joint F, simple inspection would indicate that member EF is in tension with an internal force of 20 kN and that the magnitudes of the forces CF and FH are equal.

Note that members which carry no load in one condition may carry loads when the external loading conditions change. They may also be essential to maintain the rigidity of the truss and to carry the weight of the truss.

Sample Problem 7.1

For the following truss, determine the loads on the individual members and the reactions.

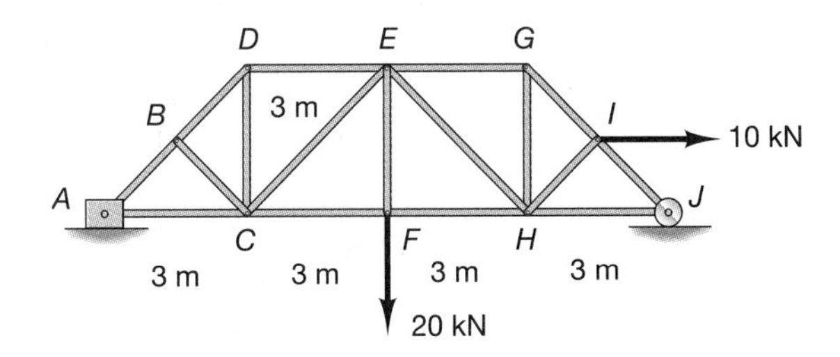

Solution Note that there is no joint with only two unknowns, and the analysis will be started by considering the structure as a rigid body. A free-body diagram for the entire structure is as follows:

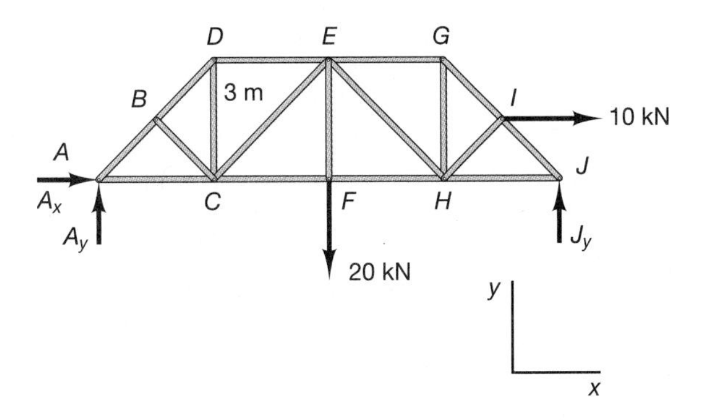

Taking moments about support A (the reactions A_x and A_y will not appear in the equation) yields

$$12J_y - 6 \times 20 - 1.5 \times 10 = 0$$

$$J_y = 11.25 \text{ kN}$$

Summing the forces in the x- and y-directions for the entire structure, we obtain

$$A_x = -10 \text{ kN (the minus sign indicates that it is in the negative } x\text{-direction)}$$

$$A_y = 20 - 11.25 = 8.75 \text{ kN}$$

Inspection of joint B shows that

$$\mathbf{F_{BC}} = 0 \quad \mathbf{F_{AB}} = \mathbf{F_{BD}}$$

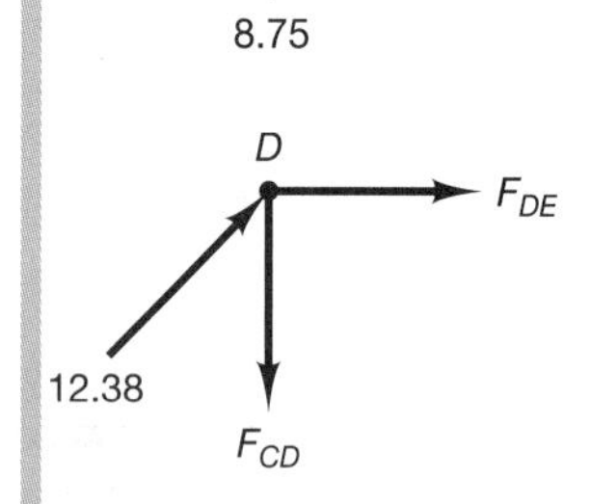

Examining joint A, allows and construction of a free-body diagram (see figure at top left) of the pin. Summing of forces in the y-direction yields

$$F_{AB} = -8.75/\sin(45) = -12.38 \text{ kN (compression)}$$

Summing of forces in the x-direction yields

$$F_{AC} = 10 - (-12.38)\cos(45) = 18.75 \text{ kN (tension)}$$

For joint D, the free-body diagram at the bottom left reveals that

$$F_{DE} = -(12.38)\sin(45) = -8.75 \text{ kN (compression)}$$

$$F_{CD} = (12.38)\cos(45) = 8.75 \text{ kN (tension)}$$

For joint C, the free-body diagram below and to the left shows that

$$F_{CE} = -8.75/\sin(45) = -12.38 \text{ kN (compression)}$$

$$F_{CF} = 18.75 - (-12.38)\cos(45) = 27.50 \text{ kN (tension)}$$

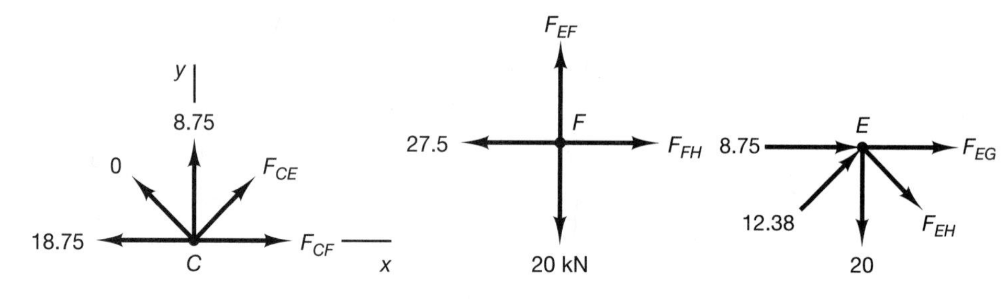

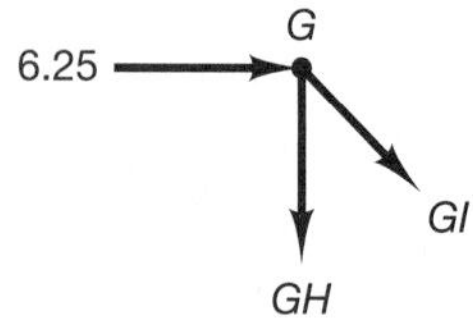

For joint F, we find, from the free-body diagram above, that

$$F_{EF} = 20 \text{ kN} \quad \text{(tension)}$$

$$F_{FH} = 27.5 \text{ kN (tension)}$$

For joint E, the free-body diagram above and to the right yields

$$F_{EH} = -[20 - 12.38\cos(45)]/\cos(45) = -15.91 \text{ kN (compression)}$$

$$F_{EG} = -[8.75 + 12.38\sin(45) - 15.91\sin(45)] = -6.25 \text{ kN (compression)}$$

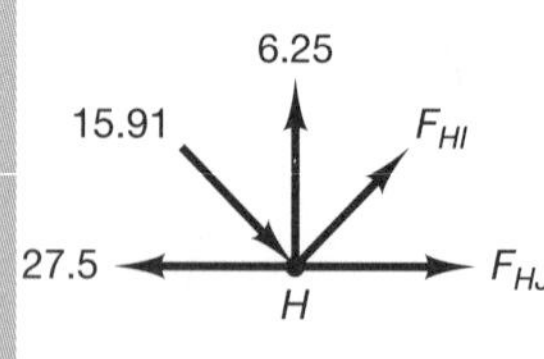

For joint G, the free-body diagram at left top gives

$$F_{GF} = -6.25/\cos(45) = -8.84 \text{ kN (compression)}$$

$$F_{GH} = -(-8.84)\sin(45) = 6.25 \text{ kN (tension)}$$

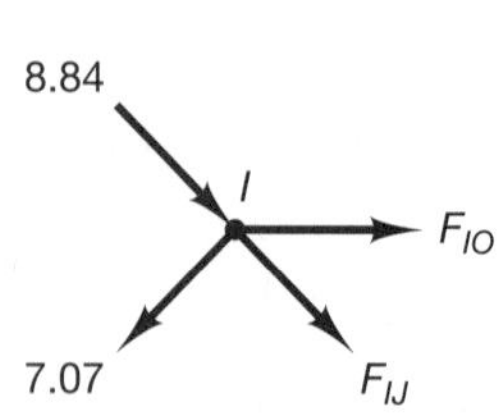

For joint H, the free-body diagram middle left results in

$$E_{HI} = [15.91\sin(45) - 6.25]/\sin(45) = 7.07 \text{ kN (tension)}$$

$$E_{HJ} = [27.5 - 15.91\cos(45) - (7.07)\cos(45)] = 11.25 \text{ kN (tension)}$$

15.91
J
11.25
11.25

And for joint I, the free-body diagram at the bottom of the previous page reveals that

$$F_{IJ} = [-8.84 - 10 \cos (45)] = -15.91 \text{ kN (compression)}$$

Since all the external and internal forces have been determined for the truss, the joint J may be used as a check: This joint is in equilibrium and indicates that the calculations are correct (see figure to the left).

For large trusses, this long and tedious procedure is not used. Practicing structural engineers and students in structural engineering courses use modern computational tools to solve these problems.

7.5 METHOD OF JOINTS USING MATRIX TECHNIQUES

In Section 7.4, we used the method of joints to determine the forces in each of the internal members within the truss. If the truss is statically determinate, the number of joints is related to the number of internal members and reactions by the equation

$$2j = m + r \tag{7.15}$$

where j is the number of joints, m is the number of internal members, and $r = 3$ is the number of external reactions.

For a plane truss, two equations of equilibrium can be written for each joint, yielding the same number of equations as the number of unknown internal member forces and external reactions. The external reactions can also be determined by treating the entire truss as a rigid body, and then the two-by-two system of equations at each joint can be solved to determine the forces in the internal members. In Sample Problem 7.1, the truss had 10 joints, and 10 systems of two linear equations each were solved. This is a tedious numerical procedure and is not used in structural engineering practice.

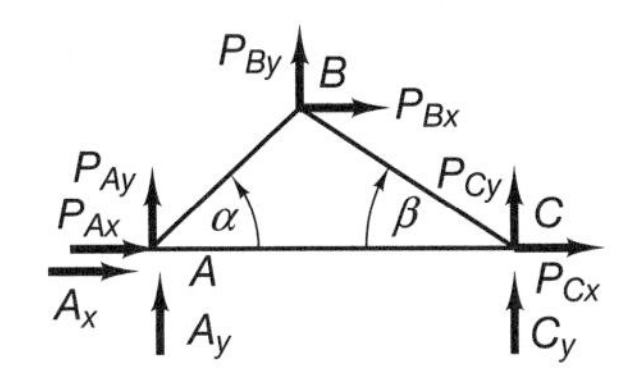

Figure 7.18

The method of joints may be written in matrix notation, and modern computational tools may be used to determine the forces in the internal members and the reactions. At each joint, the force in the internal member can be expressed as the product of the magnitude of the force and a unit vector along the midline of the member (taking into account, by the sign convention, whether the force is tensile or compressive). Consider, for example, the simple plane truss shown in Figure 7.10 and the general free-body diagram of the truss shown in Figure 7.18. External forces are shown at each joint in the free-body diagrams, and these represent applied forces. The reactions A_x, A_y, and C_y are also shown. In formulating the matrix solution, all internal forces are assumed to be tensile. The sign of the answer will indicate whether the force is in fact compressive or tensile. The internal member force in AB acting on joint A is

$$\mathbf{F}_{AB} = F_{AB}(\cos \alpha \hat{\mathbf{i}} + \sin \alpha \hat{\mathbf{j}})$$

The force in this member acting on joint B is in the opposite direction, and the unit vector is the negative of the one shown:

$$\mathbf{F}_{BA} = -\mathbf{F}_{AB} = F_{AB}(-\cos \alpha \hat{\mathbf{i}} - \sin \alpha \hat{\mathbf{j}})$$

Note that for a completely general case, external loadings F_x,F_y applied to each joint. The two equilibrium equations for each of the joints A, B, and C may be written as follows:

$$\cos\alpha(F_{AB}) + (F_{AC}) + 0(F_{BC}) + (A_x) + 0(A_y) + 0(C_y) = -P_{Ax}$$
$$\sin\alpha(F_{AB}) + 0(F_{AC}) + 0(F_{BC}) + 0(A_x) + A_y + 0(C_y) = -P_{Ay}$$
$$-\cos\alpha(F_{AB}) + 0(F_{AC}) + \cos\beta(F_{BC}) + 0(A_x) + 0(A_y) + 0(C_y) = -P_{Bx}$$
$$-\sin\alpha(F_{AB}) + 0(F_{AC}) - \sin\beta(F_{BC}) + 0(A_x) + 0(A_y) + 0(C_y) = -P_{By}$$
$$0(F_{AB}) - F_{AC} - \cos\beta(F_{BC}) + 0(A_x) + 0(A_y) + 0(C_y) = -P_{Cx}$$
$$0(F_{AB}) + 0(F_{AC}) + \sin\beta(F_{BC}) + 0(A_x) + 0(A_y) + C_y = -P_{Cy}$$

This system of equations may be written in matrix notation as: $[C][F] = [-P]$, where $[C]$ is the coefficient matrix, $[F]$ is the column matrix of the unknown internal forces and reactions, and $[-P]$ is the column matrix of the negative of the applied loads. These matrices are as follows:

$$[C] = \begin{bmatrix} \cos\alpha & 1 & 0 & 1 & 0 & 0 \\ \sin\alpha & 0 & 0 & 0 & 1 & 0 \\ -\cos\alpha & 0 & \cos\beta & 0 & 0 & 0 \\ -\sin\alpha & 0 & -\sin\beta & 0 & 0 & 0 \\ 0 & -1 & -\cos\beta & 0 & 0 & 0 \\ 0 & 0 & \sin\beta & 0 & 0 & 1 \end{bmatrix}$$

$$[F] = \begin{bmatrix} F_{AB} \\ F_{AC} \\ F_{BC} \\ A_x \\ A_y \\ C_y \end{bmatrix} \qquad [-P] = \begin{bmatrix} -P_{Ax} \\ -P_{Ay} \\ -P_{Bx} \\ -P_{By} \\ -P_{Cx} \\ -P_{Cy} \end{bmatrix}$$

All of the geometric and support characteristics of the truss are contained in the elements of the $[C]$ matrix. This matrix shows the orientation of the members and how they are connected at each joint, in addition to the location and type of reactions. The $[C]$ matrix characterizes the particular truss and would be applicable for any loading on this truss. Thus, once determined, the matrix $[C]$ can be used to calculate the internal and reaction loads contained in the matrix $[F]$ for any applied loads $[-P]$.

The matrix equation can be solved analytically to obtain the column matrix $[F]$ for any general loading, as shown in Mathematical Window 7.1. Although trusses usually are solved numerically and not analytically, the analytical solution for this simple truss can be used to see the dependence of the internal forces on the truss angles α and β.

MATHEMATICS WINDOW 7.1

$$[F] = \begin{bmatrix} \cos(\alpha) & 1 & 0 & 1 & 0 & 0 \\ \sin(\alpha) & 0 & 0 & 0 & 1 & 0 \\ -\cos(\alpha) & 0 & \cos(\beta) & 0 & 0 & 0 \\ -\sin(\alpha) & 0 & -\sin(\beta) & 0 & 0 & 0 \\ 0 & -1 & -\cos(\beta) & 0 & 0 & 0 \\ 0 & 0 & \sin(\beta) & 0 & 0 & 1 \end{bmatrix}^{-1} \begin{bmatrix} -P_{Ax} \\ -P_{Ay} \\ -P_{Bx} \\ -P_{By} \\ -P_{Cx} \\ -P_{Cy} \end{bmatrix} =$$

$$\begin{bmatrix} \dfrac{(\sin(\beta)\cdot P_{Bx} + \cos(\beta)\cdot P_{By})}{(\cos(\alpha)\cdot\sin(\beta + \sin(\alpha)\cdot\cos(\beta))} \\ \dfrac{(\sin(\alpha)\cdot\cos(\beta)\cdot P_{Bx} - \cos(\alpha)\cdot\cos(\beta)\cdot P_{By} + P_{Cx}\cdot\cos(\alpha)\cdot\sin(\beta) + P_{Cx}\cdot\sin(\alpha)\cdot\cos(\beta))}{(\cos(\alpha)\cdot\sin(\beta) + \sin(\alpha)\cdot\cos(\beta))} \\ \dfrac{-(\sin(\alpha)\cdot P_{Bx} - \cos(\alpha)\cdot P_{By})}{(\cos(\alpha)\cdot\sin(\beta) + \sin(\alpha)\cdot\cos(\beta))} \\ -P_{Ax} - P_{Bx} - P_{Cx} \\ \dfrac{-(P_{Ay}\cdot\cos(\alpha)\cdot\sin(\beta) + P_{Ay}\cdot\sin(\alpha)\cdot\cos(\beta) + \sin(\alpha)\cdot\sin(\beta)\cdot P_{Bx} + \sin(\alpha)\cdot\cos(\beta)\cdot P_{By})}{(\cos(\alpha)\cdot\sin(\beta) + \sin(\alpha)\cdot\cos(\beta))} \\ \dfrac{(\sin(\alpha)\cdot\sin(\beta)\cdot P_{Bx} - \cos(\alpha)\cdot\sin(\beta) - P_{By} - P_{Cy}\cdot\cos(\alpha)\cdot\sin(\beta) - P_{Cy}\cdot\sin(\alpha)\cdot\cos(\beta))}{(\cos(\alpha)\cdot\sin(\beta) + \sin(\alpha)\cdot\cos(\beta))} \end{bmatrix}$$

For this truss, the value of the force in each internal member and the value of the vertical reactions are inversely proportional to

$$\sin(\alpha + \beta) = \cos(\alpha)\sin(\beta) + \sin(\alpha)\cos(\beta)$$

Therefore, for the simple triangular truss, the internal member forces and reactions are a minimum if $(\alpha + \beta) = 90°$. This may vary for particular loadings, as can be seen in Mathematical Windows 7.2 and 7.3. In the first case, a unit vertical load P_{By} is applied, and the internal forces in AB and AC and the vertical reaction A_y as a function of the angle α are determined when $\alpha = \beta$. In the second case, the load is applied at B at an angle of 30° with the vertical, and the internal forces in AB, AC, and BC are examined as a function of the angle α when $\alpha = \beta$.

Modern software has rendered obsolete many of the old methods involving techniques that were developed primarily to reduce computational difficulties.

MATHEMATICS WINDOW 7.2

Consider the case when the triangle is symmetric—that is, the two angles are equal—and vary the value of the angles from 5° to 85°. (The truss will be unstable at 0° and 90°.) A vertical unit load is applied at B. We have

$$\alpha = 5°, 10°, \ldots, 85°$$

$$AB(\alpha) = \frac{\cos(\alpha)}{2\cdot\sin(\alpha)\cdot\cos(\alpha)}$$

$$AC(\alpha) = \frac{-\cos(\alpha)^2}{2\cdot\sin(\alpha)\cdot\cos(\alpha)}$$

$$A_y(\alpha) = \frac{-\sin(\alpha)\cdot\cos(\alpha)}{2\cdot\sin(\alpha)\cdot\cos(\alpha)}$$

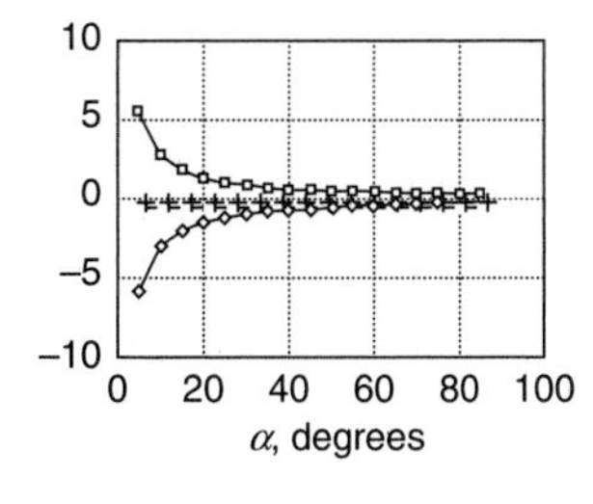

The two members AB and BC will have equal internal forces in this symmetric case, and the vertical reactions at A and C will also be equal. A unit positive vertical load is applied at B. The result is

$$AB(45°) = 0.707 \qquad AC(45°) = -0.5 \qquad A_y(45°) = -0.5$$

The graph at the left shows AB, AC, and A verses α.

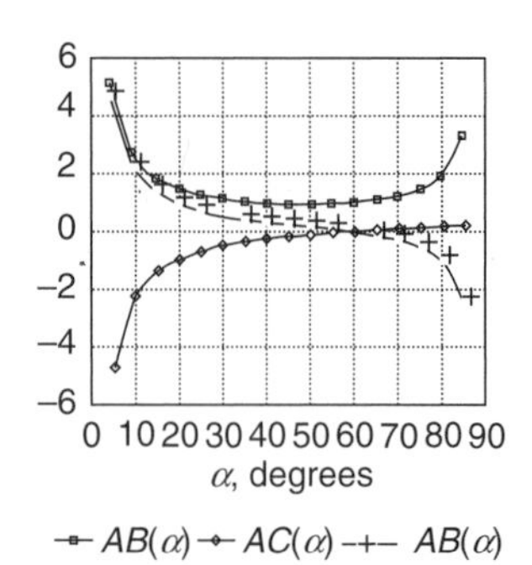

MATHEMATICS WINDOW 7.3

Now consider the case when the triangle is symmetric—that is, the two angles are equal—and vary the value of the angles from 5° to 85°. The truss will be unstable at 0° and 90°. A unit load is applied at B at an angle of 30° with the vertical. We obtain

$$\alpha = 5°, 10°, \ldots, 85° \quad P_{Bx} = 0.5 \quad P_{By} = 0.866$$

$$AB(\alpha) = \frac{\sin(\alpha)\cdot P_{Bx} + \cos(\alpha)\cdot P_{By}}{2\cdot\sin(\alpha)\cdot\cos(\alpha)}$$

$$AC(\alpha) = \frac{\sin(\alpha)\cdot\cos(\alpha)\cdot P_{Bx} - \cos(\alpha)^2\cdot P_{By}}{2\cdot\sin(\alpha)\cdot\cos(\alpha)}$$

$$BC(\alpha) = \frac{-\sin(\alpha)\cdot P_{Bx} + \cos(\alpha)\cdot P_{By}}{2\cdot\sin(\alpha)\cdot\cos(\alpha)}$$

These are internal forces of the members as a function of the truss angle. (See diagram at left.)

Problems

7.1 In Figure P7.1, use the equilibrium equations for the joint at C to prove that the forces AC and CE are equal and that the forces BC and CD are equal.

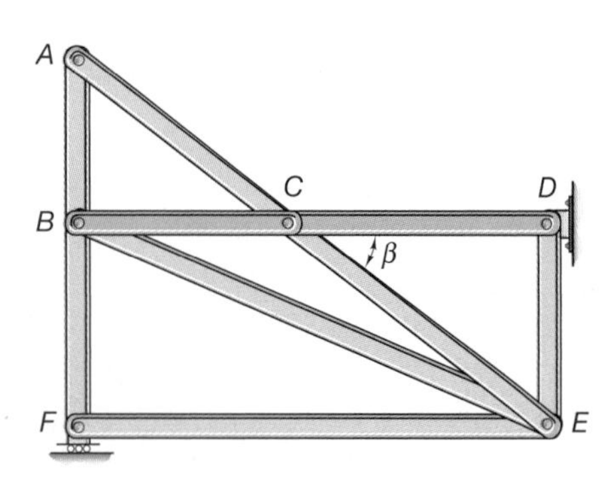

Figure P7.1

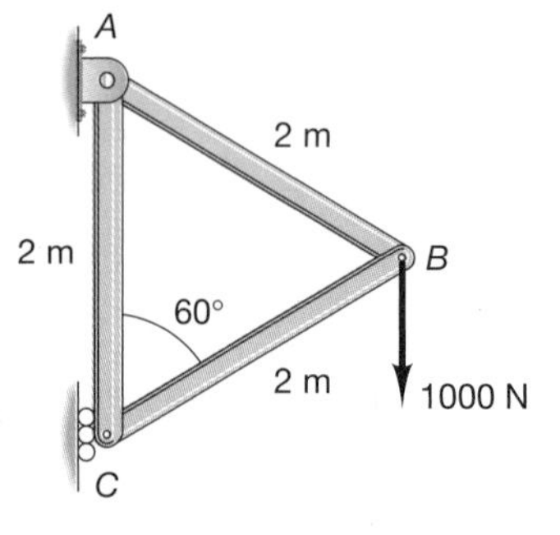

Figure P7.2

7.2 Use the method of joints to calculate the force in each member of the loaded truss illustrated in Figure P7.2, and state whether each member is in tension or in compression. Also, calculate the reaction forces at A and C.

7.3 Use the method of joints to calculate the force in each member of the truss illustrated in Figure P7.3, as well as the reaction forces at the pin B and roller C.

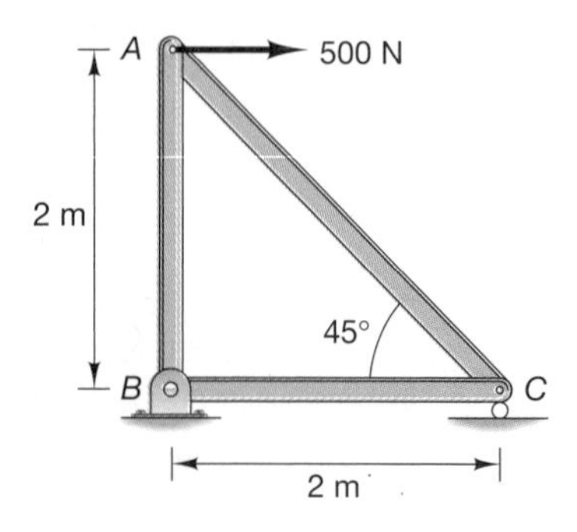

Figure P7.3

7.4 Use the method of joints to calculate the forces in each member of the truss illustrated in Figure P7.4, as well as the reaction force of the pin at B and the roller at A.

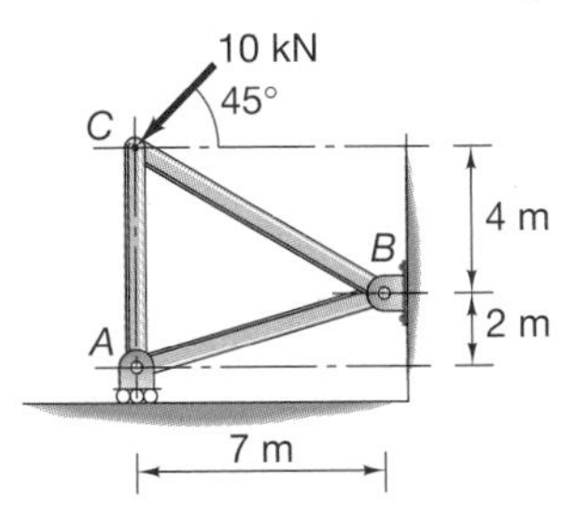

Figure P7.4

7.5 To obtain a feel for the effects of a changing load on the forces in a member, rework Problem 7.4 with a 10-kN force applied to pin A along the horizontal and pointing to the right, and with no external force at point C.

7.6 (a) Calculate the force in each member of the truss illustrated in Figure P7.6, as well as the reaction forces of the pin at C and the roller at A. (b) Add a 100-N force to the pin at D in the vertical direction and pointing downward, and recalculate the forces in the members. What changes?

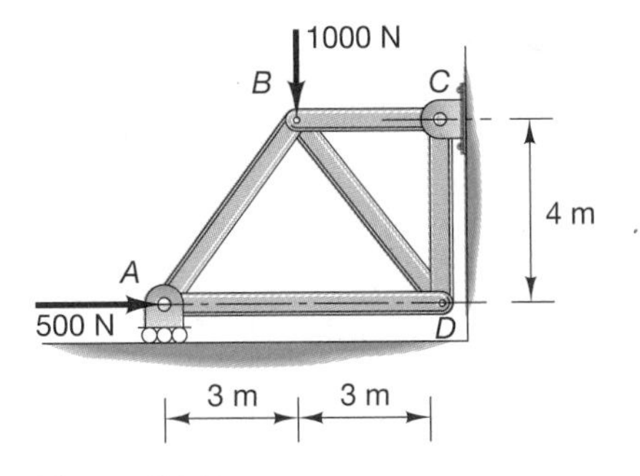

Figure P7.6

7.7 A small walkway bridge carries the load indicated in Figure P7.7. Calculate the forces in each member, and state whether they are in compression or in tension. Also, compute the reaction forces at the pin at A and the roller at point D.

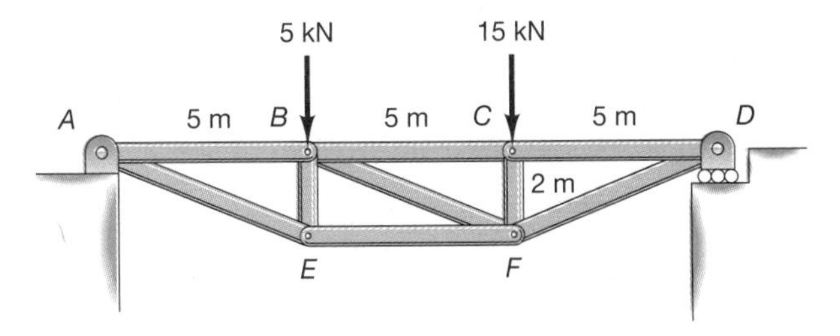

Figure P7.7

7.8 Turn the structure for the walkway of Problem 7.7 upside down, as illustrated in Figure P7.8, and compute the loads in each member. Then state whether the loads are in tension or compression, and compute the reaction forces at the pin at A and the roller at D.

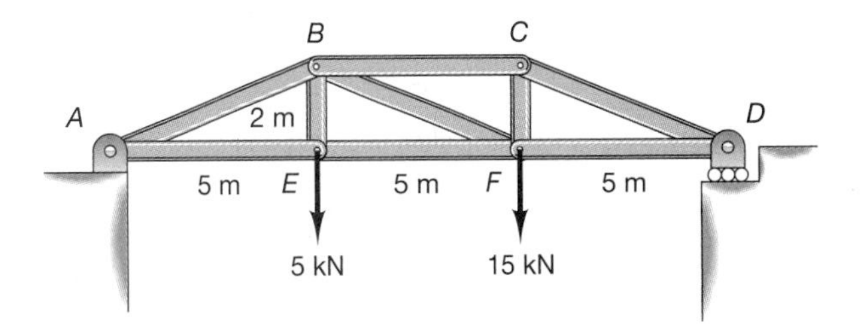

Figure P7.8

7.9 A cantilevered truss is used to support a deck represented by the loads L_1 and L_2 in Figure P7.9. The load L_3 represents a wind load on the truss. This problem, as well as the next two, examines the values of the reaction forces in the pin at A and the roller at B, and the forces in each member of the truss, as the loading changes. While such an examination does not constitute design, this is the type of question that designers must answer (i.e., "what if" the load changes?). For $L_1 = L_3 = 0$ and $L_2 = 500$ N, calculate the internal forces and the reactions.

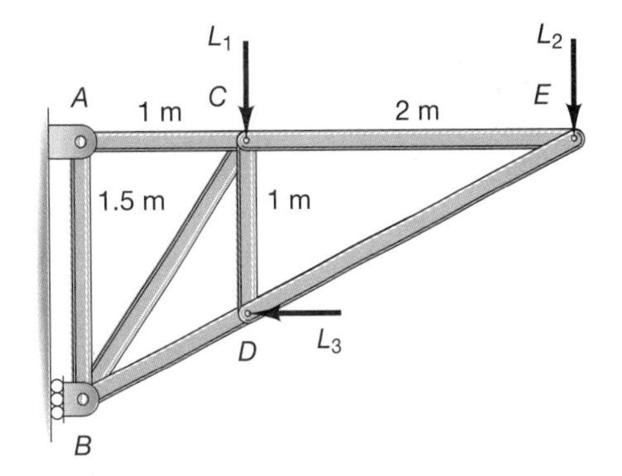

Figure P7.9

7.10 Calculate the joint forces and reactions for the system of Figure P7.9 with the load configuration $L_1 = 1000$ N, $L_2 = 500$ N, and $L_3 = 0$.

7.11 Consider the cantilevered truss of the two previous problems as sketched in Figure P7.9, and calculate the forces in each member if a wind load of approximately 1000 N is added at joint D. That is, let $L_1 = 1000$ N, $L_2 = 500$ N, and $L_3 = 1000$ N. State which members are in tension and which are in compression, and determine the reaction forces as well.

7.12 Consider a design modification of the previous three problems by making member *AB* longer, so that the point *E* remains 3 m out from the wall, but the length *AB* is changed from 1.5 m to 2 m. (See Figure P7.12). Calculate the forces in the members and the reaction forces under the load L_2 = 500 N.

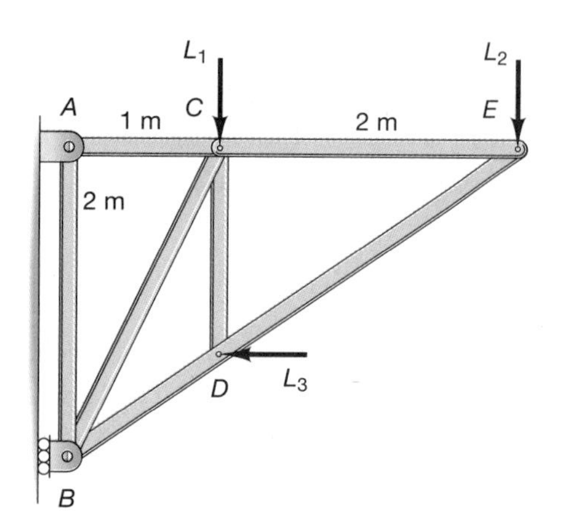

Figure P7.12

7.13 Calculate the forces in the members and the reaction forces for the system of Figure P7.12 for the load L_1 = 1000 N, L_2 = 500 N, and L_3 = 0.

7.14 Calculate the forces in the member and the reaction forces for the system of Figure P7.12 for the load given by L_1 = 1000 N, L_2 = 500 N, and L_3 = 1000 N.

7.15 A lifting mechanism is mounted on the back of a truck, as shown in Figure P7.15. Calculate the forces in each member and the reactions at *A* and *B* for (a) $\alpha = 0$, (b) $\alpha = 15°$, and (c) $\alpha = 30°$. Ignore the diameter of the pulley at point *D*.

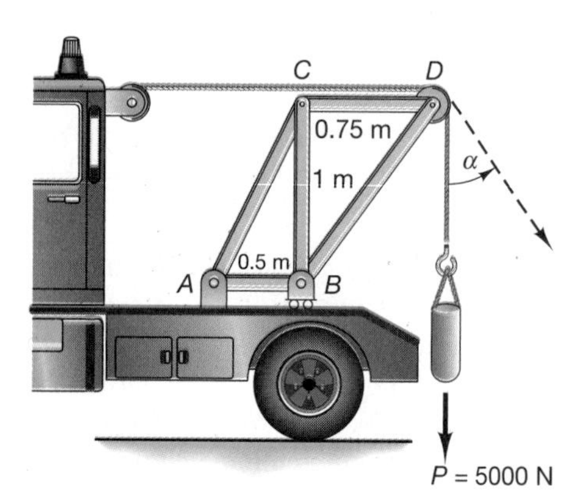

Figure P7.15

7.16 An overhanging roof-support structure is designed as a simple truss to hold a load of F = 5 kN. (See Figure P7.16.) Compute the reactions at *A* and *B* and the force in each member. Also, state whether the members are in tension or in compression.

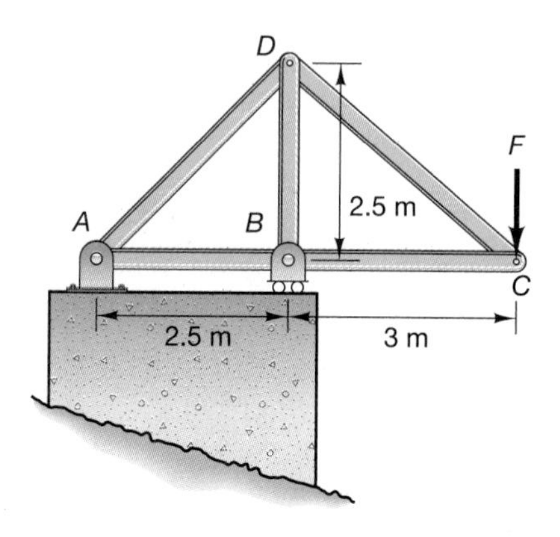

Figure P7.16

7.17 A sign of uniform density having a mass of 25 kg is supported by a five-member truss. (See Figure P7.17.) Determine the force in each member and the reactions at points *A* and *B*.

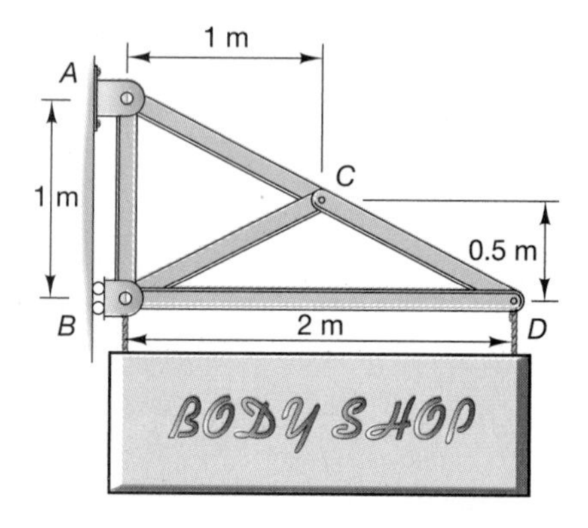

Figure P7.17

7.18 (a) Calculate the reaction forces at points *A* and *B* and the forces in the members for the frame holding the sign of Figure P7.17 if a snow load, modeled as 500 N, is acting down along the vertical at points *C* and *D*. The sign is of uniform density and has a mass of 25 kg. (b) If you also worked Problem 7.17, compare your two answers. Does any member load change drastically?

7.19 In Figure P7.19, a jack support stand for small sport vehicles is made of five pinned links and supports a weight of 1500 N. (a) Calculate the reactions at *A* and *B* and the forces in each member. (b) Is member *BD* necessary? Are there any loads for which it would be necessary? Explain your answer.

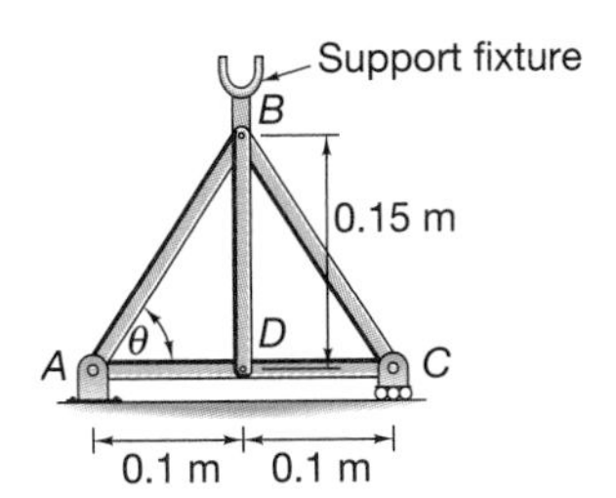

Figure P7.19

7.20 Suppose that you are asked to design the stand illustrated in Figure P7.19 by choosing the angle θ that gives the smallest values of force in each member. A constraint on your solution is that the stand must be 0.15 m high (that is, $DB = 0.15$ m) and that the width of the stand (as measured by the line AC) must be less than 0.4 m. The maximum load the stand must hold is 1500 N.

7.21 Using the approximation suggested in Figure 7.7, so that plane-truss assumptions still hold, repeat Problem 7.19 by considering the effects of the weight of the members. The mass of each member is $m_{AB} = m_{BC} = 0.5$ kg, $m_{BD} = 0.42$ kg, and $m_{AD} = m_{CD} = 0.28$ kg. Does it seem important to include the mass in this case?

7.22 Repeat Problem 7.21 for the case where the applied load is reduced to 50 N.

7.23 A crane is used for lifting a maximum load of 10^5 N. (See Figure P7.23.) Calculate the loads in each member, as well as the reaction forces at A and the tension in the cable T for $\theta = 26.6°$, $\beta = 29.7°$, and $\gamma = 68.2°$. Indicate which members are in compression and which are in tension.

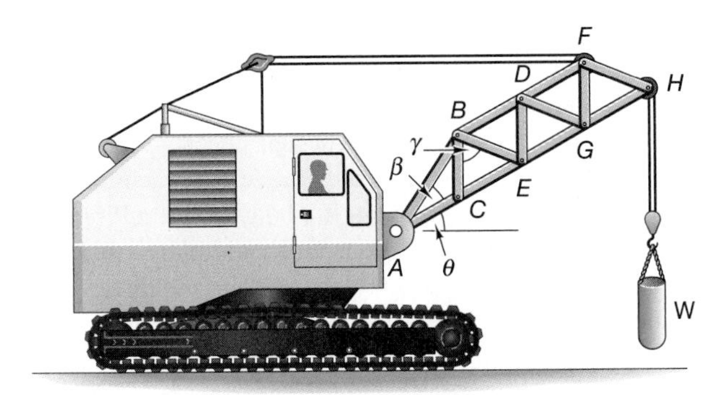

Figure P7.23

7.24 Two possible configurations of a nine-member truss are illustrated in Figure P7.24. For the given load, which configuration carries the largest force in its members?

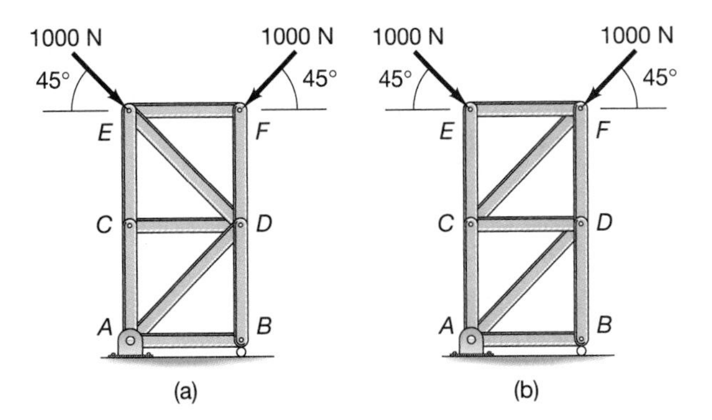

Figure P7.24

7.25 In Figure P7.25, compute the internal member forces and reactions of the three configurations of the nine-member truss for the load given. Discuss which design produces the lowest maximum member force, and state which of the three designs you would choose.

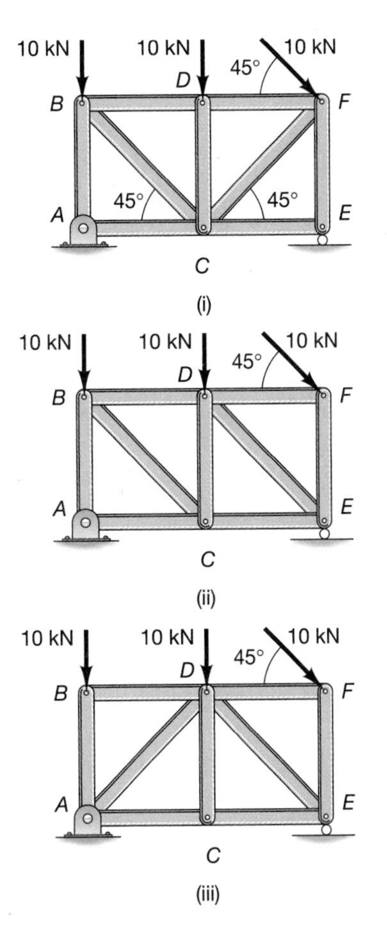

Figure P7.25

7.26 For a wind load of maximum value 100 kN, which of the configurations illustrated in Figure P7.26 produces the lowest maximum member force for values of θ between 0 and π radians?

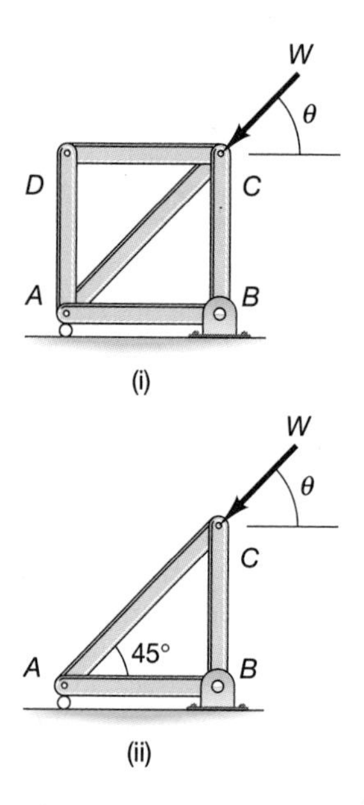

Figure P7.26

7.27 The wind load on a support structure (see Figure P7.27) is known to be a maximum value of $W = 1000$ N. However, the direction of the wind θ can vary between 0 and 180°. Neglect the weight of the member, let $\beta = 22°$, and calculate the maximum value of the magnitude of the member forces by solving for each force analytically in terms of θ, letting θ vary in increments of 5°, and plotting each member force as a function of θ. Are there any values of θ for which the structure fails to remain stationary?

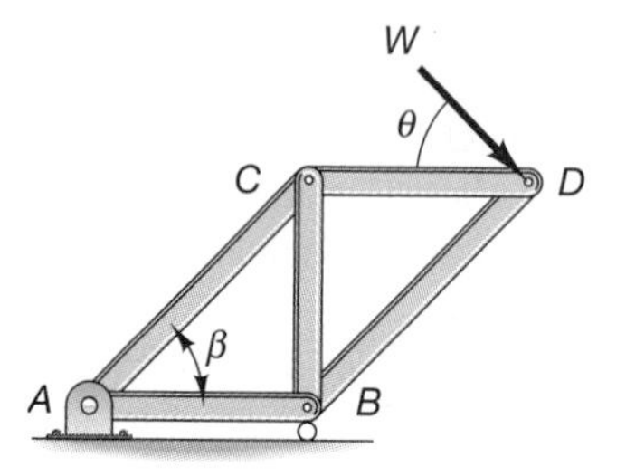

Figure P7.27

7.28 Repeat Problem 7.27 for $\beta = 45°$. If you worked Problem 7.27, compare the loads between the two cases.

7.29 Repeat Problem 7.27, including the mass of the member in the model. Approximate this by assuming that each member has a mass of 10 kg and using the method suggested in Figure 7.7.

7.30 Repeat Problem 7.29 for $\beta = 45°$. If you worked Problem 7.29, compare your answers.

7.6 METHOD OF SECTIONS

The method of joints is an efficient initial method of analysis of plane trusses when all the internal member forces are required. The geometry of the truss can be characterized by a single matrix, and the values of the reactions and internal member forces are determined using modern matrix computational software. If, however, a design check requires the determination of internal forces in only a few members, an alternative approach may be used. The ***method of sections*** is based upon the principle that, if a body is in equilibrium, all parts of that body are in equilibrium. The external reactions on a rigid plane truss can be determined by considering the entire truss as a rigid body and solving the three equilibrium equations. In a similar manner, a section of the truss can be isolated, a free-body diagram constructed, and other internal members connected to this section treated as external to the section and their effect on the isolated section represented as external forces. For a plane truss, only three equilibrium equations can be written, and the section must be chosen so that only three unknown forces act on the section. The entire truss is sectioned by passing an imaginary cut through the truss, dividing it into two separate rigid parts. Most modern structural analyses do not use a method such as this; however, the method of sections is included because it offers an easy way to determine internal forces in a few members and is of value as a prelude to some other types of analyses.

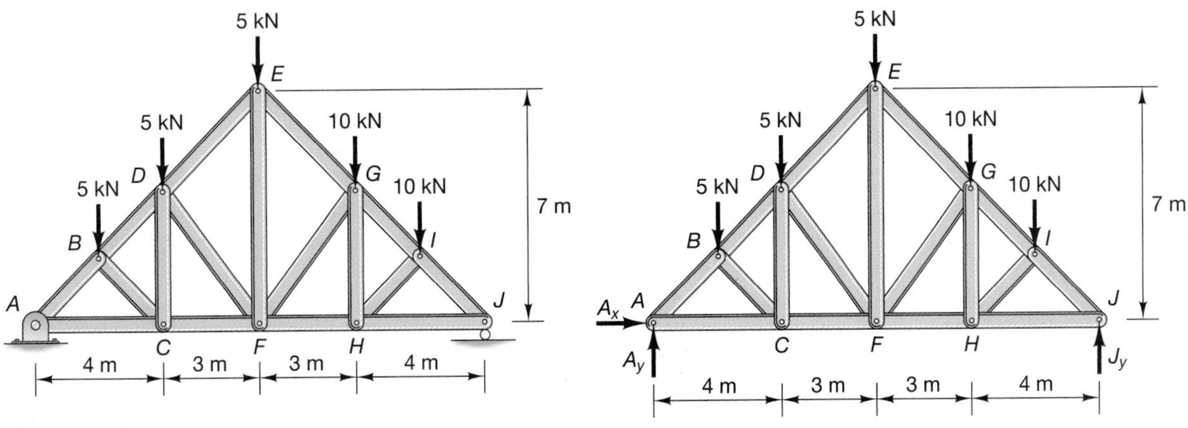

Figure 7.19

Figure 7.20

Consider the truss shown in Figure 7.19. First we will examine the free-body diagram of the entire structure as a rigid body and determine the reactions A_x and A_y at A; and the reaction J_y at J. (See Figure 7.20.)

Since there are no external forces acting in the x-direction, the reaction in that direction at A is

$$A_x = 0$$

Summing of moments about point A yields

$$14J_y - 2(5) - 4(5) - 7(5) - 10(10) - 12(10) = 0$$

$$J_y = 20.63 \text{ kN}$$

Summing of forces in the y-direction yields

$$A_y = 35 - 20.36 = 14.32 \text{ kN}$$

This truss is composed of 10 joints and 17 internal members. If the truss is analyzed by the method of joints, 20 equations are needed to determine the internal forces in the 17 members and the three external reactions. This system of 20 equations can be solved using the matrix methods introduced earlier. If, however, only the forces in members DE and DF are of interest, the method of sections yields these forces without having to solve for the forces in all 17 members. The method of sections proceeds by isolating a portion of the truss, $ABCD$, by passing a cut through members DE, DF, and CF, as shown in Figure 7.21. The following equations of equilibrium apply for the shaded rigid section of the truss shown in the figure:

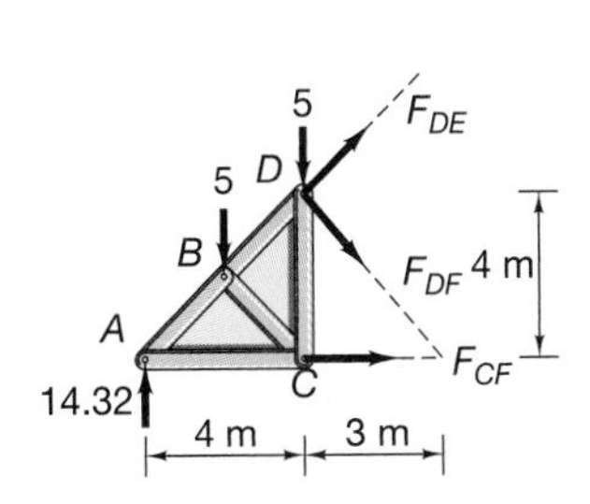

Figure 7.21

$$\sum F_x = 0 \quad F_{CF} + F_{DF}3/5 + F_{DE}\cos 45 = 0$$

$$\sum F_y = 0 \quad A_y - F_{DF}4/5 + F_{DE}\sin 45 - 10 = 0$$

$$\sum M_D = 0 \quad 4F_{CF} - 4A_y + 2(5) = 0$$

Solving these three equations for the three unknowns F_{CF}, E_{DE}, and F_{DF} yields

$$F_{CF} = 11.82 \text{ kN (tension)}$$

$$F_{DE} = -12.17 \text{ kN (compression)}$$

$$F_{DF} = -5.375 \text{ kN (compression)}$$

It should be noted that the same results would have been obtained if the section *EFGHIJ* to the right of the cut was examined. These three equations are solved using matrix methods in the Computational Supplement.

Other selected internal forces can be determined by passing cuts such that no more than three unknown forces act on the isolated portion of the truss. For example, a cut could be made through members *GI*, *GH*, and *FH* and the portion of the truss to either the right or the left of the cut isolated and treated as a rigid body.

Problems

7.31 Calculate the forces in members *BC*, *BF*, and *EF* of Problem 6.8, repeated here. (See Figure P7.31.) If you worked Problem 6.8, be sure to compare your solutions.

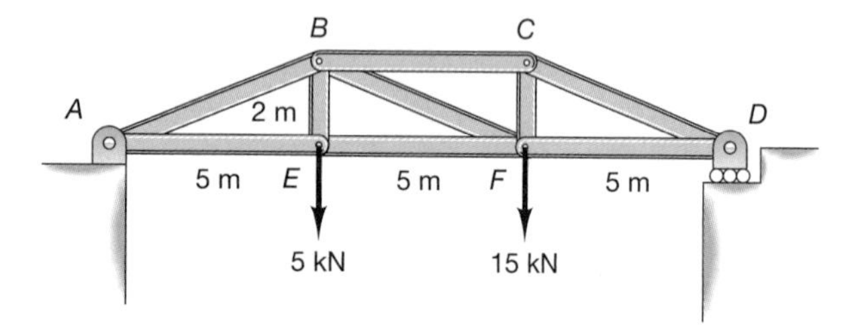

Figure P7.31

7.32 A crane supports a 1000-N load, as shown in Problem P7.32. Compute the forces in the members *CE*, *CF*, and *DF*.

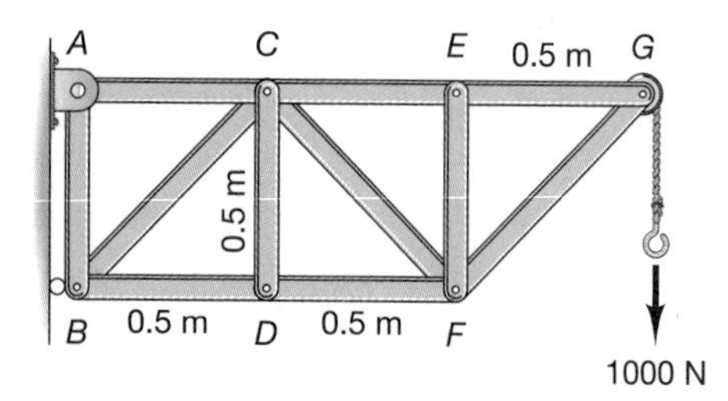

Figure P7.32

7.33 Using the method of sections, determine the forces in the members *CE*, *CF*, and *FD* of the high-voltage tower illustrated in Figure P7.33 for the case $\theta = \gamma = 0$ and $F_1 = F_2 = 1$ kN. Here, F_1 and F_2 represent the loads applied to the tower by the power lines.

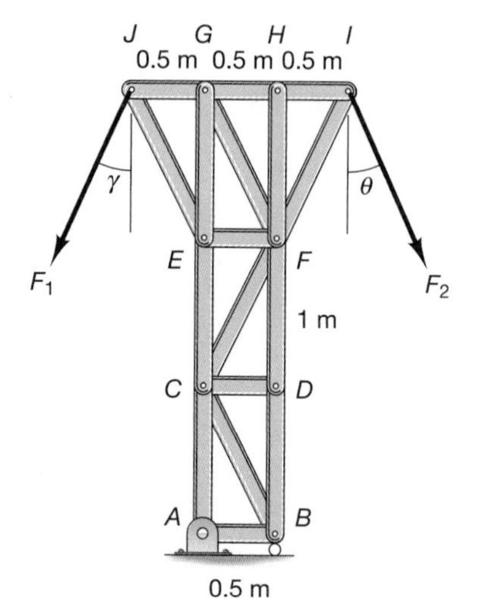

Figure P7.33

7.34 Consider the high-voltage tower of Figure P7.33. Suppose that the tower is required to hold only one line and that the wind load on the line causes the angle θ to vary between 0 and 30°. Decide whether it is better to support the load at point *J* ($F_2 = 0$, $F_1 = 1.5$ kN, $\gamma = 30°$) to the right of vertical or at point *I*, ($F_1 = 0$, $F_2 = 1.5$, $\theta = 30°$), by calculating the loads in members *CE*, *CF*, and *FD* by the method of sections.

7.35 The cross section of the roof of a storage shed is illustrated in Figure P7.35. Compute the loads in members *CE*, *CF*, and *DE*, using the method of sections for the case $P = 0$.

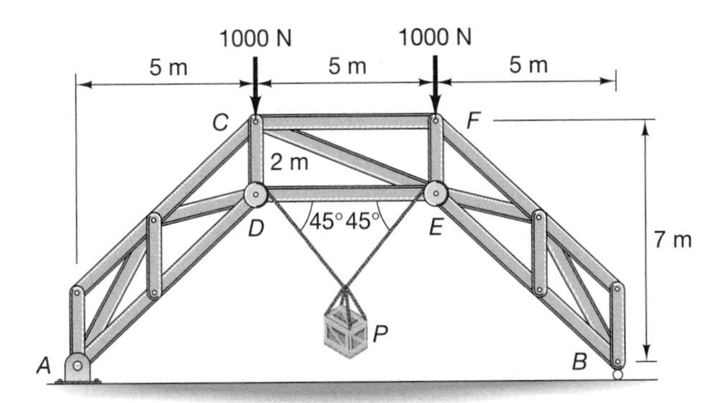

Figure P7.35

7.36 The owner of a storage shed wishes to hang a 250-kg engine from the roof, as illustrated in Figure P7.35, and wants to ensure that the roof will not collapse due to the added weight of the motor. The members are rated to hold tension and compression loads of up to 4000 N. Use the method of sections to determine the loads in *CE*, *CF*, and *DE*, and make a recommendation to the owner.

7.37 (a) Model the effects of the mass (and, hence, weight) of the members of the frame of Figure P7.32, and use the method of sections to compute the forces in the members *CE*, *CF*, and *DF*. Assume that the vertical and horizontal members have a mass of 10 kg and that the diagonal members are 41% heavier. Use the method suggested in Figure 7.7 to model the weight. (b) If you have worked Problem 7.32, compare your answer, and discuss whether it is reasonable to neglect the weight of each member.

7.38 Calculate the force in the member *BD* of the building frame shown in Figure P7.38 for a wind load of 1 kN modeled as acting at point *C*. The loads at *H* and *I* simulate roof loads. Use the concept of the method of sections to solve this problem.

7.39 Calculate the forces in the lower members (*AD*, *AB*, *BD*, and *BE*) of the building frame in Figure P7.38 in terms of the wind load *W*, of arbitrary magnitude, assumed to act on joint *C* along the horizontal direction. Plot the force in the cross member *BD* versus *W* for $0 < W < 10$ kN.

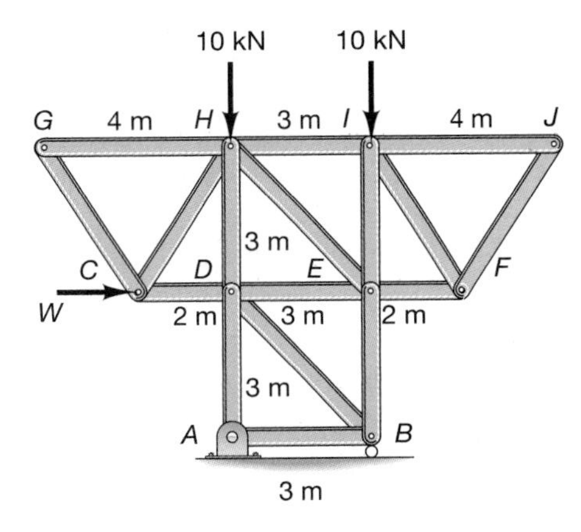

Figure P7.38

7.40 A roof truss loaded by snow and wind is modeled by the loads indicated in Figure P7.40. Here, the wind load varies between 0 and 500 N and makes an angle θ between 0 and 90°. Calculate the maximum force in the member *DJ*, and plot the force as a function of θ for $W = 500$ N. What is the maximum value of $DJ(\theta)$, and for what angle θ does this occur?

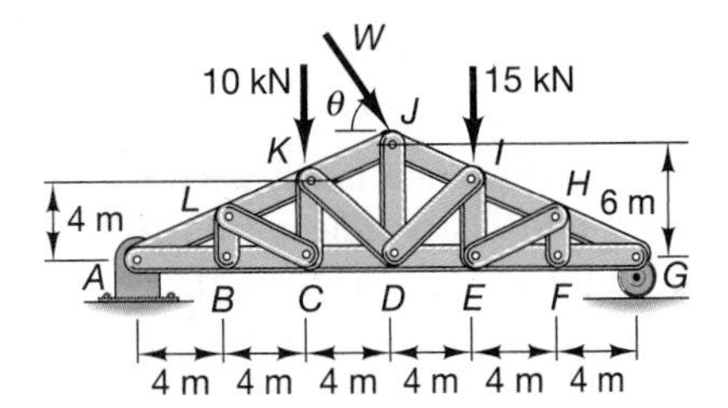

Figure P7.40

7.41 A support truss is initially designed to support a 1000-N load. (See Figure P7.41.) Compute the loads in elements *KL*, *OL*, and *ON*.

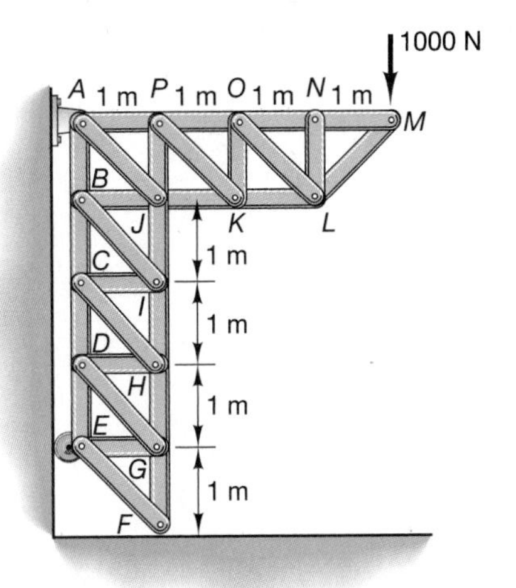

Figure P7.41

7.42 Calculate the force in member AJ of Figure P7.41 for the case that force at M is reduced to 500 N.

7.43 A roof truss is loaded by snow and wind, as modeled by the loads illustrated in Figure P7.43. Suppose that $W = 5000$ N at $\theta = 27°$, and calculate the forces in members DJ, IJ, EJ, and DE using the method of sections.

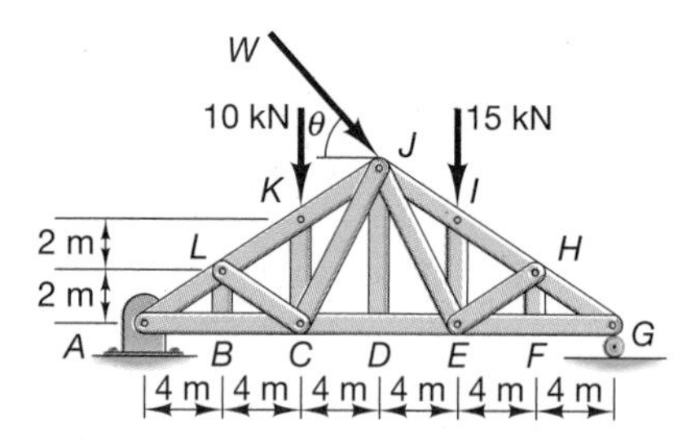

Figure P7.43

7.44 Consider the roof arrangement of Figure P7.43 and that of Figure P7.40. The two trusses are the same, except for the direction of the two diagonal elements on either side of the center member DJ. For a wind load of W = 0.5 kN at an angle of 27°, which configuraton has the largest load in the diagonal element to the right of center (i.e., EJ or DI)?

7.45 An atrium for an outdoor cafe is designed with a truss of cross section and loads illustrated in Figure P7.45. Calculate the force in the member AB.

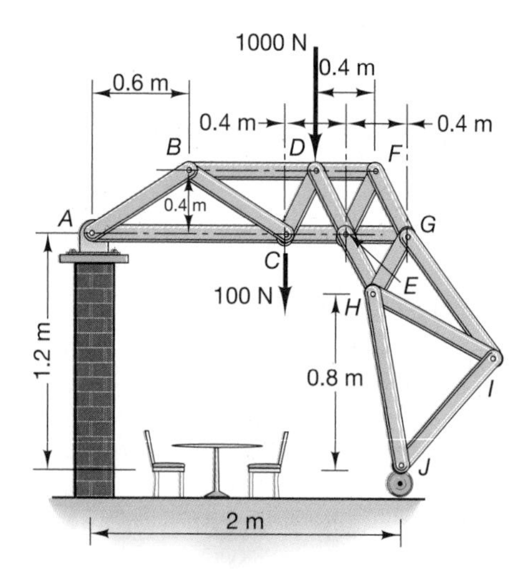

Figure P7.45

7.46 Consider the truss of Figure P7.45. Calculate the forces in the members CD, CE, and BD.

7.47 Compute the forces in members AC, BC, and BD. (See Figure P7.47.) The flowerpot has a mass of 25 kg.

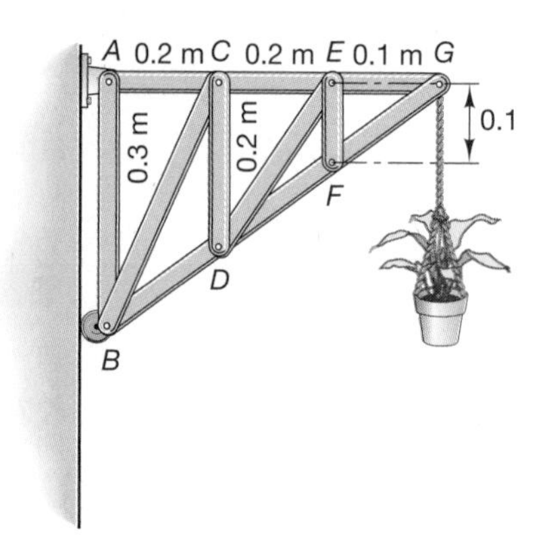

Figure P7.47

7.48 Compute the forces in members AC, AD, and BD in Figure P7.48. The flowerpot has a mass of 25 kg. Then compute the solution to Problem 7.47. Which configuration has the largest member forces in AC and BD? Which member has the larger force, BC or AD?

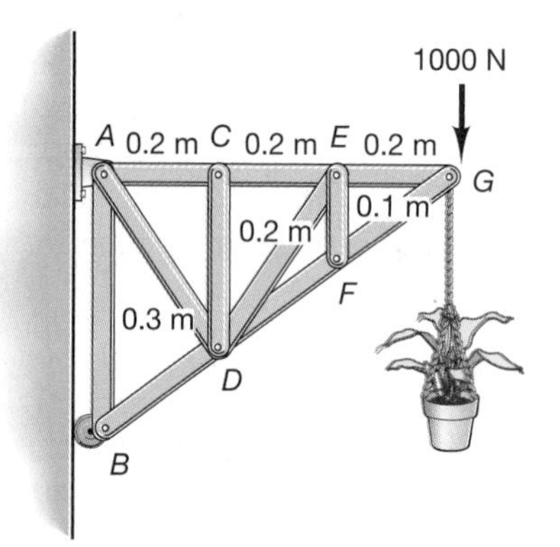

Figure P7.48

7.49 A truss supports a 1000-kg crate, as shown in Figure P7.49. Compute the forces in members JK, JL, and HL.

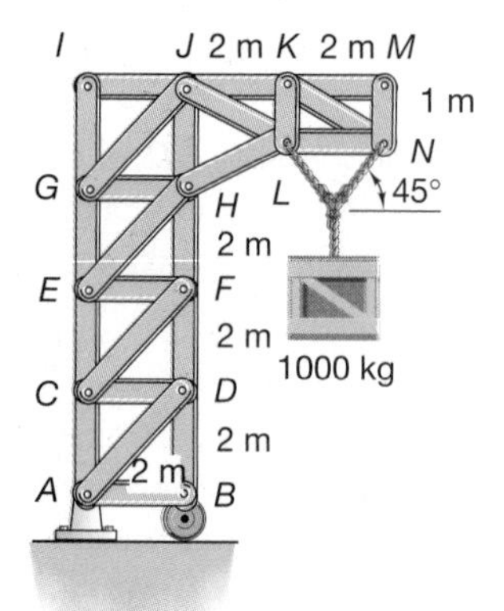

Figure P7.49

7.50 Calculate the force in member *HJ* of Figure P7.49.

7.51 Calculate the force in member *BD* under the 10-kN load illustrated in Figure P7.51.

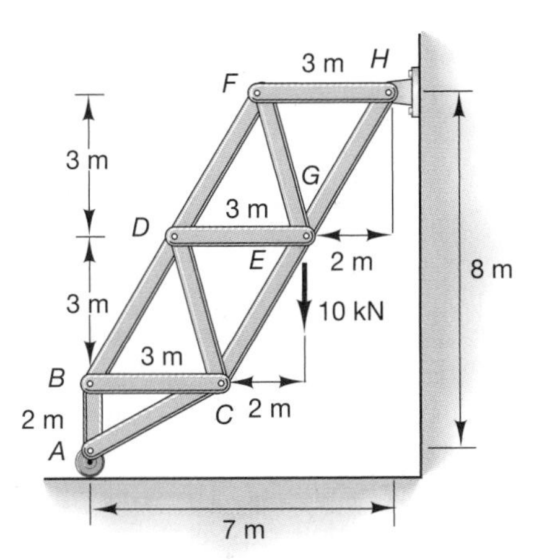

Figure P7.51

7.52 The cross section of the frame for a storage building is required to support a small aircraft engine with a maximum weight of 10 kN. (See Figure P7.52.) Calculate the load in elements *GI*, *HI*, and *HJ*. Then recompute the force on *GI* if the 10-kN load is moved from *H* to point *G*.

7.53 Calculate the force in members *LK*, *LJ*, and *MK* of Figure P7.52.

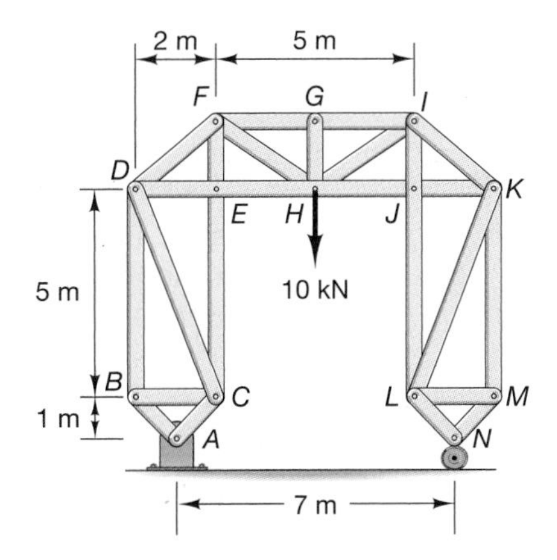

Figure P7.52

7.54 Compute the forces in members *DF* and *BE* of Figure P7.54 by using the method of sections.

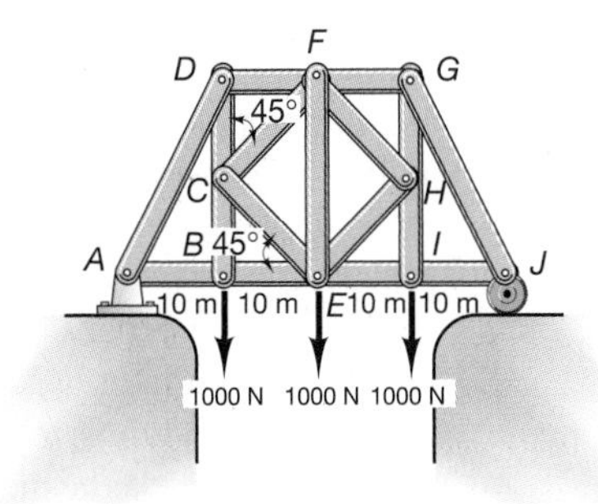

Figure P7.54

7.55 Compute the forces in members *CF* and *CE* for the truss of Figure P7.54.

7.7 SPACE TRUSSES

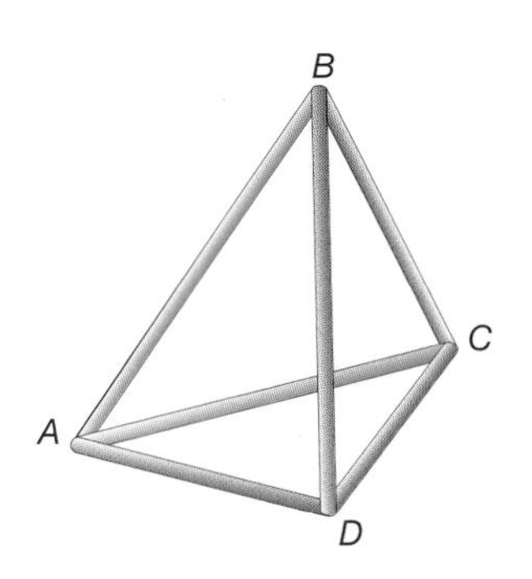

Figure 7.22

When straight members are joined in such a manner that the members do not lie in a plane, the structure is called a ***space truss***. As with plane trusses, the internal members of a space truss are modeled as two-force members, and all external loads are considered to act at the joints. The joints of the space truss are modeled as ball-and-socket connections capable of resisting forces in three directions, but not moments; that is, the members are free to rotate in the ball-and-socket joint. As with plane trusses, the weight of each member is neglected, or half the weight of each member is treated as an external load acting at the two joints at the ends of the member.

The basic element of a plane truss is a triangular arrangement of three internal members. The simple space truss, shown in Figure 7.22, is considered to be composed of tetrahedrons of six internal members. By adding three new members to this basic structure for each new joint, more complex space trusses can be developed, as shown in

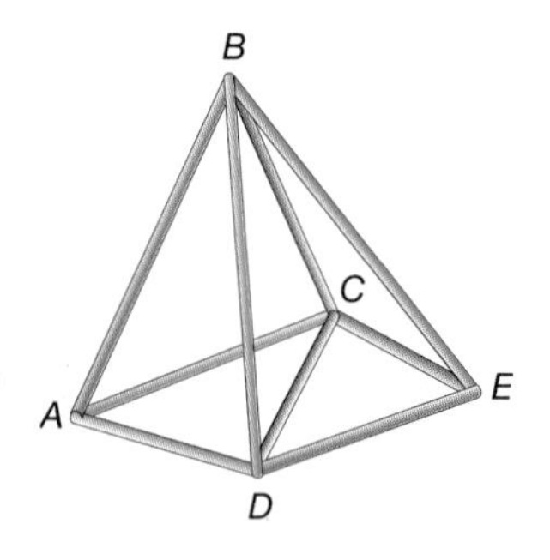

Figure 7.23

Figure 7.23. In the figure, the new joint, E, was formed by the addition of members BE, CE, and DE. If the basic tetrahedral truss shown in Figure 7.22 is considered as a single rigid body, six reactions are required to constrain the six degrees of freedom and to maintain static equilibrium. For space trusses, a relationship exists between the number of members and the number of joints. In Figure 7.22, there are four joints and six members. In the space truss shown in Figure 7.23, one new joint and three new members were added. The relationship between the number of members and the number of joints for a simple space truss is

$$3j = m + r \tag{7.16}$$

where r is the number of reactions ($r = 6$ for statically determinate structures), m is the number of members, and j is the number of joints.

If a free-body diagram of the ball at each joint is constructed, a concurrent force system exists, and equilibrium requires that the resultant of this concurrent force system be zero or that the sum of the forces acting on the ball be zero. This vector equation expressed in terms of its components yields three scalar equations at each joint. Therefore, three equations of equilibrium can be written for each joint, and a simple truss is statically determinate if three times the number of joints is equal to the number of internal forces plus the number of reactions. To constrain the truss so that it cannot move as a rigid body requires six constraints for the six degrees of freedom. If more constraints are used, the truss will be overconstrained and will be statically indeterminate. As with planar trusses, overconstraining space trusses should be avoided, as the structure can develop internal stresses due to thermal expansion or contraction.

The analysis of space trusses is accomplished using either the method of sections or the method of joints. If the forces in all the internal members must be determined, the method of joints is the more convenient. If the forces in only certain members are required, the method of sections can be used. If the method of joints is chosen, three equations of equilibrium may be written for each joint, and the solution of the system will require the solution of either groups of three equations and three unknowns or a large system of equations equal to three times the number of joints. The large system of linear equations is best solved using matrix methods and computational tools.

If the method of sections is used, the truss can be cut in a manner such that six internal forces act on the isolated section, and these internal forces are determined from the six equations of static equilibrium for a three-dimensional rigid body:

$$\begin{aligned} \sum F_x &= 0 \quad \sum F_y = 0 \quad \sum F_z = 0 \\ \sum M_x &= 0 \quad \sum M_y = 0 \quad \sum M_z = 0 \end{aligned} \tag{7.17}$$

The external reactions must be determined before the method of sections is applied.

A common three-dimensional structure that can be treated by the method of joints is called a ***uniconnected space framework***. This is a structure that results from combining plane trusses in three dimensions by replacing the pin connections with frictionless universal joints incapable of sustaining moments. Such structures are analyzed using the method of joints and the matrix methods of Section 7.5 to solve the resulting equations of equilibrium. For a statically determinate, stable system with n joints, the matrix formulation will require the inversion of a $3n \times 3n$ matrix.

Sample Problem 7.2

Determine the internal member forces and the reactions in the truss shown below, using the method of joints.

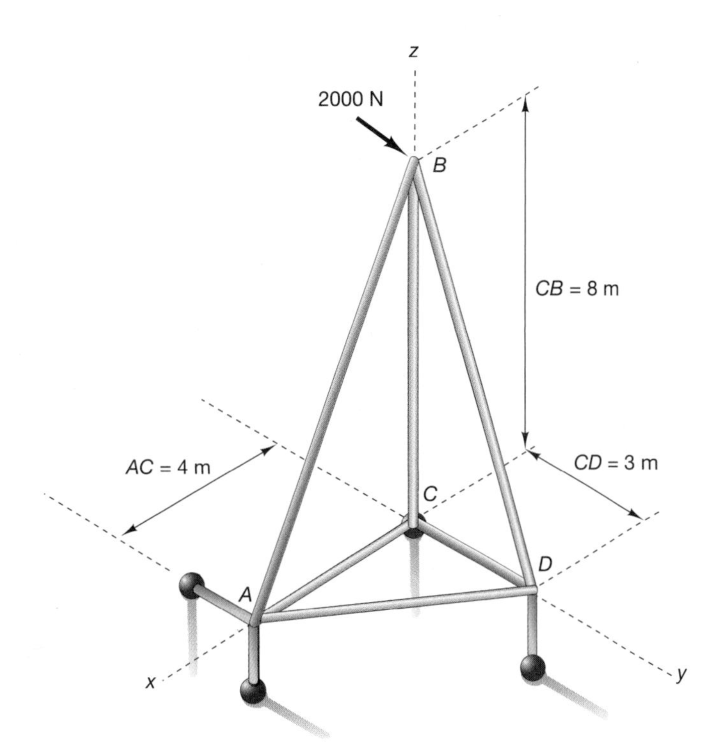

Solution A coordinate system has been chosen such that member AC lies on the x-axis, member CD lies on the y-axis, and CB lies on the z-axis. The space truss is supported by three reactions at C, two at A, and one at D, so the problem is statically determinate. Unit vectors will be determined in the AD, BD, and AB directions, and then the three equilibrium equations may be written for each joint, for a total of 12 equations for the 12 unknowns. We have

$$\mathbf{AB} = \begin{pmatrix} -4 \\ 0 \\ 8 \end{pmatrix} \quad \mathbf{AD} = \begin{pmatrix} -4 \\ 3 \\ 0 \end{pmatrix} \quad \mathbf{BD} = \begin{pmatrix} 0 \\ 3 \\ -8 \end{pmatrix}$$

so

$$\frac{\mathbf{AB}}{|\mathbf{AB}|} = \begin{pmatrix} -0.447 \\ 0 \\ 0.849 \end{pmatrix} \quad \frac{\mathbf{AD}}{|\mathbf{AD}|} = \begin{pmatrix} -0.8 \\ 0.6 \\ 0 \end{pmatrix} \quad \frac{\mathbf{BD}}{|\mathbf{BD}|} = \begin{pmatrix} 0 \\ 0.351 \\ -0.936 \end{pmatrix}$$

These are the unit vectors in the AB, AD, and BD directions. Using the method of joints, a system of 12 linear equations is obtained to determine the member forces and the reactions:

$$-0.447F_{AB} - F_{AC} - 0.8F_{AD} = 0$$

$$0.6F_{AD} + A_y = 0 \qquad \text{Joint } A$$

$$0.894F_{AB} + A_z = 0$$

$$0.447F_{AB} = 0$$

$$0.352F_{BD} + 2000 = 0 \qquad \text{Joint } B$$

$$-0.894F_{AB} - F_{CB} - 0.936F_{BD} = 0$$

$$F_{AC} + C_x = 0$$

$$F_{CD} + C_y = 0 \qquad \text{Joint } C$$

$$F_{CB} + C_z = 0$$

$$0.8F_{AD} = 0$$

$$-0.6F_{AD} - 0.351F_{BD} - F_{CD} = 0 \qquad \text{Joint } D$$

$$0.936F_{BD} + D_z = 0$$

Solving this system of equations, we obtain:

$$F_{AB} = 0 \quad F_{AC} = 0 \quad F_{AD} = 0 \quad F_{CB} = 5318\text{ N} \quad F_{BD} = -5682\text{ N} \quad F_{CD} = 1994\text{ N}$$

$$A_y = 0 \quad A_z = 0 \quad C_x = 0 \quad C_y = -1994\text{ N} \quad C_z = -5318\text{ N} \quad D_z = 5318\text{ N}$$

Note that all the forces at joint A are zero, and for this type of loading, the space truss is functioning as a planar truss in the y–z plane composed of joints B, C, and D. Although this solution involved a 12×12 system of equations, these equations carry a complete description of the geometry of the truss. By altering the loading in the equations, they may be solved for other loadings. Examples are given in the Computational Supplement.

Space trusses can also be solved using a direct vector solution, as shown in Section 2.10 and in Chapter 3. If this method is used, we will start with joint B and solve for F_{BC}, F_{BD}, and F_{BA}. We can then proceed to joint D and solve for F_{DA}, F_{DC}, and D_Z. Finally, joint A can be solved for the two reactions A_y and A_z and the force in member AC. The three reactions at joint C may then be determined by simple observation. Details of the solution are shown in the Computational Supplement.

Problems

7.56 Determine the internal member forces of the simple truss shown in Figure P7.56 by using the method of joints. The connection at C is a ball and socket, while that at B supports only a force in the vertical direction, and that at A supports one force in the vertical and one in the horizontal direction.

7.57 Determine the internal member forces of the simple truss by using the method of joints for the system of Problem 7.56, with a force of the form $F = -1000\hat{\mathbf{i}} - 1000\hat{\mathbf{j}} + 1000\hat{\mathbf{k}}$ applied at D.

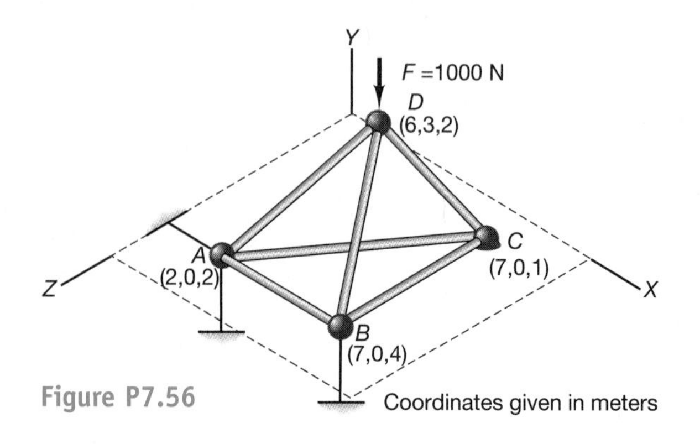

Figure P7.56

7.58 A sign is supported by a six-member space truss and a cable. (See Figure P7.58.) The truss is connected to the wall of a building by a ball and socket at point B, frictionless supports at joints A and C, and the rod between E and D. The sign causes a vertical force of 5.2 kN. Calculate the forces in each member, the reactions at A, B, and C, and the force in the rod.

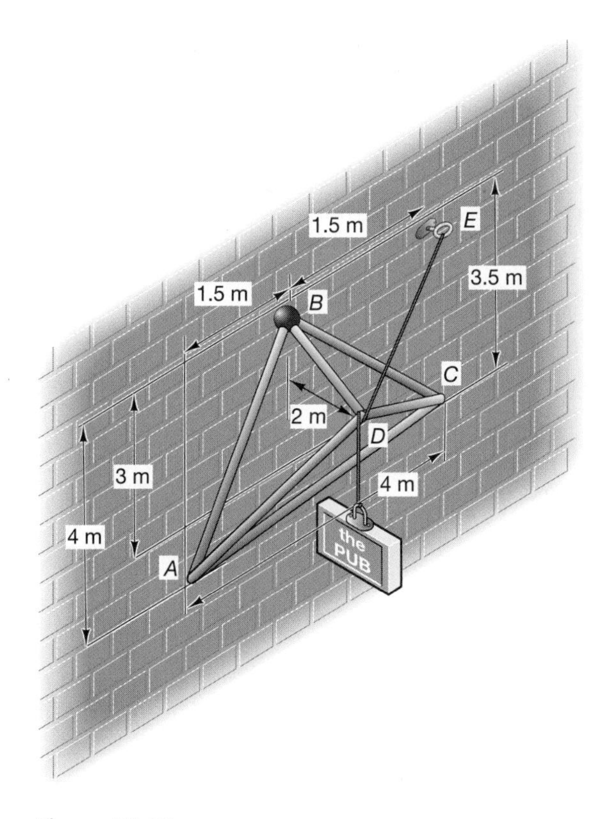

Figure P7.58

7.59 Suppose that the wind blows on the sign of Figure P7.58 such that an additional 500-N force is applied to point D in a direction parallel to the wall and the rod to the right. Calculate the forces in each member, the reactions at A, B, and C, and the force in the rod.

7.60 A window-washing scaffold is supported by two space frames, one of which is illustrated in Figure P7.60. The support at A is a ball and socket, and those at B and C are frictionless supports; C also has a pin restraint in the x-direction, to keep the frame from rotating. Joint D is centered 1 m above the midpoint of line BC, which is 1 m long. (a) How many unknowns and how many equations does this system represent if solved by the method of joints? (b) Compute the reaction forces at points and the force in each member.

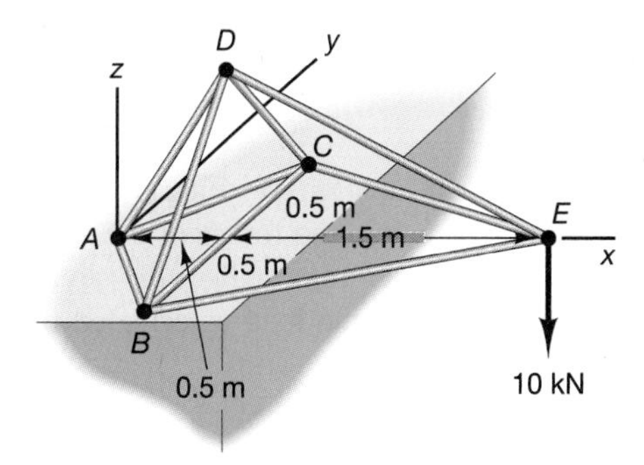

Figure P7.60

7.61 Suppose the system of Figure P7.60 has a windload of 1 kN acting on it pointed directly into the building and modeled as an additional 1-kN force applied at joint E. Calculate the reactions at A, B, and C, as well as the forces in the members.

7.62 A nine-member frame is used to connect two different-sized concrete columns, one rectangular and one cylindrical. (See Figure P7.62.) The cylindrical

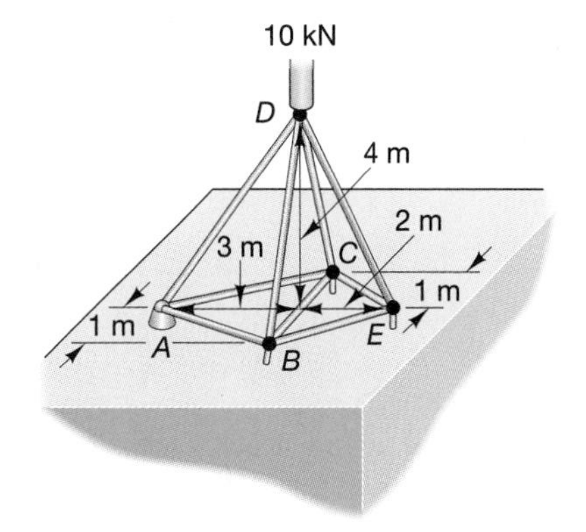

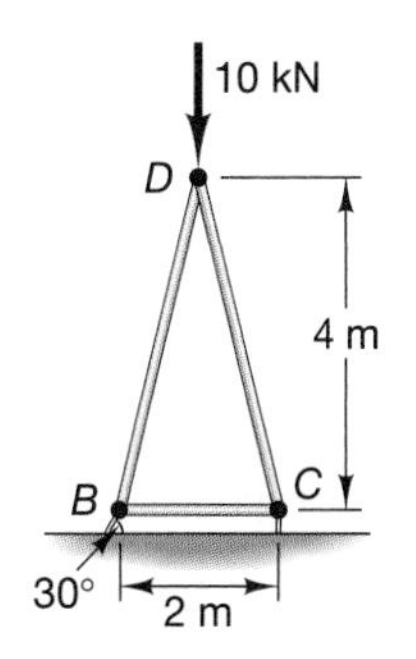

Figure P7.62

column is modeled as applying a force of 10 kN straight down at point D. The connection at A is a ball and socket; the connection at E and C are vertical pins, while the connection at B is a pin at 30° in the vertical plane of BCD. Compute the reactions at points A, B, C, and E, as well as the forces in each member.

7.63 If the cylindrical column of Figure P7.62 is misaligned during construction, the forces in the supports and members will change. To simulate this situation, resolve Problem 7.62 with the load as oriented in Figure P7.63.

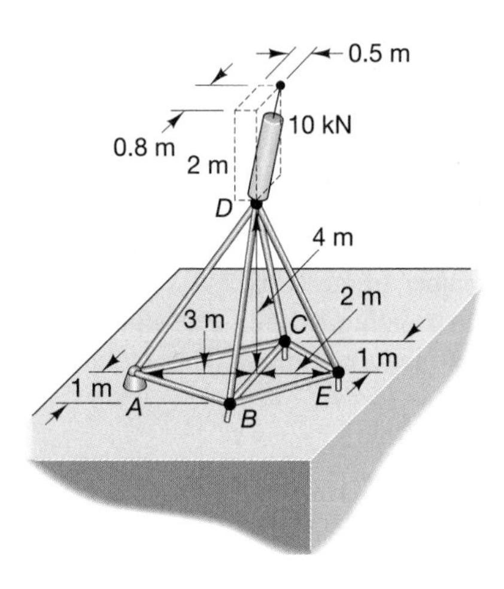

Figure P7.63

7.64 A space frame for a tow truck (see Figure P7.64) must support a maximum of 5×10^4 N. The frame is connected to the truck at point A by a ball and socket, vertical pins at points B and D that provide a reaction in only the vertical direction (+ or −), and a pin at C in the plane of ECD, making an angle of 30° with the truck bed. Compute the reactions at A, B, C, and D, plus the force in each member for this load.

7.65 Resolve Problem 7.64 for the case where the load W makes an angle of 30° from the vertical (still with magnitude of 5×10^4 N), as illustrated in the side view shown in Figure P7.65.

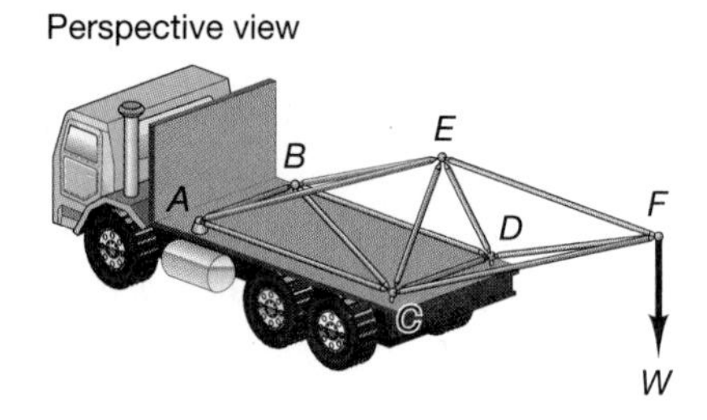

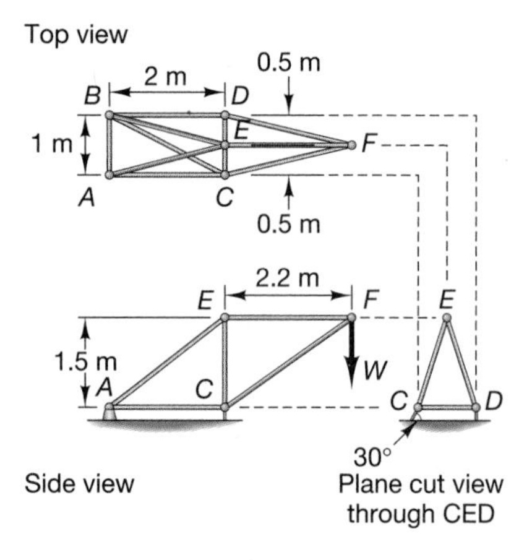

Figure P7.64

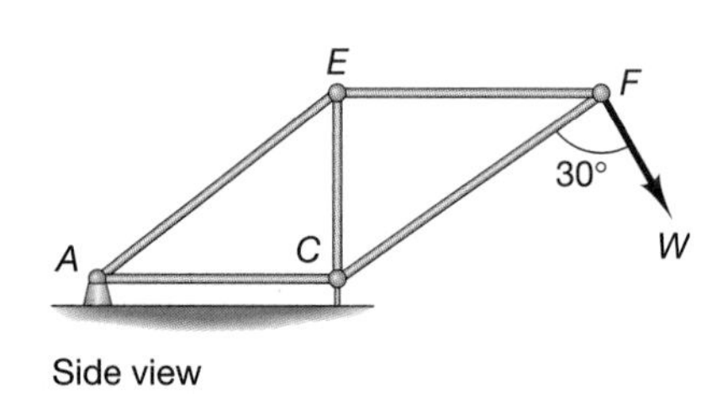

Figure P7.65

7.66 A series of metal space frames are used to support a roadway by connecting it to concrete bridge supports, as shown in Figure P7.66. If the joints at points E and F support vertical loads of 10^5 N each, compute the reaction forces at points A, B, C, and D, as well as the forces at each connection. The connection at A is a ball and socket, B is a vertical link, and the connections at C and D are slanted links, as illustrated in the figure.

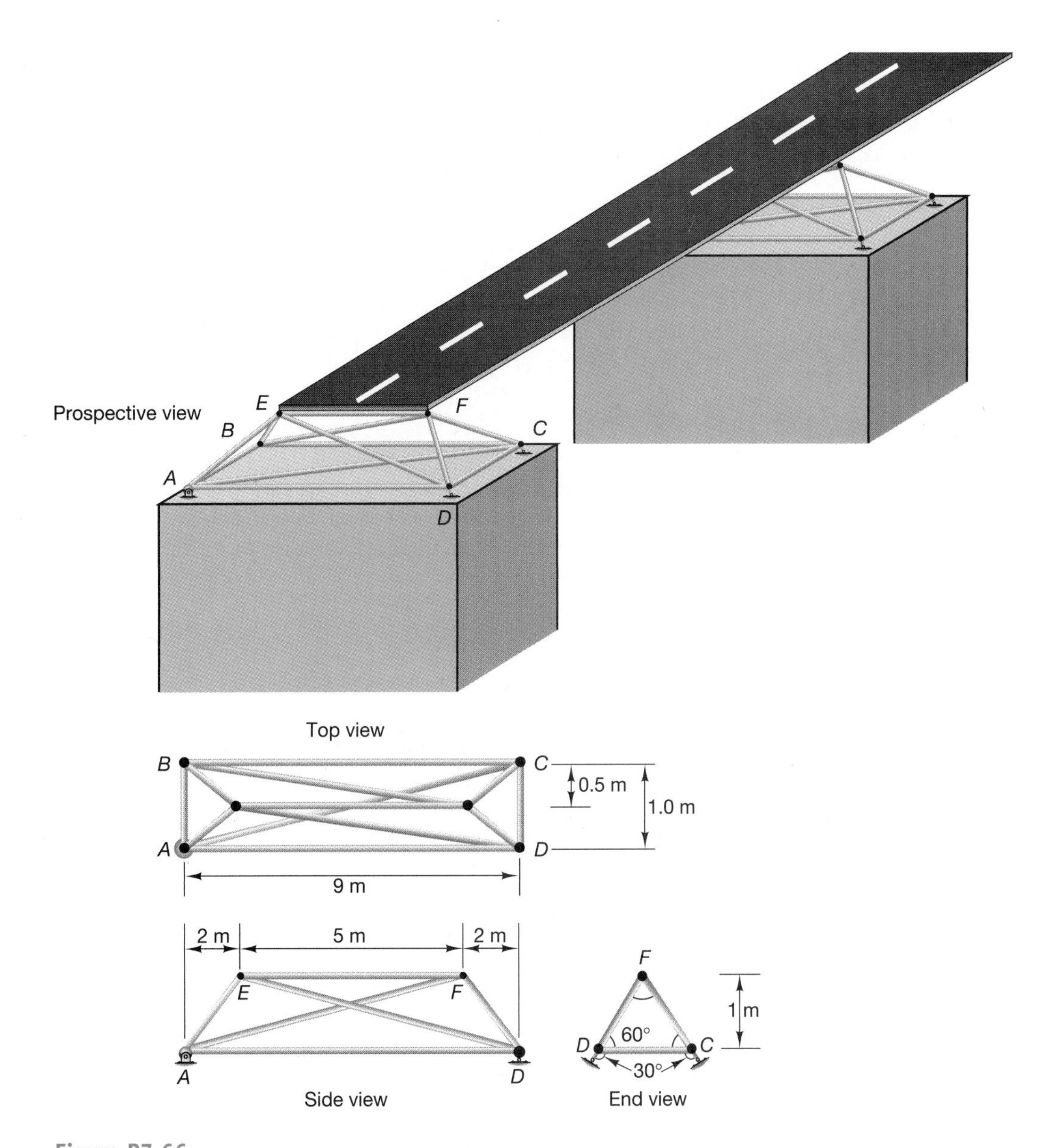

Figure P7.66

7.8 COMPOUND TRUSSES

A ***compound truss*** is a truss made by rigidly connecting several simple trusses. Consider the two simple trusses connected as shown in Figure 7.24(a). The compound truss is formed by connecting the two simple trusses *ABC* and *BDE* together at *B* and adding member *CD*. The compound truss, supported by a fixed support at *A* and a rolling support at *E*, is rigid, stable, and statically determinate. For simple planar trusses, the test for

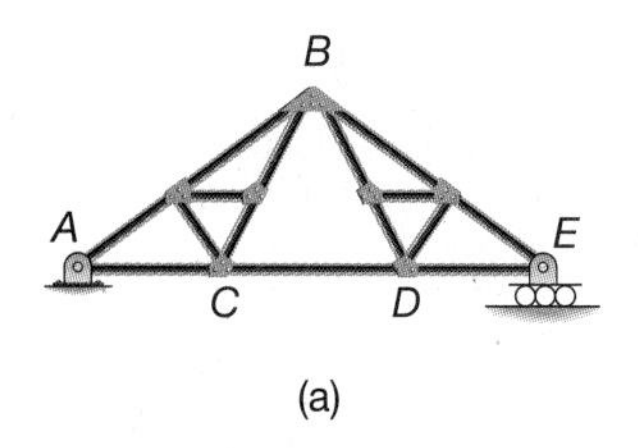

(a)

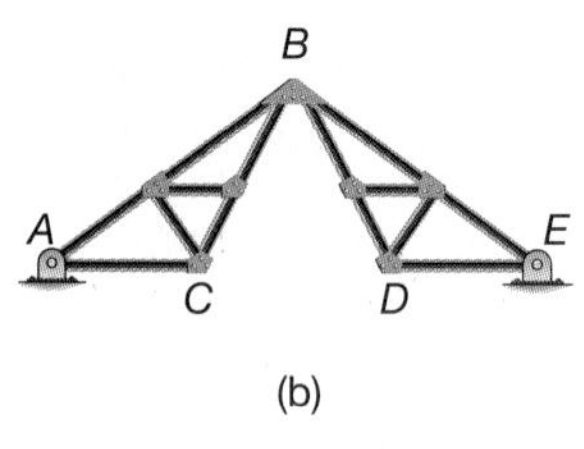

(b)

Figure 7.24

rigidity and solvability is that the number of joints be related to the number of internal members by the relationship

$$m = 2j - 3 \tag{7.18}$$

where m is the number of internal members and j is the number of joints. The compound truss in Figure 7.24(a) satisfies this relationship.

The compound truss shown in Figure 7.24(b) is formed by connecting the two simple trusses only at joint B. It is clear that this truss is not rigid and that both joints A and E must be hinged for the truss to be stable. It is also clear that this truss does not satisfy Eq. (7.18). It would, however, satisfy the relationship

$$m = 2j - r \tag{7.19}$$

where r is the number of reactions. This appears to be a more general test for determining whether the truss can be solved by equilibrium equations; however, compound trusses can even be constructed such that this equation is not a proper test. If the compound truss is statically determinate, the methods of joints and sections may be applied, as was shown earlier.

7.9 FRAMES AND MACHINES

The methods used to determine the internal member forces in trusses are dependent on the fact that the structure can be modeled as a collection of two-force members. The directions of the forces acting at each joint or connection are known. A frame or machine is characterized as having at least one multiforce member—that is, one member with three or more forces acting on it. The methods used to solve problems involving frames and those used to solve machine problems are the same. A frame may be defined as a stationary structure designed to support loads. It differs from a truss in that it contains one or more multiforce members. A machine is designed to transmit or modify forces; it contains moving parts and may not be stationary.

A frame may be a rigid structure, as shown in Figure 7.25(a), or nonrigid, as shown in Figure 7.25(b). The three external reactions on the frame in Figure 7.25(a) can be determined by treating the entire frame as a rigid structure and writing the three equations of equilibrium. The structure shown in Figure 7.25(b) is not rigid and is constrained by four reactions that cannot be determined by considering the entire structure as a rigid body. In both cases, members ABC and CDE are multiforce members. A pair of pliers is a machine that is not restrained in space, and therefore, equilibrium equations for the machine yield only the trivial result that zero is equal to zero. The only external forces are those applied to the handles and those applied to the jaws. Consequently, it is difficult to recognize whether frames and machines are statically indeterminate, underconstrained, or improperly constrained. We will begin the solution of each problem with an examination of the constraints on the frame or machine.

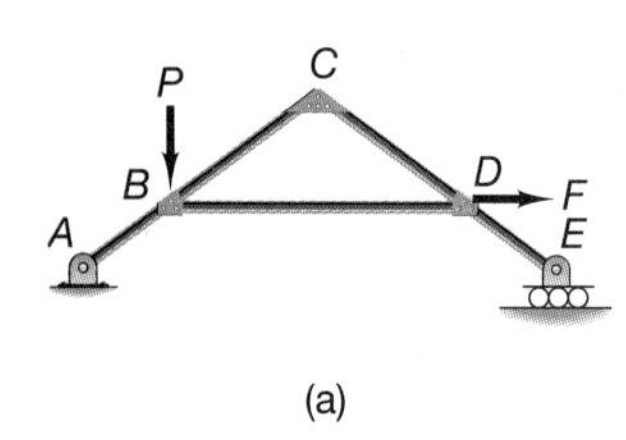

(a)

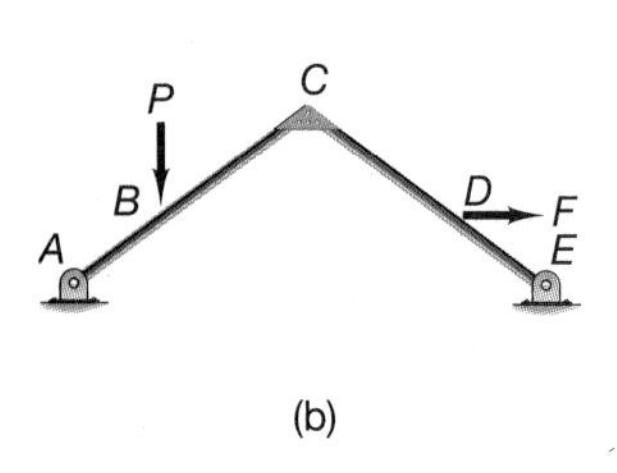

(b)

Figure 7.25

Frames and machines are analyzed by constructing a disassembled or "exploded" free-body diagram. This is done by separating each of the members from the structure and showing all the internal connecting forces. These forces will act equally and oppositely on the connecting members. Disassembled free-body diagrams for the two structures shown in Figure 7.25 are illustrated in Figure 7.26. In Figure 7.26(a), all dimensions of the structure are known, and the unknown forces due to the application of the known forces **P** and **F** can

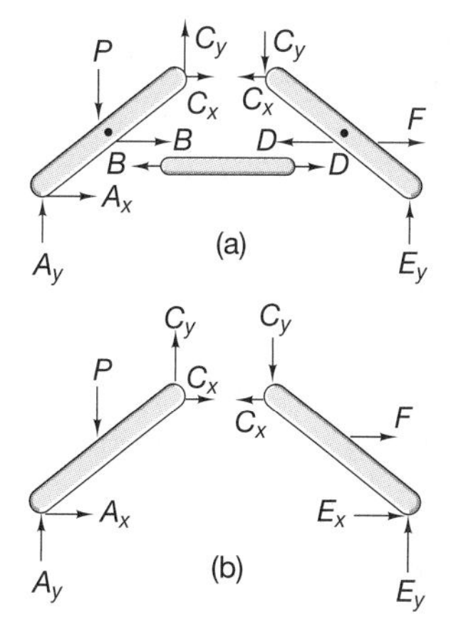

Figure 7.26

be determined. When the disassembled free-body diagram was constructed, the forces **P** and **F** were each considered to be acting on only one member. The external reactions at A and E can be determined by considering the entire structure as a rigid body. The structure is composed of three separate rigid bodies, each of which must be in equilibrium. The member BD is a two-force member, and the relationship

$$|B_x| = |D_x|$$

must be true. There are six unknown internal connecting forces and reactions on the structure: A_x, A_y, B_x, C_x, C_y, and E_y. For each member ABC and CDE, three equations of equilibrium may be written, for a total of six equations for the six unknowns. Although this frame could have been treated as a rigid body and A_x, A_y, and E_y determined from the equilibrium equations of the entire frame, these equilibrium equations are not linearly independent of the six equations obtained from the equilibrium of ABC and CDE. Therefore, it is usually better to work from the disassembled free-body diagram. Some texts recommend working with the entire structure to simplify the solution of the system of simultaneous equations. However, as stressed thoughout this text, modern computational methods eliminate that concern, and it is better to approach the analysis in a systematic manner.

The analysis of the structure shown in Figure 7.26(b) is done in exactly the same manner, and the external reactions cannot be determined by considering the entire structure as a rigid body, as it is not rigid. There are six unknown internal connecting forces and reactions: A_x, A_y, C_x, C_y, E_x, and E_y. There are three equilibrium equations for each of the two members, forming a set of six linear algebraic equations for the six unknowns. For more complex structures, there are more parts and more disassembled free-body diagrams, and the system of linear equations is larger.

Note: If the disassembled free-body diagrams are reassembled, the internal connecting forces must cancel out, leaving a free-body diagram of the entire structure with only external applied forces and external reactions. If an external force acts at a connection between members in a frame or machine, this force can be assigned to any of the members attached at that connection when they are separated. However, it can be assigned to only one of the members.

Sample Problem 7.3

In the figure on the next page, a variable torque motor is programmed to maintain a constant force P on the piston as the shaft rotates a full revolution. Determine the torque as a function of the shaft rotation angle β. Neglect the inertia of the moving parts, and solve for the quasi-static case. Graph the required torque for equilibrium for $L_1 = 500$ mm, $L_2 = 200$ mm, and $P = 1000$ N.
Note: L_1 must be greater than L_2 or the system binds.

Solution

Member L_1 is a two-force member and is in either tension or compression along its long axis. Member L_2 is attached to the motor shaft and therefore is not a two-force member. The following geometric relationship holds for all positions:

$$L_1(\sin\theta) = L_2(\sin\beta)$$

so that

$$\theta = \sin^{-1}\left(\frac{L_2}{L_1}\sin\beta\right)$$

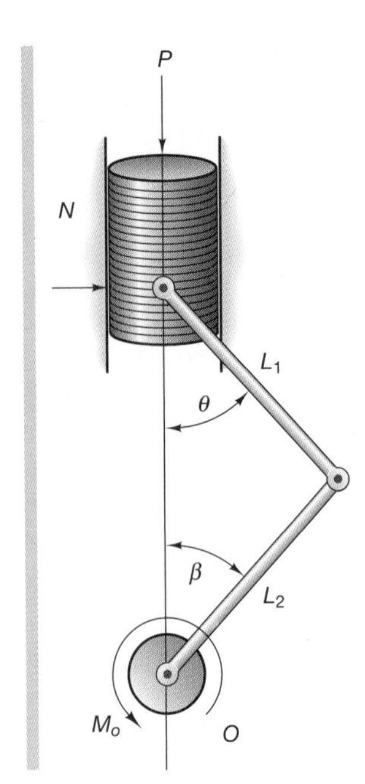

The concurrent forces acting on the piston are

$$\mathbf{N} = N\begin{pmatrix}1\\0\\0\end{pmatrix} \quad \mathbf{P} = P\begin{pmatrix}0\\-1\\0\end{pmatrix} \quad \mathbf{T}_{L_1} = T_{L_1}\begin{pmatrix}\sin\\-\cos\\0\end{pmatrix}$$

Therefore, solving these two equations, we obtain

$$N = P\tan\theta$$

Taking moments about point O for the complete assembly yields

$$M_O - (L_1\cos\theta + L_2\cos\beta)N = 0$$

$$M_O = (L_1\cos\theta + L_2\cos\beta)P\tan\theta$$

In the Computational Supplement, the required motor torque for a complete revolution is plotted for different values of the lengths L_1 and L_2.

Sample Problem 7.4

A compound-lever cutting tool is shown in the accompanying diagram. A force F is applied to each of the handles. Determine the cutting force between E and G, the pin reactions at C and D, and the tension or compression in member BH. All dimensions are in millimeters.

Solution Part BH is a two-force member, and we will not make a separate free-body diagram for this part. Free-body diagrams for parts $ACDG$, HDE, and IBC are shown in the following figures (a), (b), and (c), respectively.

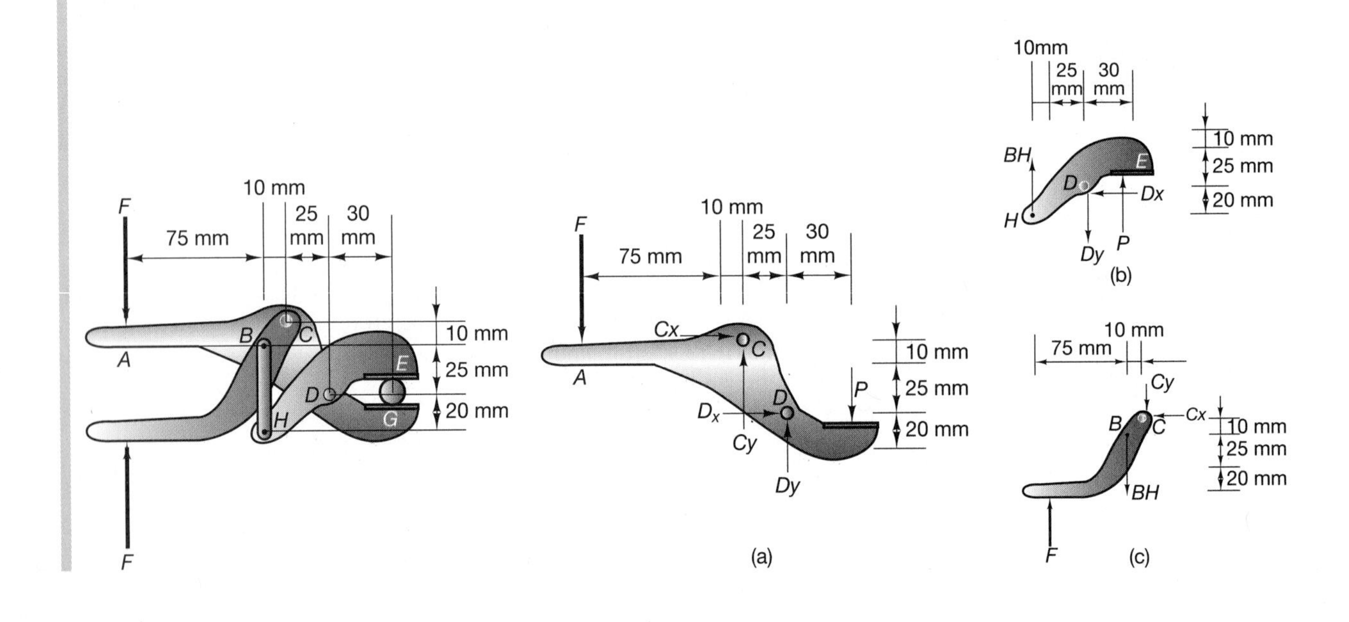

Assuming that the forces F are given, there are six unknown forces acting on the cutters: BH, C_x, C_y, D_x, D_y, and P. We can, however, write three equilibrium equations for each of the three parts of the cutting tool, thus yielding a system of nine equations for six unknowns. The cutting tool is not constrained globally, and the nine equations of equilibrium are not linearly independent. We will examine the solution in some detail. From free-body diagram (a), we obtain

$$\begin{aligned} C_x + D_x &= 0 \\ -F + C_y + D_y - P &= 0 \\ 85F + 35D_x + 25D_y - 55P &= 0 \end{aligned} \tag{a}$$

From free-body diagram (b), we have

$$\begin{aligned} -D_x &= 0 \\ BH - D_y + P &= 0 \\ -35BH + 30P &= 0 \end{aligned} \tag{b}$$

From free-body diagram (c), we obtain three more equilibrium equations:

$$\begin{aligned} C_x &= 0 \\ F - BH - C_y &= 0 \\ -75F - 10C_y &= 0 \end{aligned} \tag{c}$$

Examination of the first equation of each set shows that these equations are not linearly independent, as any two of the equations are sufficient to show that C_x and D_x are zero. Any two sets of equations may therefore be used to determine the unknown forces, which are

$$C_x = 0 \quad C_y = -15F/2 \quad D_x = 0 \quad D_y = 221F/12$$
$$BH = 17F/2 \quad P = 119F/12$$

In the Computational Supplement, the three sets of equations are solved in different combinations to demonstrate their interdependence.

Problems

7.67 A stepladder is loaded with a 600-N force, as shown in Figure P7.67. Draw the appropriate free-body diagram, and compute the forces acting at each connection. Model the contact point at A as having friction and that at E as frictionless.

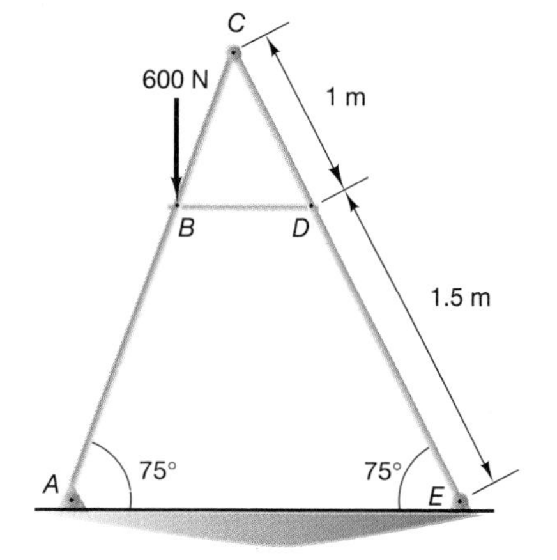

Figure P7.67

7.68 A can of paint (45 N) and a painter (810 N) are positioned on a stepladder. (See Figure P7.68.) Compute the forces at each connection. Model the stepladder as the three members shown, with friction at support A and a frictionless connection at B. The distance EG is 0.5 m and CF is 0.3 m.

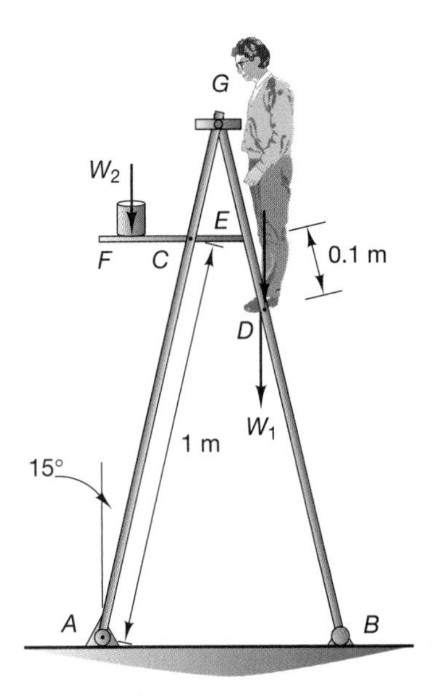

Figure P7.68

7.69 (a) In Figure 7.69, calculate the forces at each connection and the reaction forces at points A and B if the hanging mass applies a force of 40,000 N at point E, by applying the 40,000-N load to member BE. (b) Repeat (a) by applying the 40,000-N load to member AE, and show that the answers to (a) and (b) are the same. That is, show that the overall load distribution is the same.

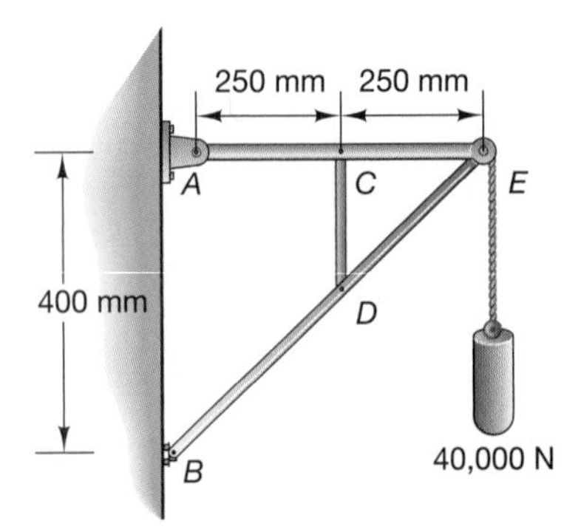

Figure P7.69

7.70 Work Problem 7.69 for the case where the hanging mass causes a downward force at E of 2200 N, which corresponds roughly to the average weight of an automobile engine.

7.71 A frame is used to support a 400-kg truck engine on a rope and pulley. (See Figure P7.71.) The rope-and-pulley system is shown as fixed, but is used to raise and lower the engine for installation. Calculate the forces at each connection and those at the supports.

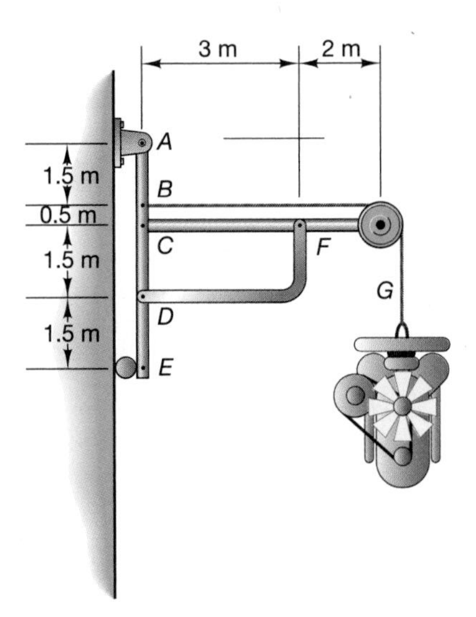

Figure P7.71

7.72 (a) A two-member frame is used to support a small overhead winch that applies a force to point D of 750 N, as shown in Figure P7.72. Calculate the reactions at A and C and the forces at each joint for $\theta = 30°$. (b) Plot the force in member AC against θ, and pick a "good design" value for θ so that the force in AC is small.

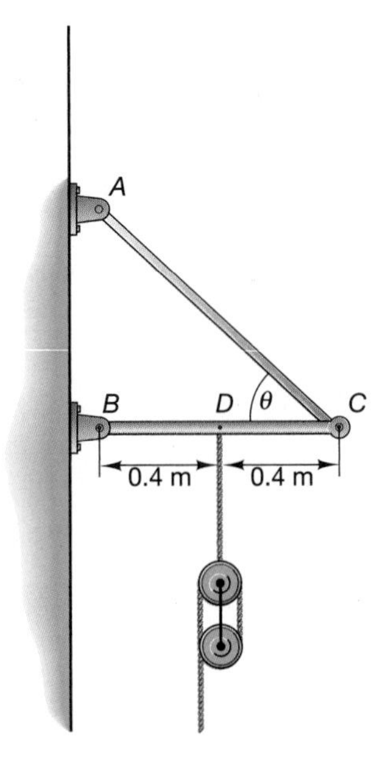

Figure P7.72

7.73 A suspension component for a racing car must support a 2000-N force configured as shown in Figure P7.73. Compute the forces at each joint.

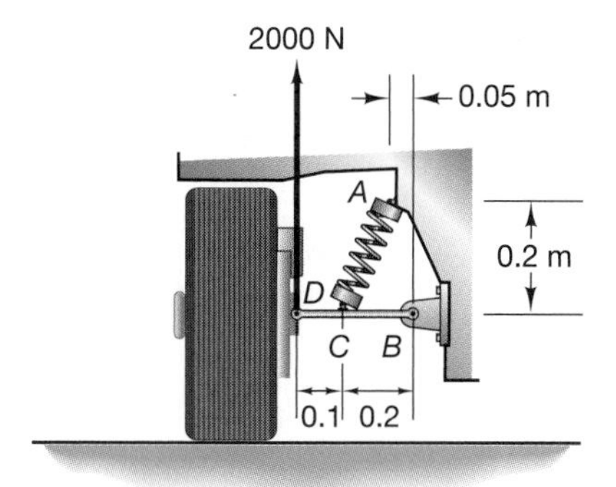

Figure P7.73

7.74 Suppose that the mass of the connecting rods of the suspension system of Figure P7.73 is given as $M_{AC} = 10$ kg and $M_{BD} = 15$ kg, respectively. Compute the forces at each joint, (a) with and (b) without the load illustrated. Assume that the center of gravity of each member is in the geometric center of the member.

7.75 Calculate the reaction forces at *A* and *C* and the forces at each connection shown in Figure P7.75.

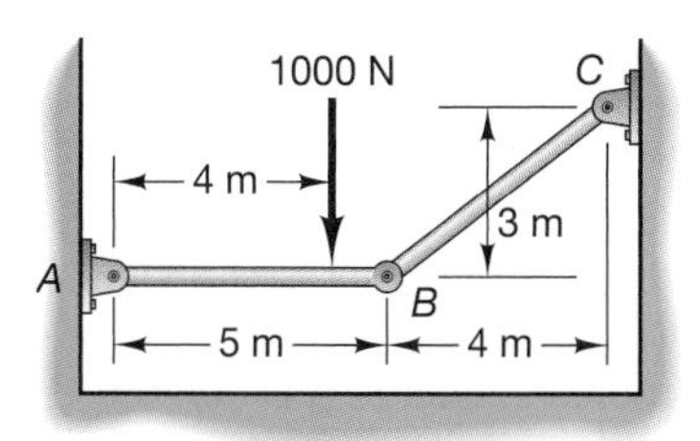

Figure P7.75

7.76 A tow-truck assembly supports a 13.5-kN load, as shown in Figure P7.76. Compute the reaction forces at *A* and *C*, as well as the forces at each connection, as the angle θ changes between 0 and 30°.

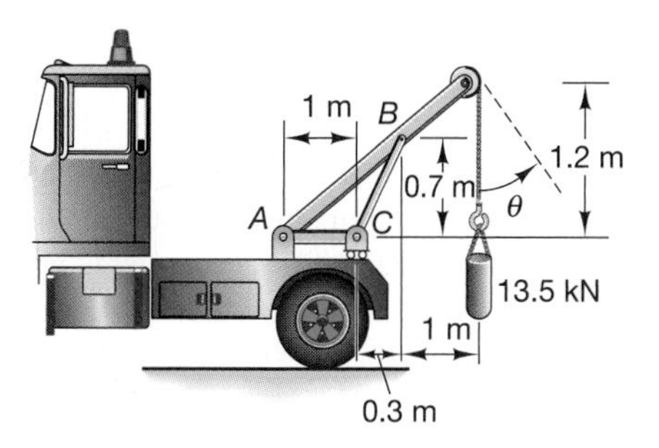

Figure P7.76

7.77 Compute the force *P* applied to the nut in Figure P7.77 by squeezing the nutcracker with 135 N of force. Also, compute the force on the support pin at point *A*.

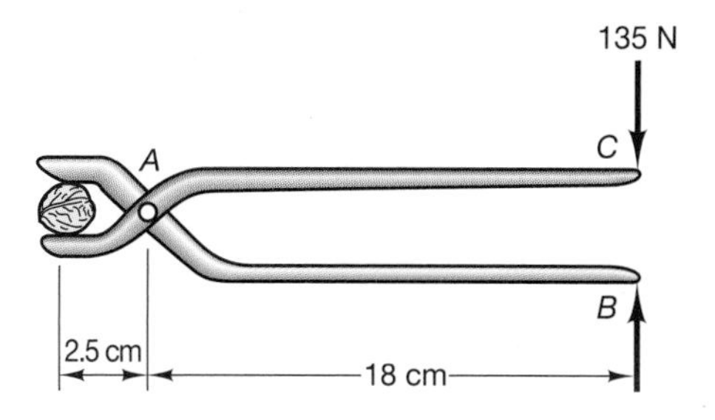

Figure P7.77

7.78 A company desires to design a pair of toy pliers for small children to use. It has been determined that children in the age group of interest can sustain 20 N of force between their thumb and fingers, and that the nut requires 100 N to maintain enough friction to turn. Complete the design of the pliers, shown in Figure P7.78, by calculating the length *a*. Also, determine the force at pin *A*.

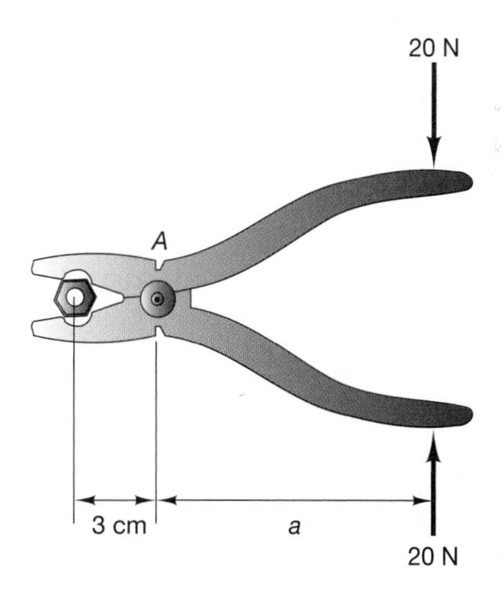

Figure P7.78

7.79 A person sitting, leaning back in a chair is modeled as applying a vertical force of 400 N and a horizontal force of 100 N to a "half-chain model" positioned, as indicated in Figure P7.79 on the next page. Assume the leg at *F* is pinned and the leg at *G* supports a vertical load only and compute the reaction forces *F* and *G* and the forces at each joint.

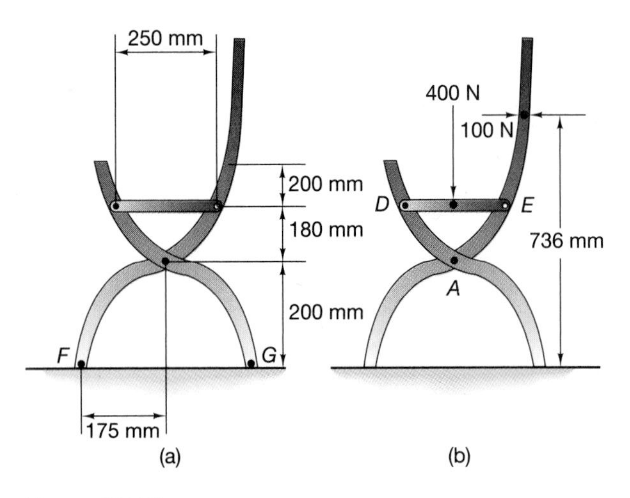

Figure P7.79

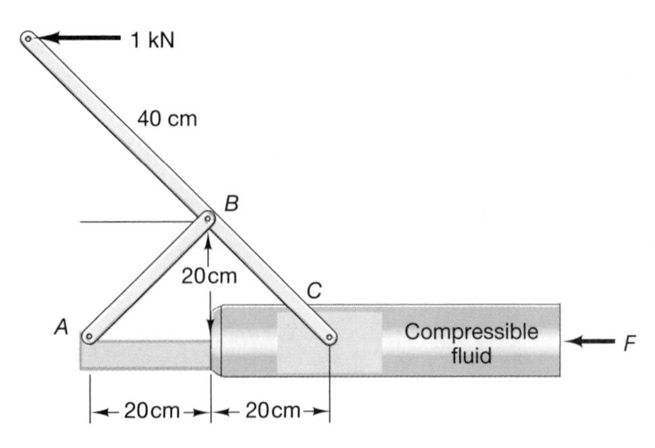

Figure P7.81

7.80 A manufacturer of exercise equipment wishes to know how much force F must be provided by the piston A shown in Figure P7.80 to balance the force of a foot pushing down with 800 N of force. Also, compute the forces in each member and the reaction force on the frictionless block.

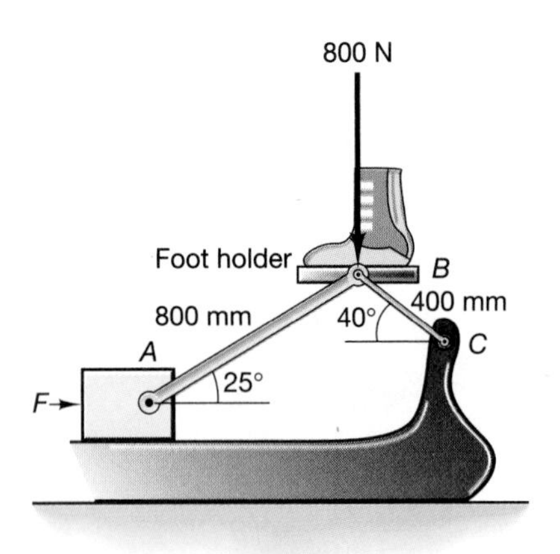

Figure P7.80

7.81 A mechanism for an exercise machine consists of two members and a piston moving against a compressible fluid. Compute the force F required for the fluid to maintain the system in equilibrium under the 1-kN load illustrated in Figure P7.81.

7.82 An airport service truck moves up and down under the action of an extending hydraulic actuator, as shown in Figure P7.82. Compute the forces at each joint if $\beta = 30°$ and the length of CE is 3.66 m. The top piece weighs 5.3 kN, and each of the 4.6-m diagonal members weighs 450 N, with offset center of gravity located 2.59 m up along each member. The weight of the telescoping actuator EC is neglected. All joints are pins except for G and B, which are sliders.

7.83 Calculate the forces at the pin A and the fixed end B on the bicycle brake lever and handlebar shown in Figure P7.83. The brake cable applies a force of 712 N to point C.

7.84 One of the purposes of most machines, such as pliers, cutters, crimpers, etc., is to provide a mechanical advantage by using one or more levers. Using Figure P7.84, show that the mechanical advantage, or amplification factor, of two levers connected by a frictionless roller at C is (ac/bd).

7.85 Compute the forces at A, B, and C, as well as the force F, for the lever system of Figure P7.84 analytically in terms of the four lengths a, b, c, and d.

7.86 The basis for the design of many hand tools (pliers, etc.) is the two-lever system of Figure P7.84. The purpose of this exercise is to look at the design process. A reasonable hand squeeze can apply 170 N of force P (about 40 lb) on the handle of a hand tool. Suppose it is required that the force F applied to a device by the hand tool needs to be 1000 N and that the total length of the hand tool must be 300 mm, such that $c + d = 200$ and $a + b = 100$ mm. Pick (i.e., design) the location of the fulcrum (points A and B) so that these requirements are met. (*Hint*: Make the calculation in Problem 7.84, and examine various values of a, b, c, and d.)

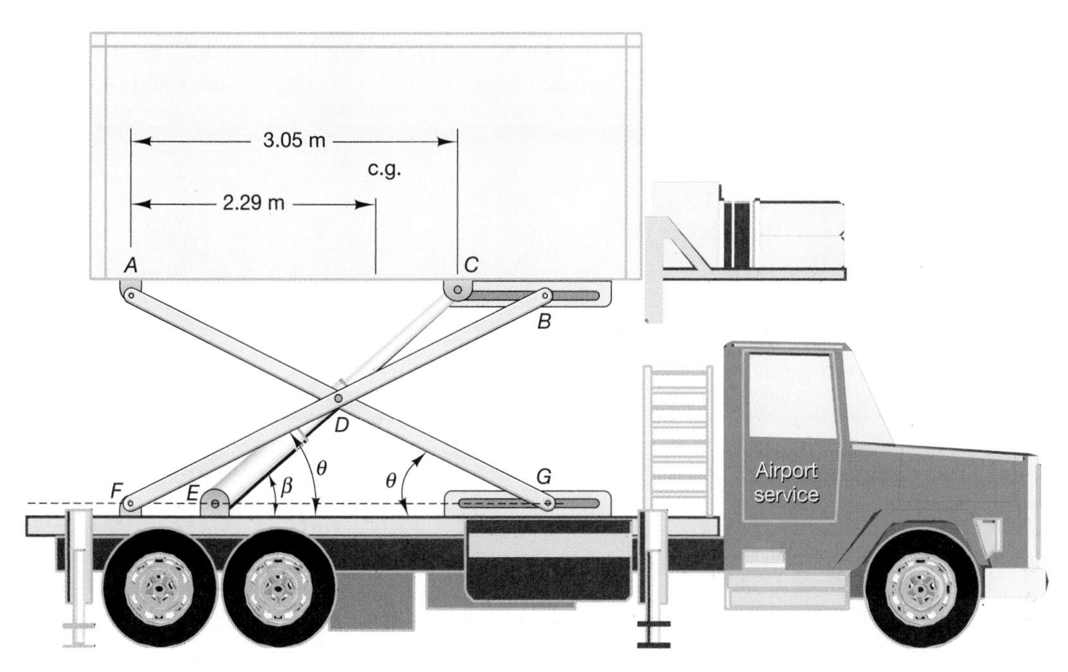

Figure P7.82

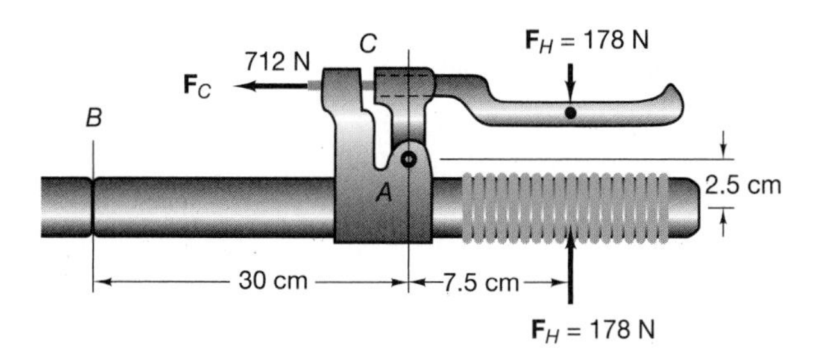

Figure P7.83

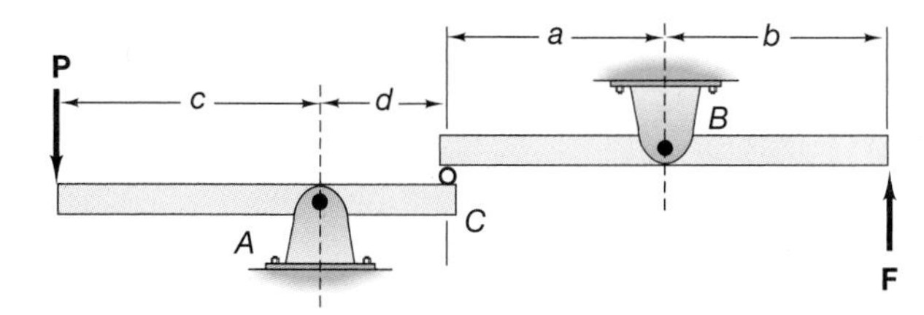

Figure P7.84

7.87 A factor in the design of hand tools is the force transmitted, as discussed in the previous three problems. Another factor is the distance that the lever must travel to apply the desired force. (a) Using the two-lever system of Figure P7.84, show that the ratio of the output deflection to the input deflections is db/ac, which is the inverse of the mechanical advantage. (b) Suppose that it is desired to design (i.e., choose a, b, c, and d) the device of Problem 7.84 such that an applied force of $P = 200$ N produces a resulting force of $F = 2000$ N. If the force F must move a distance of 10 mm, and because of the size of the hand, P can move through a distance of only 60 mm, is there a solution? If so, calculate it.

7.88 Compute the forces at the pins A, B, and C and the force F as a function of the design parameter x in Figure P7.88. What value of x should be used to cause $F = 2500$ N?

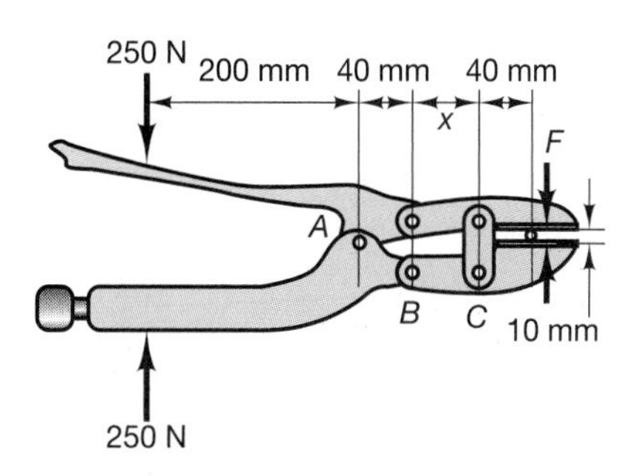

Figure P7.88

7.89 Let $x = 0.1$ m in Figure P7.88, and compute the reactions at A, B, and C and the force F.

7.90 A lifting mechanism works by a force P pulling a platform, as illustrated in Figure P7.90. Calculate the force P required to keep this system in equilibrium as a function of θ. The platform AB has a mass of 1000 kg, and the crate has a mass of 3000 kg. Also, calculate the forces at A and B. What happens at $\theta = 0$?

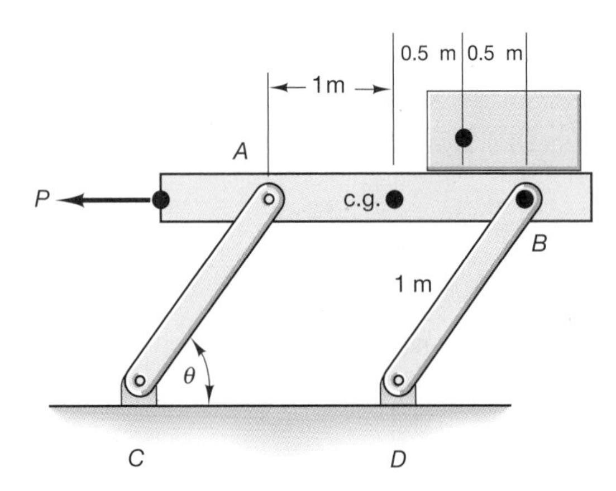

Figure P7.90

7.91 Compute the forces at the joints and the force at A and B for the motorhoist system illustrated in Figure P7.91. The force in the hydraulic piston, CE, is needed in order to design the hoist.

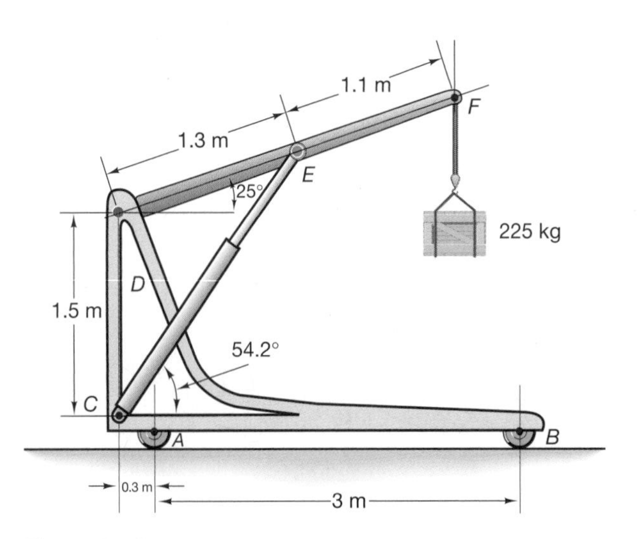

Figure P7.91

7.92 A section of a lifting mechanism is illustrated in Figure P7.92. Treat joints A and D as fixed, and compute the reaction forces at A and D and the forces at B.

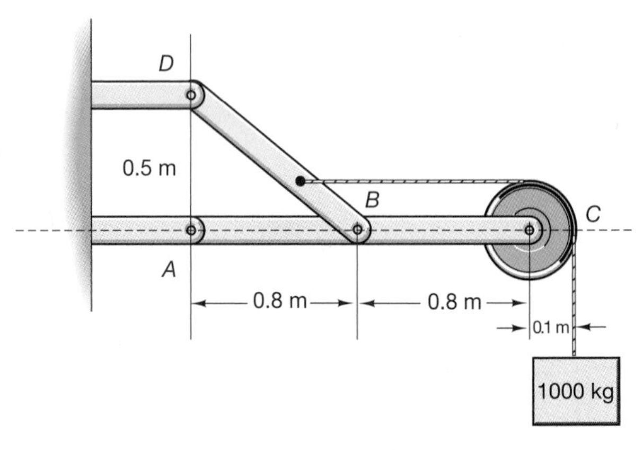

Figure P7.92

7.93 Compute the forces in the frame at each joint and connection under the load shown in Figure P7.93. Neglect the mass of the frame members.

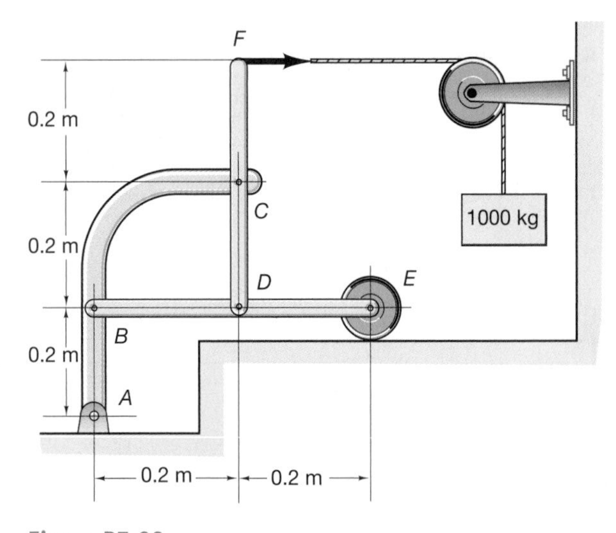

Figure P7.93

7.94 A crimping tool is squeezed to produce a force F. (See Figure P7.94.) Compute the force F as well as the forces at each joint.

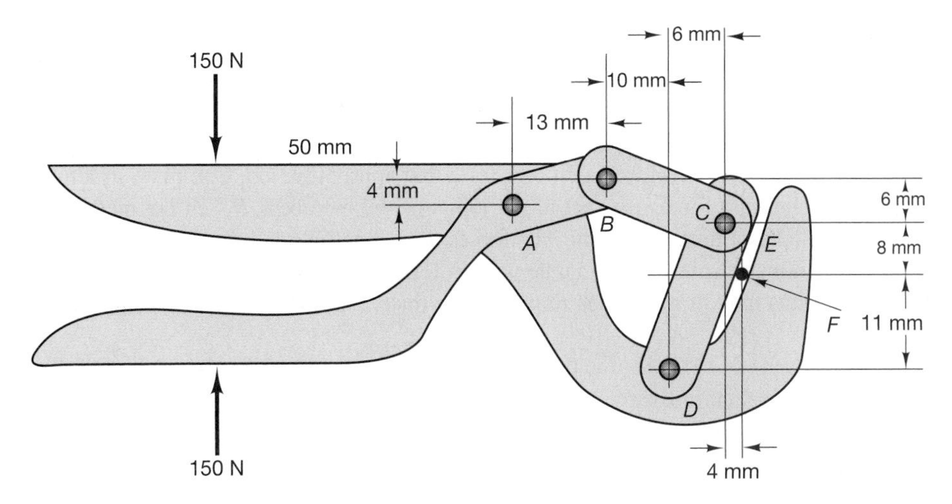

Figure P7.94

7.95 Compute the forces at the connections labeled for the plow mechanism in Figure P7.95, under the assumption that the blade of the plow provides a force of $P = 10$ kN and $W = 1.5$ kN. Neglect the weight of the other members of the machine.

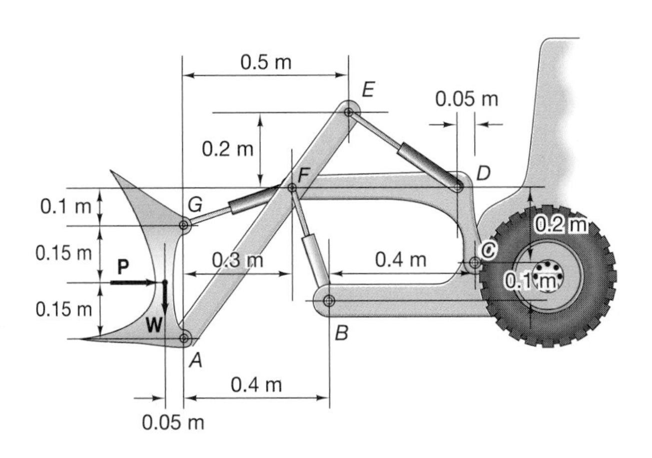

Figure P7.95

7.96 Compute the forces in the "wishbone" suspension at joints B, C, D, E, and F for a load $P = 2$ kN, for the configuration shown in Figure P7.96. The pin at G does not connect to member EH.

7.97 Compute the forces in the wishbone suspension at each joint for the configuration shown in Figure P7.97. The force $P = 2$ kN.

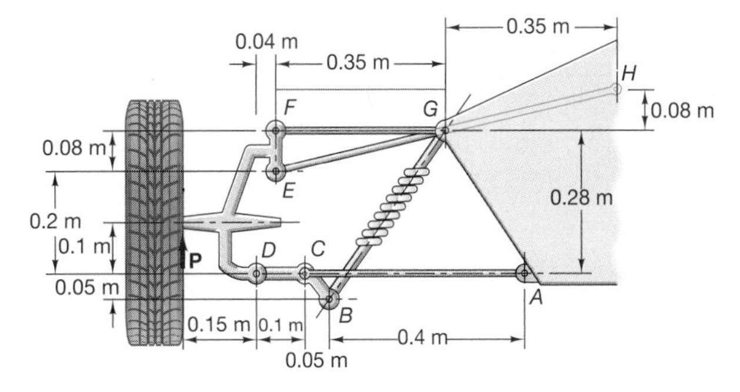

Figure P7.96

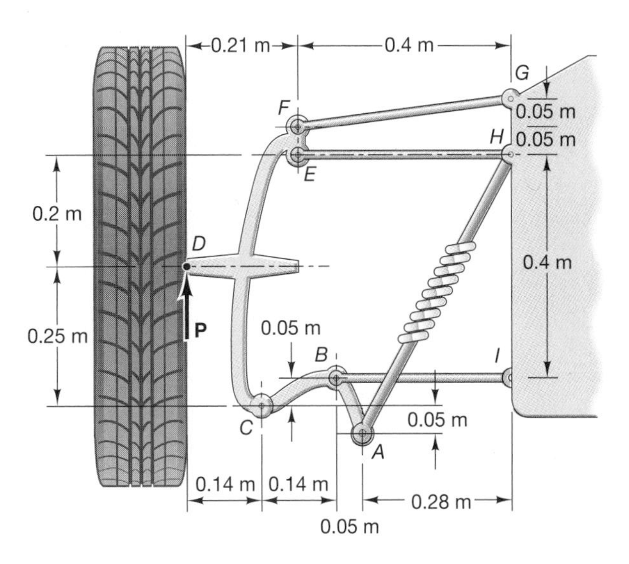

Figure P7.97

Chapter Summary

A structure is an organization of bodies arranged systematically to accomplish a particular purpose. Structures may be classified into three categories: trusses, frames, and machines.

Trusses The members of a truss may be modeled as two-force members for a first approximation. Complete structural analyses include the bending moments and shearing forces that act at the joints of the members. When the method of joints is used, each joint of the truss is modeled as a pin connection. The forces form external loads, or internal member forces act concurrently on the pin. Two equations of equilibrium can be written for each joint in a planar truss, three for each joint in a space truss. The method of joints may be employed to generate a matrix that characterizes the geometry of the truss.

Another approach—not often used, however—is to determine the forces in the truss members by the method of sections. An imaginary cut is passed through the truss, cutting three internal members. The sectioned truss is then treated as a rigid body. Although this method is not used in structural analysis, it can be employed to check results obtained by the method of joints.

Frames and Machines A frame is defined as a stationary structure designed to support loads and having at least one multiforce member. A machine is designed to transmit or modify forces, and it contains moving parts. The first step in analyzing frames or machines is the construction of a free-body diagram of each disassembled part of the structure. The connecting forces (internal forces) occur in pairs equal in magnitude, opposite in sense, and collinear. For a two-dimensional model, each part is treated as a separate rigid body, and three equations of equilibrium may be written for each part. For a three-dimensional model, six equations of equilibrium may be written for each part. The resulting system of linear equations can be solved by matrix methods.

Chapter 8

INTERNAL FORCES IN STRUCTURAL MEMBERS

The cables for traffic lights must be designed to withstand the forces due to high winds. (Courtesy of Loren Winters/Visuals Unlimited)

8.1 INTRODUCTION

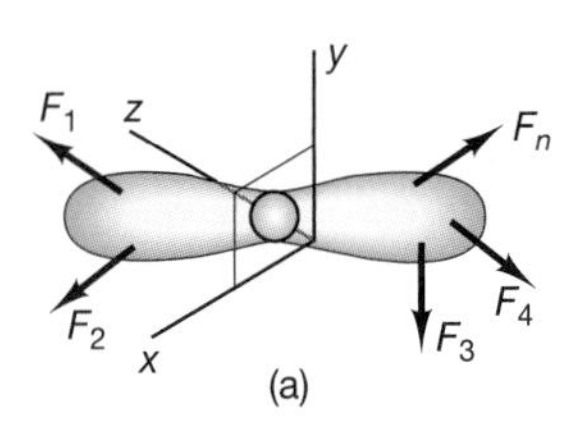

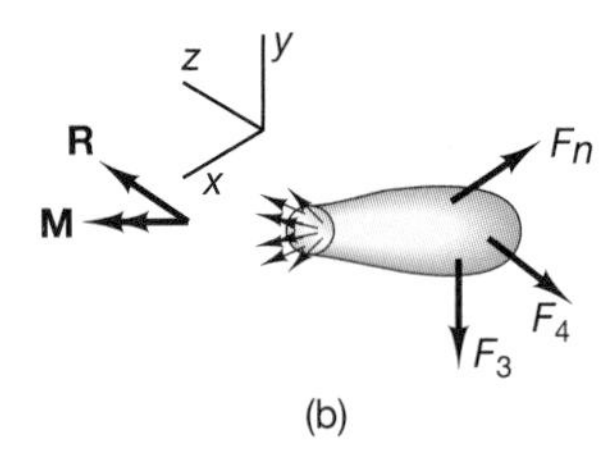

Figure 8.1

In Chapter 7, we determined the internal forces in members in a truss when those members were modeled as straight two-force members. Also, we developed methods for computing the *connecting* forces between parts of frames or machines. Engineering structural parts fail due to the *internal forces* that act within the parts. This chapter presents general methods for determining these internal forces, thus providing information about the strength of the system or the potential of the system to break or otherwise fail. The forces are distributed across any imaginary plane passing through a rigid body and are dependent upon the location, orientation, and area of such a plane, as illustrated in Figure 8.1. The distributed internal forces shown in Figure 8.1(a) are equivalent to a force system having a resultant **R** acting at a particular point on the cross section (usually the centroid of the area) of the body and a moment **M**. Distributed forces on the surface of a body were discussed in Chapters 4 and 5, and a similar approach is used to determine the internal forces. The section of the member isolated to the right of the cutting plane is in equilibrium due to **R** and **M** if the entire body is in equilibrium. The methods of deformable materials (strength of materials or mechanics of materials) are used to obtain the distributed internal forces. The intensities of these internal forces are called *stresses* and are given in terms of force per unit area. The units of stress are either pounds per square inch (psi) or newtons per square meter (N/m^2). One newton per square meter is called one *pascal*. There are many stress distributions that are equivalent to the force system made up of **R** and **M**, and the determination of the appropriate stress distribution depends upon the internal deformations of the material.

Before the internal stress distribution can be obtained, the internal force and moment of the equivalent force system must be determined for any plane passing through the body. In this chapter, we will develop the formal methods to determine this equivalent internal force system and will apply these methods to find the internal forces on long, slender members serving as bars, beams, shafts, or similar structural parts.

8.2 INTERNAL FORCES IN A MEMBER

Before computing the internal forces in a member, the external reactions should be determined. If the entire member is in equilibrium, then *any part of it is also in equilibrium.* Therefore, the internal forces may be found for any section of the member by cutting the member with an imaginary plane and noting that the separated parts are in equilibrium. Consider the long, slender rod shown in Figure 8.2 with a force of 500 N applied at the left end and an equal reaction at the right end. The rod is in equilibrium. If an imaginary cut *a–a* is made through the rod, the internal force required to hold each part of the rod in equilibrium is 500 N. This rod is said to be in uniform tension of 500 N. The rod shown in Figure 8.3 is loaded with five external axial forces acting at the points *a*, *b*, *c*, *d*, and *e*. The internal force at any point in the member is obtained by passing imaginary cuts through different sections of the rod. For the rod to be in equilibrium, the reaction on the right end must be

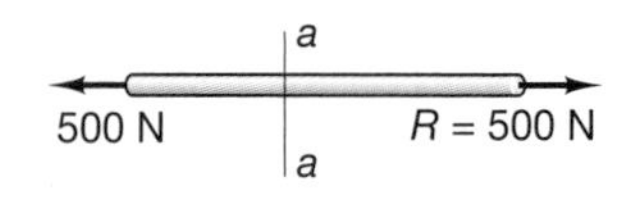

Figure 8.2

$$R = 1000 + 1500 - 500 - 500 + 500 = 2000 \text{ N}$$

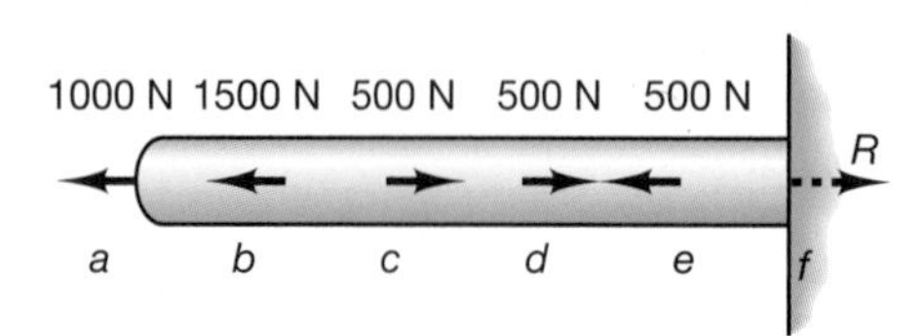

Figure 8.3

Table 8.1

Section *ab*	1000 N	Tension
Section *bc*	2500 N	Tension
Section *cd*	2000 N	Tension
Section *de*	1500 N	Tension
Section *ef*	2000 N	Tension

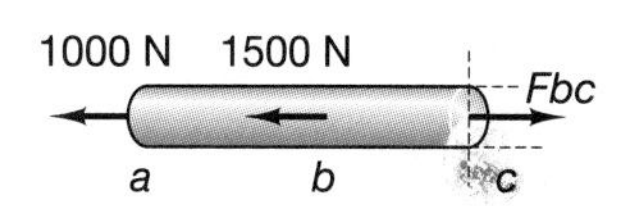

Figure 8.4

in the direction shown. To determine the internal force acting in the portion of the rod between b and c, we pass an imaginary cut through the rod anywhere in this section, as shown in Figure 8.4. Supposing the cut section to be in equilibrium, we find that the internal force is

$$Fbc - 1000 - 1500 = 0$$

or

$$Fbc = 1000 + 1500 = 2500 \text{ N}$$

The rod thus is in tension (2500 N) between points b and c. The internal force in any section of the rod can be easily determined in this manner, and the values are given in Table 8.1. The maximum internal force occurs in section bc, and the minimum internal force occurs in section ab. For this loading, the entire bar is in tension, but the magnitude of the internal tension differs in different sections.

If a bar is hung vertically and loaded by its own weight, the internal force at any section may be found by the use of imaginary cuts. Consider the bar shown in Figure 8.5. Suppose an imaginary cut is taken at a point x above the bottom of the bar. If the entire bar weighs an amount W, and the bar is of uniform constant density, the weight below the cut at x is $(W\,x/L)$, and the internal force is

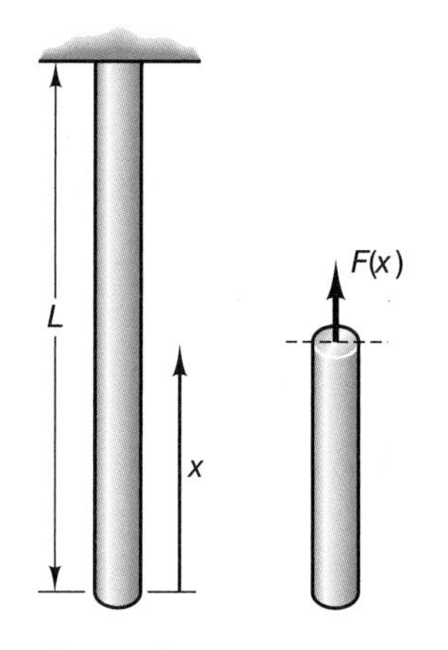

Figure 8.5

$$F(x) = Wx/L \quad \text{(Tension)}$$

Note that the internal force in the bar varies linearly from the bottom of the bar to the top.

The two bars shown in Figures 8.3 and 8.5 are axially loaded, and the internal forces in the bar are in tension or compression only. Next, consider the bar shown in Figure 8.6. The reactions at the wall are determined first, for equilibrium of the entire bar or beam. These reactions are

$$\sum F_y = 0 \qquad R = P$$

$$\sum M_{\text{wall}} = 0 \qquad M_w = PL$$

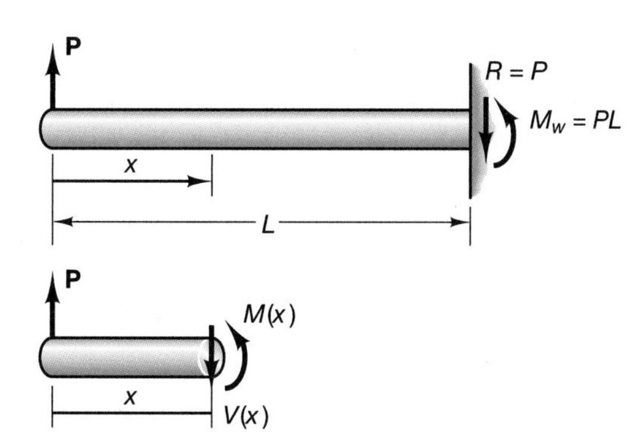

Figure 8.6

A long, slender member loaded perpendicular to its long axis is called a beam, the internal force V is called the ***shear force***, and the internal moment M is called a ***bending moment***. The internal force and moment at x are obtained by passing an imaginary cut through the beam at x and summing forces and moments at that point. Let $V(x)$ and $M(x)$ denote the shear force and bending moment, respectively, at any point x in the beam. Then

$$V(x) = P$$
$$M(x) = Px$$

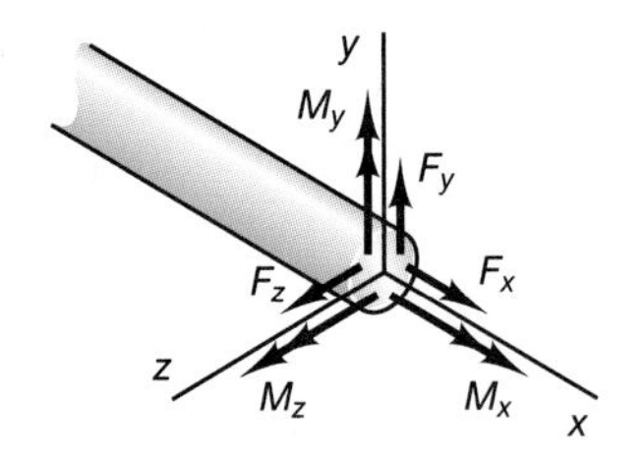

Figure 8.7

Note that the internal shear force is constant everywhere in this beam, and the bending moment increases linearly from zero at the left end to a maximum value equal to PL at the right end. The deformation and the internal stresses in the beam can now be determined at every point in it.

In general, there will be three components of the internal force and three components of the internal moment. These components are usually expressed in a coordinate system with origin at the centroid of the cross section and one coordinate axis taken perpendicular to the cross section, as shown in Figure 8.7.

The force component in the x-direction is the internal *axial force* in the bar, and the moment component in this direction is called a ***twisting or torsional moment***. The force components in the y- and z-directions are shear forces, and the moment components in these directions are bending moments.

Sample Problem 8.1

Determine the internal loading in the crane hook shown below at the leftt at any point R, θ. The crane hook can be modeled as a two-dimensional body in this case.

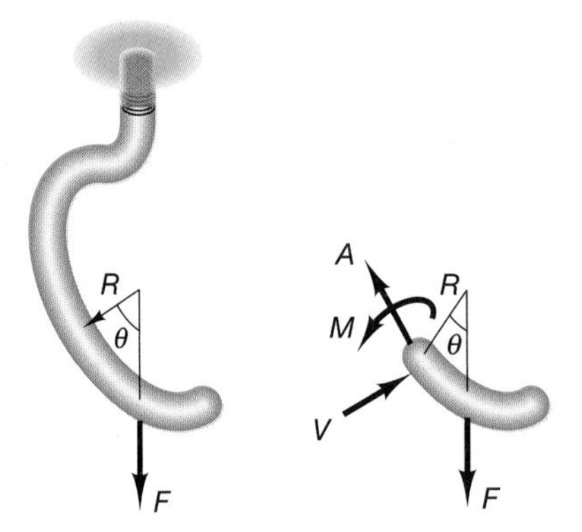

Solution We cut the hook with an imaginary plane at an angle θ and construct a free-body diagram of the part to the right of the cut. (See diagram above and to the right.) The equations of equilibrium yield

$$A = F \sin \theta$$
$$V = F \cos \theta$$
$$M = FR \sin \theta$$

The maximum shear load occurs when the angle of the cut plane is zero, and the maximum axial load and bending moment occur at an angle of 90°.

Sample Problem 8.2

Determine the internal forces and moments in the bar built into the foundation as shown in the figure at the top left and loaded with an external force P as functions of the x–y coordinates of the bar.

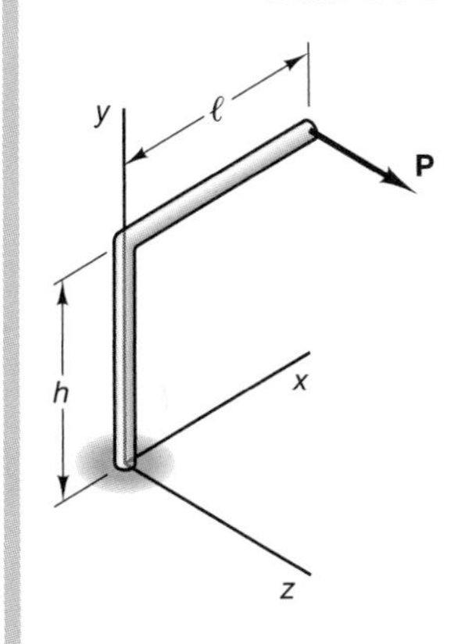

Solution We will need to make two imaginary cuts in this structure to determine the internal forces in each part. The internal forces and moments will be determined by writing the appropriate vector equations. We construct free-body diagrams of the bar after passing a section first through the horizontal portion of the bar and then through the vertical portion. (See diagram at left.) The equilibrium equations yield

$$\mathbf{R} + \mathbf{P} = \mathbf{0}$$
$$\mathbf{M} + \mathbf{r_p} \times \mathbf{P} = \mathbf{0}$$
$$\mathbf{P} = P\hat{\mathbf{k}}$$
$$\mathbf{r_p} = (l - x)\,\hat{\mathbf{i}}$$

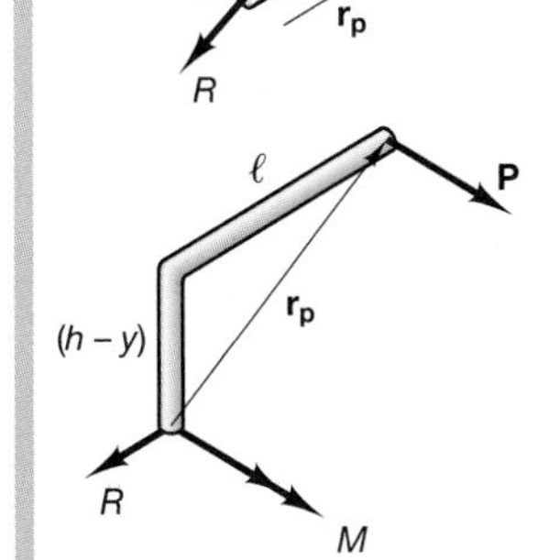

Therefore,

$$\mathbf{R} = -P\hat{\mathbf{k}}$$
$$\mathbf{M} = P(l - x)\hat{\mathbf{j}}$$

For the vertical portion of the bar;

$$\mathbf{r_p} = l\,\hat{\mathbf{i}} + (h - y)\hat{\mathbf{j}}$$

Thus,

$$\mathbf{R} = -P\hat{\mathbf{k}}$$
$$\mathbf{M} = Pl\hat{\mathbf{j}} - P(h - y)\hat{\mathbf{i}}$$

Sample Problem 8.3

Determine the internal forces and moments for any force F $= F_x\hat{\mathbf{i}} + F_y\hat{\mathbf{j}} + F_z\hat{\mathbf{k}}$ acting on the crane hook as shown at the left.

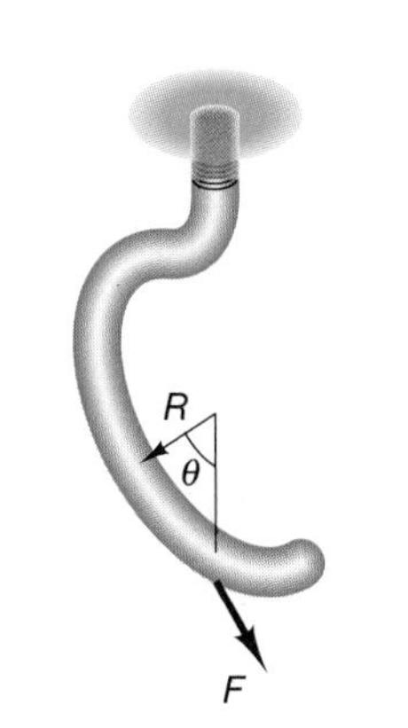

Solution We will use a vector solution and cylindrical coordinates. The cylindrical coordinate system is defined by the three unit vectors $\hat{\mathbf{e}}_\mathbf{r}$, $\hat{\mathbf{e}}_\theta$, and $\hat{\mathbf{k}}$. The unit vector $\hat{\mathbf{k}}$ acts into the paper in this case, so that a right-handed coordinate system r, θ, z is formed, as shown in the diagram at the top of the next page. The following vector relations exist between the unit vectors:

$$\hat{\mathbf{e}}_\mathbf{r} \times \hat{\mathbf{e}}_\theta = \hat{\mathbf{k}}$$
$$\hat{\mathbf{e}}_\theta \times \hat{\mathbf{k}} = \hat{\mathbf{e}}_\mathbf{r}$$
$$\hat{\mathbf{k}} \times \hat{\mathbf{e}}_\mathbf{r} = \hat{\mathbf{e}}_\theta$$

The cylindrical unit vectors may be expressed in terms of rectilinear unit vectors as

$$\hat{\mathbf{e}}_\mathbf{r} = \cos\theta\hat{\mathbf{i}} + \sin\theta\hat{\mathbf{j}}$$
$$\hat{\mathbf{e}}_\theta = -\sin\theta\hat{\mathbf{i}} + \cos\theta\hat{\mathbf{j}}$$
$$\hat{\mathbf{k}} = \hat{\mathbf{k}}$$

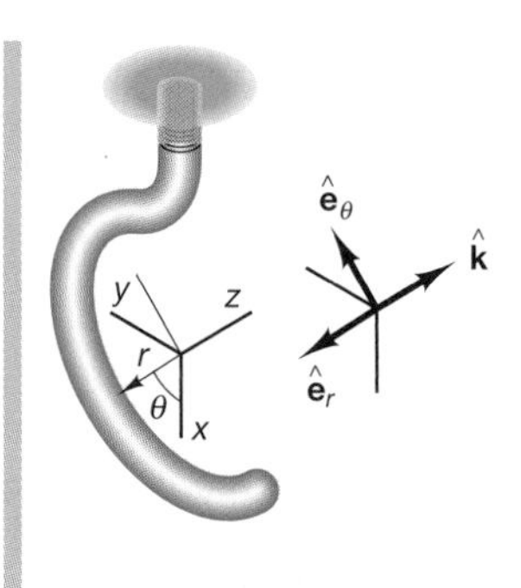

Now consider an arbitrary load applied to the crane hook where **Q** and **M** are the internal force and moment, respectively. (See diagram at left.) Summation of forces yields

$$\mathbf{Q} + \mathbf{F} = 0 \mapsto \mathbf{Q} = -\mathbf{F}$$

Summation of moments yields

$$\mathbf{M} + \mathbf{r}_{A/C} \times \mathbf{F} = 0 \mapsto \mathbf{M} = -\mathbf{r}_{A/C} \times \mathbf{F}$$

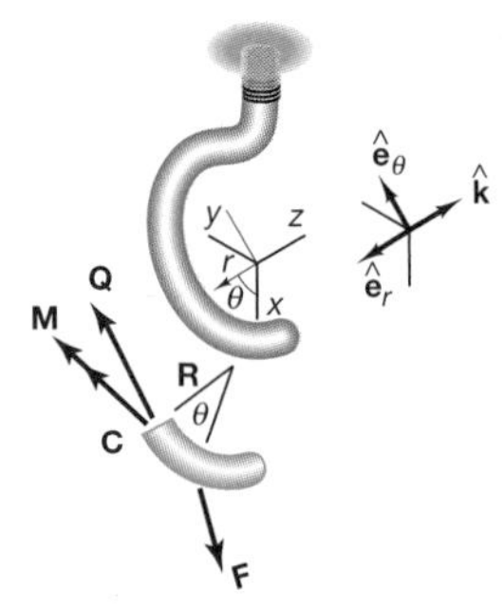

The relative position vector from C to A is

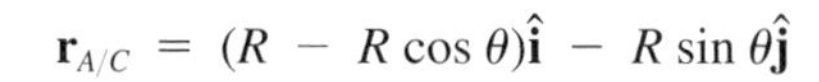

$$\mathbf{r}_{A/C} = (R - R\cos\theta)\hat{\mathbf{i}} - R\sin\theta\hat{\mathbf{j}}$$

Therefore,

$$\mathbf{Q} = -F_x\hat{\mathbf{i}} - F_y\hat{\mathbf{j}} - F_z\hat{\mathbf{k}}$$

$$\mathbf{M} = -\{(R - R\cos\theta)\hat{\mathbf{i}} - R\sin\theta\hat{\mathbf{j}}\} \times \{(F_x\hat{\mathbf{i}} + F_y\hat{\mathbf{j}} + F_z\hat{\mathbf{k}}\}$$

$$= + F_zR\sin\theta\hat{\mathbf{i}} + F_z(R - R\cos\theta)\hat{\mathbf{j}} + \{F_xR\sin\theta - F_y(R - R\cos\theta)\}\hat{\mathbf{k}}$$

These are the internal force and moment acting on the cut section. However, to interpret this force and moment, we need to express them in cylindrical coordinates. We may do so by taking the dot product with each of the cylindrical coordinate base vectors:

$$A(\theta) = \mathbf{Q}\cdot\hat{\mathbf{e}}_\theta = \{-F_x\hat{\mathbf{i}} - F_y\hat{\mathbf{j}} - F_z\hat{\mathbf{k}}\}\cdot\{-\sin\theta\hat{\mathbf{i}} + \cos\theta\hat{\mathbf{j}}\}$$
$$= F_x\sin\theta - F_y\cos\theta$$

$$V_r(\theta) = \mathbf{Q}\cdot\hat{\mathbf{e}}_\mathbf{r} = \{-F_x\hat{\mathbf{i}} - F_y\hat{\mathbf{j}} - F_z\hat{\mathbf{k}}\}\cdot\{\cos\theta\hat{\mathbf{i}} + \sin\theta\hat{\mathbf{j}}\}$$
$$= -F_x\cos\theta - F_y\sin\theta$$

$$V_z(\theta) = \mathbf{Q}\cdot\hat{\mathbf{k}} = -F_z$$

$$M_\theta(\theta) = \mathbf{M}\cdot\hat{\mathbf{e}}_\theta = \{+F_zR\sin\theta\hat{\mathbf{i}} + F_z(R - R\cos\theta)\hat{\mathbf{j}} + \{F_xR\sin\theta - F_y(R$$
$$- R\cos\theta\}\hat{\mathbf{k}}\}\cdot\{-\sin\theta\hat{\mathbf{i}} + \cos\theta\hat{\mathbf{j}}\} = F_z[-R\sin^2\theta + (R - R\cos\theta)\cos\theta]$$

$$M_r(\theta) = \mathbf{M}\cdot\hat{\mathbf{e}}_\mathbf{r} = \{F_zR\sin\theta\hat{\mathbf{i}} + F_z(R - R\cos\theta)\hat{\mathbf{j}} + \{F_xR\sin\theta - F_y(R$$
$$- R\cos\theta)\}\hat{\mathbf{k}}\}\cdot\{\cos\theta\hat{\mathbf{i}} + \sin\theta\hat{\mathbf{j}}] = F_z(R\sin\theta - 2R\sin\theta\cos\theta)$$

$$M_z(\theta) = \mathbf{M}\cdot\hat{\mathbf{k}} = \{+F_zR\sin\theta\hat{\mathbf{i}} + F_zR\hat{\mathbf{j}}$$
$$+ \{F_xR\sin\theta - F_y(R - R\cos\theta)\}\hat{\mathbf{k}}]\cdot\hat{\mathbf{k}} = \{F_xR\sin\theta - F_y(R - R\cos\theta)\}$$

The internal stresses in the crane hook can now be calculated using the mechanics of deformable bodies. Although the vector algebra in these calculations is involved, the approach is straightforward.

Problems

8.1 Compute the reaction force R and the internal forces in the bar shown in Figure P8.1 at points a and b. Indicate whether the section is in tension or compression.

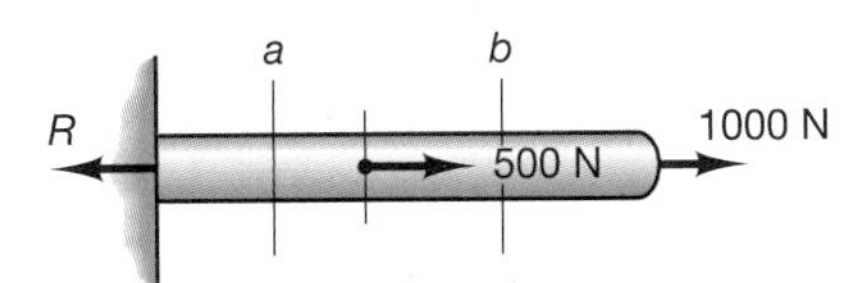

Figure P8.1

8.2 Compute the reaction force R, the internal force at a, and the internal force at b in the bar illustrated in Figure P8.2.

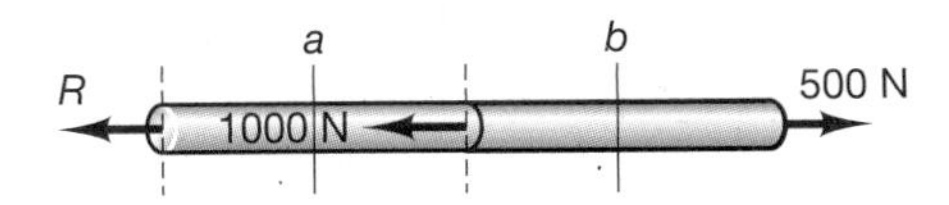

Figure P8.2

8.3 A steel bar of uniform weight of 1000 N is compressed by a 300-N force, as illustrated in Figure P8.3. Compute the reaction at the point of attachment and the internal force in the bar at every point x.

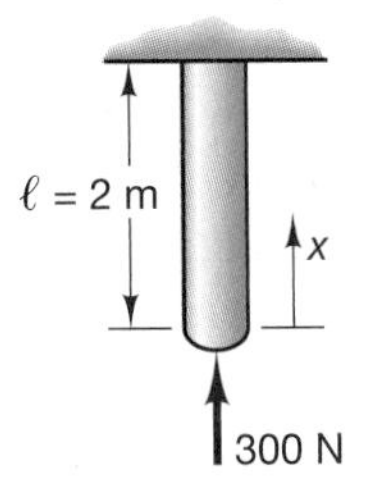

Figure P8.3

8.4 Compute the internal moment and force at points A and B for the system shown in Figure P8.4. Also, compute the reaction forces at the fixed end.

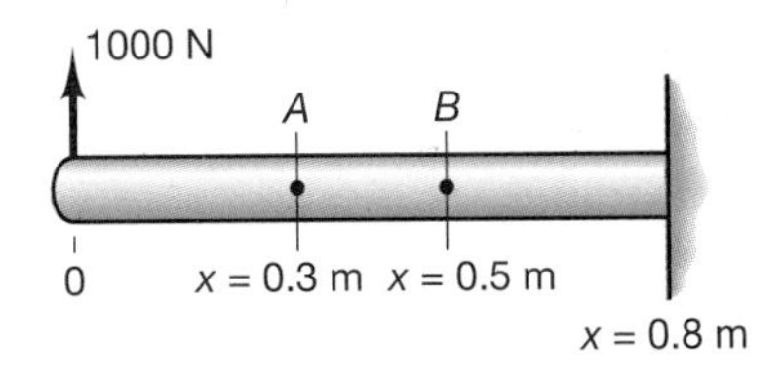

Figure P8.4

8.5 A bar is fixed at one end and is loaded as illustrated in Figure P8.5. Ignoring the weight of the bar, compute the internal force and moment at the points A and B.

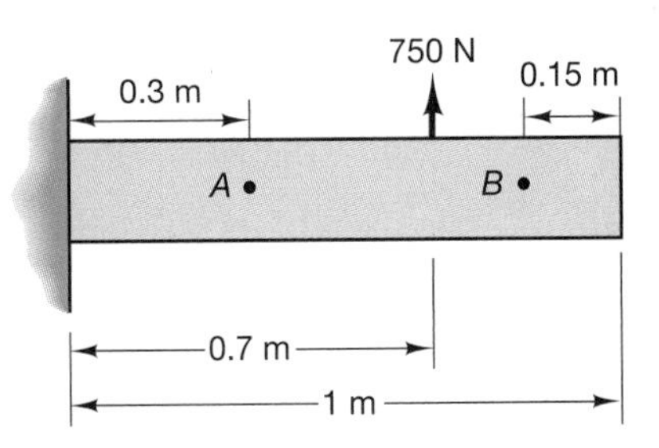

Figure P8.5

8.6 Compute the internal moment and force at the midpoint for the system of Figure P8.6.

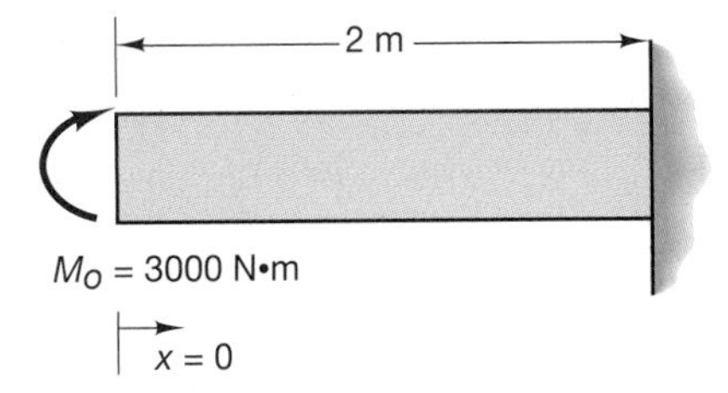

Figure P8.6

8.7 Compute the internal moment and force at a point 0.5 m from the left end for the system of Figure P8.6.

8.8 Compute the internal moment and force at a point 0.5 m from the left end for the system of Figure P8.8.

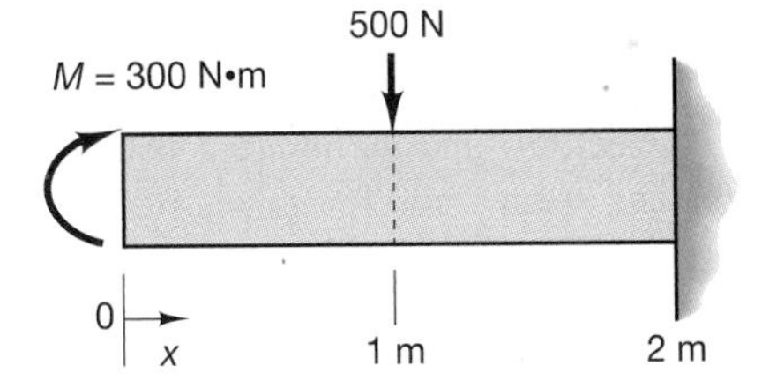

Figure P8.8

8.9 Compute the internal moment and force at a point 0.9 m from the left end for the beam of Figure P8.8.

8.10 Compute the moment and force for a point 0.9 m from the right end for the system of Figure P8.8.

8.11 Compute the moment and force at point A for the beam in Figure P8.11.

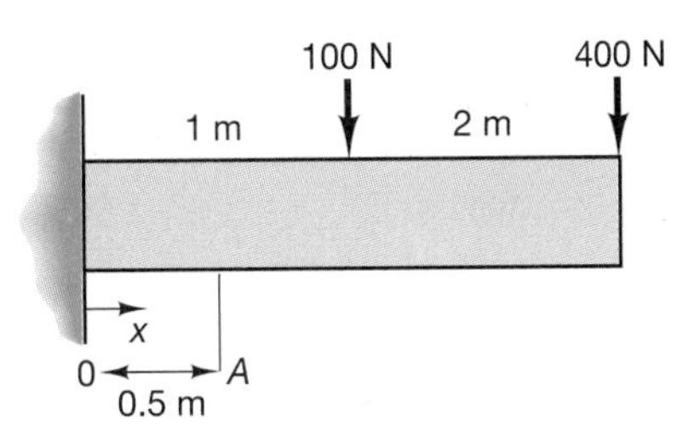

Figure P8.11

8.12 Compute the shear force, axial force, and bending moment in the hook shown in Figure P8.12 for θ between 180° and 225°.

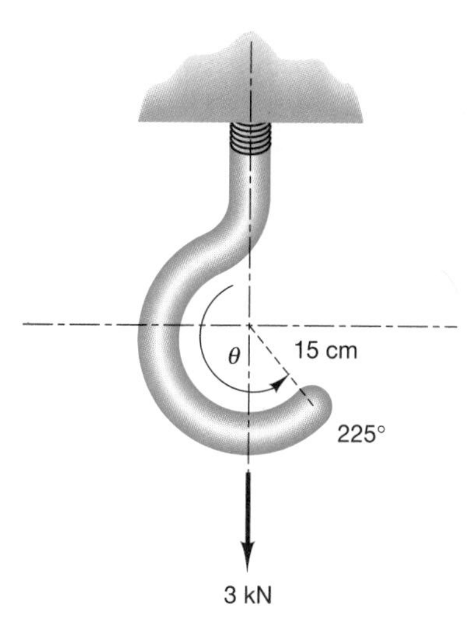

Figure P8.12

8.13 Following Sample Problem 8.1, write the value of the shear force, axial force, and bending moment for the hook of Figure P8.12 for all values of θ between 0 and 225°. Also, compute the reaction forces at the fixed connection point.

8.14 For the hook of Figure P8.14, calculate the reaction forces at the fixed connections and the internal bending moment, axial force, and shear force for values of θ between $0° < \theta < 225°$.

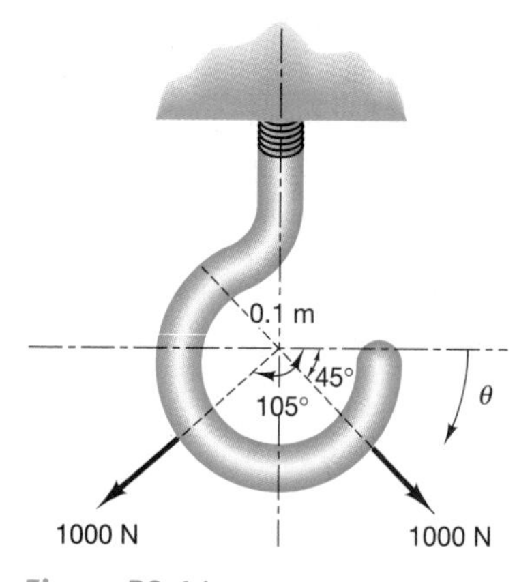

Figure P8.14

8.15 Compute the internal forces and moments at the fixed connection of the system in Figure P8.15.

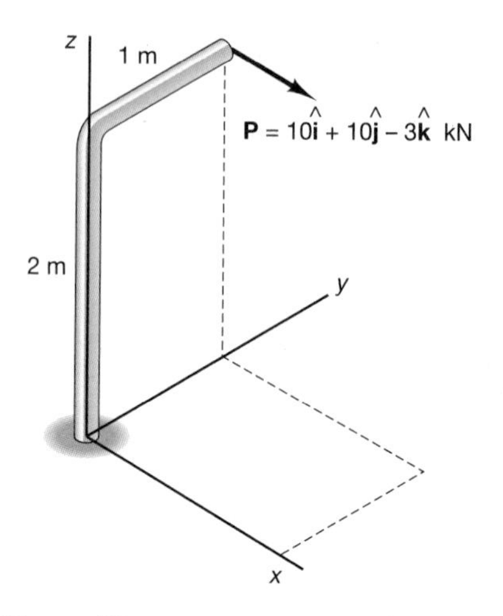

Figure P8.15

8.16 Determine the internal forces and moments in the system of figure P8.15 in terms of (a) an arbitrary point along the vertical bar and (b) an arbitrary point along the horizontal bar. (c) In each case, identify the resultant force and moment components with the terminology associated with Figure 8.7.

8.17 A rope is tied to a metal fixture, as shown in Figure P8.17. (a) Compute the reaction at the fixed connection and the internal forces in the fixture for values of θ between 0 and 170°. The rope is fixed at the point defined by $\theta = 60°$. (b) Compute the two shear components (bending) and one axial (torsional) component of the resultant force and the moment.

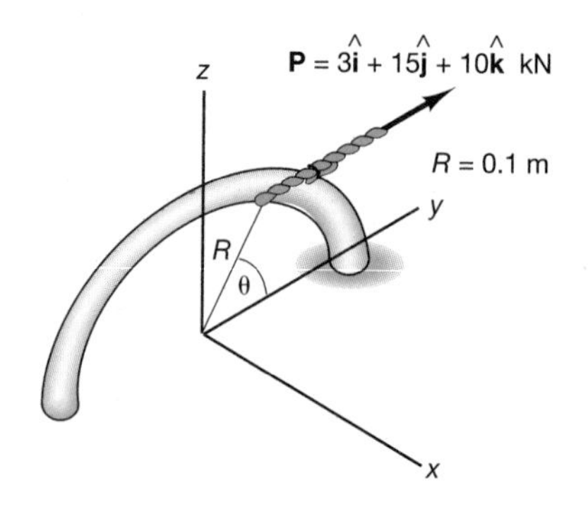

Figure P8.17

8.3 TYPES OF LOADING AND SUPPORTS IN BEAMS

A long, slender structural member that is designed to support loads at different points along its length is called a *beam.* In the simplest application, the loads are perpendicular to the long axis of the beam and parallel or perpendicular to an axis of symmetry of the cross section of the beam. (See Chapter 10 for a discussion of cross-sectional properties of beams.) Beams can support distributed loads or loads that may be treated as *concentrated* at a single point of application, as shown in Figure 8.8. Note that a concentrated load is an approximation that is made when the area over which the load is distributed is small compared to the dimensions of the beam.

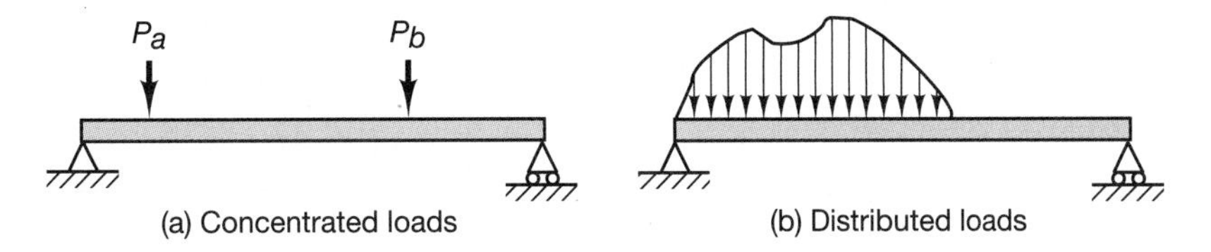

Figure 8.8

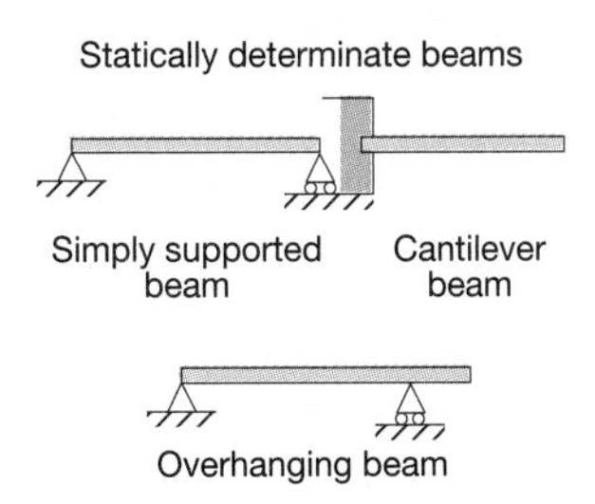

Figure 8.9

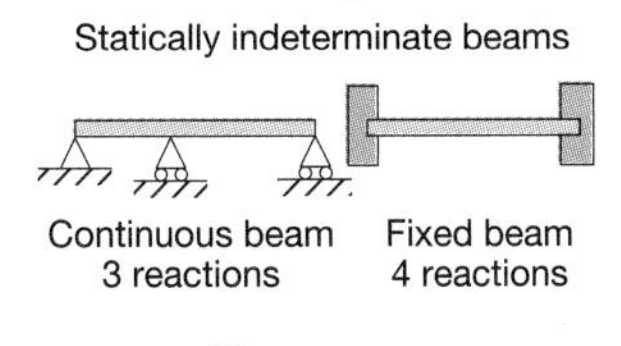

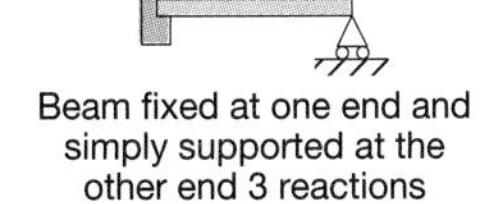

Beam fixed at one end and simply supported at the other end 3 reactions

Figure 8.10

The concentrated loads have units of newtons or pounds, or multiples of these, such as kilonewtons or kilopounds. The distributed loads are expressed in force per unit length—for example, N/m or lb/ft. Distributed loads are usually given as a function of the position on the long axis of the beam. In Chapter 5, we replaced distributed loads by equivalent force systems when we computed the reactions. When the internal forces and moments are desired, the distribution of the load must be considered. If the long axis of the beam is designated as the x-axis, with origin at the left end of the beam, the distributed load is written as $w(x)$. For example, if a beam of length L is loaded under a sinusoidal load, the distributed load function is

$$w(x) = p \sin(\pi x/L)$$

A beam under a uniform distributed load has a load distribution function

$$w(x) = \text{constant}$$

Note that $w(x)$ is treated as positive if the distributed load acts upward and y is positive upward.

Beams are classified according to the way they are supported and whether they are overconstrained and therefore are statically indeterminate. Several types of beam boundary conditions are shown in Figures 8.9 and 8.10. The beams have been modeled as two-dimensional structures with no axial loads acting on them. Therefore, only two equations of equilibrium can be written, and hence, only two reactions can be determined by static equilibrium.

8.4 SHEAR AND BENDING MOMENTS IN BEAMS

The beam shown in Figure 8.11 is loaded with a general distributed load $w(x)$ perpendicular to its long axis. The beam is sectioned at a point x measured from the left end. The internal shear force $V(x)$ can be found by integrating the distributed load from 0 to x. The internal bending moment $M(x)$ is taken about the point x, and the differential moment for

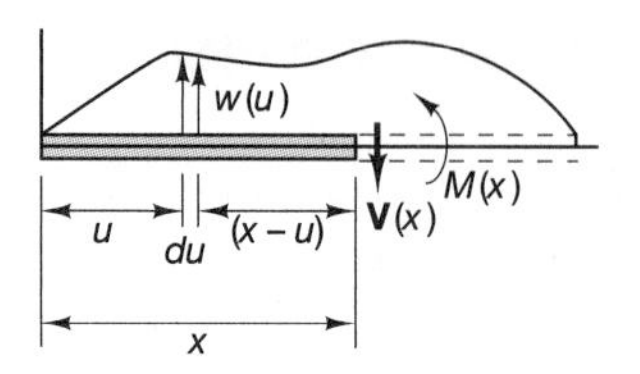

Figure 8.11

the differential load is $(x - u)\ w(u)\ du$. Integrating this differential moment from 0 to x yields the moment at the position x. Thus, we have

$$V(x) = \int_0^x w(u)du$$
$$M(x) = \int_0^x (x - u)w(u)du \tag{8.1}$$

If the loading function can be expressed as a function of x, then the integrals in Eq. (8.1) may be easily obtained. For example, consider the beams. First, a cantilever beam under a uniform load is

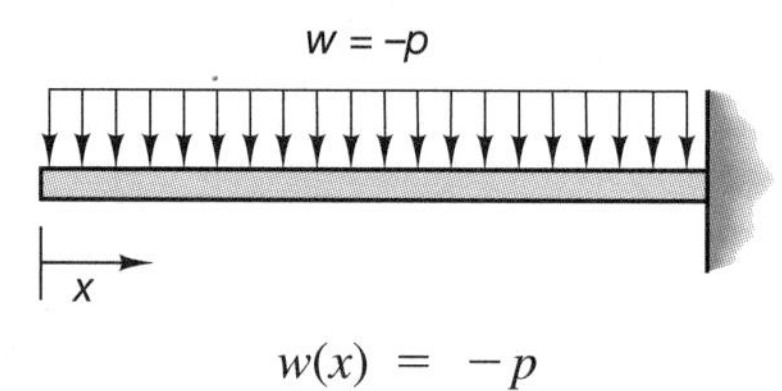

$$w(x) = -p$$

This would represent a beam loaded by its own weight or a flat deck beam loaded by snow. The shear at any position on the point of the beam is

$$V(x) = \int_0^x -pdu = -px$$

and the moment is

$$M(x) = \int_0^x (x - u)(-p)du = -px^2 + p\frac{x^2}{2} = -p\frac{x^2}{2}$$

Now, consider a beam loaded with a linearly increasing load such that

$$w(x) = kx$$

where k is the slope of the loading curve, given in N/m^2. The shear at any position on the loading curve is

$$V(x) = \int_0^x kudu = \left[\frac{1}{2}ku^2\right]_0^x = \frac{1}{2}kx^2$$

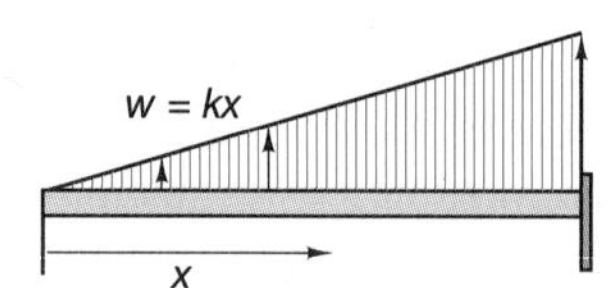

and the moment is

$$M(x) = \int_0^x (x - u)kudu = \left[\frac{1}{2}kxu^2 - \frac{1}{3}ku^3\right]_0^x = \frac{1}{6}kx^3$$

More difficult loadings may require the use of tables of integrals or analytical calculations. It is common to plot the shear and moment equations, and these graphs are called ***shear*** and ***moment diagrams***. Such diagrams were used in the past to graphically solve beam-loading problems before the advent of modern computational tools. They provide a conceptual view of the shear and moment distribution in the beam.

Sample Problem 8.4

Detemine the shear and moment equations for a ***cantilever beam*** loaded by a load function

$$\text{(a) } w(x) = kx \qquad \text{where } k = W/L$$

$$\text{(b) } w(x) = W\sin(\pi x/L)$$

where W is the maximum load intensity, given in N/m, and L is the length of the beam in meters. (See diagrams at left.)

Solution

The shear reaction at the wall is obtained from the integral relation

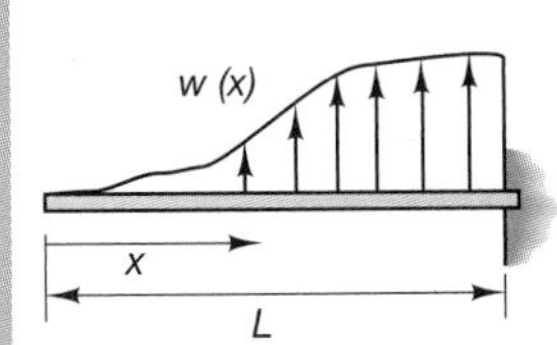

$$\text{(a) } R = \int_0^L w(u)du = \int_0^L \frac{W}{L}u\,du = \frac{Wu^2}{2L}\Big|_0^L = \frac{WL}{2}$$

$$\text{(b) } R = \int_0^L w(u)du = \int_0^L W\sin\left(\frac{\pi u}{L}\right)du = -\frac{WL}{\pi}\cos\left(\frac{\pi u}{L}\right)\Big|_0^L = \frac{2WL}{\pi}$$

The reaction moment at the wall is obtained from the integral relation

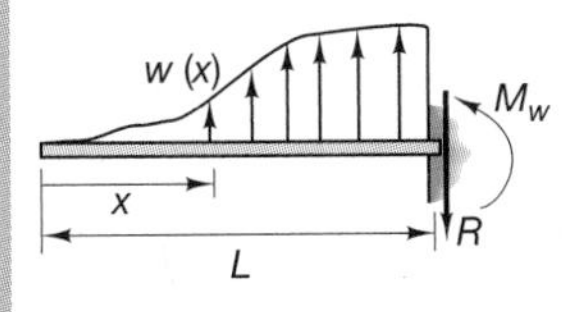

$$\text{(a) } M_w = \int_0^L (L-u)w(u)du = \int_0^L (L-u)\frac{Wu}{L}du = \frac{Wu^2}{2} - \frac{Wu^3}{3L}\Big|_0^L = \frac{WL^2}{6}$$

$$\text{(b) } M_w = \int_0^L (L-u)w(u)du$$

$$= \int_0^L (L-u)W\sin\left(\frac{\pi u}{L}\right)du = \left(\frac{WL^2}{\pi}\right)\cos\left(\frac{\pi u}{L}\right) - \frac{WL^2}{\pi^2}\sin\left(\frac{\pi u}{L}\right)$$

$$+ \frac{WL}{\pi}u\cos\left(\frac{\pi u}{L}\right)\Big)\Big|_0^L$$

$$= \frac{WL^2}{\pi}$$

The internal shear and moment are obtained using Eq. (8.1). In the discussion at the beginning of this section, these integrals were treated as definite integrals from 0 to x. They also may be treated as indefinite integrals with constants of integration that can be evaluated using the conditions that the shear and moment are zero at $x = 0$. The shear and moment equations for the two loadings are

(a)	(b)
$w(x) = kx$	$w(x) = W\sin\left(\pi\frac{x}{L}\right)$
$V(x) = k\frac{x^2}{2}$	$V(x) = -\frac{WL}{\pi}\left[\cos\left(\pi\frac{x}{L}\right) - 1\right]$
$M(x) = k\frac{x^3}{6}$	$M(x) = -\frac{WL}{\pi^2}\left[L\sin\left(\pi\frac{x}{L}\right) - \pi x\right]$

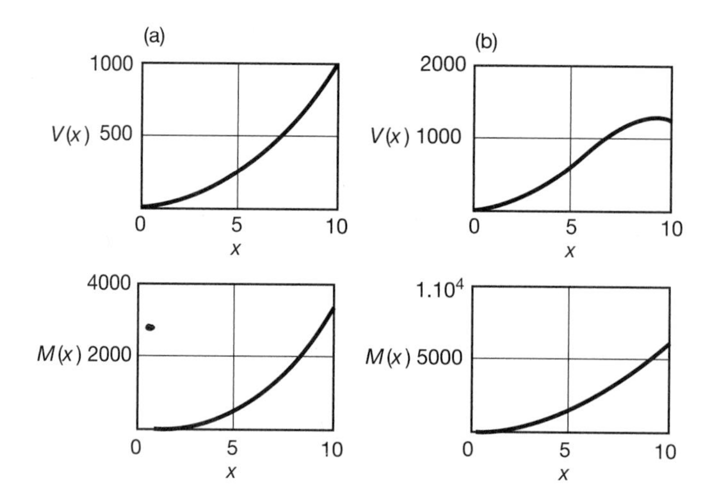

The respective shear and moment diagrams for $W = 200$ N/m and $L = 10$ m are as shown above. Graphs of this type can be obtained by using graphic calculators or computational software. Analytic and numerical solutions of Sample Problem 8.4 can be found in the Computational Supplement.

8.4.1 RELATIONSHIP BETWEEN THE LOAD DISTRIBUTION, THE SHEAR FORCE, AND THE BENDING MOMENT

Before examining the relationship between $w(x)$, $V(x)$, and $M(x)$, let us examine the differential of an integral. Consider an integral whose integrand and limits are both functions of a variable x, such as

$$\int_{a(x)}^{b(x)} F(x, u)du \tag{8.2}$$

Differentiating the integral with respect to x yields

$$\frac{d}{dx}\int_{a(x)}^{b(x)} F(x, u)du = \int_{a(x)}^{b(x)} \frac{\partial}{\partial x}[F(x, u)]du + F(x, b)\frac{db}{dx} - F(x, a)\frac{da}{dx} \tag{8.3}$$

where $\frac{\partial}{\partial x}$ denotes the partial derivative and $\frac{d}{dx}$ denotes the total derivative. If the shear function $V(x)$ is differentiated with respect to x, we obtain the relation

$$\frac{dV}{dx} = \frac{d}{dx}\int_0^x w(u)du$$

or

$$\frac{dV}{dx} = \int_0^x \frac{\partial}{\partial x}[w(u)]du + w(x)\frac{dx}{dx} - w(0)\frac{d0}{dx} = w(x) \tag{8.4}$$

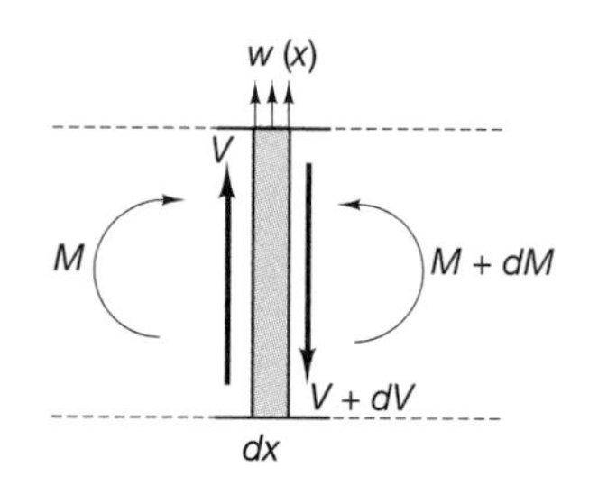

Figure 8.12

This relationship also can be derived by creating a free-body diagram of an element of the beam and writing the equilibrium equations for the element, as illustrated in Figure 8.12. Notice that the sign convention adopted for the distributed load is positive upward. The shear is positive if it is up on the left and down on the right, and the moment is positive if it is clockwise on the left and counterclockwise on the right. Summation of forces in the vertical direction yields

$$V - (V + dV) + w(x)dx = 0$$

or, after simplifying:

$$dV = w(x)dx \tag{8.5}$$

which is in agreement with Eq. (8.4). When loading and shear diagrams are drawn, it is clear that the area under the loading diagram is equal to the shear, and the slope of the shear diagram is equal to the load function. In the past, these observations were used to graphically solve for the internal shear in a beam. Since the load distribution is equal to the derivative of the shear, when the load is zero, the shear is a maximum or a minimum.

Differentiating the moment function $M(x)$ in Eq. (8.1) yields

$$\frac{dM}{dx} = \frac{d}{dx}\int_0^x (x - u)w(u)du = \int_0^x w(u)du + (x - x)w(x)$$
$$\frac{dM}{dx} = \int_0^x w(u)du = V(x) \tag{8.6}$$

The same result could have been obtained by summing moments about the right side of the element shown in Figure 8.12:

$$M - (M + dM) + V\,dx + [w(x)dx]dx/2 = 0$$

Combining terms and dropping the higher order term involving $(dx)^2$ to obtain a first-order approximation yields

$$dM = V(x)dx \tag{8.7}$$

Equation (8.7) is identical to Eq. (8.6), as expected. Equation (8.6) can be integrated to obtain the moment, and this may be used as an alternative method to Eq. (8.1). Examination of the shear and moment diagrams shows that the area under the shear diagram is equal to the moment, and the slope of the moment diagram at any point is equal to the shear. These observations were used to graphically solve for the internal moment, as was done with the shear. Recall from calculus that if the derivative of a function is zero (the slope is zero) at a point, the function is a local maximum or minimum at that point. Therefore, when the shear is equal to zero, the moment will be a local maximum or a minimum. The true maximum or minimum can also be at the endpoints.

Concentrated forces and moments are not continuous functions, and they present difficulties in creating the loading, shear, and moment equations. In general, these equations are written for different segments of the beam and are discontinuous at points of concentrated forces or moments. This discontinuity may be illustrated for the cantilever beam shown in Figure 8.13. The shear and moment equations for any point in the beam between 0 and a are

$$0 < x < a:$$
$$V(x) = P \tag{8.8}$$
$$M(x) = Px$$

For any point between a and b, the shear and moment equations are

$$a < x < b:$$
$$V(x) = P - F \tag{8.9}$$
$$M(x) = Px - F(x - a)$$

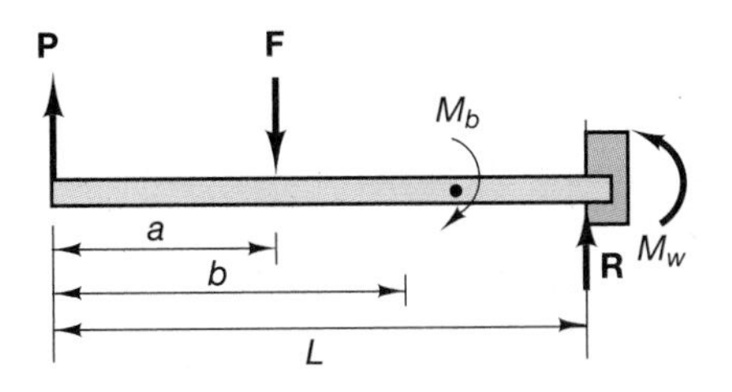

Figure 8.13

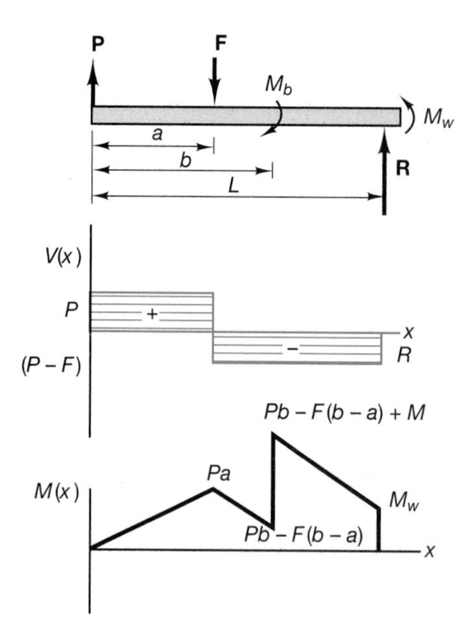

Figure 8.14

For any point between b and L, the shear and moment equations are

$$b < x < L:$$
$$V(x) = P - F \qquad (8.10)$$
$$M(x) = Px - F(x - a) + M_b$$

For points on the beam greater than L (within the wall), the internal shear and moment are assumed to be zero, and we have

$$x > L:$$
$$V(x) = P - F + R = 0 \quad \text{or} \quad R = F - P$$
$$M(x) = Px - F(x - a) - M_w + M_b + R(x - L) = 0 \qquad (8.11)$$

or

$$M_w = PL - F(L - a) + M_b$$

Equations (8.11) are the two equilibrium equations equivalent to the equation which says that the summation of forces in the vertical direction is equal to zero and the equation which says that the sum of the moments about any point $x > L$ is also equal to zero.

The shear and moment diagrams for this beam are shown in Figure 8.14. It is easily seen that the curves are discontinuous functions, and difficulties will arise if further calculations are required. These difficulties can be overcome by the introduction of special ***discontinuity functions***, discussed in Section 8.5.

Sample Problem 8.5

Write the shear and moment equations for the beam shown in the diagram at the top of the next page.

Solution

The load w N/m is assumed to be negative (acting downward) and to act on the first half of the beam. A free-body diagram of the beam is needed to determine the reactions. (See the second figure from the top on the next page.) In determining the reactions, the distributed load may be replaced by an equivalent force, as shown in Chapter 5.

Summation of forces in the vertical direction yields

$$R_L + R_R - wL/2 = 0$$

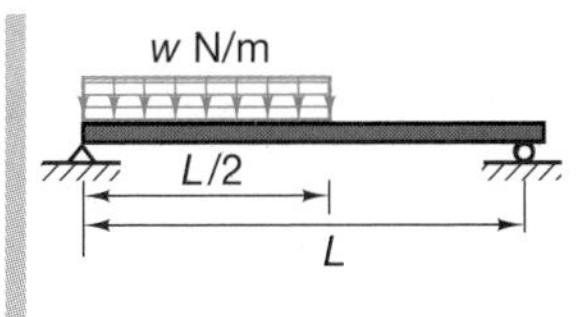

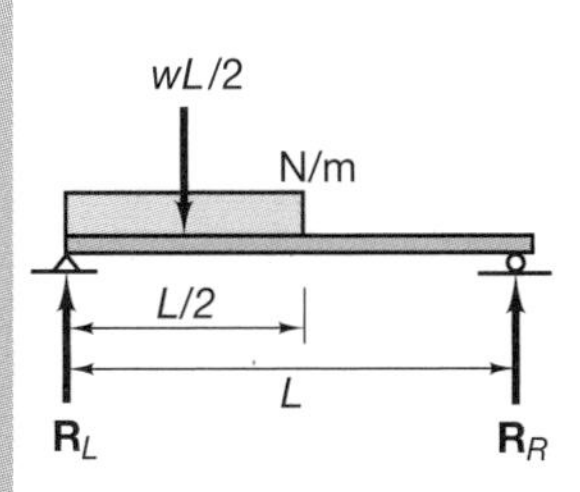

Summation of moments about the left end yields

$$R_R L - wL/2(L/4) = 0$$

Therefore,

$$R_R = +wL/8 \qquad R_L = +3wL/8 \text{ (both reactions are acting upward)}$$

Now we pass an imaginary cut through the beam for x between 0 and $L/2$ to determine the internal forces in that section:

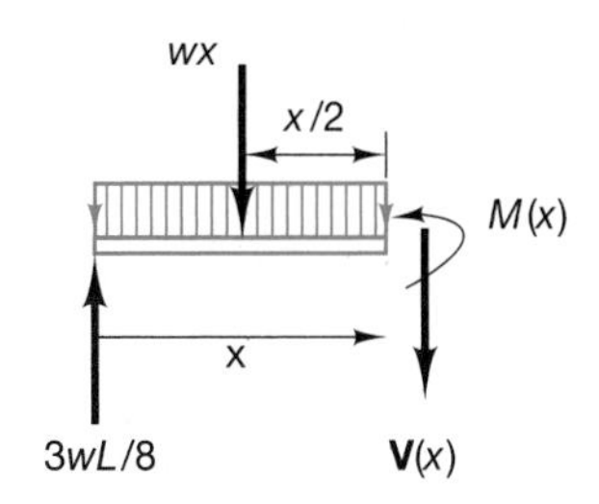

$$0 \leqslant x \leqslant \frac{L}{2}: \quad V(x) = +3wL/8 - wx$$

$$M(x) = +(3wL/8)x - wx^2/2$$

Note that the portion of the distributed load to the left of the cut is replaced by an equivalent load acting at the centroid of the distribution. A free-body diagram of the beam with a cut through the second half of the beam is needed to determine the internal shear and moment in that portion of the beam. The diagram is as follows:

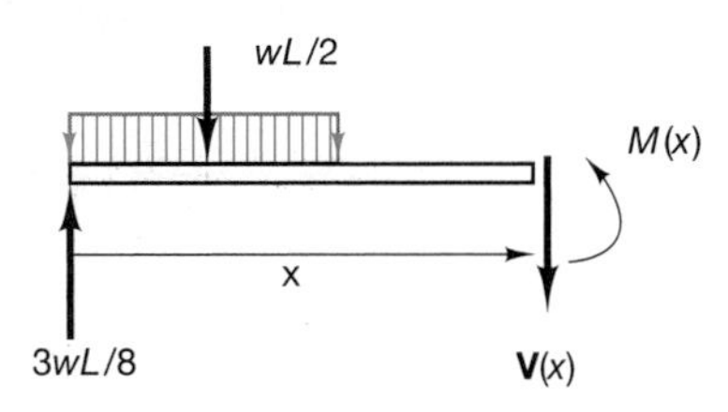

We have

$$\frac{L}{2} \leqslant x \leqslant L: \quad V(x) = +3wL/8 - wL/2$$

$$= -wL/8$$

$$M(x) = +(3wL/8)x - (wL/2)(x - L/4)$$

$$= -(wL/8)x + wL^2/8$$

Sample Problem 8.6

Write the shear and moment equations for the beam shown in the following diagram.

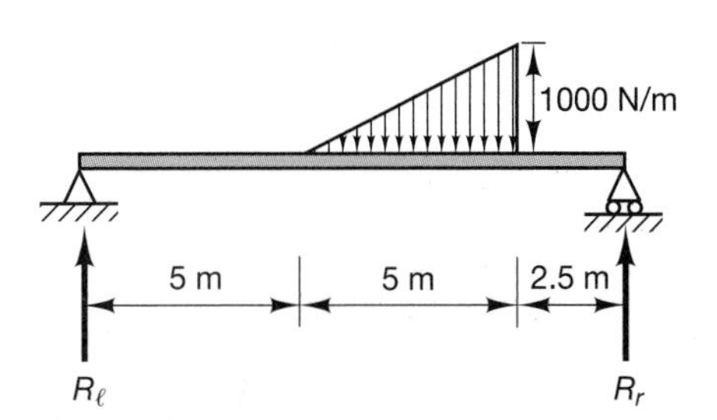

Solution The reactions should be determined first. Summing moments about the left end and noting that the distributed load can be considered to be acting through its centroid for equilibrium purposes yields

$$k = \frac{1000}{5} = 200$$

$$12.5R_r - (1/2 \times 1000 \times 5)(5 + 2/3 \times 5) = 0$$

$$R_r = 1667 \text{ N}$$

Summing forces in the vertical direction:

$$R_l + R_r - 1/2 \times 1000 \times 5 = 0$$

$$R_l = 833 \text{ N}$$

Sectioning and drawing a free-body diagram of the beam in each of three regions yields The shear and moment equations can be written in three parts:

$$0 < x < 5 \qquad V(x) = 833 \qquad M(x) = 833x$$

$$5 < x < 10{:} \qquad V(x) = 833 - 200\frac{(x-5)^2}{2} \qquad M(x) = 833x - 200\frac{(x-5)^3}{6}$$

$$10 < x < 12.5{:} \quad V(x) = 833 - 2500 = -1667 \qquad M(x) = 833x - 2500(x - 8.333) = -1667x + 20{,}833$$

(Slope of the loading curve)

The maximum moment occurs when $V(x)$ is zero. The shear is constant from zero to 5 meters and again from 10 meters to the end of the beam. Therefore, the zero value for the shear occurs in the second section of the beam. The quadratic equation for the shear must be set to zero, and the required value for x must be determined. The shear is zero at $x = 7.89$ m, and the moment at this point is 5768 N/m.

Now the moment and shear diagrams are easily drawn. However, all the necessary information is contained in the equations, and unless a graphical solution is going to be used for some other purpose, the shear and moment diagrams are not necessary. The shear and moment diagrams can be generated by using a graphic calculator or computational software and are shown in the Computational Supplement.

Problems

8.18 A pipe is simply supported at one end and effectively pinned at the other. (See Figure P8.18.) Compute the reaction forces and the internal forces, plot the shear and moment diagram, and state the maximum bending moment generated by hanging a mass at the midpoint of the pipe.

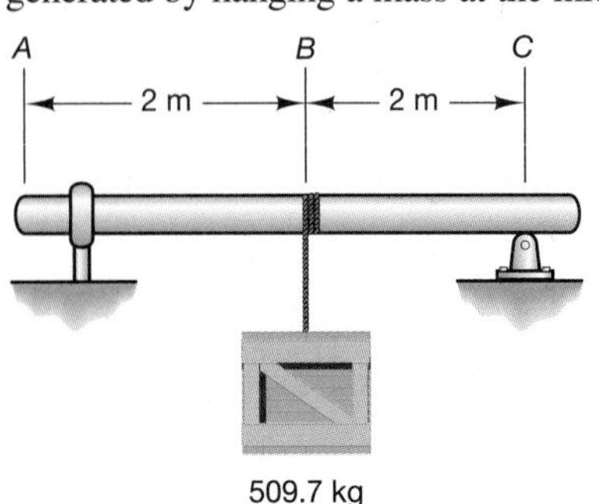

Figure P8.18

8.19 A pipe is simply supported at one end and effectively pinned at the other, as shown in Figure P8.19. Compute the reaction forces and the internal forces, plot the shear and moment diagram, and state the maximum bending moment generated by hanging a load at the point indicated.

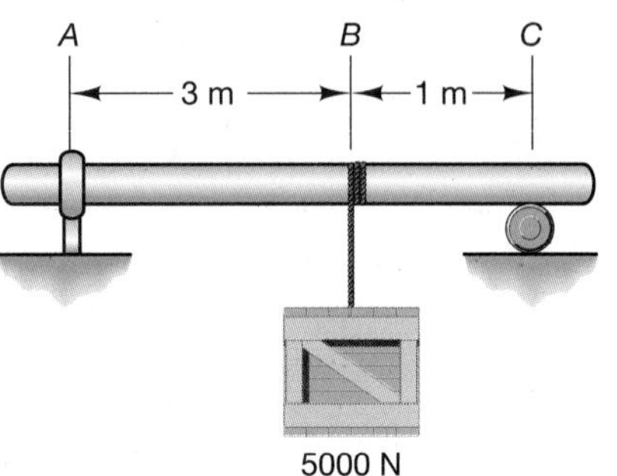

Figure P8.19

8.20 Determine the reactions at A and B and the shear and bending moment functions in terms of the unspecified, but fixed, value of the load P, position a, and length L for all points along the beam shown in Figure P8.20.

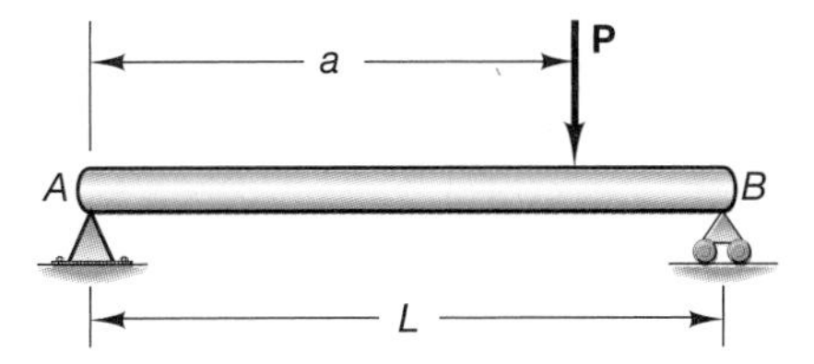

Figure P8.20

8.21 A builder wishes to use a 5-m beam to span two walls that must support the loads given in Figure P8.21. The building code states that the maximum bending moment for a beam of this construction (cross section and material) is 13 kN·m. Compute the reaction forces, shear, and moment functions, and plot the shear and moment functions. Does the beam satisfy the code?

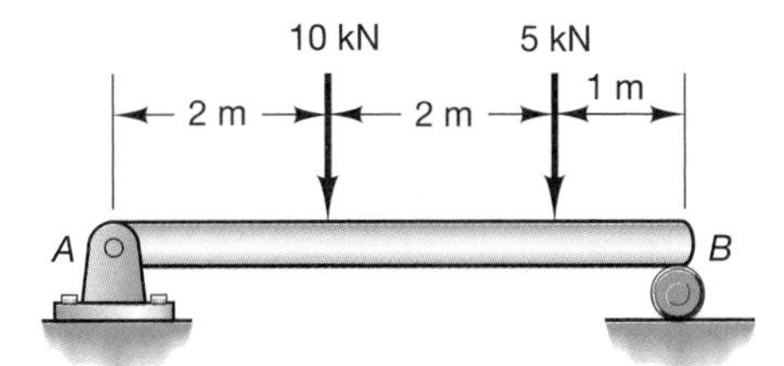

Figure P8.21

8.22 Compute the reaction forces and the shear and moments along the beam for the given load shown in Figure P8.22. What is the maximum value of the bending moment, and where in the beam does it occur?

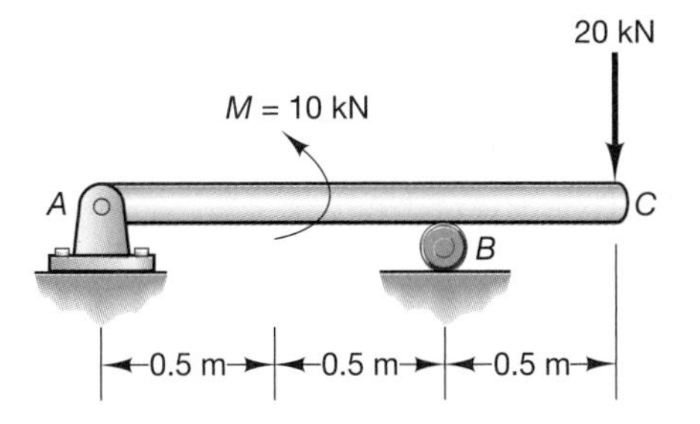

Figure P8.22

8.23 A machine located at the end of a support beam causes both a moment and a force to be applied to the beam. (See Figure P8.23.) Determine the reaction forces at A and the shear and bending moment functions in terms of the length L, applied force P_B, and applied moment M_B.

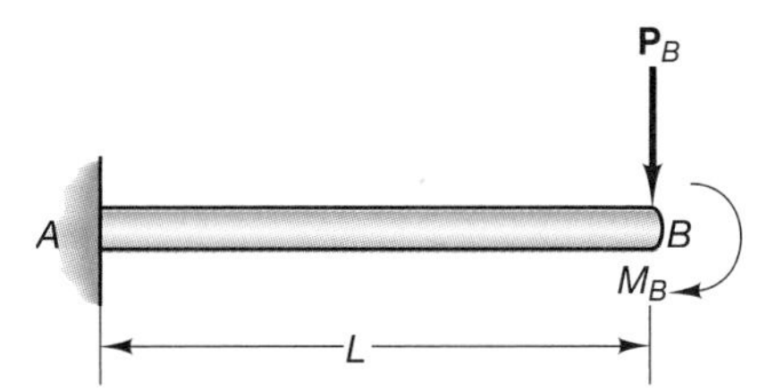

Figure P8.23

8.24 Determine the reaction forces and internal forces for the simple distributed load illustrated in Figure P8.24. Plot the shear and bending moment function against the distance x.

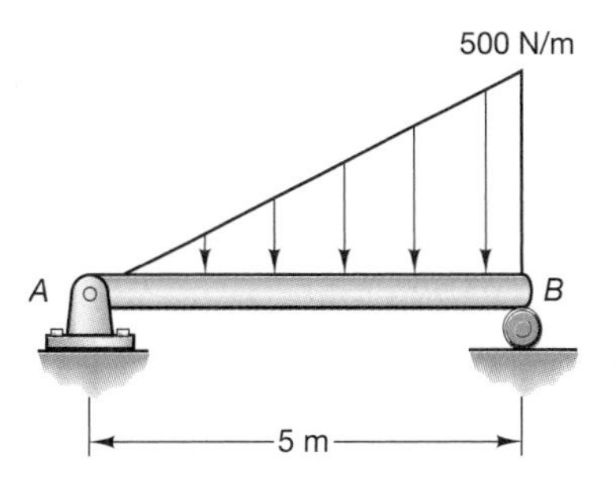

Figure P8.24

8.25 Determine the reaction forces and the internal forces and moments for the system illustrated in Figure P8.25. Plot the shear and bending moment functions against the distance x.

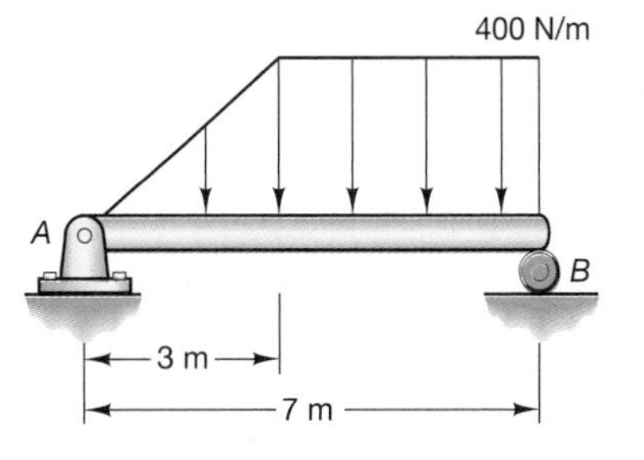

Figure P8.25

8.26 Determine the reaction forces and the internal forces and moments for the system shown in Figure P8.26. Plot the shear and bending moment function against the distance x.

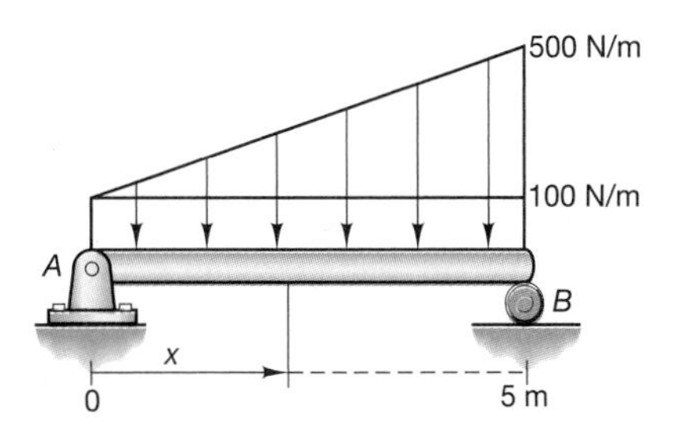

Figure P8.26

8.27 In Figure P8.27, determine the reaction forces and the internal shear force and bending moments for the load $w(x) = 10x^2$ N/m. Plot the shear and bending moment as a function of x.

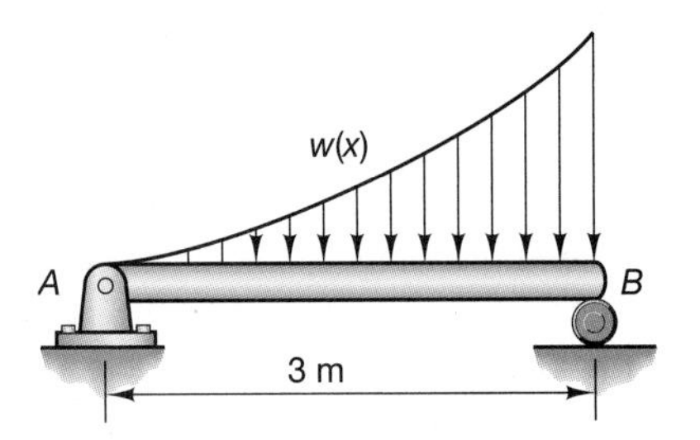

Figure P8.27

8.28 A cantilever beam is used as a crude model of a building. (See Figure P8.28.) Compute the reaction forces and the internal shear force and bending moment for a strong wind load modeled as $w(x) = (x^3 - 100x^2)\,10^{-4}$ N/m.

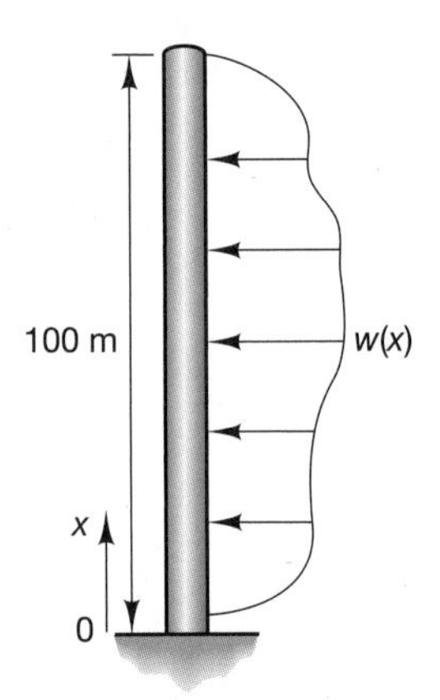

Figure P8.28

8.29 Repeat Problem 8.28 for the load $w(x) = 10(1 + \sin(\pi x/50))$ N/m.

8.30 Repeat Problem 8.28 with a wind load of $w(x) = 10(1 + \sin x)$ N/m.

8.31 A lifting load on an airplane wing is modeled by $w(x) = 1000x \sin(x)$ N/m, and the wing is modeled as a 10-m-long cantilevered beam. (See Figure P8.31.) Compute the reaction forces, as well as the internal shear force and bending moment as functions of x. Plot the shear and bending moment functions.

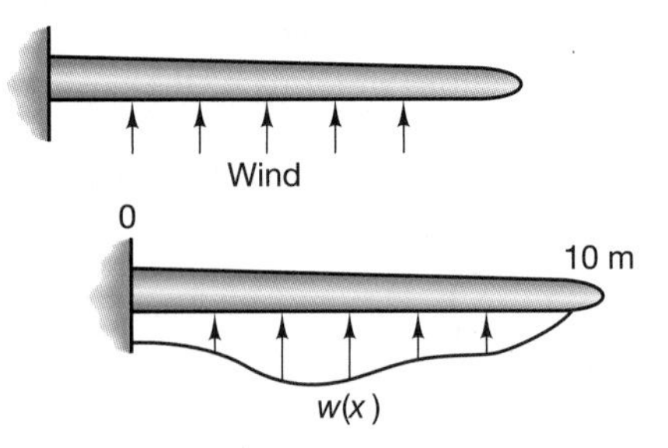

Figure P8.31

8.32 A snow load on a flat roof has the form $w(x) = 10(x \cos(\pi x/40))^2$ N/m. (See Figure P8.32.) Compute the reaction forces, and the internal shear force and bending moment, and plot the shear and bending moment as a function of x.

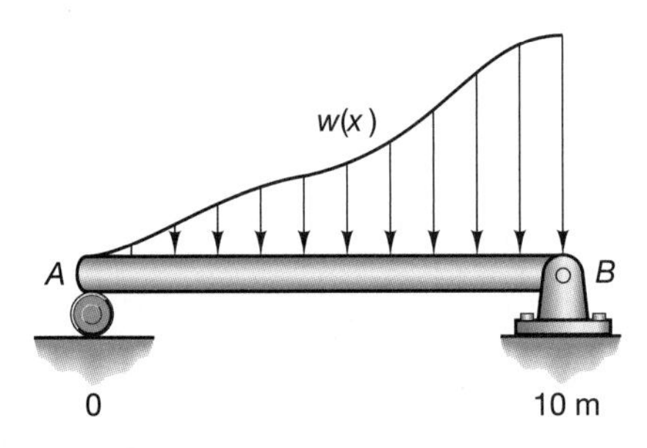

Figure P8.32

8.33 Compute the reaction forces and calculate the shear force and bending moment for each value of x along the beam shown in Figure P8.33.

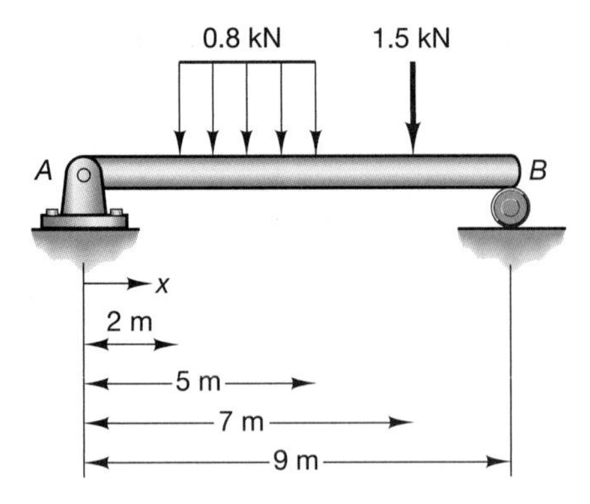

Figure P8.33

8.34 In Figure P8.34, compute the reaction forces, shear force, and bending moment for the combination of point load P and distributed load w_0, in terms of the constants a, b, L, P, and w_0 as a function of x. Assume that $a < b$.

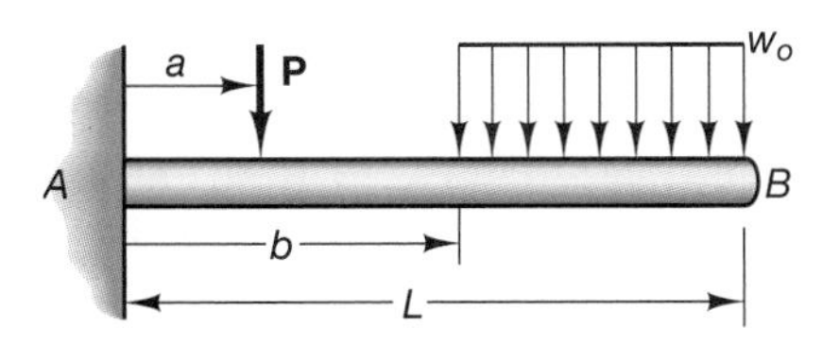

Figure P8.34

8.35 In Figure P8.35, compute the reaction forces, the shear force, and the bending moment function at each point x between the values of 0 and L, due to the applied distributed load and the applied moment M in terms of the constants a, b, L, w_0, and M. Assume that $a < b$.

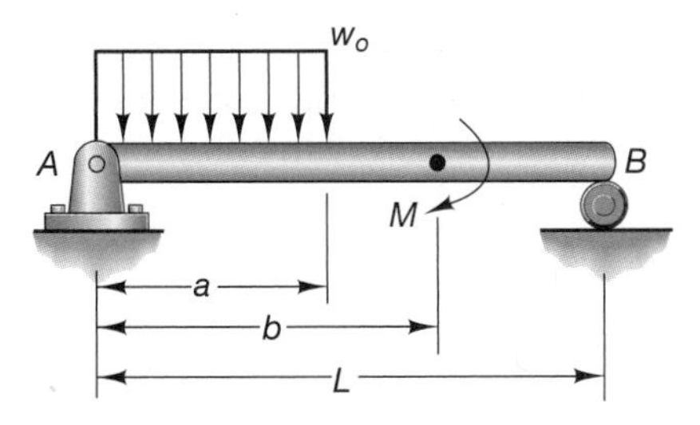

Figure P8.35

8.36 In Figure P8.36, compute the reaction forces, the shear force, and the bending moment due to the two applied point forces P and the distributed load w_0, in terms of the constants L, P, and w_0, for each value of $x = 0 < x < 6L$.

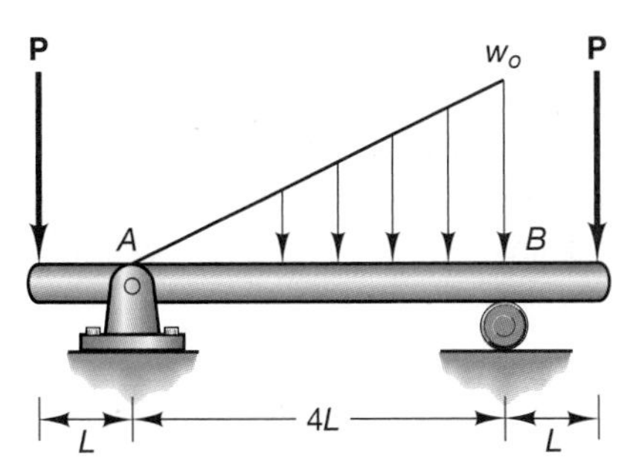

Figure P8.36

8.37 In Figure P8.37, compute the reaction forces, the shear force, and the bending moment due to the distributed load w_0 and the point load P, in terms of the constants a, b, L, P, and w_0, for each value of $x = 0 < x < L$. Assume that $a < b$.

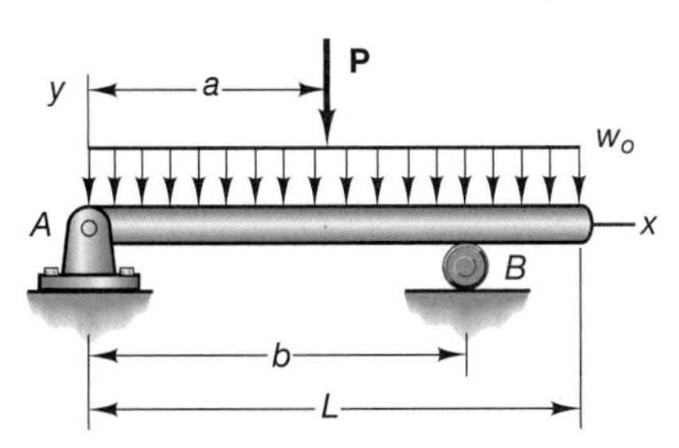

Figure P8.37

8.38 In Figure P8.38, compute the reaction forces, the shear force, and the bending moment due to the two distributed loads of strength w_0, in terms of the constants a and w_0, for all values of x between 0 and $4a$.

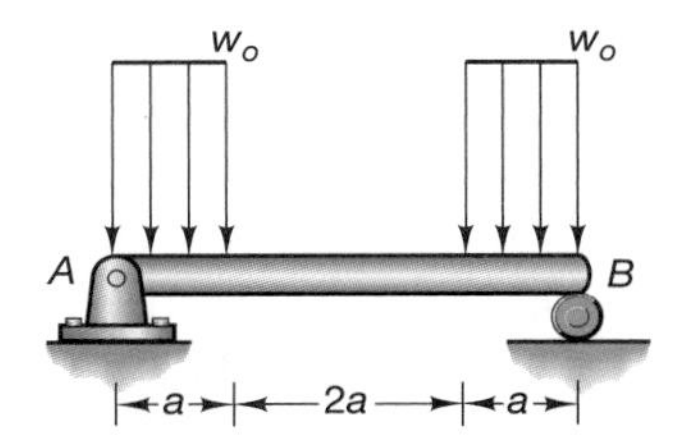

Figure P8.38

8.39 In Figure P8.39, compute the reaction forces, the shear force, and the bending moment due to the distributed load of intensity w_0, in terms of the constants L and w_0, for each value of x between 0 and L.

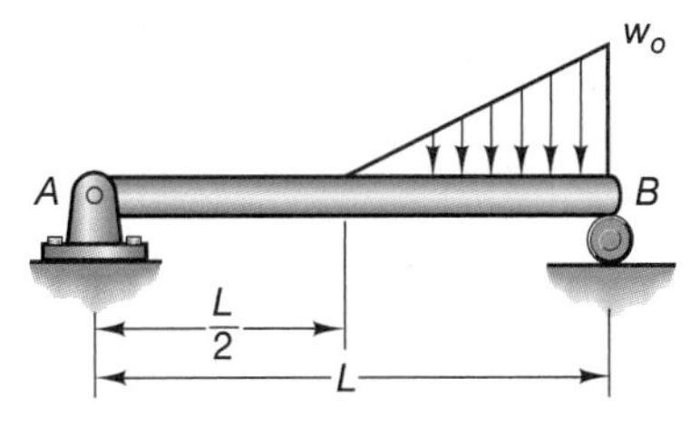

Figure P8.39

8.40 In Figure P8.40, compute the reaction forces, the shear force, and the bending moment due to the distributed load of intensity w_0, in terms of the constants L and w_0, for each value of x between 0 and L.

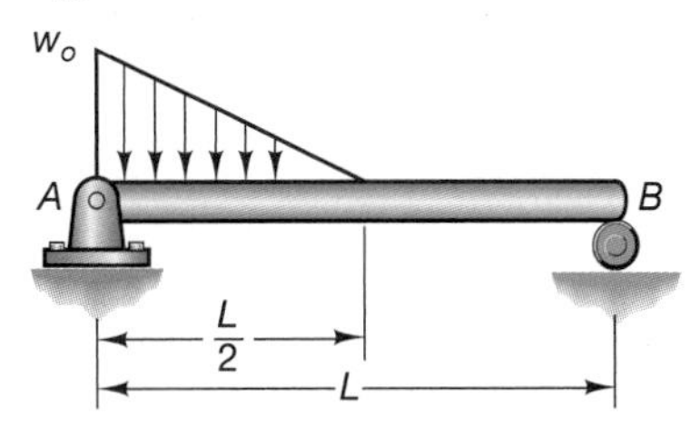

Figure P8.40

8.41 In Figure P8.41, compute the reaction forces, the shear force, and the bending moment due to the distributed load of intensity w_0, in terms of the constants L and w_0, for each value of x between 0 and L.

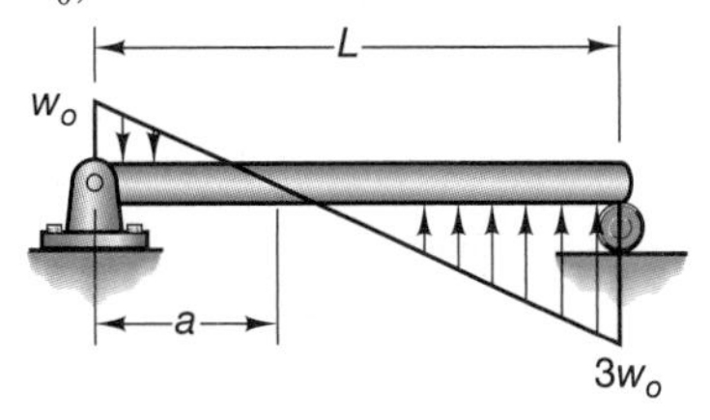

Figure P8.41

8.5 DISCONTINUITY FUNCTIONS FOR BEAM EQUATIONS

In Section 8.4, continuous load distributions could be easily expressed mathematically using Eq. (8.1), and the resulting shear and bending moment were continuous functions of x. When concentrated forces or concentrated moments are applied to a beam, discontinuities and singularities occur in the load distribution equation. In Section 8.4, the calculation of the shear force and the bending moment in the beam were obtained by handling each section of the beam separately. If the deflection of the beam is required, as in the case of deformable materials, these discontinuities will present computational difficulties.

In this section, we present an alternative way to handle discontinuous beam loadings by the use of discontinuity functions. Two different types of discontinuity functions will be used: those that are singular at the point of discontinuity (***singularity functions***) and those that are not singular at the point of discontinuity. The two singularity functions that will be used are the *unit doublet function* and the *Dirac delta function.* Nonsingular discontinuity functions such as the *Heaviside step function* or the *unit step function* will be developed from the singularity functions. The use of discontinuity functions for beams was first suggested by W. H. Macauley ("Note on the Deflection of Beams," *Messenger of Math.*, vol. 48, 129–130, 1919) and appeared in Crandal and Dahl's text, *An Introduction to the Mechanics of Solids* in 1959. These functions are therefore sometimes called Macauley discontinuity functions. The use of discontinuity functions allows all the beam loadings to be treated in an organized manner and eliminates the need to consider sections of the beam separately.

Concentrated loads have been used throughout this text, but a concentrated load must be considered to be distributed across a section of the beam of length zero. The intensity of the distributed load under a concentrated load would appear to be undefined or of infinite value. However, concentrated loads can be treated as distributed loads by the use of a singularity function called the ***Dirac delta function***. Paul A. M. Dirac defined a function that is zero everywhere except at $x = a$, and that has an area of unity. The Dirac delta function and its integral are defined as

$$\langle x - a \rangle_{-1} = \delta(x - a) = 0 \quad x \neq a$$

$$\int_{-\infty}^{x} \langle \zeta - a \rangle_{-1} \, d\zeta = \begin{cases} 0 & x < a \\ 1 & x > a \end{cases} \tag{8.12}$$

where a is the point at which the singularity occurs—that is, where the concentrated load will be applied. The Dirac delta function is not defined at $x = a$, but it is defined for all values less or greater than a. Note that two notations have been used for this function. The first, with the brackets subscripted by (-1), is commonly used in texts on the mechanics of materials, while the use of the delta to indicate this function is more common in other areas of mathematics. Figure 8.15 shows the Dirac delta function, with an infinite magnitude at point a and a zero width, but the area under the function is $\infty \cdot 0 = 1$. Although it may appear strange at first, this type of product between infinity and zero occurs elsewhere in mathematics and is usually encountered in a first calculus course. Consider, for example, the function

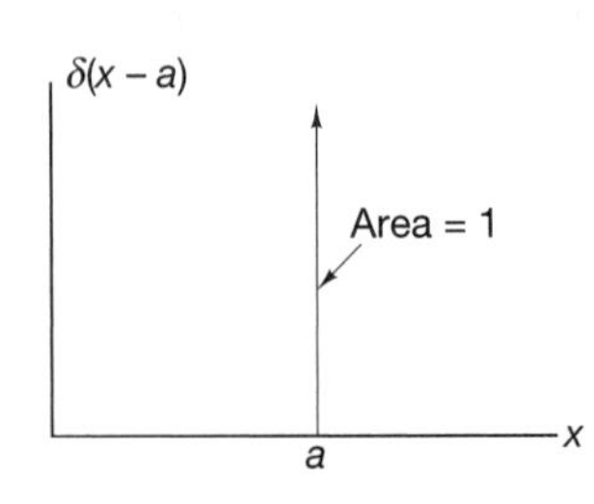

Figure 8.15

$$\frac{1}{x} \cdot \sin x$$

which, at zero, is equal to $\infty \cdot 0$, which is indeterminate. Still, the function can be evaluated by use of l'Hôpital's rule:

$$\lim_{x\to 0}\frac{f(x)}{g(x)} = \lim_{x\to 0}\frac{f'(x)}{g'(x)}$$

$$\lim_{x\to 0}\frac{\sin x}{x} = \lim_{x\to 0}\frac{\cos x}{1} = 1$$

An important property of the Dirac delta function is

$$\langle x - a\rangle_{-1} = 0 \quad x \neq a$$

$$\int_{-\infty}^{x} f(\zeta)\langle \zeta - a\rangle_{-1} d\zeta = \begin{cases} 0 & x < a \\ f(a) & x > a \end{cases} \tag{8.13}$$

Therefore, the integral of the product of a function with the Dirac delta function over a region that includes the point a is equal to the value of the function at a. The Macauley bracket notation will be used to identify discontinuity functions and their differentials and integrals.

Now consider the function shown in Figure 8.16. This is the ***Heaviside step function***, named after the English physicist and electrical engineer Oliver Heaviside (1850–1925). The step function has the value zero for all values of x less than a and has the value unity for all values of x greater than a. It is discontinuous precisely at a, stepping from the value of zero to unity at that point. This definition can be written mathematically as

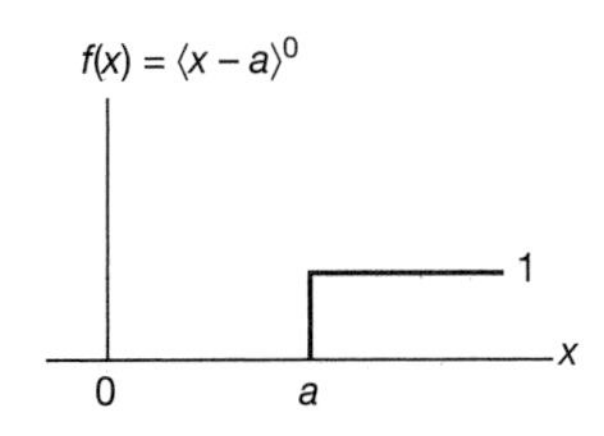

Figure 8.16

$$\langle x - a\rangle^{0} = \begin{cases} 0 & x < a \\ (x - a)^{0} = 1 & x > a \end{cases} \tag{8.14}$$

The Heaviside step function may also be specified with the notation $H(x - a)$ or $\Phi(x - a)$ in many applications, but the bracket notation will be used here for consistency with texts on the mechanics of materials. Note that, for values of x greater than a, the step function is equal to $(x - a)$ raised to the zero power, or unity. A group of polynomial-type functions having the value zero up to the point a and a value of $(x - a)^n$, $n > 0$ may now be defined:

$$\langle x - a\rangle^{n} = \begin{cases} 0 & x < a \\ (x - a)^{n} & x > a \end{cases} \tag{8.15}$$

Each of these polynomial functions has the value zero before the point a, and each begins a nonzero value at that point. Together, they make up a class of functions called discontinuity functions of order n. The Heaviside step function may be used to start any function at the point a as follows:

$$\langle x - a\rangle^{0} F(x) = \begin{cases} 0 & x < a \\ F(x) & x > a \end{cases} \tag{8.16}$$

Here, the step function has been used as a constant multiplying the function $F(x)$, causing the function to start at the desired point a. The derivative of a singularity function of order n is defined as

$$\frac{d}{dx}\langle x - a\rangle^{n} = n\langle x - a\rangle^{n-1} \text{ for } n \geqslant 1 \tag{8.17}$$

Note that Eq. (8.17) does not apply to the Heaviside step function. The derivative of the step function is more complex, and the derivative for values less than a is zero and also is zero for values greater than a, as the slope of the curve is zero. At point a, the function is discontinuous and the derivative is not defined. However, if the curve is viewed as going from a zero slope to an infinite slope and then back to a zero slope, the derivative of the step function can be seen to be the Dirac delta function. Thus,

$$\frac{d}{dx}\langle x - a\rangle^0 = \langle x - a\rangle_{-1} \tag{8.18}$$

The Dirac delta function is defined to have an area of unity at the point a. If the function is integrated, we obtain

$$\int_0^x \langle u - a\rangle_{-1}\,du = \begin{Bmatrix} 0 & \text{for } x < a \\ 1 & \text{for } x > a \end{Bmatrix} = \langle x - a\rangle^0 \tag{8.19}$$

This is consistent with Eq. (8.18), as the integral of the Dirac delta function is equal to the Heaviside step function. As was indicated earlier, the Dirac delta function is also written using the following notation:

$$\delta(x - a) = \langle x - a\rangle_{-1} \tag{8.20}$$

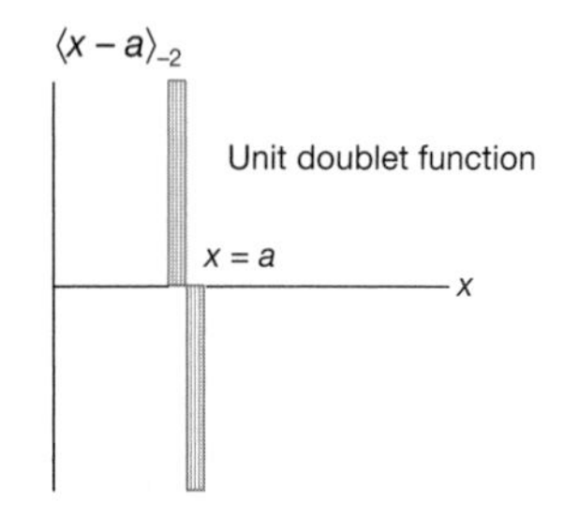

Figure 8.17

The bracket notation is consistent with that used for other singularity functions, but should not be confused to mean a notation for the inverse of $(x - a)$. The derivative of the Dirac delta function is defined and also has physical applications. Examination of the slope of the Dirac delta function shows that it has zero slope up to the point a, whereupon the slope becomes positive infinity, followed by negative infinity and then a zero slope for all values of x greater than a. A function having these characteristics is called a ***unit doublet function*** and is shown in Figure 8.17. The notation used for the unit doublet function is $\langle x - a\rangle_{-2}$ and is consistent with the notation used for other singularity functions. The unit doublet function can be considered to be a unit concentrated moment or couple. Graphically, it appears as a positive Dirac delta function and a negative Dirac delta function, separated by a distance of zero. Note that the unit doublet function is equivalent to a unit couple in a clockwise direction.

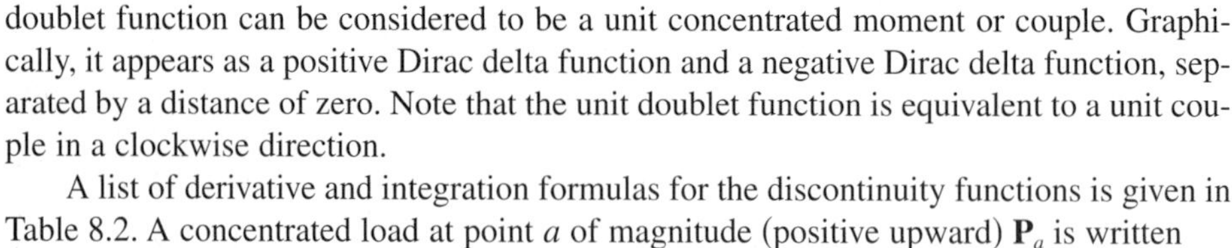

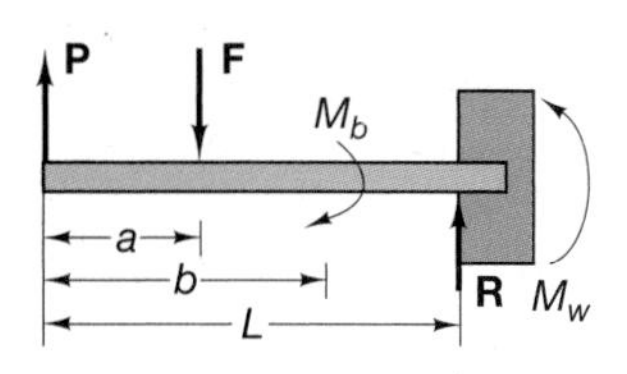

A list of derivative and integration formulas for the discontinuity functions is given in Table 8.2. A concentrated load at point a of magnitude (positive upward) $\mathbf{P}_a$ is written

$$\mathbf{P}_a = P_a\langle x - a\rangle_{-1} \tag{8.21}$$

A concentrated clockwise moment at point a of magnitude M_a is written

$$\mathbf{M}_a = M_a\langle x - a\rangle_{-2} \tag{8.22}$$

Table 8.2
Derivatives and Integrals of Discontinuity Functions

$$\frac{d}{dx}\langle x - a\rangle_{-1} = \langle x - a\rangle_{-2}$$

$$\frac{d}{dx}\langle x - a\rangle^0 = \langle x - a\rangle_{-1}$$

$$\frac{d}{dx}\langle x - a\rangle^n = n\langle x - a\rangle^{n-1} \quad \text{for } n \geqslant 1$$

$$\int_0^x \langle u - a\rangle_{-2}\,du = \langle x - a\rangle_{-1}$$

$$\int_0^x \langle u - a\rangle_{-1}\,du = \langle x - a\rangle^0$$

$$\int_0^x \langle u - a\rangle^n\,du = \frac{\langle x - a\rangle^{n+1}}{n+1} \quad \text{for } n \geqslant 0$$

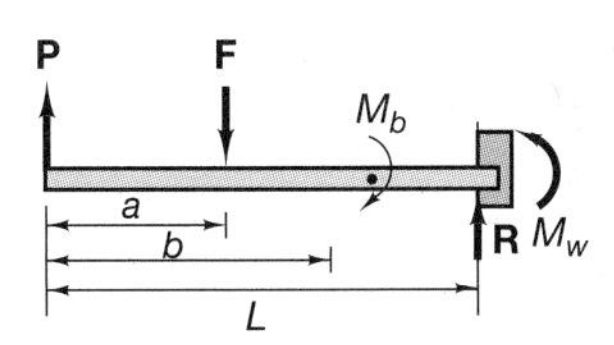

Figure 8.13

As an example of the use of the discontinuity functions, consider the beam shown in Figure 8.13 (repeated at the left). The loading equation for the beam can now be written as

$$\begin{aligned} w(x) &= P\langle x - 0\rangle_{-1} - F\langle x - a\rangle_{-1} + M_b\langle x - b\rangle_{-2} \\ &\quad + R\langle x - L\rangle_{-1} - M_w\langle x - L\rangle_{-2} \end{aligned} \tag{8.23}$$

This equation is valid for all values of x from negative infinity to positive infinity, but the beam physically exists only between $x = 0$ to $x = L$. The equation may be integrated using Table 8.2 to obtain the shear and moment equations:

$$\begin{aligned} V(x) &= P\langle x - 0\rangle^0 - F\langle x - a\rangle^0 + M_b\langle x - b\rangle_{-1} + R\langle x - L\rangle^0 \\ &\quad - M_w\langle x - L\rangle_{-1} + C_1 \\ M(x) &= P\langle x - 0\rangle^1 - F\langle x - a\rangle^1 + M_b\langle x - b\rangle^0 + R\langle x - L\rangle^1 \\ &\quad - M_w\langle x - L\rangle^0 + C_1 x + C_2 \end{aligned} \tag{8.24}$$

The two constants of integration, C_1 and C_2, can be evaluated by noting that the shear and the moment are zero at $x = 0_-$, or a point just to the left of $x = 0$. That is,

$$\begin{aligned} V(0_-) &= 0 = C_1 \\ M(0_-) &= 0 = C_2 \end{aligned} \tag{8.25}$$

The reactions R and M_w are obtained by setting V(x) and M(x) equal to zero for any value of x greater than L ($x > L$), yielding the two equations of static equilibrium:

$$V(L^+) = 0 = P - F + R \qquad R = F - P \tag{8.26}$$

$$\begin{aligned} M(L^+) &= 0 = P(L) - F(L - a) + M_b + R(L^+ - L) - M_w \\ M_w &= P(L) - F(L - a) + M_b \end{aligned} \tag{8.27}$$

Many texts on the mechanics of materials handle beam equations in this manner, as it not only simplifies the computations, but integrates loads and moments with distributed loading.

The Heaviside step function serves a very important role in modeling loading on a beam. Any loading function may be started or ended at any point on the beam by treating the step function as a constant multiplier. Therefore the function $F(x)$ may be started at a point a by writing the load function as

$$\langle x - a\rangle^0 F(x)$$

and the function may be ended at a point b by writing the load function as

$$\langle x - a\rangle^0 F(x) - \langle x - b\rangle^0 F(x)$$

Now although the function $F(x)$ is defined over the entire range of x, it only appears as a load for the range of x from a to b (actually the positive function starts at a and continues to infinity and the negative starts at b and continues to infinity but after point b, the two cancel each other). *The Heaviside step function can be used to start and stop any continuous loading function.*

The real usefulness of the use of the step function would be if we can integrate it to obtain the shear and moment distribution in a beam. Mechanics of materials courses will show that further integration will yield the slope and deflection of the beam. Let us consider the integration of

$$\int_0^x \langle \xi - a\rangle^0 F(\xi) d\xi$$

This integration may be accomplished by integration by parts.

$$\int_a^b u dv = uv\Big|_a^b - \int_a^b v du$$

We will select

$$u = \langle \xi - a \rangle^0 \text{ and } dv = F(\xi)d\xi$$

Therefore,

$$du = \langle \xi - a \rangle_{-1} d\xi \text{ and } v = \int F(\xi)d\xi = G(\xi)$$

We can now develop a general expression

$$\begin{aligned} \int_0^x \langle \xi - a \rangle^0 F(\xi)d\xi &= \Big[\langle \xi - a \rangle^0 G\xi\Big]_0^x - \int_0^x \langle \xi - a \rangle_{-1} G(\xi)d\xi \\ &= \langle x - a \rangle^0 G(x) - \langle x - a \rangle^0 G(a) \end{aligned} \tag{8.28}$$

The last term is integrated using the fact that the Dirac delta function is zero everywhere, except at the point a. If x is less than a, this integral is zero, and if x is greater than a, the integral is the value of the function $G(x)$ at a.

For example, if a beam is loaded with a partial sinusoidal loading that starts at point a (that is, the load is zero for x less than a and is $[P \sin(\pi x/L)]$ for x equal to or greater than a), the loading curve is

$$w(x) = \langle x - a \rangle^0 P \sin\left(\frac{\pi x}{L}\right)$$

$$V(x) = \langle x - a \rangle^0 P \frac{L}{\pi}\left[-\cos\left(\frac{\pi x}{L}\right) + \cos\left(\frac{\pi a}{L}\right)\right]$$

$$V(x) = \frac{PL}{\pi} \cos\frac{\pi a}{L} \cdot \langle x - a \rangle^0 - \frac{PL}{\pi} \cos\frac{\pi x}{L} \cdot \langle x - a \rangle^0$$

$$M(x) = \frac{PL}{\pi} \cos\frac{\pi a}{L} \cdot \langle x - a \rangle^1 - \frac{PL^2}{\pi^2}\left\{\sin\frac{\pi x}{L} - \sin\frac{\pi a}{L}\right] \cdot \langle x - a \rangle^0$$

Sample Problem 8.7

Write the load, shear, and moment equations for the beam shown in Sample Problem 8.6 (see the figure at the left).

Solution

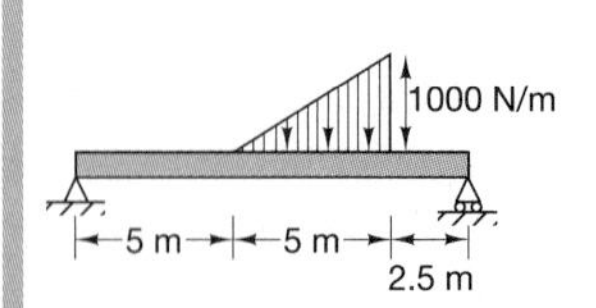

The reactions were determined when this problem was solved in Section 8.4, but finding them is not necessary when using singularity functions. However, this particular problem introduces some special considerations. The loading begins at $x = 5$ and ends at $x = 10$. Singularity functions have a definite beginning, but continue, in theory, to infinity. The distributed load is increasing at 200 (N/m)/m and ends at $x = 10$ m. This can be handled by adding the negative of the loading at $x = 10$. The negative of the loading is equal to the value of the constant distributed load at $x = 10$, plus the linearly increasing load, as shown in the figure on the next page. Therefore, the loading curve is

$$\begin{aligned} w(x) = {} & R_l\langle x - 0 \rangle_{-1} - 200\langle x - 5 \rangle^1 + 1000\langle x - 10 \rangle^0 + 200\langle x - 10 \rangle^1 \\ & + R_r\langle x - 12.5 \rangle_{-1} \end{aligned}$$

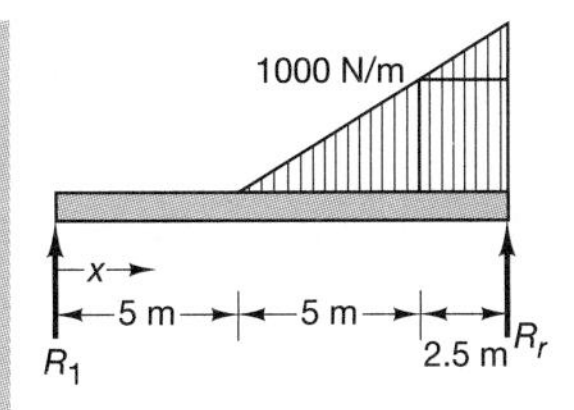

Integrating gives the shear and moment equation:

$$V(x) = R_l\langle x - 0\rangle^0 - 100\langle x - 5\rangle^2 + 1000\langle x - 10\rangle^1 + 100\langle x - 10\rangle^2$$
$$= \; + R_r\langle x - 12.5\rangle^0$$
$$M(x) = R_l\langle x - 0\rangle^1 - 33.3\langle x - 5\rangle^3 + 500\langle x - 10\rangle^2 + 33.3\langle x - 10\rangle^3$$
$$+ R_r\langle x - 12.5\rangle^1$$

The reactions may be evaluated by setting the shear and moment at L^+ equal to zero:

$$V(12.5) = R_l - 100(7.5)^2 + 1000(2.5) + 100(2.5)^2 + R_r = 0$$
$$R_l + R_r - 2500 = 0$$

This is equivalent to setting the sum of the vertical forces equal to zero. Now consider the moment at the end of the beam:

$$M(12.5) = R_l(12.5) - 33.3(7.5)^3 + 500(2.5)^2 + 33.3(2.5)^3 + R_r(0) = 0$$

This equation is equivalent to a summation of moments about the right end of the beam and yields

$$R_1 = 833 \text{ N}$$
$$R_r = 1667 \text{ N}$$

An alternate method of solving the problem can be obtained by writing the loading function as

$$w(x) = R_l\langle x - 0\rangle_{-1} - \langle x - 5\rangle^0 200(x - 5) + \langle x - 10\rangle^0 200(x - 5)$$
$$+ R_r\langle x - 12.5\rangle_{-1}$$

The second and third terms can be integrated using Eq. (8.28) to give the expression for the shear.

$$V(x) = R_l\langle x - 0\rangle^0 - \langle x - 5\rangle^0 100(x - 5)^2 + \langle x - 10\rangle^0 100(x - 5)^2$$
$$- \langle x - 10\rangle^0 2500 + R_r\langle x - 12.5\rangle^0 + C_1$$

The shear equation can now be integrated to obtain the moment using Eq. (8.28) to integrate the second and third terms.

$$M(x) = R_l\langle x - 0\rangle^1 - \langle x - 5\rangle^0 33.3(x - 3)^3 + \langle x - 10\rangle^0 33.3(x - 5)^3$$
$$- \langle x - 10\rangle^0 33.3(5)^3 - \langle x - 10\rangle^1 2500 + R_r\langle x - 12.5\rangle^1 + C_1 x + C_2$$

The boundary conditions at zero minus $(0-)$ are;

$$V(0-) = 0 \quad \Rightarrow \quad C_1 = 0$$
$$M(0-) = 0 \quad \Rightarrow \quad C_2 = 0$$

The boundary conditions at the right end of the beam $(12.5+)$ are;

$$V(12.5+) = 0 \quad \Rightarrow \quad R_l - 2500 + R_r = 0$$
$$M(12.5+) = 0 \quad \Rightarrow \quad 12.5R_l - 125(33.3) - 2500(2.5) = 0$$

Solving the two equations yields:

$$R_l = 833 \text{ N}$$
$$R_r = 1667 \text{ N}$$

In this case, either solution could have been used to integrate the equations but had the loading functions been more complex, the alternate approach must be used.

To analyze the shear and moment in the beam, each section of the shear and moment equations must be considered separately, as was done before. The advantage of discontinuity functions is their ease of integration. When the shear and moment equations are used to evaluate the slope and deflection of a beam, two more integrations are required, and the usefulness of discontinuity functions becomes apparent. Most computational software packages have the Heaviside step function, and some have the Dirac delta function; either or both may be used to solve and graph the shear and moment equations. An example is shown in the Computational Supplement.

Problems

8.42 Using singularity functions, determine the reactions at A and B, as well as the shear and bending moment functions in terms of the unspecified, but fixed, value of the load P, position a, and length L, for all points along the beam shown in Figure P8.42.

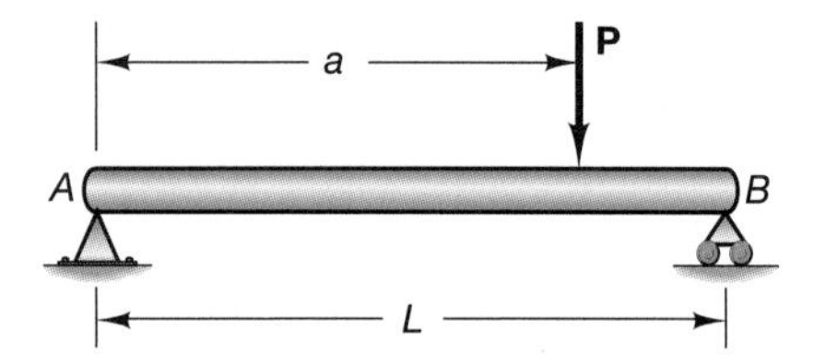

Figure P8.42

8.43 A machine located at the end of a support beam causes both a moment and a force to be applied to the beam. (See Figure P8.43.) Use discontinuity functions to determine the reaction forces at A and the shear and bending moment functions in terms of the length L, applied force P_B, and applied moment M_B.

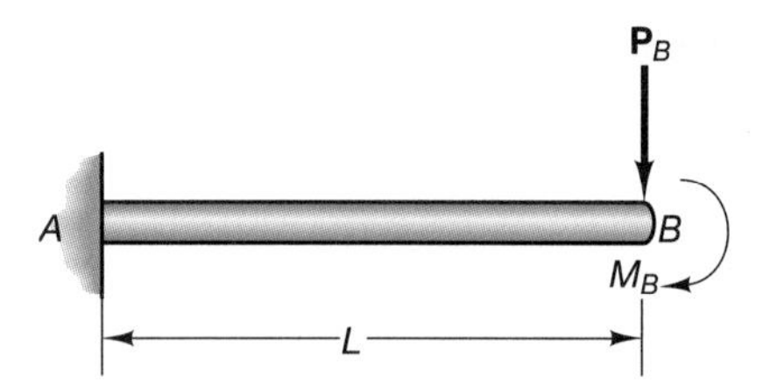

Figure P8.43

8.44 In Figure P8.44, compute the reaction forces, shear force, and bending moments for the combination point load P and distributed load w_0, in terms of the constants a, b, L, P, and w_0, as a function of x. Use the method based on discontinuity functions.

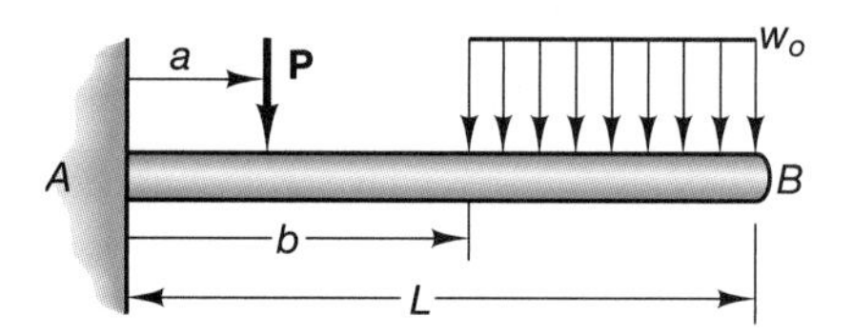

Figure P8.44

8.45 In Figure P8.45, compute the reaction forces, the shear force, and the bending moment function at each point x between 0 and L due to the applied distributed load, and the applied moment M, in terms of the constants a, b, L, w_0, and M. Assume that $a < b$, and use the discontinuity functions.

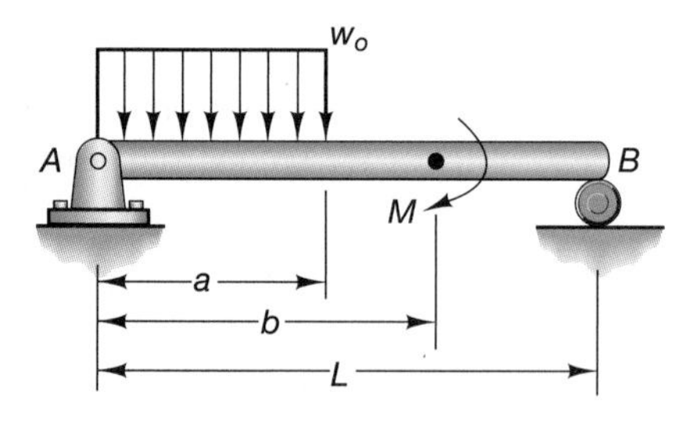

Figure P8.45

8.46 Solve Problem 8.36 using discontinuity functions.

8.47 Repeat Problem 8.37 using discontinuity functions.

8.48 Repeat Problem 8.38 using discontinuity methods.

8.49 Repeat Problem 8.39 using the discontinuity-function approach.

8.50 Resolve Problem 8.40 using discontinuity functions.

8.51 Repeat problem 8.41 with a concentrated load P at $x = a$. Use singularity functions to describe the load.

8.52 A snow load on a cantilevered deck gives the weight distribution shown in Figure P8.52 on the supporting beams. Determine the shear and the moment on the beam.

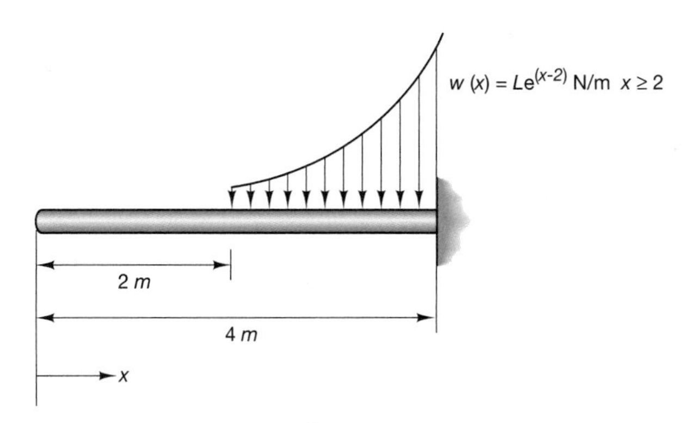

Figure P8.52

8.53 A crude model of an airplane wing under both an aerodynamic load $w(x)$ and a force P due to the weight of the engine is shown in Figure P8.53. For first approximation, $w(x) = \sin\left(\frac{\pi x}{L}\right) \frac{L}{4} \leq x \geq \frac{3L}{4}$. Determine an expression for the shear and moment for any point on the wing.

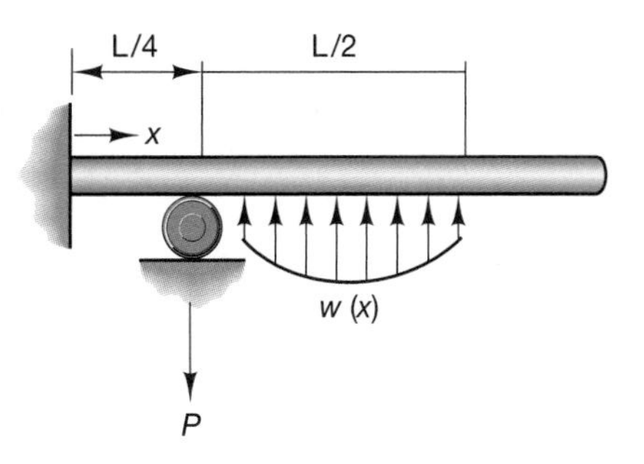

Figure P8.53

8.6 CABLES

A very important structural member is the flexible cable, which is used in suspension bridges, telephone lines, transmission lines, and many other engineering applications. The term "flexible" means that the bending resistance of the cable is negligible, and the cable is treated as a tension member. When a single load is supported by two or three cables, the forces form a concurrent force system; methods for analyzing this system were presented in Chapter 3. The structural use of a flexible cable usually involves a single cable supporting many discrete loads or distributed loads. We will consider the problem of a cable supporting discrete loads first and then examine the case of a cable resisting a distributed load.

8.6.1 CABLE SUBJECTED TO CONCENTRATED LOADS

A cable supporting a number of concentrated loads is shown in Figure 8.18. There are two unknown reactions, T_A at A and T_D at D, and these may be determined if the final geometry of the loaded cable is known. For all design specifications, the length of the

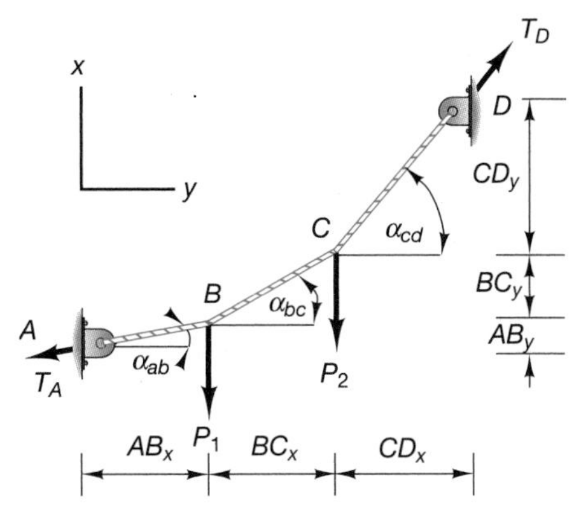

Figure 8.18

cable and the locations of the endpoints are known. The locations of the end supports add another constraint on the length of the cable—that is, the minimum. Either the individual lengths *AB*, *BC*, and *CD* may be specified, or the design may require that the span distances, AB_x, BC_x, and CD_x be specified. In the first case, the loads are applied at designated points along the length of the cable and will be separated in space according to equilibrium. In the second case, the location of attachment to the cable must be determined such that the loads will have specific locations in space.

Two equilibrium equations can be written at each of the attachment points *B* and *C*:

Point B

$$-T_{AB}\cos\alpha_{AB} + T_{BC}\cos\alpha_{BC} = 0 \qquad (8.29)$$
$$-T_{AB}\sin\alpha_{AB} + T_{BC}\sin\alpha_{BC} - P_1 = 0 \qquad (8.30)$$

Point C

$$-T_{BC}\cos\alpha_{BC} + T_{CD}\cos\alpha_{CD} = 0 \qquad (8.31)$$
$$-T_{BC}\cos\alpha_{BC} + T_{CD}\cos\alpha_{CD} - P_2 = 0 \qquad (8.32)$$

There are now four equations in six unknowns: T_{AB}, T_{BC}, T_{CD}, α_{AB}, α_{BC}, and α_{CD}. Six additional equations describe the geometry of the structure:

$$AB_x + BC_x + CD_x = AD_x \qquad (8.33a)$$
$$AB\sin\alpha_{AB} + BC\sin\alpha_{BC} + CD\sin\alpha_{CD} = AD_y \qquad (8.33b)$$
$$AB + BC + CD = L \qquad (8.33c)$$
$$AB_x = AB\cos\alpha_{AB} \qquad (8.33d)$$
$$BC_x = BC\cos\alpha_{BC} \qquad (8.33e)$$
$$CD_x = CD\cos\alpha_{CD} \qquad (8.33f)$$

Depending upon whether the span lengths or the attachment lengths are specified, these six equations may be combined to give two additional equations for the six unknowns. If AB_x, BC_x, and CD_x are specified, the two additional equations are

$$AB_x\tan\alpha_{AB} + BC_x\tan\alpha_{BC} + CD_x\tan\alpha_{CD} = AD_y \qquad (8.34a)$$
$$AB_x/\cos\alpha_{AB} + BC_x/\cos\alpha_{BC} + CD_x/\cos\alpha_{CD} = L \qquad (8.34b)$$

If the attachment lengths *AB*, *BC*, and *CD* are specified, the two additional equations are

$$AB\sin\alpha_{AB} + BC\sin\alpha_{BC} + CD\sin\alpha_{CD} = AD_y \qquad (8.35a)$$
$$AB\cos\alpha_{AB} + BC\cos\alpha_{BC} + CD\cos\alpha_{CD} = AD_x \qquad (8.35b)$$

The six equations form a system of nonlinear equations that may be solved for the six unknowns. This system of equations is not easily solved by hand, and computer software should be used. Most computer programs use an iterative approach and are sensitive to initial estimated values of the unknowns. This type of solution is shown in Sample Problem 8.8. Details of the numerical solution are found in the Computational Supplement.

The problem of a cable loaded with any number of concentrated loads can now be developed. The cable shown in Figure 8.19 is loaded with *n* loads and supported by $(n + 1)$ tensions acting at $(n + 1)$ angles. If the equilibrium equations are written for each attachment point, a system of $2n$ equations will be obtained in the form

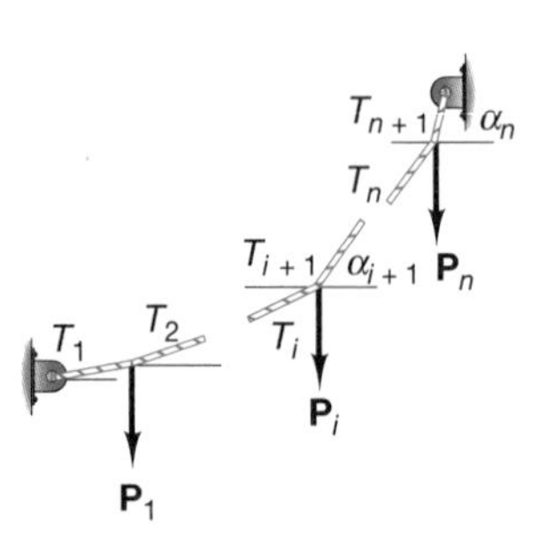

Figure 8.19

$$i = 1 \text{ to } n: \quad -T_i\cos\alpha_i + T_{i+1}\cos\alpha_{i+1} = 0 \qquad (8.36)$$
$$-T_i\sin\alpha_i + T_{i+1}\sin\alpha_{i+i} - P_i = 0 \qquad (8.37)$$

Either Eq. (8.34) or Eq. (8.35) supplies the final two equations to form a $(2n + 2)$ set of simultaneous nonlinear algebraic equations for the $(n + 1)$ tensions and the $(n + 1)$ angles. Some type of computational aid will be necessary to solve systems of equations of this type.

Sample Problem 8.8

For the cable shown in the figure at the left, $P_1 = 750$ N, $P_2 = 400$ N; and the components of the distance between the supports are $AD_x = 30$ m and $AD_y = 0$ m. The design requires that the loads be equally spaced at $AB_x = 10$ m, $BC_x = 10$ m, and $CD_x = 10$ m. Determine the tensions in the sections of the cable if a cable of length 33 m is to be used.

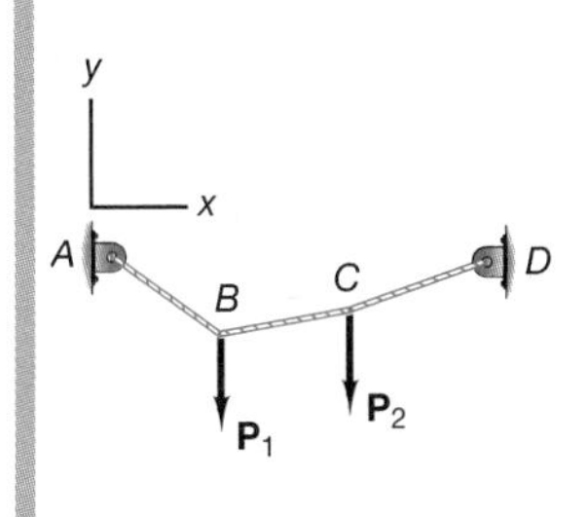

Solution

Eqs. (8.29) to (8.32) and Eqs. (8.34a) and (8.34b) form the system of nonlinear equations. The tension in the various parts of the cable decreases when the cable length increases. If the length of the cable is 33 m, the magnitudes and directions of the tensions are $T_{AB} = 1208$ N, $\alpha_{AB} = -31.63°$; $T_{BC} = 1035$ N, $\alpha_{BC} = 6.47°$; and $T_{CD} = 1151$ N, $\alpha_{AB} = 26.68°$. If the cable length is 31 m, the magnitude and directions of the tensions are $T_{AB} = 1915$ N, $\alpha_{AB} = -19.15°$; $T_{BC} = 1827$ N, $\alpha_{BC} = 3.66°$; and $T_{CD} = 1895$ N, $\alpha_{AB} = 15.82°$. These results were obtained using computational software. (See the Computational Supplement.)

If the coordinate position of a point on the cable other than the support points is known, the four unknown support reactions may be determined from the three equations of equilibrium of the structure as a whole and a fourth equation obtained by setting the moment of the reactions equal to zero for the sectioned structure at the known point. For example, if the coordinates of point B were known to be 10 m to the right of A and 6.15 m below A ($\alpha_{AB} = -31.63°$, the result of previous calculations), the reactions in the first case in Sample Problem 8.8 may be determined from the equations

$$
\begin{aligned}
&A_y + D_y - 750 - 400 = 0 \\
&A_x = D_x \\
&10 \times 750 + 20x \times 400 - 30\,D_y = 0 \\
&10\,A_y - 6.159\,A_x = 0
\end{aligned}
$$

Solution of these equations yields $A_y = 633$ N and $A_x = 1028$ N, or $T_{AB} = 1207$ N, as obtained previously. The balance of the solution may be obtained by the method of joints. Although the solution in this case can be obtained without the use of computational aids, it does not represent a practical situation.

8.6.2 CABLES SUPPORTING LOADS DISTRIBUTED UNIFORMLY ALONG A HORIZONTAL LINE

Suppose a cable is attached at two support points and carries a uniform distributed load, as illustrated in Figure 8.20. Consider the case when the load is uniformly distributed along the horizontal line, as with the main cable of a suspension bridge. The tension in the cable is tangent to the curve the cable makes in space. We construct a free-body diagram by

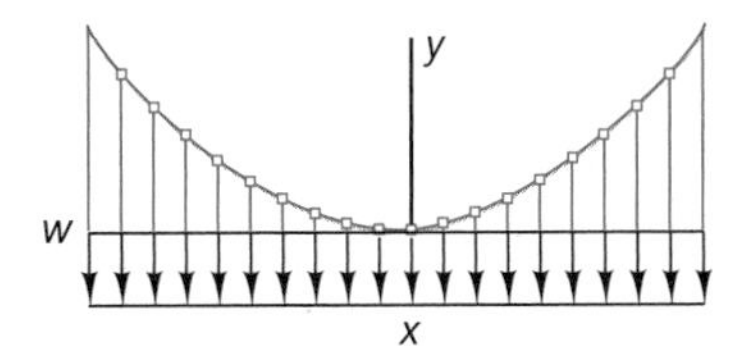

Figure 8.20

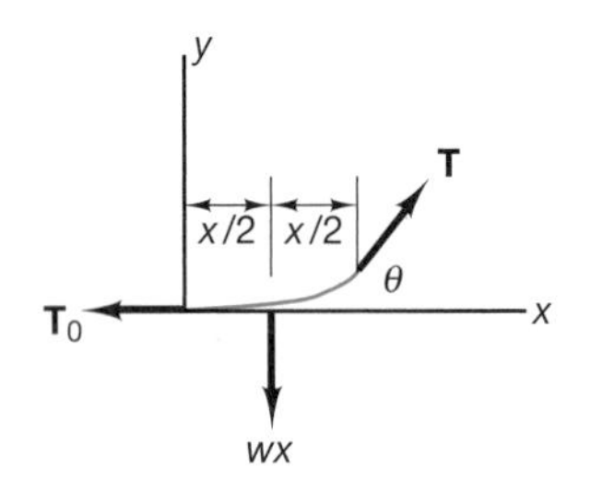

Figure 8.21

cutting the cable at its *lowest* point, as shown in Figure 8.21. If the origin of the x–y coordinate system is taken at the lowest point, the equations of equilibrium are

$$T\cos\theta = T_0 \tag{8.38}$$

$$T\sin\theta = wx \tag{8.39}$$

where w is the load per unit length along a horizontal line. Combining Eqs. (8.38) and (8.39) to form the tangent and recalling that the tangent is the slope yields

$$\tan\theta = \frac{dy}{dx} = \frac{w}{T_0}x = qx$$

$$\text{where } q = \frac{w}{T_0} \tag{8.40}$$

Integrating Eqs. (8.40) shows that the curve described by the cable is the parabola

$$y = \frac{1}{2}qx^2 \tag{8.41}$$

The tension in the cable at any point can be found in terms of q by the use of Eqs. (8.40) and (8.41), yielding

$$T = T_0\sqrt{1 + q^2x^2} \tag{8.42}$$

Sample Problem 8.9

A suspension bridge is designed to support a horizontal uniform load of 1000 N/m and span a distance of 50 meters. The distance from the lowest point on the main cable to the support on the left end is 10 meters, and that from the lowest point to the support on the right end is 12 meters. Determine the maximum tension in the cable.

Solution We establish the origin of the coordinate system at the lowest point of the cable. The vertical distances from this point to the left and right supports are, respectively,

$$y_L = 10 = \frac{1}{2}qx_L^2$$

$$y_R = 12 = \frac{1}{2}qx_R^2$$

$$\frac{x_L}{x_R} = \sqrt{\frac{10}{12}} = 0.913$$

The span of the cable is 50 meters; therefore,

$$x_L + x_R = 50$$

Thus, $x_L = 23.86$ meters and $x_R = 26.14$ meters.

The constant q may be determined using the coordinates of either end:

$$q = \frac{2 \cdot 12}{26.14^2} = 0.035$$

The maximum tension will occur where the horizontal distance is a maximum; therefore,

$$T = T_0\sqrt{1 + q^2x^2}$$

The tension at the lowest value of the cable is $T_0 = w/q = 1000/0.035 = 28{,}460$ N. So

$$T_{max} = 38{,}580 \text{ N}$$

The solution of this sample problem is shown in the Computational Supplement.

8.6.3 CABLE SUPPORTING LOADS DISTRIBUTED UNIFORMLY ALONG ITS OWN LENGTH

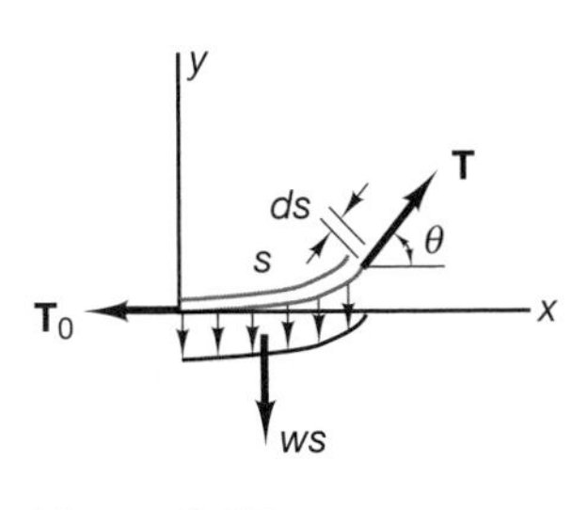

Figure 8.22

A cable that supports its own weight must support loads that are uniformly distributed along the cable, not along a horizontal line, as in the case of the suspension bridges discussed earlier. The curve that a cable loaded in this manner will describe is not a parabola, but a *catenary*. A free-body diagram of such a cable is shown in Figure 8.22. The distributed load is such that each element of length ds is acted on by a force of (wds), where w is a constant equal to the weight per unit length. As before, the cable is sectioned at its lowest point, and the tension at this point is T_0. The equilibrium equations are

$$T \sin \theta = ws \tag{8.43}$$

$$T \cos \theta = T_0 \tag{8.44}$$

Division of Eq. (8.43) by Eq. (8.44) yields

$$\tan \theta = \frac{dy}{dx} = \frac{w}{T_0} s \tag{8.45}$$

Next, we introduce a constant $q = w/T_0$ and differentiate Eq. (8.45) with respect to x. This yields

$$\frac{d}{dx}\left(\frac{dy}{dx}\right) = q \frac{ds}{dx} \tag{8.46}$$

The differential length of the cable may be written as

$$ds^2 = dx^2 + dy^2$$

$$\frac{ds}{dx} = \sqrt{1 + \left(\frac{dy}{dx}\right)^2} \tag{8.47}$$

Let $\zeta = dy/dx$, or the slope of the curve. Then, combining Eqs. (8.46) and (8.47) gives the differential equation for the slope:

$$\frac{d\zeta}{dx} = q\sqrt{1 + \zeta^2}$$

or

$$\frac{d\zeta}{\sqrt{1 + \zeta^2}} = qdx \tag{8.48}$$

The origin of the coordinate system was chosen at the lowest point of the cable, and the slope is zero at this point. Integrating both sides of Eq. (8.48) yields

$$\int_0^{\zeta} \frac{d\zeta}{\sqrt{1 + \zeta^2}} = \int_0^{x} qdx$$

$$\zeta = \frac{dy}{dx} = \frac{1}{2}(e^{qx} - e^{-qx}) = \sinh (qx) \tag{8.49}$$

The equation for the curve the cable makes in space may be found by integrating the slope equation, which results in

$$y = \frac{1}{q}[\cosh (qx) - 1] \tag{8.50}$$

This equation describes a catenary and involves the hyperbolic cosine of qx. The tension in the cable may be found by the use of Eqs. (8.44), (8.47), and (8.49), yielding

$$T = \frac{T_0}{\cos\theta} = T_0 \frac{ds}{dx}$$

or

$$T = T_0\sqrt{1 + (\sinh qx)^2} = T_0 \cosh qx \tag{8.51}$$

Here, the identity for hyperbolic functions, $\cosh^2 qx - \sinh^2 qx = 1$, has been used.

The length of the cable from its lowest point to any position x can be found from Eq. (8.45):

$$s = \zeta/q = [\sinh qx]/q \tag{8.52}$$

Note that all the relevant equations involve the tension at the lowest point in the cable, T_0, and this would not be a parameter normally specified. However, the length of the cable and the locations of the supports are known, as is the weight of the cable, and these parameters can be used to determine T_0 and the location of the lowest point in the cable.

Sample Problem 8.10

Twenty meters of telephone cable with a weight of 40 N/m is supported by two poles of equal height placed 18 meters apart. Determine how low the cable drops and the maximum tension in the cable.

Solution

The lowest point on the cable will be in the middle, and the horizontal distance to the support is $x = 9$ m. Along the cable length, $s = 10$ m. The constant q may be determined by using Eq. (8.52). We have

$$sq - \sinh(qx) = 0$$

$$10\,q - \sinh(9\,q) = 0$$

The latter is a transcendental equation for q that may be solved by graphing the equation and looking for the value of q at which the equation is zero. Most computational aids have a method to determine the root of this equation. For this case,

$$q = 0.089$$

The lowest point on the cable can be found from Eq. (8.50) by setting $x = 9$ m and determining the height of the support above the cable' s lowest point. The result of doing this is

$$y = 1/0.089\,[\cosh(0.089 \cdot 9) - 1]$$

The cable's lowest point is 3.801 beneath the supports. Using the fact that $q = w/T_0$, we find that the tension at this point is

$$T_0 = 40/0.089 = 449 \text{ N}$$

The maximum tension occurs at the supports and may be determined by Eq. (8.51):

$$T(x) = T_0 \cosh(qx) \qquad T(9) = 449 \cosh(0.089 \cdot 9)$$

Details of the computations are shown in the Computational Supplement.

Problems

8.54 Compute the tension in each segment of the rope shown in Figure P8.54.

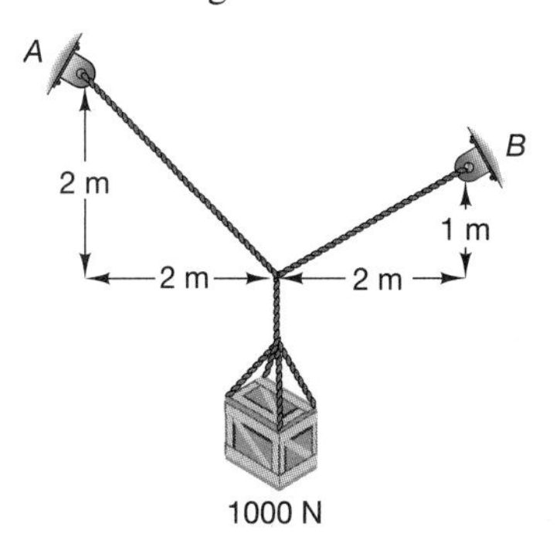

Figure P8.54

8.55 For the two cables shown in Figure P8.55, determine the tensions and the angles θ and β for the load $P = 2$ kN.

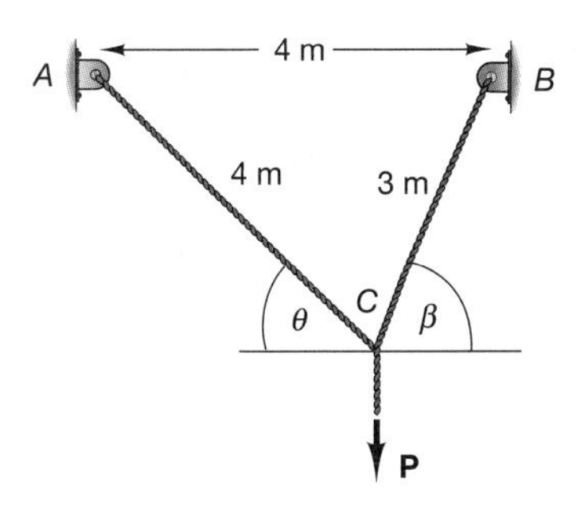

Figure P8.55

8.56 A 2-m cable is used to support a load P of 5 kN. (See Figure P8.56.) The span distance for P must be 0.6 m along the vertical from point A. (a) Determine the tension in the cables and the angles θ and β. (b) Compute the x–y components of the force at the supports A and B.

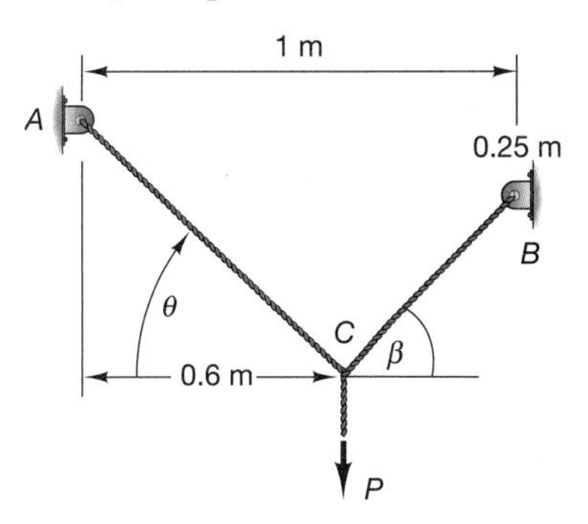

Figure P8.56

8.57 A 10.5-m cable is loaded with forces $P_1 = 1.5$ kN and $P_2 = 2$ kN. (See Figure P8.57.) (a) Compute the tensions in each segment and the angles α, β, and γ. (b) Determine the x–y components of the reaction forces at A and D and the sag at points B and C.

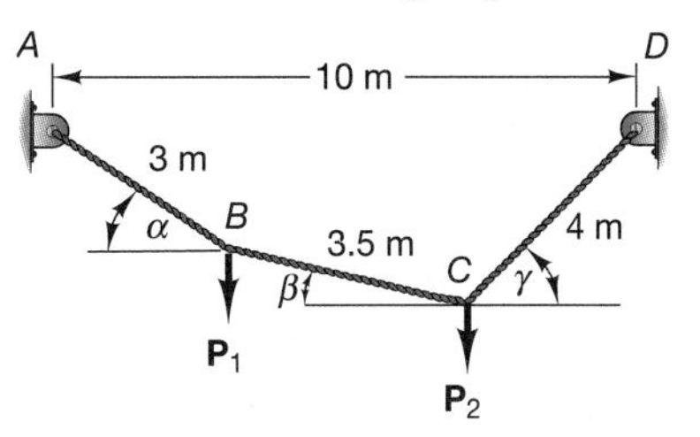

Figure P8.57

8.58 For the cable illustrated in Figure P8.58, compute the tension in each segment, the angles α, β, and γ, and the length of each segment. The load is $P_1 = 1.5$ kN and $P_2 = 5$ kN. The total length of the cable is 7.5 m.

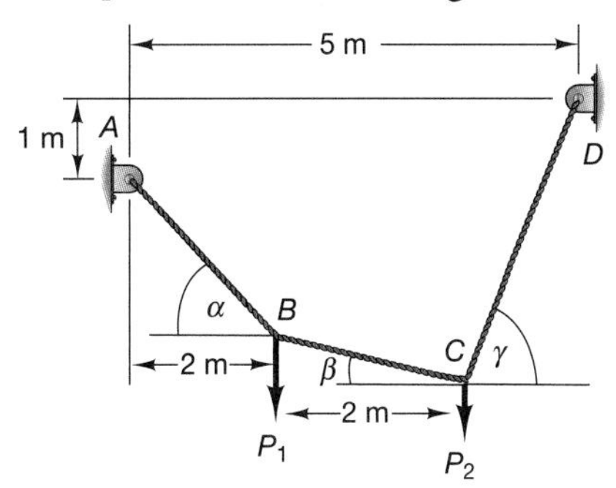

Figure P8.58

8.59 Compute the tension in each cable segment and the angles α, β, and γ for the traffic-light system shown in Figure P8.59. Design (choose) the hanger length l so that

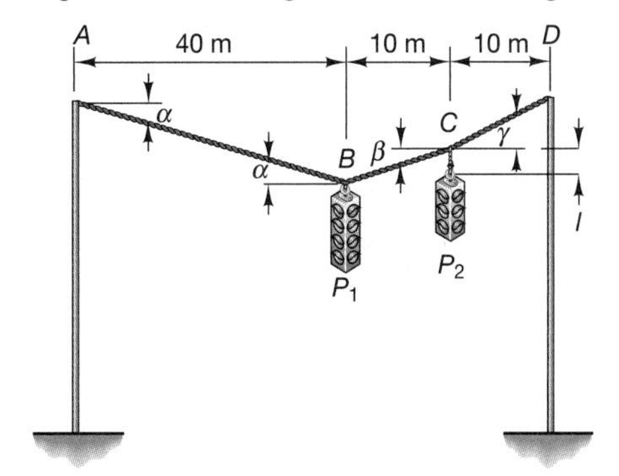

Figure P8.59

the top of the two lights hang to the same level for ease of viewing. The length of the cable is 62 m. Here, $P_1 = 500$ N and $P_2 = 400$ N.

8.60 A 16-m cable is connected between two posts that are 14 m apart. (See Figure P8.60.) Calculate the tension in each segment of the cable, the reaction forces at A and E, and the locations of points B, C, and D for the given loads.

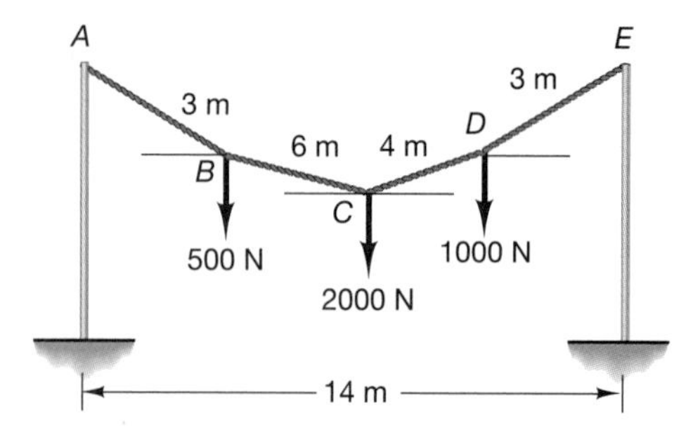

Figure P8.60

8.61 Compute the maximum tension in the cable supporting the distributed load of 1000 N/m shown in Figure P8.61.

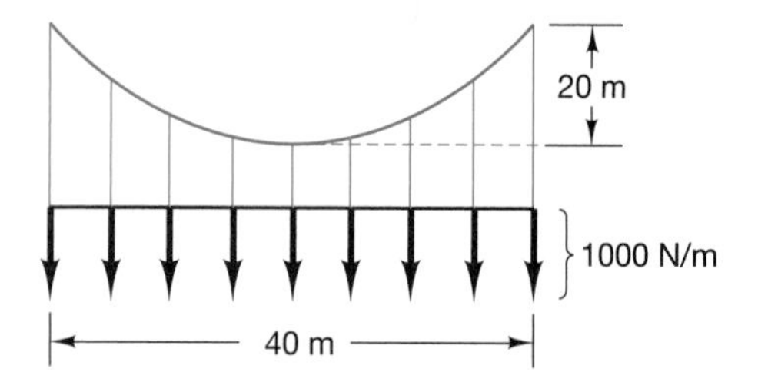

Figure P8.61

8.62 A cable supports a distributed load of 10^4 N/m. (See Figure P8.62.) Compute the maximum tension in the cable.

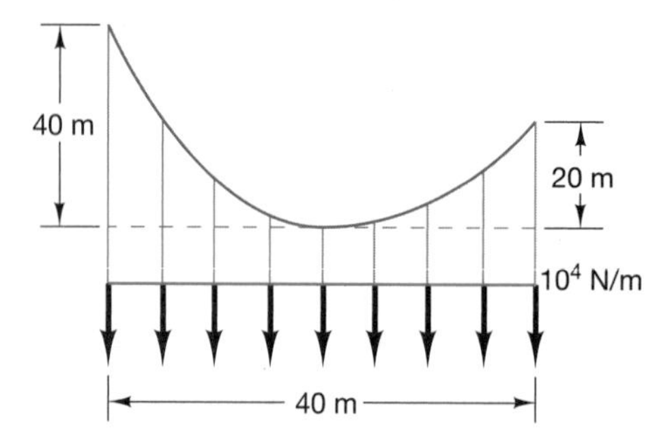

Figure P8.62

8.63 Compute the tension in the cable of the system of Figure P8.61, and plot it as a function of the tangential distance between the two endpoints.

8.64 Compute the tension in the cable of Problem 8.62, and plot the tension as a function of the horizontal distance between the two endpoints.

8.65 Compute and plot the tension in the suspension cable shown in Figure P8.65 if the uniform load on the bridge is 2000 N/m. The cable sags 10 m at the midpoint. The maximum tension the cable can sustain is 750 kN. Will the cable be able to hold the load?

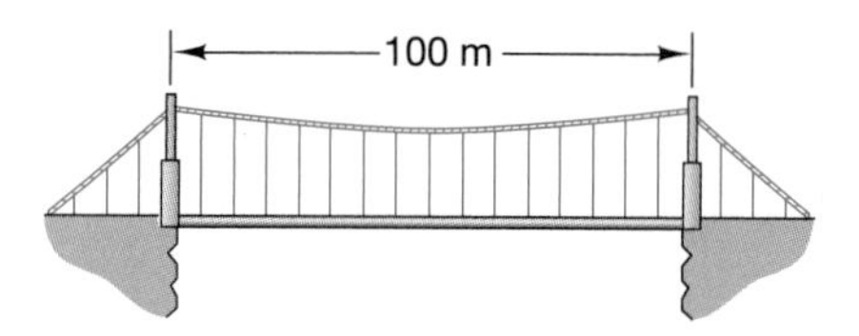

Figure P8.65

8.66 A telephone line changes from an aboveground telephone pole arrangement to a buried system at point A, as shown in Figure P8.66. The cable density is 4 kg/m, and the distance between point A and the bottom of the pole is 2.5 m. Point B is 3.5 m off the ground. Determine the equation of the cable under its own weight and the maximum tension in the cable. Plot the tension in the cable as a function of the horizontal distance from A.

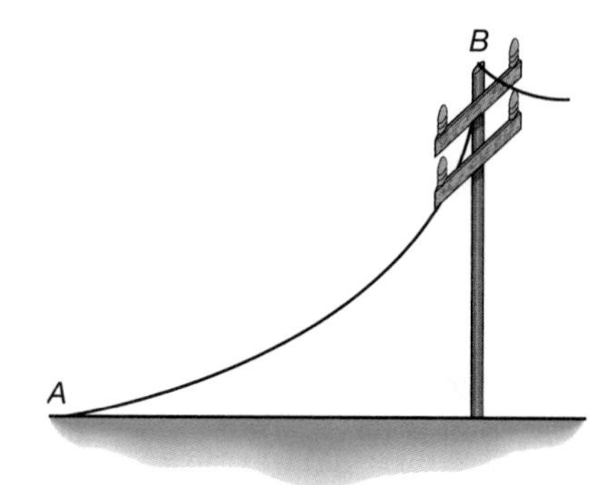

Figure P8.66

8.67 A cable is supported between two points 30 m apart, as illustrated in Figure P8.67, and has a weight per

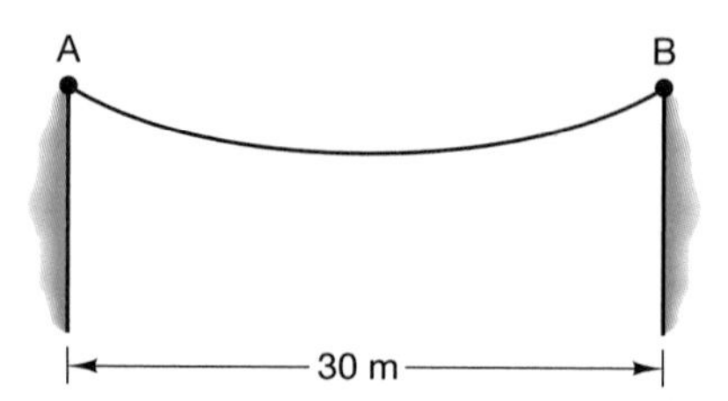

Figure P8.67

unit length of 500 N/m. Compute the magnitude of the reaction at point A due to the weight of the cable. The cable is 32 m. long. Also, plot the tension in the cable as a function of the horizontal distance from A.

8.68 A communication cable must pass across the top of a commuter-train channel. (See Figure P8.68.) The code specifies that the cable must be 2/3 m above the highest train (5 m). If 17 m of cable with a weight of 400 N is stretched across the channel, will it satisfy the code?

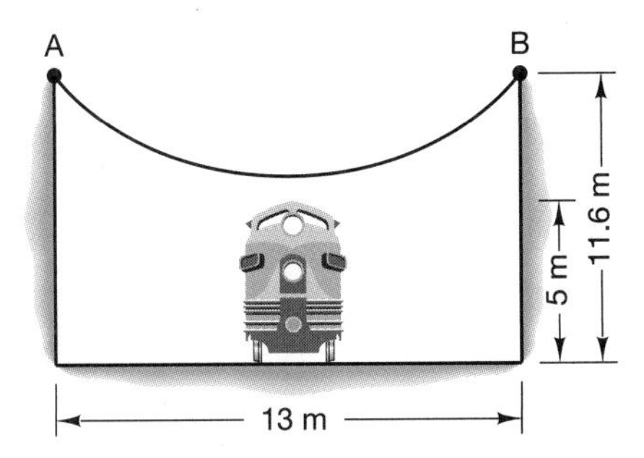

Figure P8.68

8.69 A power line runs from the top of one building to the top of the next. (See Figure P8.69.) The line has a weight per unit length of 50 N/m and is 40 m long. Determine the tension in the cable at any point and the equation for the curve describing the catenary. Plot the tension as a function of the horizontal distance, and determine the maximum tension.

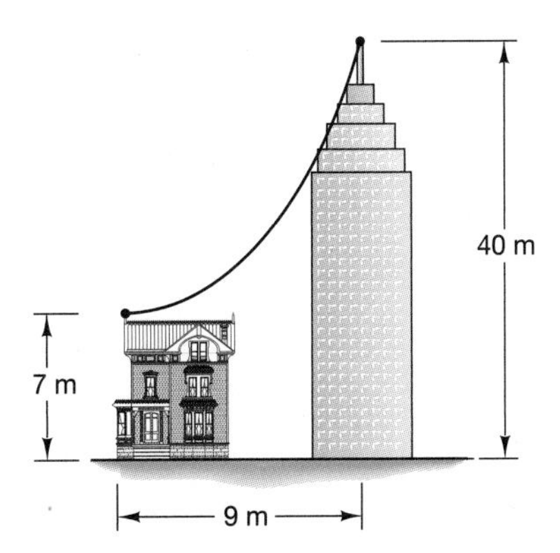

Figure P8.69

8.70 A telephone wire with a weight per unit length of 15 N/m hangs between two poles, as shown in Figure P8.70. If 45 m of cable is used between the two poles, determine both the tension in the wire as a function of the horizontal distance and the maximum tension in the wire.

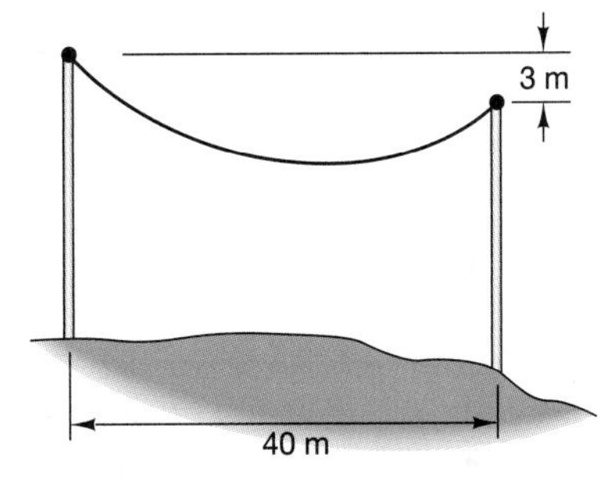

Figure P8.70

Chapter Summary

Equilibrium of an entire body implies that any part of the body is in equilibrium. Therefore, if an imaginary plane is cut through the body, internal forces acting on that plane are such that each sectioned part of the body is in equilibrium.

Beams If $w(x)$ is the distributed load acting on a beam (see figure on the right), the internal shear force and the bending moment are given by

$$V(x) = \int_0^x w(u)du$$

$$M(x) = \int_0^x (x - u)w(u)du$$

The load distribution, the shear force, and the bending moment are related as follows:

$$\frac{dV(x)}{dx} = w(x)$$

$$\frac{dM(x)}{dx} = V(x)$$

Discontinuity Functions The following singularity and discontinuity functions are useful in formulating the load distribution equations for beams (the Dirac delta function is used to represent a concentrated force at a position $x = a$):

$$\langle x - a\rangle_{-1} = \delta(x - a) = 0 \qquad x \neq a$$

$$\int_{-\infty}^{x} \langle \zeta - a\rangle_{-1} d\zeta = \int_{-\infty}^{x} \delta(\zeta - a) d\zeta = \begin{bmatrix} 0 & x < a \\ 1 & x > a \end{bmatrix}$$

$$\langle x - a\rangle_{-1} = \delta(x - a) = 0 \qquad x \neq a$$

$$\int_{-\infty}^{x} f(\zeta)\langle \zeta - a\rangle_{-1} d\zeta = \int_{-\infty}^{x} f(\zeta)\delta(\zeta - a) d\zeta = \begin{bmatrix} 0 & x < a \\ f(a) & x > a \end{bmatrix}$$

The unit doublet function is used to represent an applied moment at a position $x = a$:

$$\langle x - a\rangle_{-2} = 0 \quad x \neq a$$

$$\int \langle x - a\rangle_{-2} dx = \langle x - a\rangle_{-1}$$

The Heaviside step function has a value of zero for all values of x less than a (that is, whenever the argument is negative) and a value of unity when the argument is positive. Mathematically,

$$\langle x - a\rangle^0 = \begin{cases} 0 & x < a \\ (x - a)^0 = 1 & x > a \end{cases}$$

A group of polynomials having the value zero for negative arguments and $(x - a)^n$, $n > 0$, is called a set of discontinuity functions and is given by

$$\langle x - a\rangle^n = \begin{cases} 0 & x < a \\ (x - a)^n & x > a \end{cases}$$

Derivatives and integrals of the discontinuity function are as follows:

$$\frac{d}{dx}\langle x - a\rangle_{-1} = \langle x - a\rangle_{-2}$$

$$\frac{d}{dx}\langle x - a\rangle^0 = \langle x - a\rangle_{-1}$$

$$\frac{d}{dx}\langle x - a\rangle^n = n\langle x - a\rangle^{n-1} \qquad \text{for } n \geq 1$$

$$\int_0^x \langle u - a\rangle_{-2} du = \langle x - a\rangle_{-1}$$

$$\int_0^x \langle u - a\rangle_{-1} du = \langle x - a\rangle^0$$

$$\int_0^x \langle u - a\rangle^n du = \frac{\langle x - a\rangle^{n+1}}{n + 1} \qquad \text{for } n \geq 0$$

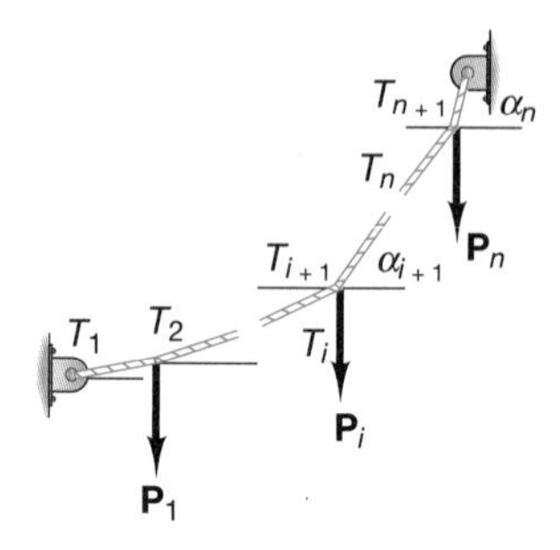

Cables The tensions in segments of a cable supporting concentrated loads can be determined by solving a set of nonlinear equations. At each point of application of a concentrated load, two equations of equilibrium may be written involving the two opposite tensions and the angles the cable makes with the horizontal. (See diagram at the right.) Therefore, a set of $2n$ simultaneous nonlinear algebraic equations is formed for the $(n + 1)$ tensions in the cable segments and the $(n + 1)$ cable segment angles with the horizontal. Two additional equations are obtained from the overall geometry of the cable attachments and the positions of the concentrated loads along the cable.

A cable supporting a load uniformily distributed along a horizontal line will make a parabolic shape in space according to the equation

$$y = \frac{1}{2}qx^2$$

where

$$q = \frac{w \text{ (uniform load)}}{T_0 \text{ (tension at the lowest point)}}$$

The tension in a point x measured horizontally from the lowest point is

$$T = T_0\sqrt{1 + q^2x^2}$$

A cable that supports its own weight will form a catenary curve in space, given by the equation

$$y = \frac{1}{q}\{\cosh (qx) - 1]$$

If s is the distance along the cable and x is the horizontal distance, q is defined by

$$s = \frac{\sinh (qx)}{q}$$

The tension at any point x measured horizontally from the lowest point is

$$T = T_0 \cosh (qx)$$

where T_0 is the tension at the lowest point and

$$q = \frac{w \text{ (weight of cable per unit length)}}{T_0 \text{ (tension at the lowest point)}}$$

Chapter 9

FRICTION

The ice climber depends upon friction between the crampons and the ice while climbing the frozen waterfall. (Courtesy of Franc Podgorsek/Shutterstock)

9.1 INTRODUCTION

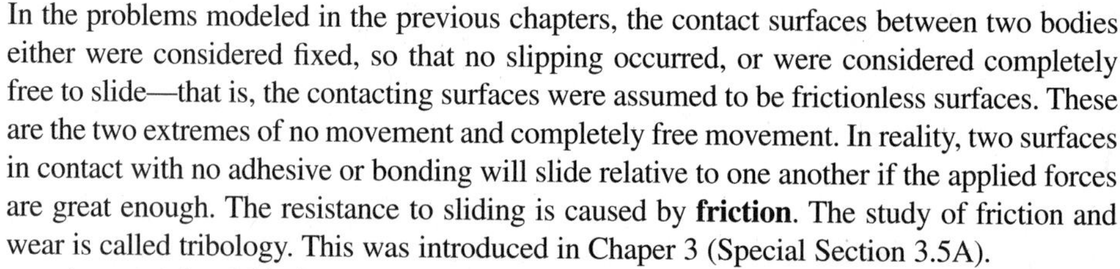

In the problems modeled in the previous chapters, the contact surfaces between two bodies either were considered fixed, so that no slipping occurred, or were considered completely free to slide—that is, the contacting surfaces were assumed to be frictionless surfaces. These are the two extremes of no movement and completely free movement. In reality, two surfaces in contact with no adhesive or bonding will slide relative to one another if the applied forces are great enough. The resistance to sliding is caused by **friction**. The study of friction and wear is called tribology. This was introduced in Chaper 3 (Special Section 3.5A).

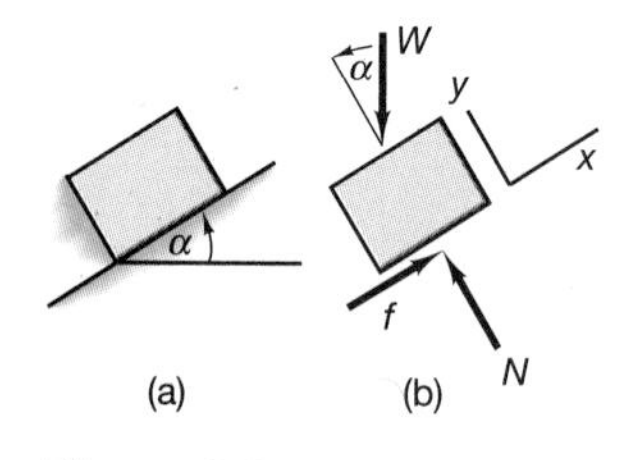

Figure 9.1

An example of friction may be seen by setting a book on an inclined plane, as shown in Figure 9.1. Figure 9.1(a) shows the book on an inclined plane of angle α, and Figure 9.1(b) is a free-body diagram of the book, with forces between the plane and the book designated as a normal force **N** and a ***friction force*** **f**. If you perform a simple experiment, you will find that the book will slide after you increase the angle of the plane to a certain value. Up to the angle at which the book slides, the magnitude of the friction force is obtained from static equilibrium equations. If we examine the free-body diagram in Figure 9.1(b), we find that the two equations of equilibrium are

$$N - W\cos\alpha = 0 \qquad (9.1)$$

$$f - W\sin\alpha = 0 \qquad (9.2)$$

Combining these two equations, we now find that the friction force is related to the normal force by

$$f = N\tan\alpha \qquad (9.3)$$

The maximum angle that the plane can be inclined before the book begins to slip represents the maximum friction force that can be generated. The ***coefficient of static friction*** is defined as the tangent of that angle:

$$\mu_s = \tan\alpha_{max} = f_{max}/N \qquad (9.4)$$

The angle α_{max} is called the ***friction angle***. The coefficient of static friction, μ_s, is equal to the maximum friction force that can be obtained between two surfaces before slipping occurs, divided by the normal force between the two surfaces.

Now consider a box on a level floor under the action of a force P, as shown in Figure 9.2(a). A free-body diagram of the box is shown in Figure 9.2(b), and if the body is in equilibrium, the following relationships hold:

$$N = W$$

$$f = P$$

$$Nb = Pa \quad \text{or} \quad b = (P/W)a$$

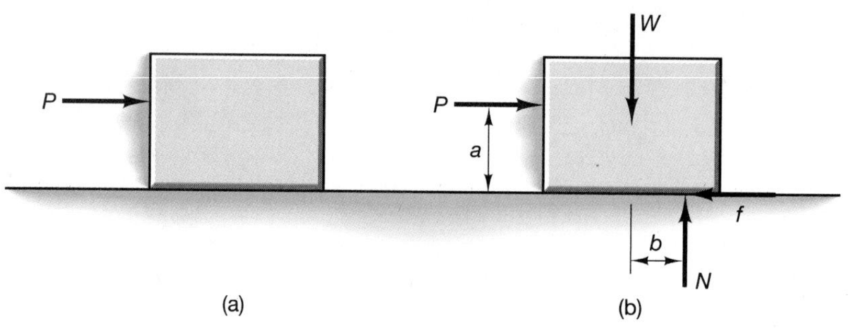

Figure 9.2

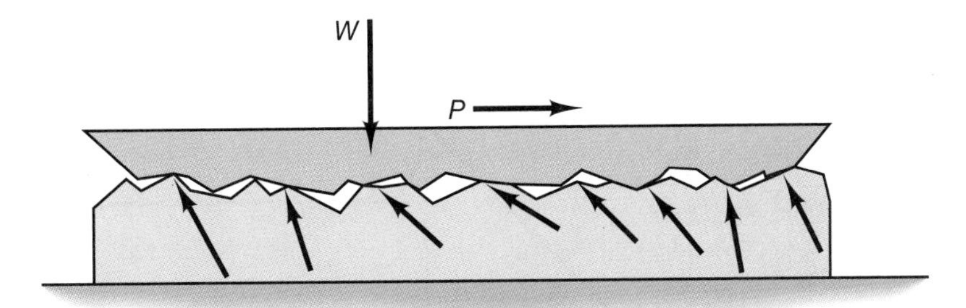

Figure 9.3

If the box were bonded to the floor, the tangential force **f** could continue to increase and would remain equal to the applied force **P**, and then the box would be in equilibrium for any value of **P**. When friction resists motion, the friction force has a limiting value given by $\mu_s N$. Therefore, the friction force in this case would increase as **P** increases, until it reached the maximum value, and the box would then be in the state of *impending slip*. The distance b is also limited and if μ_s is great enough, the box will tip when b equals a value of half the width of the box.

Two types of friction exist between contacting surfaces: fluid friction and dry friction. ***Fluid friction*** occurs when the surfaces are separated by a film of fluid, either gas or liquid. This type of friction is dependent upon the shear forces transmitted through the fluid film and is studied in detail in fluid mechanics. However, such friction is difficult to understand, as it arises from these shear forces between layers of fluids moving at different velocities. Fluid friction is the basis of most lubrication and is of great importance in the wear of machines with moving parts. The second type of friction is ***dry friction***, or ***Coulomb friction.*** The name "Coulomb friction" is derived from the fact that C. A. Coulomb studied this type of friction in 1781. Dry friction occurs between contacting bodies in the absence of a fluid film. In some situations, the distinction between the two types of friction is not clear, and the problem is usually modeled as dry friction. A car sliding on a snow-packed or icy road is such a case.

Dry friction is due to the roughness or asperities of the two contacting surfaces, as seen in Figure 9.3. As the force **P** attempts to push the upper surface to the right, the contacting forces occur at angles such that the forces oppose both **W** and **P**. The friction between two material surfaces is also influenced by temperature, the presence of foreign materials, and the molecular and atomic bonding between the surfaces.

When the surfaces begin to slip, the resisting friction decreases by as much as 25 percent, and the ratio of the friction force to the normal force is denoted by the **coefficient of kinetic friction**, μ_k, which is unitless. The change in friction force related to the applied load P is shown in Figure 9.4. The coefficient of static friction is the ratio of the maximum friction force to the normal force at the point of impending motion. When external forces are applied such that their lines of action are parallel to a surface of possible slip, the friction force will be equal in magnitude and opposite in sense to these applied forces. The line of action of the friction force is tangential to the surface. The friction force will maintain equilibrium, preventing slip until it reaches the maximum value given by the coefficient of static friction. As slip occurs, the resistance to sliding decreases such that the friction force is almost of constant magnitude, equal to the product of the coefficient of kinetic friction and the normal force between the sliding surfaces. Therefore, only at the point of

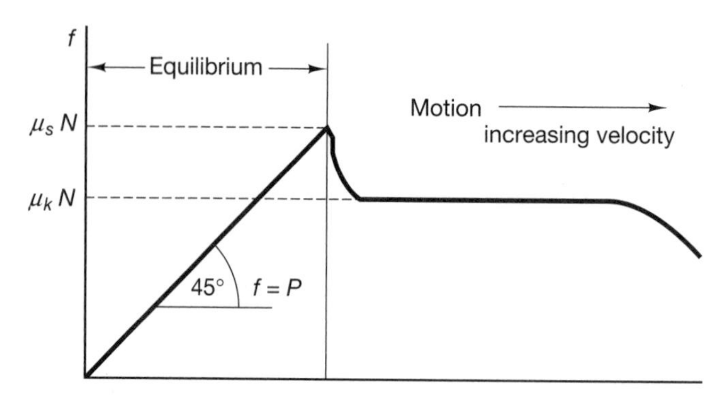

Figure 9.4

impending slip is the magnitude of the friction force equal to the product of the coefficient of static friction and the normal force.

In general, the coefficient of static friction is used only to determine whether slip will occur between the surfaces. If the required friction force for equilibrium is greater than the maximum possible between the surfaces, relative motion between the surfaces will occur. The magnitude of the friction force is then known and is equal to the product of the coefficient of kinetic friction and the normal force. If the velocity of the moving object is not constant, Newton's second law must be used to study the motion. As will all forces between contacting bodies, the internal friction forces occur in pairs equal in magnitude, opposite in sense, and collinear. Some typical values of the coefficient of static friction are given in Table 9.1. (The coefficient of kinetic friction μ_k is 20 to 25 percent lower than the corresponding coefficient of static friction.)

Problems involving rough surfaces will be examined in the next section, and kinetic friction will be considered in more detail when we study dynamics.

Table 9.1

Surface Materials	Coefficient of Static Friction
Steel on steel	0.75
Rubber on concrete	0.50–0.90
Rubber on ice	0.05–0.3
Metal on ice	0.04
Copper on copper	1.2
Metal on wood	0.20–0.60
Teflon® on Teflon®	0.04
Glass on glass	0.90
Copper on steel	0.50
Wood on wood	0.25–0.50
Aluminum on aluminum	1.10
Rubber tires on gravel	0.5
Rubber tires on dirt	0.3–0.5
Rubber tires on snow	0.1–0.3

9.2 COULOMB FRICTION

When dry friction occurs between two contacting surfaces, the friction force *always opposes the direction of the tendency to slip.* Consider the simple case of a block on a rough horizontal surface, as illustrated in Figure 9.5(a), and the corresponding free-body diagram, shown in Figure 9.5(b). The equilibrium equations are

$$P - f = 0$$
$$N - W = 0$$
$$Wb - Pa = 0$$

As the applied force P is increased in magnitude, the value of the resistive friction force f increases such that equilibrium is maintained. This balance continues until the point of impending slip, when the value of f is equal to $\mu_s N$. The maximum value of P, which can be applied before movement occurs, is then equal to the maximum value of the static friction force.

Notice that the normal force would not be directly below the center of gravity, as it must produce a moment in the opposite direction to that produced by the couple formed by the force P and the friction force f. The normal force can be considered concentrated at any point along the contacting surfaces, but the moment produced by the normal force and the weight is limited by the dimensions of the base of the block. The relevant equation is

$$b = \frac{P \cdot a}{W} \leq \left(\tfrac{1}{2}\text{ base dimension}\right) \text{ if c.g. is at center of block}$$

When the point of concentration is at the front corner of the block, the block will start to tip even if the friction force is not maximal. Therefore, one of two types of impending motion can occur: sliding or tipping.

If the friction force and the normal force are added to form a resultant contacting force, this resultant can lie anywhere within a cone formed by the sum of the maximum friction force and the normal force. This cone is illustrated in Figure 9.6. The tangent of the cone angle is given by

$$\tan \phi_s = \frac{f_{\max}}{N} = \mu_s \tag{9.5}$$

Note again that the resultant of the normal and friction force can lie anywhere within the associated cone. The cone angle is called the *angle of friction* and is equal to

$$\phi_s = \tan^{-1} \mu_s \tag{9.6}$$

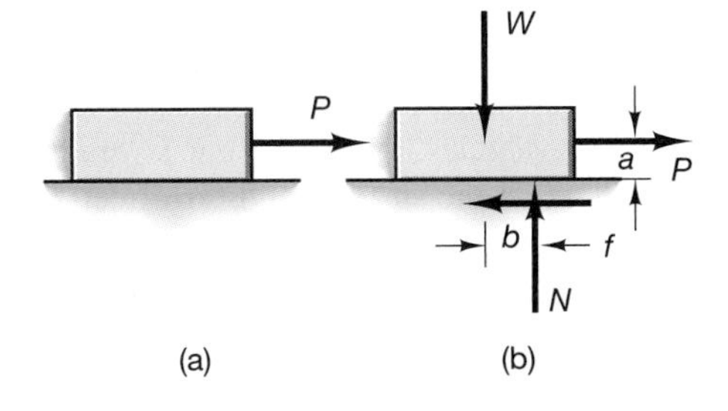

Figure 9.5

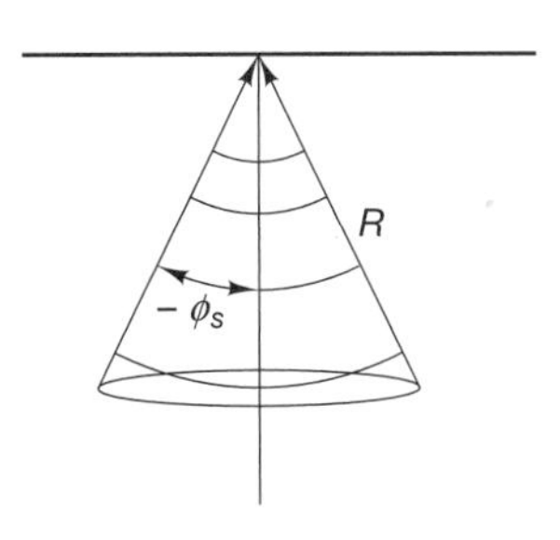

Figure 9.6

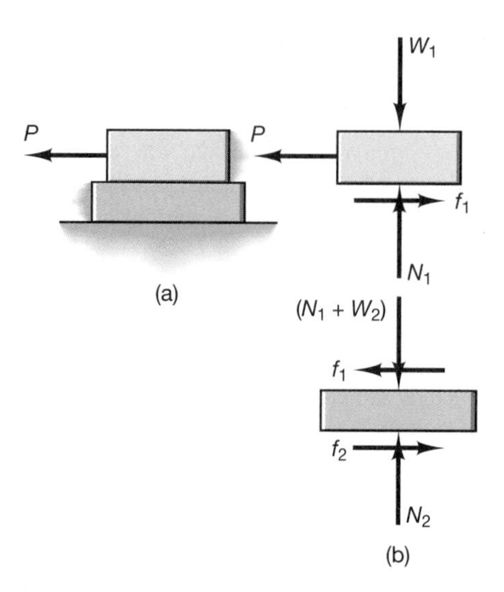

Figure 9.7

Next, consider the case of two blocks capable of sliding *relative* to one another. In Figure 9.7(a), the upper block has a tendency to slip to the left relative to the lower block, and the lower block has a tendency to slip to the right relative to the upper block. The friction force between the two blocks opposes this relative motion, and the corresponding free-body diagram is shown in Figure 9.7(b). Note that the friction force on the top block, f_1, acts to the right, and on the bottom block, f_1, is equal and opposite and acts to the left. The magnitudes of the two friction forces are equal, and their senses are indicated by the direction of the arrow. The friction force between the lower block and the surface f_2 resists motion and acts to the right. If the force P is increased, slip could occur between the upper block and the lower block, or both blocks could slide together. Note that if the coefficient of friction is the same between all surfaces, the maximum friction force between the bottom block and the floor will be greater than the friction force between the blocks as the normal force is greater between these surfaces.

Usually, the direction in which the friction force acts is easily determined by examining the direction of the possible movement. The friction force between the driving wheels of a car will depend upon whether the car is accelerating or braking. When the car is accelerating, the contact between the tire and the surface has a tendency to slip to the rear and spin through, and the friction force acts in a direction toward the front of the car. Indeed, it is this friction force that produces movement of the car. The driver gets "stuck" when the coefficient of static friction is so low that there is not enough friction force to make the car move, as in the case of tires on snow or ice. Four-wheel-drive vehicles can generate friction between all four wheels and the ground. Vehicles have approximately 60 percent of their weight on the front tires; thus, front-wheel-drive vehicles are more efficient in snow than are rear-wheel-drive vehicles. When the car is braking, the tires are not trying to rotate; the rubber is sliding on the surface, and the direction or sense of the friction force is toward the rear of the vehicle.

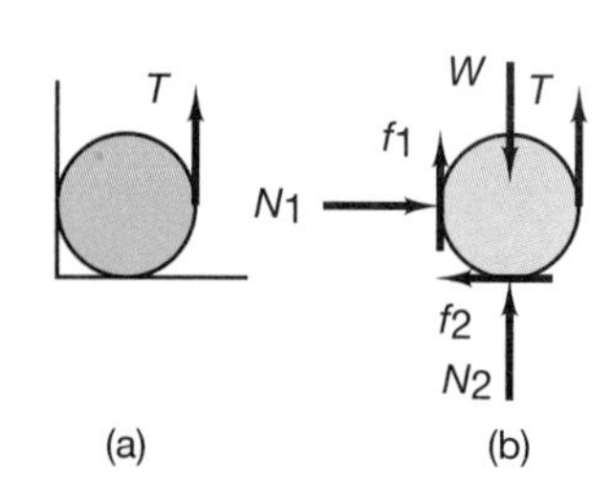

Figure 9.8

A conceptually more difficult case is the cylinder shown in Figure 9.8(a). If the cylinder is at the point of impending slip, the free-body diagram is as shown in Figure 9.8(b). The cylinder will then have the tendency to rotate counterclockwise, and the friction force at the left and the friction force at the bottom oppose this motion. If the cylinder is not at the point of impending slip, the problem is statically indeterminate, as there are four unknowns—the two normal forces and the two friction forces—and only three equations of equilibrium can be written.

Sample Problem 9.1

A worker places a ladder against a smooth wall and wishes to climb to the top of the ladder. (See the diagram at left.) If the coefficient of friction between the bottom of the ladder and the floor is μ, determine the minimum angle the ladder can make with the floor without slipping.

Solution The worker's weight acts at the top of the ladder, as that is the position for the maximum tendency to slip. From equilibrium, we have

$$N_b = W$$
$$N_a = f$$

and for the minimum value of θ, the friction force will be a maximum, given by

$$f = \mu N_b = \mu W$$

The friction force and N_a, on the one hand, and the weight and N_b, on the other, form couples that must be equal and opposite for equilibrium. Thus,

$$fL \sin\theta = WL \cos\theta$$
$$\tan\theta = 1/\mu$$

For example, if the coefficient of static friction between the floor and the base of the ladder is 0.8, the minimum angle the ladder could make with the floor for the worker to climb to the top is 51.3°.

Sample Problem 9.2

Determine the range of values of the force P required to prevent the 400-N box shown from slipping up or down the inclined plane if the coefficient of static friction between the box and the plane is 0.2.

Solution We choose a coordinate system parallel and perpendicular to the inclined plane and construct a free-body diagram of the box. (See figure at the left.) The friction force is shown acting in both directions. When P is a minimum, the tendency for slipping is down the incline, and friction will act up the incline to resist slipping. When P is a maximum, the tendency for slipping is up the plane, and friction resists by acting down the plane. Since we have specified a condition of impending slip, the magnitude of the friction force is the product of the coefficient of static friction and the normal force. Writing the equation of equilibrium, we have

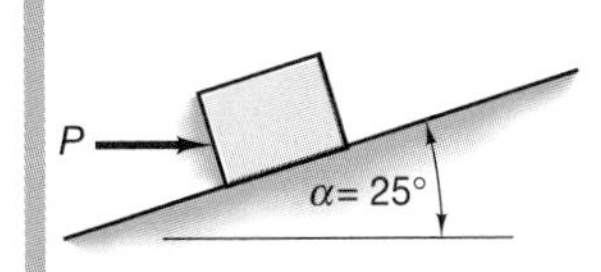

$$\mathbf{P} + \mathbf{W} + \mathbf{N} + \mathbf{f} = 0$$
$$\mathbf{P} = P(\cos 25°\hat{\mathbf{i}} - \sin 25°\hat{\mathbf{j}})$$
$$\mathbf{W} = 400(-\sin 25°\hat{\mathbf{i}} - \cos 25°\hat{\mathbf{j}})$$

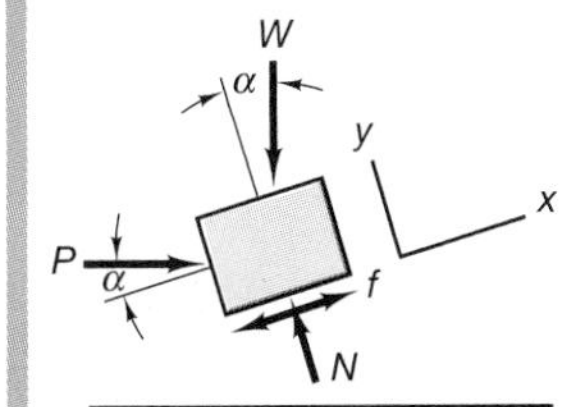

where

$$\mathbf{N} = N\hat{\mathbf{j}}$$
$$\mathbf{f} = \pm 0.2N\hat{\mathbf{i}} \ [(+) \text{ for the minimum } P \text{ and } (-) \text{ for the maximum } P]$$

Equating the $\hat{\mathbf{i}}$ and $\hat{\mathbf{j}}$ components of the vector equation gives the two scalar equations of equilibrium:

$$P_{max/min} \cos 25° - 400 \sin 25°(\mp)0.2\,N = 0$$
$$-P_{max/min} \sin 25° - 400 \cos 25° + N = 0$$

When the friction force is negative (acting down the plane), impending motion is up the plane, and P will be a maximum. The matrix equilibrium equation is

$$\begin{bmatrix} \cos 25° & -0.2 \\ -\sin 25° & 1 \end{bmatrix} \begin{bmatrix} P_{max} \\ N \end{bmatrix} = \begin{bmatrix} 300 \sin 25° \\ 400 \cos 25° \end{bmatrix}$$

Solving the matrix equation gives $P_{max} = 293.9$ N.

When the friction force is positive (acting up the plane), impending motion is down the plane, and the matrix equation is

$$\begin{bmatrix} P_{min} \\ N \end{bmatrix} = \begin{bmatrix} \cos 25° & +0.2 \\ -\sin 25° & 1 \end{bmatrix}^{-1} \begin{bmatrix} 300 \sin 25° \\ 400 \cos 25° \end{bmatrix}$$

The minimum force required to resist motion down the plane is thus $P_{min} = 97.4$ N.

Now let us solve the problem a second time with a different choice of coordinate system. The results are invariant to the choice of the coordinate system, but the equilibrium equations will have a different algebraic appearance. The free-body diagram is the same for either coordinate system. A unit vector tangent to the plane with a positive sense up the plane is

$$\hat{\mathbf{t}} = \cos 25°\hat{\mathbf{i}} + \sin 25°\hat{\mathbf{j}}$$

The unit normal vector is obtained using Eq. (5.6)—that is, the cross product of two tangential vectors:

$$\hat{\mathbf{n}} = \hat{\mathbf{k}} \times \hat{\mathbf{t}} = -\sin 25°\hat{\mathbf{i}} + \cos 25°\hat{\mathbf{j}}$$

The forces for this coordinate system are

$$\mathbf{P} = P_{max/min}\hat{\mathbf{i}}$$
$$\mathbf{W} = -400\hat{\mathbf{j}}$$
$$\mathbf{N} = N(-\sin 25°\hat{\mathbf{i}} + \cos 25°\hat{\mathbf{j}})$$
$$\mathbf{f} = (\mp)0.2N(\cos 25°\hat{\mathbf{i}} + \sin 25°\hat{\mathbf{j}})$$

The matrix equation to determine the value of P for impending motion up the plane and to resist impending motion down the plane is

$$\begin{bmatrix} P_{max/min} \\ N \end{bmatrix} = \begin{bmatrix} 1 & -\sin 25° \mp 0.2 \cos 25° \\ 0 & \cos 25° \mp 0.2 \sin 25° \end{bmatrix}^{-1} \begin{bmatrix} 0 \\ 400 \end{bmatrix}$$

The maximum and minimum magnitudes of P are

$$P_{min} = 97.4 \qquad P_{max} = 293.9$$

Sample Problem 9.3

For the block shown at the top left, determine the minimum force **P** as a function of θ for any angle of incline, α, to initiate motion.

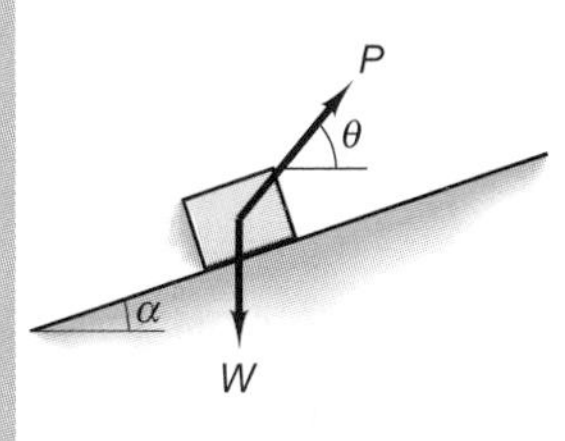

Solution

A coordinate system with x parallel to the incline and y perpendicular to the incline is selected. Then construct a free-body diagram of the block, as shown at the bottom left. Since the minimum value of P required for impending slip is sought, the magnitude of the friction force is obtained using the coefficient of static friction. The equilibrium equations become

$$P\cos(\theta - \alpha) - f - W\sin\alpha = 0$$
$$N - W\cos\alpha + P\sin(\theta - \alpha) = 0$$
$$f = \mu_s N$$

$$\begin{bmatrix} P \\ N \\ f \end{bmatrix} = \begin{bmatrix} \cos(\theta - \alpha) & 0 & -1 \\ \sin(\theta - \alpha) & 1 & 0 \\ 0 & -\mu_s & 1 \end{bmatrix}^{-1} \begin{bmatrix} W\sin\alpha \\ W\cos\alpha \\ 0 \end{bmatrix}$$

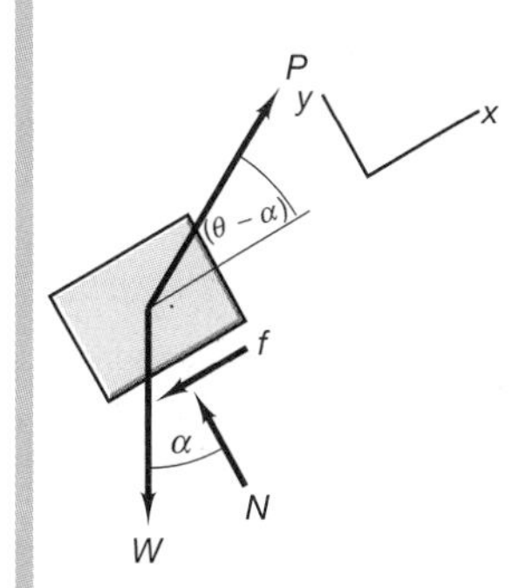

Solving these for P yields

$$P(\theta) = \frac{W(\sin\alpha + \mu_s\cos\alpha)}{\cos(\theta - \alpha) + \mu_s\sin(\theta - \alpha)}$$

It is clear that P is a function of θ, and the minimum value of P can be found by differentiating $P(\theta)$ with respect to θ and setting the differential equal to zero for a local maximum or minimum. The details of this computation are shown in the Computational Supplement. Consider a variation of this problem, pushing the block at the same angle of incline as that shown pulling the block as shown in the figure below.

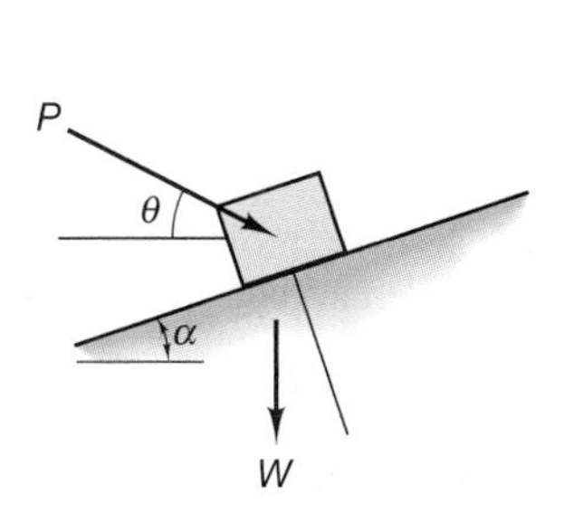

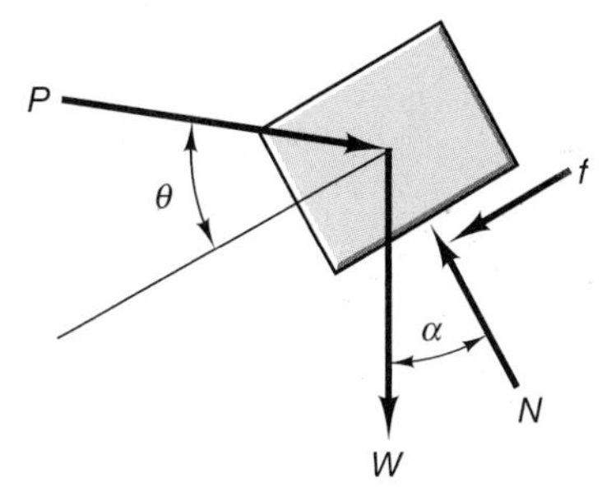

Again the equations of equilibrium are

$$P\cos(\theta + \alpha) - f - W\sin\alpha = 0$$
$$N - W\cos\alpha - P\sin(\theta + \alpha) = 0$$
$$f = \mu_s N$$

$$\begin{bmatrix} P \\ N \\ f \end{bmatrix} = \begin{bmatrix} \cos(\theta + \alpha) & 0 & -1 \\ \sin(\theta + \alpha) & 1 & 0 \\ 0 & -\mu_s & 1 \end{bmatrix}^{-1} \begin{bmatrix} W\sin\alpha \\ W\cos\alpha \\ 0 \end{bmatrix}$$

Solving for P yields

$$P(\theta) = \frac{W(\sin\alpha + \mu_s\cos\alpha)}{\cos(\theta + \alpha) - \mu_s\sin(\theta + \alpha)}$$

Note that pushing increases the normal and the minimum $P(\theta)$ required to move the block.

Sample Problem 9.4

A block of mass m rests on an inclined plane, as shown in the diagram at the top left. Determine the minimum coefficient of static friction required to keep the mass from sliding. The plane makes an angle α with the x-axis in the x–z plane and makes an angle β with the y-axis in the y–z plane.

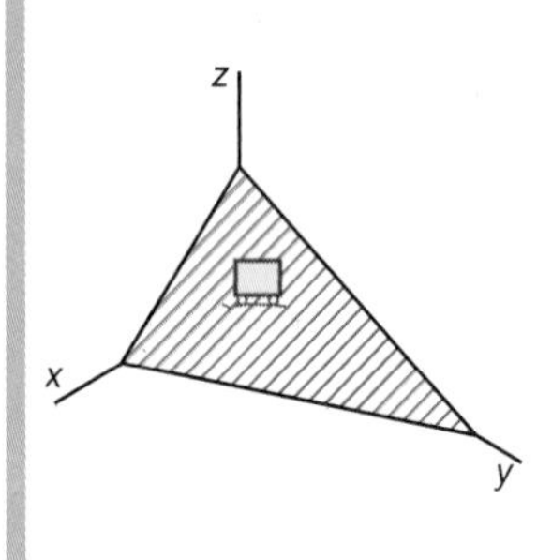

Solution The normal to the surface and the direction of the tangential force acting down the surface will be determined (see Eq. 6.6). We construct vectors along the intercepts of the surface with the x–y plane and the y–z plane:

$$\hat{\mathbf{m}} = \cos\alpha\hat{\mathbf{i}} - \sin\alpha\hat{\mathbf{k}}$$
$$\hat{\mathbf{q}} = \cos\beta\hat{\mathbf{j}} - \sin\beta\hat{\mathbf{k}}$$

The unit vector normal to the plane is defined by $\hat{\mathbf{m}}$ and $\hat{\mathbf{q}}$, as shown in the diagram at the right. Since these two vectors are not orthogonal, the unit normal is the cross product divided by its magnitude:

$$\hat{\mathbf{n}} = \frac{\hat{\mathbf{m}} \times \hat{\mathbf{q}}}{|\hat{\mathbf{m}} \times \hat{\mathbf{q}}|}$$

$$= \frac{1}{\sqrt{\cos^2\alpha + \sin^2\alpha\cos^2\beta}}[\sin\alpha\cos\beta\hat{\mathbf{i}} + \cos\alpha\sin\beta\hat{\mathbf{j}} + \cos\alpha\cos\beta\hat{\mathbf{k}}]$$

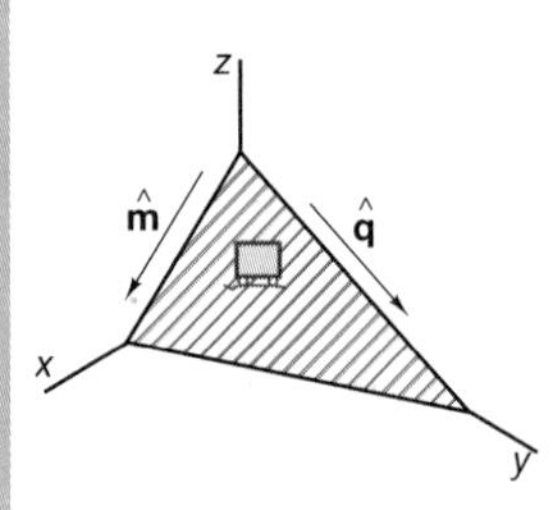

The only force—other than the normal force and the friction force—acting on the mass is gravitational attraction, which results in the block's weight.

$$W = -mg\hat{\mathbf{k}}$$

The components of $\mathbf{W}$ in the normal and tangential directions are, respectively,

$$W_n = (\mathbf{W} \cdot \hat{\mathbf{n}})\hat{\mathbf{n}}$$
$$W_t = W - W_n$$

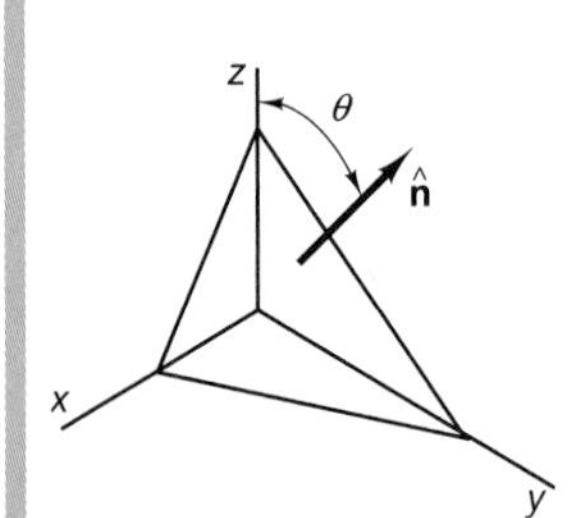

The required friction force must be equal in magnitude and opposite in sense to $\mathbf{W}_t$, and the normal force is equal in magnitude and opposite in sense to $\mathbf{W}_n$ so that

$$N = -W_n \quad f = -W_t$$

The minimum coefficient of static friction is

$$\mu_{s\,\min} = \frac{|\mathbf{f}|}{|\mathbf{N}|}$$

Since the unit vector $\hat{\mathbf{n}}$ is normal to the surface, the angle θ between the z-axis and the normal vector is the angle of incline of the plane. (See figure at left.) This angle is

$$\theta = \cos^{-1}(\hat{\mathbf{n}} \cdot \hat{\mathbf{k}})$$

We have shown earlier that the coefficient of static friction can be defined by the tangent of the angle of incline corresponding to impending slip. Therefore, the minimum coefficient of static friction for any inclined surface is

$$\mu_s = \tan\theta = \tan[\cos^{-1}(\hat{\mathbf{n}} \cdot \hat{\mathbf{k}})]$$

The vector calculations can be carried out for specific numerical cases, and an example is shown in the Computational Supplement.

Sample Problem 9.5

The 3-m-tall 100-kg crate shown at the top left is loaded such that its center of mass is 2 m above the 1-m base. If the coefficient of static friction between the base and the inclined surface is 0.4, determine the force required to initiate motion, and determine the minimum and maximum height h where the force can be applied.

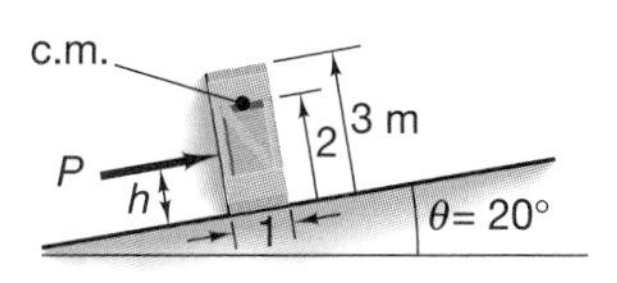

Solution The crate can slide up the incline, or, if the force **P** is applied too high, it will tip forward; if the force **P** is applied too low, it will tip back. First, we construct a free-body diagram of the crate. (See figure at the bottom left.) The x-axis is chosen to be parallel to the inclined plane. We have:

$$\sum F_x = P - W\sin\theta - f = 0$$

$$\sum F_y = -W\cos\theta + N = 0$$

$$\sum M_{\text{c.m.}} = P(2 - h) - 2f - Nd = 0$$

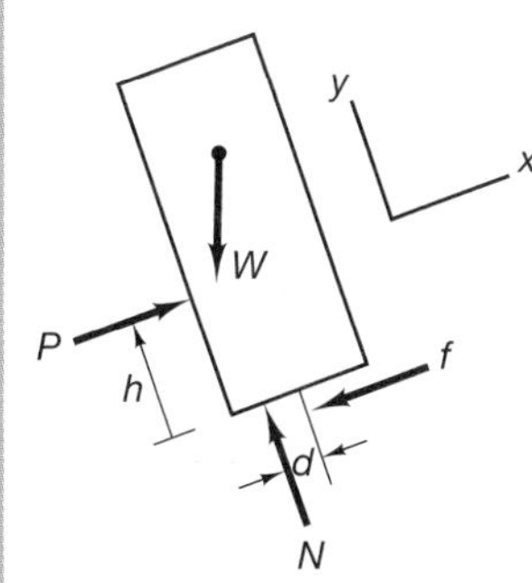

Therefore,

$$N = W\cos\theta$$
$$P = W\sin\theta + \mu_s W\cos\theta$$

The f maximum value d can have is $+0.5$, to prevent tipping backward, and the minimum is -0.5, to prevent tipping forward. Therefore, the minimum value of h required to prevent tipping backward can be found from the moment equation:

$$(W\sin\theta + \mu_s W\cos\theta)(2 - h) - (\mu_s W\cos\theta)(2) - (W\cos\theta)(0.5) = 0$$

$$h = \frac{2\sin\theta - \cos\theta}{\sin\theta + \mu_s\cos\theta}$$

For the given incline and coefficient of static friction, the minimum value of $h = 0.3$ m. The maximum value of h required to prevent tipping forward is found by letting $d = -0.5$ in the moment equation:

$$h = \frac{2\sin\theta + \cos\theta}{\sin\theta + \mu_s\cos\theta}$$

The maximum value of h is $h = 1.6$ m.

Sample Problem 9.6

A yo-yo is held by its string on an inclined plane, as shown in the figure at the top left. Determine the minimum value of the coefficient of static friction required for equilibrium.

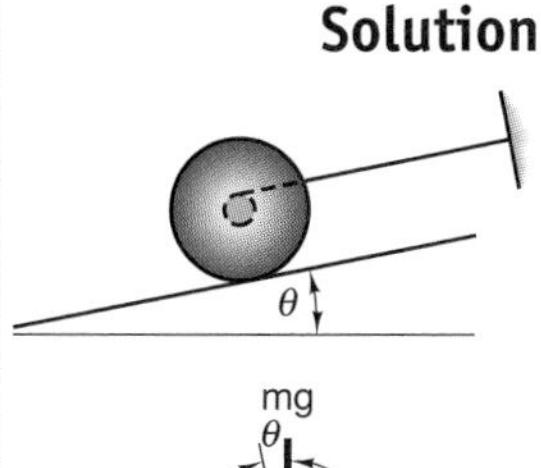

Solution We construct a free-body diagram as illustrated at the bottom left, from which we obtain the following equations:

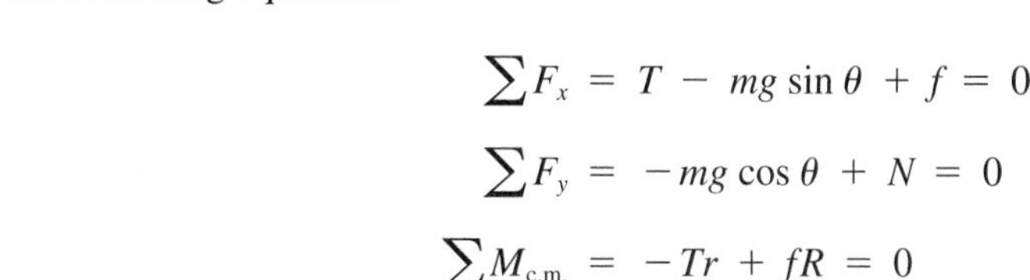

$$\sum F_x = T - mg\sin\theta + f = 0$$

$$\sum F_y = -mg\cos\theta + N = 0$$

$$\sum M_{\text{c.m.}} = -Tr + fR = 0$$

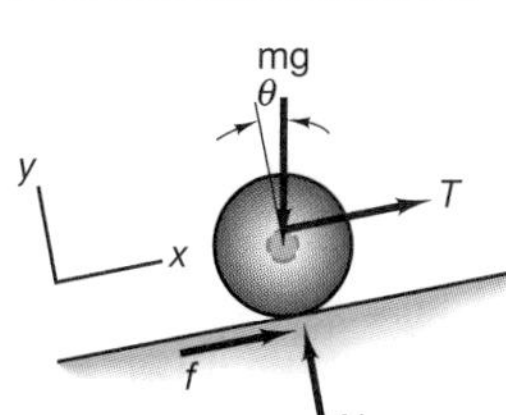

Solving for f and N yields

$$\mu_{s\,\min} = \frac{f}{N} = \frac{r}{r + R}\tan\theta$$

Problems

9.1 A 100-N force pulls a 200-N block, as shown in Figure P9.1. The coefficient of static friction between the block and the floor is $\mu_s = 0.6$, and the coefficient of kinetic friction is $\mu_k = 0.4$. (a) What is the friction force between the block and the floor? (b) Will the block move?

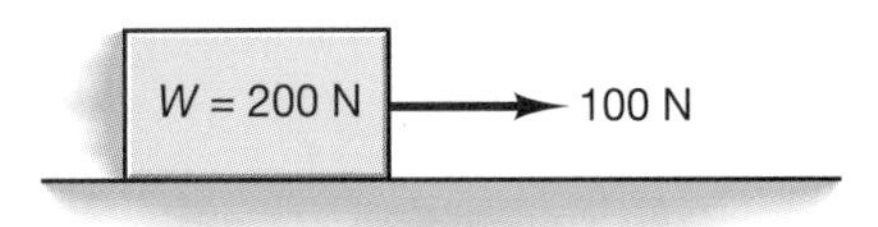

Figure P9.1

9.2 (a) A 200-kg block is resting on a 30° incline. The coefficient of static friction between the block and the plane is $\mu_s = 0.8$. Will the block remain at rest? (b) A 50-kg mass is resting on a 45° incline. The coefficient of static friction between the block and the plane is $\mu_s = 0.6$. Will the block remain at rest?

9.3 What is the maximum angle of incline for a 50-kg block to rest on an inclined plane if the coefficient of friction between the block and the plane is $\mu_s = 0.35$?

9.4 A 10-N force is pushing a 40-N block, as shown in Figure P9.4. The angle of incline of the plane is $\alpha = 40°$. The coefficient of static friction between the block and the incline is $\mu_s = 0.75$, and the coefficient of kinetic friction is $\mu_k = 0.65$. Will the block slide on the plane? If it does, will it slide up or down the plane? What is the friction force between the block and the plane?

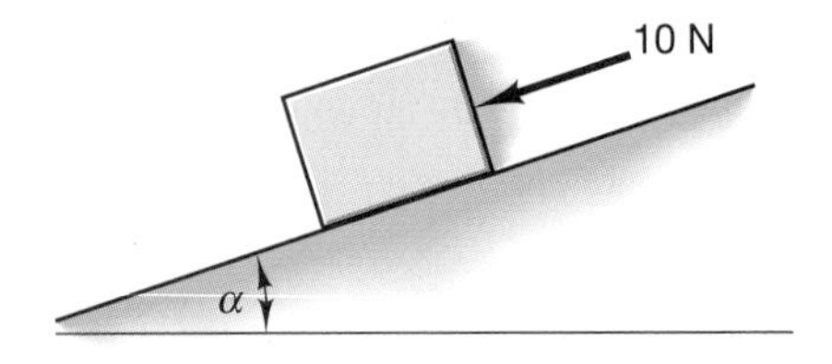

Figure P9.4

9.5 A 60-N force is pushing a 10-kg mass down the plane as shown in Figure P9.5. The angle of incline of the plane is $\alpha = 20°$. The coefficient of static friction between the block and the plane is $\mu_s = 0.6$, and the coefficient of kinetic friction is $\mu_k = 0.5$. Will the block slide on the plane? If it does, will it slide up or down the plane? What is the friction force between the block and the plane?

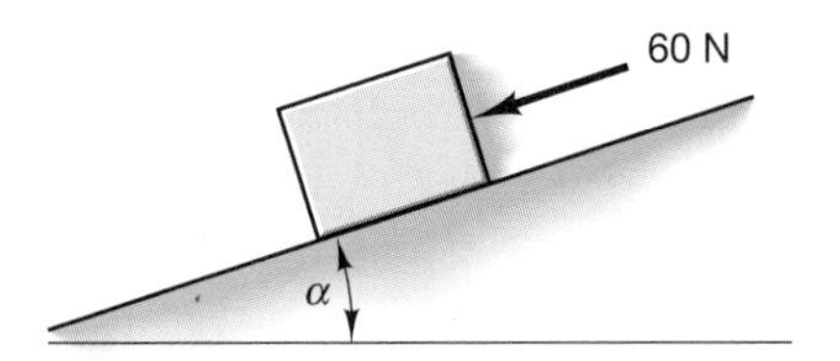

Figure P9.5

9.6 For Problem 9.4, what is the maximum angle of incline, α, if static equilibrium is to be maintained?

9.7 A 100-N force is pulling a 50-kg block as shown in Figure P9.7. The coefficient of static friction between the block and the incline is $\mu_s = 0.4$, and the coefficient of kinetic friction is $\mu_k = 0.3$. The angle of incline is $\alpha = 35°$. (a) Will the block move? (b) What is the friction force between the block and the plane?

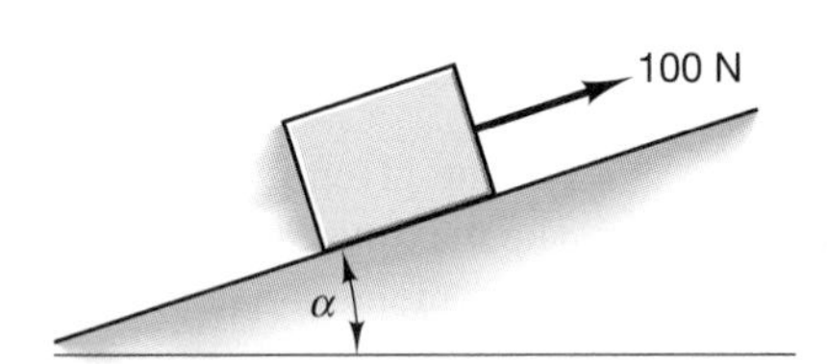

Figure P9.7

9.8 A 75-N force, inclined at an angle $\alpha = 45°$ as shown in Figure P9.8, pulls a 100-N block. The coefficient of static friction between the block and the floor is $\mu_s = 1.3$, and the coefficient of kinetic friction is $\mu_k = 1.1$. (a) Will the block move? (b) What is the friction force between the block and the plane?

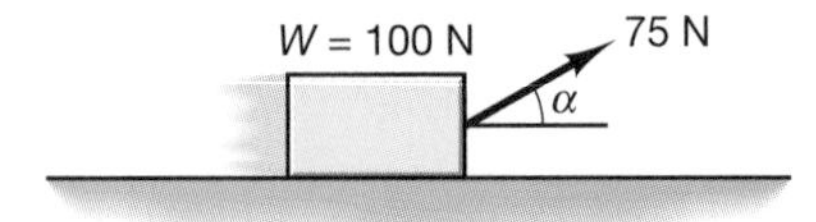

Figure P9.8

9.9 A 5-kg block rests on a rail with a 100-kg mass hanging from it. (See Figure P9.9.) (a) What force is required to begin moving the block along the rail? The coefficient of static friction for the block on the rail is $\mu_s = 0.6$, and the coefficient of kinetic friction is $\mu_k = 0.5$. The force is inclined at 60°, as shown. (b) For what angles α can the force F slide the block to the right?

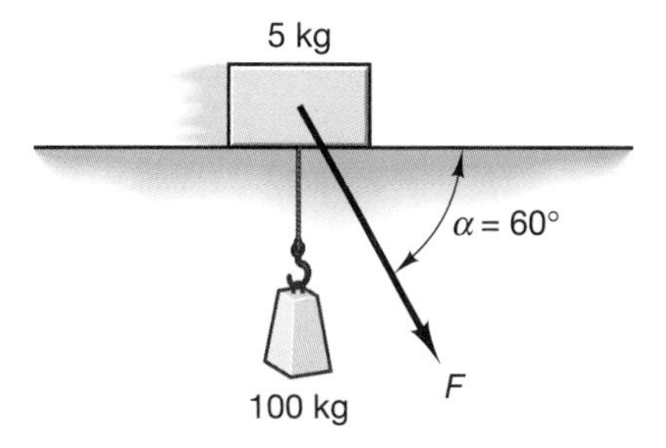

Figure P9.9

9.10 A 1000-N force, inclined at an angle $\beta = 35°$, pulls a 100-kg block. (See Figure P9.10.) The coefficient of static friction between the block and the incline is $\mu_s = 0.25$, and the coefficient of kinetic friction is $\mu_k = 0.15$. The angle of incline α, is 40°.(a) What is the friction force between the block and the plane? (b) Will the block move?

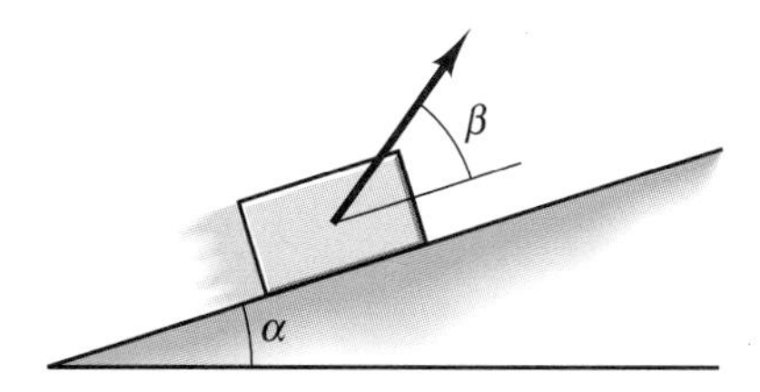

Figure P9.10

9.11 A 650-N force, inclined at an angle $\beta = 25°$, pushes a 1500-N block, as shown in Figure P9.11. The coefficient of static friction between the block and the floor is $\mu_s = 1.05$, and the coefficient of kinetic friction is $\mu_k = 0.95$. The angle of incline, α, is 30°. (a) Will the block move? (b) What is the friction force between the block and the plane?

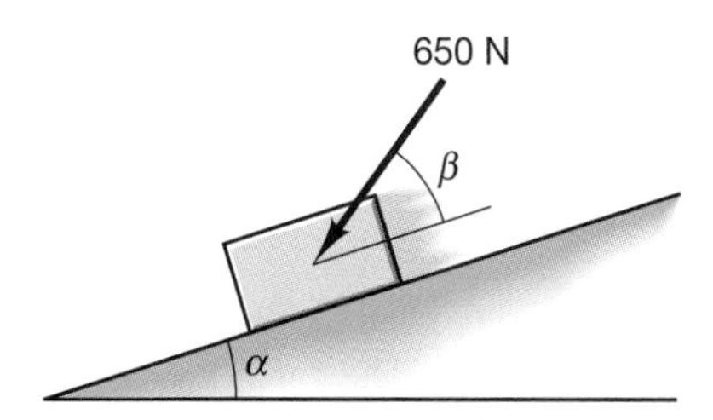

Figure P9.11

9.12 A rope pulls a disk, as shown in Figure P9.12. What is the maximum tension in the rope prior to slipping? The coefficient of static friction for both interfaces is $\mu_s = 0.6$, the coefficient of kinetic friction is $\mu_k = 0.5$, the radius is 1 m, and the weight of the disk is 20 N.

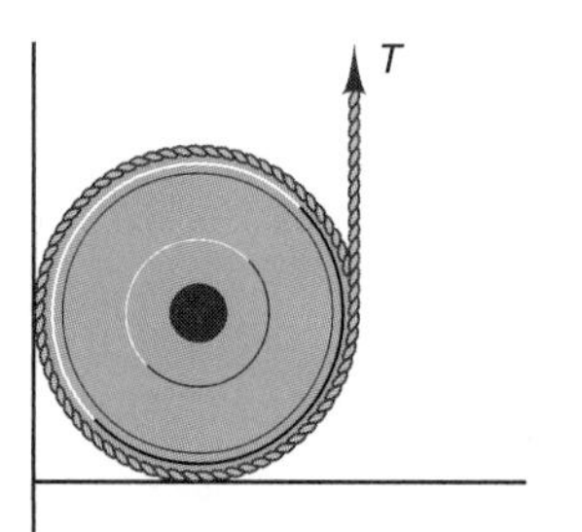

Figure P9.12

9.13 Solve for the maximum moment M, in terms of μ_s, r, and W, that can be applied to the wheel shown in Figure P9.13 prior to the wheel slipping. The coefficient of static friction, μ_s, is the same for both interfaces.

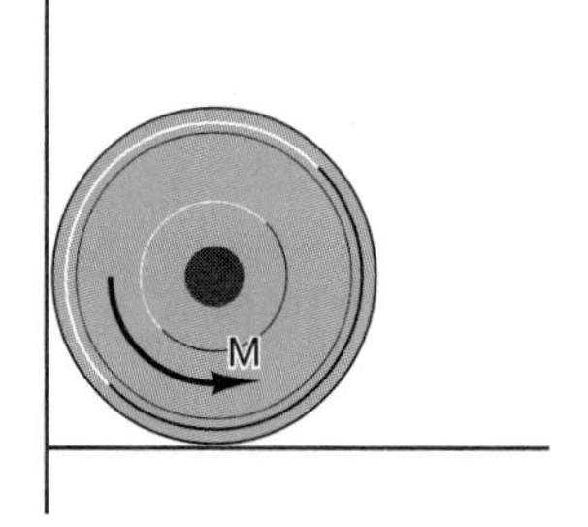

Figure P9.13

9.14 Two blocks are on an inclined plane, as shown in Figure P9.14. Block A has a mass of 100 kg and block B has a mass of 10 kg. The angle of incline is 35°. (a) What is the minimum coefficient of friction required to prevent the blocks from slipping? Assume that the coefficient of friction is the same between the plane and the block and between the two blocks. (b) What is the tension in the rope?

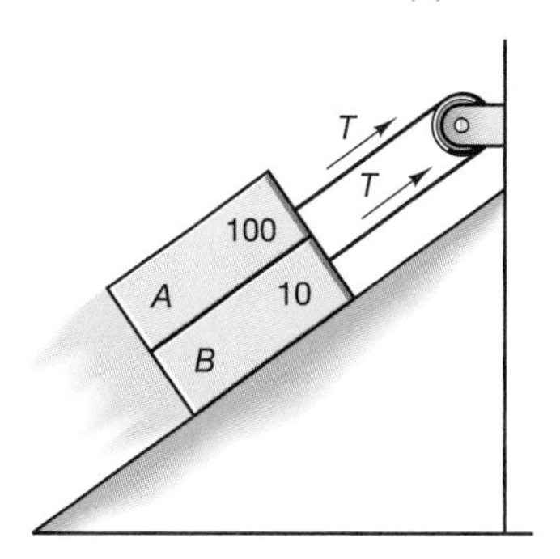

Figure P9.14

9.15 Two blocks are on an inclined plane, as shown in Figure P9.15. Block A has a mass of 10 kg and block B has a mass of 25 kg. The angle of incline is 20°. (a)What is the minimum coeffiecient of friction required to prevent the blocks from slipping? Assume that the coefficient of friction between the blocks is four times the coefficient of friction between the block and the plane. (b) What is the tension in the rope?

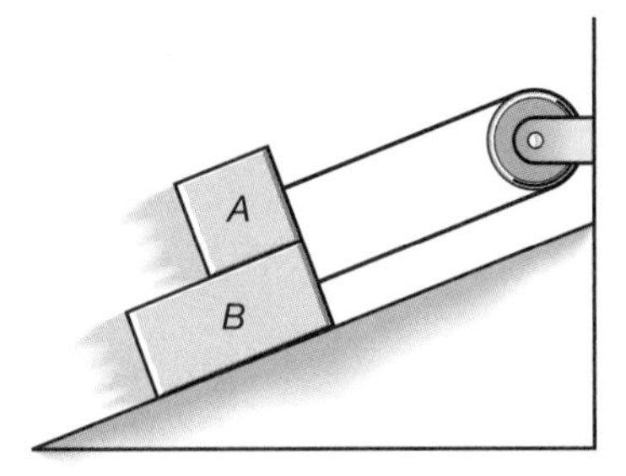

Figure P9.15

9.16 A plane is inclined 30° from the x-axis. A 50-kg mass is resting on the plane. The coefficient of static friction between the plane and the block is $\mu_s = 0.5$, and the coefficient of kinetic friction is $\mu_k = 0.4$. A force $\mathbf{F} = 20\hat{\mathbf{i}} + 40\hat{\mathbf{j}}$ N is applied to the block. Assume that gravity acts in the y-direction. Will the block slide?

9.17 A roofer is standing on a roof. The roofer has a mass of 80 kg. The coefficient of friction between the roofer's boots and the plywood is $\mu_s = 0.4$. What is the maximum incline on the roof required to prevent the roofer from slipping?

9.18 A conveyor belt is used in an airport to move baggage from the ground to the cargo hold of an airplane. (See Figure P9.18.) The angle of incline of the belt is 40°. What minimum coefficient of friction between the baggage and the belt is required to prevent the baggage from slipping?

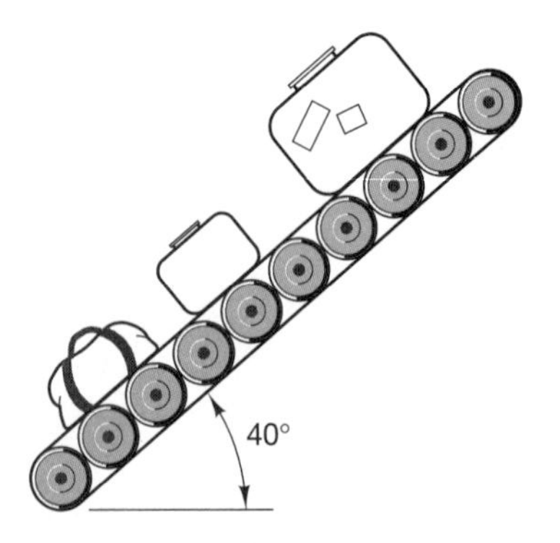

Figure P9.18

9.19 A truck, shown in Figure P9.19, is going down the road at 100 km/hr. The truck has a mass of 1100 kg. The coefficient of rolling friction between the wheels of the truck and the ground is $\mu_r = 0.35$. What is the friction force between each wheel and the ground? Assume that the center of gravity of the truck is at the center of the truck.

Figure P9.19

9.20 A hoist is used in a garage to lift the engine out of a car and move it to a stand. (See Figure P9.20.) The engine has a mass of 600 kg. Assume that the hoist has negligible mass. The coefficient of static friction between the hoist and the rail is $\mu_s = 0.4$, and the coefficient of kinetic friction is $\mu_k = 0.3$. What is the maximum force that must be applied to start the engine moving?

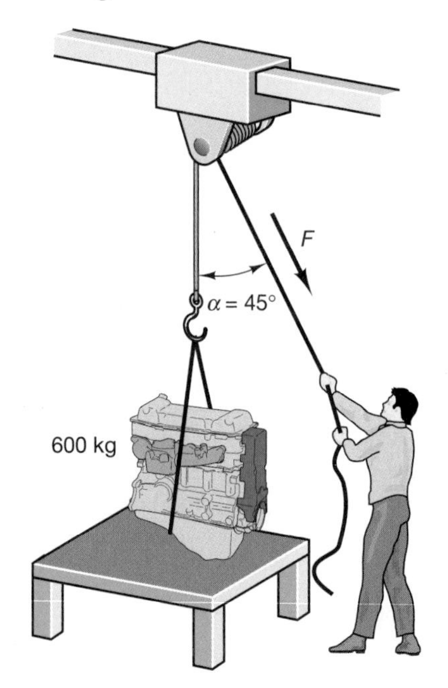

Figure P9.20

9.21 A person is trying to roll an 8-kg log along the ground, as shown in Figure P9.21. The force the person exerts is at 40° to the ground. The coefficient of static friction between the log and the ground is $\mu_s = 0.85$, and the coefficient of kinetic friction is $\mu_k = 0.7$. How hard must the person push the log (i.e., how much force must she exert) to start it rolling?

Figure P9.21

9.22 A skier wishes to start down a hill from rest at the top of the hill. (See Figure P9.22.) She has a mass of 54 kg, the coefficient of friction between her skis and the snow is $\mu_s = 0.04$, and the coefficient of kinetic friction is $\mu_k = 0.03$. The slope of the hill is 60°. Her poles form a 45° angle with the ground. How much force must she exert on the poles? Assume that she can exert force straight down the pole.

Figure P9.22

9.23 A sled is used to move a lathe from one end of a factory to the other as shown in Figure P9.23. How much force must the tractor supply to start the machine moving? The lathe weighs 1000 N and the sled weighs 300 N. The coefficient of static friction between the sled and the concrete is $\mu_s = 0.7$, and the coefficient of kinetic friction is $\mu_k = 0.6$. The horizontal distance between the sled and the tractor is 3 m, and the vertical distance between the eyebolt on the sled and the attachment to the tractor is 1 m.

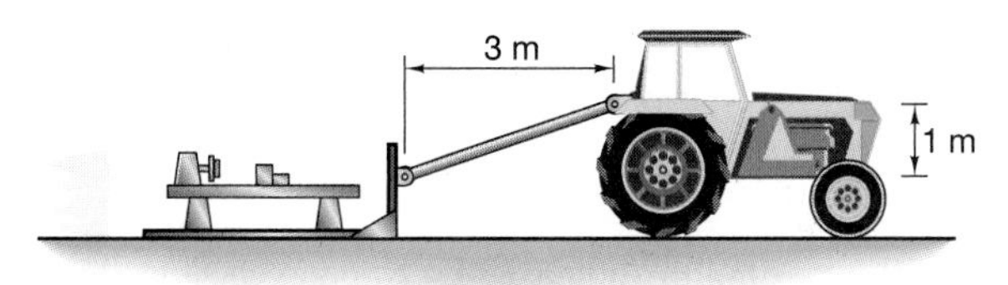

Figure P9.23

9.24 Solve Problem 9.7; however, vary the angle of incline from 0 to 80° in 5° increments.

9.25 Solve Problem 9.8; however, vary the angle of incline from 0 to 80° in 5° increments.

9.26 A contractor needs to choose the slope of a driveway so that a car may safely park on it. (See Figure P9.26.) With the rear wheel (emergency) brake set, compute the largest angle that the driveway may make with the horizontal. The car has a mass of 1800 kg, and the coefficient of static friction for a tire on wet pavement is about 0.6.

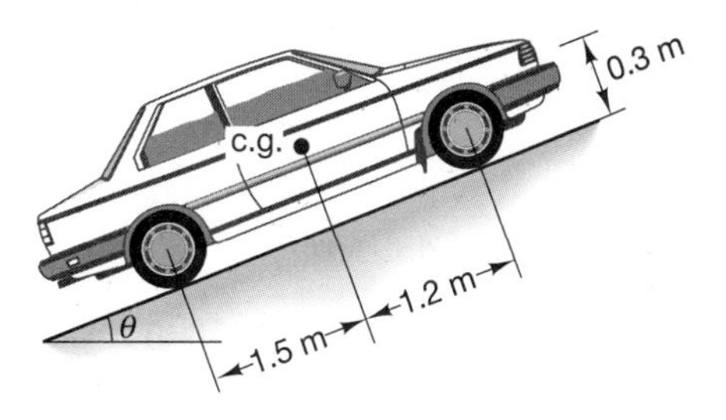

Figure P9.26

9.27 Repeat Problem 9.26 for the case where both of the brakes are set, so that the front *and* rear wheels apply friction. They will do so when the brake pedal is pushed.

9.28 Consider the problem of finding the maximum angle of a driveway, stated in Problem 9.26. For a car with just its parking brake on, compare the maximum angle for a snow-covered driveway ($\mu_s = 0.3$), a rain-covered driveway ($\mu_s = 0.6$), and a dry driveway ($\mu_s = 0.75$).

9.29 Compute the largest force allowed and the largest height of application for a block to simultaneously slip and tip up where the interface between the block and the surface is (a) steel on steel, (b) copper on copper, and (c) aluminum on aluminum. (See Figure P9.29.) The center of mass of the block is at its geometric center. Does the largest height depend on the weight?

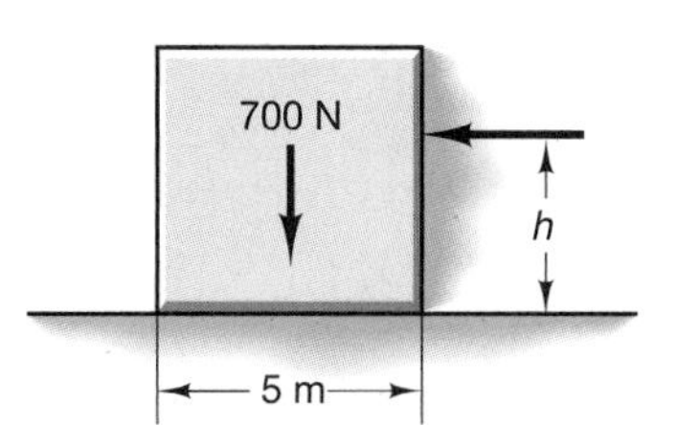

Figure P9.29

9.30 What values of mass m can be held in equilibrium in the steel container shown resting on a steel incline in Figure P9.30 for the case where $\theta = 30°$?

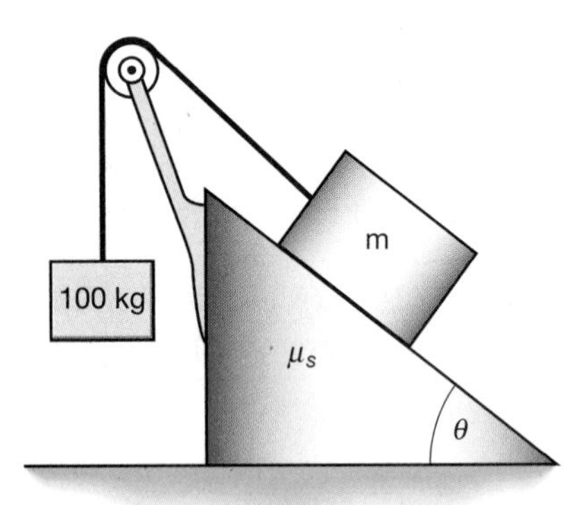

Figure P9.30

9.31 Repeat Problem 9.30 for $\theta = 60°$.

9.32 Is the crate shown in Figure P9.32 in equilibrium? The center of gravity of the crate is at its geometric center, and $\mu_s = 0.6$.

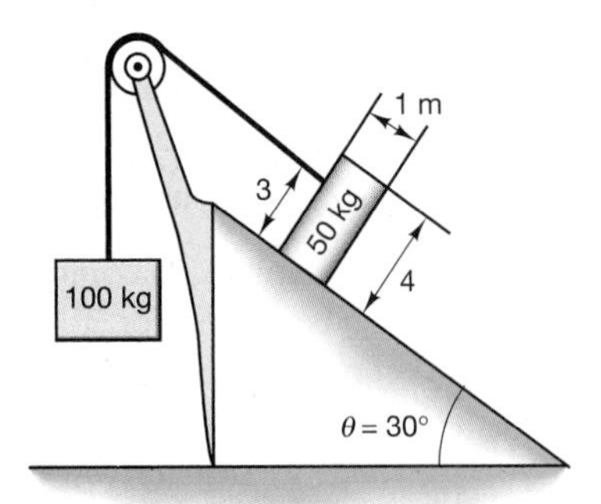

Figure P9.32

9.33 If the rope is removed from the crate of Figure P9.32, will the crate tip, slide, or stay put?

9.34 Is the box illustrated in Figure P.9.34 in equilibrium? Assume that the center of gravity of the box is at its geometric center.

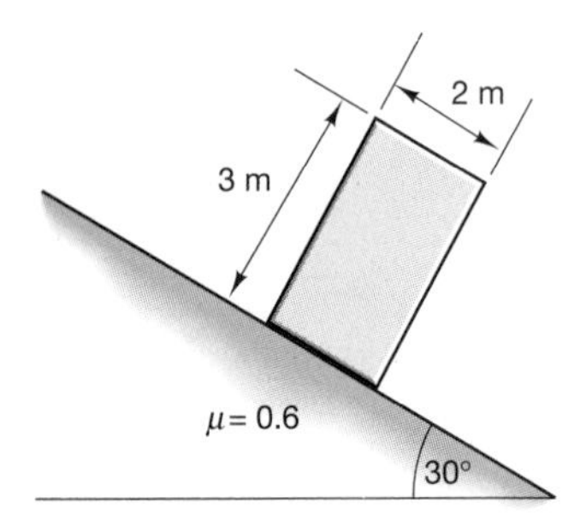

Figure P9.34

9.35 Suppose that the box in Figure P9.34 has its center of gravity 1 m down and 1 m over from the upper left-hand corner. Is the box in equilibrium?

9.36 A crate with an off-center center of gravity must be held on an incline, as shown in Figure P9.36. Assuming that the worker can provide a maximum force P of 750 N at a height h of no more than 1.5 m, can she hold the crate in equilibrium? Assume that the mass of the crate is 500 kg.

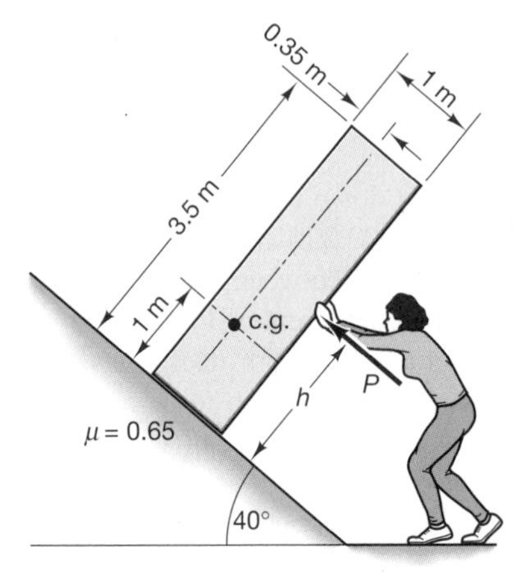

Figure P9.36

9.37 It is desired to push a wooden chest of drawers across a wood floor without tipping the chest. (See Figure P9.37.) If the chest has a mass of 75 kg, how much force should be applied, and where should it be applied, so that the chest slides without tipping?

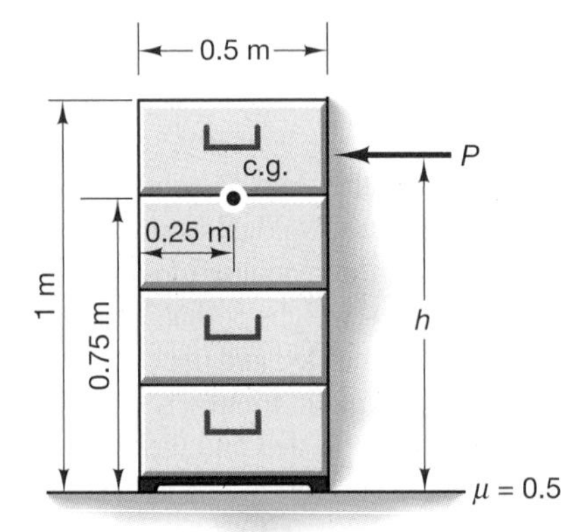

Figure P9.37

9.38 Repeat Problem 9.37 for the case where the floor is polished so that the coefficient of static friction is reduced to 0.3.

9.39 It is desired to slide a 75-kg wooden chest of drawers across a wood floor without tipping the chest. (See Figure P9.39.) (a) How much force should be applied, and where should it be applied, so that the chest slides without tipping? (b) If the chest is turned around so that the force is applied on the opposite side of the center of gravity, how much force should be applied, and where should it be applied, so that the chest slides without tipping?

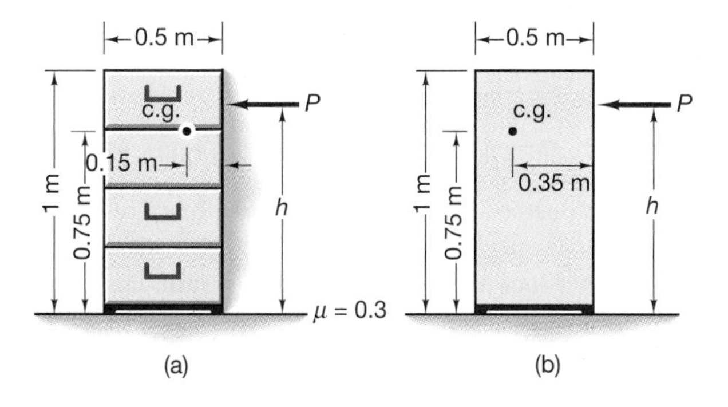

Figure P9.39

9.40 Determine the value of the applied force P in terms of the weight W of the box shown in Figure P9.40, and the coefficient of static friction μ_s at the condition of impending slip. Derive an expression for the impending slip, and show that it is independent of W and b.

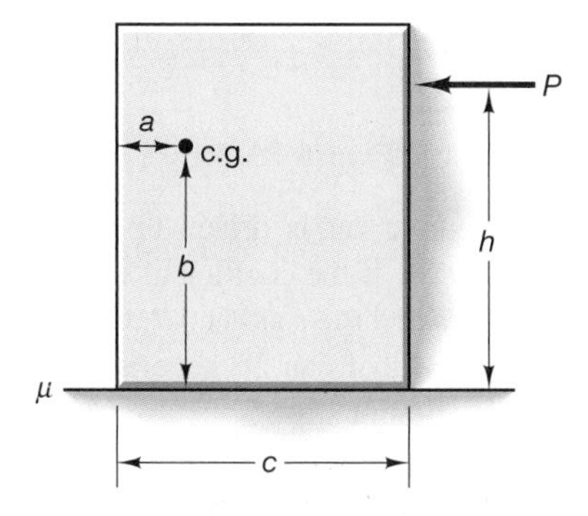

Figure P9.40

9.41 A man attempts to slide a 100-kg box by applying a 600-N force at the angle shown in Figure P9.41. The coefficient of static friction is 0.5. Does the box move (slide or tip)? The center of gravity is at the geometric center of the box.

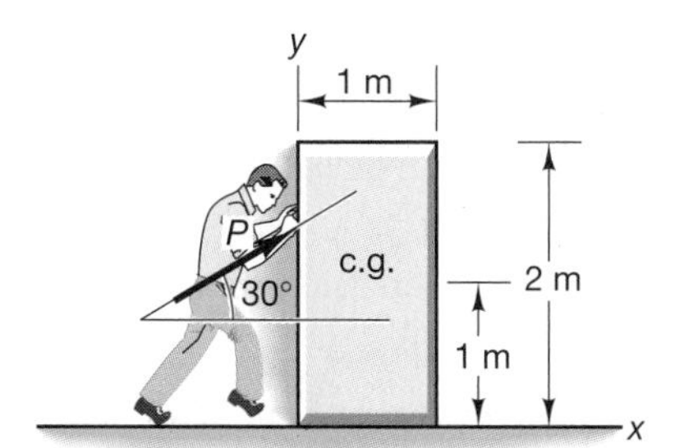

Figure P9.41

9.42 Consider the man pushing on the box in Figure P9.41. Suppose that he pushes at an angle of 45°, that the box has a mass of 100 kg, and that the coefficient of static friction is $\mu_s = 0.8$. Does the box move, and, if so, does it slide or tip?

9.43 A 55-N force is applied to the top box in Figure P9.43, which weighs 100 N. The bottom box weighs 50 N, and the coefficient of static friction between the ground and the bottom box is $\mu_s = 0.3$, while that between the two boxes is $\mu_s = 0.6$. Assuming that the boxes are wide enough so that no tipping will occur, is there any sliding, or do the boxes stay put?

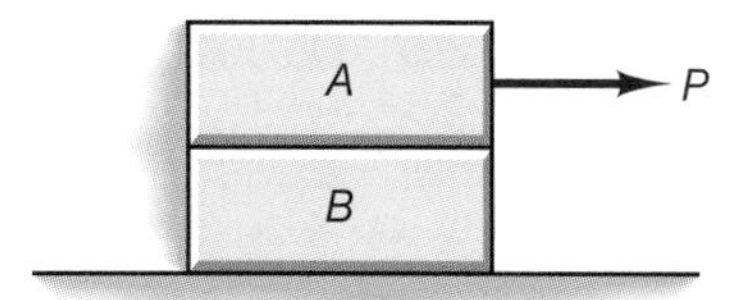

Figure P9.43

9.44 Consider again the two boxes of Figure P9.43. Compute the value of the maximum force P that may be applied to box A so that no sliding occurs at either surface, in terms of the weight of the two boxes and the two coefficients of static friction at the two surfaces.

9.45 In Figure P9.45, compute the value of the maximum force P that may be applied to box B so that no sliding occurs at either surface, in terms of the weight of the two boxes and the two coefficients of static friction at the two surfaces.

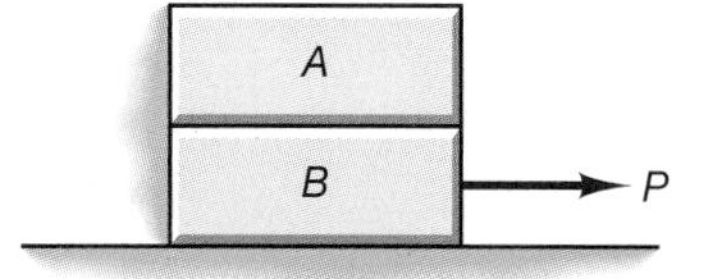

Figure P9.45

9.46 Compute the value of the maximum force P that may be applied to the top box in Figure P9.46 so that no sliding occurs at any surface, in terms of the weight of the three boxes and the three coefficients of static friction at the three surfaces.

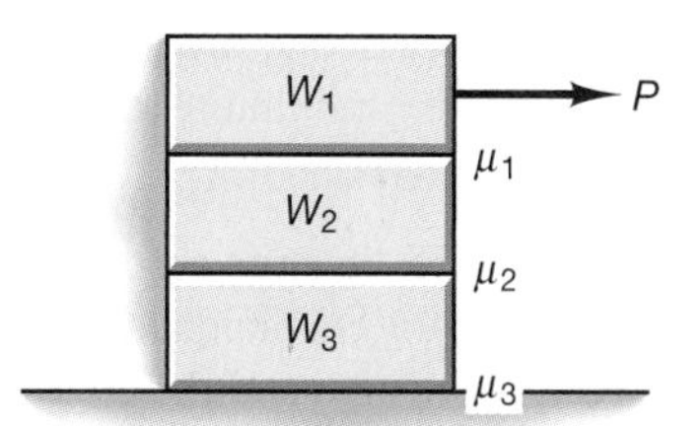

Figure P9.46

9.47 What is the state of pending motion for $P = 0.25\,W$ for the system of boxes shown in Figure P9.47? The coefficients of static friction are the same at $\mu_s = 0.5$, the weight of each box is the same (W), and the center of gravity of each box is at its geometric center.

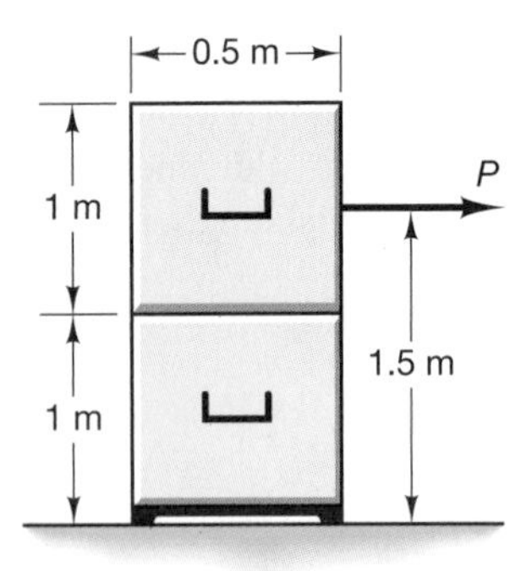

Figure P9.47

9.48 A yo-yo is placed on the floor against a baseboard. (See Figure P9.48.) If the coefficient of static friction is the same for all contacting surfaces, determine the minimum force P necessary to initiate motion.

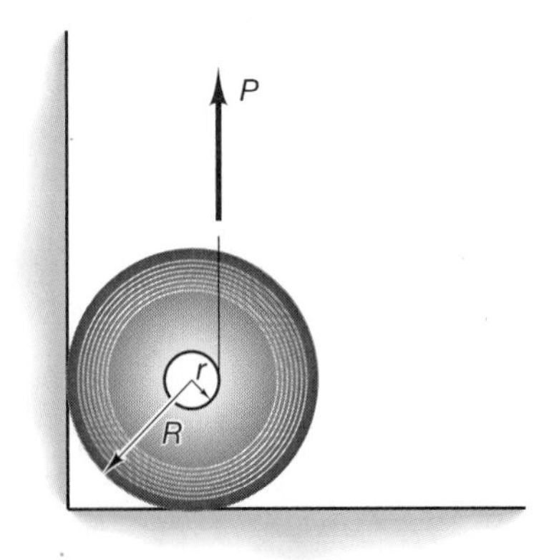

Figure P9.48

9.49 If the wall in Figure P9.48 is smooth, and friction acts only at the floor, determine the minimum force P necessary to initiate motion of the yo-yo.

9.50 If the floor in Figure P9.48 is smooth, and friction acts only at the wall, determine the minimum force P required to initiate motion of the yo-yo.

9.51 A person attempts to push a 75-kg crate across the floor, as shown in Figure P9.51. If the coefficient of friction between the crate and the floor is 0.6 and the crate's center of mass is at its geometric center, how much force must be exerted to move the crate?

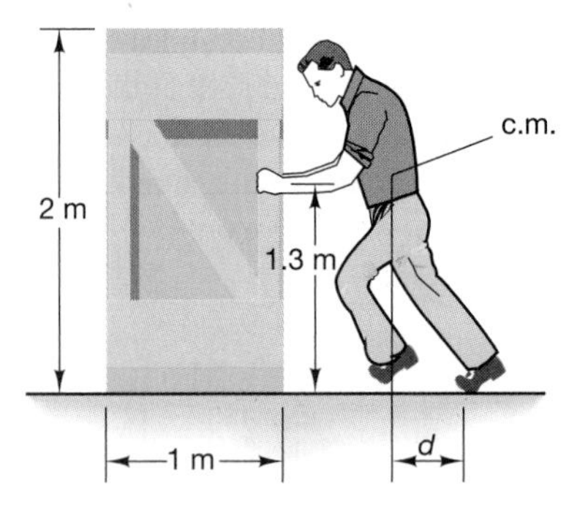

Figure P9.51

9.52 If the person in Figure P9.51 has a mass of 73 kg, what must be the minimum coefficient of static friction between his foot and the ground to move the crate, and what distance d must his foot be behind his center of mass?

9.53 Three identical 10-kg cylinders are stacked as shown in Figure P9.53. Determine the minimum coefficients of static friction between the cylinders and between the cylinder and the floor necessary to maintain equilibrium.

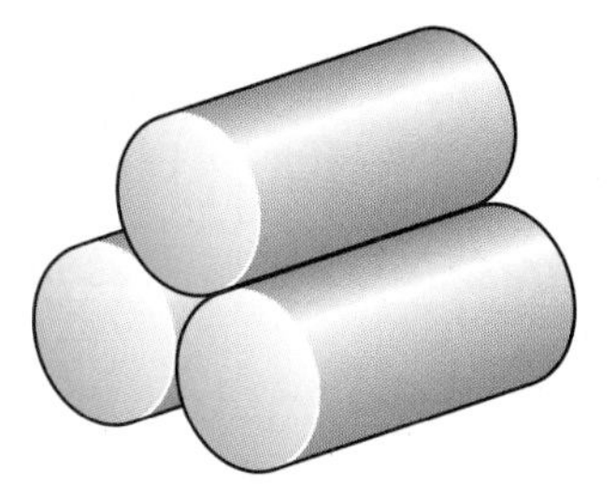

Figure P9.53

9.54 A 1360-kg rear-wheel-drive car is driven up a 15° incline as shown in Figure P9.54. If the coefficient of friction between the two rear tires and the road is 0.7, what is the maximum propulsion force that can be generated on the two rear tires so that they do not slip?

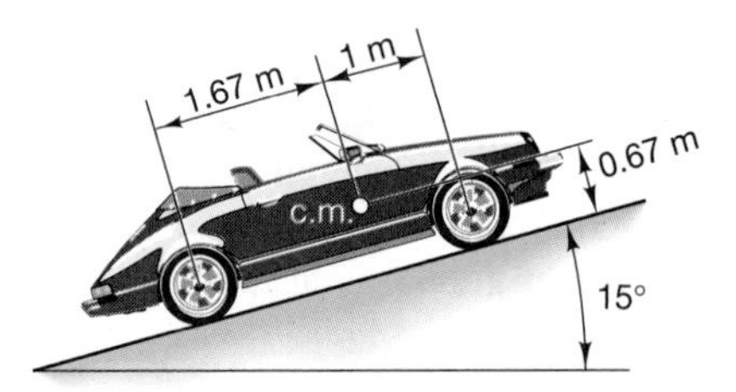

Figure P9.54

9.55 If the car in Problem 9.54 is a front-wheel-drive car, what is the maximum propulsion force that can be generated on the two front tires if they do not slip?

9.56 If the car in Problem 9.54 is a four-wheel-drive vehicle, what is the maximum propulsion force that can be generated without slip?

9.57 If the wheel radius of the rear-wheel-drive car in Problem 9.54 is 0.3 m, what is the maximum drive axle torque that can be applied to car?

9.58 A 90-kg man pulls an 40-kg boy on a sled, as shown in Figure P9.58. Determine the tension in the rope required for the sled to slip if the coefficient of static friction between the sled and the snow is 0.4. Neglect the weight of the sled.

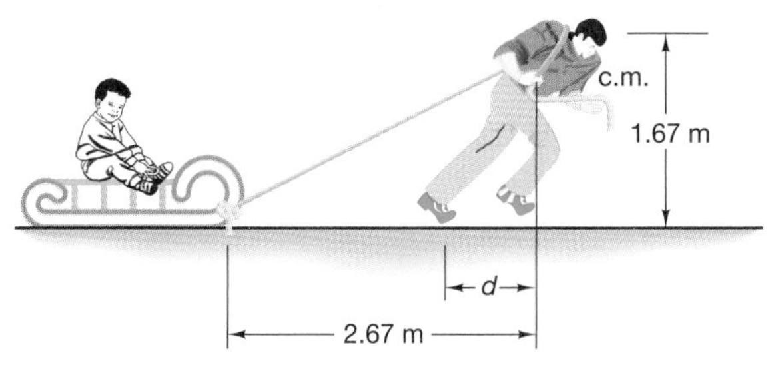

Figure P9.58

9.59 Determine the distance d the man must plant his feet behind the center of mass of his body to pull the sled in Problem 9.58 if he is not falling forward or backward.

9.60 What is the minimum coefficient of static friction that the tong–log interface shown in Figure P9.60 can have to hold a log in this position without the log slipping?

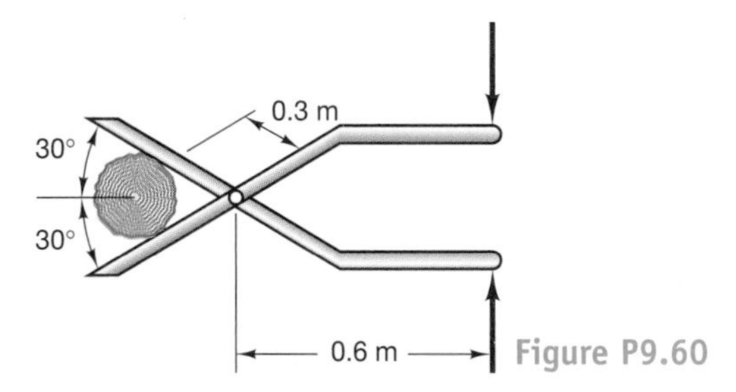

Figure P9.60

9.61 Compute the maximum value of the force P before the steel wrench slips on the steel pipe illustrated in Figure P9.61.

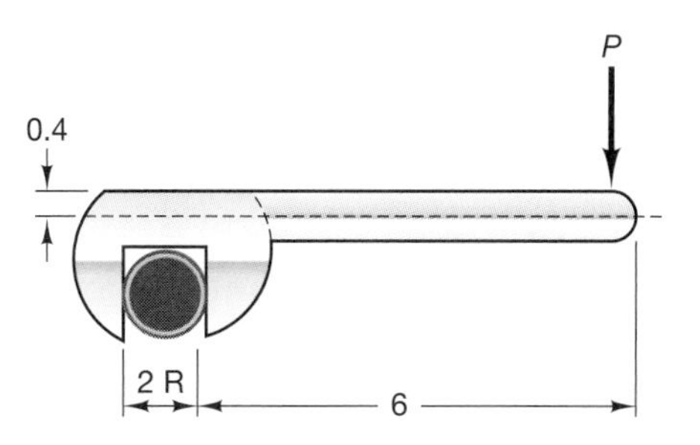

Figure P9.61

9.62 A 100-kg climber crosses an ice face using crampons. Determine the minimum coefficient of static friction between the crampons and the ice face required for the climber to traverse the wall. (Note that the climber can obtain a high equivalent coefficient of static friction, as the crampons actually dig into the ice surface. A climber would also use one or two ice axes for such a climb.) The coordinates of points A, B, and C in Figure P9.62 are A (0,0,600), B (100,0,0), and C (0,200,0).

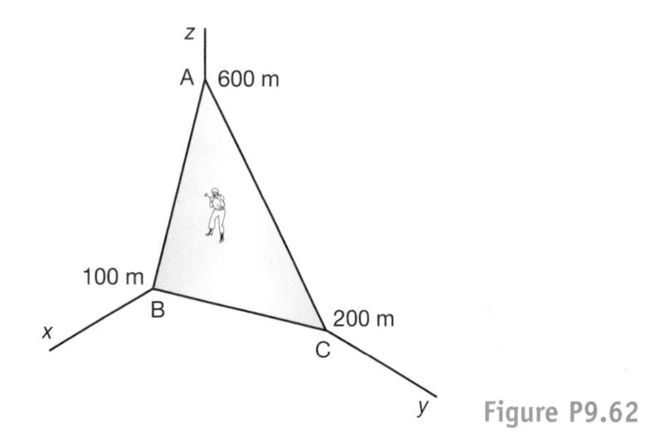

Figure P9.62

9.3 WEDGES

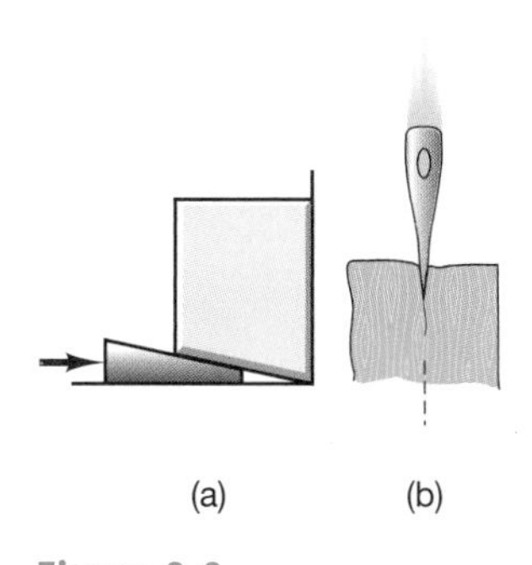

Figure 9.9

The ***wedge*** is one of the simplest and oldest machines or tools known to humankind. It consists of two faces set at a small acute angle to each other, as illustrated in Figure 9.9. When a wedge in Figure 9.9(a) is pushed forward, it will exert a large force normal to its inclined face, so it is used to lift heavy objects. The concept of a wedge is also employed in axes, to split wood, as shown in Figure 9.9(b). The wedge is an efficient lifting device because the load applied to it is usually considerably smaller than the forces normal to its faces. The friction that exists between the contacting surfaces resists driving the wedge in, but may be used to advantage in holding the wedge in place after the wedge is driven into its intented location. (The wedge is therefore said to be self-locking.) The wedge not only can lift a weight, but can hold it in place, allowing for adjustment in height. Carpenters use special wedges, called shims, to make walls and cabinets level. Wedges are frequently used in pairs, as shown in the system illustrated in Figure 9.10(a). The angle θ is called the wedge angle, and it determines the amount of force P required to raise the weight W, as well as the change in height of the

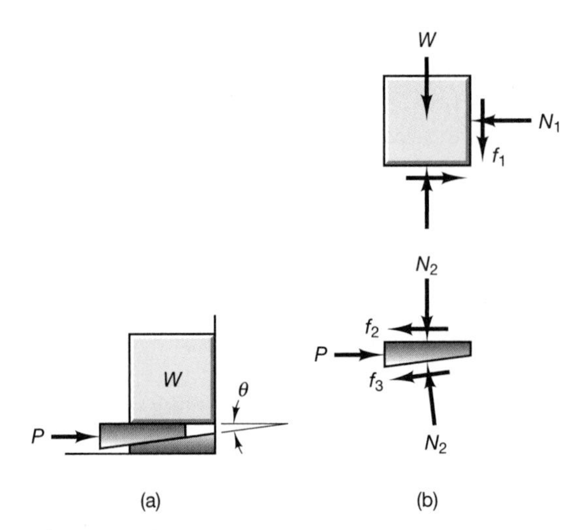

Figure 9.10

weight as the wedge is driven into place. The smaller the angle, the less is the required force, but also, the less the weight is lifted. Free-body diagrams of the wedge and the weight are shown in Figure 9.10(b). Friction is assumed between all contacting surfaces, and at the point of impending motion, the friction forces will be a maximum; that is, they will equal the respective coefficients of static friction, multiplied by the normal force. In general, we wish to determine the value of the force P required to initiate lifting a given weight, and the coefficient of static friction is used. If the weight is lifted at a constant rate, all friction forces are determined using the coefficient of kinetic friction. There are four unknowns in the problem: the force P and the three normal forces. Two scalar equations of equilibrium can be written for the weight and the wedge when both are modeled as particles, giving the required four equations. The equations of equilibrium for the weight and wedge, assuming impending slip, are

$$\mu_{s2}N_2 - N_1 = 0$$
$$N_2 - \mu_{s1}N_1 - W = 0$$
$$P - u_{s2}N_2 - \mu_{s3}N_3 \cos\theta - N_3 \sin\theta = 0$$
$$N_3 \cos\theta - \mu_{s3}N_3 \sin\theta - N_2 = 0$$

For a known weight, the dependency of the applied force **P** on the wedge angle can now be examined. A general solution of the preceding four equations can be obtained by using symbolic operators found in most computational software programs. Details of such a solution are in the Computational Supplement. Solving this system of linear equations yields

$$P = \frac{(\mu_{s2} + \mu_{s3})\cos\theta + (1 - \mu_{s2}\mu_{s3})\sin\theta}{(1 - \mu_{s1}\mu_{s2})\cos\theta - \mu_{s3}(1 - \mu_{s1}\mu_{s2})\sin\theta} W$$
$$N_1 = \frac{\mu_{s2}}{1 - \mu_{s1}\mu_{s2}} W$$
$$N_2 = \frac{1}{1 - \mu_{s1}\mu_{s2}} W$$
$$N_3 = \frac{1}{(1 - \mu_{21}\mu_{s2})\cos\theta - \mu_{s3}(1 - \mu_{21}\mu_{s2})\sin\theta} W$$

In general, design problems are approached in just this manner; that is, we first determine the dependence upon the parameters that can be controlled in the design. In the case of the wedge, the wedge angle can be chosen, and the selection of the material will control the coefficient of friction. If a high coefficient of friction is desired to lift a metal object, the wedge would be made of a similar metal. This would produce a self-locking wedge, but would require a greater force to lift it. A wooden wedge would give a lower coefficient of static friction. The weight of the wedge was neglected in this analysis, as it usually is small compared to the weight being lifted. It may be included without any significant increase in difficulty, as it enters into only the fourth equation.

Sample Problem 9.7

Determine the minimum coefficient of static friction required to ensure that the wedge system shown in Figure 9.10(a) is self-locking. The solution will express the coefficient of static friction as a function of the wedge angle.

Solution

The weight will not bear against the wall in this case and is free to move with the wedge. The free-body diagram will be of the block and wedge combined, as shown to the left.

The equation of equilibrium becomes

$$f\cos\theta - N\sin\theta = 0$$

Therefore,

$$\mu_s = \tan\theta$$

Thus, the problem is equivalent to that of a block on an inclined plane, and for a self-locking wedge, the coefficient of friction must be equal to or greater than the tangent of the wedge angle.

Sample Problem 9.8

Determine the minimum force P required to overcome static friction and begin to lift the 100-N block A shown in the diagram at the top left. The weight of the 20° wedge is 5 N, and the coefficient of static friction between all surfaces is 0.4.

Solution

We draw free-body diagrams of both the block and the wedge, as shown at the left. For the block,

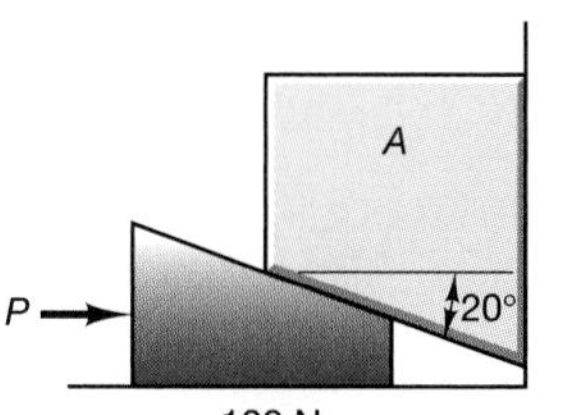

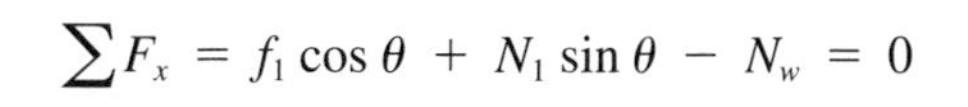

$$\sum F_x = f_1\cos\theta + N_1\sin\theta - N_w = 0$$

$$\sum F_y = N_1\cos\theta - f_1\sin\theta - f_w - 100 = 0$$

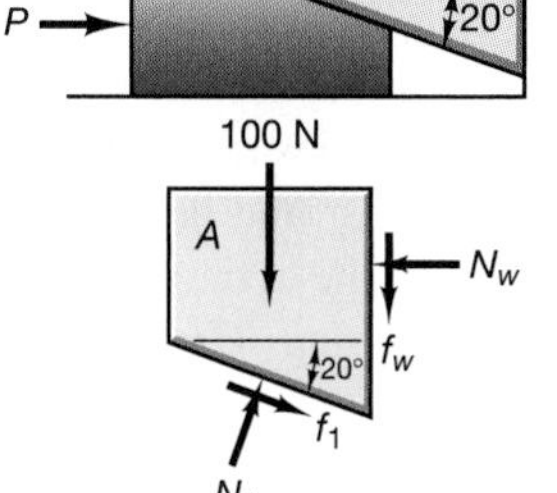

For the wedge,

$$\sum F_x = P - f_1\cos\theta - N_1\sin\theta - f_2 = 0$$

$$\sum F_y = N_2 - N_1\cos\theta + f_1\sin\theta - 5 = 0$$

For impending motion, the friction forces may be set to $f = \mu_s N$, and we have a system of four linear equations in four unknowns: N_1, N_2, N_w, and P. Written in matrix form, these equations become

$$\begin{bmatrix} -1 & \mu_s\cos\theta + \sin\theta & 0 & 0 \\ -\mu_s & \cos\theta - \mu_s\sin\theta & 0 & 0 \\ 0 & -\mu_s\cos\theta - \sin\theta & -\mu_s & 1 \\ 0 & -\cos\theta + \mu_s\sin\theta & 1 & 0 \end{bmatrix} \begin{bmatrix} N_w \\ N_1 \\ N_2 \\ P \end{bmatrix} = \begin{bmatrix} 0 \\ 100 \\ 0 \\ 5 \end{bmatrix}$$

Solving the system of equations for $\mu_s = 0.4$ and $\theta = 20°$ yields

$$\begin{bmatrix} N_w \\ N_1 \\ N_2 \\ P \end{bmatrix} = \begin{bmatrix} 139.2 \\ 193.9 \\ 160.7 \\ 203.5 \end{bmatrix}$$

The force required to lift the 100-N block is 203.5 N, and this wedge system is not too efficient. Note that almost all wedge problems will involve the solution of systems of linear equations, and matrix methods should be used to reduce the numerical work.

Sample Problem 9.9

Detemine the minimum force P required to initiate upward motion of the 50-kg cylinder shown in the diagram at the top left. The coefficients of static friction are 0.4 between the wall and the cylinder, 0.3 between the cylinder and the wedge, and 0.2 between the wedge and the floor. Neglect the weight of the wedge. The wedge angle is 20°.

Solution

First construct free-body diagrams of the wedge and the cylinder, and then select a coordinate system. The equilibrium equations for the cylinder are

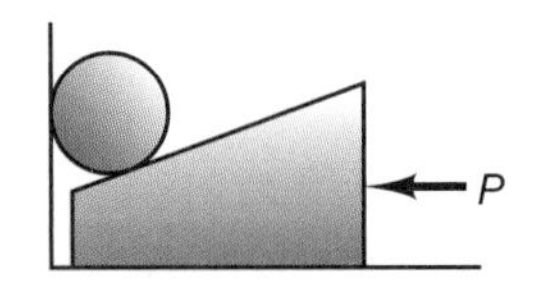

$$\sum F_x = N_1 - f_2 \cos\theta - N_2 \sin\theta = 0$$

$$\sum F_y = -f_1 - f_2 \sin\theta + N_2 \cos\theta - mg = 0$$

$$\sum M_{\text{c.m.}} = f_1 r - f_2 r = 0$$

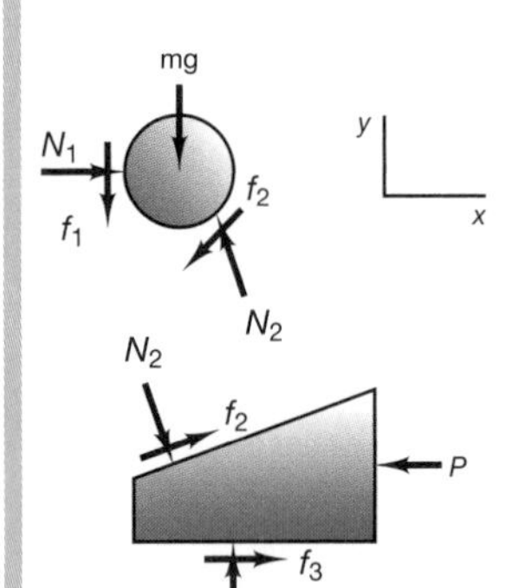

The equilibrium equations for the wedge are

$$\sum F_x = f_3 + f_2 \cos\theta + N_2 \sin\theta - P = 0$$

$$\sum F_y = N_3 + f_2 \sin\theta - N_2 \cos\theta = 0$$

Note that when the weight of the wedge is neglected, the wedge is a three-force member, and the forces will be concurrent. The summation of moments only verifies this fact. If the system is to move, there must be slip between the wedge and the floor and between the cylinder and the wall, or between the cylinder and the wedge—that is, the cylinder can roll up the wall without slipping, or it can roll on the wedge without slipping. The condition for slip is either

$$f_3 = \mu_{s3} N_3 \quad f_1 = \mu_{s1} N_1 \quad f_2 \leq \mu_{s2} N_2 \quad \text{(slip between cylinder and wall)}$$

or

$$f_3 = \mu_{s3} N_3 \quad f_2 = \mu_{s2} N_2 \quad f_1 \leq \mu_{s1} N_1 \quad \text{(slip between clyinder and wedge)}$$

We now have a system of seven equations and one inequality that must be satisfied for the wedge–cylinder system to move. The seven unknowns are the three normal forces, the three friction forces, and the force P. These equations may be solved by hand or by using computational aids. We must assume no slip on one of the surfaces and then check this assumption. In other words, is the inequality satisfied? The results for condition 1—that is, slipping at the wall and rolling on the wedge—are

$$N_1 = 416\,N \quad N_2 = 760\,N \quad N_3 = 657\,N \quad f_1 = f_2 = 167\,N$$

$$f_3 = 131N \quad P = 548\,N$$

Note that the inequality condition is satisfied: The friction force f_2 is less than the maximum allowable static friction.

However, no solution was found that would cause the cylinder to roll up the wall, as slip occurred. If the coefficient of static friction between the wedge and the cylinder is reduced to 0.2, the only solution has the cylinder slipping on the wedge and rolling up the wall. If the wedge angle is increased to 30°, the only solution has the cylinder rolling up the wall. We can determine the transition wedge angle from rolling on the wedge to rolling up the wall by setting all frictions to their maximum (adding another equation) and solving for the required wedge angle. For the original coefficients of friction, the transition wedge angle is 29.22°. We saw that we could change the form of motion from rolling on the wedge to rolling up the wall by reducing the coefficient of friction between the wedge and the cylinder. The coefficient of friction for pending slip on all surfaces is $\mu_{s2} = 0.219$; that is, if the coefficient of friction between the cylinder and the wedge is this value or lower, the cylinder will slip on the wedge and roll up the wall. You can see how an engineer can design for a particular type of motion by choice of wedge angle or coefficient of friction.

Problems

9.63 Determine the minimum force P required to overcome static friction and begin to lift block A in Figure P9.63. Block A weighs 100 N and the 20° wedge weighs 5 N. The coefficient of static friction between all surfaces is 0.4.

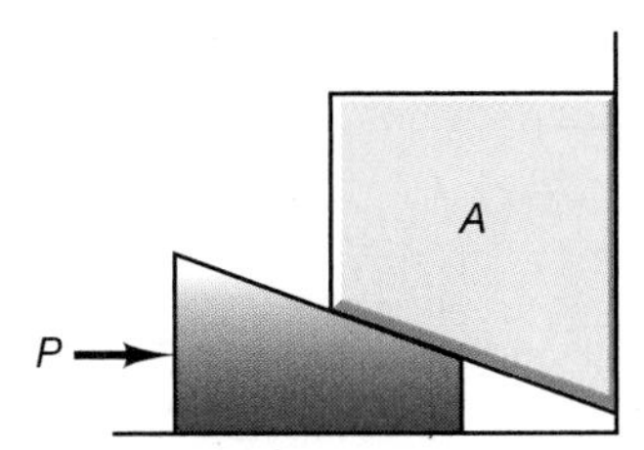

Figure P9.63

9.64 If, in Figure P9.63, the coefficient of kinetic friction between all surfaces is 0.35, determine the force P necessary to raise the block at a constant speed.

9.65 A double-wedge system is used to position the 500-kg crate shown in Figure P9.65. Neglect the mass of the wedges, and determine the minimum force P required to initiate movement. The coefficient of static friction between the 10° wedges and all other surfaces is 0.2, and between the crate and the floor it is 0.6.

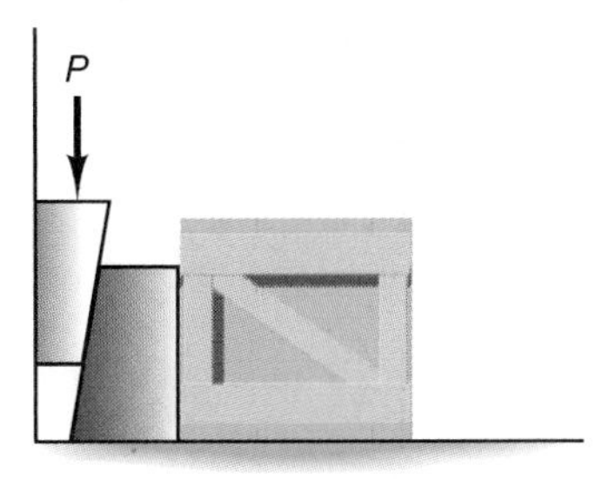

Figure P9.65

9.66 If the wedges in Problem 9.65 were coated with Teflon®, reducing their coefficient of static friction to 0.02, what would be the required force P to initiate motion?

9.67 What is the effect of increasing the wedge angles from 10° to 15° in Problem 9.65?

9.68 What is the effect of decreasing the wedge angles from 10° to 5° in Problem 9.65?

9.69 The wedges in Problem 9.65 are reversed, as shown in Figure P9.69. If the coefficients of static friction are the same as given in Problem 9.65, what is the minimum force necessary to initiate motion of the crate?

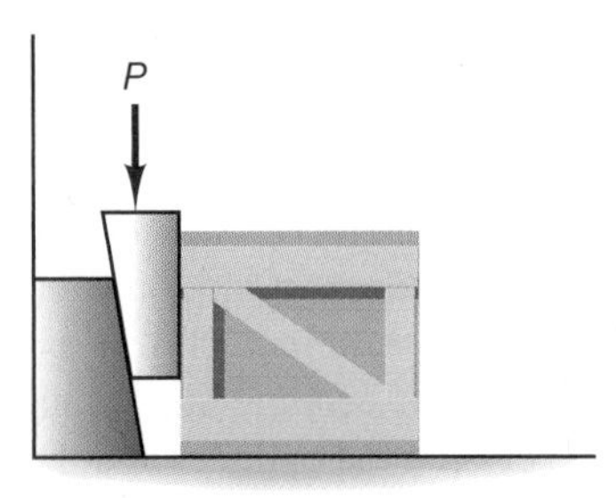

Figure P9.69

9.70 If a three-wedge system is used to move the crate illustrated in Figure P9.70, what is the minimum force necessary to initiate motion? The coefficient of static friction between the wedges and all surfaces is 0.2, and it is 0.6 between the 500-kg crate and the floor. Neglect the weight of the wedges.

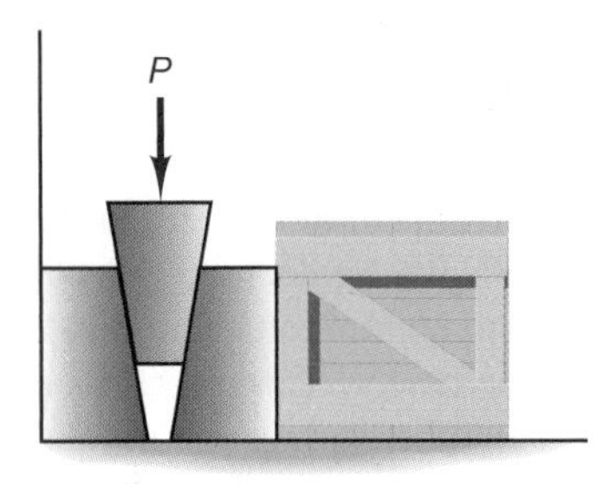

Figure P9.70

9.71 If the coefficients of kinetic friction are 0.16 between the wedges and all contacting surfaces and 0.48 between the crate and the floor in the three-wedge system of Problem 9.70, what is the minimum force P necessary to move the crate at a constant speed?

9.72 Determine the minimum force P required to hold in place the wedge and the 50-kg cylinder shown in Sample Problem 9.9. (See Figure P9.72.) The coefficient of friction between the wedge and the floor is 0.2, between the wedge and the cylinder is 0.3, and between the cylinder and the wall is 0.4. The wedge angle is 20°. Is the impending motion rolling on the wedge or rolling on the wall?

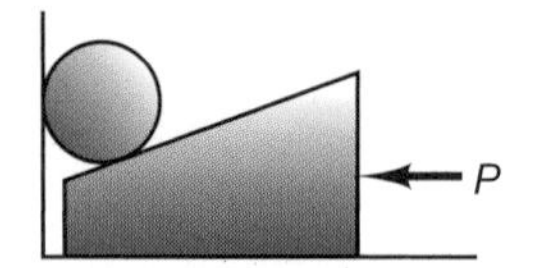

Figure P9.72

9.73 If the wall is considered to be a smooth surface in Problem 9.72, determine the minimum force required to raise the cylinder. Will the cylinder roll on the wall or on the wedge?

9.74 If the coefficient of static friction is 0.4 between all surfaces in Problem 9.72, determine the minimum force required to raise the cylinder. Will the cylinder roll on the wall or on the wedge?

9.75 If the cylinder in Problem 9.72 is smooth, having no friction between the wall and the wedge, (a) determine the minimum force necessary to hold the cylinder in place, and (b) determine the minimum coefficient of static friction between the floor and the wedge for the wedge to be self-locking.

9.76 A 5-kg, 15° wedge is used to move a 150-kg cylinder to the right, as shown in Figure P9.76. Determine the minimum force P required to move the cylinder if the coefficient of static friction of the wedge and all contacting surfaces is 0.2 and the coefficient of static friction between the cylinder and the floor is 0.5. Will the cylinder roll on the floor or on the wedge?

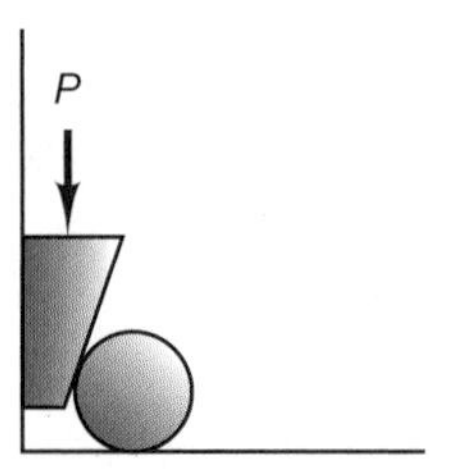

Figure P9.76

9.77 Determine the wedge angle and the force P such that there will be slip at all surfaces of the wedge–cylinder system in Problem 9.76.

9.78 Determine the minimum coefficient of friction between the wedge and its contacting surfaces and the force P such that there will be slip at all surfaces of the wedge–cylinder system in Problem 9.76.

9.79 It is necessary to level the refrigerator shown in Figure P9.79 with a 5° wedge system. The weight distribution is such that the weights on all feet of the 800-kg refrigerator are equal. The coefficient of friction between the feet and their contacting surface is 0.8, the coefficient of static friction between the wedges is 0.2, and the coefficient of static friction between the wedge and the floor is 0.4. What is the minimum force P required to initiate leveling of the refrigerator?

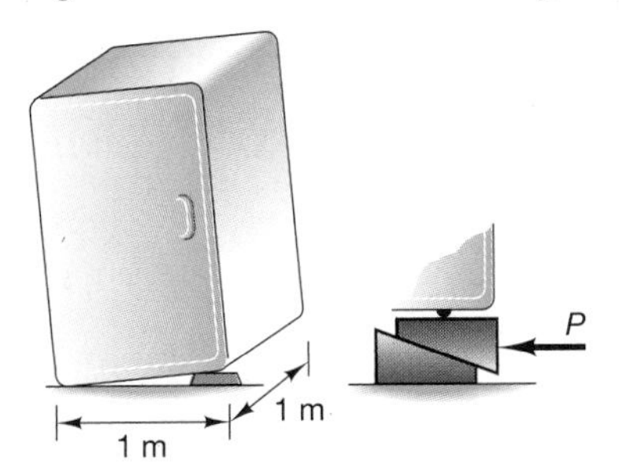

Figure P9.79

9.80 Is the wedge system in Problem 9.79 self-locking?

9.81 If a 10° wedge system is used in Problem 9.79, what is the minimum force P required to initate leveling of the refrigerator?

9.4 SQUARE-THREADED SCREWS

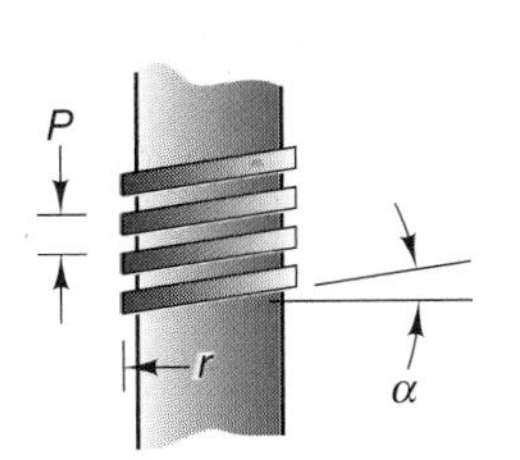

Figure 9.11

Square-threaded ***screws*** are efficient devices used to transmit power or motion. This type of screw is used in vices, shafts, clamps, and adjustable supports. The friction force on the threads determines the action of the screw. The coefficient of kinetic friction for a well-lubricated power screw is approximately 0.15. Although a similar analysis could be developed for the V-threaded screw, only the square-threaded screw will be considered here. A section of such a screw is shown in Figure 9.11. The lead or pitch of the screw, p, is the distance of advancement of the screw for each revolution. The slope of the thread, or lead angle, is designated α. If r is the mean radius of the threads, the pitch is

$$p = 2\pi r \tan(\alpha)$$

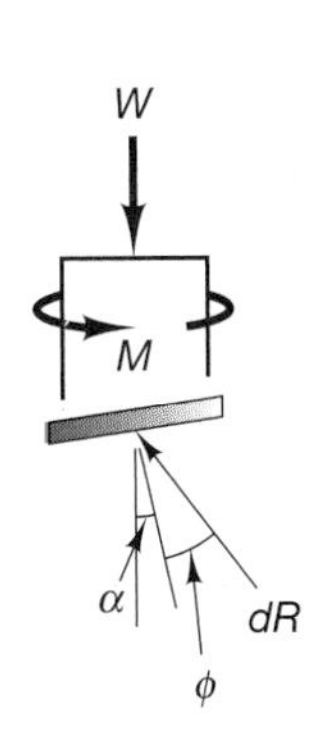

Figure 9.12

The screw thread may be thought of as an inclined plane wrapped around a cylinder. As the screw is turned in its frame, the screw will slide around and up the thread in the frame. The friction force will be related to the normal force by the coefficient of kinetic friction, and the resultant of the friction and the normal forces will make an angle of $(\alpha + \phi)$ with the vertical axis, as is shown in the free-body diagram of Figure 9.12. The angle f is the friction angle given in Eq. (9.6). For equilibrium to exist, the vertical component of the total reaction R must equal the weight being lifted (W), and the applied moment M must equal the moment of the horizontal component about the screw axis. If the length of the screw thread in contact at any position is given by L, the equations of equilibrium are

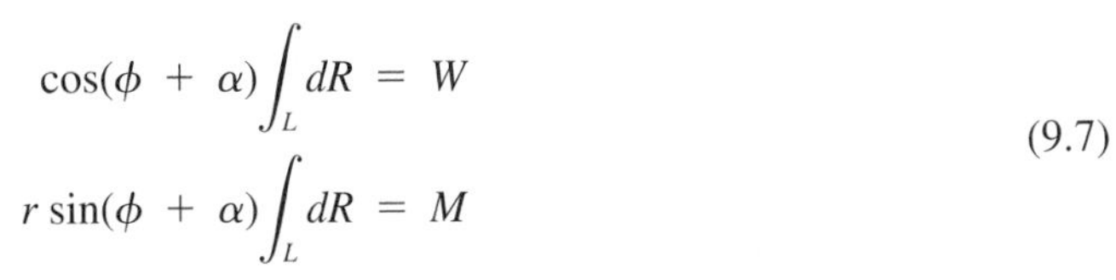

$$\cos(\phi + \alpha)\int_L dR = W$$
$$r\sin(\phi + \alpha)\int_L dR = M \tag{9.7}$$

Eliminating the integral for R yields

$$M = Wr\tan(\phi + \alpha) \tag{9.8}$$

This is the moment required for the screw to be on the verge of rotating and moving in the direction to lift W. Many machines design references will express Eq. (9.8) in a different form. The tangent of the sum of two angles can be written as

$$\tan(\phi + \alpha) = \frac{\tan\phi + \tan\alpha}{1 - \tan\phi\tan\alpha} \tag{9.9}$$

The coefficient of friction is equal to the tangent of ϕ ($\mu = \tan\phi$), and the tangent of the thread angle can be related to the lead or the pitch of the screw and the mean thread radius by

$$\tan\alpha = \frac{p}{2\pi r} \tag{9.10}$$

so that Eq. (9.8) may be written as

$$M = Wr\left(\frac{p + 2\pi\mu r}{2\pi r - \mu p}\right) \tag{9.11}$$

The moment to initiate movement is determined using the coefficient of static friction in Eq. (9.11). The moment required to continue movement at a constant rate is obtained using the coefficient of kinetic friction. Design specifications usually give the pitch and the mean diameter or mean radius of the screw, and Eq. (9.11) is referenced. This equation can be developed directly without introducing the friction angle ϕ. To see how, consider one circumference of the thread, as shown in the free-body diagram of Figure 9.13. For equilibrium, we must have

$$\sum F_x = F - \mu N\cos\alpha - N\sin\alpha = 0$$

$$\sum F_y = -W - \mu N\sin\alpha + N\cos\alpha = 0$$

$$\tan\alpha = \frac{p}{2\pi r}$$

Solving the equilibrium equations yields

$$F = W\frac{(\mu\cos\alpha + \sin\alpha)}{(\cos\alpha - \mu\sin\alpha)}$$

Eliminating α in terms of the mean radius and the pitch results in

$$F = W\frac{(\mu 2\pi r + p)}{(2\pi r - \mu p)}$$

The required moment is

$$M = Fr = W\frac{(\mu 2\pi r + p)}{(2\pi r - \mu p)}r$$

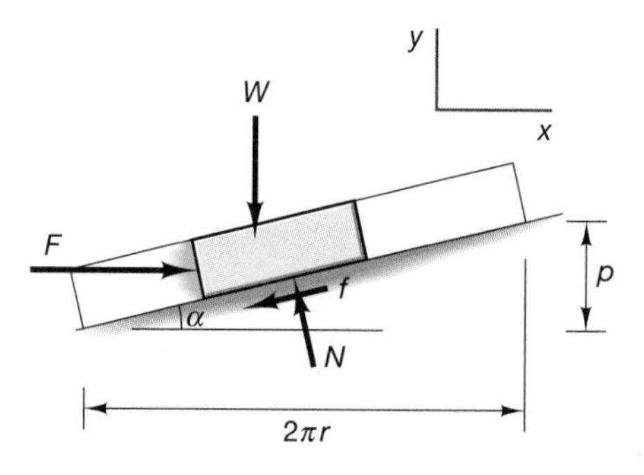

Figure 9.13

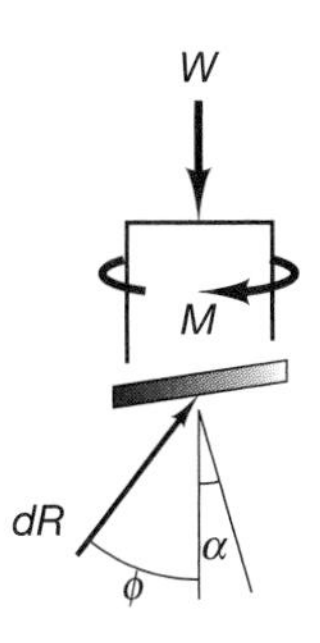

Figure 9.14

This agrees with Eq. (9.11). Again, remember that the coefficient of static friction is used to determine the moment required to initiate movement, and the coefficient of kinetic friction is used to determine the moment required to sustain constant movement.

When the weight is lowered, the friction force on the threads acts in the opposite direction, as shown in Figure 9.14. The equations of equilibrium now become

$$\cos(\phi - \alpha)\int_L dR = W$$
$$r\sin(\phi - \alpha)\int_L dR = M \tag{9.12}$$

The relationship between the moment M required to lower the weight and the weight W is

$$M = Wr\tan(\phi - \text{a}) \tag{9.13}$$

Notice that if the coefficient of static friction is such that ϕ is equal to or less than α, the weight will not remain stationary without a resisting moment, and the screw will turn due to the weight. Again, machine design references will express the moment required to lower the weight as

$$M = Wr\left(\frac{2\pi\mu r - p}{2\pi r + \mu p}\right) \tag{9.14}$$

where, again, p is the pitch of the screw and r is the mean radius of the thread. Equation (9.14) can also be developed without reference to the screw angle, as was done for Eq. (9.11). The screw will be self-locking as long as the moment needed to lower the screw is greater than or equal to zero. Therefore, the coefficient of static friction needed for a self-locking screw is

$$\mu_s \geq \frac{p}{2\pi r} \tag{9.15}$$

For a double-threaded screw (one with two separate threads intertwined), the pitch would be replaced with $2p$. Therefore, for an n-threaded screw, the pitch would be replaced with np.

The efficiency of a power screw is defined as the ratio of the output work to the input work. Work will be formally defined in Chapter 11, but for the computation of the efficiency of a power screw, the output work is the product of the weight W and the distance moved in one revolution, p. The input work is the product of the lifting moment M and the angle in radians for one revolution, or 2π. Therefore, the efficiency of the power screw is

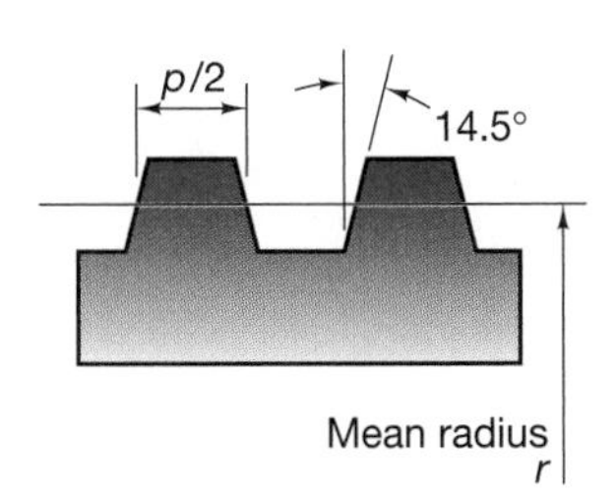

$$e + \frac{Wp}{2\pi M} = \frac{p(2\pi r - \mu_k p)}{r(p + \mu_k 2\pi r)} \tag{9.16}$$

Using Eq. (9.10), we may write the efficiency as

$$e = \frac{1 - \mu_k\tan\alpha}{1 + \mu_k\cot\alpha} \tag{9.17}$$

The square-threaded screw provides high efficiency, but is more difficult to machine because of its perpendicular face. Therefore, the threads on most screws are cut at an angle, as shown for the Acme thread in the diagram above and to the left.

The moment required to raise an Acme-threaded power screw is

$$M = Wr\left(\frac{p\cos\beta + 2\pi\mu r}{2\pi r\cos\beta - \mu p}\right) \tag{9.18}$$

where β is the screw thread angle (14.5° for an Acme thread). The moment required to lower an Acme-threaded power screw is

$$M = Wr\left[\frac{2\pi\mu r - p\cos\beta}{2\pi r\cos\beta + \mu p}\right] \tag{9.19}$$

The efficiency of an Acme-threaded power screw is

$$e = \frac{\cos\beta - \mu_k \tan\alpha}{\cos\beta + \mu_k \cot\alpha} \tag{9.20}$$

Sample Problem 9.10

A square-threaded power screw is used to lift and hold a 500-kg mass. The mean radius of the screw is 30 mm, the lead angle is 3°, the coefficient of kinetic friction is 0.15, and the coefficient of static friction is 0.19.

a. Determine the moment required to lift the mass at a constant rate.
b. Is the screw self-locking?
c. What is the efficiency of the screw?
d. What is the moment required to initiate downward movement of the mass?

Solution

a. The pitch of the screw is

$$p = 2\pi r \tan(\alpha) = 2\pi(0.030)\tan(3°) = 0.010 \text{ m}$$

The moment to lift the mass at a constant rate is

$$M = Wr\left(\frac{p + 2\pi\mu_k r}{2\pi r - \mu_k p}\right)$$

$$= (500)(9.81)(0.03)\left(\frac{0.01 + 2\cdot\pi\cdot 0.15\cdot 0.03}{2\cdot\pi\cdot 0.03 - 0.15\cdot 0.01}\right) = 29.90 \text{ N}\cdot\text{m}$$

b. For the screw to be self-locking, Eq. (9.15) requires that

$$\mu_s \geq \frac{p}{2\pi r}$$

so we have

$$0.18 \geq \frac{0.01}{2\cdot\pi\cdot 0.03}$$

$$0.18 \geq 0.053$$

Therefore, the screw is self-locking.

c. The efficiency of the screw is obtained by Eq. (9.17):

$$e = \frac{1 - \mu_k \tan\alpha}{1 + \mu_k \cot\alpha}$$

$$= \frac{1 - 0.15\cdot\tan(3°)}{1 + 0.15\cdot\cot(3°)} = 0.257 \text{ or } 25.7\%$$

d. The moment required to initiate downward movement is

$$M = Wr\left(\frac{2\pi\mu_s r - p}{2\pi r + \mu_s P}\right)$$

$$= (500)(9.81)(0.03)3\left(\frac{2\cdot\pi\cdot 0.18\cdot 0.03 - 0.01}{2\cdot\pi\cdot 0.03 + 0.18\cdot 0.01}\right) = 18.50 \text{ N}\cdot\text{m}$$

Problems

9.82 A square-threaded power jack is used to lift a 500-kg load as shown in Figure P9.82. If a screw with a mean diameter of 75 mm and a pitch of 1.0 mm with a coefficient of friction of 0.15 is used, what force must be applied to the 360-mm jack handle to raise the load? (Note that, for a well-lubricated power screw, the coefficient of kinetic friction is approximately equal to the coefficient of static friction.)

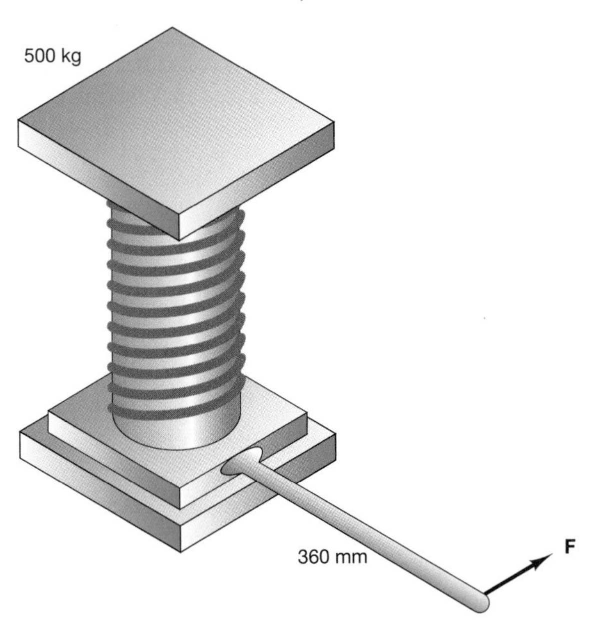

Figure P9.82

9.83 Is the jack in Problem 9.82 self-locking? What force must be applied to the jack handle to lower the load?

9.84 What is the efficiency of the square-threaded power jack in Problem 9.82?

9.85 If an Acme-threaded power jack with the same mean diameter and pitch as that in Problem 9.82 were used, what force must be applied to the jack handle to raise the load?

9.86 What is the efficiency of the Acme-threaded power jack in Problem 9.85?

9.87 Two identical C-clamps are used to hold the two 50-mm by 100-mm boards together during gluing. (See Figure P9.87.) If an average pressure of 3×10^3 Pa is necessary for the glue to set properly, what moment must be applied to each of the C-clamps? The clamps have square threads with 250 mm mean diameter, a pitch of 50 mm, and a coefficient of static friction of 0.2. The clamps are placed such that their distances from the ends of the boards are $a = b = 250$ mm.

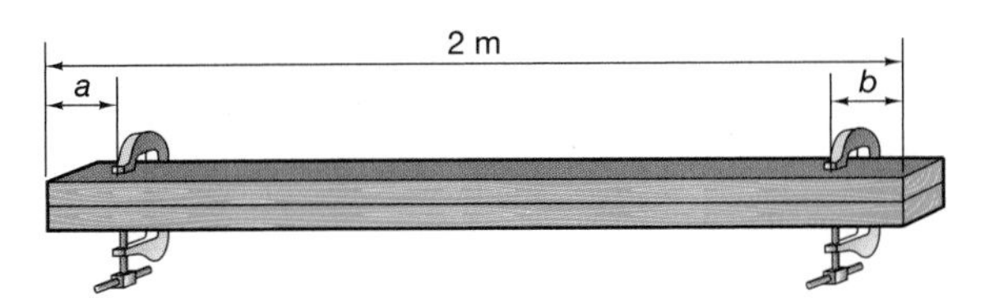

Figure P9.87

9.88 If the C-clamp on the right end in Problem 9.87 is placed such that $b = 50$ cm, what moment must be applied to the two clamps to have a pressure of 3×10^3 Pa between the boards?

9.89 What moment must be applied to the C-clamps in Problem 9.87 to remove them?

9.90 If a screw is non-self-locking, it can be "back driven" by the load W; that is, pushing on the nut will cause the screw to rotate. A Yankee screwdriver (see Figure P9.90) is an application of a back-driven screw. A high-pitch angle is used so that when a force is placed on the handle, the screwdriver will rotate, turning the wood screw into place. Develop the relationship for the handle axial load and the moment applied to the wood screw.

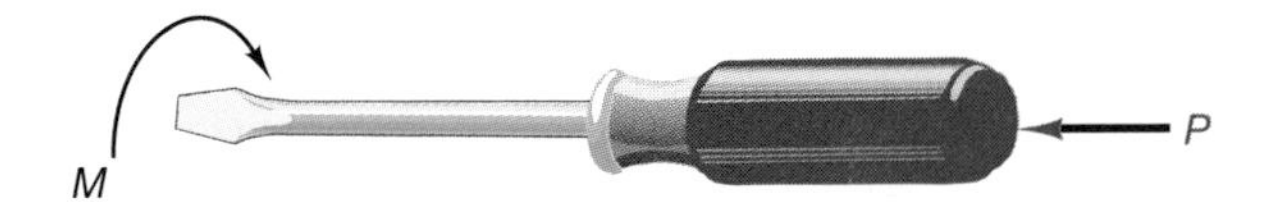

Figure P9.90

9.91 Derive Eq. (9.18) for an Acme-threaded power screw.

9.92 A square-screw power jack is used to level a house floor (See Figure P9.92.) If the screw has a mean diameter of 80 mm and a pitch of 15 mm with a coefficient of static friction of 0.3, determine the moment that must be used to level the I-beam. Each floor joist spaced at 250 mm applies a load of 4.5 kN to the beam. Neglect the weight of the beam.

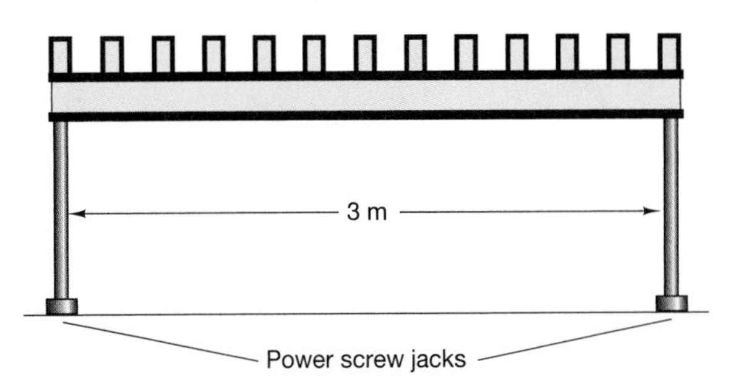

Figure P9.92

9.93 An automoble jack (see Figure P9.93) is used to raise the rear of an automobile. The jack is run by a square-threaded screw (right-handed thread at A and left-handed thread at B). The mean diameter of the screw is 25 mm and the pitch is 5 mm. Assume that the coefficient of kinetic friction is 0.15. Determine the required moment on the jack as a function of the load W and the angle θ required to raise the car.

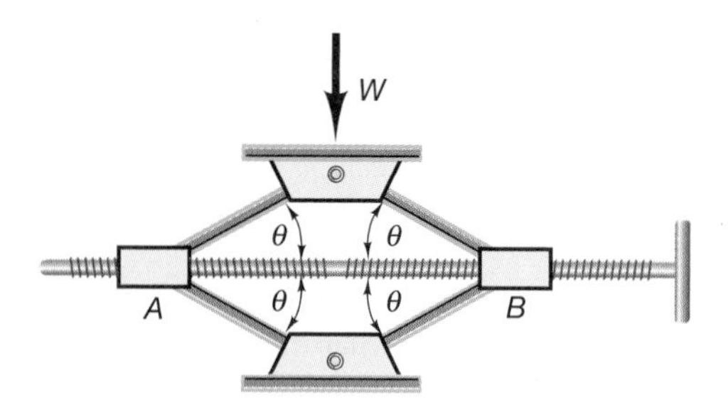

Figure P9.93

9.94 Determine the moment required to lower the automobile in Problem 9.93.

9.95 A freezer has 25-mm-diameter Acme-threaded leveling screws on each corner. The freezer has a mass of 300 kg, which is evenly distributed on the four leveling screws. If the screw pitch is 5 mm and the coefficient of static friction is 0.3, determine the moment required to raise a corner of the freezer.

9.96 Determine the moment required to lower a corner of the freezer in Problem 9.95.

9.5 BELT FRICTION

When a ***belt*** or rope passes over a drum, friction is developed between the belt and the drum. This friction is of great importance in the design of belt drives used in many machines, from cars to vacuum cleaners. The flat belt illustrated in Figure 9.15 passes over a fixed curved surface such that the contact angle between the belt and the surface is β.

Suppose that the resisting tension T_1 is known and that it is desired to determine the tension T_2 that will cause the belt to slip on the drum. This tension is great enough to overcome the friction between the belt and the drum and the force T_1, so that impending motion of the belt is counterclockwise on the drum. Consider the element of the belt in contact with the drum denoted by $d\alpha$, as shown in Figure 9.16(a). A free-body diagram of a

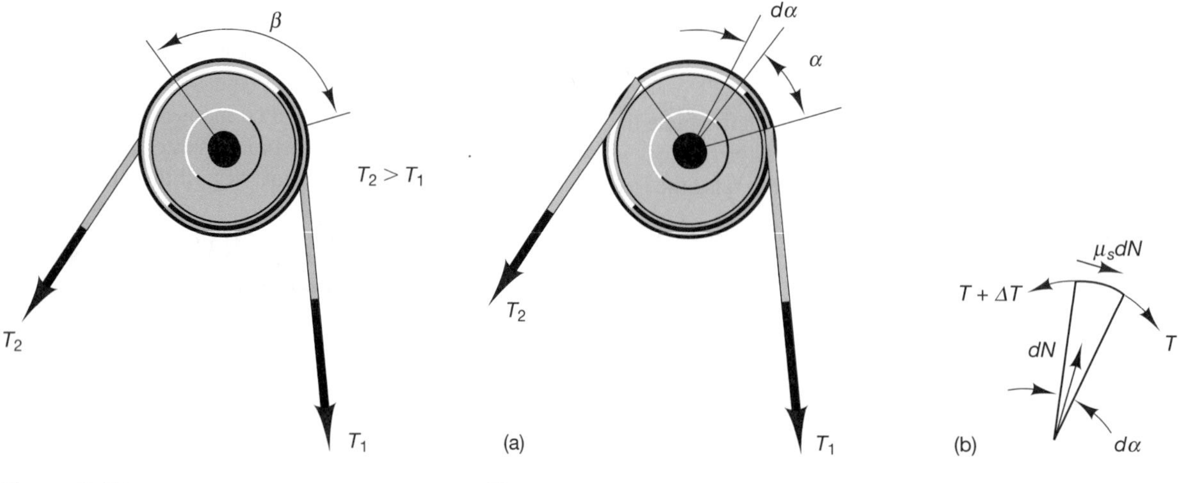

Figure 9.15

Figure 9.16

section of the belt is shown in Figure 9.16(b). The equilibrium equations in the normal and tangential directions are

$$\begin{aligned}\sum F_{\text{tangential}} &= \mu_s dN + T\cos\frac{d\alpha}{2} - (T + dT)\cos\frac{d\alpha}{2} = 0 \\ \sum F_{\text{normal}} &= dN - T\sin\frac{d\alpha}{2} - (T + dT)\sin\frac{d\alpha}{2} = 0\end{aligned} \tag{9.21}$$

The following small-angle approximations are applicable:

$$\begin{aligned}\cos\frac{d\alpha}{2} &\approx 1 \\ \sin\frac{d\alpha}{2} &\approx \frac{d\alpha}{2}\end{aligned} \tag{9.22}$$

If the higher order term ($dT\, d\alpha/2$) is neglected, Eqs. (9.21) become

$$\begin{aligned}\mu_s dN - dT &= 0 \\ dN - T d\alpha &= 0\end{aligned} \tag{9.23}$$

Eliminating dN yields the differential equation

$$\frac{dT}{T} = \mu_s d\alpha \tag{9.24}$$

Integrating both sides between corresponding limits results in

$$\begin{aligned}\int_{T_1}^{T_2} \frac{dT}{T} &= \int_0^{\beta} \mu_s d\alpha \\ \ln\frac{T_2}{T_1} &= \mu_s \beta \\ T_2 &= T_1 e^{\mu_s \beta}\end{aligned} \tag{9.25}$$

Note that the ratio T_2/T_1 does not depend on the radius of the drum. The angle β must be expressed in radians. Replacing the coefficient of static friction with the coefficient of kinetic friction determines the force required to cause the belt to slide over the drum at a constant rate. If the belt is wrapped around the drum such that the contact angle β is 90°, the sensitivity to the value of the coefficient of friction can be seen in a graph of the ratio of the tension T_2 to the tension T_1 versus the coefficient of friction. (See Figure 9.17.)

When mountain climbers are crossing an ice field, they are trained to move cross roped to one another. But if one climber falls into a crevasse, the friction between the rope and the

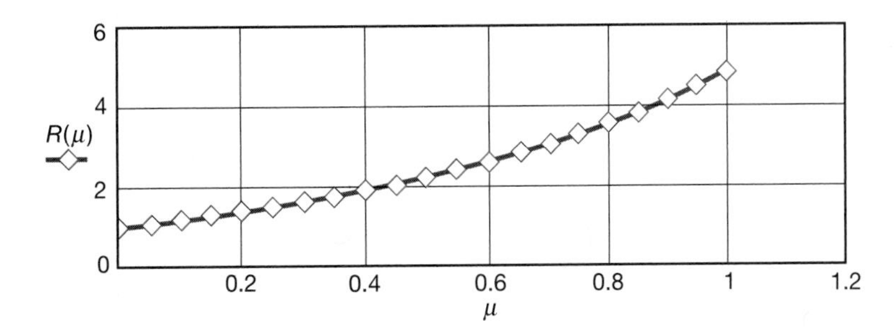

Figure 9.17

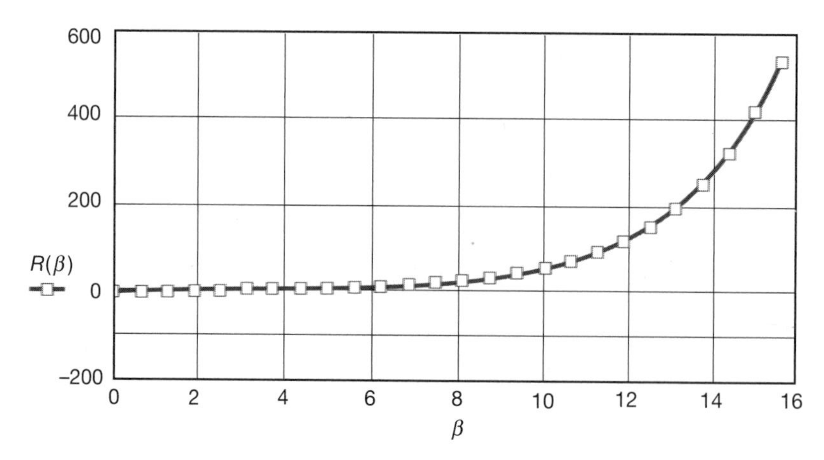

Figure 9.18

edge of the crevasse is such that it requires a force greater than twice the body weight to pull the fallen climber out of the crevasse. The climber must climb out himself, or, if he is injured, a rescuer must climb down and bring him up. A coefficient of static friction greater than 0.5 occurs between the rope and the ice lip of the crevasse.

It is worthwhile to examine the dependency of the ratio of T_2 to T_1 on the angle of contact, β, by plotting the one against the other. (See Figure 9.18.)

Wrapping the belt around the drum two and a half times would require a force of 500 times the resisting tension. Cowboys secure horses by wrapping the reins around the hitching post.

9.5.1 V-BELTS

The analysis shown in this section has assumed that the belt is flat and that the normal force is directed in the direction of the radius of curvature of the drum. Many applications in cars, farm equipment, and other machines, however, use belts that fit into grooves in the pulleys. This does not alter the analysis in any significant manner, as will be shown. Impending slip is assumed, as it was before, but the belt must be considered in cross section, as well as in the tangential direction, as shown in Figure 9.19. The side-view free-body diagram is the same as that shown in Figure 9.16(b). Writing the equations of equilibrium in the tangential and the normal directions yields

$$\sum F_{\text{tangential}} = 2\mu_s dN + T_{\cos}\frac{d\alpha}{2} - (T + dT)\cos\frac{d\alpha}{2} = 0$$
$$\sum F_{\text{normal}} = 2dN\sin\frac{\phi}{2} - T\sin\frac{d\alpha}{2} - (T + dT)\sin\frac{d\alpha}{2} = 0 \tag{9.26}$$

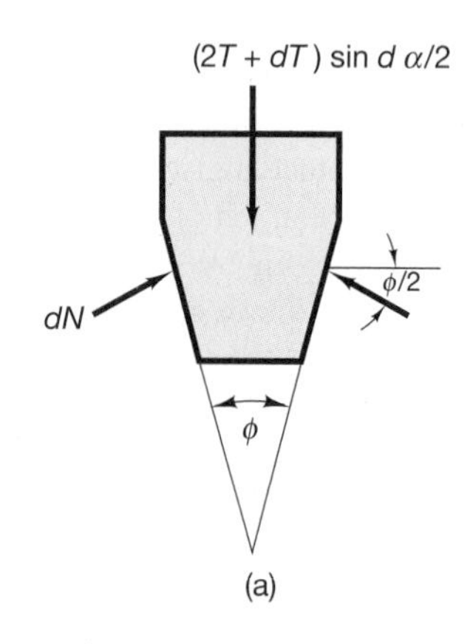

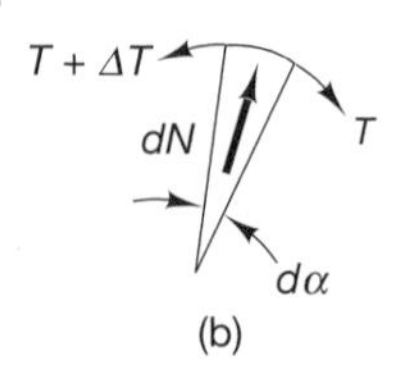

Figure 9.19

Using the small-angle approximations for $d\alpha$, we proceed with the analysis proceeds as for a flat belt, and the relationship between T_2 and T_1 becomes

$$T_2 = T_1 e^{\frac{\mu_s \beta}{\sin\frac{\phi}{2}}} \tag{9.27}$$

where β and ϕ are in radians. The use of ***V-belts*** increases the effective coefficient of friction greatly, as can be seen by the division by the sine of the half-angle. If all chance of slip of the belt must be eliminated, a chain and sprocket wheel should be used.

Sample Problem 9.11

A external brake-band assembly is shown in the figure below. The coefficient of static friction between the flat belt and the drum is 0.3. Determine the minimum force F necessary to prevent the drum from rotating when subjected to a counterclockwise torque of 100 N · m.

Solution We establish free-body diagrams for the drum and the brake lever:
The contact angle between the belt and the drum is π, and Eq. (9.25) yields

$$\frac{T_2}{T_1} = e^{\mu_s \beta} = e^{0.3\pi} = 2.566$$

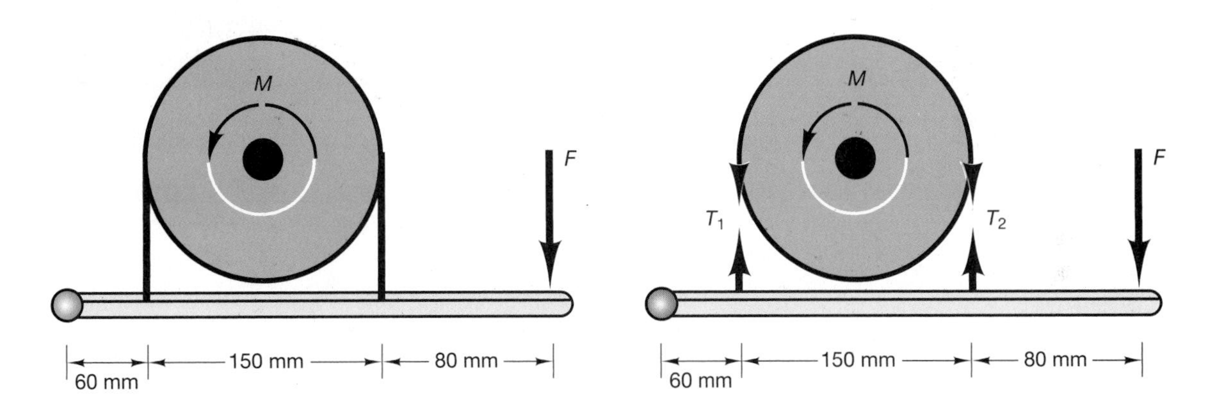

Summing moments about the center of the drum, we obtain

$$100 - (T_2 - T_1)0.075 = 0$$

Summing moments about point 0 on the brake lever yields

$$0.060T_1 + 0.210T_2 - 0.290F = 0$$

We now have three equations for the three unknowns F, T_1, and T_2. Solving the equations results in

$$F = 1758 \text{ N} \qquad T_1 = 851 \text{ N} \qquad T_2 = 2185 \text{ N}$$

9.6 BEARINGS

The two major types of bearings used in machines are rolling-contact bearings and lubrication or journal bearings. The terms "rolling-contact bearing", "antifriction bearing", and "***rolling bearing***" are used interchangeably to describe a bearing in which loads are transferred by rolling contact instead of sliding contact. The static friction before motion is about twice the kinetic friction in these bearings, but is still neglected in most analyses. A rolling-contact bearing is illustrated in Figure 9.20. Since friction is neglected in analyzing of these types of bearings, only journal bearings will be considered in this section.

Journal bearings consist of a shaft or journal rotating inside a sleeve or bearing and the relative motion is sliding. In general, this type of bearing is dependent upon lubrication to reduce friction, wear, and the heat generated during rotation of the shaft.

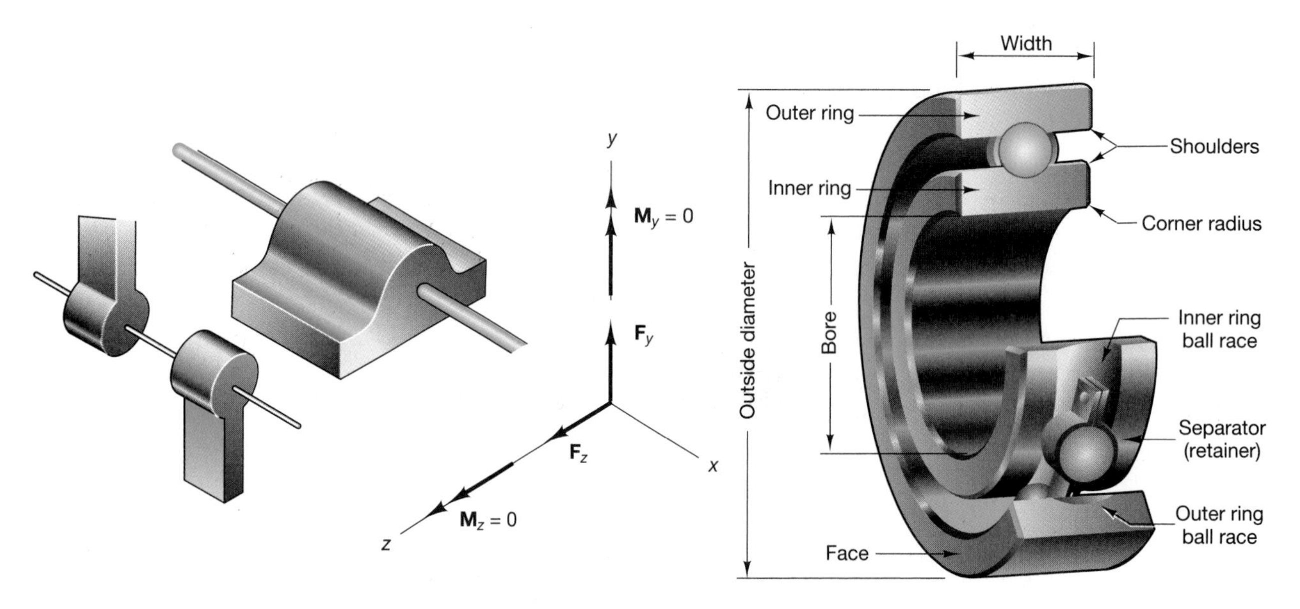

Figure 9.20

The applications of journal bearings are many, from crankshafts in cars to steam turbines. There are also many applications in which the loads are light, maintenance is unimportant, and a simple, unlubricated bearing is a satisfactory design. Powder-metallurgy bearings have built-in lubrication, and many other bearings require only a light grease lubricant. Bearings that are dependent upon full lubrication need to be analyzed using techniques from the field of fluid mechanics and will not be discussed here.

An unlubricated bearing, shown in Figure 9.21, consists of a shaft and a bearing of a slightly larger diameter. As a moment is applied to the shaft, the shaft rides up the bearing side until slipping occurs. After that, the shaft slides down a small amount and settles into

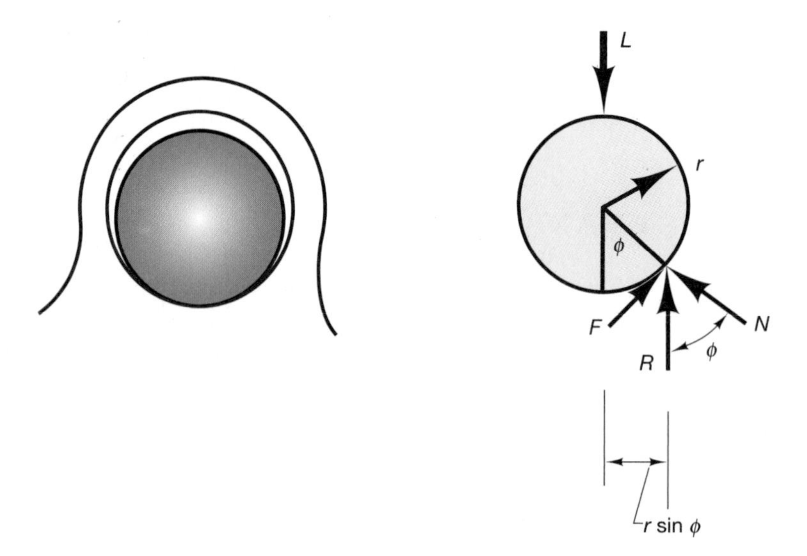

Figure 9.21

a stable position. The couple, developed by the resultant of the friction and the normal forces and the load on the shaft, will equal the driving moment. For equilibrium, the following relations must hold:

$$R = L \tag{9.28}$$
$$M = Rr \sin \phi \tag{9.29}$$

Here, ϕ is the friction angle such that the coefficient of friction is equal to the tangent of this angle. The distance $r \sin \phi$ is called the radius of the *friction circle.* For small angles, the sine may be set equal to the tangent, and Eq. (9.29) may be approximated as

$$M = Rr\mu_k \tag{9.30}$$

This approximation will give reasonable results for the frictional moment. Values of the coefficient of kinetic friction and the angle ϕ are shown with the value of the sine of the angle in the following table:

μ_k	$\phi = \tan^{-1} \mu_k$	$\sin \phi$
0.05	2.86°	0.05
0.10	5.71	0.10
0.15	8.53	0.148
0.20	11.31	0.196
0.25	14.04	0.243
0.30	16.70	0.287

For a coefficient of friction of 0.3, the approximation has an error of less than 5 percent. You can always make the judgment as to whether or not the approximation should be made. The additional labor required to calculate the friction angle and to use Eq. (9.29) is minimal. Well-lubricated journal bearings have low coefficients of friction to reduce wear on the bearing.

Sample Problem 9.12

Determine the torque required to maintain a constant speed for a system consisting of a 10-kg shaft of radius 0.05 m and length 1 m. The coefficient of kinetic friction for the rolling-contact bearing is 0.15. The following diagram shows the system:

Solution The free-body diagram of the shaft is

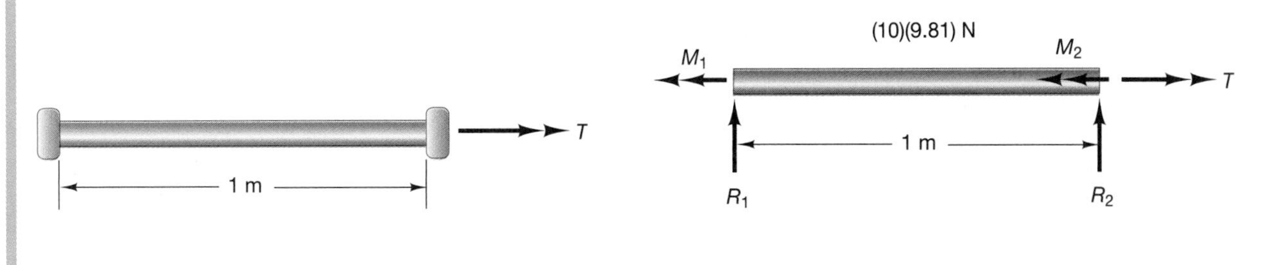

We have

$$R_1 = R_2 = 1/2(10)(9.81) = 49 \text{ N}$$
$$T = M_1 + M_2$$
$$T = R_1 r\mu_k + R_2 r\mu_k = (49)(0.05)(0.15)2 = 0.736 \text{ Nm}$$

9.7 THRUST BEARINGS, COLLARS, AND CLUTCHES

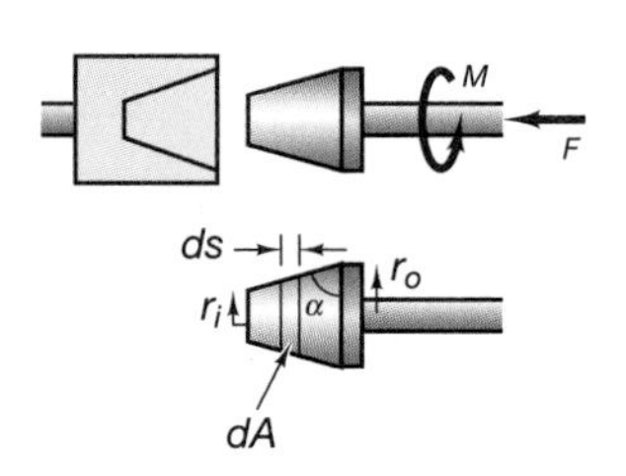

Figure 9.22

In the design of machines that rotate, clutches, collars, and thrust bearings are often used to control axial forces while transmitting torque and rotation. A ***clutch*** is a two-part device that is often used to transmit a motor's circular motion to a driveshaft in such a way that the driveshaft can be easily connected and disconnected. The mating surfaces of a clutch are made of materials with high relative coefficients of friction. ***Collars*** are fixtures used to restrict the axial motion of rotating shafts. Thrust bearings are used in similar applications to restrict axial motion. The mechanics of all three of these devices are similar and will be covered in this section.

A ***thrust bearing*** is capable of resisting an axial load on a shaft, but friction exists between the sliding surfaces. A thrust bearing is illustrated in Figure 9.22. A differential element of area on the conical contact surface is

$$dA = 2\pi r ds = 2\pi r \frac{dr}{\cos\alpha} \tag{9.31}$$

Integrating from r_i to r_o, we find that the entire contact area is

$$a + \int_{r_1}^{r_o} 2\pi r \frac{dr}{\cos\alpha} = \frac{\pi(r_o^2 - r_1^2)}{\cos\alpha} \tag{9.32}$$

Let p be the contact pressure between the surfaces. Then for equilibrium,

$$pA\cos\alpha = F \tag{9.33}$$

and, assuming that the surfaces are in uniform contact, the pressure is

$$p = \frac{F}{A\cos\alpha} = \frac{F}{\pi(r_o^2 - r_i^2)} \tag{9.34}$$

Note that the pressure does not depend upon the cone angle. The moment due to friction between the contacting surfaces is

$$M_f = \int_A \mu_k prdA = \int_{r_i}^{r_o} \mu_k r \frac{F}{\pi(r_o^2 - r_i^2)} \frac{2\pi rdr}{\cos\alpha}$$

or

$$M_f = \frac{2\mu_k F}{3\cos\alpha}\left[\frac{r_o^3 - r_i^3}{r_o^2 - r_i^2}\right] \tag{9.35}$$

For a thrust bearing, the coefficient of friction is the kinetic coefficient of friction. For a clutch, the inner radius is zero, indicating full contact across the mating clutch faces, and the cone angle is zero, as shown in Figure 9.23. The pressure acting on the clutch face is

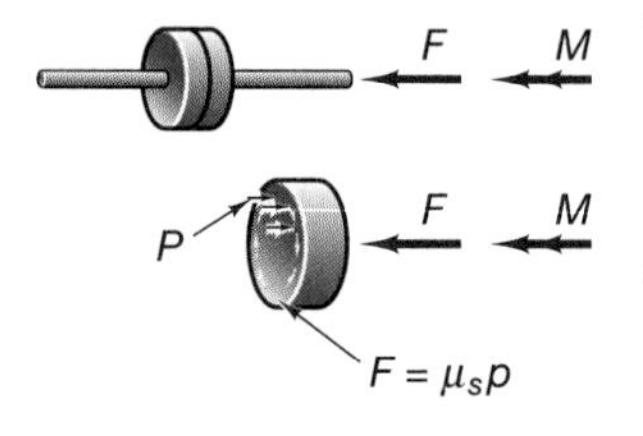

Figure 9.23

$$p = \frac{F}{\pi r_o^2}$$

The maximum moment that can be transmitted by the clutch without slipping is

$$M_f = \int_A \mu_s prdA = \int_{r_i}^{T_o} \mu_k r \frac{F}{\pi r_o^2} 2\pi rdr$$

or

$$M_f = \frac{2\mu_s F r_o}{3}$$

Sample Problem 9.13

A radial sander uses a 125-mm-radius disk rotating at a constant angular velocity. If the sander is pressed on a surface made of wood with a total force of 100 N and the coefficient of friction between the sander and the wood surface is 0.8, what motor torque is required to maintain a constant angular velocity?

Solution The sander is sliding relative to the wood surface, and therefore the coefficient of kinetic friction is used. Equation (9.35) may be used to determine the friction moment the motor must overcome. For this case, $\alpha = 0$ and $r_i = 0$; therefore,

$$M_f = 2/3\mu_k F r_o$$

$$M_f = 2/3(0.8)(100)125 = 6.666\ \text{N}\cdot\text{m}$$

9.8 ROLLING RESISTANCE

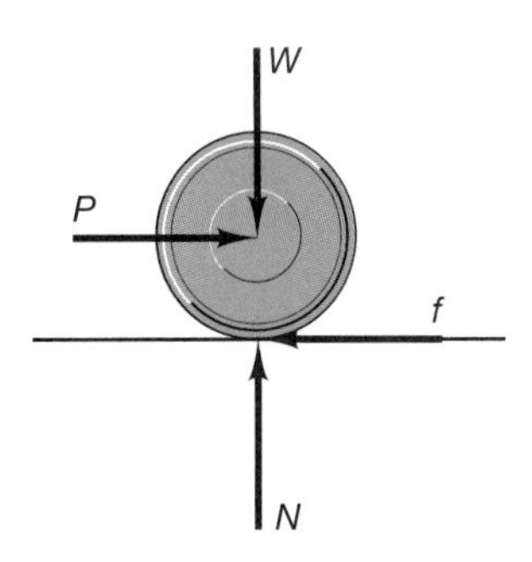

Figure 9.24

"Rolling resistance" refers to the forces acting on a rolling wheel that cause it to slow down and stop, even when axle friction and air resistance are negligible. If the wheel is rolling without sliding, there is no relative motion at the point of contact between the wheel and the horizontal surface. Consider the free-body diagram of a wheel rolling along a level surface under the action of a horizontal force P and supporting a weight W. (See Figure 9.24.) Summation of moments about the center of the wheel shows that the friction force f is zero; regardless of the value of the coefficient of friction between the surfaces, no friction would exist. Examination of equilibrium in the horizontal direction indicates that P is also zero. Therefore, the wheel would continue to roll without any propelling force. Common observation shows that, although the wheel is one of humankind's greatest accomplishments, it is not a perpetual-motion machine. If there were axle friction, as has been seen with bearings, the couple formed by P and f would be equal and opposite to the bearing friction moment, and a force P would be required to continue the motion of the wheel. This, however, is not the full explanation of why the rolling wheel slows and stops.

To understand rolling resistance, one must examine the deformations of the two contacting surfaces. This is especially true if one material is much stiffer than the other, as would be the case with a tire rolling in mud or a metal wheel rolling on a plastic surface. The latter occurs in total knee joint replacements, where a metal component rolls on a polyethylene wear surface. (See Figure 9.25.) As the wheel rolls, it deforms the material in front of it and pushes it forward. The resistance to this deformation is Nd. Although there is some restoring force Nr from the material behind the wheel, it is always less than the deformation force.

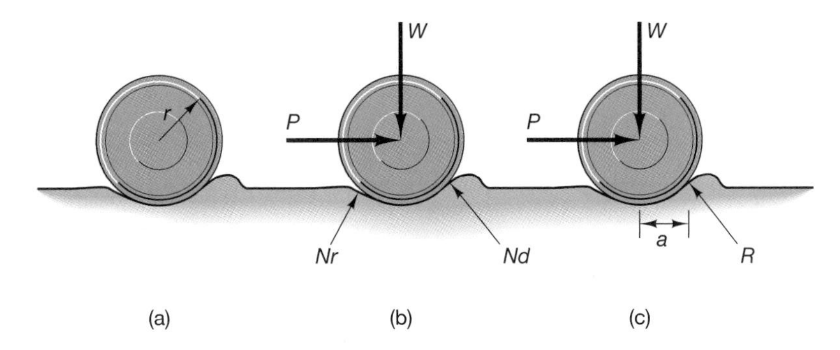

Figure 9.25

The equivalent free-body diagram is shown in Figure 9.25(c). If moments are taken around the point of application of R, the force necessary to overcome the rolling resistance is

$$P = W(a/r) \tag{9.36}$$

P is therefore equal to the rolling resistance force, and Eq. (9.36) may be written as

$$P = \mu_r N \tag{9.37}$$

where N is equal to the normal force between the wheel and the undeformed surface and μr has the same form as a coefficient of friction and is termed the *coefficient of rolling resistance.* It must be emphasized that there is no sliding friction between the wheel and the surface, and the rolling resistance is due to a complex deformation and, sometimes, plastic flow of the material. Many references call the distance a the coefficient of rolling resistance, as experiments have shown that rolling resistance is not very sensitive to the radius of the wheel. However, experiments of this type are extremely difficult to perform, and not much credence is given to the numbers obtained. If only the value a is reported, it will be in units of length, such as inches or millimeters. Some reported values of a are shown in Table 9.2.

Table 9.2

Materials	a in Millimeters	a in Inches
Steel on steel	0.01	0.0004
Steel on wood	2	0.08
Steel on soft ground	130	5
Tire on pavement	0.6	0.025
Tire on dirt	1.3	0.05

Problems

9.97 If the coefficient of friction between leather and wood is 0.6, how many wraps of the reins around a hitching post must a cowboy make to resist a pull by the horse of 3.2 kN? Assume that the free end of the reins weighs 8 N.

9.98 In Figure P9.98, if the coefficient of friction between the rope and the safety ring is 0.4, how much force must support the 80-kg fallen climber?

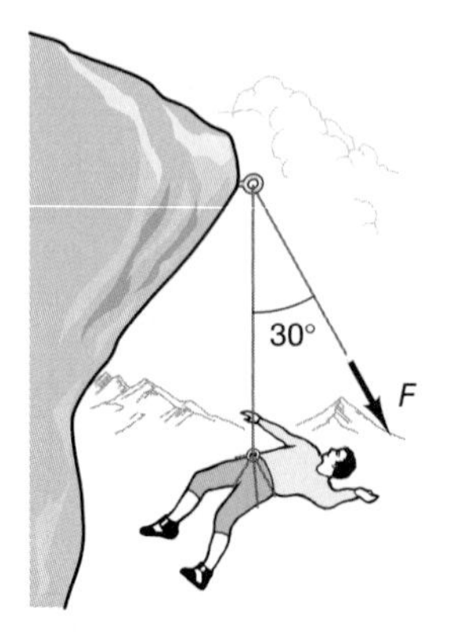

Figure P9.98

9.99 How much force would be required to overcome friction and to begin to pull up the climber in Problem 9.98?

9.100 In Figure P9.100, the drive wheel A is used to transmit a torque to wheel B, attached to a drilling machine. Wheel C is used to apply controlled tension to the flat belt, but is free to rotate. The mass of C is 30 kg, and its weight is fully supported by the belt. Determine the maximum torque that can be transmitted from A to B if the coefficient of static friction between the belt and wheel is 0.4.

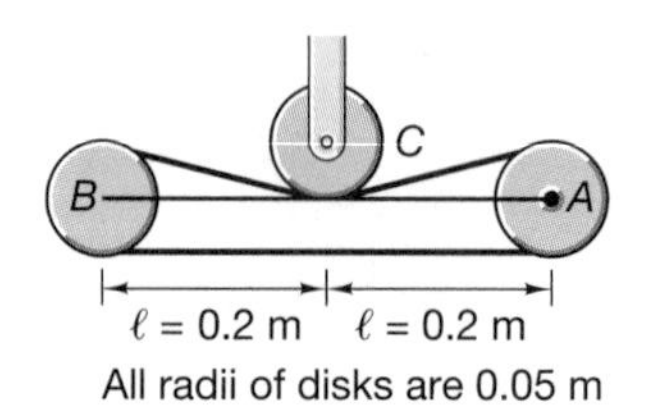

Figure P9.100

9.101 In Problem 9.100, determine the maximum coefficient of friction between the belt and wheels necessary to transmit a torque of 2500 N · m.

9.102 A strap wrench is used to loosen a pipe connection without damage to the surface of the pipe. (See Figure P9.102.) Determine the minimum coefficient of static friction between the strap and the pipe so that the wrench will not slip. Determine the amount of torque applied to the pipe.

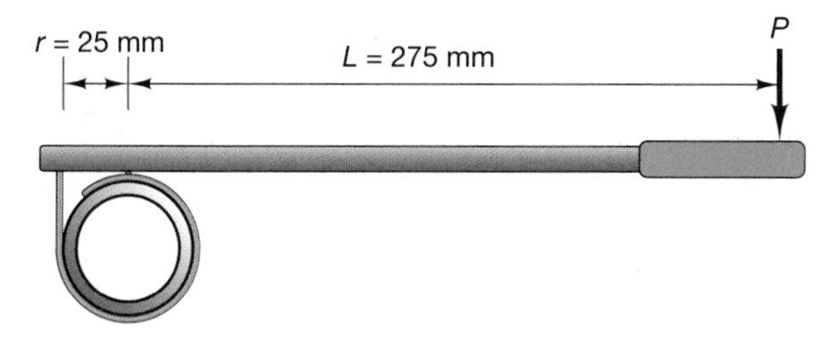

Figure P9.102

9.103 Show for the strap wrench that the minimum coefficient of static friction required to prevent slip depends on only the ratio $L/(L + r)$. Investigate the dependency on this ratio.

9.104 Two 50-kg flywheels are supported by two journal bearings and driven at a constant angular velocity by a motor. (See Figure P9.104.) Determine the torque necessary to overcome bearing friction if the coefficient of kinetic friction is 0.2 and the connecting 4-kg shaft has a diameter of 80 mm.

Figure P9.104

9.105 If a moment of 4 N · m is required to maintain the constant speed of the flywheels in Problem 9.104, what is the kinetic coefficient of friction at the bearings?

9.106 If the coefficient of friction at the bearings is 0.15, determine the torque required to raise the 50-N weight shown in Figure P9.106 at constant speed. The 20-cm-diameter pulley weighs 20 N and the 3-cm-diameter shaft weighs 5 N.

Figure P9.106

9.107 Determine the torque required to lower the weight at a constant speed in Problem 9.106.

9.108 Negligence on the part of service personnel results in poor lubrication of the bearing in the pulley system of Problem 9.106, and the coefficient of friction between the shaft and the bearing rises to 0.9. Determine the torque required to raise the weight and the friction wear force acting between the shaft and the bearing.

9.109 Total knee replacement (total knee arthroplasty) consists of implanting a cobalt–chromium femoral condyle component that slides on an ultra-high-molecular weight polyethylene wear surface on the tibial plate. (See Figure P9.109.) The cobalt–chromium component acts like a shaft, and the polyethylene acts like a bearing. If the coefficient of friction between the metal and the polyethylene is 0.05, and the radius of the condyle component is 25 mm, determine the flexion moment required to overcome friction when the knee is loaded by a force of 900 N during single-stance phase while walking.

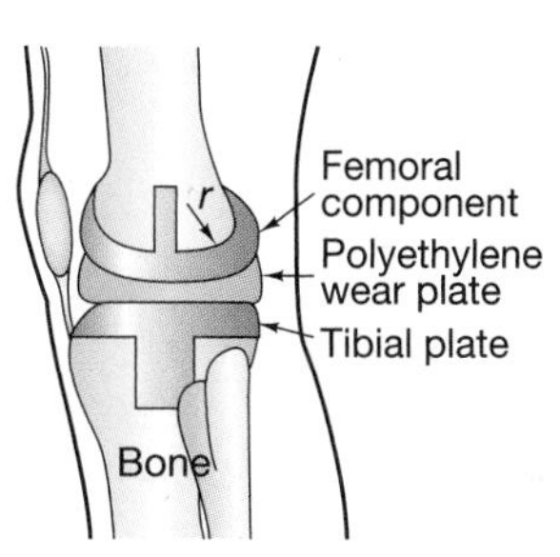

Figure P9.109

9.110 A lever of negligible weight is attached with clearance to a shaft of radius 20 mm. (See Figure P9.110.) If the coefficient of friction between the shaft and the lever is 1.1, determine the force F necessary to start to lift a 30-kg block.

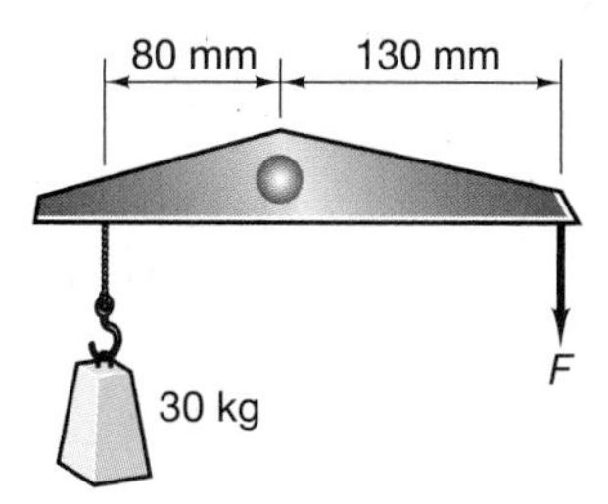

Figure P9.110

9.111 Determine the minimum force F required to prevent the lever in Problem 9.110 from rotating counterclockwise.

9.112 Determine the maximum torque that can be transmitted across the spring-loaded clutch of radius 100 mm shown in Figure P9.112. The coefficient of static friction between the clutch faces is 0.8. The unstretched length of the spring is 500 mm and the spring constant is 1000 N/m.

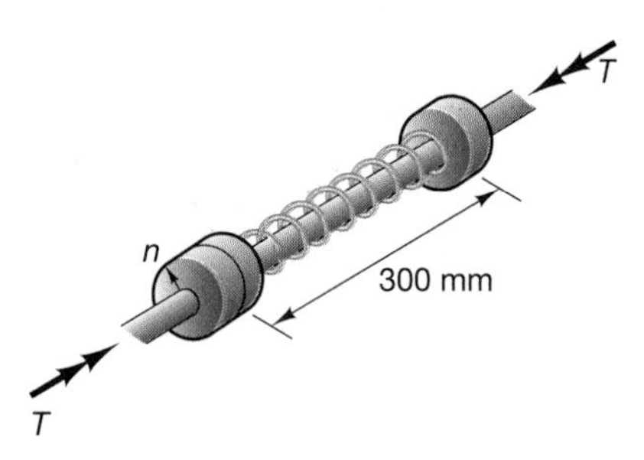

Figure P9.112

9.113 Determine the torque required to overcome the kinetic friction $\mu_k = 0.2$ in a thrust bearing carring an axial thrust of 1000 N. (See Figure P9.113.)

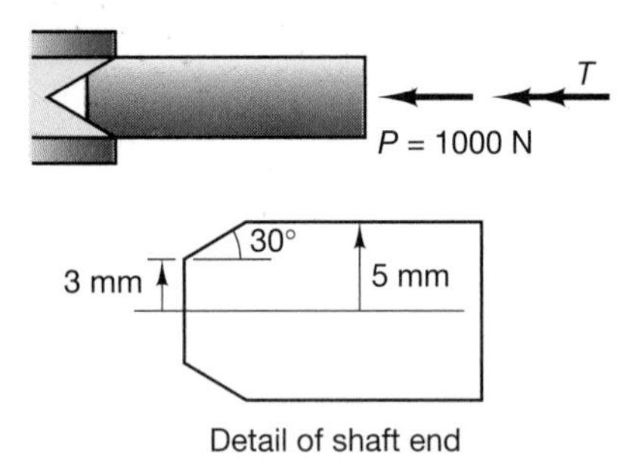

Figure P9.13

9.114 In Figure P9.114, two friction pads form the contacting surface of a thrust bearing supporting a vertical shaft and pulley weighing 2000 N. If the coefficient of kinetic friction between the pads and the shaft is 0.4, determine the torque required to overcome friction.

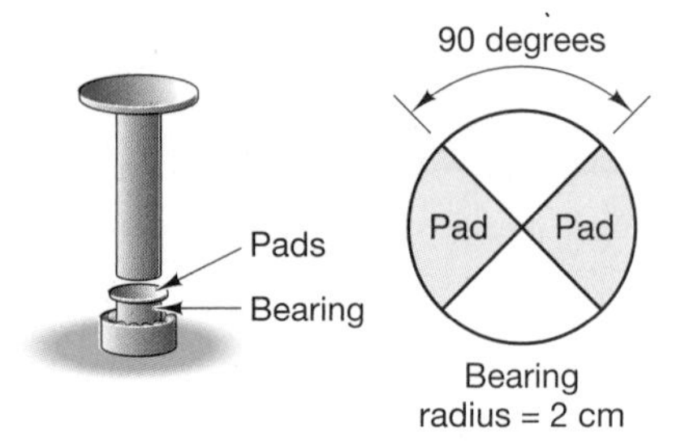

Figure P9.114

9.115 In Problem 9.109, if the coefficient of rolling resistance $a = 2$ mm, determine the horizontal force necessary to make the femoral component roll on the polyethylene wear plate.

9.116 If a tire radius is 330 mm, how much force is necessary to push a 1100-kg car over a dirt surface (with coefficient of rolling resistance $a = 1.3$ mm)? Over a muddy surface ($a = 40$ mm)?

9.117 How much force is required to push the same car in Problem 9.116 on pavement ($a = 0.6$ mm)?

Chapter Summary

Friction is the resistance to sliding between two contacting surfaces. Friction always acts to resist motion. The maximum friction force that can be obtained before slipping occurs is defined by the coefficient of static friction,

$$\mu_s = f_{\max}/N$$

where $f_{\max}$ is the maximum friction force and N is the normal force between the contacting surfaces. When sliding occurs, the friction force decreases by about 20 to 25 percent and is defined by the coefficient of kinetic friction,

$$\mu_k = f_{\text{sliding}}/N$$

If an object is on an inclined surface, and $\hat{\mathbf{m}}$ and $\hat{\mathbf{q}}$ are two nonparallel unit vectors tangent to the surface, the unit vector normal to the surface is

$$\hat{\mathbf{n}} = \frac{\hat{\mathbf{m}} \times \hat{\mathbf{t}}}{|\hat{\mathbf{m}} \times \hat{\mathbf{t}}|}$$

Square-threaded screws are efficient devices used to transmit power or motion. Such screws convert angular motion into linear motion or linear motion into angular motion. The lead or pitch of a screw is the distance of advancement of the screw for each full revolution. The moment required to initiate motion or to continue movement at a constant rate to lift a weight is

$$M = Wr\left(\frac{p + 2\pi\mu r}{2\pi r - \mu p}\right)$$

where r is the mean radius of the screw, p is the pitch of the screw, and μ is the coefficient of friction between the screw thread and the thread in the frame. The coefficient of static friction is used to determine the moment to initiate movement, and the coefficient of kinetic friction is used to determine the moment for constant movement.

If the screw is self-locking—that is, if the screw will support the weight without a resisting moment—the coefficient of static friction must be

$$\mu_s \geq \frac{p}{2\pi r}$$

The moment required to lower the weight is

$$M = Wr\left(\frac{2\pi\mu r - p}{2\pi r + \mu p}\right)$$

where the coefficient of static friction is used to determine the moment needed to initiate movement and the coefficient of kinetic friction is used during movement at a constant rate. The efficiency of a power screw is

$$e = \frac{Wp}{2\pi M} = \frac{p(2\pi r - \mu_k p)}{r(p + \mu_k 2\pi r)}$$

Belt Friction When a belt or rope passes over a drum, friction will cause a difference in the tensions at the two ends of the belt such that

$$T_2 = T_1 e^{\mu b}$$

where β is the contacting angle in radians.

Roller Bearings The friction moment resisting rotation of a shaft in a dry or rolling contact bearing is

$$M \approx Rr\mu_k$$

where R is the force acting normal to the shaft surface at the bearing and r is the radius of the shaft.

Thrust Bearings, Collars, and Clutches Thrust bearings, collars, and clutches are used to control axial forces while transmitting torque and rotation. For a thrust bearing in the shape of a truncated cone, the friction moment is

$$M_f = \int_A \mu_k pr\,dA = \int_{r_i}^{r_o} \mu_k r \frac{F}{\pi(r_o^2 - r_i^2)} \frac{2\pi r dr}{\cos\alpha}$$

$$M_f = \frac{2\mu_k F}{3\cos\alpha}\left[\frac{r_o^3 - r_i^3}{r_o^2 - r_i^2}\right]$$

The coefficient of static friction is used to determine the moment required to initiate movement, and the coefficient of kinetic friction is used to determine the resisting friction during movement.

For a clutch, the angle α and the inner radius r_i are zero. The maximum moment that can be transmitted through an engaged clutch is

$$M_f = \int_A \mu_s pr\, dA = \int_{r_i}^{r_o} \mu_k r \frac{F}{\pi r_o^2} 2\pi r dr$$

$$M_f = \frac{2\mu_s F r_o}{3}$$

where F is the axial force and r_o is the radius of the clutch.

Chapter 10 MOMENTS OF INERTIA

A gymnast changes the moment of inertia of her body by tucking and increasing the rate of rotation. (*Tumblers, 1942* by Harold Edgerton. Reprinted by permission of Palm Press, Inc. and the Edgerton Foundation.)

10.1 INTRODUCTION

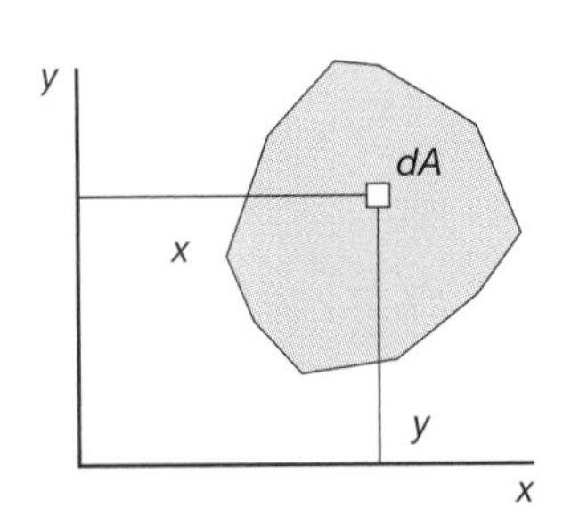

Figure 10.1

In this chapter, we will introduce a specific mathematical property of an area called the **second moment of an area**, or the *area moment of inertia.* The latter name used for this property is based upon its mathematical similarity to the *mass moment of inertia,* which is the resistance to rotation of a rigid body. This property of a rigid body will be discussed in detail in the study of dynamics. The second moment of an area is also called the *transverse moment of inertia.*

The second moment of an area characterizes the way the area is distributed about the centroid and is directly related to the resistance to bending of a beam. The second moment of an area has the mathematical properties of a second-order tensor. In Chapter 5, the first moment of an area was introduced and used to find the centroid of the area. In Figure 10.1, the first moment of the area about the x- and y-axes is defined to be

$$\begin{aligned}\text{First moment of the area about } x &= \int_A y\,dA = y_c A \\ \text{First moment of the area about } y &= \int_A x\,dA = x_c A\end{aligned} \tag{10.1}$$

where x_c and y_c are the x-and y-coordinates of the centroid of the area, which can be considered to be the average of the coordinates of each element of the area, or a point at which the area can be considered to be concentrated.

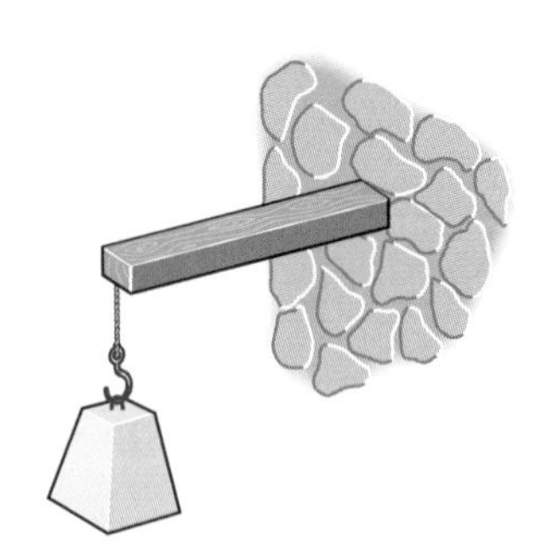
Figure 10.2

We will study the second moment of an area in detail for two reasons: First, the second moment of the cross-sectional area of a beam is directly related to its resistance to bending; second, this quantity is a second-order tensor, and tensors of this order will be encountered as stresses, strains, and mass moments of inertia, and their mathematical properties need to be understood. Examination of a 5×10 cm board will show that the board bends more easily when the 10-cm side is down than when the 5-cm side is down. The cross-sectional area of the board is the same, so that the resistance to bending is not the area. In the 1500s, Galileo examined the resistance to the bending of a beam by anchoring one end in a stone wall and hanging loads on the other end, as shown in Figure 10.2. He determined that the resistance to the bending of a beam with a rectangular cross section was proportional to the base times the cube of the height, but he assumed the wrong proportionality constant.

10.2 SECOND MOMENT OF AN AREA

The second moments of an area with respect to the x- and y-reference axes are defined as

$$\begin{aligned}I_{xx} &= \int_A y^2 dA \\ I_{yy} &= \int_A x^2 dA\end{aligned} \tag{10.2}$$

The units of the second moment of the area are length to the fourth power—that is, mm^4, in^4, m^4, or ft^4. Note that the integrand is a squared quantity, so that, regardless of the reference axes, the second moment of the area is positive. The second moment of the area increases as the square of the distance from the reference axis to the element of area.

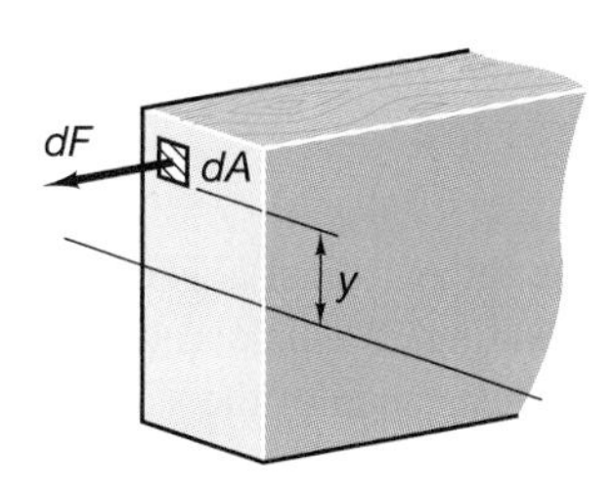

Figure 10.3

It will be shown in the study of deformable solids that the distributed internal forces in a beam in pure bending (equal and opposite moments on each end of the beam) vary linearly about an axis passing through the centroid of the cross-sectional area—that is, a centroidal axis of the beam. This relationship is illustrated in Figure 10.3. The force on the element of area dA is

$$dF = kydA \quad \text{where } k \text{ is a constant} \tag{10.3}$$

The total force acting on the cross section of the beam is

$$R = \int dF = \int_A kydA = ky_cA \tag{10.4}$$

The total force acting upon the cross section is zero, as the distance from the centroidal axis to the centroid y_c is zero. The total moment of the internal forces about the centroidal axis is

$$M = \int ydF = \int_A ky^2dA = kI_{xx} \tag{10.5}$$

The internal moment is then proportional to the second moment of the area about the centroidal axis.

10.2.1 DETERMINATION OF THE SECOND MOMENT OF AN AREA BY INTEGRATION

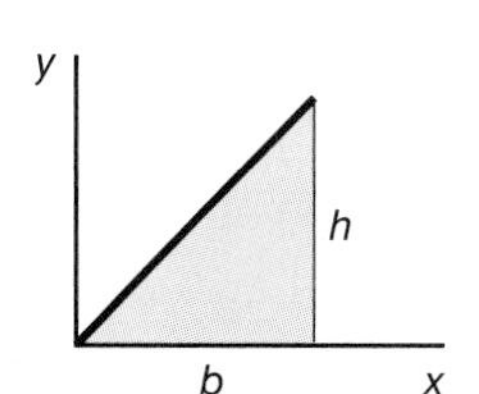

Figure 10.4

The most straightforward manner of obtaining the second moment of areas by integration is to use repeated or double integration. Consider, for example, the triangular area shown in Figure 10.4. The equation of the line forming the top of the triangle is $y = (h/b)\,x$, and the second moment of the area about the x-axis is

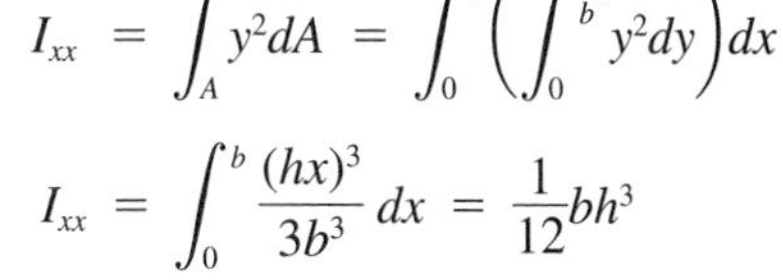

$$I_{xx} = \int_A y^2dA = \int_0^b \left(\int_0^{\frac{h}{b}x} y^2dy \right) dx$$

$$I_{xx} = \int_0^b \frac{(hx)^3}{3b^3}\,dx = \frac{1}{12}bh^3$$

Note that the inner integral has functional limits and must be integrated first.

In a similar manner, the second moment of the area about the y-axis is

$$I_{yy} = \int_A x^2dA = \int_0^b x^2 \left(\int_0^{\frac{h}{b}x} dy \right) dx$$

$$I_{yy} = \int_0^b \frac{h}{b}\,x^3dx = \frac{1}{4}b^3h$$

10.3 POLAR MOMENT OF INERTIA

The resistance of a shaft to torsion is given by the ***polar moment of inertia*** or the *second moment of the area about the z-axis*. The polar moment of inertia is defined as

$$J_{0z} = \int_A r^2dA \tag{10.6}$$

In polar coordinates, $r^2 = x^2 + y^2$, and therefore,

$$J_{0z} = I_{xx} + I_{yy} \tag{10.7}$$

Figure 10.5

The second moment of a cross-sectional area of a beam about a centroidal axis of the cross-sectional area is a measure of the resistance of the beam to bending. The polar moment of inertia of a circular shaft about its centroid is a measure of the resistance of the shaft to torsion. The first moment of the area was used to determine the centroid of the area. Realize that there are an infinite number of axes that pass through the centroid. In Section 10.8, we will define a particular set of orthogonal centroidal axes called the ***principal axes***.

Consider the circular area shown in Figure 10.5, wherein the centroid is easily identified to be the center of the circle. The polar moment of inertia is

$$J_{0z} = \int_A r^2 dA = \int_0^{2\pi}\left[\int_0^R r^3 dr\right] d\theta = \frac{\pi R^4}{2} \tag{10.8}$$

The polar moment of inertia can also be obtained from the second moments of the area about the x- and y-axes. For a circular area, the second moment of the area about any centroidal axis is the same. The polar moment of inertia can be written as

$$J_{0z} = I_{xx} + I_{yy} = 2I_{xx}$$

$$I_{xx} = \int_A y^2 dA \quad y = r\sin\theta$$

$$I_{xx} = \int_0^{2\pi}\left[\int_0^R r^3 dr\right] = \sin^2\theta d\theta = \int_0^{2\pi}\frac{R^4}{4}\sin^2\theta d\theta$$

$$I_{xx} = \frac{R^2}{4}\left[\frac{\theta}{2} - \frac{\sin\theta}{4}\right]_0^{2\pi} = \frac{\pi R^4}{4}$$

Therefore,

$$J_{0z} = \frac{\pi R^4}{2} \tag{10.9}$$

As expected, the result is the same as Eq. (10.8). Note that the resistance of a circular shaft to bending is one-half its resistance to torsion.

Integrals of this type can be evaluated using mathematical software that is available in commercial packages. (See the computational Supplement.)

10.4 SECOND MOMENT OF AN AREA ABOUT CENTROIDAL AXES FOR SPECIFIC AREAS

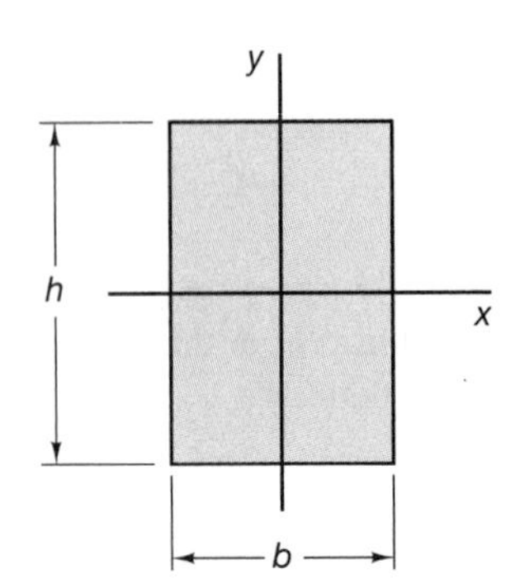

The second moment of regular areas may be easily obtained by multiple integration. The diagrams at the left and on the top of the next page show a rectangle, a triangle, and a circle. For these figures, we have the following second moments:

Rectangle:

$$I_{xx} = \int_{-\frac{b}{2}}^{\frac{b}{2}}\int_{-\frac{h}{2}}^{\frac{h}{2}} y^2 dy\, dx = \frac{1}{12}\cdot h^3\cdot b$$

$$I_{yy} = \int_{-\frac{b}{2}}^{\frac{b}{2}}\int_{-\frac{h}{2}}^{\frac{h}{2}} x^2 dy\, dx = \frac{1}{12}\cdot b^3\cdot h$$

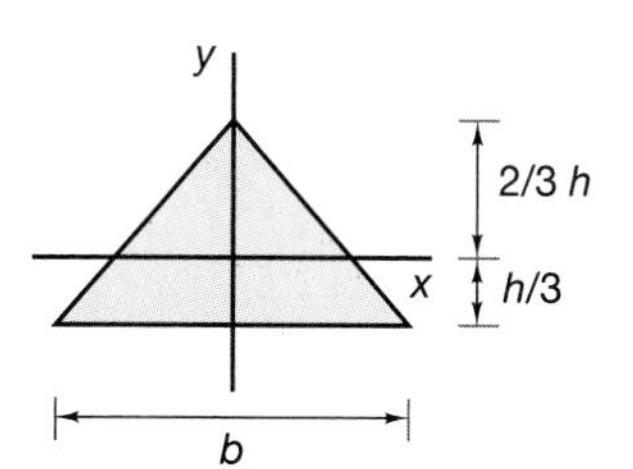

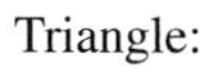
Triangle:

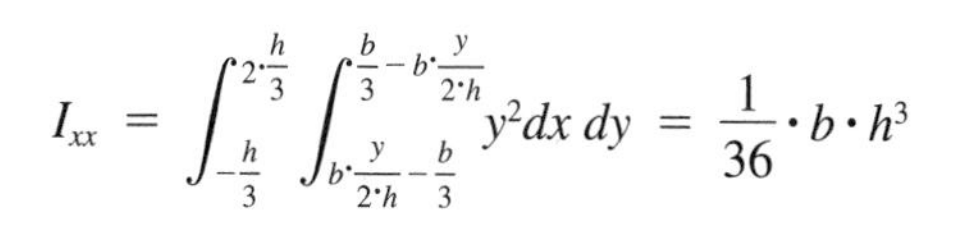

$$I_{xx} = \int_{-\frac{h}{3}}^{2\cdot\frac{h}{3}} \int_{b\cdot\frac{y}{2\cdot h}-\frac{b}{3}}^{\frac{b}{3}-b\cdot\frac{y}{2\cdot h}} y^2 dx\, dy = \frac{1}{36}\cdot b\cdot h^3$$

$$I_{yy} = \int_{-\frac{h}{3}}^{\frac{2\cdot h}{3}} \int_{\frac{b\cdot y}{2\cdot h}-\frac{b}{3}}^{\frac{b}{3}-\frac{b\cdot y}{2\cdot h}} x^2 dx\, dy = \frac{1}{48}\cdot b^3\cdot h$$

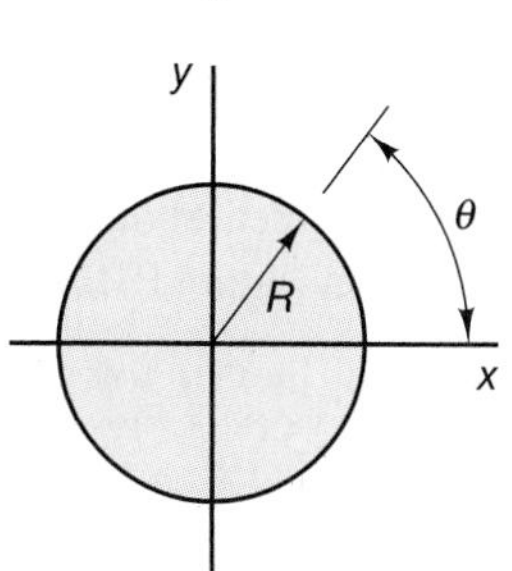

Circle:

$$I_{xx} = I_{yy} = \int_0^{2\cdot\pi}\left(\int_0^R r^3 dr\right)\cdot \sin^2(\theta)\, d\theta = \frac{1}{4}\cdot\pi\cdot R^4$$

Each of these second moments of area has been obtained by double integration.

Problems

10.1 Determine the location of the centroid in Figure P10.1.

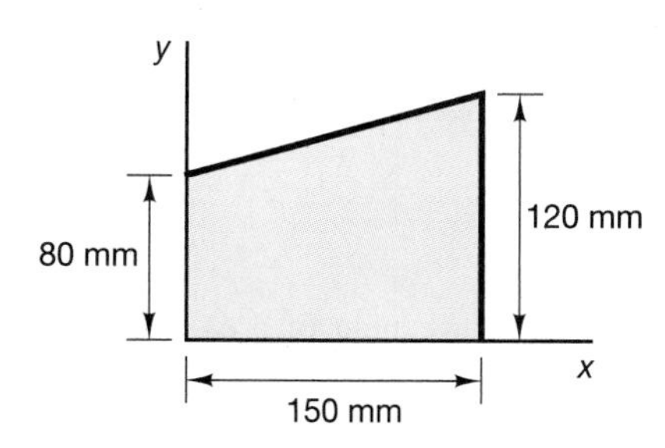

Figure P10.1

10.2 Determine the second moment I_{xx} of the area about the coordinate axis in Figure P10.1.

10.3 Determine the second moment I_{yy} of the area about the coordinate axis in Figure P10.1.

10.4 Determine I_{xx} for the ring shown in Figure P10.4.

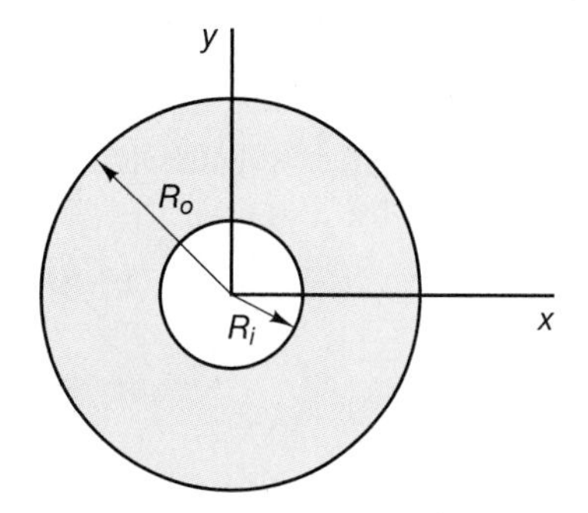

Figure P10.4

10.5 In Figure P10.4, determine J_{0z}, and verify that it is equal to the sum $(I_{xx} + I_{yy})$.

10.6 Determine I_{xx} about the reference axis for the rectangle shown in Figure P10.6.

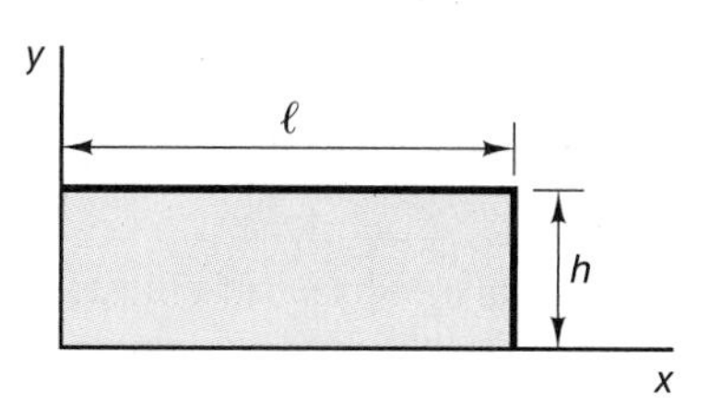

Figure P10.6

10.7 Determine I_{yy} about the reference axis in Figure P10.6.

10.8 Determine I_{xx} for the area shown in Figure P10.8.

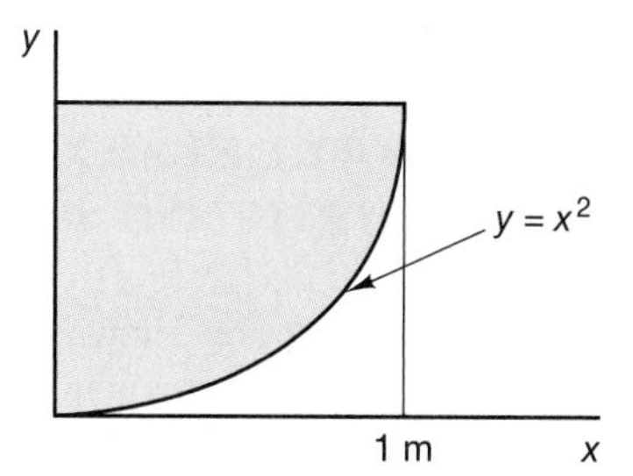

Figure P10.8

10.9 Determine I_{yy} for the area shown in Figure P10.8.

10.10 Determine the location of the centroid for the area shown in Figure P10.8.

10.11 Determine the second moments of the area and the polar moment of inertia about the point 0 in Figure P10.11.

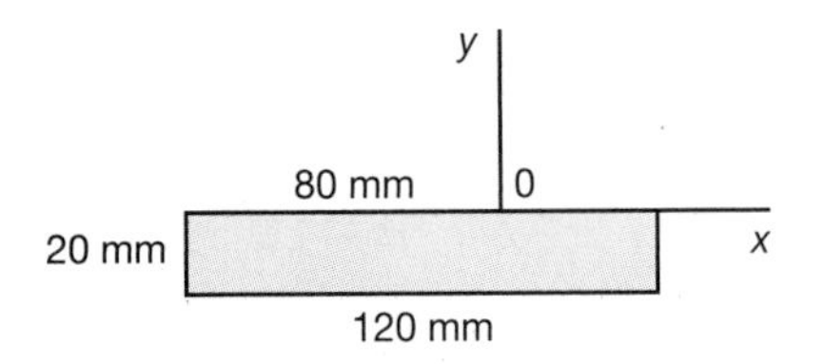

Figure P10.11

10.12 The equation for the ellipical cross section in Figure P10.12 is

$$\frac{x^2}{a^2} + \frac{y^2}{b^2} = 1 \quad \text{where } a = 3 \text{ and } b = 2$$

Determine the second moment of area about the x-axis.

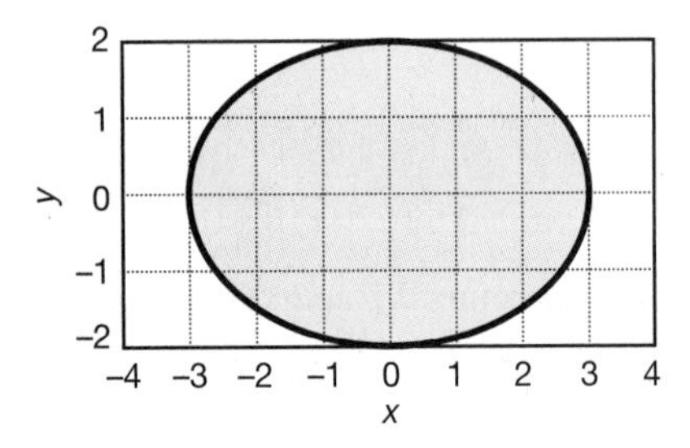

Figure P10.12

10.13 Determine the second moment of area about the y-axis for the ellipse in Figure P10.12.

10.14 Determine the polar moment of inertia of the ellipse in Figure P10.12.

10.15 A cross section is approximated by a half-ellipse, as shown in Figure P10.15. Determine the second moment of area about the x-axis. Note that

$$-a \leqslant x \leqslant a$$

$$y = b\sqrt{1 - \frac{x^2}{a^2}}$$

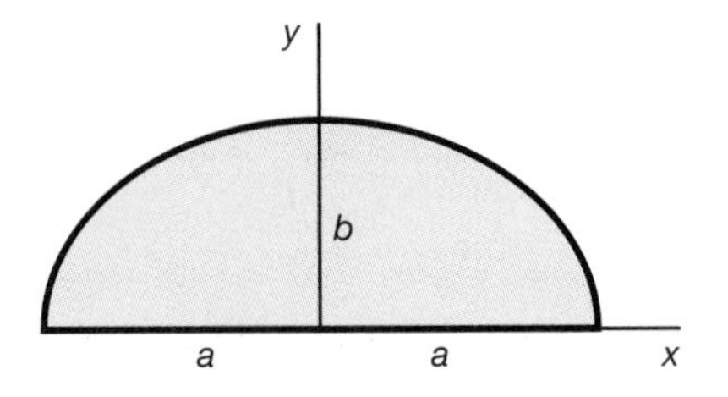

Figure P10.15

10.16 Determine the second moment of area about the y-axis for the half-ellipical cross section in Figure P10.15.

10.17 Determine the polar moment of inertia for the half-ellipical cross section in Figure P10.15.

10.18 Determine the second moment of area about the x-axis for a cross-sectional area defined by the half-sine curve $y = h \sin\left(\frac{\pi x}{\ell}\right)$, $0 \leqslant x \leqslant \ell$, in Figure P10.18.

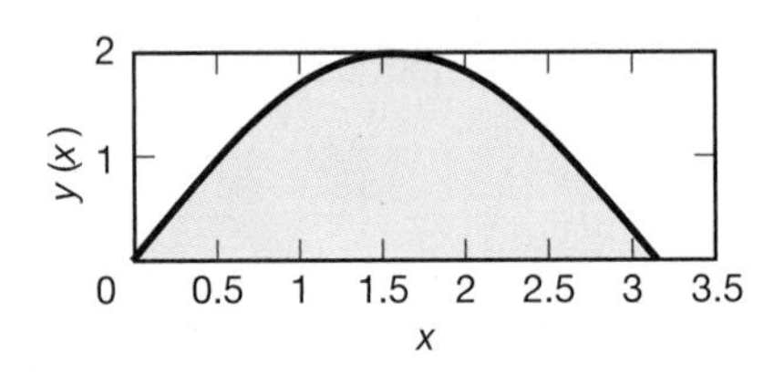

Figure P10.18

10.19 Determine the second moment of area about the y-axis in Figure P10.18.

10.20 Determine the polar moment of area in Figure P10.18.

10.21 Determine the y-centroidal axis of the half-sine area and the second moment of area about this axis in Figure P10.18.

10.5 PARALLEL-AXIS THEOREM FOR THE SECOND MOMENT OF AREA

Consider the area shown in Figure 10.6 with two parallel reference axes, x–y and x'–y'. The y-coordinate of any point on the area is related to the y'-coordinate by

$$y = y' + y_0 \tag{10.10}$$

The second moment of the area about the x-axis is

$$I_{xx} = \int_A y^2\, dA$$

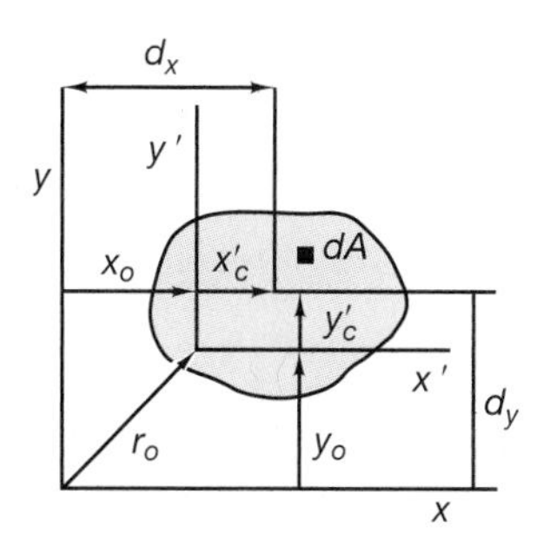

Figure 10.6

or

$$I_{xx} = \int_A (y' + y_0)^2\, dA = \int_A (y'^2 + 2y'y_0 + y_0^2)\, dA \tag{10.11}$$

The first term in the integral in Eq. (10.11) is the second moment of the area about the x'-axis and may be written as I'_{xx}. The second term is the integral of the product of y' and dA and is equal to the distance from the x'-axis to the centroid, times the area. The third term is equal to y_0^2 times the area. Therefore, Eq. (10.11) can be written as

$$I_{xx} = I'_{xx} + 2y_0 y'_c A + y_0^2 A \tag{10.12}$$

If the x'-axis is a centroidal axis, the distance y'_c is zero, and Eq. (10.12) can be written as

$$I_{xx} = I_{xx_c} + Ad_y^2 \tag{10.13}$$

where I_{xx_c} is the second moment of the area about the x'-centroidal axis and d_y is the distance bewteen the reference x-axis and the parallel centroidal axis x_c.

This is the parallel-axis theorem, which allows the computation of the second moment of area about any axis if the second moment of area about a parallel centroidal axis is known. The theorem permits complex cross-sectional areas to be treated as a composite of simple geometric shapes.

The parallel-axis theorem applies equally to the second moment of an area about the y-axis, which may be written as

$$I_{yy} = I_{yy_c} + Ad_x^2 \tag{10.14}$$

where I_{yy_c} is the second moment of the area about the y-centroidal axis and d_x is the distance between the reference y-axis and the parallel centroidal axis y_c.

Adding these two forms of the parallel-axis theorem together yields a similar result for the polar moment of inertia, or the second polar moment of the area, which is

$$J_0 = J_{0_c} + Ar_0^2 \tag{10.15}$$

where J_{0_c} is the second polar moment of the area about the centroid and r_0 is the distance from the reference origin to the centroid.

In this case, the parallel axes are the z-axes, and the second moment of the area is taken about those axes.

Sample Problem 10.1

Using the parallel-axis theorem, compute the second moment of a semicircular area about its centroidal axes.

Solution

We will first compute the second moment of area about a reference set of axes x, y. Note that the y-axis is an axis of symmetry and therefore is a centroidal axis. The coordinates of the centroid are $\left(0, \dfrac{4R}{3\pi}\right)$, and the area of the cross section is $\pi R^2/2$. Thus, for a semicircle,

$$I_{xx} = \int_0^{\pi}\left(\int_0^R r^3 dr\right) \cdot \sin^2(\theta)\, d\theta = \frac{1}{8}\cdot \pi \cdot R^4$$

$$I_{yy} = \int_0^{\pi}\left(\int_0^R r^3 dr\right) \cdot \cos^2(\theta)\, d\theta = \frac{1}{8}\cdot \pi \cdot R^4$$

I'_{yy} is the second moment of area about the centroidal axis. The second moment of area about the x-centroidal axis may be found using the parallel-axis theorem:

$$
\begin{aligned}
I'_{xx} &= I_{xx} - Ay_c^2 \\
&= \left(\frac{\pi}{8} - \frac{8}{9\pi}\right)R^4
\end{aligned}
$$

10.6 RADIUS OF GYRATION OF AN AREA

A useful manner of presenting the second moment of area about an axis is the **radius of gyration of an area**. "Gyration" means "rotation about a point or axis," and the radius of gyration may be perceived to be the distance from the point or axis to a point at which the area is considered to be concentrated. The radii of gyration are defined as

$$
\begin{aligned}
k_x &= \sqrt{\frac{I_{xx}}{A}} \\
k_y &= \sqrt{\frac{I_{yy}}{A}} \qquad (10.16) \\
k_z &= \sqrt{\frac{J_{0z}}{A}}
\end{aligned}
$$

The second moments of the area can be expressed in terms of the radii of gyration as

$$
\begin{aligned}
I_{xx} &= k_x^2 A \\
I_{yy} &= k_y^2 A \qquad (10.17) \\
J_{0z} &= k_z^2 A
\end{aligned}
$$

Examining the relationship between the polar second moment of the area and the second moments of area about the x- and y-axes, we obtain the following equations:

$$
\begin{aligned}
J_{0z} &= I_{xx} + I_{yy} \\
k_z^2 A &= k_x^2 A + k_y^2 A
\end{aligned} \qquad (10.18)
$$

Therefore,

$$k_z^2 = k_x^2 + k_y^2$$

The parallel-axis theorem may be written in terms of the radius of gyration as

$$I_{xx} = I_{xx_c} + Ad_y^2$$

so that

$$k_x^2 = k_{xc}^2 + d_y^2 \qquad (10.19)$$

where k_{xc} is the radius of gyration about the x-centroidal axis.

Sample Problem 10.2 Determine the radius of gyration of the rectangular area shown at the left, where

$$I_{xx_c} = \frac{1}{12}bh^3$$

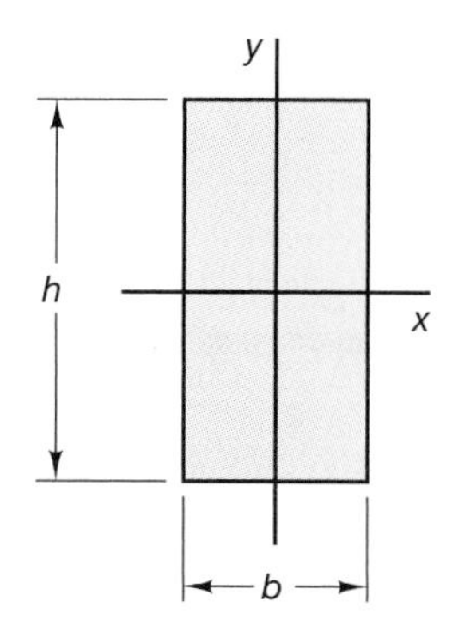

Solution We have

$$k_{xc} = \sqrt{\frac{I_{xx_c}}{A}} = \sqrt{\frac{\frac{1}{12}bh^3}{bh}} = \sqrt{\frac{1}{12}}h$$

If the radius of gyration about an axis at the base of the rectangle had been requested, it could be found by the parallel-axis theorem for radii of gyration:

$$k_x^2 = k_{xc}^2 + d_y^2$$

$$k_x^2 = \frac{1}{12}h^2 + \left(\frac{h}{2}\right)^2 = \frac{1}{3}h^2$$

$$k_x = \sqrt{1/3}h$$

Problems

10.22 Determine the centroid of the cross section shown in Figure P10.22 and the second moment of area about the x- and y-axes. Use the parallel-axis theorem to find the radii of gyration about the centroidal axes.

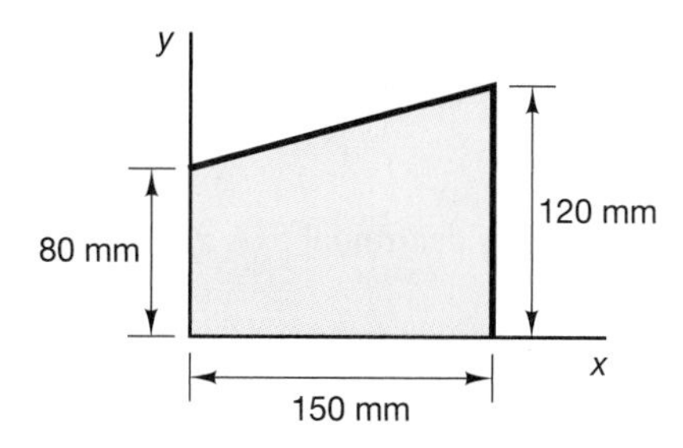

Figure P10.22

10.23 Use the parallel-axis theorem to determine the second moment of the area about x- and y-axes with origin at A and B in Figure P10.23.

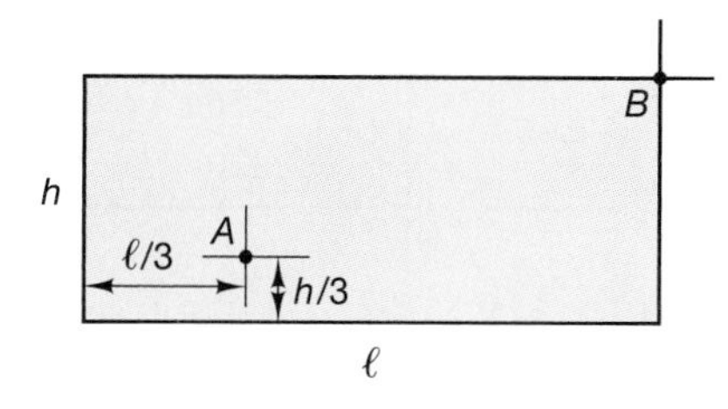

Figure P10.23

10.24 Determine the second moment of a semicircle of radius r about the centroidal axes.

10.25 Determine the second moment of the area about parallel centroidal axes for the area shown in Problem 10.8.

10.26 Determine the second moment of the area about parallel centroidal axes for the half-ellipse area shown in Problem 10.15.

10.27 Determine the centroidal radii of gyration for the half-sine area in Problem 10.18.

10.28 Determine the centroidal polar moment of inertia of the half-sine area in Problem 10.18.

10.29 For a circular cross section of radius r (see Figure P10.29), determine the second moment of the area about an axis x' located $r/2$ above the center of the circle.

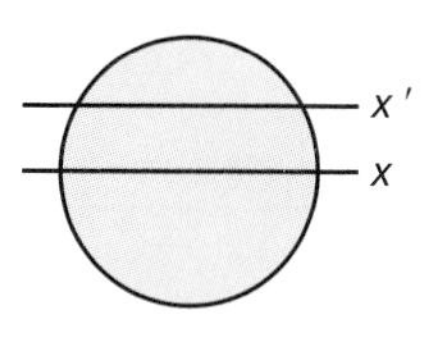

Figure P10.29

10.30 Determine the radius of gyration about an axis tangent to the top of the circle shown in Figure P10.29.

10.31 Determine the second moment of area about the centroidal axes of the triangular cross section illustrated in Figure P10.31.

10.32 Determine the second moment of the triangle about a pair of axes parallel to the base and height with origin at the top corner of the triangle in Figure P10.31.

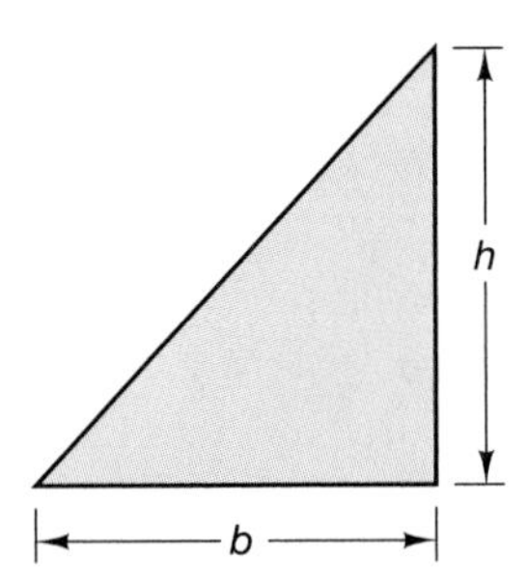

Figure P10.31

10.7 SECOND MOMENTS OF COMPOSITE AREAS

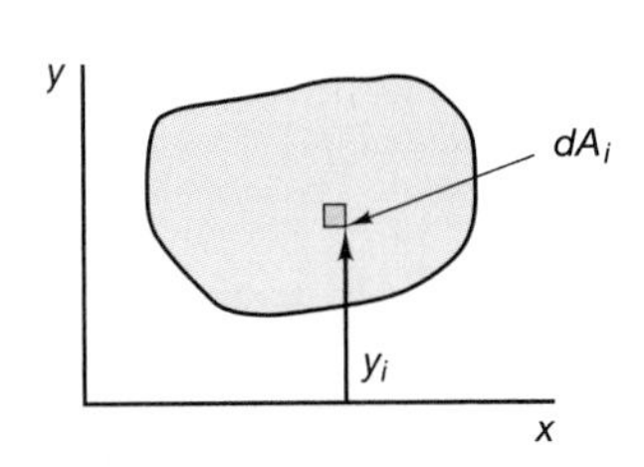

Figure 10.7

The increased use of computers in the solution of engineering problems has introduced major improvements in numerical integration. Second moments of regular areas were established by the use of multiple integrals in Section 10.4. However, an area can be divided into a large number of small areas and the second moment of the area obtained by summation (numerical integration) of the second moments of area of the divided areas, as shown in Figure 10.7. The second moment of area about the x-axis is

$$I_{xx} = \sum_i y_i^2 A_i \tag{10.20}$$

The accuracy of the numerical integration will increase as the area is divided into a greater number of subareas. This means that the size of the subarea decreases, and, in the limit, this process will approach integration.

An area can be considered as a composite of a finite number of regular areas of known properties (area, second moment of area about the centroidal axis, or radius of gyration). The parallel-axis transfer theorem can then be used to determine the second moment of the entire composite area. The centroids and second moments of area for common geometries are shown in Table 10.1 on page 380.

Sample Problem 10.3 Find the second moment of the area shown at the left about its centroidal axis x_c.

Solution The location of the x'-centroidal axis first must be established by using the methods for composites presented in Chapter 5. We divide the cross section into two parts: the 2 cm × 6 cm rectangle at the top and the 6 cm × 2 cm rectangle beneath it. We then have the following table:

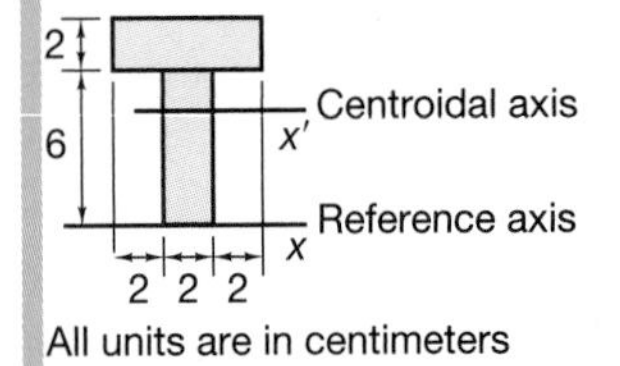

Section	Area, cm^2	y Distance to Centroid, cm	$A \cdot y$ cm^3
1	12	7	84
2	12	3	36
Σ	24		120

The distance from the reference axis to the x'-centroidal axis is

$$y_c = 120/24 = 5 \text{ cm}$$

The second moment of the area of a rectangle about its own centroidal x-axis is

$$I_{xx_c} = \frac{1}{12}bh^3$$

where b is the base and h is the height. Therefore, the second moment of the area about its centroidal axis may be found by the use of the composite values shown in the following table:

Section	Ixx_c, cm^4	Area, cm^2	dy, cm	Ady^2, cm^4
1	4	12	2	48
2	36	12	−2	48
Σ	40			96

The second moment of the area about the x-centroidal axis is

$$Ixx_c = 40 + 96 = 136 \text{ cm}^4$$

Sample Problem 10.4

Determine the second moment of area about the x-centroidal axis of the area shown to the left.

Solution

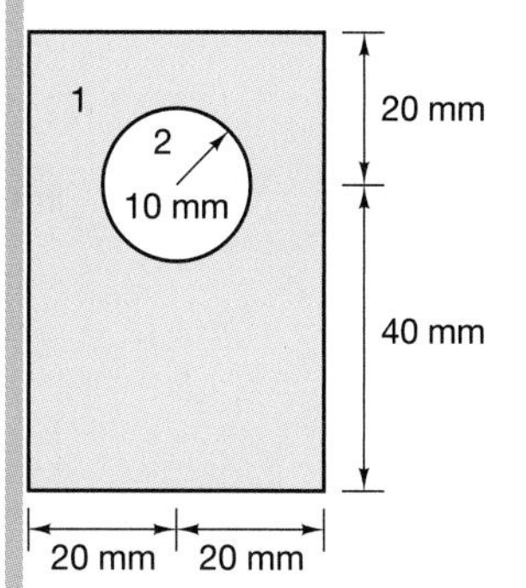

We divide the area into a rectangular solid area A_1 and a circular hole A_2 and establish a reference axis at the base of the rectangle. The hole will be treated as a negative area. First, we determine the centroid of the entire composite area, making use of the following table:

Segment	A_i, mm^2	y_{ci}, mm	$A_i y_{ci}$, mm^3
1	2400	30	72,000
2	−314	40	−12,560
Σ	2086		59,440

The centroid is

$$y_c = \frac{\Sigma A_i y_i}{\Sigma A_i} = \frac{59440}{2086} = 28.5 \text{ mm}$$

Noting the second moment of a circular and a rectangular area about their centroidal axes, we determine the second moment of the composite area in a similar manner:

Segment	A_i, mm^2	$D_{yi} = (y_{ci} - y_c)$	I_{xxci}, mm^4	$A_i d_{yi}^2$, mm^4
1	2400	−1.5	720,000	5400
2	−314	11.5	−7854	−41,526
Σ			712,146	−36,126

$$I_{xxc} = \sum_i (I_{xxc_i} + A_i d_{yi}^2) = 676,020 \text{ mm}^4$$

Table 10.1 *Geometric Properties of Line and Area Elements*

Centroid location

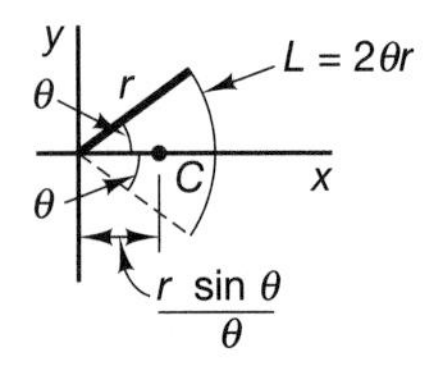

Circular arc segment

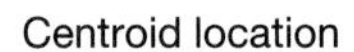

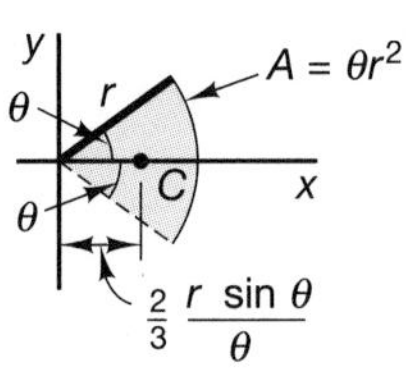

Circular sector area

Area moment of Inertia

$I_x = \frac{1}{4} r^4 (\theta - \frac{1}{2} \sin 2\theta)$

$I_Y = \frac{1}{4} r^4 (\theta - \frac{1}{2} \sin 2\theta)$

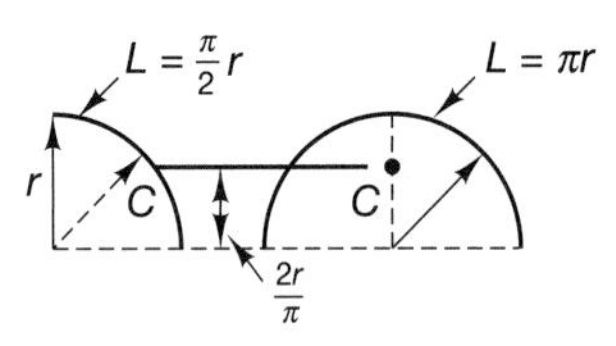

Quarter and semicircular arcs

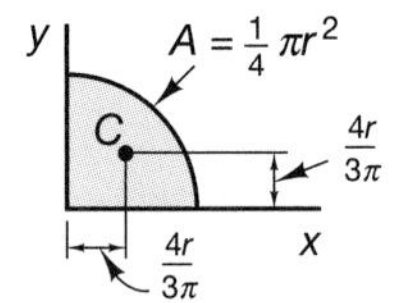

Quarter circular area

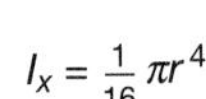

$I_y = \frac{1}{16} \pi r^4$

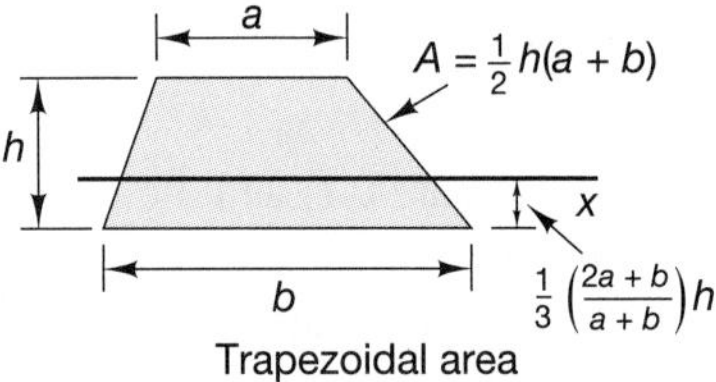

Trapezoidal area

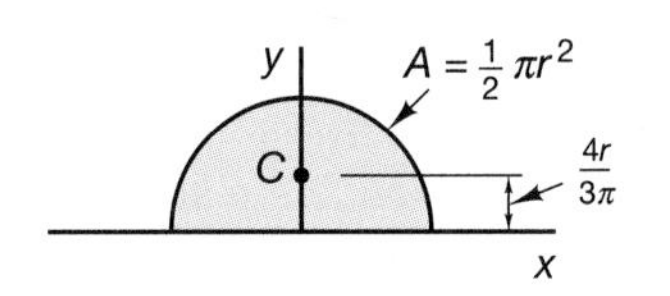

Semicircular area

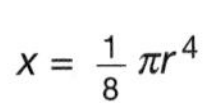

$I_y = \frac{1}{8} \pi r^4$

Semiparabolic area

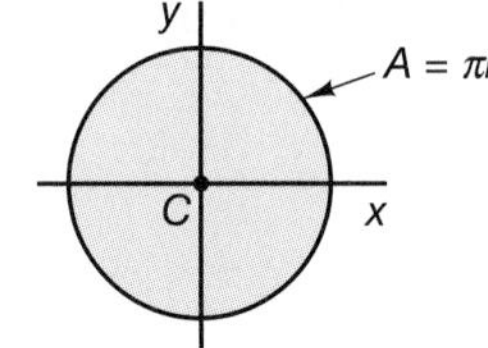

Circular area

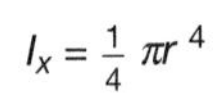

$I_y = \frac{1}{4} \pi r^4$

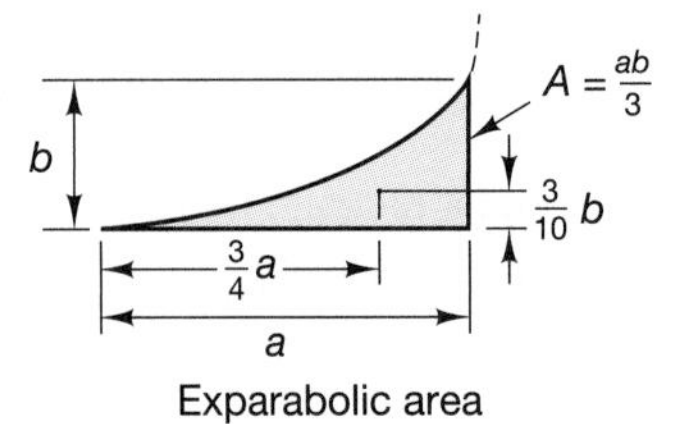

Exparabolic area

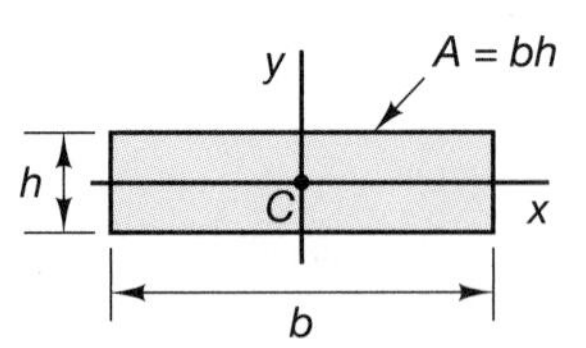

Rectangular area

$I_x = \frac{1}{12} bh^3$

$I_y = \frac{1}{12} hb^3$

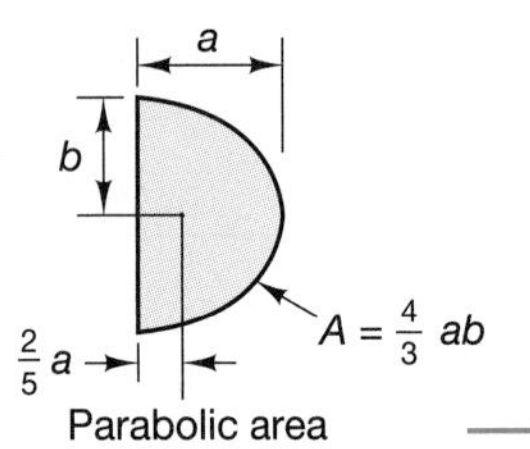

Parabolic area

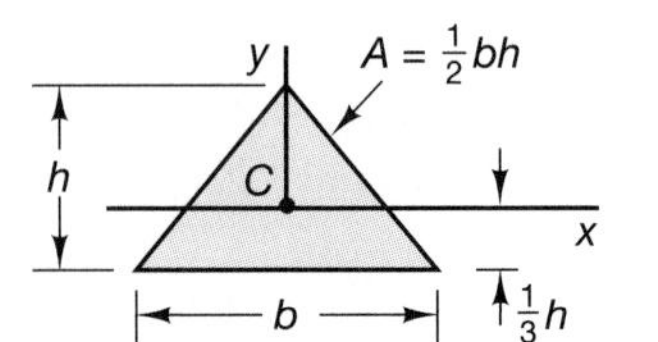

Triangular area

$I_x = \frac{1}{36} bh^3$

Problems

10.33 Determine the second moment of area about the centroidal axes of the cross-sectional area shown in Figure P10.33.

10.34 Find the centroid of the area shown in Figure P10.34, and determine the second moment of area about the centroidal axes. $b = \frac{\ell}{2}$.

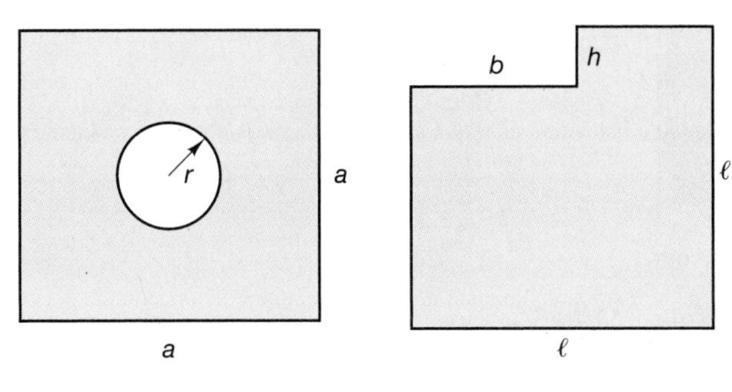

Figure P10.33

Figure P10.34

10.35 Determine the second moment of area about horizontal and vertical centroidal axes for the z-section shown in Figure P10.35.

10.36 Determine the radius of gyration for the centroidal polar moment of inertia for the z-section in Figure P10.35.

10.37 Determine the second moment of area about the centroidal axes of the channel cross section with base b, height h, and thickness t in Figure P10.37.

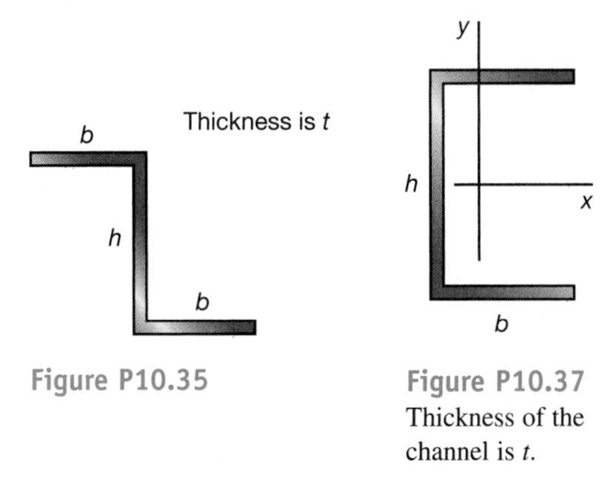

Figure P10.35

Figure P10.37 Thickness of the channel is t.

10.38 Determine the second moment of area about the centroidal axes of the I-beam cross section with base b, height h, and thickness t in Figure P10.38.

10.39 Determine the second moment of area and the polar moment of inertia for a circular cross section with a square hole in the middle, as shown in Figure P10.39.

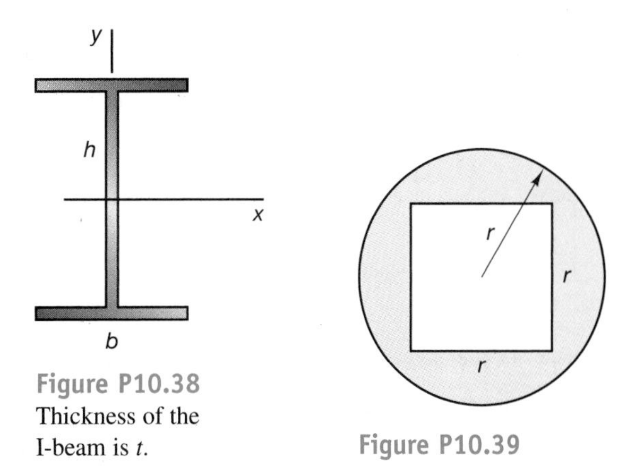

Figure P10.38 Thickness of the I-beam is t.

Figure P10.39

10.40 Determine the increase in torsional stiffness (the polar moment of inertia) of the cross section shown in Figure P10.39 if the square hole is reduced in size from $(r \times r)$ to $(r/2 \times r/2)$.

10.41 Determine the increase in bending stiffness (second moment of the area) when the square hole in the cross section shown in Figure P10.39 is reduced from $(r \times r)$ to $(r/3 \times r/3)$. What would be the increase in the polar moment of inertia?

10.42 Determine the centroidal second moments of area for the cross section shown in Figure P10.42.

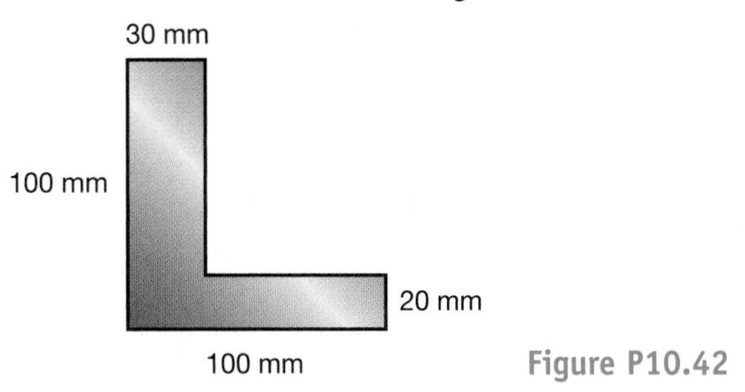

Figure P10.42

10.43 Determine the centroidal second moments for the area shown in Figure P10.43.

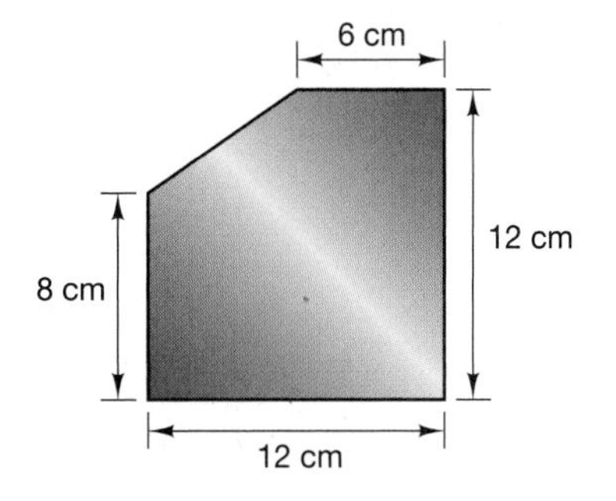

Figure P10.43

10.44 Find the centroidal second moments of area for the T-section shown in Figure P10.44.

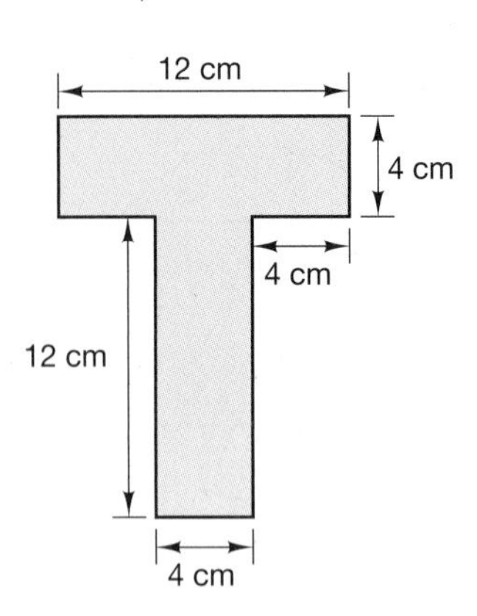

Figure P10.44

10.45 Determine the centroidal second moments of area for the cross-sectional area shown in Figure P10.45.

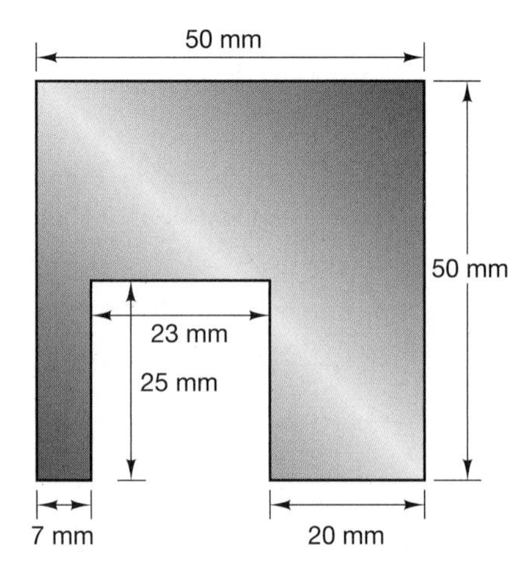

Figure P10.45

10.8 PRINCIPAL SECOND MOMENTS OF AREA

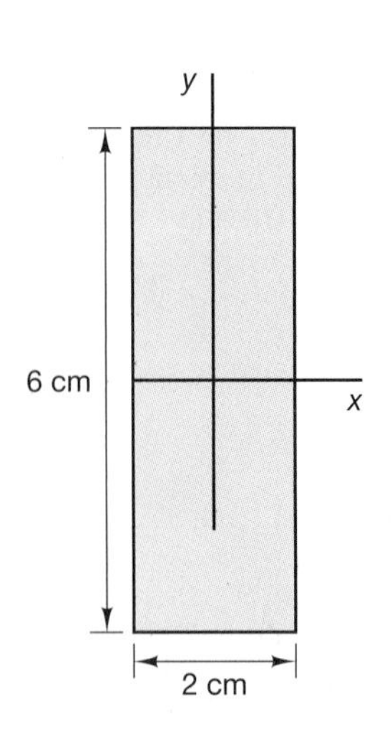

Figure 10.8

An infinite number of axes pass through the centroid of an area. Therefore, an infinite number of second moments of area that can be computed. Of this number, two orthogonal axes, called the *principal axes,* are of immense physical importance. When a beam is bent, it will always respond as if it were bent by loads perpendicular to the principal centroidal axes. The analysis of stresses and deflection of beams requires that the loads perpendicular to the long axis of the beam be resolved into components perpendicular to the principal axes, which have the following characteristics:

1. They are orthogonal axes.
2. The second moment of area about one of these axes will be the maximum value of the second moment of area for all centroidal axes, and the second moment of area about the other principal axes will be a minimum value.
3. A property called the *product moment of area* will be zero about these axes.

It is not surprising that, when a beam is bent, it responds as if it were bent about the axis of maximum resistance to bending and about the axis of minimum resistance to bending, independently. If the 2×6 rectangular area shown in Figure 10.8 is examined using the formulas presented in Section 10.4, it can be seen that the second moments of area about the x- and y-centroidal axes are

$$\begin{aligned} I_{xx} &= \frac{1}{12}bh^3 = \frac{2(6)^3}{12} = 36 \text{ cm}^4 \\ I_{yy} &= \frac{1}{12}hb^3 = \frac{6(2)^3}{12} = 4 \text{ cm}^4 \end{aligned} \tag{10.21}$$

It will be shown that the x- and y-axes are the principal axes of the area, that 36 is the maximum value of all possible second moments of area, and that 4 is the minimum value.

10.8.1 PRODUCT MOMENT OF AN AREA, OR PRODUCT MOMENT OF INERTIA

A mathematical property called the *product moment of an area* is related to any set of orthogonal axes and is defined by

$$I_{xy} = -\int_A xy\,dA \tag{10.22}$$

The minus sign in Eq. (10.22) is introduced for continuity with the formal tensor definitions of second moments of areas and mass moments of inertia. Different authors may or may not use the minus sign. The product moment of an area is a measure of antisymmetry and is necessary to determine the principal axes of the area in question. As has been pointed out earlier, one characteristic of the principal axes is that the product moment of the area about these axes is zero. By contrast, the second moment of area about an axis must always be positive, as its integrand is either x^2 or y^2, which is nowhere negative within the area. The product moment of the area can be either positive, negative, or zero, as its integrand can be positive or negative at different points within the area.

Even and odd functions were defined when centroids were discussed in Chapter 5, and an axis of symmetry was shown to be a centroidal axis owing to the fact that the integral of an odd function between symmetric limits is always zero. The integrand of the product moment of area is an odd function in both x and y, and therefore, *the product moment of area will be zero about any set of orthogonal axes such that one of these axes is an axis of symmetry.* Therefore, any axis of symmetry of an area is a principal axis of that area.

A parallel-axis theorem can also be developed for the product moment of area, as was done for the second moment of area. Referring to Figure 10.6, we can relate the product moment of area about the x–y axes to the product moment of area about the x'–y' axes by the equations

$$\begin{aligned} x &= x' + x_0 \qquad y = y' + y_0 \\ I_{xy} &= -\int_A xy\,dA = -\int_A (x' + x_0)(y' + y_0)\,dA \\ I_{xy} &= -\int_A x'y'\,dA - x_0\int_A y'\,dA - y_0\int_A x'\,dA - x_0y_0\int_A dA \\ I_{xy} &= I_{x'y'} - x_0y_cA - y_0x_cA - x_0y_0A \end{aligned} \tag{10.23}$$

where x_c and y_c are the distances from the x'–y' axes to the centroid of the area. If the x'–y' axes are centroidal axes, these distances are zero, and the parallel-axis theorem for the product moment of area is

$$I_{xy} = I_{x'y'} - x_0y_0A \tag{10.24}$$

10.8.2 ROTATION OF AXES

Consider the two sets of axes x–y and x'–y' with common origin, as shown in Figure 10.9. The x'–y' coordinate axes are rotated through an angle β counterclockwise from the x–y axes. Introducing a notation similar to that used for direction cosines in Chapter 2, let

$$\begin{aligned} \theta_{x'x} &= \text{the angle between the } x'\text{-axis and the } x\text{-axis} \\ \theta_{x'y} &= \text{the angle between the } x'\text{-axis and the } y\text{-axis} \\ \theta_{y'x} &= \text{the angle between the } y'\text{-axis and the } x\text{-axis} \\ \theta_{y'y} &= \text{the angle between the } y'\text{-axis and the } y\text{-axis} \end{aligned} \tag{10.25}$$

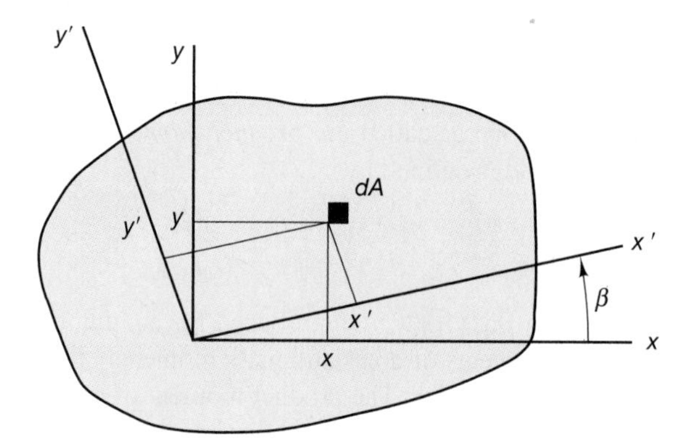

Figure 10.9

The x'–y' coordinates of any point may be related to the x–y coordinates of that point by

$$\begin{aligned} x' &= x\cos\theta_{x'x} + y\cos\theta_{x'y} \\ y' &= x\cos\theta_{y'x} + y\cos\theta_{y'y} \end{aligned} \tag{10.26}$$

These are called the transformation equations for rotation of a coordinate system about the z-axis. The matrix of the cosines, $[R]$, that define this rotation is called the *orthogonal rotation transformation matrix* and is written as

$$\begin{aligned} x' &= x\cos\theta_{x'x} + y\cos\theta_{x'y} \\ y' &= x\cos\theta_{y'x} + y\cos\theta_{y'y} \\ \begin{bmatrix} x' \\ y' \end{bmatrix} &= \begin{bmatrix} \cos\theta_{x'x} & \cos\theta_{x'y} \\ \cos\theta_{y'x} & \cos\theta_{y'y} \end{bmatrix} \begin{bmatrix} x \\ y \end{bmatrix} \\ [x'] &= [R][x] \end{aligned} \tag{10.27}$$

where

$$[R] = \begin{bmatrix} \cos\theta_{x'x} & \cos\theta_{x'y} \\ \cos\theta_{y'x} & \cos\theta_{y'y} \end{bmatrix}$$

For a counterclockwise rotation β, the cosines become

$$\begin{aligned} &\cos\theta_{x'x} = \cos\beta \quad \cos\theta_{x'y} = \cos(90° - \beta) = \sin\beta \\ &\cos\theta_{y'x} = \cos(90° + \beta) = -\sin\beta \quad \cos\theta_{y'y} = \cos\beta \end{aligned} \tag{10.28}$$

Therefore, the transformation between x'–y' coordinates and x–y coordinates becomes

$$\begin{aligned} x' &= x\cos\beta + y\sin\beta \\ y' &= -x\sin\beta + y\cos\beta \end{aligned} \tag{10.29}$$

The second moment of area about the x'-axis is

$$I_{x'x'} = \int_A (y')^2\, dA \tag{10.30}$$

Using Eq. (10.29), we can write Eq. (10.30) as

$$\begin{aligned} I_{x'x'} &= \int_A (-x\sin\beta + y\cos\beta)^2\, dA \\ &= \int_A (x^2\sin^2\beta - 2xy\sin\beta\cos\beta + y^2\cos^2\beta)\, dA \end{aligned} \tag{10.31}$$

or

$$I_{x'x'} = I_{xx}\cos^2\beta + I_{xy}(2\sin\beta\cos\beta) + I_{yy}\sin^2\beta$$

If the double-angle relationships

$$\begin{aligned} 2\sin\beta\cos\beta &= \sin 2\beta \\ \cos^2\beta &= \frac{1 + \cos 2\beta}{2} \\ \sin^2\beta &= \frac{1 - \cos 2\beta}{2} \end{aligned} \tag{10.32}$$

are introduced, the relationship between the second moments of area of the two coordinate systems becomes

$$I_{x'x'} = \frac{I_{xx} + I_{yy}}{2} + \frac{I_{xx} - I_{yy}}{2}\cos 2\beta + I_{xy}\sin 2\beta \tag{10.33}$$

In a similar manner, the product moment of the area in the primed coordinate system may be related to the unprimed coordinate system by

$$I_{x'y'} = \frac{I_{xx} - I_{yy}}{2}\sin 2\beta + I_{xy}\cos 2\beta \tag{10.34}$$

The changes in $I_{x'x'}$ and $I_{x'y'}$ with β can be seen in Figure 10.10, in which $I_{xx} = 10\text{ mm}^4$, $I_{yy} = 5\text{ mm}^4$, and $I_{xy} = 0$. Graphs of this type are easily obtained on a graphical computer or in most computational software. (See the Computational Supplement.)

Examining Figure 10.10, we can see that the second moment of the area, $I_{x'x'}$, reaches a maximum or minimum when $I_{x'y'}$ is zero. Differentiation of Eq. (10.33) with respect to 2β yields

$$\frac{d}{d(2\beta)} I_{x'x'} = -\frac{I_{xx} - I_{yy}}{2}\sin 2\beta + I_{xy}\cos 2\beta = I_{x'y'} \tag{10.35}$$

Therefore, setting the derivative of the second moment of the area equal to zero is equivalent to setting $I_{x'y'}$ equal to zero, and *the product moment of the area is zero about the principal axes where the second moment of area is either a maximum or a minimum.*

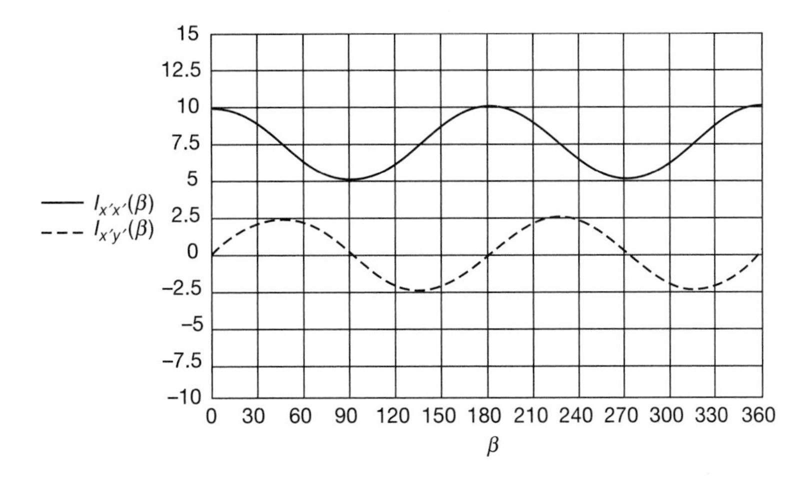

Figure 10.10 $I_{x'x'}$ and $I_{x'y'}$ versus the angle β

The angle to the principal axes from any reference set of axes x–y can be found by setting Eq. (10.35) equal to zero. Therefore, the angle β is

$$\tan 2\beta = \frac{2I_{xy}}{I_{xx} - I_{yy}}$$
$$\beta = \frac{1}{2}\tan^{-1}\left(\frac{2I_{xy}}{I_{xx} - I_{yy}}\right) \tag{10.36}$$

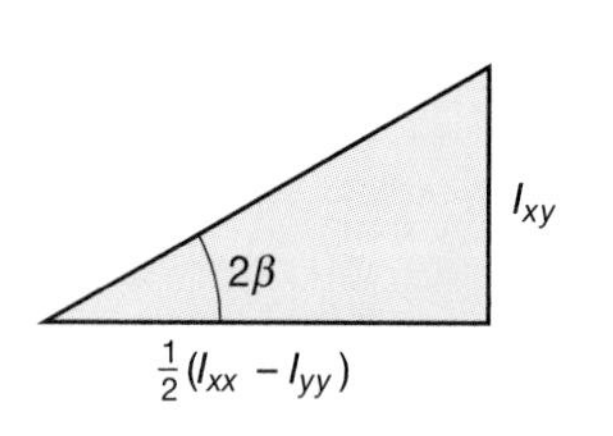

There will be two solutions to Eq. (10.36), separated by 90°. These solutions represent the x' and y' axes, which are principal axes. The tangent of 2β can be examined by considering the right triangle shown at the left. We have

$$\sin 2\beta = \frac{I_{xy}}{\sqrt{\left(\frac{I_{xx} - I_{yy}}{2}\right)^2 + I_{xy}^2}}$$
$$\cos 2\beta = \frac{\frac{I_{xx} - I_{yy}}{2}}{\sqrt{\left(\frac{I_{xx} - I_{yy}}{2}\right)^2 + I_{xy}^2}}$$

Substitution into Eq. (10.33) yields

$$I_{max/min} = \frac{I_{xx} + I_{yy}}{2} \pm \sqrt{\left(\frac{I_{xx} - I_{yy}}{2}\right)^2 + I_{xy}^2} \tag{10.37}$$

Equations (10.36) and (10.37) yield the values of the principal second moments of area and the angles from the reference set of axes.

Sample Problem 10.5

For the area shown to the left, find the centroid, and the principal centroidal axes, and the value of the second moment of area about these axes.

All dimensions are in centimeters

Solution

The X–Y axes will be taken as reference axes, and the x–y axes will be centroidal axes, but not necessarily principal. First, we determine the centroid of the area, considering it as a composite area. We have the following table:

	A_i, cm²	x_{ic}, cm	y_{ic}, cm	$A x_{ic}$, cm³	$A y_{ic}$, cm³
1	12	3	7	36	84
2	12	1	3	12	36
Σ	24			48	120

Therefore, the coordinates of the centroid are $x = 48/24 = 2$ cm, $y = 120/24 = 5$ cm.

The second moment of the area about the centroidal axes will be found by using the parallel-axis theorem. We obtain the following table:

	A, cm²	d_x, cm	d_y, cm	I_{xx}, cm⁴	I_{yy}, cm⁴	I_{xy}, cm⁴	Ad_x^2, cm⁴	Ad_y^2, cm⁴	-1^* Ad_xd_y, cm⁴
1	12	1	2	4	36	0	12	48	−24
2	12	−1	−2	36	4	0	12	48	−24
Σ	24			40	40	0	24	96	−48

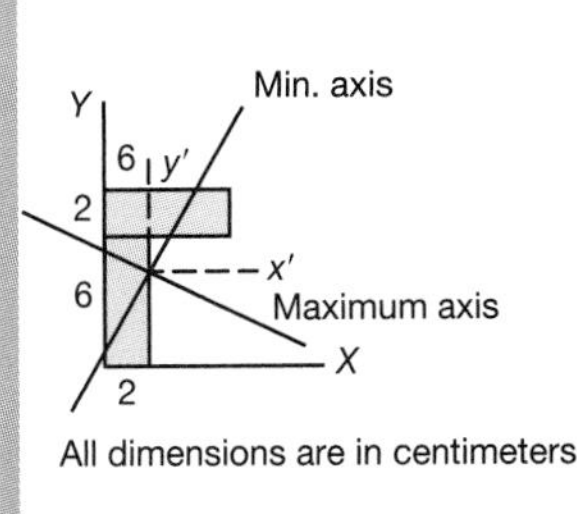

The second moments of the area and the product moment of the area about the $x-y$ centroidal axes are

$$I_{xx} = \sum(I_{xx} + Ad_y^2) = 40 + 96 = 136 \text{ cm}^4$$
$$I_{yy} = \sum(I_{yy} + Ad_x^2) = 40 + 24 = 64 \text{ cm}^4$$
$$I_{xy} = \sum(I_{xy} - Ad_xd_y) = 0 - 48 = -48 \text{ cm}^4$$

The angle to the principal axes is

$$\beta = \frac{1}{2}\tan^{-1}\left(\frac{2I_{xy}}{I_{xx} - I_{yy}}\right) = \frac{1}{2}\tan^{-1}\left(\frac{-96}{72}\right) = -26.6° \text{ or } -116.6°$$

The two angles represent the angles to the x' and y' principal axes, respectively. The values of the principal second moments of area are $I_{max} = 160 \text{ cm}^4$ and $I_{min} = 40 \text{ cm}^4$. The principal axes are shown in figure to the left.

Problems

For each of the following cross-sectional areas, determine the principal axes and the principal second moments of area using Eqs. (10.36) and (10.37).

10.46 The z-section shown in Problem 10.35. (See Figure P10.46.) $b = \frac{h}{2}$

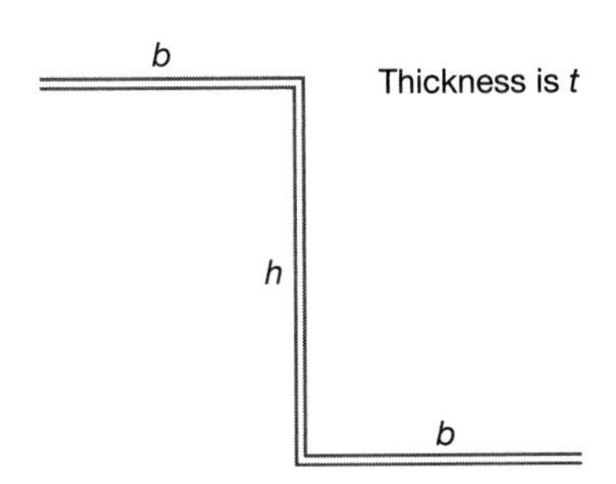

Figure P10.46

10.47 The cross section shown in Problem 10.22. (See Figure P10.47.)

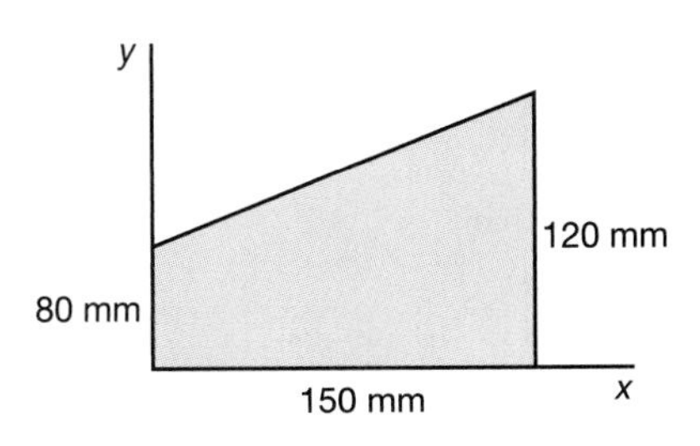

Figure P10.47

10.48 The cross section shown in Problem 10.42. (See Figure P10.48.)

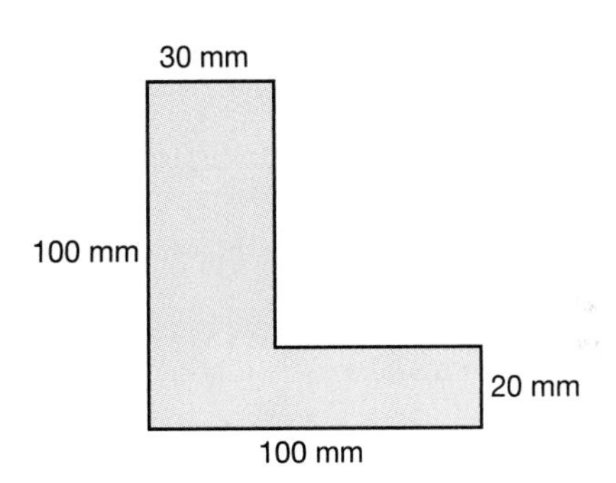

Figure P10.48

10.49 The cross-sectional area shown in Problem 10.43. (See Figure P10.49.)

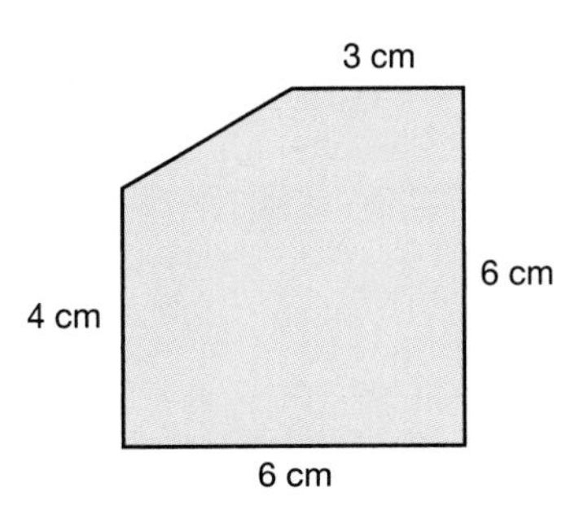

Figure P10.49

10.50 The cross-sectional area shown in Problem 10.45. (See Figure P10.50.)

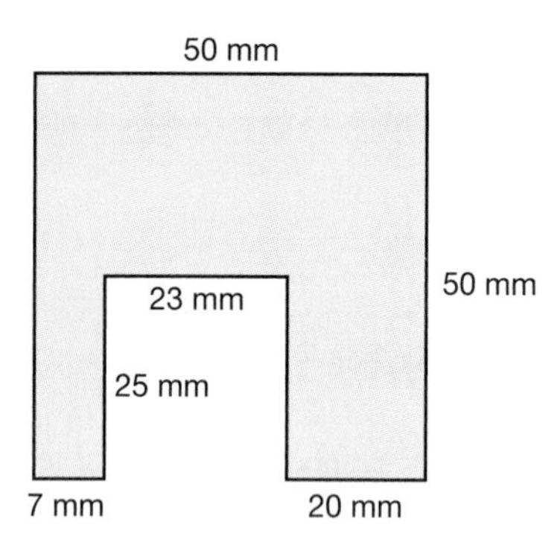

Figure P10.50

10.51 Determine the centroidal principal axes and principal second moments of area for the triangle shown in Figure P10.51.

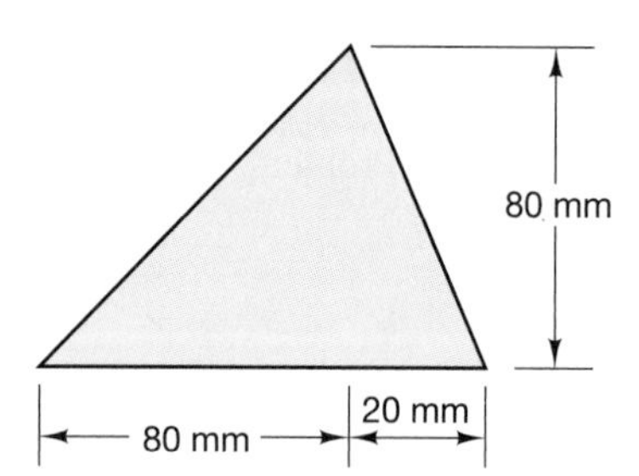

Figure P10.51

10.52 Prove that every centroidal axis is a principal axis for a square cross-sectional area.

10.53 Prove that every centroidal axis is a principal axis for an equilateral triangle.

10.54 Prove that every centroidal axis is a principal axis for a hexagon.

10.55 Determine the centroidal principal axes and the principal second moments of area for the cross-sectional area in Figure P10.55.

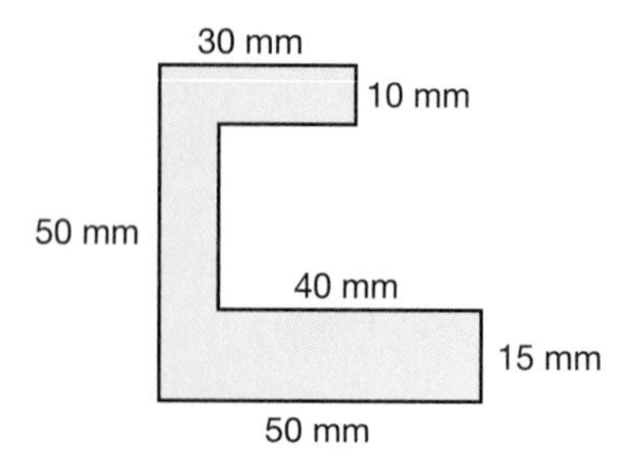

Figure P10.55

10.56 Determine the centroidal principal axes and principal second moments of area for the square cross-sectional area with the 10-mm-radius hole in the upper left quadrant in Figure P10.56.

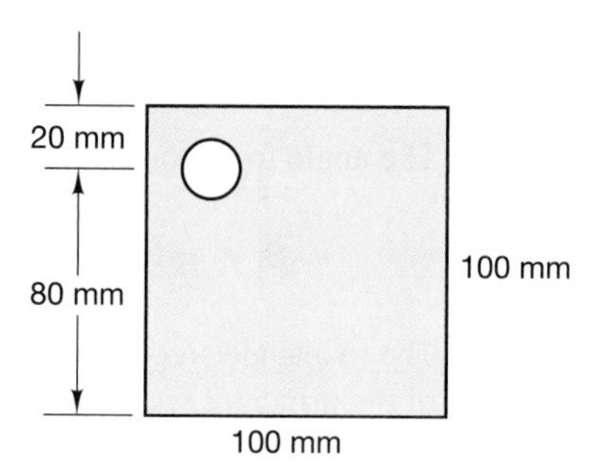

Figure P10.56

10.57 Determine the centroidal principal axes and the principal second moments of area for an angular cross-sectional beam made of stock of thickness t in Figure P10.57.

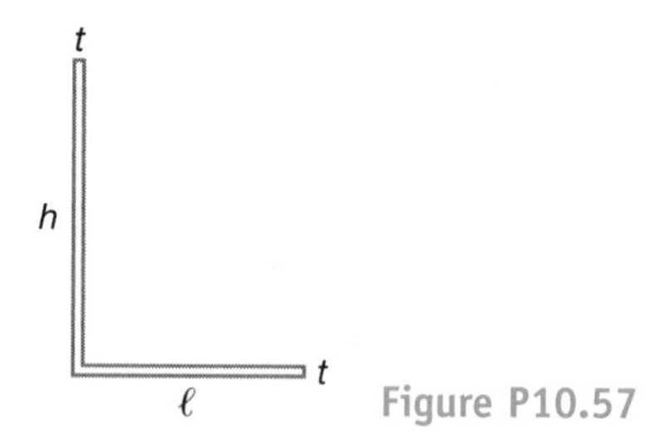

Figure P10.57

10.58 In Figure P10.57, show that if the legs of the angle are equal ($h = \ell$), the axis of symmetry is a principal axis.

10.59 Determine the centroidal principal axes and the principal second moments of area for the cross-sectional area shown in Figure P10.59.

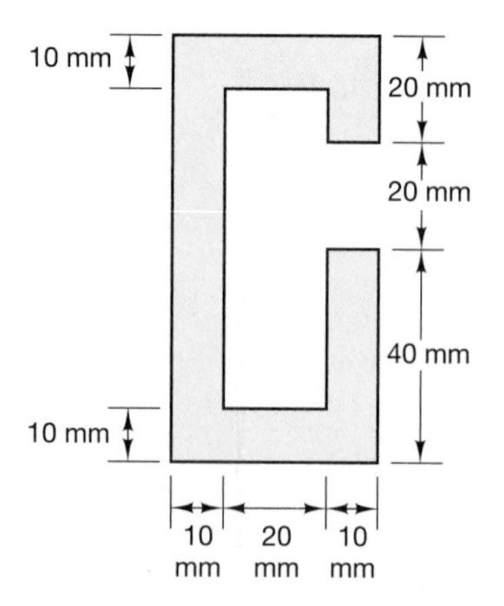

Figure P10.59

10.9 MOHR'S CIRCLE TO DETERMINE PRINCIPAL SECOND MOMENTS OF AREA

Given the second moments and the product moment of area about any pair of orthogonal axes, Eqs. (10.33) and (10.34) allow the determination of these properties about any other set of axes x'–y' having the same origin. A graphical representation of the transformation equations was developed by the German engineer Otto Mohr in 1882. Mohr showed that the loci of all possible values of the second moment of area and the product moment of area lie on a circle. If the ***Mohr's circle*** is accurately drawn to an appropriate scale, it can be used as a graphical tool to determine the second moment of an area about any axis. As with other graphical techniques, this one is combined with trigonometry and used as a semigraphical tool or as a conceptual tool for the transformation equations.

The transformation Eqs. (10.33) and (10.34) are dependent upon trigonometric functions of β, or *circular functions.* It is not surprising that Mohr looked for some relationship that involved a circle. The transformation equations will yield the equation of a circle if they are squared and added together. This result may be easily obtained by writing the transformation equations as

$$I_{x'x'} - \frac{I_{xx} - I_{yy}}{2} = \frac{I_{xx} - I_{yy}}{2} \cos 2\beta + I_{xy} \sin 2\beta$$

$$I_{x'y'} = \frac{I_{xx} - I_{yy}}{2} \sin 2\beta + I_{xy} \cos 2\beta$$

Squaring and adding the left and right sides yields

$$\left(I_{x'x'} - \frac{I_{xx} + I_{yy}}{2}\right)^2 + (I_{x'y'})^2 = \left(\frac{I_{xx} - I_{yy}}{2}\right)^2 + (I_{xy})^2 \tag{10.38}$$

If this equation is considered to be an equation in $I_{x'x'}$, $I_{x'y'}$ space instead of x, y space, it can be seen to be the equation of a circle:

$$(x - x_0)^2 + y^2 = R^2 \tag{10.39}$$

The circle in second-moment space has its origin at $\left(\frac{I_{xx} + I_{yy}}{2}, 0\right)$ and radius

$$R = \sqrt{\left(\frac{I_{xx} - I_{yy}}{2}\right)^2 + I_{xy}^2} \tag{10.40}$$

Plotting Mohr's circle requires some careful consideration of the transformation equations. Note first that the double angle 2β appears in the equation, so that all angles on the circle are double the angles between the coordinate systems on the area. A positive sign convention must be selected for I_{xy}, as the product moment of the area can have either a positive or a negative sign. The second moment of the area is always positive, so Mohr's circle will appear as shown in Figure 10.11.

Note that the x- and y-axes become points on the circle with coordinates $(I_{xx}, -I_{xy})$ and $(I_{yy}, +I_{xx})$, respectively. The choice to plot I_{xy} positive downward is made so that rotation of the axes on the area coincides with the direction of rotations on Mohr's circle. Note that angles on the circle are double those on the area.

The principal second moments of the area can be found by noting that

$$I_{max} = (I_{xx} + I_{yy})/2 + R$$

$$I_{min} = (I_{xx} + I_{yy})/2 - R$$

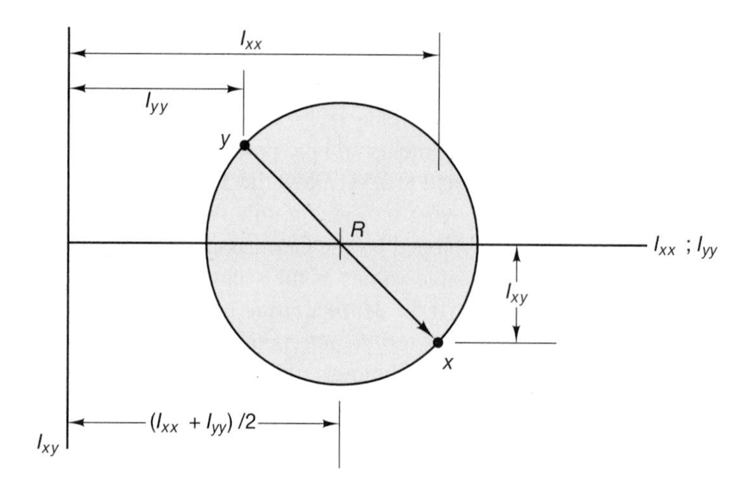

Figure 10.11

where $(I_{xx} + I_{yy})/2$ is the distance to the center of the circle. Therefore,

$$I_{\text{max/min}} = \frac{I_{xx} + I_{yy}}{2} \pm \sqrt{\left(\frac{I_{xx} - I_{yy}}{2}\right)^2 + I_{xy}^2} \tag{10.41}$$

Sample Problem 10.6 Sketch the Mohr's circle for Sample Problem 10.5.

Solution We set up a second moment–product moment of the area coordinate system. (See the figure at left below.) Then we plot the coordinates of the x-moments and the y-moments, remembering that I_{xy} is plotted with opposite sign when plotting the y-moments. Connecting the x- and y-points with a line forms the diameter of the circle and locates its center. (See the two figures on the right below.)

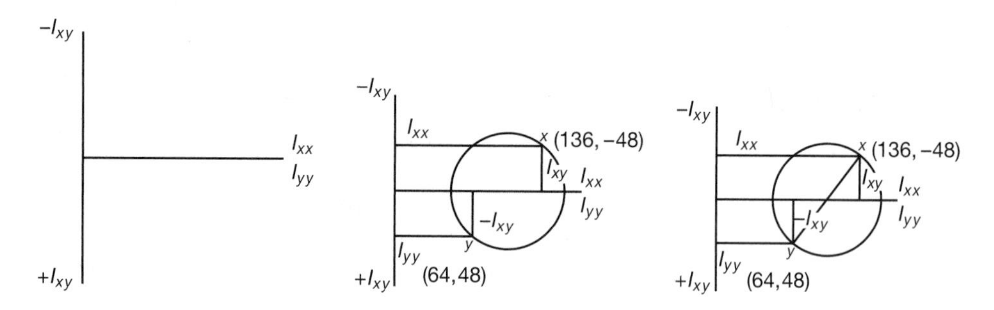

The double angle between the x-axis and the maximum value can easily be determined from the circle and in this case is found to be

$$\tan^{-1}(48/\{[136 - 64]/2\}) = 53.13°$$

Therefore, the principal axis about which the second moment of area is a maximum is 26.6° clockwise, rotated from the x-axis. The center of the circle is at $(136 + 64)/2$, or 100 cm^4 on the $I_{x'x'}$ axis, and the radius of the circle is

$$R = \sqrt{\left(\frac{I_{xx} - I_{yy}}{2}\right)^2 + I_{xy}^2}\ \text{cm}^4$$
$$= \sqrt{36^2 + 48^2} = 60$$

The maximum second moment of the area is 160 cm^4, and the minimum is 40 cm^4. Note that there is no savings in computational effort using Mohr's circle, but it does provide a conceptual tool to see the effects of changing centroidal axes.

Problems

For each of the following cross-sectional areas, determine the principal axes and the principal second moments of area, using Mohr's circle.

10.60 The z-section shown in Problem 10.35. (See Figure P10.60.) $b = \frac{h}{2}$

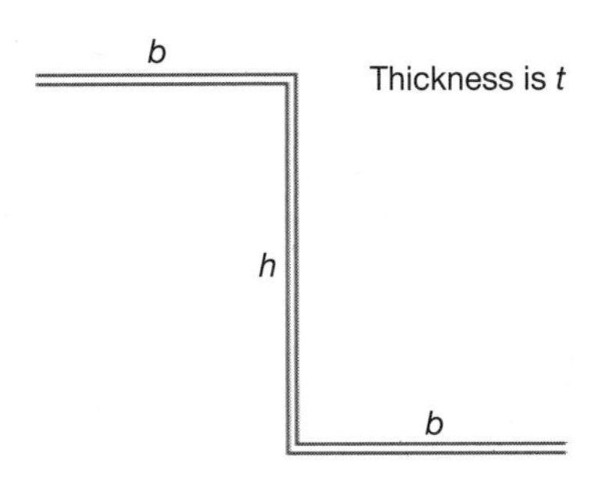

Figure P10.60

10.61 The cross section shown in Problem 10.22. (See Figure P10.61.)

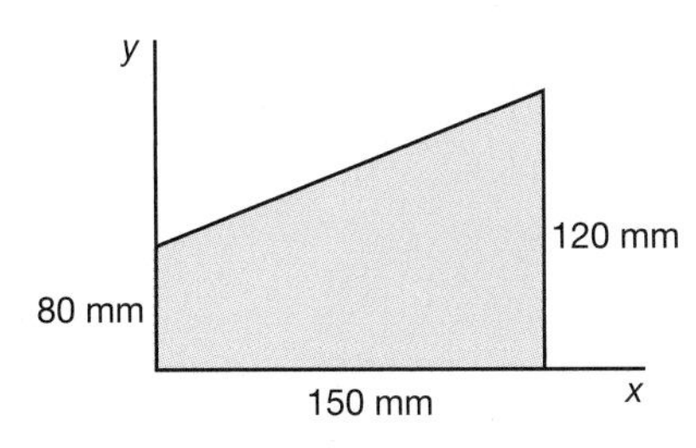

Figure P10.61

10.62 The cross section shown in Problem 10.42. (See Figure P10.62.)

10.63 The cross-sectional area shown in Problem 10.43. (See Figure P10.63.)

10.64 The cross-sectional area shown in Problem 10.45. (See Figure P10.64.)

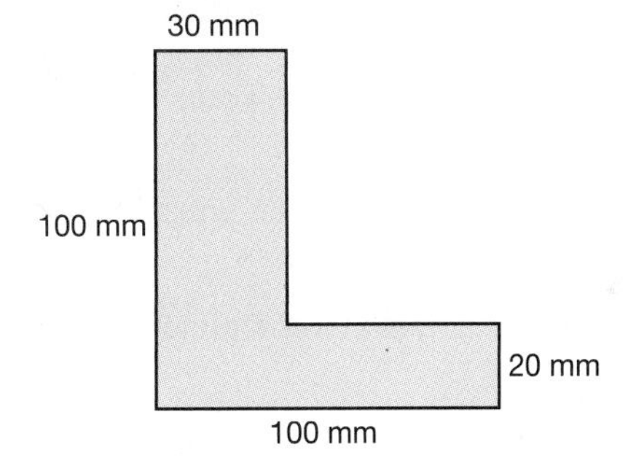

Figure P10.62

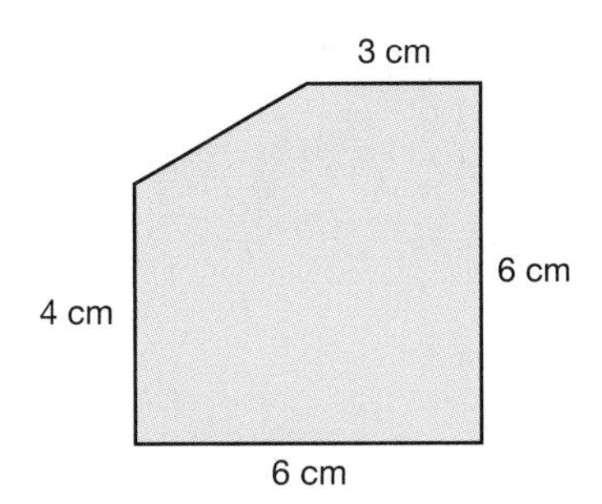

Figure P10.63

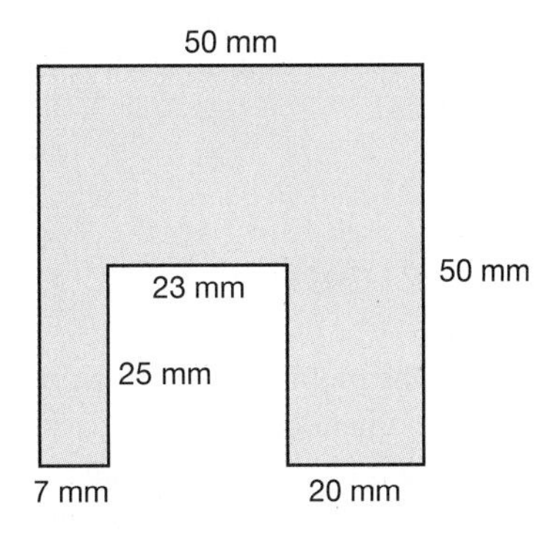

Figure P10.64

10.65 Determine the centroidal principal axes and principal second moments of area for the triangle shown in Figure P10.65.

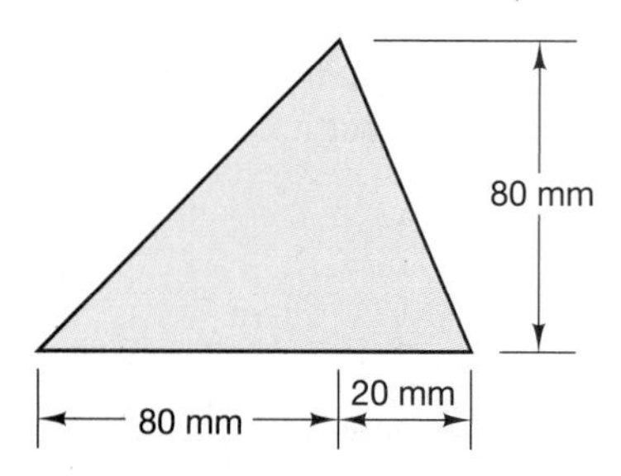

Figure P10.65

10.66 Prove that every centroidal axis is a principal axis for a square cross-sectional area.

10.67 Prove that every centroidal axis is a principal axis for an equilateral triangle.

10.68 Prove that every centroidal axis is a principal axis for a hexagon.

10.69 Determine the centroidal principal axes and the principal second moments of area for the cross-sectional area in Figure P10.69.

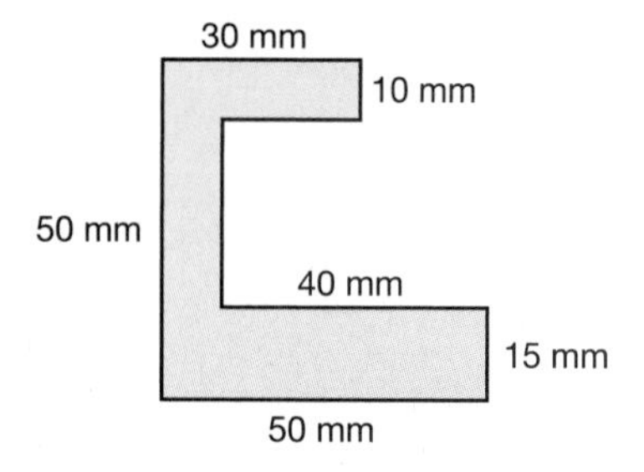

Figure P10.69

10.70 Determine the centroidal principal axes and principal second moments of area for the square cross-sectional area with the 10-mm-radius hole in the upper left quadrant in Figure P10.70.

10.71 Determine the centroidal principal axes and the principal second moments of area for an angular cross-sectional beam made of stock of thickness t in Figure P10.71.

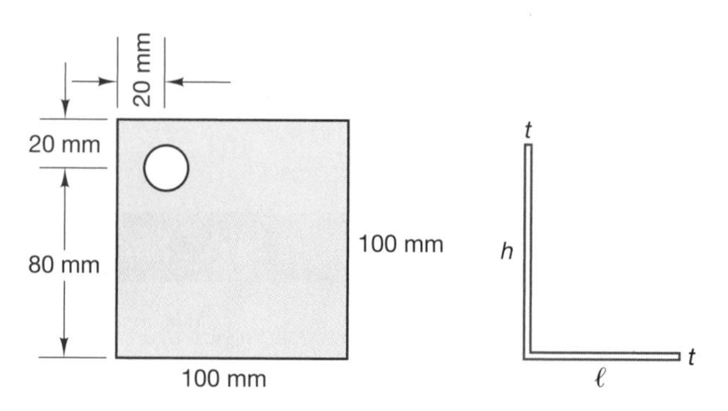

Figure P10.70

Figure P10.71

10.72 In Figure P10.71, show that if the legs of the angle are equal ($h = \ell$), the axis of symmetry is a principal axis.

10.73 Determine the centroidal principal axes and the principal second moments of area for the cross-sectional area shown in Figure P10.73.

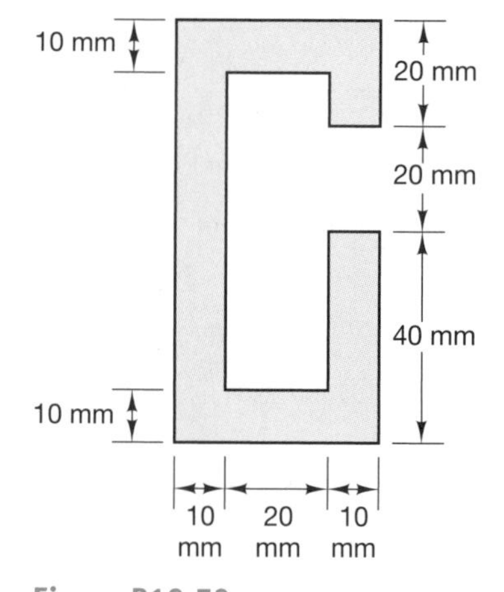

Figure P10.73

10.10 EIGENVALUE PROBLEM

The problem of finding principal axes and the corresponding principal values of the second moment of the area arises in many analogous situations in physics and mathematics. It is therefore not surprising that the problem has been examined in detail. The principal values have been given the name "characteristic values" or *eigenvalues,* and the unit vectors along the principal axes are called "characteristic vectors" or *eigenvectors.* The eigenvalue problem arises in many areas of applied mathematics, including linear algebra, tensor analysis, and differential equations. In Section 10.8, an orthogonal rotation transformation matrix was introduced. This matrix transformed the x,y coordinates into the x',y' coordinates and was used to determine the second moment of an area and the product moment of area about the x'-axis. The transformation can be generalized by the relationship

$$[I'] = [R][I][R]^T \tag{10.42}$$

where $[R]^T$ is the transpose of the matrix $[R]$. (Details can be found in linear algebra texts.)

Expanding Eq. (10.42) yields

$$\begin{bmatrix} I'_{xx} & I'_{xy} \\ I'_{yx} & I'_{yy} \end{bmatrix} = \begin{bmatrix} \cos\beta & \sin\beta \\ -\sin\beta & \cos\beta \end{bmatrix} \begin{bmatrix} I_{xx} & I_{xy} \\ I_{yx} & I_{yy} \end{bmatrix} \begin{bmatrix} \cos\beta & -\sin\beta \\ \sin\beta & \cos\beta \end{bmatrix} \tag{10.43}$$

Performing the matrix multiplication in Eq. (10.43) gives the values for the second moments of area and the product moment of area in the primed coordinate system:

$$\begin{bmatrix} I'_{xx} & I'_{yx} \\ I'_{xy} & I'_{yy} \end{bmatrix} = \begin{bmatrix} I_{xx}\cos^2\beta + 2I_{xy}\sin\beta\cos\beta + I_{yy}\sin^2\beta & -(I_{xx} - I_{yy})\sin\beta\cos\beta + I_{xy}(\cos^2\beta - \sin^2\beta) \\ -(I_{xx} - I_{yy})\sin\beta\cos\beta + I_{xy}(\cos^2\beta - \sin^2\beta) & I_{yy}\cos^2\beta - 2I_{xy}\sin\beta\cos\beta + I_{xx}\sin^2\beta \end{bmatrix} \tag{10.44}$$

Using the double-angle relationships in Eq. (10.32) gives Eqs. (10.43) and (10.44).

Now consider that the (2×2) matrix $[I]$ was multiplied by the (2×1) column matrix $[u]$ yielding the (2×1) column matrix $[v]$. That is,

$$[I][u] = [v] \tag{10.45}$$

The column matrices $[u]$ and $[v]$ can be considered as vectors in two dimensions. Now we ask whether there is a nonzero vector $[u]$ that, when multiplied by $[I]$, will yield a vector **parallel** to itself; that is, we ask whether

$$[I][u] = \lambda[u] \tag{10.46}$$

This is the eigenvalue problem. Eq. (10.46) can be written as

$$\begin{bmatrix} (I_{xx} - \lambda) & I_{xy} \\ I_{xy} & (I_{yy} - \lambda) \end{bmatrix} \begin{bmatrix} u_x \\ u_y \end{bmatrix} = 0 \tag{10.47}$$

If Eq. (10.47) is considered to be a system of linear equations to determine u_x and u_y, it will have a solution *if and only* if the determinant of the coefficient matrix is zero; that is, we must have

$$\lambda^2 - (I_{xx} + I_{yy})\lambda + (I_{xx}I_{yy} - I_{xy}^2) = 0 \tag{10.48}$$

The eigenvalues λ are the principal values of the second moment of the area, and the eigenvectors $[u]$ are along the principal axes. Using the quadratic formula, we see that Eq. (10.48) yields

$$\lambda_{1/2} = \frac{I_{xx} + I_{yy}}{2} \pm \sqrt{\frac{(I_{xx} + I_{yy})^2 - 4(I_{xx}I_{yy} - I_{xy}^2)}{4}}$$

$$\lambda_{1/2} = \frac{I_{xx} + I_{yy}}{2} \pm \sqrt{\frac{(I_{xx} - I_{yy})^2}{4} + I_{xy}^2} \tag{10.49}$$

Equation (10.49) agrees with Eq. (10.41), obtained by the use of Mohr's circle.

All computational software programs have routines for solving eigenvalue problems, as can be seen in the Computational Supplement. Eigenvalue problems will arise when principal axes of mass moments of inertia, principal stresses or principal strains, and natural frequencies are to be determined.

Sample Problem 10.7

Determine the principal values of the second moment of area and the principal axes for the cross-sectional area shown in Sample Problem 10.5, using the eigenvalue method.

Solution The second-moment-of-area tensor may be written as

$$[I] = \begin{bmatrix} 136 & -48 \\ -48 & 64 \end{bmatrix} \text{cm}^4$$

The eigenvalues are

$$\lambda = \begin{bmatrix} 160 \\ 40 \end{bmatrix} \text{cm}^4$$

The eigenvector for the eigenvalue $\lambda = 160\ \text{cm}^4$ is

$$n_1 = \begin{bmatrix} 0.894 \\ -0.447 \end{bmatrix}$$

$$\beta_1 = \tan^{-1}\left(\frac{n_{1y}}{n_{1x}}\right) = -26.565°$$

The eigenvector for the minimum eigenvalue $\lambda = 40\ \text{cm}^4$ is

$$n_2 = \begin{bmatrix} 0.447 \\ 0.894 \end{bmatrix}$$

$$\beta_2 = \tan^{-1}\left(\frac{n_{1y}}{n_{1x}}\right) = 63.435°$$

The orthogonal rotation transformation matrix is

$$[R] = \begin{bmatrix} 0.894 & -0.447 \\ 0.447 & 0.894 \end{bmatrix}$$

$$[R][I][R]^T = \begin{bmatrix} 160 & 0 \\ 0 & 40 \end{bmatrix}$$

Problems

For each of the following cross-sectional areas, determine the principal axes and the principal second moments of area, using the eigenvalue–eigenvector method.

10.74 The z-section shown in Problem 10.35. (See Figure P10.74.) $b = \frac{h}{2}$

10.75 The cross section shown in Problem 10.22. (See Figure P10.75.)

10.76 The cross section shown in Problem 10.42. (See Figure P10.76.)

10.77 The cross-sectional area shown in Problem 10.43. (See Figure P10.77.)

10.78 The cross-sectional area shown in Problem 10.45. (See Figure P10.78.)

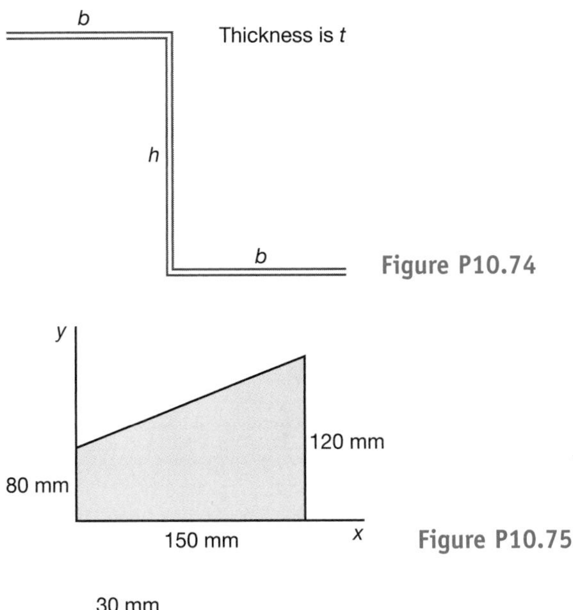

Figure P10.74

Figure P10.75

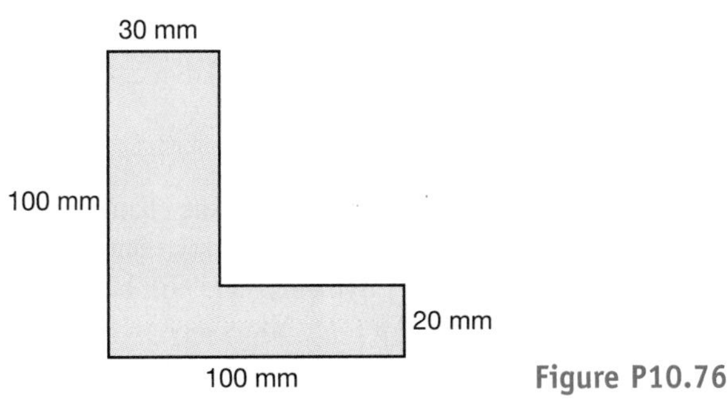

Figure P10.76

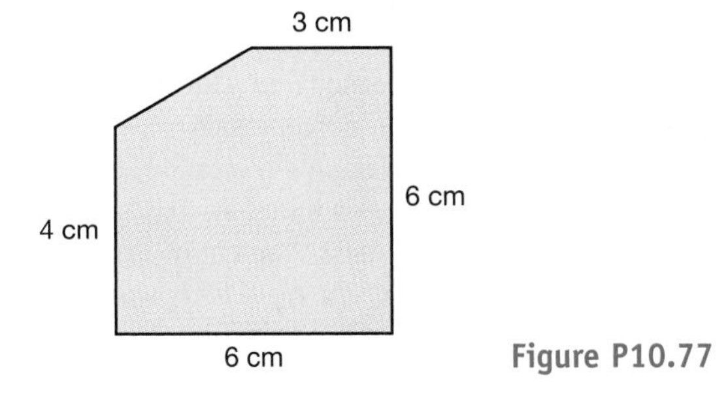

Figure P10.77

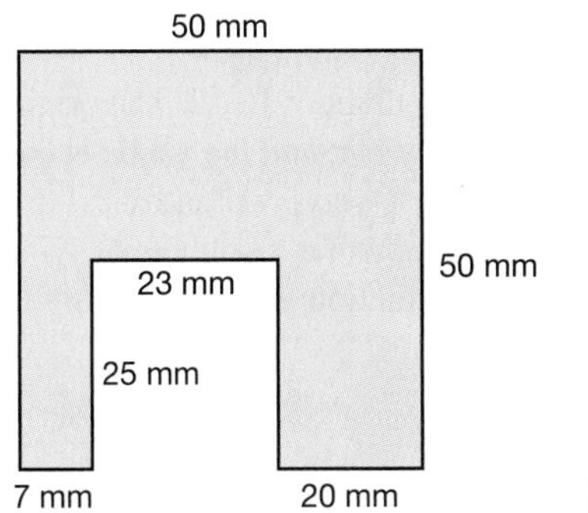

Figure P10.78

10.79 Determine the centroidal principal axes and principal second moments of area for the triangle shown in Figure P10.79.

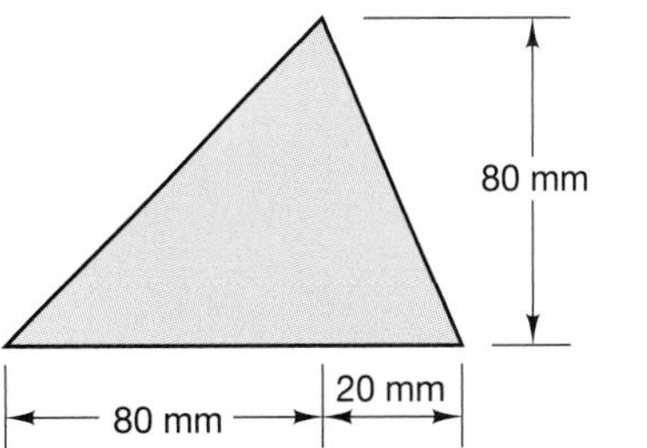

Figure P10.79

10.80 Prove that every centroidal axis is a principal axis for a square cross-sectional area.

10.81 Prove that every centroidal axis is a principal axis for an equilateral triangle.

10.82 Prove that every centroidal axis is a principal axis for a hexagon.

10.83 Determine the centroidal principal axes and the principal second moments of area for the cross-sectional area in Figure P10.83.

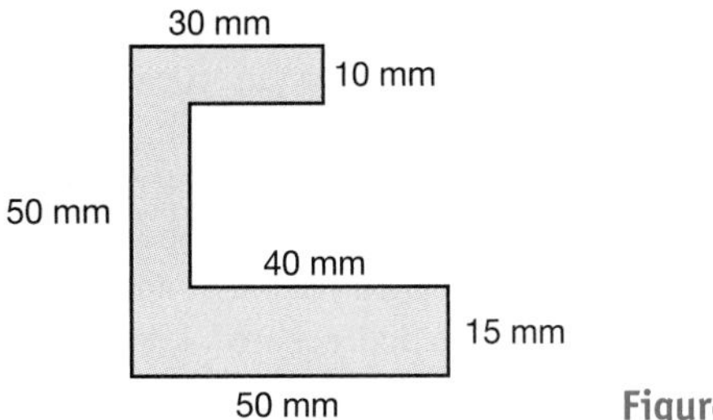

Figure P10.83

10.84 Determine the centroidal principal axes and principal second moments of area for the square cross-sectional area with the 10-mm-radius hole in the upper left quadrant in Figure P10.84.

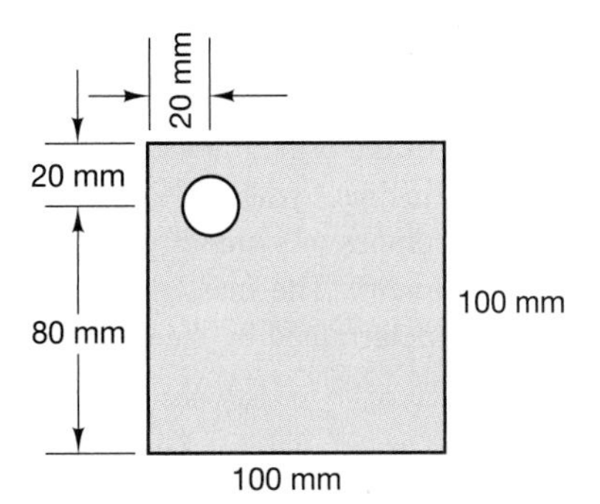

Figure P10.84

10.85 Determine the centroidal principal axes and the principal second moments of area for an angular cross-sectional beam made of stock of thickness t in Figure P10.85.

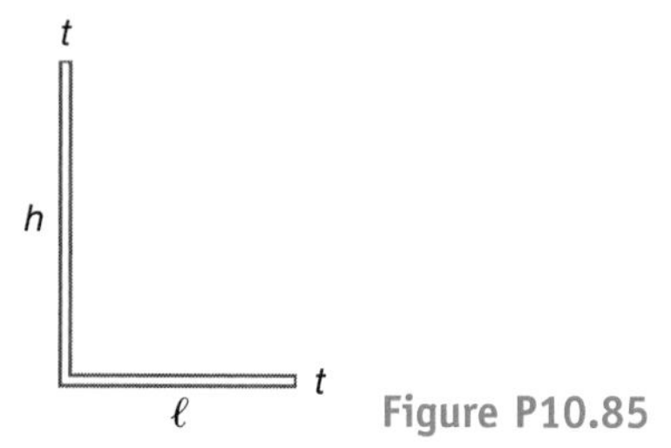

Figure P10.85

10.86 In Figure P10.85, show that if the legs of the angle are equal ($h = l$), the axis of symmetry is a principal axis.

10.87 Determine the centroidal principal axes and the principal second moments of area for the cross-sectional area shown in Figure P10.87.

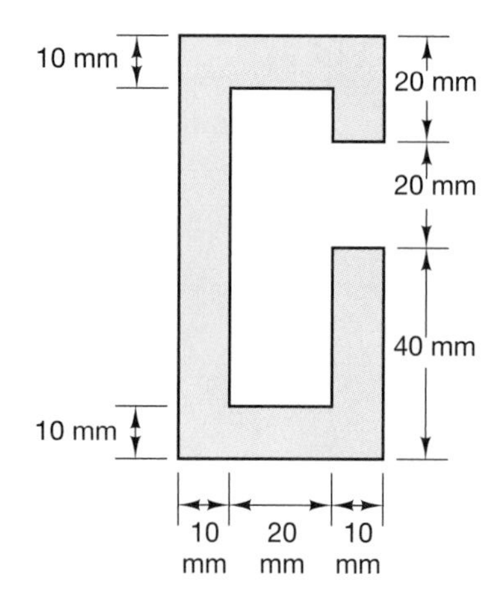

Figure P10.87

10.11 MASS MOMENTS OF INERTIA

Newton's second law relates the unbalanced forces acting on a particle to the change in the particle's linear momentum mv. The mass of a particle is a measure of its resistance to linear acceleration. This principle is fundamental to the field of dynamics and will be considered in detail when we examine the subject. The moment of a force about a point is a measure of the turning effect of the force about that point. If a rigid body is subjected to an unbalanced moment, the body will undergo an angular acceleration. The resistance to angular acceleration is determined by the mass distribution within the body. This resistance is called the *mass moment of inertia*. The units of the mass moment of inertia are mass times length squared, kilograms times meters squared, pounds times feet squared (slugs times feet squared). This property of a rigid body is considered in detail in dynamics. Here, we will examine only rigid bodies that have a plane of mass symmetry, delaying a detailed study of mass moments until we analyze dynamics problems. It can be shown that an axis perpendicular to a body's plane of symmetry is a principal axis of the mass moment of inertia: *If a rigid body has a plane of mass symmetry, the center of mass of the rigid body will lie in the plane of mass symmetry, and axes perpendicular to this plane will be principal axes.* This is the basis of any plane dynamics that models rigid bodies two dimensionally and assumes that the body has only three degrees of freedom—two translations in the plane of symmetry and rotation about an axis perpendicular to the plane of symmetry.

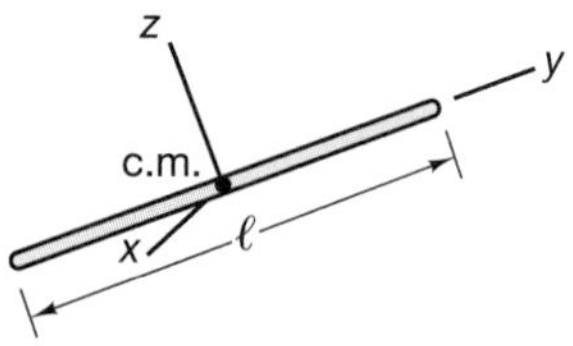

Figure 10.12

Consider the simple, slender bar of uniform density shown in Figure 10.12. The origin of the orthogonal coordinate system is at the center of mass of the bar, and the y-axis is coincident with the long axis of the bar. All three coordinate planes—x–y, y–z, and x–z—are planes of symmetry, and therefore, each coordinate axis is perpendicular to a plane of symmetry. The mass moment of inertia—that is, the resistance to rotation about the axis—is determined by the distribution of the mass about the axis; that is,

$$I = \int_M r^2\, dm \tag{10.50}$$

where M is the total mass and r is the perpendicular from the axis to the mass element. If ρ is the mass density and A is the cross-sectional area of the bar, the mass moment of inertia about the x- or z-axis is

$$I_{xx} = I_{zz} = \int_{-\frac{l}{2}}^{\frac{l}{2}} y^2 \rho A \, dy = \frac{1}{12}(\rho A l) l^2 = \frac{1}{12} M l^2 \quad (10.51)$$

A radius of gyration of the mass is defined so as to be consistent with the radius of gyration of an area, Eq. (10.16) in Section 10.6:

$$k = \sqrt{\frac{I}{M}} \quad (10.52)$$

The mass moment of inertia can be expressed in terms of the radius of gyration as

$$I = k^2 M \quad (10.53)$$

Therefore, the radius of gyration of the long, slender rod about its center of mass is

$$k_x = k_z = \sqrt{\frac{1}{12}}\, l \quad (10.54)$$

10.11.1 PARALLEL-AXIS THEOREM

We can develop a parallel-axis theorem for the mass moment of inertia similar to that used for the second moment of area. This theorem relates the mass moment of inertia about a point to the mass moment of inertia about the center of mass. Consider a rigid body having a mass plane of symmetry designated by the x–y plane, as shown in Figure 10.13. The x'–y' axes are axes through the center of mass, and the position vector to the mass element is

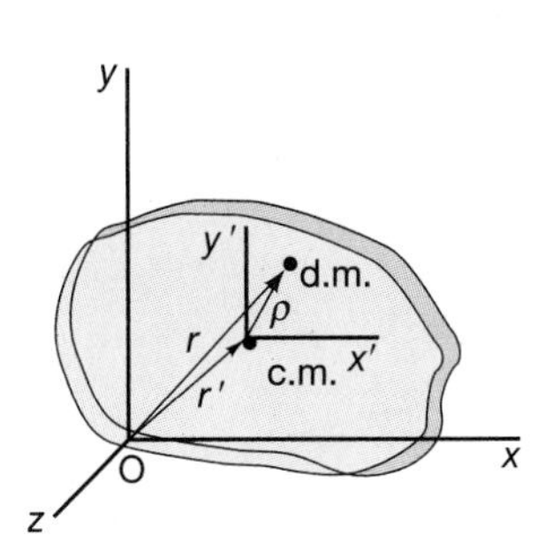

Figure 10.13

$$r = r' + \rho \quad (10.55)$$

The mass moment of inertia about the z-axis through point O is

$$I_{Oz} = \int_M r \cdot r \, dm \quad (10.56)$$

Substituting Eq. (10.55) into Eq. (10.56) yields

$$I_{0z} = \int_M [r' \cdot r' + 2r' \cdot \boldsymbol{\rho} + \boldsymbol{\rho} \cdot \boldsymbol{\rho}] \, dm \quad (10.57)$$

$$I_{0z} = (r')^2 M + 2r' \cdot \int_M \boldsymbol{\rho} \, dm + \int_M \rho^2 \, dm$$

The integral in the second term in Eq. (10.57) is the definition of the distance from a reference point to the center of mass, and since ρ is measured from the center of mass, this distance is zero. The mass moment of inertia about the center of mass is

$$I_{\text{c.m.}} = \int_M \rho^2 \, dm \quad (10.58)$$

Therefore, the parallel-axis theorem for mass moment of inertia is

$$I_{0z} = (r')^2 M + I_{\text{c.m.}z} \quad (10.59)$$

This parallel-axis theorem can be used to determine the mass moment of inertia of composite bodies. The center of mass and the mass moment of inertia for homogeneous solids are given in Table 10.2.

Table 10.2 *Center of Mass and Mass Moment of Intertia of Homogenous Solids*

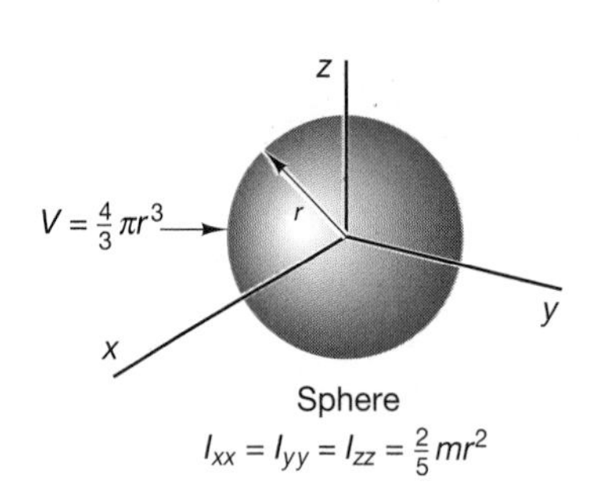

Sphere

$I_{xx} = I_{yy} = I_{zz} = \frac{2}{5}mr^2$

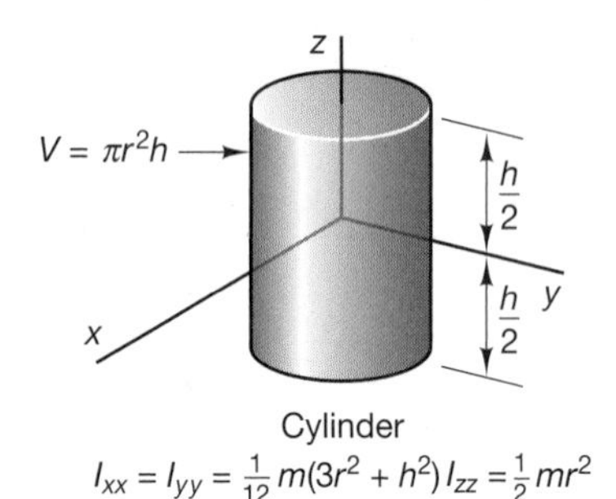

Cylinder

$I_{xx} = I_{yy} = \frac{1}{12}m(3r^2 + h^2)\ I_{zz} = \frac{1}{2}mr^2$

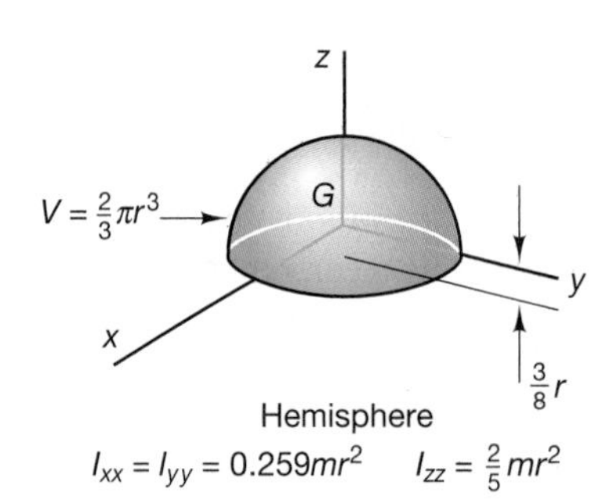

Hemisphere

$I_{xx} = I_{yy} = 0.259mr^2 \quad I_{zz} = \frac{2}{5}mr^2$

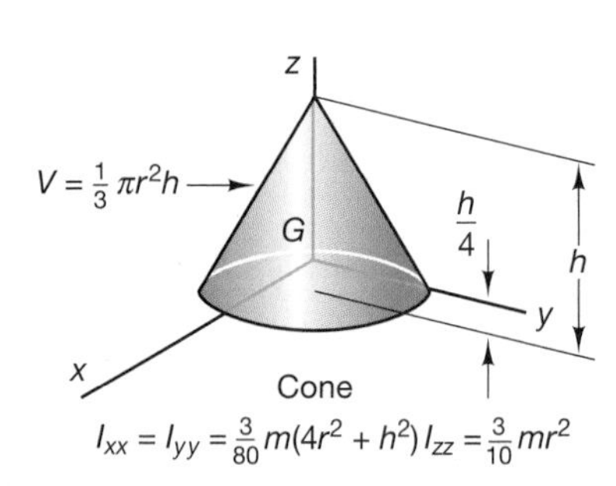

Cone

$I_{xx} = I_{yy} = \frac{3}{80}m(4r^2 + h^2)\ I_{zz} = \frac{3}{10}mr^2$

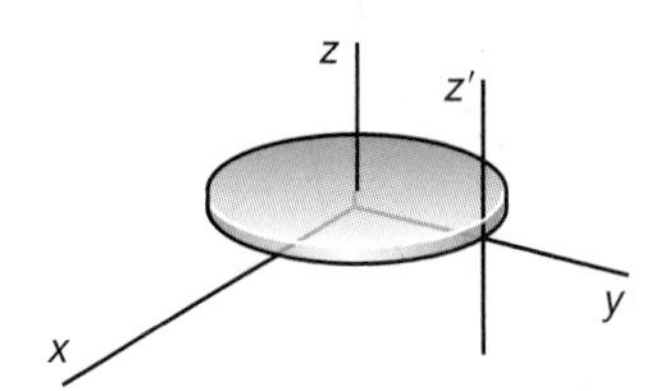

Thin circular disk

$I_{xx} = I_{yy} = \frac{1}{4}mr^2\ I_{zz} = \frac{1}{2}mr^2\ I_{zz} = \frac{3}{2}mr^2$

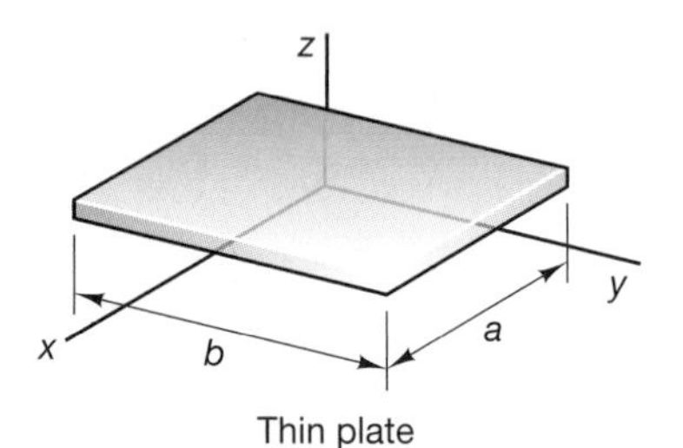

Thin plate

$I_{xx} = \frac{1}{12}mb^2\ I_{yy} = \frac{1}{12}ma^2\ I_{zz} = \frac{1}{12}m(a^2 + b^2)$

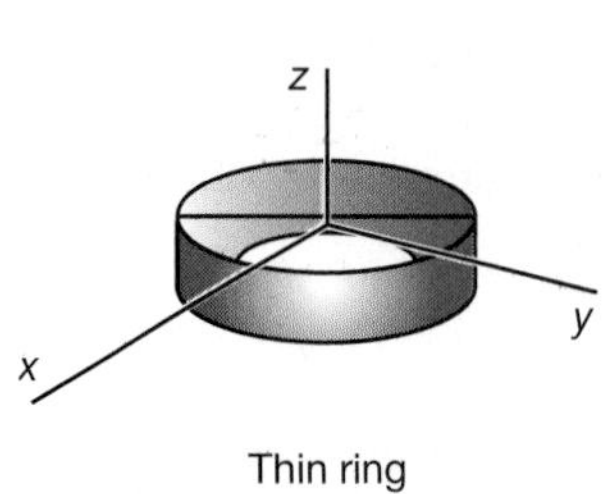

Thin ring

$I_{xx} = I_{yy} = \frac{1}{2}mr^2\ I_{zz} = mr^2$

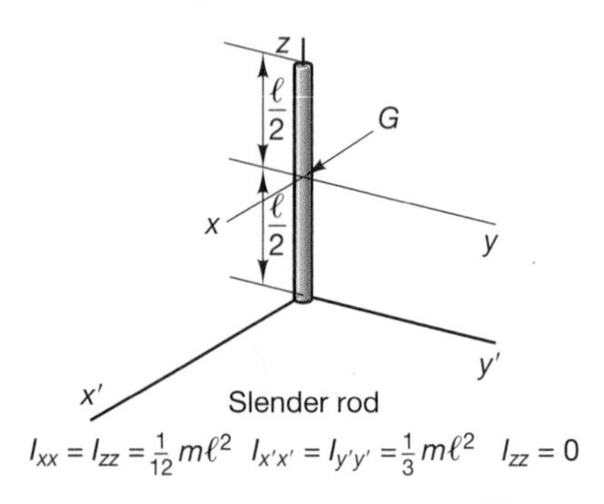

Slender rod

$I_{xx} = I_{zz} = \frac{1}{12}m\ell^2\ I_{x'x'} = I_{y'y'} = \frac{1}{3}m\ell^2\ I_{zz} = 0$

Sample Problem 10.8

Determine the mass moment of inertia of the pendulum shown at the left about point O. The pendulum is composed of a slender bar of mass m and a disk of mass M.

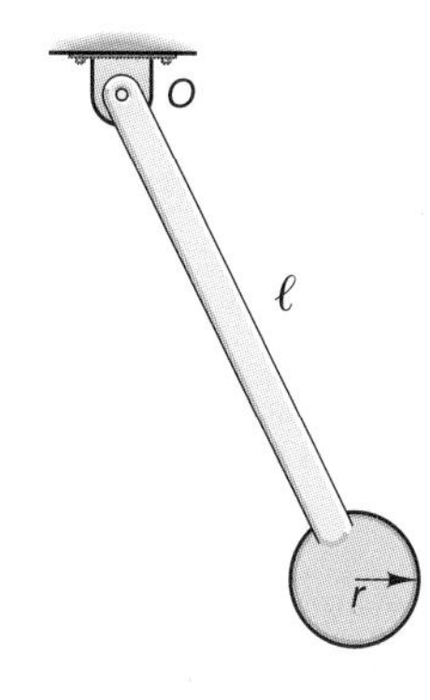

Solution

We need first to determine the mass moment of inertia of the disk about its center of mass. If the disk is of uniform density ρ and has a thickness t, the mass moment of inertia about a z-axis passing through the center of mass is

$$I_{\text{c.m.}} = \int_{-\frac{1}{2}t}^{\frac{1}{2}t}\int_{0}^{2\pi r}\int_{0}^{r} \rho[(r^2 r\, dr)\, d\theta]\, dz$$

$$I_{\text{c.m.}} = \frac{r^2}{2}(\rho\pi r^2 t) = \frac{1}{2}Mr^2$$

The mass moment of area of the bar about its center of mass is

$$I_{\text{c.m.}} = \frac{1}{12}m\ell^2$$

Using the parallel-axis theorem, we find that the mass moment of inertia about point O is

$$I_O = \frac{1}{12}m\ell^2 + m\left(\frac{1}{2}\right)^2 + \frac{1}{2}Mr^2 + M(\ell + r)^2$$

Therefore,

$$I_O = \frac{1}{3}m\ell^2 + M\left(l^2 + 2\ell r + \frac{2}{3}r^2\right)$$

Problems

10.88 In Figure P10.88, determine the mass moment of inertia of the rectangular plate of mass m, assuming that the plate is of uniform density.

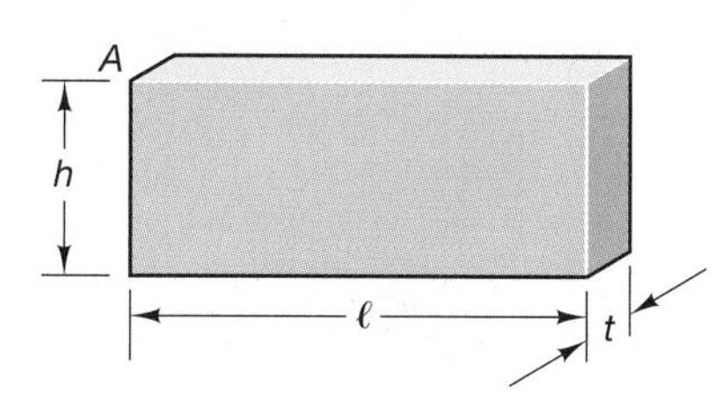

Figure P10.88

10.89 Determine the mass moment of inertia of the plate about the upper-left corner A in Figure P10.88.

10.90 Determine the mass moment of inertia for the car modeled as symmetric about a midplane in Figure P10.90. Approximate the car as four rectangles, three making up the body of the car and the fourth being the engine. Neglect the mass of the tires and the mass of the other components. The mass of the three components of the body is 360 kg, and the engine block mass 200 kg. Assume the center of mass of each rectangle to be at its geometric center.

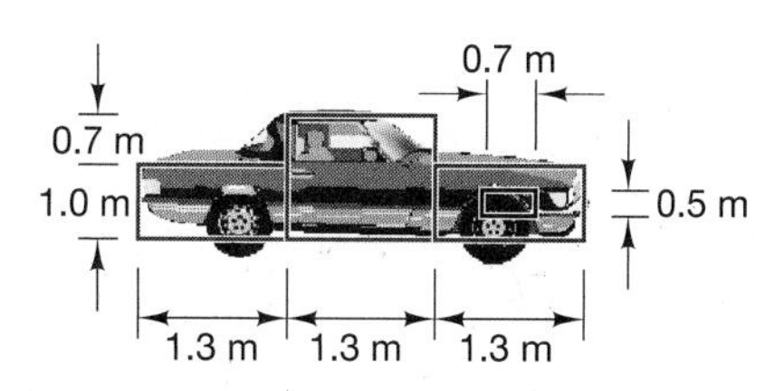

Figure P10.90

10.91 Determine the mass moment of inertia of the car in Problem 10.90 when two 75-kg passengers modeled as 0.3-m by 1-m rectangles are added to the front of the passenger compartment. (See Figure P10.91 on the next page.)

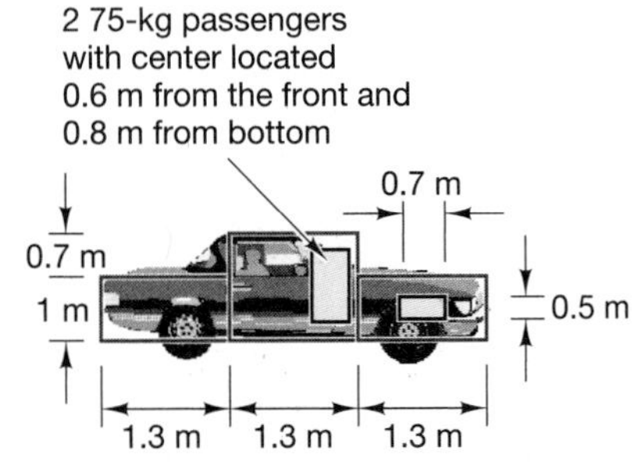

Figure P10.91

10.92 How would the center of mass and the mass moment of inertia change in Figure P10.92 if two more 75-kg passengers were in the back seat of the passenger compartment?

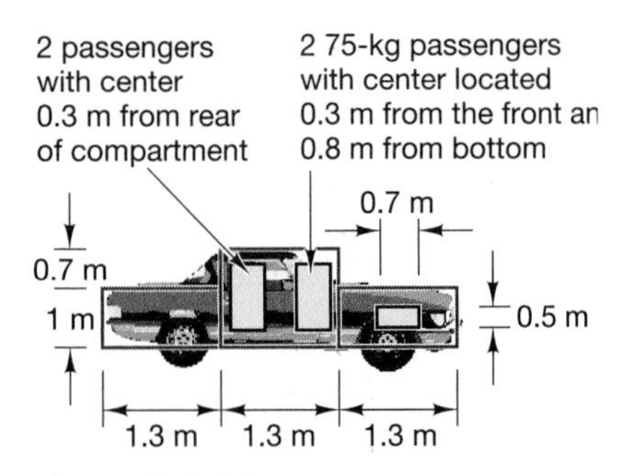

Figure P10.92

10.93 In Figure P10.93, determine the center of mass and the mass moment of inertia about point O of the triangular 60-mm-thick plate if the mass density is 0.8×10^{-6} kg/mm^3.

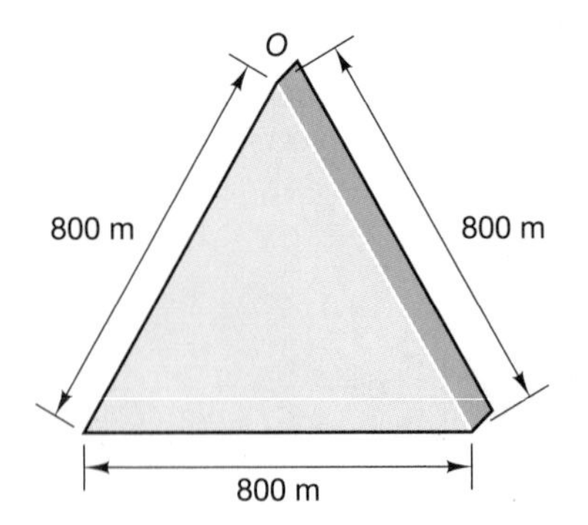

Figure P10.93

10.94 In Figure P10.94, determine the mass moment of inertia of the barbell about the z-axis through the center of mass. (*Hint:* You must first determine the moment of inertia of a disk about an axis along its diameter.)

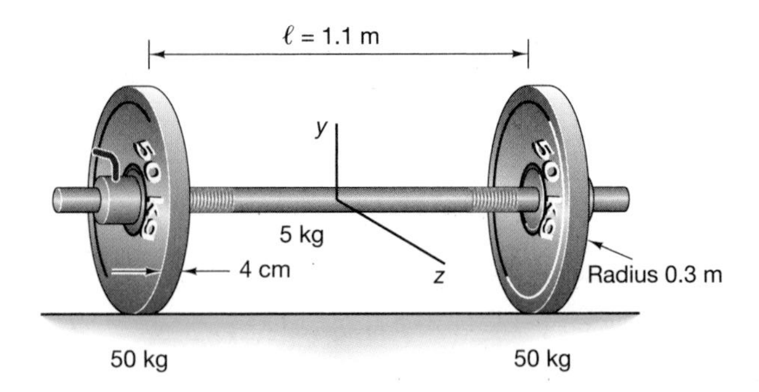

Figure P10.94

10.95 Determine the mass moment of inertia of the barbell about the long axis of the bar in Figure P10.94.

10.96 Determine the mass moment of inertia about point O of the pendulum composed of a sphere of mass 5 kg and radius 0.2 m and a 1-kg bar of length 1.5 m in Figure P10.96.

10.97 In Figure P10.97, determine the mass moment of inertia of an 8-mm-thick, 500-mm-square plate with a 100-mm-diameter hole at the center. The mass density is 250 kg/m^3.

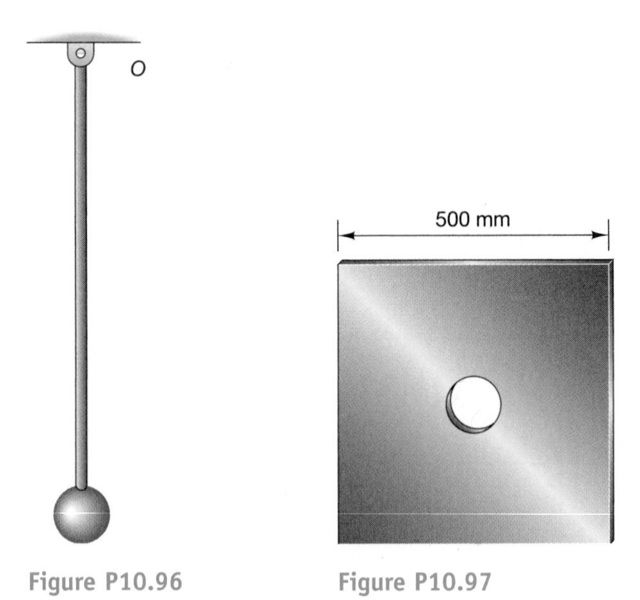

Figure P10.96 Figure P10.97

10.98 Determine the mass moment of inertia of the plate in Figure P10.97 about its upper-left corner.

10.99 Determine the mass moment of inertia of the plate in Figure P10.97 about a midpoint along an edge.

10.100 A long, slender bar of mass m and length l is bent at its midpoint by an angle θ in Figure P10.100. Develop a general expression for the mass moment of inertia about the center of mass of the bar. Show that, for an angle $\theta = 0$, the expression is the same as for a straight bar of mass m and length l.

10.101 Develop a general expression for the mass moment of inertia for the bent bar about point O in Figure P10.100.

10.102 A shovel can be modeled as a bar and a plate. In Figure P10.102, determine the mass moment of inertia about the handle point O. The mass of the bar is 2 kg and the mass of the plate is 3 kg.

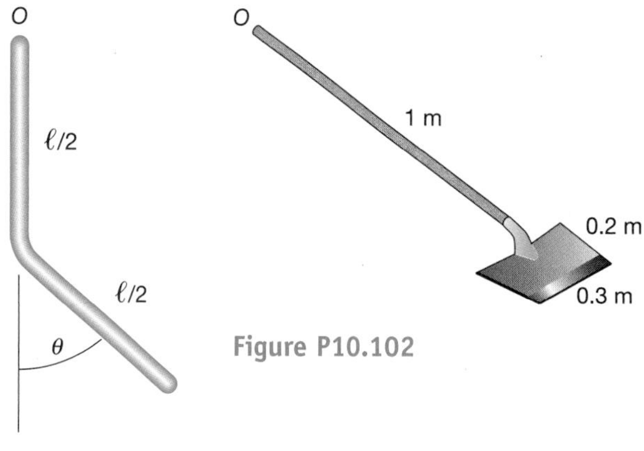

Figure P10.100

Figure P10.102

Chapter Summary

The first moment of area was defined in Chapter 5 as

$$\text{First moment of the area about } x = \int_A y\, dA = y_c A$$

$$\text{First moment of the area about } y = \int_A x\, dA = x_c A$$

The second moment of area is defined as

$$I_{xx} = \int_A y^2\, dA$$

$$I_{yy} = \int_A x^2\, dA$$

The polar moment of inertia, or the second moment of the area about the z-axis, is defined as

$$J_{0z} = \int_A r^2\, dA$$

In polar coordinates, $r^2 = x^2 + y^2$; therefore,

$$J_{0z} = I_{xx} + I_{yy}$$

The parallel-axis theorem for the second moment of the area is

$$I_{xx} = I_{xx_c} + Ad_y^2 \tag{10.13}$$

where I_{xx_c} is the second moment of the area about the x'-centroidal axis and d_y is the distance between the reference x-axis and the parallel centroidal axis x_c.

The parallel-axis theorem allows the computation of the second moment of area about any axis if the second moment of area about a parallel centroidal axis is known. The theorem also allows complex cross-sectional areas to be treated as a composite of simple geometric shapes.

The parallel-axis theorem applies equally to the second moment of the area about the y-axis in which case it is written as

$$I_{yy} = I_{yy_c} + Ad_x^2 \tag{10.14}$$

where I_{yy_c} is the second moment of the area about the y'-centroidal axis and d_x is the distance between the reference y-axis and the parallel centroidal axis y_c.

Addition of these two forms of the parallel-axis theorem yields a similar result for the polar moment of inertia, or the second polar moment of the area:

$$J_0 = J_{0_c} + Ar_0^2 \tag{10.15}$$

where J_{0_c} is the second polar moment of the area about the centroid and r_0 is the distance from the reference origin to the centroid.

In this case the parallel axes are the z-axes, and the second moment of the area is taken about those axes.

The radius of gyration is another method of specifying the second moments of area:

$$k_x = \sqrt{\frac{I_{xx}}{A}}$$

$$k_y = \sqrt{\frac{I_{yy}}{A}}$$

$$k_z = \sqrt{\frac{J_{0z}}{A}}$$

Principal Second Moment of Area The principal moments of area are the second moments of area about the principal centroidal axes. The principal axes have the following characteristics:

1. They are orthogonal axes.
2. The second moment of area about one of these axes will be the maximum value of the second moment of area for all centroidal axes, and the second moment of area about the other principal axes will be a minimum value.
3. A property called the *product moment of area* will be zero about these axes.

The product moment of area is defined as

$$I_{xy} = -\int_A xy\,dA$$

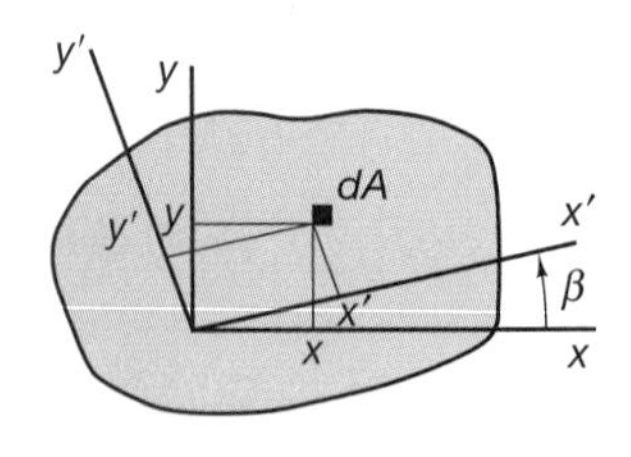

If the second moments of area and the product moment of area are known about one set of rectangular axes, the second moment of area about a set of rotated axes x'–y' (see diagram to the left) are

$$I_{x'x'} = \frac{I_{xx} + I_{yy}}{2} + \frac{I_{xx} - I_{yy}}{2}\cos 2\beta + I_{xy}\sin 2\beta$$

$$I_{x'y'} = \frac{I_{xx} - I_{yy}}{2}\sin 2\beta + I_{xy}\cos 2\beta$$

The angle to the principal axis is given by

$$\tan 2\beta = \frac{2I_{xy}}{I_{xx} - I_{yy}}$$

or

$$\beta = \frac{1}{2}\tan^{-1}\left(\frac{2I_{xy}}{I_{xx} - I_{yy}}\right)$$

The product moment of inertia about the principal axes is zero. The second moments of area about these axes are maximum and minimum, and are given by

$$I_{\text{max/min}} = \frac{I_{xx} + I_{yy}}{2} \pm \sqrt{\left(\frac{I_{xx} - I_{yy}}{2}\right)^2 + I_{xy}^2}$$

Mohr's circle is a graphical representation of the rotation transformation equations of the second moments of area. It is a valuable conceptual tool for examining of second-order tensors.

The principal second moments of area and the principal axes are the eigenvalues and eigenvectors, respectively, of the second moment of area matrix (tensor)

$$[I] = \begin{bmatrix} I_{xx} & I_{xy} \\ I_{xy} & I_{yy} \end{bmatrix}$$

Mass Moments of Inertia The resistance to angular acceleration of a rigid body about an axis is the mass moment of inertia of the body. The units of the mass moment of inertia are mass times length squared. If a rigid body has a plane of mass symmetry, the center of mass of the rigid body will lie in that plane, and the axes normal to the plane will be principal axes. The mass moment of inertia about these axes is

$$I = \int_M r^2\, dm$$

The radius of gyration is

$$k = \sqrt{\frac{I}{M}}$$

The parallel-axis theorem for the mass moment of inertia is

$$I_{0z} = (r')^2 M + I_{\text{c.m.}z}$$

where r' is the distance from the center of mass to the reference axis and $I_{\text{c.m.}z}$ is the mass moment of inertia about the principal perpendicular to the plane of mass symmetry that passes through the center of mass.

Chapter 11 VIRTUAL WORK

The principles of virtual work can be used to analyze the forces on the backhoe. (Photo courtesy of ronfromyork/Shutterstock)

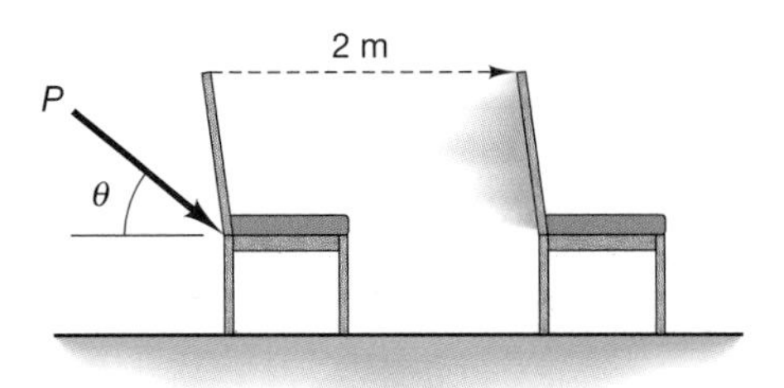

Figure 11.1

11.1 INTRODUCTION

The method of *virtual work* is based on a branch of mathematics called the calculus of variations. The development of this subject was begun by Newton in 1686 and further refined by Johann and Jakob Bernoulli (1696), Euler (1744), Legendre (1786), Lagrange (1788), Hamilton (1833), and Jacobi (1837). The method was formally presented by Lagrange in 1788, with the publication of *Mécanique Analytique* as an alternative to Newton's laws of equilibrium. The methods of virtual work and *potential energy* will be used in this chapter to examine the equilibrium of a series of connected rigid bodies and to investigate the *stability* of these systems.

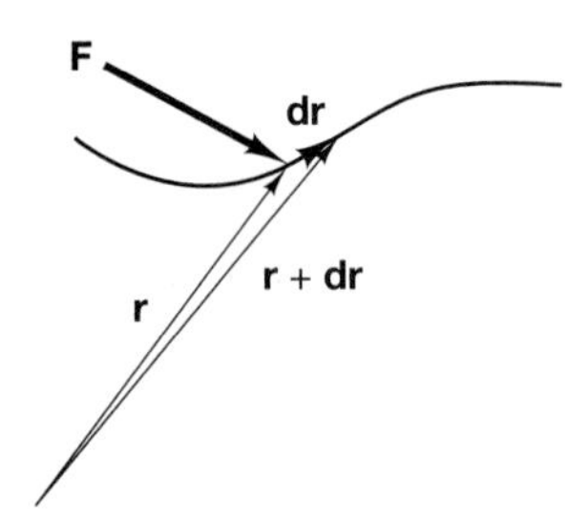

Figure 11.2

11.1.1 WORK—BY A FORCE OR MOMENT

Work may be defined as "the production of results" or "the performance of a function." Consider the simplest example of the work done by a force P, as illustrated in Figure 11.1. The desired result is to push the chair two meters across the floor. It is obvious that the vertical component of the force does nothing to help produce the desired result and, therefore, does no work. The force P works only when the chair undergoes a displacement in the direction of the force. With this motivation, consider a particle moving from position $\mathbf{r}$ to position $\mathbf{r} + \mathbf{dr}$ under the action of a force F, as shown in Figure 11.2. The formal mathematical definition of the work done by the force F undergoing a displacement $\mathbf{dr}$ is

$$dU = \mathbf{F} \cdot \mathbf{dr} \tag{11.1}$$

In words, the differential work is equal to the dot or scalar product of the force with the differential change in the position vector. Work is therefore a scalar quantity, and we can never refer to the work done in some direction. The units of work are (force) × (length). If U.S. customary units are used, work is expressed in ft • lb or in. · lb. If SI units are used, work is expressed in N • m, and one N • m is equal to one joule (J). Although a moment has units of force times length, it is not a measure of work or energy and is not expressed in joules. If the angle between the force and the displacement vector is greater than 90°, the dot product is negative, and the force does negative work. From a conceptual view, this is reasonable, as, in that case, the force would inhibit the desired result. Certain forces, such as friction, always do negative work and, as will be shown, dissipate energy.

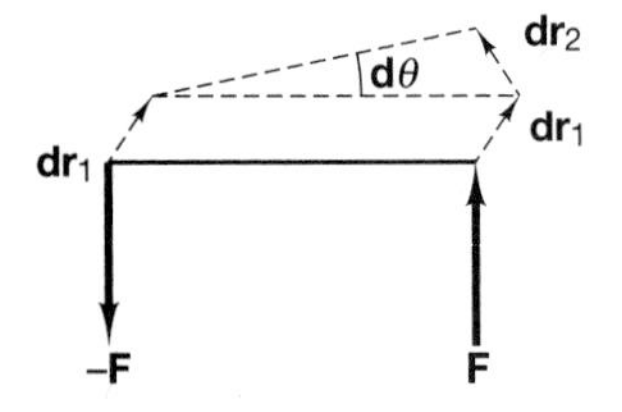

Figure 11.3

The work done by a moment is best examined by considering the work done by two forces forming a couple, as shown in Figure 11.3. When the couple is displaced $\mathbf{dr}_1$, the moment of the couple does no work, but when the force $\mathbf{F}$ is displaced by an amount $\mathbf{dr}_2$, net work is done. If a differential angle vector is defined by the right-hand rule as $\mathbf{d\theta}$, the displacement is

$$\mathbf{dr}_2 = \mathbf{d\theta} \times \mathbf{r} = d\theta\, \hat{\mathbf{n}} \times \mathbf{r} \tag{11.2}$$

where $\hat{\mathbf{n}}$ is a unit vector along the axis of the rotation $d\theta$ with sense and direction defined by the right-hand rule. The work done is

$$dU = \mathbf{F} \cdot (d\theta\, \hat{\mathbf{n}} \times \mathbf{r}) \tag{11.3}$$

The scalar triple product identity allows Eq. (11.3) to be written as

$$dU = d\theta\, \hat{\mathbf{n}} \cdot (\mathbf{r} \times \mathbf{F}) = d\theta\, \hat{\mathbf{n}} \cdot \mathbf{M} \tag{11.4}$$

When the moment of the couple is rotated through an angle $d\theta$, the moment does work equal to

$$dU = \mathbf{M} \cdot \mathbf{d\theta} \tag{11.5}$$

where $\mathbf{M} = \mathbf{r} \times \mathbf{F}$. Although finite rotations are not vectors, as was discussed in Chapter 2, infinitesimal rotations can be treated mathematically as vectors.

Certain constraining forces do no work. For example, the reactions at a frictionless ball-and-socket joint or a hinged joint do no work, as they are not displaced during movement of the structure. Internal forces occur in pairs that are equal, opposite, and collinear, and therefore, considered as pairs, they do no work.

11.2 VIRTUAL WORK

Thus far, we have defined work in terms of actual movements expressed in terms of differential changes in the position vector $\mathbf{dr}$. An imagined displacement or variation from the equilibrium position consistent with the constraints to the movement is called a ***virtual displacement***, and the work done by the external forces during such a virtual displacement is called ***virtual work***. These virtual displacements do not actually take place, but are imagined in order to analyze different equilibrium positions. Virtual displacements are assumed to be first-order differentials and are designated by a special symbol, $\delta\mathbf{r}$. The differential virtual displacement vector can be written in component notation as

$$\delta\mathbf{r} = \delta x\hat{\mathbf{i}} + \delta y\hat{\mathbf{j}} + \delta z\hat{\mathbf{k}} \tag{11.6}$$

In a similar manner, the virtual work will be

$$\delta U = \mathbf{F} \cdot \delta\mathbf{r} + \mathbf{M} \cdot \delta\theta \tag{11.7}$$

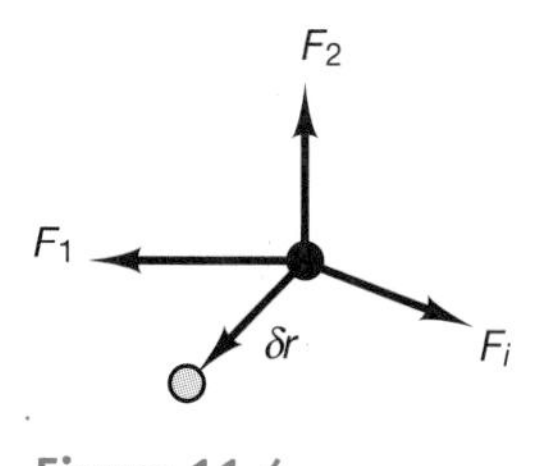

Figure 11.4

Now consider a particle in equilibrium under the action of several forces when it is displaced through a virtual displacement $\delta\mathbf{r}$, as shown in Figure 11.4. The virtual work done during this virtual displacement is

$$\delta U = \mathbf{F}_1 \cdot \delta\mathbf{r} + \mathbf{F}_2 \cdot \delta\mathbf{r} + \mathbf{F}_3 \cdot \delta\mathbf{r} = \left(\sum \mathbf{F}_i\right) \cdot \delta\mathbf{r} \tag{11.8}$$

Since the particle is in equilibrium, the sum of the forces is zero, and the virtual work is equal to zero. In a similar manner, the virtual work of a three-dimensional body or system is zero if it is in equilibrium and undergoes a virtual displacement from the equilibrium configuration. Therefore, *for a body or system of bodies in equilibrium, the virtual work associated with any virtual displacement is zero.* Symbolically,

$$\delta U = 0 \text{ for equilibrium.} \tag{11.9}$$

The forces involved are assumed to remain constant during the virtual displacement. If changes in these forces are considered, the product of the changes and the infinitesimal virtual displacement consists of higher order terms that are neglected. Clearly, there are only limited advantages to using of the concept of virtual work to solve problems of equilibrium of a single rigid body, as the equations of equilibrium can be written directly and then solved. The method of virtual work is most suitable for solving equilibrium problems involving a system of connected rigid bodies. This type of application will be discussed in the next section.

11.3 PRINCIPLE OF VIRTUAL WORK FOR A SYSTEM OF CONNECTED RIGID BODIES

Before outlining how solutions can be obtained for a system of connected rigid bodies using virtual work, some other concepts need to be reviewed. The number of degrees of freedom of a single rigid body has been defined to be the number of coordinate positions that must be specified to locate a body in three-dimensional space. In general, a single rigid body has six degrees of freedom: three translations and three rotations. For example, the bar shown in Figure 11.5 has coordinates specified at two points, so that six coordinates are specified. The position vectors to point A and B on the rigid bar are given by

$$\begin{aligned} \mathbf{r}_A &= x_A\hat{\mathbf{i}} + y_A\hat{\mathbf{j}} + z_A\hat{\mathbf{k}} \\ \mathbf{r}_B &= x_B\hat{\mathbf{i}} + y_B\hat{\mathbf{j}} + z_B\hat{\mathbf{k}} \end{aligned} \tag{11.10}$$

where (x,y,z) are the coordinates of the points A and B. Since six coordinates are specified, the body has six degrees of freedom. So, initially, we would think that these six coordinates would be sufficient to locate the body in space. A simple examination of the body, however, indicates that it is free to rotate about an axis through A and B and that the six coordinates in Eq. (11.10) do not specify this rotation. The reason they do not is because the six coordinates are not independent. To see this, recall that the body has been defined to be a rigid body, and therefore, the magnitude of a vector from A to B must be constant. This relationship may be expressed mathematically as

$$\begin{aligned} \mathbf{r}_{B/A} &= (x_B - x_A)\hat{\mathbf{i}} + (y_B - y_A)\hat{\mathbf{j}} + (z_B - z_A)\hat{\mathbf{k}} \\ |\mathbf{r}_{B/A}| &= \sqrt{(x_B - x_A)^2 + (y_B - y_A)^2 + (z_B - z_A)^2} = l_{AB} \end{aligned} \tag{11.11}$$

Equation (11.11) is a constraint equation for the coordinates of points A and B; hence, there are only five independent coordinates for the bar.

The principle of virtual work depends upon the specification of a set of ***independent coordinates***, or ***generalized coordinates***, that are equal in number to the number of degrees of freedom of the system. The generalized coordinates are usually denoted by q_i, where the index i has the values 1 to n, with n being the number of degrees of freedom. A one-degree-of-freedom system is shown in Figure 11.6. If point B is constrained to remain

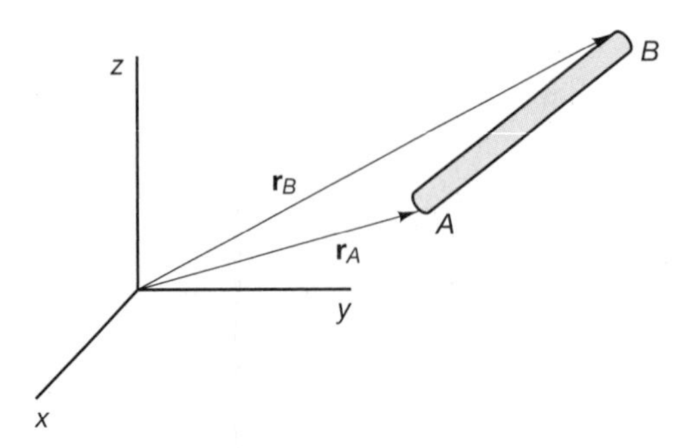

Figure 11.5

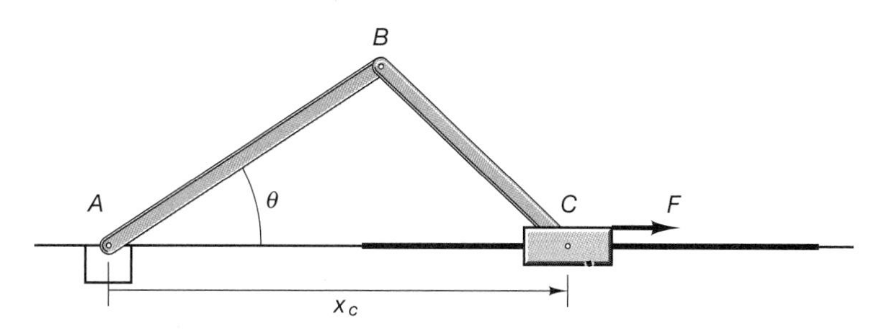

Figure 11.6

above the horizontal line AC, the configuration of the system is completely specified if either x_c or θ is given, and the system has only one degree of freedom. Therefore, either x_c or θ could be used as the independent, or generalized, coordinate:

$$q_1 = x_c \qquad \text{or} \qquad q_1 = \theta \tag{11.12}$$

The x- and y-coordinates of points B and C can be specified in terms of the generalized coordinate q_1.

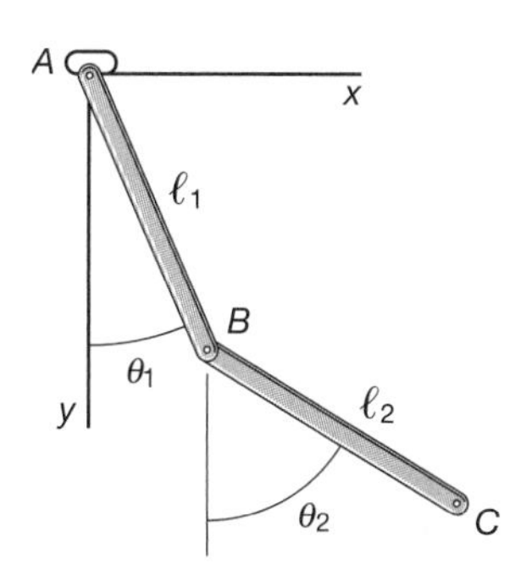

Figure 11.7

The planar double pendulum illustrated in Figure 11.7 is an example of a two-degree-of-freedom system. Since the length of the two segments of the pendulum are constant and the pendulum is constrained to move in a plane, the two generalized coordinates are

$$\begin{aligned} q_1 &= \theta_1 \\ q_2 &= \theta_2 \end{aligned} \tag{11.13}$$

The number of degrees of freedom of a system of n rigid bodies is equal to

$$\text{degrees of freedom} = 6\,n - m \tag{11.14}$$

where m is the number of constraints. For the double pendulum shown in Figure 11.7, let B_1 be the point B on link 1 and B_2 be the point B on link 2. Then the 10 constraint equations are

$$\begin{aligned} x_A &= 0 \\ y_A &= 0 \\ z_A &= 0 \\ z_{B_1} &= 0 \\ z_{B_2} &= 0 \\ z_C &= 0 \\ x_{B_1} &= x_{B_2} \\ y_{B_1} &= y_{B_2} \\ (x_{B_1} - x_A)^2 &+ (y_{B_1} - y_A)^2 = \ell_1^2 \\ (x_C - x_{B_2})^2 &+ (y_C - y_{B_2})^2 = \ell_2^2 \end{aligned} \tag{11.15}$$

The four nonzero coordinates can be written in terms of the two generalized coordinates as follows:

$$\begin{aligned} x_B &= \ell_1 \sin\theta_1 \\ y_B &= \ell_1 \cos\theta_1 \\ x_C &= \ell_1 \sin\theta_1 + \ell_2 \sin\theta_2 \\ y_C &= \ell_1 \cos\theta_1 + \ell_2 \cos\theta_2 \end{aligned} \tag{11.16}$$

Equations (11.16) are called the ***transformation equations*** from a set of coordinates (x, y, z) to the generalized coordinate ($q_1 = \theta_1$ and $q_2 = \theta_2$).

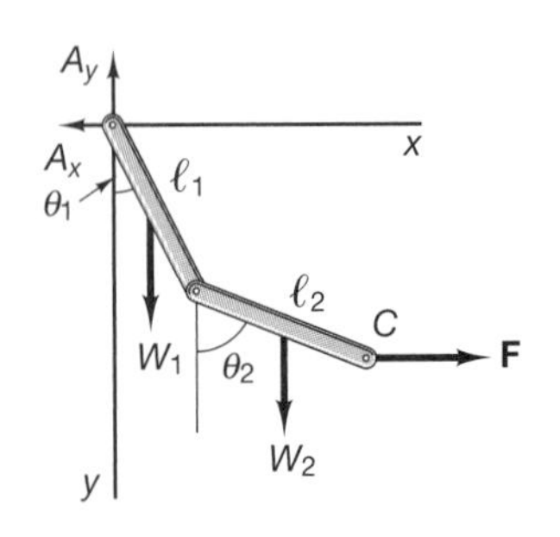

Figure 11.8

Next, consider the case when the two links forming the double pendulum are uniform, and therefore, their weight can be considered to be acting at the midlength point. The coordinates of the midpoints of the two bars are designated (x_1, y_1) and (x_2, y_2). The system is held in equilibrium by a horizontal force **F** applied at point *C*, as shown in the free-body diagram in Figure 11.8. Equilibrium requires that, for given weights and a given force **F**, the angles that the links make with the vertical coordinate have specific values. Although these angles can be determined by the methods shown in Chapter 7, the principle of virtual work offers an efficient alternative to that approach. To see this, consider virtual displacements $\delta\theta_1$ and $\delta\theta_2$ and the resulting virtual work

$$\delta U = W_1 \delta y_1 + W_2 \delta y_2 + F \delta x_C$$

The virtual displacements are obtained by taking the partial derivatives of the transformation equations with respect to the generalized coordinates:

$$y_1 = \frac{\ell_1}{2} \cos\theta_1$$

$$y_2 = \ell_1 \cos\theta_1 + \frac{\ell_2}{2}\cos\theta_2$$

$$x_c = \ell_1 \sin\theta_1 + \ell_2 \sin\theta_2$$

Therefore,

$$\delta y_1 = -\frac{\ell_1}{2}\sin\theta_1 \delta\theta_1$$

$$\delta y_2 = -\ell_1 \sin\theta_1 \delta\theta_1 - \frac{\ell_2}{2}\sin\theta_2 \delta\theta_2$$

$$\delta x_c = \ell_1 \cos\theta_1 \delta\theta_1 + \ell_2 \cos\theta_2 \delta\theta_2$$

The **principle of virtual work** for a system of rigid bodies connected with *frictionless* connections is as follows:

If a rigid body or a system of frictionless connected rigid bodies is in equilibrium, the total virtual work of the external forces acting on the body or system is zero for any virtual displacement consistent with the constraints on the body or system.

Sample Problem 11.1

Determine the equilibrium angles for the double pendulum shown in Figure 11.8 by both the principle of virtual work and the solution of the equilibrium equations.

Solution

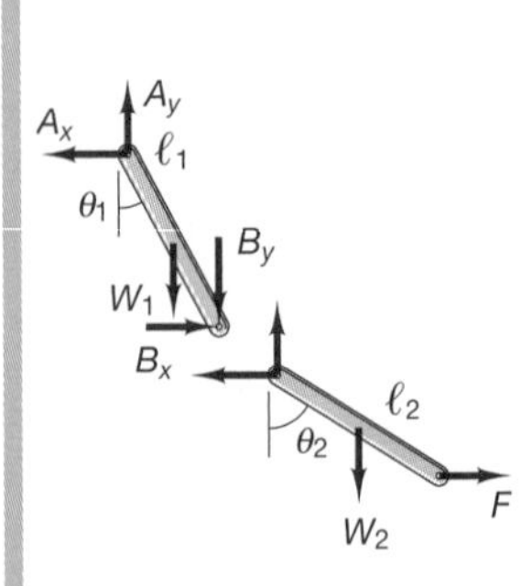

The total virtual work of the system is

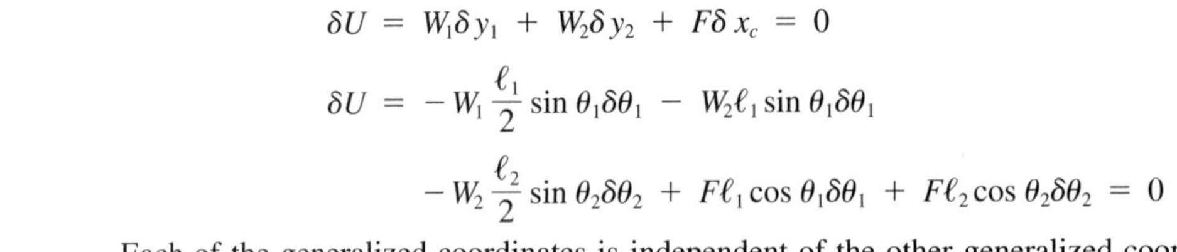

$$\delta U = W_1 \delta y_1 + W_2 \delta y_2 + F \delta x_c = 0$$

$$\delta U = -W_1 \frac{\ell_1}{2}\sin\theta_1 \delta\theta_1 - W_2 \ell_1 \sin\theta_1 \delta\theta_1$$

$$- W_2 \frac{\ell_2}{2}\sin\theta_2 \delta\theta_2 + F\ell_1 \cos\theta_1 \delta\theta_1 + F\ell_2 \cos\theta_2 \delta\theta_2 = 0$$

Each of the generalized coordinates is independent of the other generalized coordinates, and the virtual displacements can be considered separately. Therefore, the virtual work of each virtual displacement is equal to zero, and the equations of equilibrium are

$$\left(-W_1 \frac{\ell_1}{2}\sin\theta_1 - W_2 \ell_1 \sin\theta_1 + F\ell_1 \cos\theta_1\right)\delta\theta_1 = 0$$

$$\left(-W_2 \frac{\ell_2}{2}\sin\theta_2 + F\ell_2 \cos\theta_2\right)\delta\theta_2 = 0$$

It follows that:

$$\tan\theta_1 = \frac{2F}{W_1 + 2W_2}$$
$$\tan\theta_2 = \frac{2F}{W_2}$$

This problem can be solved by using the equations of equilibrium and creating free-body diagrams for each of the links, as is shown in the figure at the bottom of the previous page. Summing forces on the second link gives

$$B_x = F \quad \text{and} \quad B_y = W_2$$

Taking moments about point B results in

$$F\ell_2 \cos\theta_2 - W_2\frac{\ell_2}{2}\sin\theta_2 = 0$$
$$\tan\theta_2 = \frac{2F}{W_2}$$

This equation agrees with the result obtained by the method of virtual work. Taking moments about point A for the first link gives

$$F\ell_1 \cos\theta_1 - W_2\ell_1 \sin\theta_1 - W_1\frac{\ell_1}{2}\sin\theta_1 = 0$$

Therefore,

$$\tan\theta_1 = \frac{2F}{W_1 + 2W_2}$$

This equation also agrees with the result obtained by the method of virtual work. If the weights are equal and the force F is equal to the weight of one link, the equilibrium angles are

$$\theta_1 = 33.7^\circ$$
$$\theta_2 = 63.4^\circ$$

Note that the angles are independent of the lengths of the links and depend only upon the external forces applied to the system.

Problems

11.1 Determine the angle θ in terms of the applied force F corresponding to equilibrium of the two-bar linkage illustrated in Figure P11.1. The bars are identical, each of mass m and length ℓ. The center of mass of each link is at its geometric center.

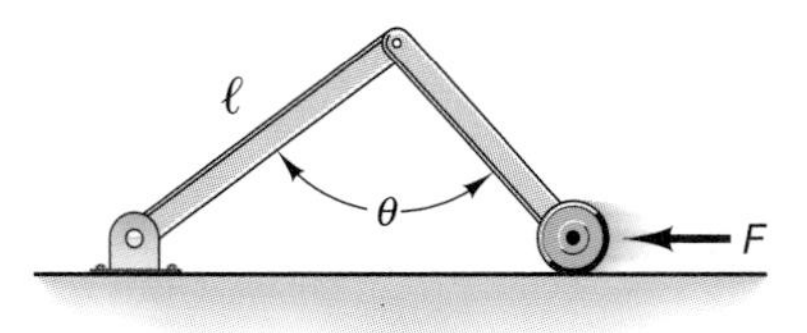

Figure P11.1

11.2 A mechanism consists of two levers used to change the direction of the applied force. Compute the force required to keep the system in the equilibrium position shown in Figure P11.2.

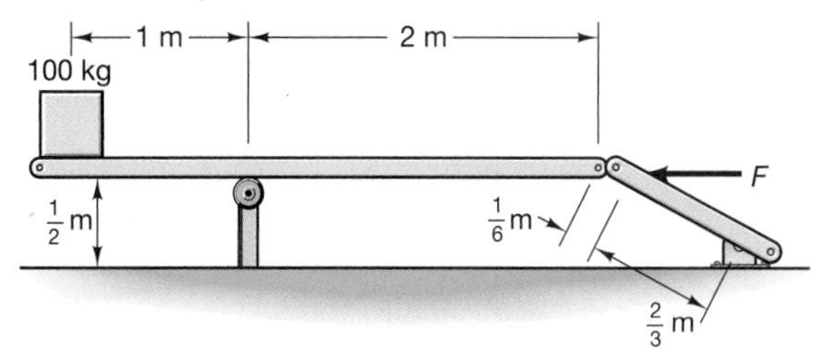

Figure P11.2

11.3 A motor applies a moment M to the two-bar linkage illustrated in Figure P11.3 in order to keep the indicator arm a at the position shown. The short rod has a mass m, and the long rod has a mass $4m/3$, with center of gravity at the geometric center of each link. Compute the required value of M as a function of θ, m, and ℓ for equilibrium.

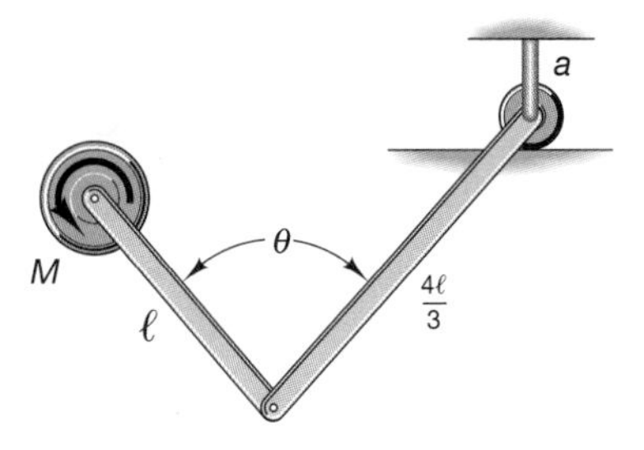

Figure P11.3

11.4 Compute the force **F** as a function of mass m, length a, and angle θ for equilibrium of the system shown in Figure P11.4.

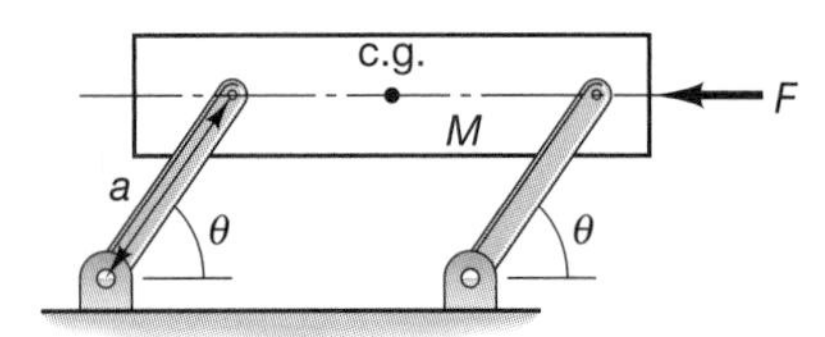

Figure P11.4

11.5 Repeat Problem 11.4 for the case where the support links are not aligned with the center of gravity, as illustrated in Figure P11.5.

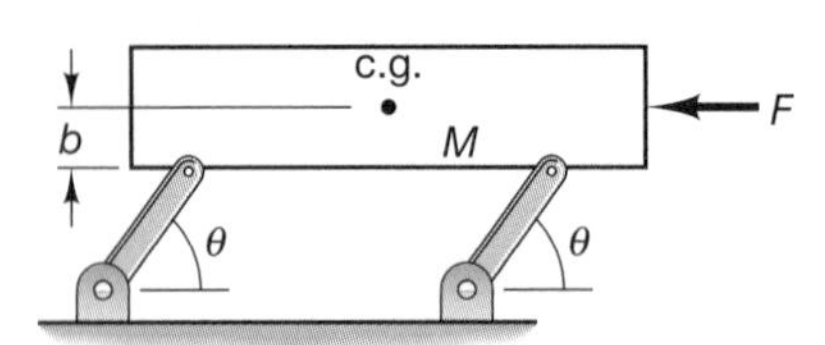

Figure P11.5

11.6 An apple press is constructed of two equal links and a cup, as illustrated in Figure P11.6. An applied force F is generated at joint B by a screw-and-crank mechanism. Compute the force P at the apple (point A) in terms of the angle θ and the force F required for equilibrium. Neglect the mass of the links.

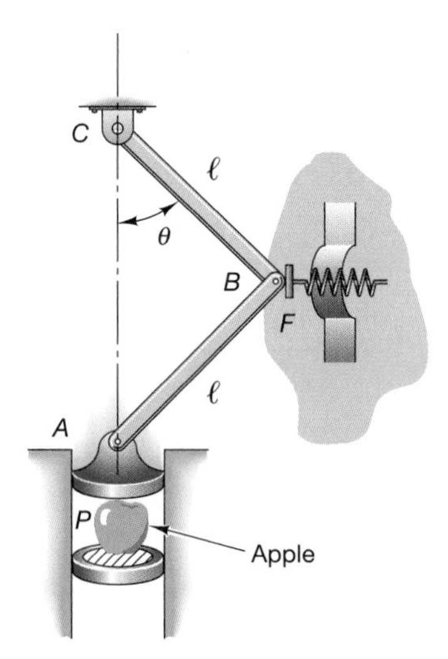

Figure P11.6

11.7 Repeat Problem 11.6 for the apple press, but include the weight of the links (mg), assuming that the center of mass of each link is at its midpoint. Solve for the force F in terms of mg, θ, ℓ, and P.

11.8 A motor is used through a linkage to position a part for machining. Ignore the mass of each part, and compute the moment M required to hold the force **F** in equilibrium at the position shown in Figure P11.8. The force F may be provided by a spring or hydraulic component not shown in the figure.

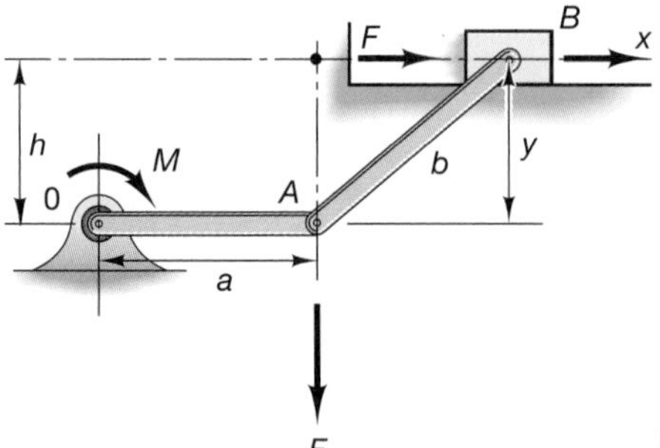

Figure P11.8

11.9 In Figure P11.9, compute the force **F** required by the piston at B to hold the mass W (500 kg) at an angle of $\theta = 30°$. Use $\ell = 2.5$ m.

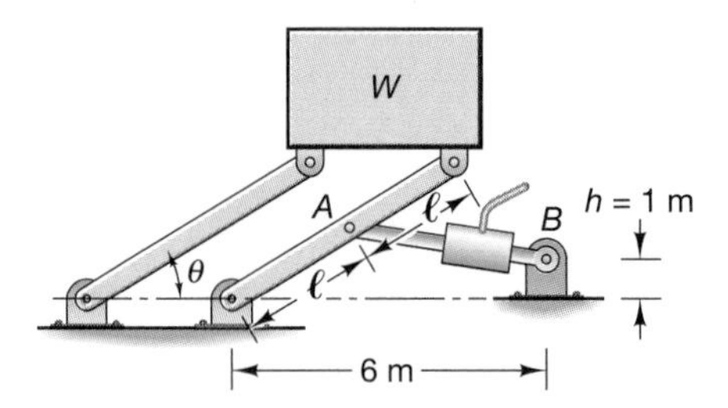

Figure P11.9

11.10 Compute the moment M required of the motor at A to keep the system shown in Figure P11.10 in equilibrium, and evaluate M at $\theta = 30°$, $W = 1000$ N, and $\ell = 2$ m.

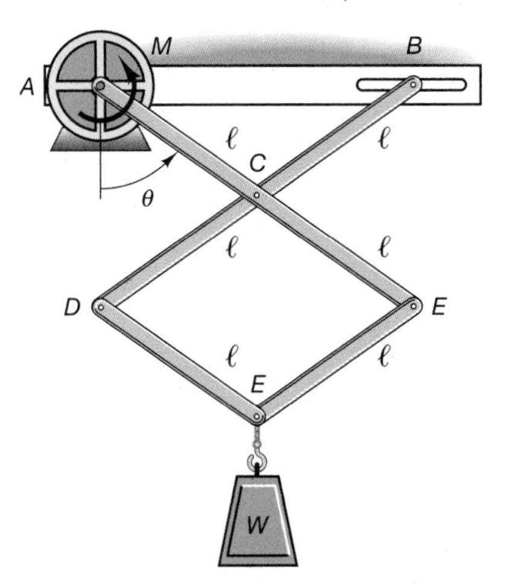

Figure P11.10

11.11 Compute the weight W required to hold the system illustrated in Figure P11.11 in equilibrium at $\theta = 30°$.

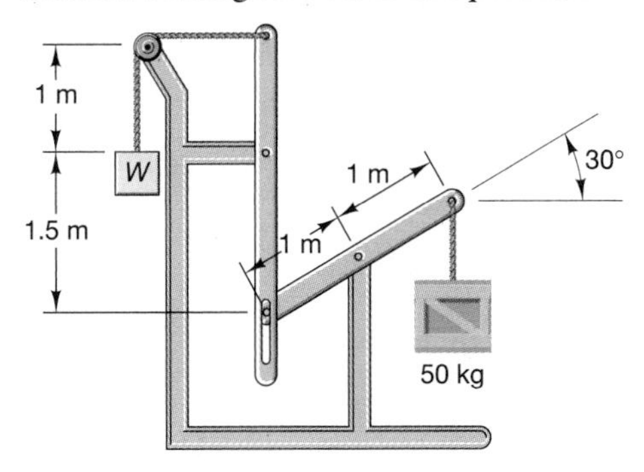

Figure P11.11

11.12 A hydraulic foot pump is held in equilibrium by applying a 250-N force to counteract the force of a piston. (See Figure P11.12.) Plot the required piston force as a function of θ for θ between 5° and 85°.

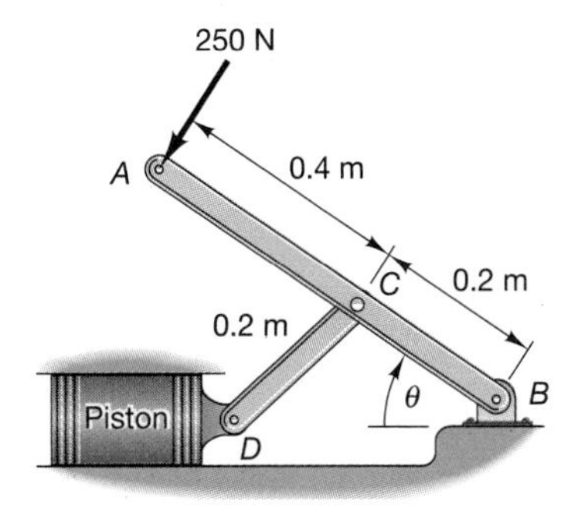

Figure P11.12

11.13 Repeat Problem 11.12, this time including the mass of the two links. The force of gravity on link AB is 25 N acting on its center, and that of DC is 15 N acting at its center.

11.14 Figure P11.14 shows a lifting mechanism operated by applying a force **F** to the roller by some drive system (not shown). Compute the force needed to keep this mechanism in equilibrium, in terms of θ, a, and b. Then discuss how to design the mechanism so that the required force is as small as possible for a given W. (In other words, what are good choices of a and b?)

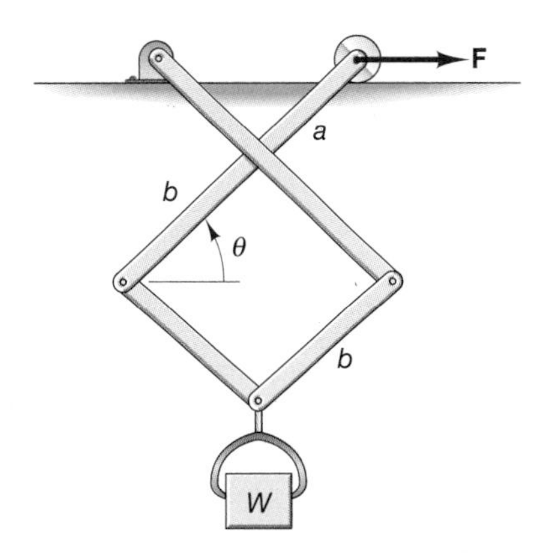

Figure P11.14

11.15 Compute the force at E resulting from the applied 150-N forces shown in Figure P11.15.

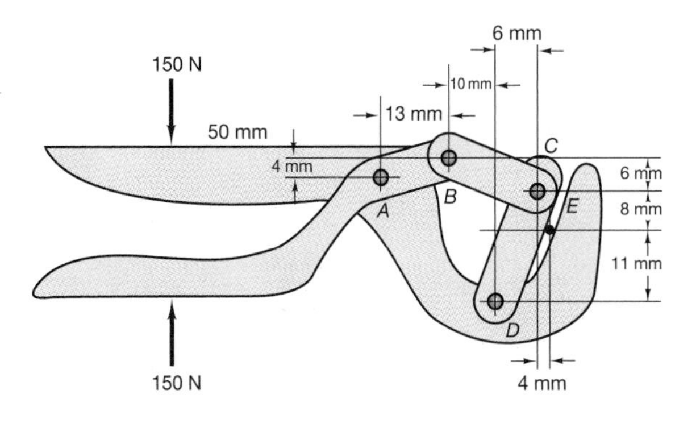

Figure P11.15

11.16 Calculate the force P in terms of θ if the platform shown in Figure P11.16 has a mass of 1000 kg and the crate has a mass of 3000 kg. Assume that the masses of the two links AC and DB are negligible.

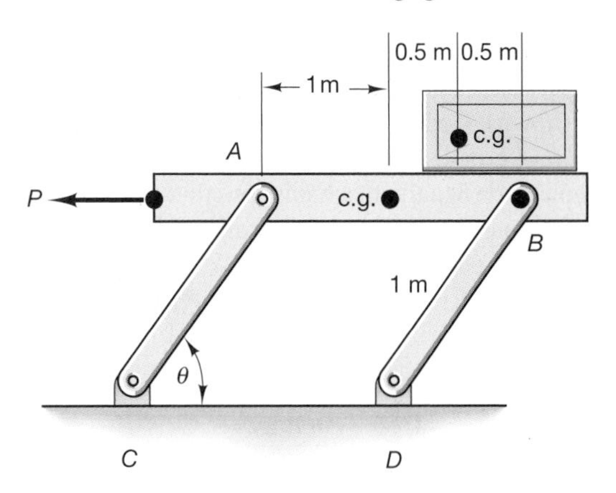

Figure P11.16

11.17 A dump-truck lifting mechanism is used to hold the load W in the position shown in Figure P11.17. Calculate the required force F as a function of a, b, c, d, and θ.

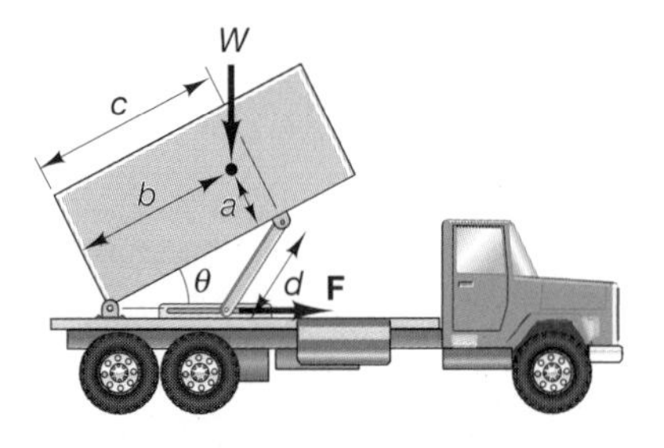

Figure P11.17

11.18 A precision machining tool uses a motor to apply and hold a machine part against a cutting tool that requires a force P. (See Figure P11.18.) Compute the moment M in terms of the link length a, applied force P, and cutting distance h.

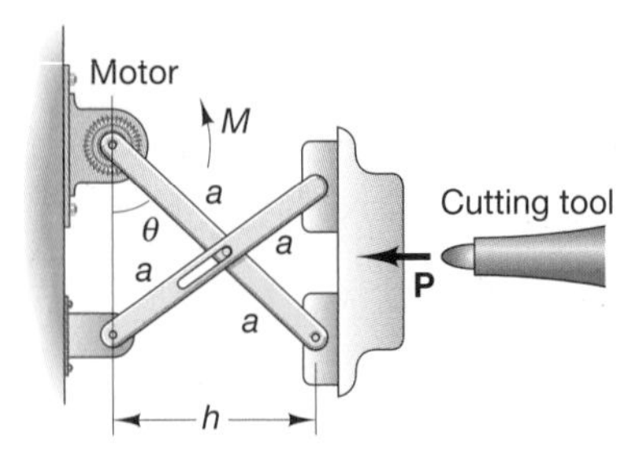

Figure P11.18

11.19 A section of an amusement-park ride is illustrated in Figure P11.19. Compute the moment M required of the motor at point C to hold the passenger in the position shown as a function of the mass m, g, R, and θ. Ignore the mass of the components.

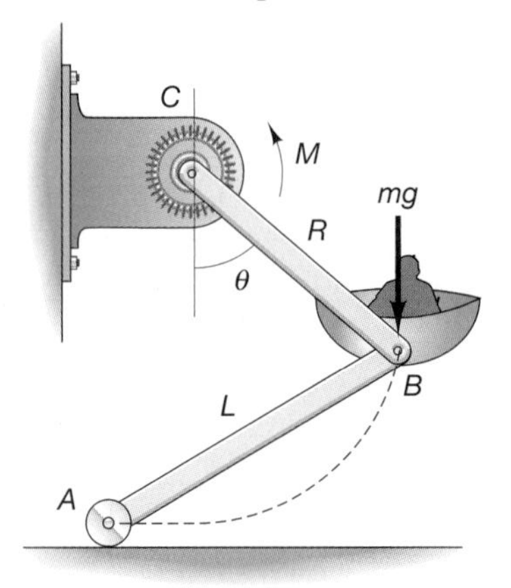

Figure P11.19

11.20 Repeat Problem 11.19, but this time include the weight of each member. Denote the mass of link CB as M_R and that of link AB as M_L. The center of gravity of each link is at its geometric center. Compute the required torque from the motor in terms of m, g, M_R, M_L, θ, and R.

11.21 Compute the weight (W) required to hold the four uniform, 100 N bars shown in Figure P11.21 at the angle θ.

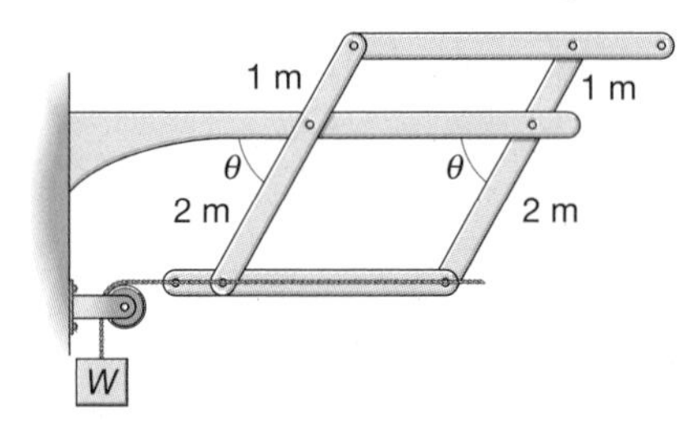

Figure P11.21

11.22 Compute the angles made by each link illustrated in Figure P11.22 in terms of L_1, L_2, W_1, W_2, W_B, W_C, and the force F. Assume that the weights of the links act at their geometric centers.

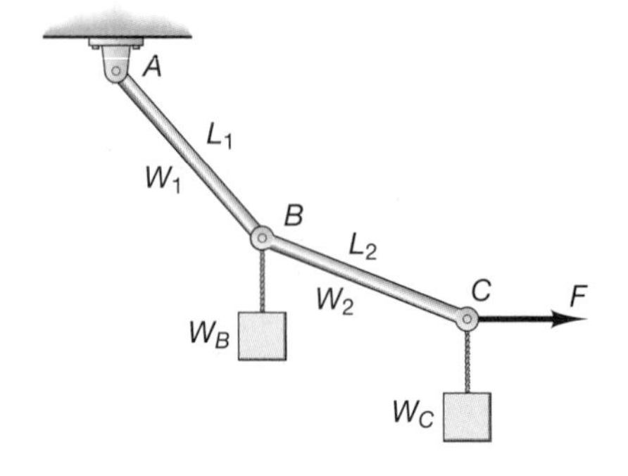

Figure P11.22

11.23 Compute the angles made by each link if a force of 100 N is applied horizontally to the end of the double pendulum shown in Figure P11.23. Each link has a weight force of 20 N acting at its geometric center.

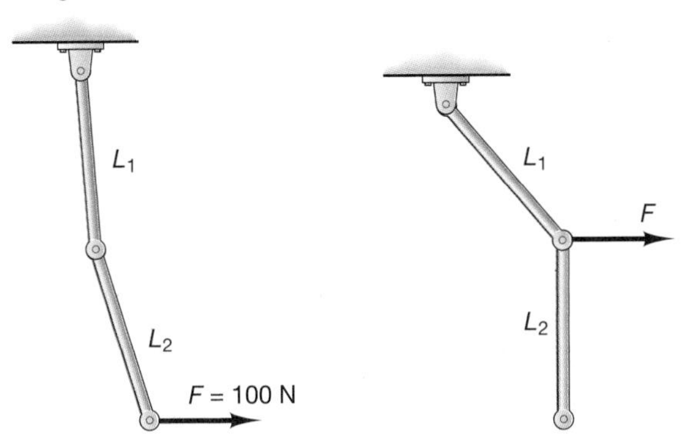

Figure P11.23 **Figure P11.24**

11.24 Compute the angles made by each link if a force of F is applied horizontally to the middle of the double pendulum shown in Figure P11.24. The links have a weight force of W_1 and W_2 acting at their geometric centers. Show that the lower link remains vertical.

11.25 Compute the angles made by each link if a force of 100 N is applied horizontally to the middle of the double pendulum of Figure P11.24. Each link has a weight force of 20 N acting at its geometric center.

11.26 In Figure P11.26, compute the angles made by each link in terms of L_1, L_2, L_3, W_1, W_2, W_3, and the force F. Assume that the weights of the links act at their geometric centers.

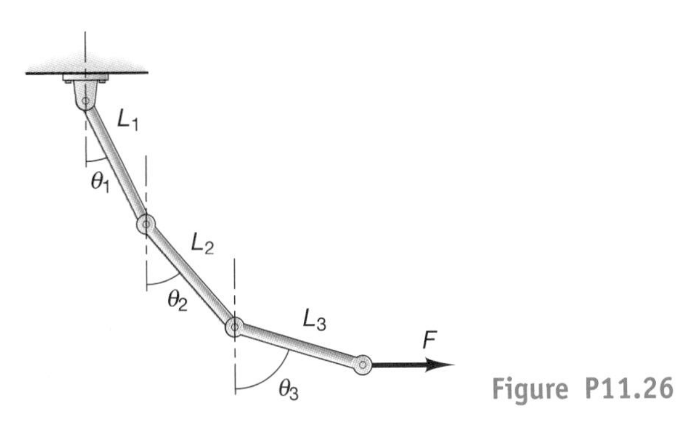

Figure P11.26

11.27 Compute the angles made by each link of Figure P11.26 if a 10-N force is applied horizontally to the last link and each link has a mass of 10 kg.

11.28 If a 10-N force is applied horizontally to the end of link 1 in Figure P11.26, what angles will result? (Each link has a mass of 10 kg.)

11.4 FINITE WORK OF A FORCE AND MOMENT

In Section 11.1, we found that the differential work done by a force moving through a differential change in the position vector was given by

$$dU = \mathbf{F} \cdot \mathbf{dr} \tag{11.17}$$

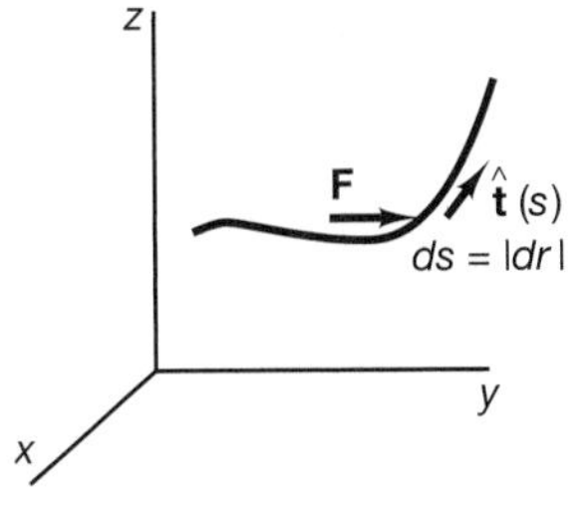

Figure 11.9

A curve in space can be described by a parameter along its length s, as shown in Figure 11.9. Both two-dimensional and three-dimensional curves may be expressed in terms of a parameter. (See Section 5.3.) The change in position vector can be written in terms of ds and the unit vector tangent to the curve as

$$\mathbf{dr} = \hat{\mathbf{t}}(s)ds \tag{11.18}$$

The force $F(s)$ may also be a function of the path of movement, and therefore, the work done moving from position s_1 to s_2 is

$$U_{1\to2} = \int_{s_1}^{s_2} \mathbf{F}(s) \cdot \hat{\mathbf{t}}(s)ds \tag{11.19}$$

In general, it is difficult to express the force as a function of position along a particular path in space using a unit tangent vector that also is a function of that path.

The work that a force does moving along a path in two dimensions can be computed in a general manner if the tangent along the path can be expressed mathematically as a function of the path. (See Figure 11.10.) For planar curves, this unit vector can be related to a reference x–y coordinate system as

$$\hat{\mathbf{t}}(s) = \cos\theta(s)\hat{\mathbf{i}} + \sin\theta(s)\hat{\mathbf{j}} \tag{11.20}$$

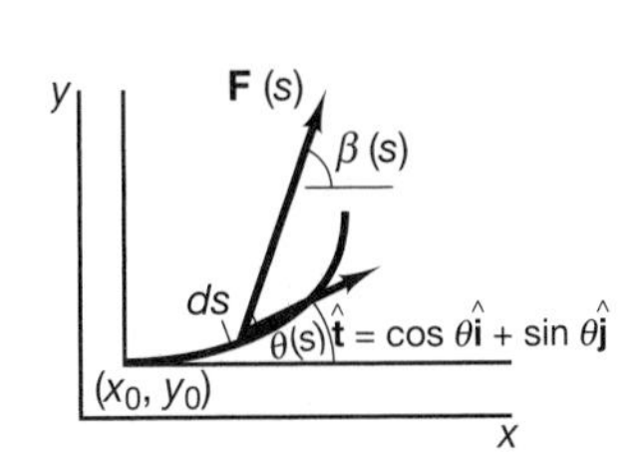

Figure 11.10

The position of the particle in the plane may be found by using the expression

$$\mathbf{r}(s) = \int_0^s \hat{\mathbf{t}}(u)du \tag{11.21}$$

The work done by a force acting on a particle moving along a path s is given by

$$U_{s_1 \to s_2} = \int_{s_1}^{s_2} \mathbf{F}(s) \cdot \hat{\mathbf{t}}(s)ds \tag{11.22}$$

where

$$\hat{\mathbf{t}}(s) = \cos\theta(s)\hat{\mathbf{i}} + \sin\theta(s)\hat{\mathbf{j}}$$
$$\mathbf{F}(s) = f(s)[\cos\beta(s)\hat{\mathbf{i}} + \sin\beta(s)\hat{\mathbf{j}}]$$

Both the force and the unit tangent vector have been expressed in a completely general form. The magnitude of the force is expressed as a function of the position s, and the unit vector for the force makes an angle $\beta(s)$ with the x-axis. Substituting into the work equation yields

$$\begin{aligned}
U_{s_1 \to s_2} &= \int_{s_1}^{s_2} \mathbf{F}(s) \cdot \hat{\mathbf{t}}(s)ds \\
U_{s_1 \to s_2} &= \int_{s_1}^{s_2} \{f(s)[\cos\beta(s)\hat{\mathbf{i}} + \sin\beta(s)\hat{\mathbf{j}}] \cdot [\cos\theta(s)\hat{\mathbf{i}} + \sin\theta(s)\hat{\mathbf{j}}]\}ds \\
U_{s_1 \to s_2} &= \int_{s_1}^{s_2} \{f(s)[\cos\beta(s)\cos\theta(s) + \sin\beta(s)\sin\theta(s)]\}ds \\
U_{s_1 \to s_2} &= \int_{s_1}^{s_2} \{f(s)\cos[\beta(s) - \theta(s)]\}ds
\end{aligned} \tag{11.23}$$

If the variation in the magnitude of the force and the angle the force vector forms with the x-axis are known, then the work done by the force moving along a path defined by $\theta(s)$ can be obtained from Eq. (11.23).

Sample Problem 11.2

Determine the work done by a force acting in the x-direction on a particle moving along a circular path of radius R from 0 to 90° when

(a) The magnitude of the force is constant

(b) The magnitude of the force is given by θ; $F(\theta) = F\sin\theta$

Solution

The vector tangent to the path is $\hat{\mathbf{t}} = \hat{\mathbf{i}}_\theta = -\sin\theta\hat{\mathbf{i}} + \cos\theta\hat{\mathbf{j}}$. The force in both (a) and (b) can be written as

$$\begin{aligned}
\mathbf{F}(\theta) &= |\mathbf{F}|f(\theta)\hat{\mathbf{i}} \\
\text{a.}\ f(\theta) &= 1 \\
\text{b.}\ f(\theta) &= \sin\theta
\end{aligned}$$

The work done by the force is

$$U_{0 \to 90^\circ} = F\int_0^{\frac{\pi}{2}} -f(\theta)\sin\theta R d\theta$$

Thus, we have:

$$a.\ U_{0\to 90^\circ} = FR[\cos\theta]_0^{\frac{\pi}{2}} = -FR$$

$$b.\ U_{0\to 90^\circ} = -FR\left[\frac{\theta}{2} - \frac{\sin 2\theta}{4}\right]_0^{\frac{\pi}{2}} = -\frac{FR\pi}{2}$$

Sample Problem 11.3

Consider a path defined by $\hat{\mathbf{t}}[\theta(s)]$, where $\theta(s) = 1 + s^2$. It can be shown that this path is a spiral in the plane. Plot the path of the particle in the plane and determine the work done by a constant force in the x-direction of 50 N as the particle moves 4 meters along the path.

Solution The curve in space can be obtained from Eq. (11.21). The components of the position vector at any value of s are

$$x(s) = \int_0^s \cos(1 + \varsigma^2)\delta\varsigma$$

$$y(s) = \int_0^s \sin(1 + \varsigma^2)d\varsigma$$

These integrals have to be evaluated numerically in terms of Fresnel integrals. A plot of the curve in $x-y$ space is a spiral. (See diagram below at left.) The work done by the 50-N force as it moves 4 meters along the spiral path is obtained from Eq. (11.23). Since the force is constant in the x-direction, β is zero, and the work is given by the integral

$$U(s) = \int_0^s 50 \cos[\theta(u)]du$$

Again, this integral must be evaluated numerically, and details are shown in the Computational Supplement. A graph of the total work done as a function of s is shown below at right.

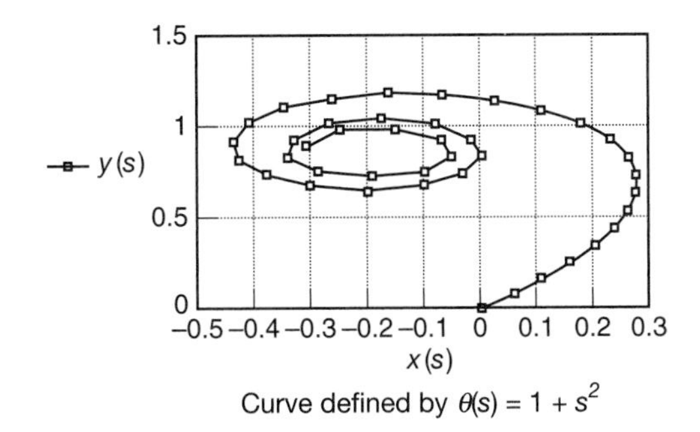

Curve defined by $\theta(s) = 1 + s^2$

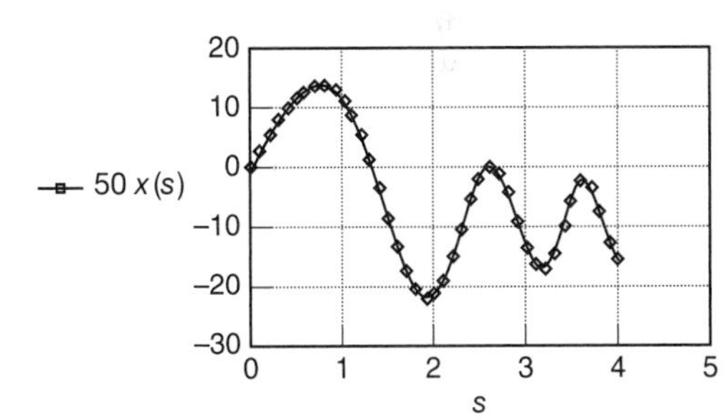

Total work as a function of position

Sample Problem 11.4

Consider a child of 20-kg mass sliding down a slide in a playground with a coefficient of friction of 0.3. The equation for the slide path can be considered to be parabolic in nature; the initial angle of the slide is −60°, and the final angle is 0°. If the length along the slide is taken to be 1 unit, the total work can be determined. *Note that, since the child, modeled as a particle, is moving along a curved path in space, the normal force would also contain an inertial term equal to the product of the mass and the velocity squared, divided by the radius of curvature of the path in space. The problem becomes nonlinear if this inertial term is included; the problem then becomes one of dynamics, not statics. Here, we consider the inertial term negligible, and the normal force will balance the gravitational attraction.*

Solution The angle of the slide can be written as

$$\theta(s) = -60(1 - s^n)$$

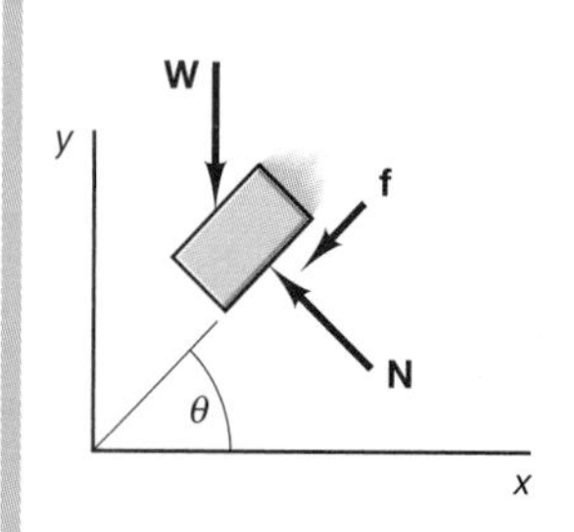

where n will determine the shape of the parabola. We draw a free-body diagram of the child modeled as a particle. (See figure at left.) It is best to sketch the angle θ as positive and let the equation take care of the possibility that it is negative. In this manner, sign errors can be avoided. The vectors are

$$\hat{\mathbf{t}} = \cos\theta(s)\hat{\mathbf{i}} + \sin\theta(s)\hat{\mathbf{j}}$$
$$\overline{\mathbf{W}} = -W\hat{\mathbf{j}}$$
$$\mathbf{N} = W\cos\theta(s)[-\sin\theta(s)\hat{\mathbf{i}} + \cos\theta(s)\hat{\mathbf{j}}]$$
$$\bar{\mathbf{f}} = \mu W\cos\theta(\text{s})[-\cos\theta(s)\hat{\mathbf{i}} - \sin\theta(s)\hat{\mathbf{j}}]$$

The integrand of the work integral is

$$\overline{\mathbf{F}}(s)\cdot\hat{\mathbf{t}}(s) = -W\{\sin\theta(s) + \mu\cos\theta(s)\}$$

The work can be obtained by numerical integration for any value of s. (See the Computational Supplement for details.)

Problems

11.29 Determine the work done for the system of Sample Problem 11.2 for the case where $f(\theta) = 10\cos(\theta)$ N and $R = 2$ m.

11.30 Determine the work done for the system of Sample Problem 11.2 for the case where $f(\theta) = F\theta\,\hat{\mathbf{i}}$ N and $R = 1$ m.

11.31 In Figure P11.31, determine the work done moving from A to B by the force F acting in the negative x-direction for the case where $F(\theta) = 10$ N and $R = 2$ m.

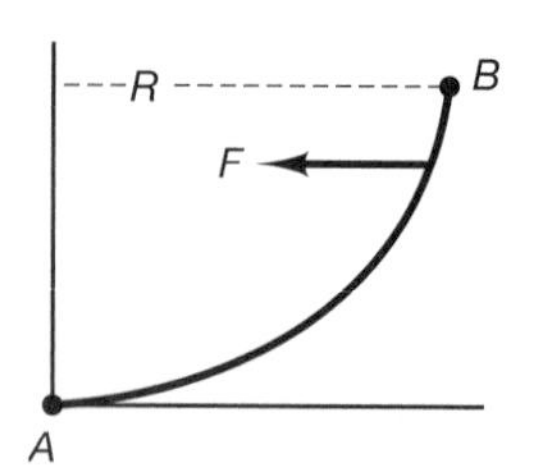

Figure P11.31

11.32 In Figure P11.31, determine the work done by the force F acting in the x-direction for the case where $F(\theta) = 5\sin(\theta)$ N and $R = 2$ m.

11.33 In Figure P11.31, determine the work done by the force F acting in the x-direction for the case where $F(\theta) = 2\,\theta$ N and $R = 5$ m.

11.34 A 32-kg child slides down a slide shaped as a quarter-section of a circle. (See Figure P11.34.) (a) Using Eq. (11.19), compute the work done by the force of gravity. Assume no friction. (b) Compute the weight of the child times the height.

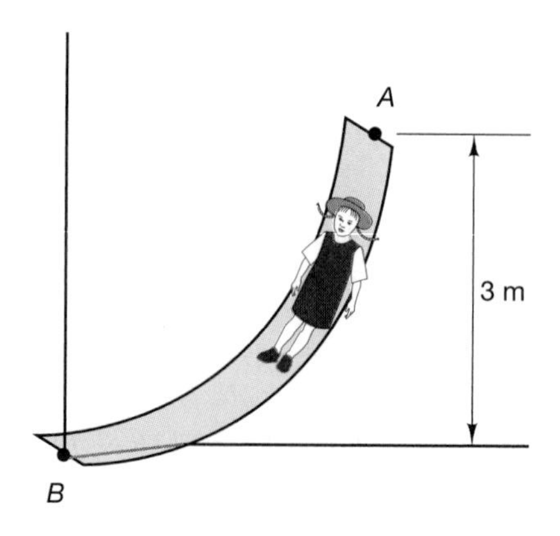

Figure P11.34

11.5 CONSERVATIVE FORCES AND POTENTIAL ENERGY

In Section 11.4, the work done by a force acting on a particle was determined using the component of the force tangent to the path of movement. Certain forces depend upon only their position in space, and the work done by these forces is independent of the actual path of movement, depending only on the initial and final positions of the particle. For these forces, if the particle upon which they act is moved along a closed path (having the same initial and final positions), the work done by the forces is zero:

$$\oint_c \mathbf{F} \cdot \mathbf{dr} = 0 \tag{11.24}$$

Forces that satisfy the preceding conditions are called ***conservative forces***. If the work depends on only the initial and final positions, then the integrand must be an exact differential of some function; that is,

$$\mathbf{F} \cdot \mathbf{dr} = -dV \tag{11.25}$$

The minus sign is introduced to simplify later equations. The function V is called a *potential function.* An example of a conservative force was given in Sample Problem 11.2. When the constant force acting in the x-direction was moved in a circular path, the work done was

$$U_{1\rightarrow 2} = \int_{s_1}^{s_2} \mathbf{F}(s) \cdot \hat{\mathbf{t}}(s) ds = F_x \int_{\theta_1}^{\theta_2} [-R \sin \theta d\, \theta]$$

$$U_{1\rightarrow 2} = F_x \int_{\theta_1}^{\theta_2} d(R\cos\theta) = F_x \int_{x_1}^{x_2} dx = F_x(x_2 - x_1) \tag{11.26}$$

Also, note that

$$\oint_c d(R \cos \theta) = R \cos \theta \Big|_{\theta_1}^{\theta_1 + 2\pi} = 0$$

The potential function for a conservative force can be obtained by observing that

$$dV = \frac{\partial V}{\partial x} dx + \frac{\partial V}{\partial y} dy + \frac{\partial V}{\partial z} dz \tag{11.27}$$

The change in the position vector and the force vector can be respectively written as

$$\mathbf{dr} = dx\hat{\mathbf{i}} + dy\hat{\mathbf{j}} + dz\hat{\mathbf{k}}$$

$$\mathbf{F} = F_x\hat{\mathbf{i}} + F_y\hat{\mathbf{j}} + F_z\hat{\mathbf{k}} \tag{11.28}$$

The integrand of the work integral then becomes

$$\mathbf{F} \cdot \mathbf{dr} = F_x dx + F_y dy + F_z dz = -\left(\frac{\partial V}{\partial x} dx + \frac{\partial V}{\partial y} dy + \frac{\partial V}{\partial z} dz\right)$$

Therefore,

$$F_x = -\frac{\partial V}{\partial x}$$

$$F_y = -\frac{\partial V}{\partial y} \tag{11.29}$$

$$F_z = -\frac{\partial V}{\partial z}$$

These equations may be solved for V. The work done by the force moving from point 1 to point 2 is

$$U_{1\rightarrow 2} = V_1 - V_2 \tag{11.30}$$

In vector calculus, a vector operator for the directed derivative is defined as

$$\vec{\nabla} = \hat{\mathbf{i}}\frac{\partial}{\partial x} + \hat{\mathbf{j}}\frac{\partial}{\partial y} + \hat{\mathbf{k}}\frac{\partial}{\partial z} \tag{11.31}$$

Equations (11.29) can now be written in vector form as

$$\mathbf{F} = -\vec{\nabla}V \tag{11.32}$$

The symbol $\vec{\nabla}$ is called the ***del operator***. The term $\vec{\nabla}V$ is called the ***gradient*** of V, or "grad" V, and is equal to the directed derivative of V in each of the coordinate directions. Some other properties of the del operator are given in Mathematics Window 11.1.

MATHEMATICS WINDOW 11.1

The del vector operator: $\vec{\nabla} = \hat{\mathbf{i}}\frac{\partial}{\partial x} + \hat{\mathbf{j}}\frac{\partial}{\partial y} + \hat{\mathbf{k}}\frac{\partial}{\partial z}$

The gradient of a scalar function φ(x,y,z,):

$$\vec{\nabla}\varphi = \hat{\mathbf{i}}\frac{\partial \varphi}{\partial x} + \hat{\mathbf{j}}\frac{\partial \varphi}{\partial y} + \hat{\mathbf{k}}\frac{\partial \varphi}{\partial z}$$

The divergence of a vector: $\vec{\nabla}\cdot\mathbf{u} = \frac{\partial u_x}{\partial x} + \frac{\partial u_y}{\partial y} + \frac{\partial u_z}{\partial z}$

The curl of a vector:

$$\vec{\nabla}\times\mathbf{u} = \hat{\mathbf{i}}\left(\frac{\partial u_z}{\partial y} - \frac{\partial u_y}{\partial z}\right) + \hat{\mathbf{j}}\left(\frac{\partial u_x}{\partial z} - \frac{\partial u_z}{\partial x}\right) + \hat{\mathbf{k}}\left(\frac{\partial u_y}{\partial x} - \frac{\partial u_x}{\partial y}\right)$$

A special scalar operator, the Laplacian:

$$\nabla^2 = \vec{\nabla}\cdot\vec{\nabla} = \frac{\partial^2}{\partial x^2} + \frac{\partial^2}{\partial y^2} + \frac{\partial^2}{\partial z^2}$$

It can be shown that the ***curl*** of the gradient of any function is zero; that is,

$$\vec{\nabla}\times\vec{\nabla}\varphi = 0 \tag{11.33}$$

Therefore, another definition of a conservative force is a force whose curl is zero:

$$\vec{\nabla}\times\mathbf{F} = \vec{\nabla}\times(-\vec{\nabla}V) = 0 \tag{11.34}$$

Thus, to determine whether a force is conservative, we take the curl of the force and check to see whether it is zero.

Although all the operations involving the del operator may be new to you at this time, probably no function is more common in engineering, and the operator appears in almost all branches of physics and engineering, from mechanics to electrical engineering.

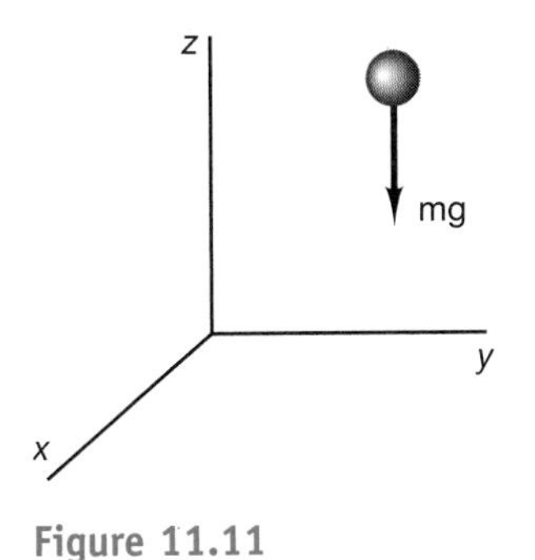

Figure 11.11

The most common conservative force is the force due to gravitational attraction, or the weight of a body. This force is illustrated in Figure 11.11 and can be written as

$$\mathbf{F} = -mg\hat{\mathbf{k}} \tag{11.35}$$

Equations (11.29) become

$$\begin{aligned} \frac{\partial V}{\partial x} &= 0 \Rightarrow V = f(y,z) \\ \frac{\partial V}{\partial y} &= 0 \Rightarrow \frac{\partial f}{\partial y} = 0 \Rightarrow V = f(z) \\ \frac{\partial V}{\partial z} &= mg \Rightarrow \frac{df}{dz} = mg \\ V &= f = mgz + C \end{aligned} \tag{11.36}$$

where C is a constant of integration.

V is usually written as V_g and is called the gravitational potential. V_g is the ***potential energy*** of a body due to the gravitational attraction force. As the body is raised, the potential energy increases, and the body has the capability to perform positive work when released. The gravitational force is equal to the negative of the gradient of the potential function, and therefore, the work it performs is independent of the constant C in Eq. (11.36). The reference datum plane (the $x-y$ plane) or reference height can be arbitrarily chosen.

The force developed in a spring when it is compressed or stretched is another example of a conservative force. Consider the spring shown in Figure 11.12. The spring force for this spring can be written as

$$\mathbf{F}(x) = -kx\hat{\mathbf{i}} \tag{11.37}$$

The work done by the spring force when the particle is moved from x_1 to x_2 is

$$U_{1\to 2} = \int_{x_1}^{x_2} -kxdx = -\frac{k}{2}(x_2^2 - x_1^2) \tag{11.38}$$

The work done by the spring force is independent of the path of motion, depending only upon the initial and final positions of the spring. (Note that the position is measured from the unstretched length of the spring.) The spring force is therefore a conservative force and can be derived from a potential function. The potential energy function is

$$V_e = \frac{1}{2}kx^2 \tag{11.39}$$

As the spring is stretched, it gains potential energy, or the potential to do work. This is the potential energy due to the elasticity of the spring. Since all elastic materials deform when forces are applied to them, these materials acquire potential energy due to the deformation.

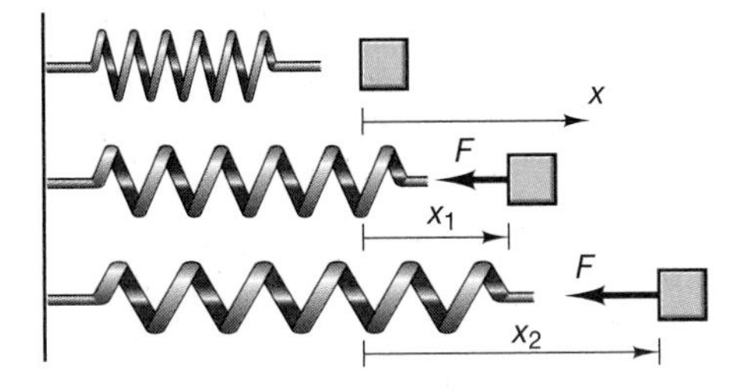

Figure 11.12

Sample Problem 11.5

Determine the total potential energy of a weight supported by a spring as a function of its position relative to the unstretched length of the spring. (See diagram below.)

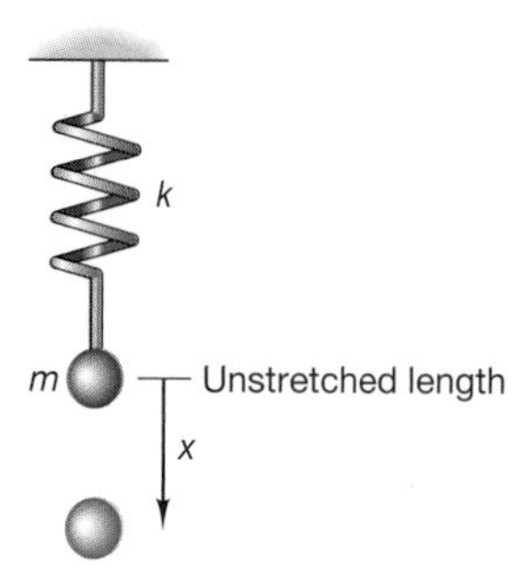

Solution The total potential energy is the sum of the gravitational potential energy and the spring deformation potential energy:

$$V = V_g + V_s = -mgx + (1/2)\,kx^2$$

11.6 POTENTIAL ENERGY AND EQUILIBRIUM

Using the theorem of virtual work, we note that if a body or system is in equilibrium, the virtual work of the external forces acting on the body or system is zero for any virtual displacement consistent with the constraints on the body or system. If all the external forces that do work are conservative, the virtual work and the virtual change in the potential energy are

$$\delta U = -\delta V = 0 \tag{11.40}$$

This principle applies to a system of interconnected bodies, as long as the external forces that do work are conservative and the interconnecting forces either do no work or are conservative forces. A system such as this is called a ***conservative system***. In Section 11.3, the concept of generalized coordinates was introduced, and each of the generalized coordinates is independent. *For a conservative system in equilibrium, the derivative of the potential energy is zero.* Therefore, the condition of equilibrium for a conservative system is

$$\delta V = \sum_i \frac{\partial V}{\partial q_i}\delta q_i = 0 \tag{11.41}$$

Since the generalized coordinates are independent, the condition for equilibrium of a conservative system can be written as

$$\frac{\partial V}{\partial q_i} = 0 \tag{11.42}$$

Equation (11.42) can be used to determine positions for the equilibrium of a conservative system.

Sample Problem 11.6

Consider the double pendulum shown at the top of the next page in the reproduction of Figure 11.7. The masses of the two links are m_1 and m_2. Determine positions of equilibrium for the system.

Solution If a zero reference plane for the potential energy is selected when the pendulum is hanging down vertically, the potential energy of the system for any position is

$$V = m_1 g\frac{\ell_1}{2}(1 - \cos\theta_1) + m_2 g\left[\ell_1(1 - \cos\theta_1) + \frac{\ell_2}{2}(1 - \cos\theta_2)\right]$$

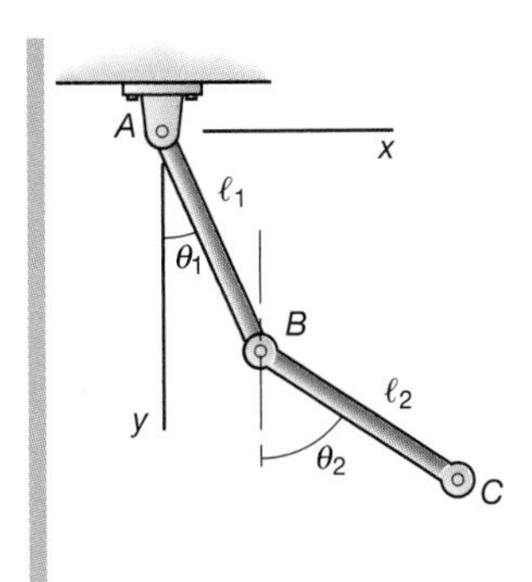

For equilibrium:

$$\frac{\partial V}{\partial \theta_1} = \left[g\ell_1\left(\frac{m_1}{2} + m_2\right)\right]\sin\theta_1 = 0$$

$$\frac{\partial V}{\partial \theta_2} = m_2 g\frac{\ell_2}{2}\sin\theta_2 = 0$$

Therefore, the system is in equilibrium when

$$\theta_1 = 0 \text{ or } 180°$$

$$\theta_2 = 0 \text{ or } 180°$$

11.7 STABILITY OF EQUILIBRIUM

In Sample Problem 11.6, there is a fundamental difference in the equilibrium positions. If both θ_1 and θ_2 are zero and the double pendulum is displaced slightly from the equilibrium position and released, the pendulum will return to the equilibrium position. This equilibrium position is said to be stable, and the system is said to be in a state of ***stable equilibrium***. If, on the other hand, either of the angles is 180° and the system is perturbed, the system will not return to the initial position, and these ***equilibrium*** positions would be ***unstable***.

If the derivative of the function of a single variable is zero at a point, then the value of the function is a ***relative extremum***—that is, a relative maximum or minimum. This property is used to determine a "peak" or a "valley" in the graph of that function. If $f(x)$ is a relative maximum at a point c, then

$$f(c) \geqslant f(x) \text{ for all } x \text{ in an open interval containing } c \tag{11.43}$$

If $f(x)$ is a relative minimum at a point c, then

$$f(c) \leqslant f(x) \text{ for all } x \text{ in an open interval containing } c \tag{11.44}$$

For a function of a single variable, the second-derivative test for a relative extremum is given in Mathematics Window 11.2.

MATHEMATICS WINDOW 11.2

Let $f(x)$ be a function such that $f'(c) = 0$ and the second derivative exists on an open interval containing c.

If $f''(c) > 0$, there is a relative minimum at $x = c$.

If $f''(c) < 0$, there is a relative maximum at $x = c$. (11.45)

If $f''(c) = 0$, then the second derivative test fails and gives no information.

Using this theorem from the calculus of a single variable, we can show that if the potential energy is a function of a single generalized coordinate, then if

$$\frac{dV}{dq} = 0 \quad \text{and} \quad \frac{d^2V}{dq^2} > 0, \text{ we have a position of stable equilibrium} \tag{11.46a}$$

If, however,

$$\frac{dV}{dq} = 0 \quad \text{and} \quad \frac{d^2V}{dq^2} < 0, \text{ we have a position of unstable equilibrium} \quad (11.46b)$$

If both the first and second derivatives are zero, it is necessary to examine higher derivatives to determine whether the position of equilibrium is stable. If all the derivatives of V are zero, the position of equilibrium is said to be a ***neutral position of equilibrium***, and the value of potential energy does not change when the system is perturbed. Stable equilibrium corresponds to a state of minimum energy, and the system will always tend to return to this state.

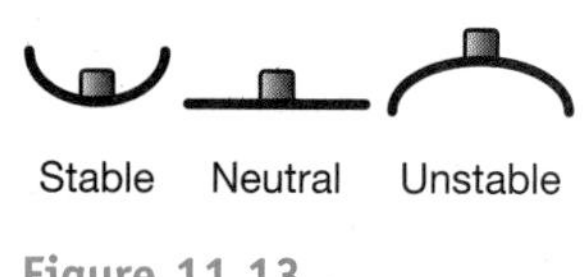

Figure 11.13

The three states of equilibrium are illustrated in Figure 11.13. If the system has more than one degree of freedom and, therefore, the potential energy is a function of more than one generalized coordinate, the test for stable, unstable, and neutral equilibrium is more complex. Theorems of multivariable calculus must be used in this case. These theorems, given without proof, are as follows:

Theorem 1.

Let $f(x,y)$ be defined on a domain that includes the point (a,b) and all points within some circle centered at (a,b). If $f(x,y)$ has a relative extremum at (a,b) and the partial derivatives of $f(x,y)$ with respect to x and y exist, then these partial derivatives are both zero at (a,b). That is,

$$\frac{\partial f(a,b)}{\partial x} = 0 = \frac{\partial f(a,b)}{\partial y} \quad (11.47)$$

Points where the first derivatives are zero are called critical points.

It might be expected that if the second partial derivatives of $f(x,y)$ were both positive at (a,b), the function would be a relative minimum at (a,b) and, conversely, if the two second partial derivatives of $f(x,y)$ were negative at (a,b), the function would be a relative maximum at (a,b). Examination of the function for a relative extremum is not this simple, however, as is shown in Sample Problem 11.7.

Theorem 2.

The second-partial-derivative test for $f(x,y)$. The partial derivatives will be represented by

$$f_x = \frac{\partial f}{\partial x}, \quad f_y = \frac{\partial f}{\partial y}, \quad f_{xx} = \frac{\partial^2 f}{\partial x^2}, \quad f_{yy} = \frac{\partial^2 f}{\partial y^2}, \quad f_{xy} = \frac{\partial^2 f}{\partial x \partial y}$$

Let (a,b) be a critical point of $f(a,b)$. Assume that the partial derivatives are continuous at and near (a,b). Let

$D = f_{xx}(a,b)f_{yy}(a,b) - [f_{xy}(a,b)]^2$

If $D > 0$ and $f_{xx}(a,b) > 0$ or $f_{yy}(a,b) > 0$, *then $f(a,b)$ is a relative minimum.*

If $D > 0$ and $f_{xx}(a,b) < 0$ or $f_{yy}(a,b) < 0$, *then $f(a,b)$ is a relative maximum.*

If $D < 0$, then $f(x,y)$ *has neither a relative minimum nor a relative maximum at (a,b). {There is a saddle point at (a,b).}*

The term D is called the *discriminant* of $f(x,y)$. It is easier to express the fact that the discriminant must be greater than zero in the following manner:

$$[f_{xy}(a,b)]^2 < f_{xx}(a,b)f_{yy}(a,b) \quad (11.48)$$

For the function to have an extremum at the point, the mixed partial derivative must not be too large. If the discriminant is zero, the test of the second derivative fails and gives no information.

Sample Problem 11.7

Find the critical points of $f(x,y) = x^2 + 3xy + y^2$, and determine whether there is an extremum at these critical points.

Solution Taking the partial derivatives of $f(x,y)$ with respect to x and y and setting them equal to zero yields

$$2x + 3y = 0$$

$$3x + 2y = 0$$

The only solution is $(x,y) = (0,0)$.

A surface plot of $f(x,y)$ in the region of the origin shows the general shape of the surface. (See diagram below.) Although the second partial derivative of $f(x,y)$ with respect to either x or y is positive, the surface plot shows that the function has a saddle at the point (0,0) and is not a minimum. The discriminant is constant and negative for this function $[D = (2 \star 2) - (3)^2 = -5]$, and the point (0,0) is a saddle point.

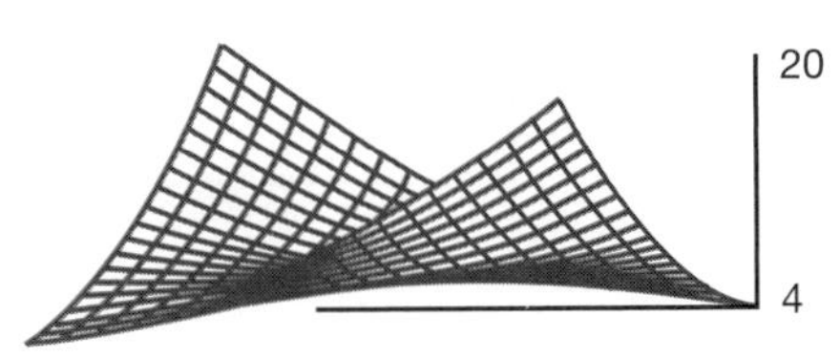

Surface plot of $f(x,y) = x^2 + 3xy + y^2$

Problems

11.35 A mass is suspended from a spring. (See Figure P11.35.) (a) Compute and plot the potential energy of the system for x between zero and 0.2 m. (b) Show that the minimum value of the potential occurs at the rest or equilibrium position of the system. Choose $V = 0$ at $x = 0$ to correspond to the unstretched position of the spring, as illustrated in the figure.

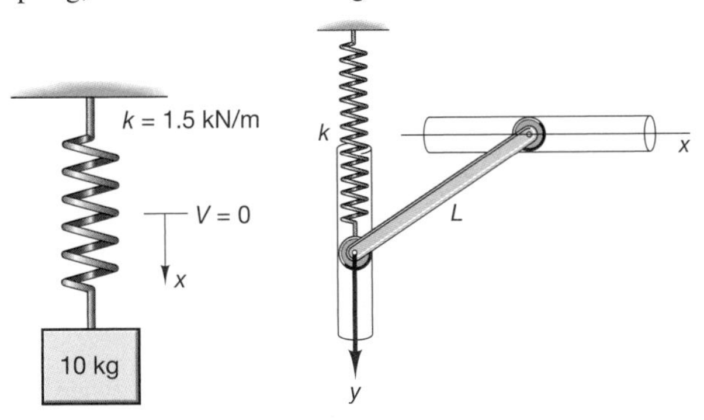

Figure P11.35 **Figure P11.36**

11.36 Determine the potential energy of the spring-and-link system shown in Figure P11.36. The ends of the link are on rollers, and the spring is stretched by the weight of the link. The unstretched position of the spring is at the line $y = 0$. Use m to denote the mass of the link and k to denote the stiffness of the spring.

11.37 Determine the equilibrium positions for the system of Figure P11.36 and decide whether or not they are stable.

11.38 A 100-kg mass is supported by two links, each 0.5 m long, as shown in Figure P11.38. Determine the relationship between the spring stiffness k and the angle the link makes for the system to be in equilibrium. The unstretched position of the spring is when both links are horizontal. Assume that the mass of the links is negligible, and plot the stiffness versus the angle. What happens near $\theta = 0°$?

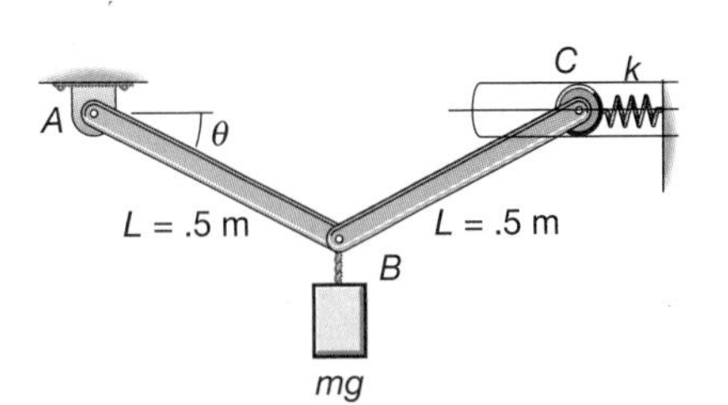

Figure P11.38

11.39 Consider the system of Figure P11.38 as described in Problem 11.38. Rather than neglecting the weight of the two links, assume each length to have a mass of 10 kg, and determine the relationship between the spring constant k and the angle for the system to be in equilibrium. Determine the position and the stability of the equilibrium if the spring has stiffness $k = 10{,}000$ N/m.

11.40 A 150-kg mass is placed on the lever mechanism illustrated in Figure P11.40. Compute the angle the lever AB makes with the x-axis in equilibrium, and show that this position is stable. Assume that the unstretched position of the spring is that shown in the figure, and neglect the mass of the lever.

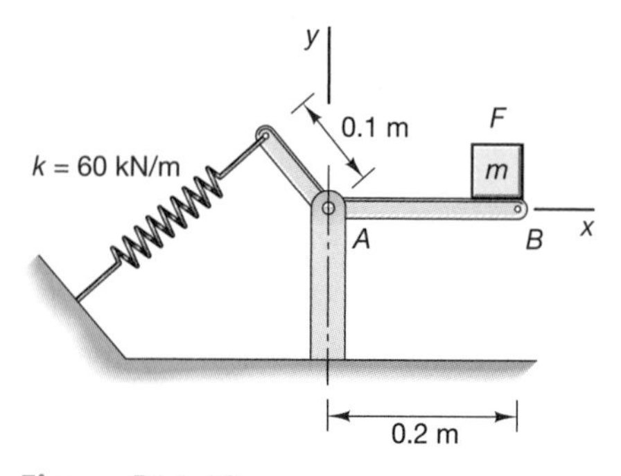

Figure P11.40

11.41 A mass is hung from a rope-and-pulley system to deflect a spring. (See Figure P11.41.) Determine the distance at which the mass will hang in equilibrium in terms of the spring stiffness and the weight of the mass.

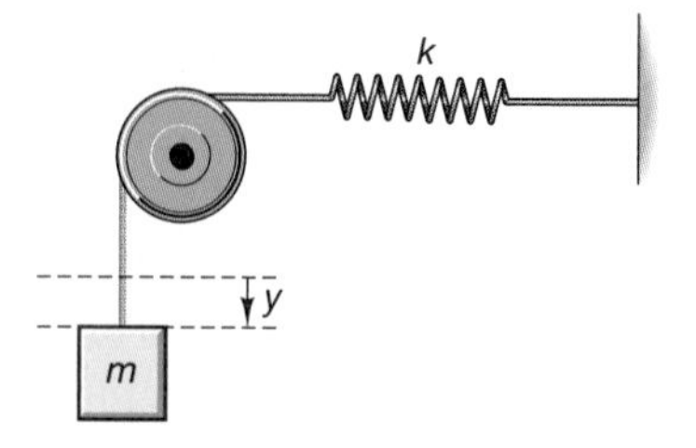

Figure P11.41

11.42 The device shown in Figure P11.42 is held in place by a spring. Assume that the unstretched position of the spring corresponds to the lower link being in the horizontal position and that the mass of the links may be neglected. Compute the angle(s) between 0 and 360° for the system to be in equilibrium, and discuss the stability of each angle for the case where $a = 0.3$ m, $m = 125$ kg, and $k = 1$ kN.

11.43 Suppose that it is desired to design the system of Figure P11.42 such that the mass hangs in a prescribed position. This amounts to choosing the value ofthe spring constant k so that equilibrium is satisfied. Calculate k so that the mass hangs 0.2 m down from the spring.

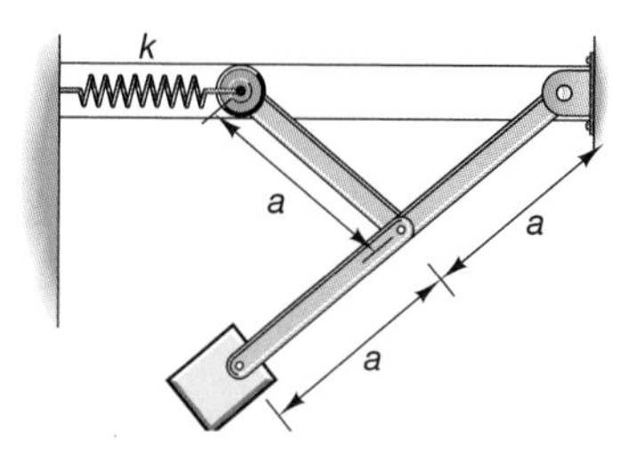

Figure P11.42

11.44 A small balance system consists of a cup and a lever attached to a spring, as illustrated in Figure P11.44. If the cup has a mass of 10 kg and the lever has a mass of 5 kg acting at its center, compute the spring constant required to keep the system in equilibrium at the horizontal position shown in the figure. Assume that the unstretched position of the spring is 0.01 m below the line corresponding to $\theta = 0°$.

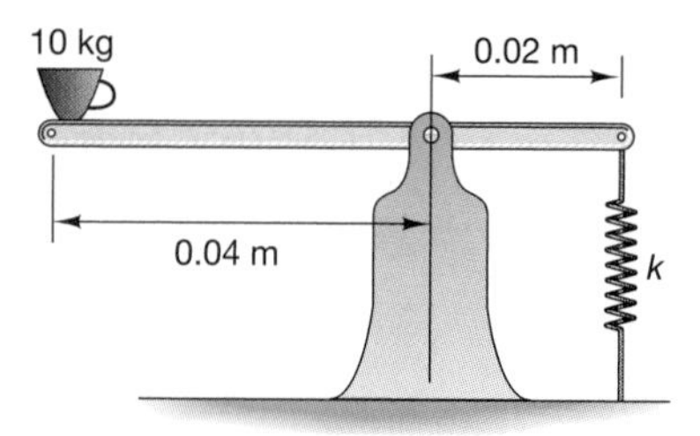

Figure P11.44

11.45 The device shown in Figure P11.45 is held in place by two springs. Assume that the unstretched position of the springs corresponds to the link being in the vertical position and that the mass of the link may be neglected. Compute the angle necessary for the system to be in equilibrium, and discuss the stability for the case where $a = 0.3$ m, $m = 125$ kg, and $k = 1$ kN.

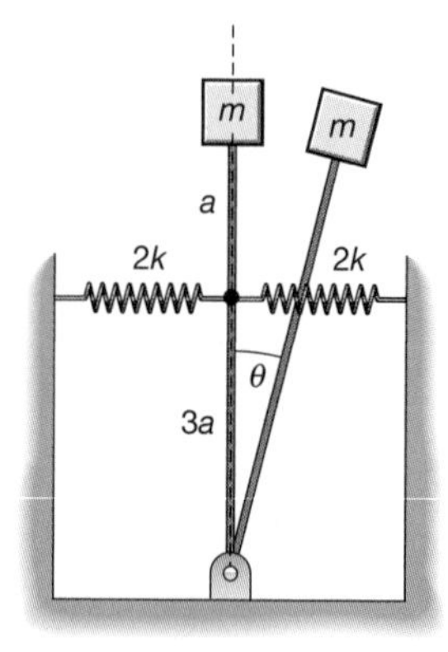

Figure P11.45

11.46 In Figure P11.46, consider the effect of where the spring is attached to the inverted pendulum on the equilibrium position by plotting the angle θ corresponding to the equilibrium position as a function of the parameter α, where α varies between 0 and 1. Set the constant

$(mg/kL) = 10^{-4}$, and assume that the unstretched position of the spring is at $\theta = 0°$. Can $\alpha = 0$?

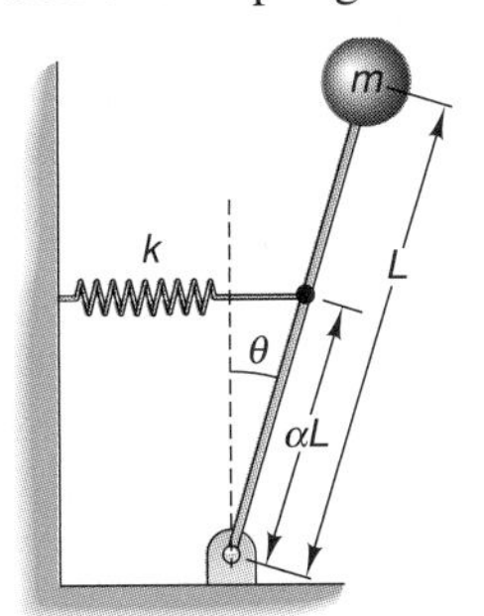

Figure P11.46

11.47 Determine the angle θ in Problem 11.46 for $\alpha = 0.8$ m, $L = 1$ m, $m = 1000$ kg, and $k = 20$ kN/m. Will the spring hold the mass and rod up with a stiffness of 1 kN/m?

11.48 Compute the distance h corresponding to equilibrium of the system shown in Figure P11.48. Here, $m = 100$ kg, $k = 10$ kN, and the unstretched position of the spring corresponds to the link being in the horizontal position. Show that this position is, in fact, stable.

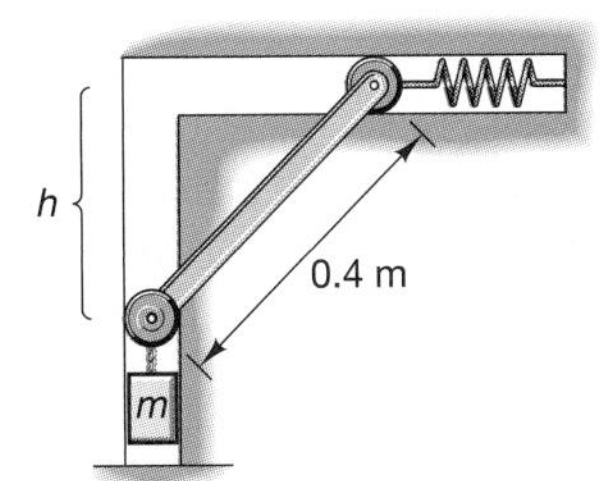

Figure P11.48

11.49 Two rods of equal length L are connected to two equal springs of stiffness k that support a weight W. (See Figure P11.49.) Assume that the spring forces remain horizontal, calculate the equilibrium positions as a function of L, k, and W, and discuss the stability of each position. The unstretched position of the spring corresponds to the link being in the vertical position. For what values of these parameters is the system stable? (Assume the springs remain horizontal.)

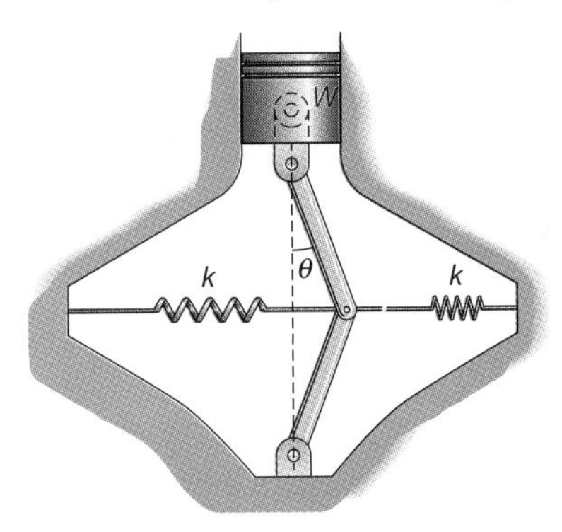

Figure P11.49

11.50 Repeat Problem 11.49, but include the weight of the rods ($0.25\ W$).

11.51 Compute the equilibrium positions for the device shown in Figure P11.51, and discuss the stability of each position. Assume that the unstretched position of the spring is at $\theta = 0°$ and that the center of mass of the link is at its geometric center.

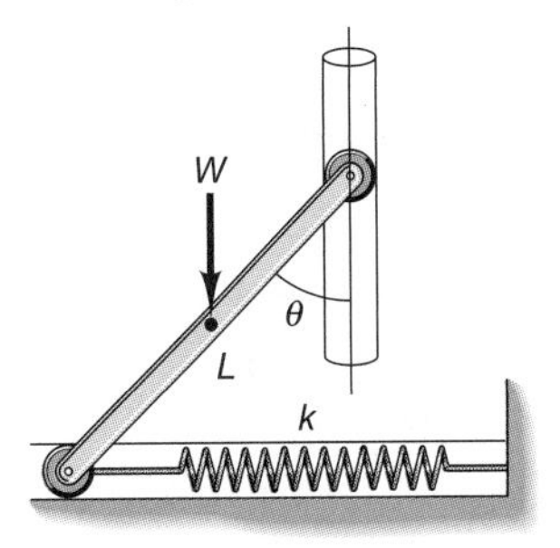

Figure P11.51

11.52 Repeat Problem 11.51 for the case where the unstretched position of the spring is at $\theta = 20°$, and $W = 100$ N, $L = 1$, and $k = 200$ N/m.

11.53 A force $\mathbf{F} = x^2\hat{\mathbf{i}} + xy\hat{\mathbf{j}} + g\hat{\mathbf{k}}$, where g is a constant, is applied to a particle. Is this force conservative?

11.54 A force $\mathbf{F} = x^2\hat{\mathbf{i}} + y^2\hat{\mathbf{j}} + z^2\hat{\mathbf{k}}$ is applied to a particle. Is this force conservative?

11.55 A force $\mathbf{F} = f_1(x)\hat{\mathbf{i}} + f_2(y)\hat{\mathbf{j}} + f_3(z)\hat{\mathbf{k}}$ is applied to a particle. Is this force conservative?

11.56 A force $\mathbf{F} = x^2\hat{\mathbf{i}} + z^2\hat{\mathbf{j}} + 2yz\hat{\mathbf{k}}$ is applied to a particle. Is this force conservative?

11.57 A potential energy function for a particle is determined to be $V = x^2 + y^2 + z^2$. Calculate the force that caused this potential.

11.58 A potential energy function for a particle is determined to be $V = yx^2 + zy^2 + xz^2$. Calculate the force that caused this potential.

11.59 A potential energy function for a particle is determined to be $V = (x - 1)^2 + y^2 + z^2$. Calculate the force that caused this potential. Is this a conservative force?

11.60 A potential energy function for a particle is determined to be $V(x, z) = (1 - x)^2 + \sin z$. Calculate the force that caused this potential, and determine whether it is a conservative force. If the force is conservative, compute the equilibrium positions and discuss the stability of each.

Chapter Summary

The differential work done by a force F undergoing a displacement $\mathbf{dr}$ is

$$dU = \mathbf{F} \cdot \mathbf{dr}$$

The work done by a moment rotated through a differential angle $\mathbf{d}\theta$ is

$$dU = \mathbf{M} \cdot \mathbf{d}\,\theta$$

The differential virtual displacement, written in component notation, is

$$\delta \mathbf{r} = \delta x \hat{\mathbf{i}} + \delta y \hat{\mathbf{j}} + \delta z \hat{\mathbf{k}}$$

and the virtual work is

$$\delta U = \mathbf{F} \cdot \delta \mathbf{r} + \mathbf{M} \cdot \delta \boldsymbol{\theta}$$

For a body or system of bodies in equilibrium, the virtual work associated with any displacement is zero.

The principle of virtual work depends upon the specification of a set of independent, or generalized, coordinates q_i that are equal in number to the number of degrees of freedom of the system. The number of degrees of freedom of a system of n rigid bodies is equal to

$$\text{degrees of freedom} = 6\,n - m$$

where m is the number of constraints on the system.

The equations relating a set of orthogonal coordinates to the generalized coordinates are called the transformation equations. The virtual displacements are obtained from the variations in the transformation equations with respect to the generalized coordinates.

The principle of virtual work is as follows:

If a rigid body or a system of frictionless connected rigid bodies is in equilibrium, the total virtual work of the external forces acting on the body or system is zero for any virtual displacement consistent with the constraints on the body or system.

A curve in space can be described by a parameter s and a unit vector tangent to the curve as shown in Figure 11.14:

$$\hat{\mathbf{t}}(s) = \cos\theta(s)\hat{\mathbf{i}} + \sin\theta(s)\hat{\mathbf{j}}$$

The position of a particle is

$$\mathbf{r}(s) = \int_0^s \hat{\mathbf{t}}(u)du$$

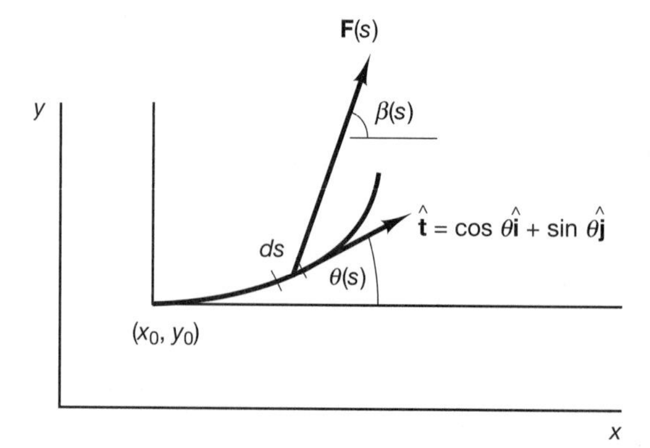

Figure 11.14

The work done by a force acting on a particle moving along a path s is

$$U_{s_1 \to s_2} = \int_{s_1}^{s_2} \mathbf{F}(s) \cdot \hat{\mathbf{t}}(s) ds$$

where

$$\hat{\mathbf{t}}(s) = \cos \theta(s)\hat{\mathbf{i}} + \sin \theta(s)\hat{\mathbf{j}}$$
$$\mathbf{F}(s) = f(s)[\cos \beta(s)\hat{\mathbf{i}} + \sin \beta(s)\hat{\mathbf{j}}]$$

Conservative Forces If the work done by a force acting on a particle moving in a closed path is zero, the force is called a conservative force. Mathematically,

$$\oint_c \mathbf{F} \cdot \mathbf{dr} = 0$$

The work done by a conservative force is independent of the actual path of movement of the particle and depends upon only the particle's initial and final positions. Conservative forces are derivable from a potential function as follows:

$$\mathbf{F} \cdot \mathbf{dr} = -dV$$
$$\mathbf{F} \cdot \mathbf{dr} = F_x dx + F_y dy + F_z dz = -\left(\frac{\partial V}{\partial x} dx + \frac{\partial V}{\partial y} dy + \frac{\partial V}{\partial z} dz\right)$$

Therefore,

$$F_x = -\frac{\partial V}{\partial x}$$
$$F_y = -\frac{\partial V}{\partial y}$$
$$F_z = -\frac{\partial V}{\partial z}$$
$$\mathbf{F} = -\vec{\nabla} V$$

Since the curl of the gradient of any function is zero, the curl of a conservative force is zero:

$$\vec{\nabla} \times \mathbf{F} = \vec{\nabla} \times (-\vec{\nabla} V) = 0$$

If all the forces that do work on a system are conservative, the virtual change in the potential energy is zero:

$$\delta V = \sum_i \frac{\partial V}{\partial q_i} \delta q_i = 0$$

The condition for equilibrium can be written as

$$\frac{\partial V}{\partial q_i} = 0$$

The equilibrium position can be stable, unstable, or neutral. The stability of a position can be determined by examining the relative extrema of the potential energy:

If $\dfrac{dV}{dq} = 0$ and $\dfrac{d^2V}{dq^2} > 0$, then we have a position of stable equilibrium

If $\dfrac{dV}{dq} = 0$ and $\dfrac{d^2V}{dq^2} < 0$, then we have a position of unstable equilibrium

Appendix SOLUTION OF SYSTEMS OF LINEAR EQUATIONS

In the text, matrix methods were examined to solve a system of linear equations. The system of equations were written in matrix notation as:

$$[A][x] = [c]$$

where $[A]$ is the coefficient matrix, $[x]$ is the unknown column matrix and $[c]$ is the matrix of the known right-hand side of each equation. The coefficient matrix is a square matrix of size $(n \times n)$ for the n equations and n unknowns. The inverse of this matrix can be determined using computational software and the solution expressed in the form:

$$[x] = [A]^{-1}[c] \qquad \text{where}[A]^{-1} \text{ is the inverse of } [A]$$

The inverse of the coefficient matrix exists unless the determinant of the coefficient matrix is zero in which case the coefficient matrix is said to be singular. When computational software is available, a matrix solution of the system of equations is the preferable method. Most other solutions are based on a form of the Gauss–Jordan reduction method.

GAUSS–JORDAN REDUCTION

Consider a system of equation in the following form.

$$\begin{aligned} a_{11}x_1 + a_{12}x_2 + \cdots\cdots + a_{1n}x_n &= c_1 \\ a_{21}x_1 + a_{22}x_2 + \cdots\cdots + a_{2n}x_n &= c_2 \\ \cdots\cdots\cdots\cdots\cdots\cdots\cdots \\ a_{m1}x_1 + a_{m2}x_2 + \cdots\cdots + a_{mn}x_n &= c_m \end{aligned}$$

The problem is: Given $a_{11}, a_{12}, \cdots a_{mn}$ and $c_1, c_2, \cdots c_m$, determine $x_1,\ x_2, \cdots x_n$. A systematic approach to the solution of such a system of equations is given by the Gauss-Jordan reduction. The Gauss–Jordan reduction proceeds as follows for the system of equations. First order the equations such that a_{11} is not zero. Divide the first equation by a_{11} so that the coefficient of the x_1 term is one. Multiply the equation by a_{21} and subtract from the second equation, thereby eliminating x_1 from this equation. Repeat this process for the balance of equations so the system is reduced to the first equation and the $m - 1$ equations which do not contain the x_1 term. Repeat the process on the second equation, thereby reducing the system such that the balance of the equations do not contain either x_1 or x_2. This process is repeated until the last equation contains only x_n and may be easily solved for this unknown.

A simple example of this solution is as follows:

$$\begin{aligned} 2x + 3y - z &= 6 \\ 3x - y + z &= 8 \\ -x + y - 2z &= 0 \end{aligned}$$

After the first reduction, the equations are

$$\begin{aligned} x + 1.5y - 0.5z &= 3 \\ -5.5y + 2.5z &= -1 \\ 2.5y - 2.5z &= 3 \end{aligned}$$

Repeating the process to eliminate y from the last equation, the reduced equations become

$$\begin{aligned} x + 1.5y - 0.5z &= 3 \\ y - 0.455z &= 0.182 \\ -1.363z &= 2.545 \end{aligned}$$

Solving for z yields: $z = -1.867$
Back substituting into the first two equations yields: $y = -0.667$ and $x = 3.067$

We have proposed in the text that matrix methods be used to solve systems of linear equations. Important to the determination as to whether a solution of this system of equations exists is the value of the determinant of the coefficient matrix. If this determinant is non-zero, the coefficient matrix is non-singular and a solution exists. The *determinant* of a square matrix is denoted by

$$|\mathbf{A}| = |A_{ij}| \tag{A.1}$$

and is a number obtained as the sum of all possible products in each of which there appears one and only one element from each row and each column, each such product being assigned a plus or minus sign according to the following rule: Let the elements involved in a given product be joined in pairs by line segments. If the total number of such segments sloping upward to the right is even, prefix a plus sign to the product. Otherwise, prefix a negative sign.

The following properties of determinants are helpful when working with systems of equations.

1. *If all elements of any row or column of a square matrix are zeros, its determinant is zero.*
2. *The value of the determinant is unchanged if the rows and columns of the matrix are interchanged.*
3. *If two rows (or columns) of a square matrix are interchanged, the sign of its determinant is changed.*
4. *If all elements of one row (or one column) of a square matrix are multiplied by a number α, the determinant is multiplied by α.*
5. *If corresponding elements of two rows (or two columns) are equal or in a constant ratio, then the determinant is zero.*
6. *If each element in one row (or one column) is expressed as the sum of two terms, then the determinant is equal to the sum of two determinants, in each of which one of the two terms is deleted in each element of that row (or column).*
7. *If to the elements of any row (column) are added α times the corresponding elements of any other row (column), the determinant is unchanged.*

The properties of determinants are useful when *Cramer's rule* is applied to the solution of a system of linear equations. Let $[\mathbf{a}]$ be a square $(n \times n)$ matrix of coefficients of n linear equations and let $[\mathbf{x}]$ be the $(n \times 1)$ matrix of unknowns. The right-hand side of the equations is written as the $(n \times 1)$matrix $[\mathbf{c}]$. The system of equations may be written as

$$[\mathbf{a}][\mathbf{x}] = [\mathbf{c}] \tag{A.2}$$

Cramer's rule states that: *If the determinant* $|\mathbf{a}|$ *is not zero, the system of linear algebraic equations has a unique solution for the* n *unknown* x*'s. The expression for any* x_j *is the ratio of two determinants, the denominator being the determinant of the matrix of coefficients* $[\mathbf{a}]$*, and the numerator being the determinant of the matrix obtained by replacing* *the jth column of coefficient matrix with the elements of the column matrix* [**c**] *of the right-hand side of the equations.*

As an example of the use of Cramer's rule, examine the system of three equations and three unknowns solved by the Gauss–Jordan reduction method earlier.

$$\begin{aligned} 2x + 3y - z &= 6 \\ 3x - y + z &= 8 \\ -x + y - 2z &= 0 \end{aligned}$$

The determinant of the coefficient matrix is

$$|\mathbf{a}| = \begin{vmatrix} 2 & 3 & -1 \\ 3 & -1 & 1 \\ -1 & 1 & -2 \end{vmatrix} = (4 - 3 - 3 + 1 - 2 + 18) = 15$$

$$x = \begin{vmatrix} 6 & 3 & -1 \\ 8 & -1 & 1 \\ 0 & 1 & -2 \end{vmatrix} /15 = (12 + 0 - 8 + 0 - 6 + 48)/15 = 46/15 = 3.067$$

In a similar manner, $y = -0.667$ and $z = -1.867$, which agree with the previous solution.

A matrix whose determinant vanishes (equal to zero) is called a *singular* matrix. Examination of Cramer's rule would indicate that if the determinant of the coefficient matrix of a system of linear equations were zero, this matrix would be singular and the system could not be solved. This usually arises when two rows of the matrix are equal or in a constant ratio and the equations are not linearly independent. Note this is property number 5 of determinants and is illustrated for (2×2) and (3×3) systems in the previous section. As the determinant of this matrix approaches zero, the equations become ill-conditioned. Certain numerical methods lead to ill-conditioned equations in their limit.

Another useful matrix manipulation in mechanics is the concept of the *transpose* of a matrix. The transpose of a $(m \times n)$ matrix will be an $(n \times m)$ matrix formed by the interchange of rows and columns. For example, the transpose of a matrix [**a**] denoted by $[\mathbf{a}]^T$ is

$$[\mathbf{a}] = \begin{bmatrix} 1 & 2 & 3 \\ 4 & 5 & 6 \end{bmatrix} \qquad [\mathbf{a}]^T = \begin{bmatrix} 1 & 4 \\ 2 & 5 \\ 3 & 6 \end{bmatrix} \tag{A.3}$$

INVERSE OF A MATRIX

Consider the (2×2) matrix [**A**] defined by

$$[\mathbf{A}] = \begin{bmatrix} a & b \\ c & d \end{bmatrix} \tag{A.4}$$

The inverse of a square matrix [**A**] is a matrix of the same dimension, denoted by $\mathbf{A}^{-1}$, such that

$$[\mathbf{A}][\mathbf{A}]^{-1} = [\mathbf{A}]^{-1}[\mathbf{A}] = [\mathbf{I}] \tag{A.5}$$

where [**I**] is the identity matrix. In this case [**I**] has the form

$$[\mathbf{I}] = \begin{bmatrix} 1 & 0 \\ 0 & 1 \end{bmatrix} \tag{A.6}$$

The inverse matrix for a general (2 × 2) matrix is

$$[\mathbf{A}]^{-1} = \frac{1}{|A|}\begin{bmatrix} d & -b \\ -c & a \end{bmatrix} \tag{A.7}$$

provided that $|A|$, the determinant of the matrix [**A**] does not equal zero. The determinant of the matrix has the value

$$|A| = ad - bc \tag{A.8}$$

To see that Equation (A.7) is in fact the inverse, note that

$$[\mathbf{A}]^{-1}[\mathbf{A}] = \frac{1}{ad - bc}\begin{bmatrix} d & -b \\ -c & d \end{bmatrix}\begin{bmatrix} a & b \\ c & c \end{bmatrix}$$

$$= \frac{1}{ad - bc}\begin{bmatrix} ad - bc & bd - bd \\ ac - ac & ad - bc \end{bmatrix} = \begin{bmatrix} 1 & 0 \\ 0 & 1 \end{bmatrix} \tag{A.9}$$

It is important to realize that the matrix [**A**] has an inverse if and only if $|A| \neq 0$. If the determinant of [**A**] is zero, [**A**] does not have an inverse and the matrix is called a singular matrix.

The determinant of a matrix [**A**] is defined by the formula

$$|A| = \sum_{j=1}^{n}(-1)^{1+j}a_{1j}\,|A_{1j}| \tag{A.10}$$

where a_{1j} is the element in the $(1, j)$ postion of the matrix [**A**] and [$\mathbf{A}_{1j}$] is the $(n - 1) \times (n - 1)$ matrix formed from the matrix [**A**] by deleting the first row and the jth column of the matrix [**A**]. Note that the determinant is a scalar.

The matrix [$\mathbf{A}_{ij}$] formed by deleting the ith row and the jth column of the matrix [**A**] is called a minor of [**A**]. Let α_{ij} denote the scalar found by taking the determinant of [$\mathbf{A}_{ij}$] with a particular sign.

$$\alpha_{ij} = (-1)^{i+j}\,|A_{ij}| \tag{A.11}$$

Then the matrix defined by forming the elements α_{ij} into a matrix is called the ***adjoint*** of [**A**] denoted by adj [**A**]. The inverse $[\mathbf{A}]^{-1}$ can be calculated from the adjoint by

$$[\mathbf{A}]^{-1} = \frac{\text{adj}[\mathbf{A}]}{|A|} \tag{A.12}$$

Note again that the matrix [**A**] does not have an inverse if its determinant is zero and the matrix would be singular. A matrix that has an inverse is called a nonsingular matrix.*

Some properties of the transpose, inverse, and determinant of a matrix are shown in the following mathematics window.

*(This discussion of the inverse of a matrix is from ***Engineering Vibration***, *Daniel J. Inman; Prentice-Hall, Inc. Englewood Cliffs, N. J. 1994.*)

MATHEMATICS WINDOW A.1:

Properties of Transpose, Inverse, and Determinants of Matrices

$$([\mathbf{a}] + [\mathbf{b}])^{\mathrm{T}} = [\mathbf{a}]^{\mathrm{T}} + [\mathbf{b}]^{\mathrm{T}}$$
$$(\alpha [\mathbf{a}])^{\mathrm{T}} = \alpha [\mathbf{a}]^{\mathrm{T}}$$
$$([\mathbf{a}][\mathbf{b}])^{\mathrm{T}} = [\mathbf{b}]^{\mathrm{T}}[\mathbf{a}]^{\mathrm{T}}$$
$$([\mathbf{a}]^{\mathrm{T}})^{\mathrm{T}} = [\mathbf{a}]$$
$$([\mathbf{a}]^{-1})^{\mathrm{T}} = ([\mathbf{a}]^{\mathrm{T}})^{-1}$$
$$([\mathbf{a}][\mathbf{b}])^{-1} = [\mathbf{b}]^{-1}[\mathbf{a}]^{-1}$$
$$\text{determinant: } \det([\mathbf{a}][\mathbf{b}]) = \det([\mathbf{a}])\det([\mathbf{b}])$$
$$\det([\mathbf{a}]^{\mathrm{T}}) = \det([\mathbf{a}])$$
$$\det(\alpha [\mathbf{a}]) = \alpha \det([\mathbf{a}]) \text{ where } \alpha \text{ is a scalar.}$$

SOLUTION OF VECTOR EQUATIONS

A problem that arises frequently in statics is the three dimensional problem where the directions of the supporting cables are known and the magnitude of the tension in each cable is unknown. This problem takes the mathematical form:

$$T_A\hat{\mathbf{a}} + T_B\hat{\mathbf{b}} + T_C\hat{\mathbf{c}} = \mathbf{P} \tag{A.13}$$

The unit vectors express the direction of the unknown tensions and **P** is the known applied load. We will show three separate solutions to this problem, each of which, involve the solution of simultaneous scalar equations that can be written in matrix notation. First let us use the traditional approach of expressing these equations in a reference Cartesian coordinate system. We will form the scalar or dot product of Eq. (A.13) with each of the unit base vectors yielding three scalar equations.

$$\begin{aligned} T_A\hat{\mathbf{a}}\cdot\hat{\mathbf{i}} + T_B\hat{\mathbf{b}}\cdot\hat{\mathbf{i}} + T_C\hat{\mathbf{c}}\cdot\hat{\mathbf{i}} &= \mathbf{P}\cdot\hat{\mathbf{i}} \\ T_A\hat{\mathbf{a}}\cdot\hat{\mathbf{j}} + T_B\hat{\mathbf{b}}\cdot\hat{\mathbf{j}} + T_C\hat{\mathbf{c}}\cdot\hat{\mathbf{j}} &= \mathbf{P}\cdot\hat{\mathbf{j}} \\ T_A\hat{\mathbf{a}}\cdot\hat{\mathbf{k}} + T_B\cdot\hat{k} + T_C\hat{\mathbf{c}}\cdot\hat{\mathbf{k}} &= \mathbf{P}\cdot\hat{\mathbf{k}} \end{aligned} \tag{A.14}$$

Eq. (A.14) can be written in the familiar matrix notation to complete the solution.

$$\begin{bmatrix} \hat{\mathbf{a}}\cdot\hat{\mathbf{i}} & \hat{\mathbf{b}}\cdot\hat{\mathbf{i}} & \hat{\mathbf{c}}\cdot\hat{\mathbf{i}} \\ \hat{\mathbf{a}}\cdot\hat{\mathbf{j}} & \hat{\mathbf{b}}\cdot\hat{\mathbf{j}} & \hat{\mathbf{c}}\cdot\hat{\mathbf{j}} \\ \hat{\mathbf{a}}\cdot\hat{\mathbf{k}} & \hat{\mathbf{b}}\cdot\hat{\mathbf{k}} & \hat{\mathbf{c}}\cdot\hat{\mathbf{k}} \end{bmatrix} \begin{bmatrix} T_A \\ T_B \\ T_C \end{bmatrix} = \begin{bmatrix} \mathbf{P}\cdot\hat{\mathbf{i}} \\ \mathbf{P}\cdot\hat{\mathbf{j}} \\ \mathbf{P}\cdot\hat{\mathbf{k}} \end{bmatrix} \tag{A.15}$$

This matrix equation can now be solved using computational software or a calculator. The advantage of generating the equations in this manner is that the chance of error is greatly reduced.

The second method is similar to the first but uses the vectors $\hat{\mathbf{a}}$, $\hat{\mathbf{b}}$, and $\hat{\mathbf{c}}$ as base vectors and forms the three scalar equations by taking the dot product of Eq. (A.13) with each of these base vectors. We also use the fact that the dot product of any unit vector with itself is equal to one.

$$\begin{aligned} T_A + T_B\hat{\mathbf{b}}\cdot\hat{\mathbf{a}} + T_C\hat{\mathbf{c}}\cdot\hat{\mathbf{a}} &= \mathbf{P}\cdot\hat{\mathbf{a}} \\ T_A\hat{\mathbf{a}}\cdot\hat{\mathbf{b}} + T_B + T_C\hat{\mathbf{c}}\cdot\hat{\mathbf{b}} &= \mathbf{P}\cdot\hat{\mathbf{b}} \\ T_A\hat{\mathbf{a}}\cdot\hat{\mathbf{c}} + T_B\hat{\mathbf{b}}\cdot\hat{\mathbf{c}} + T_C &= \mathbf{P}\cdot\hat{\mathbf{c}} \end{aligned} \tag{A.16}$$

Eq. (A.16) can now be written in matrix notation as follows:

$$\begin{bmatrix} 1 & \hat{\mathbf{b}}\cdot\hat{\mathbf{a}} & \hat{\mathbf{c}}\cdot\hat{\mathbf{a}} \\ \hat{\mathbf{a}}\cdot\hat{\mathbf{b}} & 1 & \hat{\mathbf{c}}\cdot\hat{\mathbf{b}} \\ \hat{\mathbf{a}}\cdot\hat{\mathbf{c}} & \hat{\mathbf{b}}\cdot\hat{\mathbf{c}} & 1 \end{bmatrix} \begin{bmatrix} T_A \\ T_B \\ T_C \end{bmatrix} = \begin{bmatrix} \mathbf{P}\cdot\hat{\mathbf{a}} \\ \mathbf{P}\cdot\hat{\mathbf{b}} \\ \mathbf{P}\cdot\hat{\mathbf{c}} \end{bmatrix} \tag{A.17}$$

This matrix equation has the advantage that the coefficient matrix is symmetric with diagonal elements equal to one. A review of the matrix literature will show that these equations are easier to solve numerically as the diagonal elements are all equal to one.

The third approach to the solution of Eq. (A.13) will develop a second set of base vectors that are biorthogonal to the original base vectors $\hat{\mathbf{a}}$, $\hat{\mathbf{b}}$, and $\hat{\mathbf{c}}$. Realize that the original base vectors are not orthogonal and this required the solution in the form of a system of scalar equations. The process we will use now is a modified Gram-Schmidt orthogonalization developed by the Danish mathematician Jorgen Pederson Gram. We will determine a new set of base vectors $\hat{\mathbf{r}}$, $\hat{\mathbf{s}}$, and $\hat{\mathbf{t}}$ each of which will be orthogonal to two of the original base vectors. This new set of independent base vectors will not be orthogonal to each other but will be biorthogonal with the original set. The new set of base vectors will be formed using cross products of the original base vectors as follows:

$$\begin{aligned} \hat{\mathbf{r}} &= \frac{\hat{\mathbf{b}}\times\hat{\mathbf{c}}}{|\hat{\mathbf{b}}\times\hat{\mathbf{c}}|} \quad \hat{\mathbf{r}} \text{ is a unit vector orthogonal to } \hat{\mathbf{b}} \text{ and } \hat{\mathbf{c}}. \\ \hat{\mathbf{s}} &= \frac{\hat{\mathbf{c}}\times\hat{\mathbf{a}}}{|\hat{\mathbf{c}}\times\hat{\mathbf{a}}|} \quad \hat{\mathbf{s}} \text{ is a unit vector orthogonal to } \hat{\mathbf{c}} \text{ and } \hat{\mathbf{a}}. \\ \hat{\mathbf{t}} &= \frac{\hat{\mathbf{a}}\times\hat{\mathbf{b}}}{|\hat{\mathbf{a}}\times\hat{\mathbf{b}}|} \quad \hat{\mathbf{t}} \text{ is a unit vector orthogonal to } \hat{\mathbf{a}} \text{ and } \hat{\mathbf{b}}. \end{aligned} \tag{A.18}$$

We will now form three scalar equations from Eq. (A.13) by taking the dot product with each of these biorthogonal vectors.

$$\begin{aligned} T_A\hat{\mathbf{a}}\cdot\hat{\mathbf{r}} + T_B\hat{\mathbf{b}}\cdot\hat{\mathbf{r}} + T_C\hat{\mathbf{c}}\cdot\hat{\mathbf{r}} &= \mathbf{P}\cdot\hat{\mathbf{r}} \\ T_A\hat{\mathbf{a}}\cdot\hat{\mathbf{s}} + T_B\hat{\mathbf{b}}\cdot\hat{\mathbf{s}} + T_C\hat{\mathbf{c}}\cdot\hat{\mathbf{s}} &= \mathbf{P}\cdot\hat{\mathbf{s}} \\ T_A\hat{\mathbf{a}}\cdot\hat{\mathbf{t}} + T_B\hat{\mathbf{b}}\cdot\hat{\mathbf{t}} + T_C\hat{\mathbf{c}}\cdot\hat{\mathbf{t}} &= \mathbf{P}\cdot\hat{\mathbf{t}} \end{aligned} \tag{A.19}$$

Using the biorthogonal properties of the two sets of base vectors, Eq. (A.19) can be written in terms of a diagonal matrix as follows:

$$\begin{bmatrix} \hat{\mathbf{a}}\cdot\hat{\mathbf{r}} & 0 & 0 \\ 0 & \hat{\mathbf{b}}\cdot\hat{\mathbf{s}} & 0 \\ 0 & 0 & \hat{\mathbf{c}}\cdot\hat{\mathbf{t}} \end{bmatrix} \begin{bmatrix} T_A \\ T_B \\ T_C \end{bmatrix} = \begin{bmatrix} \mathbf{P}\cdot\hat{\mathbf{r}} \\ \mathbf{P}\cdot\hat{\mathbf{s}} \\ \mathbf{P}\cdot\hat{\mathbf{t}} \end{bmatrix} \tag{A.20}$$

Since the coefficient matrix is in a diagonal form, the solution for the tensions can be obtained without solving a system of simultaneous equations. The solution is

$$T_A = \frac{\mathbf{P}\cdot\hat{\mathbf{r}}}{\hat{\mathbf{a}}\cdot\hat{\mathbf{r}}}$$
$$T_B = \frac{\mathbf{P}\cdot\hat{\mathbf{s}}}{\hat{\mathbf{b}}\cdot\hat{\mathbf{s}}} \qquad (A.21)$$
$$T_C = \frac{\mathbf{P}\cdot\hat{\mathbf{t}}}{\hat{\mathbf{c}}\cdot\hat{\mathbf{t}}}$$

This solution is a form of a direct vector solution of Eq. (A.13).

An example of solution of Eq. (A.13) using each of these methods is shown in the Computational Window A.1.

COMPUTATIONAL WINDOW A.1

Consider a mass of 100 kg supported by three cables. We will take the origin of the coordinate system to be at the mass and the cable attachments are A (4, −2, 10), B (−2, −3, 10), and C (0, 5, 10). The vector equation to be solved is:

$$\mathbf{T}_A + \mathbf{T}_B + \mathbf{T}_C + \mathbf{W} = 0$$

$$\mathbf{A} := \begin{pmatrix} 4 \\ -2 \\ 10 \end{pmatrix} \quad \mathbf{B} := \begin{pmatrix} -2 \\ -3 \\ 10 \end{pmatrix} \quad \mathbf{C} := \begin{pmatrix} 0 \\ 5 \\ 10 \end{pmatrix} \quad \mathbf{W} := \begin{pmatrix} 0 \\ 0 \\ -100 \cdot 9.81 \end{pmatrix}$$

We will form unit vectors along each of the cables and also designate unit vectors in the cartesian coordinate directions:

$$\hat{\mathbf{a}} := \frac{\mathbf{A}}{|\mathbf{A}|} \quad \hat{\mathbf{b}} := \frac{\mathbf{B}}{|\mathbf{B}|} \quad \hat{\mathbf{c}} := \frac{\mathbf{C}}{|\mathbf{C}|} \quad \hat{\mathbf{i}} := \begin{pmatrix} 1 \\ 0 \\ 0 \end{pmatrix} \quad \hat{\mathbf{j}} := \begin{pmatrix} 0 \\ 1 \\ 0 \end{pmatrix} \quad \hat{\mathbf{k}} := \begin{pmatrix} 0 \\ 0 \\ 1 \end{pmatrix}$$

The biorthogonal base vectors are:

$$\hat{\mathbf{r}} := \frac{\hat{\mathbf{b}} \times \hat{\mathbf{c}}}{|\hat{\mathbf{b}} \times \hat{\mathbf{c}}|} \quad \hat{\mathbf{s}} := \frac{\hat{\mathbf{c}} \times \hat{\mathbf{a}}}{|\hat{\mathbf{c}} \times \hat{\mathbf{a}}|} \quad \hat{\mathbf{t}} := \frac{\hat{\mathbf{a}} \times \hat{\mathbf{b}}}{|\hat{\mathbf{a}} \times \hat{\mathbf{b}}|}$$

The three solution forms are:

$$\mathrm{T1} := \begin{pmatrix} \hat{\mathbf{a}}\cdot\hat{\mathbf{i}} & \hat{\mathbf{b}}\cdot\hat{\mathbf{i}} & \hat{\mathbf{c}}\cdot\hat{\mathbf{i}} \\ \hat{\mathbf{a}}\cdot\hat{\mathbf{j}} & \hat{\mathbf{b}}\cdot\hat{\mathbf{j}} & \hat{\mathbf{c}}\cdot\hat{\mathbf{j}} \\ \hat{\mathbf{a}}\cdot\hat{\mathbf{k}} & \hat{\mathbf{b}}\cdot\hat{\mathbf{k}} & \hat{\mathbf{c}}\cdot\hat{\mathbf{k}} \end{pmatrix}^{-1} \cdot \begin{pmatrix} \mathbf{W}\cdot\hat{\mathbf{i}} \\ \mathbf{W}\cdot\hat{\mathbf{j}} \\ \mathbf{W}\cdot\hat{\mathbf{k}} \end{pmatrix} \qquad \mathrm{T1} = \begin{pmatrix} 233.616 \\ 453.399 \\ 381.493 \end{pmatrix}$$

$$\mathrm{T2} := \begin{pmatrix} \hat{\mathbf{a}}\cdot\hat{\mathbf{a}} & \hat{\mathbf{b}}\cdot\hat{\mathbf{a}} & \hat{\mathbf{c}}\cdot\hat{\mathbf{a}} \\ \hat{\mathbf{a}}\cdot\hat{\mathbf{b}} & \hat{\mathbf{b}}\cdot\hat{\mathbf{b}} & \hat{\mathbf{c}}\cdot\hat{\mathbf{b}} \\ \hat{\mathbf{a}}\cdot\hat{\mathbf{c}} & \hat{\mathbf{b}}\cdot\hat{\mathbf{c}} & \hat{\mathbf{c}}\cdot\hat{\mathbf{c}} \end{pmatrix}^{-1} \cdot \begin{pmatrix} \mathbf{W}\cdot\hat{\mathbf{a}} \\ \mathbf{W}\cdot\hat{\mathbf{b}} \\ \mathbf{W}\cdot\hat{\mathbf{c}} \end{pmatrix} \qquad \mathrm{T2} = \begin{pmatrix} 233.616 \\ 453.399 \\ 381.493 \end{pmatrix}$$

$$\mathbf{T}_A := \frac{\mathbf{W}\cdot\hat{\mathbf{r}}}{\hat{\mathbf{a}}\cdot\hat{\mathbf{r}}} \qquad T_A\ 5\ 233.616$$

$$\mathbf{T}_B := \frac{\mathbf{W}\cdot\hat{\mathbf{s}}}{\hat{\mathbf{b}}\cdot\hat{\mathbf{s}}} \qquad T_B\ 5\ 453.399$$

$$\mathbf{T}_C := \frac{\mathbf{W}\cdot\hat{\mathbf{t}}}{\hat{\mathbf{c}}\cdot\hat{\mathbf{t}}} \qquad T_C\ 5\ 381.493$$

STATICS INDEX DICTIONARY

associative: A rule of combining mathematical quantities two at a time is associative if the result of the combination of three quantities (order being preserved) does not depend on the way in which the objects are grouped. Vector addition. (19)

beam: A structural member, generally straight whose length is much greater than its height or width and is loaded perpendicular to the long axis. (196)

bearing: A support which permits rotation of a shaft while restricting translations normal to the shaft axis and resists bending of the shaft at that point. (212)

> **journal bearing:** (lubrication bearing) A bearing that consists of a shaft (journal) and a sleeve or bearing and the relative motion between the two is sliding motion. (359)
>
> **rolling bearing:** (Antifriction bearing) A bearing which allows rotation of the shaft while transmitting loads by rolling contact instead of sliding contact. (359)
>
> **thrust bearing:** A rolling or journal bearing that resists axial motion of the shaft. (362)

belt: A flat, flexible rope like device used to transmit forces and motion. (356)

> **V-belt:** A belt whose cross-section is shaped like a wedge, or the letter V, used to increase the friction between the belt and the pulley. (358)

bending moment: The component of the internal moment in a beam that causes a bending deformation of the beam. (292)

body force: A force due to the attraction between the distributed mass of the body and a separated body. This force is distributed throughout the volume of the body. (14)

buoyancy: The force of buoyancy is the force that a fluid exerts on a body and is equal to the weight of the displaced fluid, Archimedes principle. (201)

cable: A strong rope, usually made of metal, designed to have great tensile strength and to be used in structures. (315)

> **cable subjected to concentrated loads.** (315–317)
>
> **cables supporting loads distributed uniformly along a horizontal line, parabola shape.** (317–318)
>
> **cables supporting loads distributed uniformly along its own length, catenary.** (319–320)

cantilever beam: A beam that is built in at one end such that the rotation and translation of that end is zero. (297)

Cartesian coordinate: The distance of a point from either of two intersecting straight-line axes measured parallel to the other axis. The distance form any of three intersecting coordinate planes measured parallel to that one of three straight-line axes that is the intersection of the other two planes. (20)

center of gravity: The point within a body where the resultant of the total gravitational attraction forces may be considered to be acting. (174)

center of mass: The point in a system of masses where the total mass may be considered to be concentrated. Mathematically equal to the first moment of the mass divided by the total mass. (172)

center of pressure: The point where the resultant force of a distributed pressure acts on a surface. (200)

centroid: of a line, area or volume. The point defined as where the line, area or volume may be considered to be concentrated. Mathematically equal to the first moment of the line, area, or volume divided by the length of the line, the area or the volume. (171)

clutch: A device used to connect and disconnect two rotating members. (362)

coefficient of kinetic friction: μ_k The ratio of the friction force to the normal force during sliding between two surfaces. (392)

coefficient of rolling resistance: μ A measure of the resistance of one material to roll over another material. This resistance is due to the deformations of the two materials in the contact region. (364)

coefficient of static friction: μ_s The ratio of the maximum friction force to the normal force at the point of impending slip. (109)

collar: A band, usually circular used as a support of a shaft to restrict motion along the axis of the shaft. (362)

collinear: Lying on or passing through the same straight line. (26)

commutative: A mathematical rule between two quantities is commutative if the result is independent of the order of the combination. (17)

components of a vector: components of a vector: Any set of two non collinear, coplanar or three non collinear, non coplanar vectors whose sum is a given vector are components of that vector. The set of vectors is then said to span the space (two or three dimensions). (25)

computational software: Commercial software packages to perform numerical and symbolic computations with graphical capabilities. (12)

concentrated force: A force which may be represented as a single vector such that the force is considered to be distributed across zero area. (170)

concurrent: Converging, occupying the same point in space. (17)

concurrent force systems: A force system where all the forces are applied at a common point on the body or having their lines of action with a common intersection point. (153)

coplanar: Lying in a common plane in space, two-dimensional. (17)

coplanar force systems: A system of forces acting in a single plane on the body or in space. (153)

Coulomb friction: Friction in the absence of fluids. (329)

couple: Two parallel forces which are equal in magnitude, opposite in direction and non-collinear. (144)

Cramer's rule: Solution of a system of linear equations by use of determinants. (432)

cross-product: See vector product.

curl: The curl of a vector is equal to the cross product of the del operator with the vector and is written as $\vec{\nabla} \times \overline{F}$. The curl of a conservative force is equal to zero. (420)

degrees of freedom: The number of translations and rotations that must be specified to locate a rigid body in space. A single rigid body can have six degrees of freedom (translation in the x, y, and z direction and rotation about three axes. (210)

del operator: $\vec{\nabla} = \hat{\mathbf{i}}\frac{\partial}{\partial x} + \hat{\mathbf{j}}\frac{\partial}{\partial y} + \hat{\mathbf{k}}\frac{\partial}{\partial z}$ The del operator is a directed derivative. (420)

determinant: (something that determines) An operation applied to a square matrix which may be used to determine the inverse of the matrix and is a measure of the singularity of the matrix. (432)

Dirac delta function: A function having a value of 1 at a particular value of the variable and equal to zero for all other values;

$$\delta(x - a) = 0 \qquad x \neq a$$

$$\int_{-\infty}^{x} \delta(\zeta - a)d\zeta = \begin{bmatrix} 0 & x < a \\ 1 & x > a \end{bmatrix} \tag{308}$$

direct vector method: A method to solve the vector equation $\mathbf{A} + \mathbf{B} + \mathbf{C} = \mathbf{D}$ when the unit vectors $\hat{\mathbf{a}}$, $\hat{\mathbf{b}}$, $\hat{\mathbf{c}}$ and the vector **D** are known and the magnitude of **A**, **B** and **C** are unknown. This equation has a unique solution when the three unknown vectors are non-coplanar. This equation may be solved directly without expanding it into scalar components and the method is very useful in solution of particle equilibrium problems. (75)

direct vector solution: A method for determining either **A** or **B** in the vector equation $\mathbf{A} \times \mathbf{B} = \mathbf{C}$, if the three vectors are mutually orthogonal. The solution is obtained without expanding the equation into scalar component form. This method was first used by Rodrigue for determining angular velocity of a rigid body in plane motion. (129)

direction cosine: The cosine of the angle between a vector and a coordinate axis. The direction cosine is equal to the ratio of the orthogonal scalar component of a vector to the magnitude of the vector. (22)

distributed force: A force which is distributed across a line, area or volume and may not be represented as concentrated at a single point. (170)

dot product: See scalar product.

dry friction: See Coulomb friction.

equilibrium: Static equilibrium. A condition where there are no net external forces acting upon a particle or rigid body and the body remains at rest or continues at a constant velocity. (87)

equivalent: Corresponding or virtually identical in effect or function. Equivalent force system produce the same resultant force and moment about any point on a body or in space. (148)

fluid friction: The resistance to sliding between two surfaces separated by a layer of fluid or gas. (329)

force: The action of one body on another body. (3) A surface force is due to the contact of one body with another. A body force is due to the attraction of one body to another body when the bodies are separated. (14)

> **conservative force:** A force which depends only upon its position in space and the work done by this force is independent of the path of movement and depends only upon initial and final positions. (419)

frame: A structure designed to support loads and to prevent any motion and containing at least one member which has more than two forces acting on it. (251)

free-body diagram: A diagram used to isolate a body from its environment, showing all external constraints and forces acting upon it and all geometric measurements necessary to model the body. (82)

friction: The resistance to sliding between two surfaces. (108)

friction force: The force generated by friction to resist motion between two contacting surfaces. (109)

Gauss–Jordan reduction: Organized method of solution of a system of linear equations. (431)

generalized coordinates: A set of independent coordinates equal to the degrees of freedom of a system. (408)

geometric instability: The condition that results from supporting a body in a manner such that motion in some direction can result. (238)

gradient: The gradient of a function is equal to the del operator acting on that function and is equal to the directed derivative of that function. (420)

gravitational acceleration: A term that is sometimes applied to the constant,

$$g = \frac{GM}{R^2}$$

where G is the universal gravitational constant, M is the mass of the Earth and R is the radius of the Earth. g equals 9.81 $\mathrm{m/s^2}$ in SI units and 32.2 $\mathrm{ft/s^2}$ in U. S. customary units. (4)

gravitational system of units: A system of units based upon weight such as U.S. customary units. (4)

Heavyside step function or unit step function: A singularity function that is equal to zero for all values of the variable less than a specific value and equal to one for all values greater than the specified value.

$$\langle x - a\rangle^0 = \begin{cases} 0 & x < a \\ (x - a)^0 = 1 & x > a \end{cases} \tag{309}$$

hydrostatic pressure: The pressure that a fluid at rest excerts at any point which is equal in all directions according to Pascal's law. (199)

internal forces: Forces which act internal to parts of structures and are needed to determine the material stresses and strains. (122)

inverse: The inverse of a square nonsingular matrix is equivalent to its reciprocal and is used in the solution of systems of linear equations. (56)

inversion of coordinates: transformation that carries a right-handed coordinate system into a left-handed coordinate system or visa versa. (35)

irrational number: A number not represented by an integer or a fraction. (11)

kilogram: The basic unit of mass in the SI units. $1\ \mathrm{N} = 1\ \mathrm{kg} \cdot \mathrm{m/s^2}$ (5)

line of action: The line in space along which a force vector acts. (16)

machine or mechanism: An assemblage of parts that transmit forces, motion and energy one to another in a predetermined manner. Machines contain moving parts and have at least one multiforce member. (251)

mass: A quantity of matter related to the volume and density of a body. Gravitational mass is that property of a body that determines the attraction force to another body. This force is proportional to the product of the masses and the universal gravitational constant and inversely proportional to the square of the distance between the bodies. Inertial mass is the property of matter that resists acceleration when forces to a body. Gravitational mass and inertial mass are equivalent. (3)

matrix: A rectangular array of terms called elements with dimension or order ($\mathrm{m} \times \mathrm{n}$) used to facilitate the study of problems in which the relation between these elements is fundamental, as in the study of simultaneous linear equations. (41)

metacenter: The point of intersection of the vertical line through the center of buoyancy with the centerline of the cross-section of the hull of a ship. (201)

method of joints: A method of analysis of trusses which constructs free-body diagrams of each joint and determines the forces acting on that joint by considering equilibrium of the joint pin. (254)

using matrix methods: The solution of a truss by the method of joints but employing matrix notation to describe the geometry of the truss. (259)

method of sections: A method of analysis of trusses which passes an imaginary cut through the truss cutting no more than three members, constructing a free-body diagram of the isolated section of the truss and treating this section as if it were a rigid body in equilibrium. (266)

Mohr's circle: A semi-graphical method to determine the principle axes of an area developed by Otto Mohr in 1882. The circle is the loci of the second moments of area obtained by rotating the axes and is a conceptual representation of the transformation equations, (389)

moment arm: The perpendicular distance from a reference point or axis to the line of action of the force. (124)

moment diagram: A plot of the internal moment in a beam versus position along the axial length of the beam. In the past, used to solve beam problems in a semi-graphical manner. Currently used only as a conceptual aid in most engineering practice. (298)

moment of a force:

1. about a point - The turning effect of a force about a point equal to the magnitude of the force times the perpendicular distance from the point to the line of action of the force, $M = Fd$. It may be computed as the vector product of the position vector from the point to any point on the line of action of the force and the force vector, $\mathbf{M} = \mathbf{r} \times \mathbf{F}$. (127)
2. about a line or an axis - The turning effect of a force about a line or axis in space equal to the component of the moment vector parallel to the line or axis. (140)

neutral equilibrium: If the potential energy of a system is a function of only one generalized coordinate and all derivatives of that function are zero, the system is said to be in a position of neutral equilibrium. (424)

newton: A unit of measure of force in the SI system equal to the force required to accelerate 1 kg of mass 1 m/s^2. (5)

Newton's laws:

1. Every body or particle continues at a state of rest or of uniform motion in a straight line, unless it is compelled to change that state by forces acting upon it. $\Sigma\mathbf{F} = \mathbf{0}$
2. The change of motion of the body is proportional to the net force imposed on the body and is in the direction of the net force. $\Sigma\mathbf{F} = m\mathbf{a}$
3. If one body exerts a force on a second body then the second body exerts a force on the first body which is equal in magnitude, opposite in direction and collinear. $\mathbf{F}_{12} - \mathbf{F}_{21}$
4. Any two particles are attracted toward one another with a force whose magnitude is proportional to the product of their gravitational masses and inversely proportional to the square of the distance between them.

$$F = \frac{Gm_1m_2}{r^2} \tag{3}$$

orthogonal: Mutually perpendicular, mathematically independent. (20)

parallel axes theorem for the second moment of area: A mathematical relationship yielding the second moment of area about any axis parallel to a centroidal axis about which the second moment of area is known. This may be written mathematically as $I_{xx} = I_{xx_c} + Ad_y^2$. (374)

parallel force system: A force system composed of forces with parallel lines of action. (155)

parallelogram law: The law of vector addition in which two vectors are added by placing the origin of one vector at the terminus of another vector and forming a parallelogram to determine the sum. (17)

particle: The model of a body considered to occupy only a point in space and having no orientation. The particle may be considered a point mass. (14)

Pascal: A unit of measure of force per unit area. $1 \text{ Pa} = 1 \text{ N}/1 \text{ m}^2$. (30)

Pascal's law: A fluid at rest exerts a pressure, the hydrostatic pressure, at any point in the fluid which is equal in all directions. (199)

polar moment of inertia: A property of the cross-sectional area of a shaft related to its resistance to torsion. Mathematically this may be written as:

$$J_{0z} = \int_A \bar{r}^2 dA \tag{371}$$

position vector: A vector from the origin of a coordinate system to a point in space. The scalar components of the vector are the coordinates of the point. (39)

potential energy: The potential function of a conservative force, V and related to the force by $\bar{\mathbf{F}} \cdot d\bar{\mathbf{r}} = -dV$. The potential energy is a scalar measure of the force's capability to do work. Potential energy is usually the result of the position of a mass in the gravitational field or stored by elastic deformation of a body or spring. (421)

principal axes of an area: A pair of orthogonal centroidal axes about which the second moment of area will be a maximum or minimum and the product moment of area is zero. (372)

principle of transmissibility: If a body is modeled as a rigid body, forces acting on that body may be treated as sliding vectors and therefore the force may be considered to be acting a any point on its line of action. (123)

principle of virtual work: If a rigid body or a system of frictionless connected rigid bodies is in equilibrium, the total virtual work of the external forces acting on the body or system is zero for any virtual displacement consistent with the constraints on the body or system. (410)

product moment of an area: (product moment of inertia). A mathematical property of an area related to any set of orthogonal axes and defined as

$$I_{xy} = -\int_A xydA \tag{383}$$

radius of gyration of an area: A distance equivalent to placing all the area at this point when computing the second moment of the area. The radius of gyration may be written as:

$$k_x = \sqrt{\frac{I_{xx}}{A}} \tag{376}$$

relative extrema: If the derivative of the function of a single variable is zero at a point, then the value of the function is a relative extrema, that is, a relative maximum or minimum. (423)

relative position vector: A position vector relating the position of a point B to a point A and written as: $\mathbf{r}_{B/A} = \mathbf{r}_B - \mathbf{r}_A$ where $\mathbf{r}_A$ and $\mathbf{r}_B$ are the position vectors of A and B. (40)

resultant force: The resultant of a system of forces is the vector sum of all the forces. (31)

right-handed coordinate system: A coordinate system such that the x, y, and z coordinates may be aligned with the thumb, index and middle fingers of the right hand, respectively. (35)

rigid body: A body which is considered to be of a defined geometry but which will not deform or if deformed, the changes in dimensions are negligible. (123)

scalar: A physical quantity that has magnitude only. A tensor of order 0. Examples are mass, temperature, volume, etc. (15)

scalar product: or dot product. Multiplication of two vectors such that the result is a scalar. Defined as $\mathbf{A} \cdot \mathbf{B} = |\mathbf{A}||\mathbf{B}| \cos \theta$, where θ is the angle between 0 and 180 degrees between the vectors. (61)

scalar triple product: or mixed vector product. The vector operation between three vectors which produces a scalar result and is defined as $\mathbf{A} \cdot (\mathbf{B} \times \mathbf{C})$ (70)

screw: A mechanical device consisting of a helical rib projected from a cylindrical or conical shaft used to support loads, raise or lower loads or to fasten objects together. (351)

second moment of the area: (area moment of inertia) A property of an area related to the resistance of bending and equal to the integral

$$\int_A y^2\, dA \quad \text{or} \quad \int_A x^2\, dA \tag{370}$$

sense: The sense of a vector designates the positive direction in which the vector acts. (16)

shear diagram: A plot of the shear force in a beam versus the position along the axis of the beam. Previously used in semi-graphical solutions. (298)

shear force: The internal force usually in a beam which acts in the plane of the cross-section of the beam. (30)

SI units (Systeme International d'Unites): A system of units based on length, mass and time adopted as an international absolute system of units. The unit of length is meter (m), mass is measured in kilograms (kg) and time is expressed in seconds (s). (4)

singular matrix: A square matrix whose determinant is equal to zero. (56)

singularity functions: A function that has the value of zero for all values of the variable less that a specified value and is a defined function for all values of the variable greater than the specified value of the variable.

$$\langle x - a \rangle^n = \begin{cases} 0 & x < a \\ (x - a)^n & x > a \end{cases} \tag{308}$$

space: The boundless extent in which objects and events occur and have relative position and direction. The concept of length, area and volume. Space is considered to be infinite, homogeneous, isotropic and absolute. (2)

spring constant: The slope of the force deflection curve of a linear spring, that is, the ratio of the force to the deformation. (101)

stable equilibrium: If the potential energy is a function of a single generalized coordinate, the system is in stable equilibrium if the first derivative of the potential energy is zero and the second derivative is greater than zero. (423)

statically indeterminate: A structure or body which is over-constrained such that there are more unknown supports than there are equations of static equilibrium. The deformations of the supports or members must be considered in addition to equilibrium to determine the unknown forces. (88)

stresses: The intensities of the internal distributed forces in a body. (30)

structure: Something made up of interdependent parts in a definite pattern of organization, such as trusses, frames or machines. (250)

supports: Connections of a rigid body with other bodies used to restrict movement of the rigid body. (207)

surface force: The force acting on the surface of a body due to contact with another body. (14)

system of linear equations: A system of algebraic equations usually n equations and n unknowns all of which are linear. (53)

Theorems of Pappus and Guldinus:

1. The area A of a surface of revolution generated by revolving a plane curve of length L about any nonintersecting axis in its plane is equal to the product of the length of the curve and the length of the path traveled by the centroid of the curve. (186)
2. The volume V of the solid of revolution generated by revolving a plane area A about any non-intersecting axis in its plane is equal to the product of the area and the length of the path traveled by the centroid of the area. (187)

three-force body: A body in equilibrium under the action of three forces. The forces must be coplanar and be either concurrent or parallel. (219)

time: The concept used to order the flow of events; the measured or measurable period during which an action, process or condition exists or continues. (3)

transpose: The transpose of a matrix is obtained by interchanging rows and columns. The transpose is denoted by $[\mathbf{a}]^T$. (433)

truss: A structure designed to support loads and to prevent any motion and whose parts or members may all be modeled as two-force members. (251)

compound truss: A truss constructed by connecting several simple trusses. (277)

planar truss: A truss that may be modeled such that all of its members will be coplanar. (251)

simple truss: A planar truss that has its members organized such that they form triangles. A space truss comprised of tetrahedrons of six internal members. (254)

space truss: A truss constructed in such a manner that its straight members do not lie in a single plane. (272)

twisting or torsion moment: The component of a moment which causes twisting or torsion of a shaft or long member. (292)

two-force body: A rigid body in equilibrium under the action of only two forces. The two forces must be equal in magnitude, collinear and have opposite sense. (219)

unit matrix: A square matrix whose diagonal elements equal one and whose off diagonal elements are zero. (57)

unit vector: A vector of magnitude one, having a specific direction in space and having no units. Used as base vectors of a coordinate system. (36)

units: Determinate quantities (of length, time, mass, etc.) adopted as a standard of measurement. (4)

universal gravitational constant: $G = 66.73 \times 10^{-12}\ m^3/kg\ s^2$. Constant relating the gravitational attraction between two masses;

$$F = \frac{Gm_1m_2}{r^2} \qquad (3)$$

unstable equilibrium: If the potential energy is a function of a single generalized coordinate, the system is in unstable equilibrium if the first derivative of the potential energy is zero and the second derivative is less than zero. (423)

U.S. Customary units: A system of units based on the old English system for length, force and time; foot, pound, and second, respectively. (4)

Varignon's theorem: or principle of moments. The moment of a force about a point is equal to the sum of the moments of the force's components about the point. (Application of the distributive property of the vector product) (131)

vector: A mathematical quantity that has magnitude, direction and sense and that satisfies the law of vector addition and multiplication by a scalar. A tensor of order one. (14)

vector addition: The process of adding two vectors by use of the parallelogram law. (16)

vector calculator: a computer software program that allows vector calculations to be performed similar to ordinary arithmetic calculations. (43)

vector product: or cross product. A vector operation that multiplies two vectors to produce a third vector perpendicular to these two vectors and having a magnitude equal the product of the magnitudes of the two vectors times the sine of the angle between them. The sense of the resulting vector is determined by the right-hand rule. $\mathbf{C} = \mathbf{A} \times \mathbf{B}$ where $|\mathbf{C}| = |\mathbf{A}||\mathbf{B}| \sin\theta$. (61)

vector triple product: A vector operation which involves the successive vector product between three vectors resulting in another vector and defined as: $\mathbf{A} \times (\mathbf{B} \times \mathbf{C})$ (71)

vector type:

fixed - a vector describing a physical quantity that has a fixed point of application (origin or terminus) and line of action on a body or in space. (23)

free - a vector describing a physical quantity that is independent of its point of application (origin, terminus or line of action) on a body or in space. (23)

sliding - a vector describing a physical quantity that has a specific line of action on a body or in space but is independent of its point of application on this line of action. (23)

virtual displacement: An imagined displacement or variation from the equilibrium position consistent with the constraints on the system. (407)

virtual work: The work performed by a force when moved through an imagined or virtual displacement. (407)

wedge: A simple machine consisting of a rigid body which tapers to a thin edge and is used to separate or force objects apart. (345)

weight: The force exerted upon a body due to gravitational attraction to a planet. On the surface of the Earth $W = mg$ (3)

work: To produce results or to perform a function. Mathematically defined as the scalar product between a force and the displacement through which the force moves. $dU = \mathbf{F} \cdot \mathbf{dr}$ (406)

wrench: A force system composed of a single force and a parallel moment. (160)

Answers to Selected Problems

Chapter 1

1.1 d

1.2 a

1.3 b

1.4 c

1.5 Yes

1.6 6 ft

1.7 622.7 N

1.8 63.48 kg

1.9 4.35 slugs

Chapter 2

2.1 $x = x' \cos 30°$
$y = y' \cos 30°$

2.3 $\lambda_x = .8660, \lambda_y = .5$
$\lambda_{x'} = 1, \lambda_{y'} = 0$

2.5 $R = 77.9$ N, 59.5° from the x-axis.

2.7 $R = 49.8$ m/s, 34.74° from the $+x$-axis.

2.9 $F_{AC} = 331.2$ N, $F_{AB} = 244.0$ N

2.11 $F_{AB} = 202.32$ N, $F_{CB} = 418.28$ N

2.13 $F_x = 707.1$ N, $F_y = 707.1$ N
$\lambda_x = 0.707, \lambda_y = 0.707$

2.15 $v_{x'} = v_x \cos\theta + v_y \sin\theta$
$v_{y'} = -v_x \sin\theta + v_y \cos\theta$

2.17 $\beta = 24.74°$, $F_{b\text{-}b'} = 57.98$ N

2.19 $R = 138.5$ N, 79.2° from the $+x$-axis.

2.21 $R = 1761$ N, 20.5° from the $+x$-axis.

2.23 $|\mathbf{A}| = 7$, $\mathbf{a} = (3\hat{\mathbf{i}} + 2\hat{\mathbf{j}} + 6\hat{\mathbf{k}})/7$

2.25 $F_y = \pm 120$ N

2.27 $2\mathbf{U} - \mathbf{V} = 7\hat{\mathbf{i}} - 4\hat{\mathbf{j}} + 10\hat{\mathbf{k}}$
$\theta_x = 56.99°$ (with $+x$-axis)
$\theta_y = 108.14°$ (with $+y$-axis)
$\theta_z = 38.90°$ (with $+z$-axis)

2.29 $\theta_x = 105.54°$, $\theta_y = 16.94°$, $\theta_z = 83.41°$
$F_x = -268$ N, $F_y = 956.6$ N, $F_z = -114.8$ N

2.31 $\dfrac{\mathbf{AB}}{|\mathbf{AB}|} = \dfrac{2}{3}\hat{\mathbf{i}} - \dfrac{1}{3}\hat{\mathbf{j}} + \dfrac{2}{3}\hat{\mathbf{k}}$

2.33 $\mathbf{r}_{\mathbf{b/a}} = -\hat{\mathbf{i}} - 4\hat{\mathbf{j}} - 2\hat{\mathbf{k}}$
$\hat{\mathbf{e}}_{\mathbf{b/a}} = (-\hat{\mathbf{i}} - 4\hat{\mathbf{j}} - 2\hat{\mathbf{k}})/\sqrt{21}$

2.35 $T_1 = 366$ N, $T_2 = 259$ N

2.37 $|\mathbf{R}| = 324.8$ N, $-1.9°$ from the $+x$-axis.

2.39 $|\mathbf{R}| = 1759.8$ N, 20.7° from the $+x$-axis.

2.41 $\mathbf{B} = -\hat{\mathbf{i}} + 2\hat{\mathbf{j}} - \hat{\mathbf{k}}$
$\lambda_x = -0.408, \lambda_y = 0.816,$
$\lambda_z = -0.408$

2.43 $\mathbf{B} = -3\hat{\mathbf{i}} - 16\hat{\mathbf{j}} - 2\hat{\mathbf{k}}$
$|\mathbf{B}| = 16.401, \lambda_x = -0.183,$
$\lambda_y = -0.976, \lambda_z = -0.123$

2.45 $F_x = 272.1$ N, $F_y = 127.3$ N, $F_z = 169.7$ N

2.47 $\mathbf{F} = 1278\hat{\mathbf{i}} + 634\hat{\mathbf{j}} + 464\hat{\mathbf{k}}$ N

2.49 $\mathbf{F} = 400\hat{\mathbf{i}} + 565.6\,\hat{\mathbf{j}}$ N $+ 400\hat{\mathbf{k}}$ N

2.51 $\mathbf{F} = 604\hat{\mathbf{i}} - 220\hat{\mathbf{j}} + 766\hat{\mathbf{k}}$ N

2.53 $\mathbf{A} = 25.561\hat{\mathbf{i}} + 14.758\hat{\mathbf{j}}$
$\mathbf{B} = 9.459\hat{\mathbf{i}} + 20.242\hat{\mathbf{j}}$

2.57 **A**: $\theta_x = 57.69°$ (with $+x$-axis), $\theta_y = 143.30°$ (with $+y$-axis), $\theta_z = 74.50°$ (with $+z$-axis)
B: $\theta_x = 54.74°$ (with $+x$-axis), $\theta_y = 54.74°$ (with $+y$-axis), $\theta_z = 54.74°$ (with $+z$-axis)
C: $\theta_x = 33.69°$ (with $+x$-axis), $\theta_y = 90°$ (with $+y$-axis), $\theta_z = 123.69°$ (with $+z$-axis)

2.59 $|\mathbf{A}| = 6736$ N
$|\mathbf{B}| = 13{,}428$ N

2.61 $R_a = -29.74$, $R_b = 64.74$,
$R_c = 26.24$

2.63 $R_a = 47.17$, $R_b = -81.70$,
$R_c = 47.17$

2.65 $P_A = 244.79$, $P_B = 229.48$
$P_C = -162.86$

2.67 $|\mathbf{A}| = 272.46$ N
$\hat{\mathbf{c}} = -0.635\hat{\mathbf{i}} + 0.454\hat{\mathbf{j}} + 0.625\hat{\mathbf{k}}$

2.69 $F_A = -63.64$, $F_B = 218.18$,
$F_c = 59.09$

2.71 $T_1 = -81.06$, $T_2 = 8.27$, $T_3 = 46.67$,
$T_4 = -24.66$, $T_5 = -8.36$

2.73

$$AB = \begin{bmatrix} 4 & 8 & 8 \\ -7 & 1 & -3 \\ -5 & 1 & 5 \end{bmatrix},$$

$$BA = \begin{bmatrix} 2 & 7 & 8 \\ -4 & 6 & 0 \\ -9 & 3 & 2 \end{bmatrix}$$

2.75

$$A^{-1} = \begin{bmatrix} 0.3750 & 0.2500 & -0.2500 \\ 0.2500 & 0.500 & -0.5000 \\ -0.1875 & -0.1250 & 0.6250 \end{bmatrix}$$

$$B^{-1} = \begin{bmatrix} 0.0769 & -0.2692 & -0.3077 \\ 0.2308 & 0.1923 & 0.0769 \\ 0.2308 & -0.3077 & 0.0769 \end{bmatrix}$$

2.77 $F_1 = 126.92$, $F_2 = 34.62$,
$F_3 = -7.69$, $F_4 = -23.08$

2.79 $\mathbf{A}\cdot\mathbf{B} = 2$, $\mathbf{A}\cdot\hat{\mathbf{i}} = 3$,
$\mathbf{A}\cdot\hat{\mathbf{j}} = -2$, $\mathbf{A}\cdot\hat{\mathbf{k}} = 1$

2.81 $\hat{\mathbf{e}}_n = \dfrac{1}{\sqrt{5}}(2\hat{\mathbf{j}} + \hat{\mathbf{k}})$
$\mathbf{F}\cdot\hat{\mathbf{e}}_n = 2.2361$

2.83 $\mathbf{R}_v = 1.049\hat{\mathbf{i}} + 1.817\hat{\mathbf{j}}$
$\mathbf{R}_\perp = -2.049\hat{\mathbf{i}} + 1.183\hat{\mathbf{j}}$

2.85 $B_z = 1$

2.87 $\mathbf{F} = 4.402\hat{\mathbf{e}}_a + 2.319\hat{\mathbf{e}}_b$
$\mathbf{F}\cdot\hat{\mathbf{e}}_a = 6.25$
$\mathbf{F}\cdot\hat{\mathbf{e}}_b = 5.82$

2.89 $\mathbf{F} = -1.4\hat{\mathbf{e}}_a - 2\hat{\mathbf{e}}_b$
$\mathbf{F}\cdot\hat{\mathbf{e}}_a = 0$
$\mathbf{F}\cdot\hat{\mathbf{e}}_b = -1$

2.91 $\mathbf{F} = 70.71\,\hat{\mathbf{e}}_a + 50\hat{\mathbf{e}}_b$ N
$\mathbf{F}\cdot\hat{\mathbf{e}}_a = 106.07$ N
$\mathbf{F}\cdot\hat{\mathbf{e}}_b = 100$ N

2.93
$F_a = -135.4$ N, $F_b = -178.9$ N,
$\mathbf{F}\cdot\hat{\mathbf{e}}_a = -281.96$ N, $\mathbf{F}\cdot\hat{\mathbf{e}}_b = -289.8$ N

2.95 **a.** $l = 0.922$ m
b. $\theta = 45°$

2.97 $\theta = 14°$

2.99 $\mathbf{A} \times (\mathbf{B} + \mathbf{C}) = -2\hat{\mathbf{j}} + 2\hat{\mathbf{k}}$
$(\mathbf{A} \times \mathbf{B}) + (\mathbf{A} \times \mathbf{C}) = -2\hat{\mathbf{j}} + 2\hat{\mathbf{k}}$

2.101 $r_x = 1$
$\mathbf{M} = 4\hat{\mathbf{i}} - 4\hat{\mathbf{j}}$
$\mathbf{M}\cdot\mathbf{F} = 0$, $\mathbf{M}\cdot\mathbf{r} = 0$

2.103
$(\mathbf{r}_1 + \mathbf{r}_2) \times \mathbf{F} = 30\hat{\mathbf{i}} - 9\hat{\mathbf{j}} + 61\hat{\mathbf{k}}$
$(\mathbf{r}_1 \times \mathbf{F}) + (\mathbf{r}_2 \times \mathbf{F}) = 30\hat{\mathbf{i}} - 9\hat{\mathbf{j}} + 61\hat{\mathbf{k}}$

2.105 $\hat{\mathbf{n}} = .707\hat{\mathbf{i}} - .707\hat{\mathbf{j}}$

2.107 $\mathbf{B}_\perp = -\hat{\mathbf{i}} + 2\hat{\mathbf{j}} + 4\hat{\mathbf{k}}$

2.109
$\mathbf{W}_n = -39.51\hat{\mathbf{i}} - 69.14\hat{\mathbf{j}} - 39.51\hat{\mathbf{k}}$ N
$\mathbf{W}_t = 39.51\hat{\mathbf{i}} + 69.14\hat{\mathbf{j}} - 160.49\hat{\mathbf{k}}$ N

2.111 **a.** $\mathbf{r}_{B/A} = \mathbf{r}_B - \mathbf{r}_A$
$\mathbf{r}_{C/A} = \mathbf{r}_C - \mathbf{r}_A$

$$\hat{\mathbf{i}} = \frac{\mathbf{r}_{B/A}}{|\mathbf{r}_{B/A}|}$$

$$\hat{\mathbf{k}} = \frac{\mathbf{r}_{B/A} \times \mathbf{r}_{C/A}}{|\mathbf{r}_{B/A} \times \mathbf{r}_{C/A}|}$$

$$\hat{\mathbf{j}} = \hat{\mathbf{k}} \times \hat{\mathbf{i}}$$

b.

$\hat{\mathbf{i}} = -0.159\hat{\mathbf{I}} + 0.379\hat{\mathbf{J}} - 0.912\hat{\mathbf{K}}$
$\hat{\mathbf{j}} = 0.411\hat{\mathbf{I}} + 0.865\hat{\mathbf{J}} + 0.288\hat{\mathbf{K}}$
$\hat{\mathbf{k}} = 0.898\hat{\mathbf{I}} - 0.329\hat{\mathbf{J}} - 0.293\hat{\mathbf{K}}$

2.113 $A = 5.66$
$B = -7$

2.115 $A = 59.72$
$B = 66.67$
$C = -16.72$

2.117 $A = 77.1$
$B = 111.32$
$C = 118.79$

2.119 $A = 100$
$B = 49.5$
$C = 77.79$

2.121 $A = 77.47$
$B = 30.71$
$C = -21.37$

Chapter 3

3.9 $\hat{\mathbf{e}}_1 = -0.577\hat{\mathbf{i}} + 0.577\hat{\mathbf{j}} + 0.577\hat{\mathbf{k}}$
$\hat{\mathbf{e}}_2 = -0.577\hat{\mathbf{i}} + 0.577\hat{\mathbf{j}} - 0.577\hat{\mathbf{k}}$
$\hat{\mathbf{e}}_3 = 0.447\hat{\mathbf{i}} + 0.894\hat{\mathbf{j}}$
$\mathbf{T}_1 = T_1\hat{\mathbf{e}}_1, \mathbf{T}_2 = T_2\hat{\mathbf{e}}_2, \mathbf{T}_3 = T_3\hat{\mathbf{e}}_3$

3.11 $T_1 = 7181$ N, $T_2 = 8795$ N, $T_3 = 9810$ N

3.13 $T_A = 1212$ N, $T_B = 2948$ N

3.15 **a.** $T_1 = 3328$ N, $T_2 = 29.56$ N, $T_3 = 59.09$ N, $T_4 = 108$ N
b. 2.5 m:
$T_1 = 28.66$ N, $T_2 = 27.62$ N, $T_3 = 55.27$ N, $T_4 = 108$ N
0.25 m:
$T_1 = 100.95$ N, $T_2 = 66.08$ N
$T_3 = 132.16$ N, $T_4 = 108$ N

3.17 $T_1 = 2230$ N, $T_2 = 3717$ N, $T_3 = W_2 = 3539$ N, $T_4 = 3433$ N

3.19 $T_1 = 208.14$ N, $T_2 = 348.40$ N, $T_3 = 232.56$ N

3.21 $\mathbf{F}_2 = -800\hat{\mathbf{i}} + 64{,}290\hat{\mathbf{j}}$ N

3.23 $\mathbf{T}_1 = -100\hat{\mathbf{i}} + 100\hat{\mathbf{j}} - 50\hat{\mathbf{k}}$ N
$\mathbf{T}_2 = -100\hat{\mathbf{j}} + 100\hat{\mathbf{k}}$ N
$\mathbf{T}_3 = 100\hat{\mathbf{i}} - 100\hat{\mathbf{j}} - 50\hat{\mathbf{k}}$ N

3.25 $T = (m + M/2)g/2$

3.27 $N_1 = 8g/(\sin 45°/\tan\beta + \cos 45°)$
$N_2 = 8g/(\sin\beta/\tan 45° + \cos\beta)$

3.29 $T = 85.52$ N, $\alpha = 6.59°$

3.31 $\tan\alpha = \dfrac{m_A \tan\beta - m_B \tan\theta}{m_A + m_B}$

$T = g(m_A + m_B)\dfrac{\sqrt{1 + \tan^2\alpha}}{\tan\theta + \tan\beta}$

3.33 $a = 4.42$ m, $b = 3.64$ m
$\hat{\mathbf{I}}_\mathbf{B}\cdot(\mathbf{B} - \mathbf{A}) = 0$

3.35
$\hat{\mathbf{I}}_\mathbf{A} = -0.23\hat{\mathbf{i}} + 0.69\hat{\mathbf{j}} - 0.69\hat{\mathbf{k}}$
$\hat{\mathbf{I}}_\mathbf{B} = -0.89\hat{\mathbf{i}} + 0.45\hat{\mathbf{j}}$
$\mathbf{A} = 3\hat{\mathbf{k}} + 2.27\hat{\mathbf{I}}_\mathbf{A}$
$\mathbf{B} = 4\hat{\mathbf{i}} + 2.18\hat{\mathbf{I}}_\mathbf{B}$

3.37
$\mathbf{W}_t = -95.2\hat{\mathbf{i}} - 63.45\hat{\mathbf{j}} + 687.46\hat{\mathbf{k}}$ N

3.39 **a.** $f = k_1c - mg$
b. $f = 2k_1c - mg$

3.41 $T_1 = 1.652 \times 10^3$ N
$T_2 = 1.607 \times 10^3$ N
$x = 0.098$ m

3.43 $\delta = 0.136$ m

3.45 $\delta = 0.42$ m

3.47 **a.** $f = 100$ N
b. The block will not move.

3.49 19.29°

3.51 The block slides down the incline.
$f = 46.09$ N

3.53 **a.** The block slides down the incline.
b. $f = 120.55$ N

3.55 **a.** There is no such force.
b. $0° < \alpha < 59°$

3.57 **a.** The block will not move.
b. $f = 1339.1$ N

3.59 **a.** $\mu_s = 0.047$
b. $T = 50.88$ N

3.61 21.8°

3.63 $\mu_s = 0.84$

3.65 Answers given by (3.5B.8) and (3.5B.9).

Chapter 4

4.1 $M_0 = 6.92$ N m

4.3 $d = 0.3$ m

4.5 $M_0 = 386.6$ N m, $M_p = 86.6$ N m

4.7 $M_{\min} = 0$, $M_{\max} = 100$ N m
Counterclockwise.
It makes a big difference what angle you push on a wrench.

4.9 $M_A = M_B = 70.71$ N m

4.11 $\mathbf{M}_0 = -386.6\hat{\mathbf{k}}$ N m

4.13 $\mathbf{M}_A = 3.6\hat{\mathbf{k}}$ N cm
$\mathbf{M}_B = 0$
$\mathbf{M}_C = -1.2\hat{\mathbf{k}}$ N cm
$\mathbf{M}_D = -1.2\hat{\mathbf{k}}$ N cm

4.15 $\mathbf{M}_0 = 7\hat{\mathbf{i}} - 14\hat{\mathbf{j}} - 7\hat{\mathbf{k}}$ kN m

4.17 $\mathbf{M}_0 = -1679\hat{\mathbf{k}}$ N m
$\mathbf{M}_B = -779.4\hat{\mathbf{k}}$ N m

4.19 **a.** $\mathbf{M}_0 = -28.7\hat{\mathbf{k}}$ N cm

4.21
$\mathbf{M}_0 = -0.8\hat{\mathbf{i}} + 1.9\hat{\mathbf{j}} - 0.7\hat{\mathbf{k}}$ kN m

4.23 $\mathbf{M}_0 = -141.4\hat{\mathbf{i}} + 84.64\hat{\mathbf{j}}$
Same moment using both methods.

4.25 $\mathbf{M}_0 = -1.60\hat{\mathbf{i}} - 0.40\hat{\mathbf{j}} - 4.8\hat{\mathbf{k}}$ N m

4.27 $\mathbf{M}_0 = 63.6\hat{\mathbf{i}} - 178.7\hat{\mathbf{k}}$ kN m

4.29 a. $\mathbf{F}_1$ provides the greatest moment about point 0 in the $\hat{\mathbf{j}}$ direction.
b. $\mathbf{M}_0 = -15\hat{\mathbf{i}} + 20.4\hat{\mathbf{j}}$ N m

4.31 $\mathbf{r} = -3\hat{\mathbf{k}}$

4.33 $\mathbf{F} = -1.44\hat{\mathbf{i}} + 1.56\hat{\mathbf{j}} + 0.22\hat{\mathbf{k}}$ N

4.35 $\mathbf{F} = 3.95\hat{\mathbf{i}} - 3.42\hat{\mathbf{j}} + 2.63\hat{\mathbf{k}}$ N

4.37 $\mathbf{P} = 1.57\hat{\mathbf{i}} - 3.39\hat{\mathbf{j}} + 2.43\hat{\mathbf{k}}$ N

4.39 $c_x = 0.25$, $c_y = 1.57$
$\mathbf{F} = 553$ N (compression)

4.41
$\mathbf{M}_A = 3.12\hat{\mathbf{i}} - 1.44\hat{\mathbf{j}} + 2.52\hat{\mathbf{k}}$ N m
$\mathbf{M}_{AB} = 2.52\hat{\mathbf{k}}$ N m

4.43 $\mathbf{M}_{AB} = -48.0\hat{\mathbf{i}} - 24.0\hat{\mathbf{j}}$ N m

4.45 $\mathbf{M}_0 \cdot \hat{\mathbf{k}} = -848.5$ N m

4.47 $\mathrm{M}_P \cdot \hat{\mathbf{j}} = -94$ N m

4.49 $\mathbf{F} = R\hat{\mathbf{i}} + (0.8R - 120)\hat{\mathbf{j}} + 60\hat{\mathbf{k}}$ N for any value of R.

4.51 $M_{\max} = 5.66$ N m at $\theta = 90°$

4.53 $\mathbf{C} = -8.84\hat{\mathbf{k}}$ N m

4.55
$\mathbf{C} = -2.2 \times 10^4\hat{\mathbf{k}} - 5 \times 10^3\hat{\mathbf{i}}$ N m
$|\mathbf{C}| = 22{,}561$ N m

4.57 $\mathbf{C}_A = -60\hat{\mathbf{k}}$ Nm
$\mathbf{C}_B = 40.44\hat{\mathbf{k}}$ Nm
$\mathbf{C}_A + \mathbf{C}_B = -19.56\hat{\mathbf{k}}$ N m

4.59 $\mathbf{C} = -116.4\hat{\mathbf{i}} + 89.1\hat{\mathbf{j}}$
$\hat{\mathbf{e}}_C = -0.794\hat{\mathbf{i}} + 0.608\hat{\mathbf{j}}$
$\mathbf{C} = 146.6\,\hat{\mathbf{e}}_C$ N m

4.61 $\mathbf{C} = 122.474\hat{\mathbf{i}} - 61.237\hat{\mathbf{k}}$ N m

4.63 $\mathbf{F}_A = 433\hat{\mathbf{i}} + 250\hat{\mathbf{j}}$ N
$\mathbf{C} = -91.5\hat{\mathbf{k}}$ N m

4.65 $\mathbf{C} = -5\hat{\mathbf{j}} + 5\hat{\mathbf{k}}$ N m

4.67 $\mathbf{R} = (\hat{\mathbf{i}} + 20\hat{\mathbf{j}})$ kN
$\mathbf{M}_A = -0.3\hat{\mathbf{k}}$
$x = 0.27$ m

4.69 $\mathbf{P}_{AO} = 0.269\hat{\mathbf{i}} - 0.013\hat{\mathbf{j}}$ m

4.71 $\mathbf{M}_0 = -237.28\hat{\mathbf{k}}$ N m
$\mathbf{R} = 66.86\hat{\mathbf{i}} + 132.43\hat{\mathbf{j}}$ N
$\mathbf{r}_{ro} = -1.43\hat{\mathbf{i}} + 0.721\hat{\mathbf{j}}$ m

4.73 a. $\mathbf{R} = -15\hat{\mathbf{i}} - 47.3\hat{\mathbf{j}}$ kN
$\mathbf{r}_{R/O} = 6.738\hat{\mathbf{i}} - 2.136\hat{\mathbf{j}}$ m
b. $3.155x - y = 23.39$ (7.1414, 0)

4.75 $\mathbf{C}_{\parallel} = 8.69\hat{\mathbf{i}} - 2.35\hat{\mathbf{j}} + 4.35\hat{\mathbf{k}}$ N m
$\mathbf{C}_{\perp} = 41.31\hat{\mathbf{i}} + 52.35\hat{\mathbf{j}} - 54.35\hat{\mathbf{k}}$ N m

4.77 $\mathbf{r} = 1.072\hat{\mathbf{i}} + 0.24\hat{\mathbf{j}} + 1.764\hat{\mathbf{k}}$ m
$\mathbf{R} = 20\hat{\mathbf{i}} - 825\hat{\mathbf{j}} + 100\hat{\mathbf{k}}$ N
$\mathbf{C}_{\parallel} = 0.729\hat{\mathbf{i}} - 30.09\hat{\mathbf{j}} + 3.65\hat{\mathbf{k}}$ N m

4.79
$\mathbf{r} = 0.571\hat{\mathbf{i}} + 0.122\hat{\mathbf{j}} - 1.224\hat{\mathbf{k}}$ m
$\mathbf{M}_{\parallel} = 4.408\hat{\mathbf{i}} + 8.816\hat{\mathbf{j}} + 2.939\hat{\mathbf{k}}$ N m

4.81 $\mathbf{T} = 3.6\hat{\mathbf{i}}$ N m
$\mathbf{p}_0 = -0.89\hat{\mathbf{i}} - 0.165\hat{\mathbf{j}} + 0.189\hat{\mathbf{k}}$ m

4.83 $\mathbf{p}_0 = 0.05(-\hat{\mathbf{i}} + \hat{\mathbf{j}})$ m
$\mathbf{T} = -5.5(\hat{\mathbf{i}} + \hat{\mathbf{j}})$ N m

Chapter 5

5.1 $x_c = 0.5$ m, $y_c = 0.5$ m

5.3 $x_c = 6.62$ m, $y_c = 49.61$ m

5.5 $x_c = 0.73$ mm, $y_c = 0.6$ mm

5.7 $x_c = 6.36$ cm, $y_c = 0$

5.9 $x_c = 0$, $y_c = 0$

5.11 $x_c = 2b/3$, $y_c = h/3$

5.13 $x_c = (2R/3\alpha)\sin\alpha$, $y_c = 0$

5.15 $x_c = 4a/3\pi$, $y_c = 4b/3\pi$

5.17 $x_c = 3b/5$, $y_c = 3h/8$

5.19 $A = 3ab/4$, $x_c = 4a/7$, $y_c = 2b/5$

5.21 $V = \pi hR^2$, $\bar{x} = 0$, $\bar{y} = 0$, $\bar{z} = h/2$

5.23 $V = \pi hR^2/3$, $\bar{x} = 3h/4$, $\bar{y} = 0$, $\bar{z} = 0$

5.25 $V = 2\pi R^3/3$, $\bar{x} = 5R/8$, $\bar{y} = 0$, $\bar{z} = 0$

5.27 $A = 4\pi R^2$

5.29 $V = \pi ab^2$

5.31 $V = 2\pi ab^2/3$

5.33 $\bar{x} = 0.1$ m, $\bar{y} = 0.11$ m, $\bar{z} = 0.25$ m

5.35 $x_c = 0.72$ m
$y_c = 2.2$ m
$z_c = 2.48$ m

5.37 $x_c = 2$ m, $y_c = 2$ m
$x_c = 2$ m, $y_c = 1.33$ m (solid triangle)

5.39 $x_c = 0.41$ m, $y_c = 0.2$ m
$x_c = 0.4$ m, $y_c = 0.2$ m (without holes)

5.41 $\bar{x} = 2.06$ m, $\bar{y} = 3.58$ m
Centroid is *on* the structure.

5.43 $\bar{x} = 44.5$ mm, $\bar{y} = 34.6$ mm
Centroid is *not on* the object.

5.45 $\bar{x} = 3.06$ cm, $\bar{y} = 55.35$ cm
Centroid is *not on* the cane.

5.47 $\bar{x} = 5$ m, $\bar{y} = 4.81$ m, $\bar{z} = 1.5$ m

5.49 $V = 80.59$ cm^3, $\bar{x} = 0$, $\bar{y} = 4.65$ cm, $\bar{z} = 0$

5.51 At $l = 6$ m, $\bar{x} = 6.94$ m, $\bar{y} = 0$, $\bar{z} = 0$

5.53 1500 N at $x_c = 1.91$ m

5.55 $W = 90$ N at $x_c = 2.25$ m

5.57 $W = 603$ N at $x_c = 19$ m

5.59 $W = 3111$ N at $x_c = 1.86$ m

5.61 $W = 13{,}390$ N at $x_c = 5.83$ m

5.63 $R = 6898$ kN, $y_c = 5$ m

5.65 $R = 12{,}844$ kN

5.67 $R_{bot} = 58.8$ N at the center of the bottom.
$R_{end} = 58.8$ N at 2/3 of the way down from the top (13.3 cm).
$R_{side} = 19.7$ N at 2/3 of the way down from the top (13.3 cm).

5.69 $W = 9418$ N

Chapter 6

6.11 $F_{Ax} = 0$, $F_{Ay} = 391$ N, $F_{By} = 492$ N

6.13 $F_{Ax} = 0$, $F_{Ay} = -265$ N, $F_{By} = 1588$ N

6.15 Lincoln, $d_f = 1.04$ m, $d_r = 1.73$ m
BMW, $d_f = 1.31$ m, $d_r = 1.39$ m
Porsche, $d_f = 1.135$ m, $d_r = 1.135$ m
Blazer, $d_f = 1.20$ m, $d_r = 1.52$ m

6.17 $F_A = 491$ N, $F_B = 180$ N, $F_C = 670$ N

6.19 $T = 981$ N, $F_{Ax} = 0$, $F_{Ay} = 981$ N

6.21 $F_{Ax} = 0$, $F_{Ay} = 250$ N, $F_{By} = 450$ N

6.23 $F = 39$ N, $T_1 = 53$ N, $T_2 = 43$ N

6.25 At $\theta = 0°$, T cannot be determined.
At $\theta = 10°$, $T = 115.2$ lb, $F_x = 113.4$ lb, $F_y = 20$ lb
At $\theta = 90°$, $T = 20$ lb, $F_x = 0$, $F_y = 20$ lb

6.27 **a.** $F_{Ax} = -55{,}220$ N, $F_{Bx} = 55{,}220$ N, $F_{By} = 49{,}050$ N
b. $F_{By} = 49{,}050$ N, $F_{Bx} = -F_{Ax} = 67{,}970$, 59,470, 50,970, 42,480, 33,980, 25,490 N

6.29 **a.** $F_{Ax} = 1.5mg \cos\beta \sin\beta/l$, $F_{Ay} = mg(1 - 1.5\cos^2\beta/l)$, $F_{By} = 1.5mg\cos\beta/l$
b. $F_{Ax} = 173.2$ N, $F_{Ay} = 100$ N, $F_{By} = 346.4$ N

6.31 $\alpha = \sin^{-1}\dfrac{9mg}{2l(k_1 + 9k_2)}$

6.33 $T = D(BW + W)/2d$

6.35 $A_x = -133.33$ N
$A_y = -162.62$ N
$B_y = -108.09$ N
$B_z = 66.67$ N
$C_x = -66.67$ N
$C_z = -137.38$ N

6.37 $F = 125.5$ N
$A_x = 71.5$ N
$A_y = 80.6$ N
$A_z = 130.3$ N
$B_y = -22.8$ N
$B_z = -19.6$ N

6.39 $A_x = 452.7$ N
$A_y = 942.9$ N
$A_z = 1811.1$ N
$T_{BD} = 769.8$ N
$T_{BC} = 1281.9$ N

6.41 $\theta = 13.66°$

6.43 $T = 133.28$ N
$A_x = 48.58$ N
$A_y = 50$ N
$A_z = 161.8$ N
$B_x = 48.58$ N
$B_z = -35.16$ N

6.45 $S_x = -664$ N
$S_y = 0$
$S_z = 4607.6$ N
$T_x = -167.1$ N
$T_z = -167.1$ N
$M = -3943.6$ N

6.47 $\mathbf{M} = 1177.2\,\hat{\mathbf{i}} - 1569.6\,\hat{\mathbf{k}}$ N
M_x is the component of the total moment that could cause an ankle sprain.

6.49 The plate carries 75% of the load, the bone 25%.

6.51 $\Delta = \dfrac{100}{k_B}$

6.53 $F_A = F_C = 350$ N, $F_D = 450$ N, $F_B = 250$ N

6.55 $4F = 1000$ N (symmetry)

6.57 $F_A = F_D = 250$ N, $F_B = F_C = F_E = 166.67$ N

Chapter 7

7.1

$\sum F_y = 0$: $CD \sin\beta - CB \sin\beta = 0$

$\sum F_x = 0$:

$CE - CA + CD\cos\beta - CB\cos\beta = 0$

7.3 $AC = -707$ N (compression)
$AB = 500$ N (tension)
$BC = 500$ N (tension)
$C_y = 500$ N (up)
$B_x = -500$ N (left)
$B_y = -500$ N (down)

7.5 $AB = -10.4$ kN
$AC = 0, BC = 0$
$B_x = -10$ kN
$B_y = 2.9$ kN
$A_y = 2.9$ kN

7.7 $D_y = 11{,}670$ N, $A_x = 0, A_y = 8330$ N
$AE = 22{,}430$ N, $AB = -20{,}830$ N,
$EF = 20{,}830$ N,
$BE = -8330$ N, $DF = 31{,}420$ N,
$CD = -29{,}180$ N,
$BC = -29{,}180$ N, $BF = 8990$ N,
$CF = -15{,}000$ N

7.9 $B_x = 1000$ N, $A_x = 1000$ N,
$A_y = 500$ N,
$AC = 1000$ N, $AB = 500$ N,
$BC = 0$,
$CE = 1000$ N, $CD = 0$,
$DE = BD = -1119$ N

7.11

Force	
A_x	1331 N
A_y	1500 N
B_x	2331 N
AC	1331 N
AB	1500 N
BC	−600 N
BD	−2235 N
CE	999 N
CD	−500 N
DE	−1117 N

7.13 $A_x = -1250$ N
$A_y = 1500$ N
$B_x = 1250$ N
$AC = 1250$ N
$DE = -901.4$ N
$AB = 1500$ N
$CE = 750$ N
$BC = -1117.8$ N
$BD = -901.4$ N
$CD = 0$

7.15 a., b., c.

	$\alpha = 0$	$\alpha = 15^\circ$	$\alpha = 30^\circ$
A_x	5000 N(R)	3706 N(R)	2500 N(R)
A_y	2500 N(U)	167 N(U)	1495 N(U)
B_y	2500 N(U)	4662 N(U)	5825 N(U)
AB	3750 N(C)	3622 N(C)	3248 N(C)
AC	2795 N(C)	187 N(C)	1671 N(T)
BC	2500 N(T)	167 N(T)	1495 N(C)
BD	6250 N(C)	6037 N(C)	5413 N(C)
CD	1250 N(C)	84 N(C)	747 N(T)

R = right, U = up, C = compression,
T = tension

7.17 $A_x = -245.3$ N
$A_y = 245.3$ N
$B_x = 245.3$ N
$AC = 274.2$ N
$CD = 274.2$ N
$AB = 122.65$ N
$BD = -245.3$ N
$BC = 0$

7.19 a. $A_x = 0$, $AB = -901$ N,
$BD = 0$, $A_y = 750$ N,
$AD = 500$ N, $CD = 500$ N
$C_y = 750$ N, $BC = -901$ N,

b. The force in BD is zero unless joint D is loaded directly. Therefore, BD is not necessary.

7.21 $A_x = 0$, $AB = -908.5$ N,
$BD = 4.78$ N, $A_y = 759.7$ N,
$AD = 503.9$ N, $CD = 503.9$ N,
$C_y = 759.7$ N, $BC = -908.5$ N
It is not necessary to consider the weights of the members.

7.23 $A_x = 21{,}960$ N, $AB = 1799$ N
$BE = 1192$ N (C), $A_y = 10{,}000$ N
$AC = -25{,}680$ N (C), $BD = 2334$ N
$T = 7890$ N, $BC = 0$,
$CE = -25{,}680$ N (C), $DE = 997$ N,
$EG = -26{,}910$ N (C),
$GH = -12{,}420$ N (C), $DF = 3592$ N
$FH = 11{,}960$ N, $DG = -1192$ N (C)
$GH = -6049$ N (C),
C = compression

7.25 i.

$A_x = -7.071$ kN $AC = 7.071$ kN
$CD = -10$ kN $DF = -1.464$ kN
$A_y = 11.464$ kN $BC = 2.071$ kN
$CE = 0$ $E_y = 15.607$ kN
$AB = -11.464$ kN $BD = -1.464$ kN
$CF = 12.01$ kN $EF = -15.689$ kN

ii.

$A_x = -7.07$ kN $AC = 7.07$ kN
$CD = -1.464$ kN $DF = 7.07$ kN
$A_y = 11.464$ kN $BC = 2.01$ kN
$CE = 8.536$ kN $E_y = 15.607$ kN
$AB = -11.464$ kN $BD = -1.464$ kN
$DE = -12.071$ kN $EF = -7.07$ kN

iii.

$A_x = -7.071$ kN $AC = 8.536$ kN
$CD = 0$ $DF = 7.071$ kN
$A_y = -8.536$ kN $AD = -2.071$ kN
$CE = 8.536$ kN $E_y = 15.607$ kN
$AB = 10$ kN $BD = 0$
$DE = -12.071$ kN $EF = -7.07$ kN

7.27

$$A_x = -W\cos\theta$$
$$B_y = W\frac{\sin\beta\cos\theta + 2\sin\theta\cos\beta}{\cos\beta}$$
$$A_y = -W\frac{\sin\beta\cos\theta + \sin\theta\cos\beta}{\cos\beta}$$
$$BC = -W\frac{\sin\beta\cos\theta + \sin\theta\cos\beta}{\cos\beta}$$
$$AB = -W\sin\theta\cos\beta/\sin\beta$$
$$BD = -W\sin\theta/\sin\beta$$
$$AC = W\frac{\sin\beta\cos\theta + \sin\theta\cos\beta}{\sin\beta\cos\theta}$$
$$CD = W\frac{\sin\beta\cos\theta + \sin\theta\cos\beta}{\sin\beta}$$

Structure not in equilibrium for $168.5 < \theta < 180°$.

7.29

$$A_x(\theta) = -W\cos(\theta)$$
$$B_y(\theta) = (490.5\cos\beta + W\sin\beta\cos\theta + 2W\sin\theta\cos\beta)/\cos\beta$$
$$A_y(\theta) = -W\frac{\sin\beta\cos\theta\sin\theta\cos\beta}{\cos\beta}$$
$$BC(\theta) = -(245.25\cos\beta + W\sin\beta\cos\theta + W\sin\theta\cos\beta)/\cos\beta$$
$$AB(\theta) = \frac{\cos\beta}{\sin\beta}(W\sin\theta + 98.1)$$
$$BD(\theta) = -(W\sin\theta + 98.1)/\sin\beta$$
$$AC(\theta) = \frac{W(\sin\beta\cos\theta + \sin\theta\cos\beta) + 98.1\cos\beta}{\cos\beta\sin\beta}$$
$$CD(\theta) = \frac{W(\sin\beta\cos\theta + \sin\theta\cos\beta) + 98.1\cos\beta}{\sin\beta}$$

Structure always in equilibrium.

7.31 $BC = 29.2$ kN
$BF = 8.97$ kN
$EF = 20.9$ kN

7.33 $FD = -1$ kN
$EC = -1$ kN
$CF = 0$

7.35 $CE = 0$
$CF = -2500$ N (compression)
$DE = 2500$ N (tension)

7.37 a. $CF = 2123.8$ N
$CE = 1118.2$ N
$DF = -2620$ N (compression)

b. Including the weight of the members makes a significant difference.

7.39 $AD = -W - 10$,
$BD = \sqrt{2}\,W$, $BE = -10$

7.41 $OL = 1414$ N
$ON = 1000$ N
$KL = -2000$ N

7.43 $DJ = 0$
$JI = -34.789$ kN (tension)
$EJ = 18.028$ kN (compression)
$DE = 21.116$ kN (compression)

7.45 $AB = -409$ N

7.47 $AC = 490.5$ N
$BD = -548.34$ N
$BC = 0$

7.49 $JK = 9.81$ kN, $HL = -16.45$ kN,
$JL = 5.48$ kN

7.51 $BD = -34.3$ kN

7.53 $MK = -2.5$ kN, $LK = 0$,
$LJ = -2.5$ kN

7.55 $-CE = CF = 354$ N

7.57 $A_x = 0.988$ N $F_{BD} = -1329$ N
$A_y = 799.543$ N $F_{BC} = -596.261$ N
$F_{AB} = 382$ N $C_x = 999.0122$ N
$F_{AC} = 723.9$ N $C_y = -865.6$ N
$F_{AD} = -1332.6$ N $C_z = -1000$ N
$B_y = 1333$ N $F_{dc} = -956.464$ N

7.59 $A_x = 1.71$ kN, $AD = -2.30$ kN
$B_x = -2.05$ kN, $C_x = 1.02$ kN
$AB = 776$ N, $BC = 469$ N
$B_y = 0$, $CD = -1.66$ kN
$AC = 1.01$ kN, $BD = 3.72$ kN
$B_z = 4.20$ kN, $DE = 1.30$ kN

7.61 $A_x = 1$ kN, $AD = 33.5$ kN,
$C_z = 19.9$ kN, $A_y = 0$, $B_z = 19.9$ kN,
$CD = -22.4$ kN, $A_z = -29.9$ kN,
$BC = 20.7$ kN, $CE = -8.4$ kN,
$AB = -11.3$ kN, $BE = 8.4$ kN,
$DE = 18.0$ kN, $AC = -11.3$ kN,
$BD = -22.4$ kN, $C_x = 0$

7.63 $A_x = 2.26$ kN, $B = 4.18$ kN,
$C_z = -122.3$ kN, $A_y = 0$,
$BC = -5.14$ kN, $CD = -12.76$ kN,
$A_Z = 9.54$ kN, $BD = -2.15$ kN,
$CE = 2.74$ kN, $AB = 2.58$ kN,
$BE = 2.74$ kN, $AC = 2.58$ kN,
$DE = -10.95$ kN, $AD = -11.93$ kN,
$E_z = 4.79$ kN

7.65 $A_x = -25{,}000$ N, $B_z = -95{,}032$ N,
$C = 7217$ N, $A_y = -6250$ N,
$BC = -100{,}727$ N, $CD = 39{,}999$ N,
$A_z = 21{,}357$ N, $BD = -36{,}616$ N,
$CE = 22{,}513$ N, $AB = 13{,}369$ N,
$BE = 161{,}523$ N, $CF = 45{,}092$ N,
$AC = 53{,}476$ N, $EF = 98{,}233$ N,
$AE = -36{,}300$ N

7.67 $A_x = 0$, $B = -120$, $C_y = 180$
$A_y = 420$, $C_x = -120$,
$E_y = 180$ $(B = -D)$

7.69 **a.** Same as (b) except
$E_y = 40{,}000$ N.

b.
$A_x = -50{,}000$ N, $C = D = 80{,}000$ N
$A_y = -40{,}000$ N, $E_x = 50{,}000$ N
$B_x = 50{,}000$ N, $E_y = 80{,}000$ N

7.71
$A_x = 4316$ N, $C_x = 9156$ N
$D_x = 10{,}080$ N, $E_x = 4316$ N
$A_y = 3924$ N, $C_y = 2616$ N
$D_y = 6540$ N, $F_y = 6540$ N
$F_x = 1380$ N, $G_x = G_y = B = 3924$

7.73
$A_x = 2250$ N, $B_x = 2250$ N, $C_x = 2250$ N,
$A_y = 3000$ N, $B_y = 1000$ N, $C_y = 3000$ N

7.75 $A_x = -1067$ N, $B_x = -1067$ N,
$C_x = -1067$ N, $A_y = 200$ N,
$B_y = 800$ N, $C_y = 800$ N

7.77 $A_x = 0$, $A_y = 1187$ N, $P = -972$ N

7.79 $A_x = 538$ N,
$A_y = 210$ N, $D_x = E_x = -538$ N,
$D_y = E_y = 200$ N, $F_x = -100$ N,
$F_y = -10$ N, $G_y = 410$ N

7.81 $F = 0.207$ kN

7.83 $A_x = 712$ N, $A_y = 180$ N,
$B_x = 712$ N, $B_y = 0$,
$M_B = 3150$ N cm

7.85 $A_y = \dfrac{d + c}{d} P$, $B_y = \dfrac{(a + b)c}{db} P$,
$C_y = \dfrac{c}{d} P$, $F = \dfrac{ac}{db} P$

7.87 **b.** No solution.

7.89 $A_x = B_x = C_x = 0$, $A_y = -1500$ N,
$B_y = -1250$ N, $C_y = 4375$ N,
$F = 3125$ N

7.91 $A_y = 214$ N, $B = 5967$ N,
$CE = 7569$ N, $D_x = 4428$ N,
$D_y = 3932$ N

7.93 $A_x = 9810$ N, $A_y = -14{,}715$ N,
$B_x = 9810$ N, $B_y = 14{,}715$ N,
$C_x = 19{,}620$ N, $C_y = 29{,}430$ N,
$D_x = 9810$ N, $D_y = -29{,}430$ N,
$E = 14{,}715$ N

7.95 $A_x = -5.25$ kN, $A_y = 3.083$ kN,
$B = 3.582$ kN, $C_x = -8.867$ kN,
$C_y = -1.898$ kN, $D = -10.907$ kN,
$F_x = -17.942$ kN, $F_y = -4.152$ kN,
$G = -5.007$ kN

7.97 $E = -15.54$ kN, $F = 14.92$ kN,
$A = -0.171$ kN, $B = -0.651$,
$C_x = -735$ kN,
$C_y = -0.149$ kN

Chapter 8

8.1 $R = 1500$ N (tension)
$F_a = 1500$ N (tension)
$F_b = 1000$ N (tension)

8.3 $R = 700$ N
$F(x) = 500x - 300$ N

8.5 $V(0.3) = 750$ N
$M(.3) = 300$ N m
$V(.85) = 0$
$M(.85) = 0$

8.7 $V(0.5) = 0$
$M(0.5) = 3000$ N m

8.9 $V(0.9) = 0$
$M(0.9) = 300$ N m

8.11 $V(0.5) = 500$ N
$M(0.5) = 1050$ N m

8.13

$$V(\theta) = \begin{cases} 3\cos\theta \text{ kN} & 0 \le \theta \le 180° \\ 0 & 180° \le \theta \le 225° \end{cases}$$

$$M(\theta) = \begin{cases} 450\sin\theta \text{ N m} & 0 \le \theta \le 180° \\ 0 & 180° \le \theta \le 225° \end{cases}$$

$$A(\theta) = \begin{cases} 3\sin\theta \text{ kN} & 0 \le \theta \le 180° \\ 0 & 180° \le \theta \le 225° \end{cases}$$

8.15 $R = -10\hat{\mathbf{i}} - 10\hat{\mathbf{j}} + 3\hat{\mathbf{k}}$ kN
$M(0) = 23\hat{\mathbf{i}} - 20\hat{\mathbf{j}} + 10\hat{\mathbf{k}}$ kN m

8.17 a.
$\mathbf{R}_0 = -3\hat{\mathbf{i}} - 15\hat{\mathbf{j}} - 10\hat{\mathbf{k}}$ kN
$\mathbf{M}_0 = 1.799\hat{\mathbf{i}} - 0.2598\hat{\mathbf{j}} - 0.15\hat{\mathbf{k}}$ N m
For $0° \le \theta < 60°$,
$\mathbf{Q}(\theta) = 3\hat{\mathbf{i}} + 15\hat{\mathbf{j}} + 10\hat{\mathbf{k}}$,
$\mathbf{M}(\theta) = (1.5\sin\theta - \cos\theta - 0.799)\hat{\mathbf{i}}$
$+ (0.2598 - 0.3\sin\theta)\hat{\mathbf{j}}$
$+ (0.3\cos\theta - 0.15)\hat{\mathbf{k}}$,
For θ between 60º and up to 170º, **M** and **Q** are zero.

b.
$\hat{\mathbf{i}}\cdot\mathbf{Q}(\theta) = 3$ kN
$\hat{\mathbf{e}}_r\cdot\mathbf{Q}(\theta) = 15\cos\theta + 10\sin\theta$ kN
$\hat{\mathbf{e}}_\theta\cdot\mathbf{Q}(\theta) = -15\sin\theta + 10\cos\theta$ kN
$\hat{\mathbf{i}}\cdot\mathrm{M}(\theta) = (1.5\sin\theta - \cos\theta$
$- 0.799)$ kN m
$\hat{\mathbf{e}}_r\cdot\mathbf{M}(\theta) = 0.2598\cos\theta$
$- 0.15\sin\theta$ kN m
$\hat{\mathbf{e}}_\theta\cdot\mathbf{M}(\theta) = -0.2598\sin\theta - 0.15\cos\theta$
$+ 0.3$ kN m

8.19 $C_y = 3.75$ kN
$A_y = 1.25$ kN
$V(x) = 1.25$ kN, $0 \le x < 3$
$M(x) = 1.25x$ kN m, $0 \le x < 3$
$V(x) = -3.75$ kN, $3 < x \le 4$
$M(x) = 15 - 3.75x$ kN m,
$3 < x \le 4$
The maximum value of the moment is 3.75 kN m.

8.21 $A_y = 7$ kN
$B_y = 8$ kN
$V(x) = 7$ kN, $0 \le x < 2$ (down)
$M(x) = 7x$ kN m, $0 \le x < 2$
$V(x) = -3$ kN, $2 < x < 4$ (up)
$M(x) = 20 - 3x$ kN m, $2 < x < 4$
$V(x) = -8$ kN, $4 < x < 5$ (up)
$M(x) = 40 - 8x$, $4 < x \le 5$
The maximum value of the bending moment is 14 kN m.

8.23 $R_A = P_0$
$M_A = P_0L + M_0$
$V = P_0$ (up)
$M(x) = P_0(x - L) - M_0$

8.25 $A_y = 885.7$ N
$B_y = 1314.3$ N
$V(x) = 885.7 - 66.67x^2$ N,
$0 \le x < 3$ m
$M(x) = 885.7x - 22.22x^3$,
$0 \le x < 3$
$V(x) = -400x + 1485.7$ N,
$3 < x \le 7$
$M(x) = -200x^2 + 1485.7x - 600$ N m,
$3 < x \le 7$

8.27 $B_y = 67.5$ N, $A_y = 22.5$ N,

$$V(x) = \frac{-10x^3}{3} + 22.5 \text{ N},$$

$$M(x) = \frac{-5x^4}{6} + 22.5x \text{ N m}$$

8.29 $R_y = 1000$ N
$M_0 = 34{,}000$ N/m

$$V(x) = \frac{10}{\pi}\left(x\pi - 50\cos\frac{\pi x}{50} + 50\right) - 1000 \text{ N}$$

$$M(x) = 5x^2 - \frac{25{,}000}{\pi^2}\sin\frac{\pi x}{50} + \frac{500}{\pi}x - 1000x + 3400 \text{ N m}$$

8.31
$R_y = 7847$ N
$M_0 = 69{,}349$ N m
$V(x) = 1000(\sin x - x\cos x) - 7847$ N
$M(x) = 71{,}349 - 7847x - 2000\cos x$
$- 1000x\sin x$ N m

8.33
$B_y = 2.1$ kN, $A_y = 1.8$ kN
$V(x) = 1.8$, $0 \le x < 2$
$M(x) = 1.8x$, $0 \le x < 2$
$V(x) = 3.4 - 0.8x$ kN, $2 < x \le 5$
$M(x) = -0.4x^2 + 3.4x - 1.6$ kN m,
$2 < x \le 5$

$V(x) = -0.6$ kN, $5 < x \leq 7$
$M(x) = 8.4 - 0.6x$ kN m
$5 < x \leq 7$
$V(x) = -2.1$ kN, $7 < x \leq 9$
$M(x) = 18.9 - 2.1x$ kN m,
$7 < x \leq 9$

8.35

$$R_B = \frac{a^2}{2L}w_0 + \frac{M}{L},$$

$$R_A = w_0 a\left(1 - \frac{a}{2L}\right) - \frac{M}{L}$$

For $0 \leq x < a$,

$$V(x) = w_0 a\left(1 - \frac{a}{2L}\right) - \frac{M}{L} - w_0 x,$$

$$M(x) = \left(w_0 a\left(1 - \frac{a}{2L}\right) - \frac{M}{L}\right)x - \frac{w_0}{2}x^2$$

For $a \leq x < b$,

$$V(x) = -\frac{a^2}{2L}w_0 - \frac{M}{L},$$

$$M(x) = -\frac{a^2}{2L}w_0 x - \frac{M}{L}x + \frac{a^2 w_0}{2}$$

For $b \leq x \leq L$, $V(x) = -\frac{a^2}{2L}w_0 - \frac{M}{L}$,

$$M(x) = -\frac{a^2}{2L}w_0 x + M\left(1 - \frac{x}{L}\right) + \frac{a^2 w_0}{2}$$

8.37

$$R_B = \frac{a}{b}P + \frac{w_0}{2}\left(\frac{L^2}{b}\right),$$

$$R_A = P\left(1 - \frac{a}{b}\right) + w_0 L\left(1 - \frac{L}{2b}\right)$$

For $0 \leq x < a$,

$$V(x) = -w_0 x + P\left(1 - \frac{a}{b}\right) + w_0 L\left(1 - \frac{L}{2b}\right),$$

$$M(x) = -w_0\frac{x^2}{2} + \left(P\left(1 - \frac{a}{b}\right) + w_0 L\left(1 - \frac{L}{2b}\right)\right)x$$

For $a \leq x < b$,

$$V(x) = -w_0 x + w_0 L\left(1 - \frac{L}{2b}\right) - \frac{a}{b}P,$$

$$M(x) = -w_0\frac{x^2}{2} + w_0 L\left(1 - \frac{L}{2b}\right)x + Pa\left(1 - \frac{x}{b}\right)$$

For $b \leq x < L$, $V(x) = -w_0 x + w_0 L$,

$$M(x) = -\frac{(L - x)^2}{2}w_0$$

8.39

$$R_A = \frac{1}{24}Lw_0,\ R_B = \frac{5}{24}Lw_0$$

For $0 \leq x < L/2$

$$V(x) = \frac{1}{24}Lw_0,\ M(x) = \frac{1}{24}Lw_0 x$$

For $L/2 \leq x < L$,

$$V(x) = -\frac{1}{L}w_0\left(x - L/2\right)^2 + \frac{1}{24}Lw_0,$$

$$M(x) = -\frac{w_0}{3L}\left(x - \frac{L}{2}\right)^3 + \frac{1}{24}Lw_0 x$$

8.41

$a = L/4$,

$$R_A = -\frac{w_0 L}{6},\ R_B = -\frac{5}{6}w_o L$$

For $0 \leq x < a$,

$$V(x) = -w_0 x\left(1 - \frac{2x}{L}\right) - \frac{w_0 L}{6},$$

$$M(x) = -\frac{w_0 x^2}{2}\left(1 - \frac{4x}{3L}\right) - \frac{w_0 L}{6}x$$

For $a < x \leq L$, same as above.

8.43 $R_A = P_B$

$M_A = (M_B + P_B L)$

$$V(x) = M_B\langle x - L\rangle_{-1} - P_B\langle x - L\rangle^0 - M_A\langle x - 0\rangle_{-1} + R_A\langle x - 0\rangle^0$$

$$M(x) = M_B\langle x - L\rangle^0 - P_B\langle x - L\rangle^1 - M_A\langle x - 0\rangle^0 + R_A\langle x - 0\rangle^1$$

8.45 $R_A = w_0 a\left(1 - \frac{a}{2L}\right) - \frac{M}{L}$

$$R_B = \frac{a^2}{2L}w_0 + \frac{M}{L}$$

$$V(x) = -w_0\langle x - 0\rangle^1 + w_0\langle x - a\rangle^1 + M\langle x - b\rangle^{-1} + R_A\langle x - 0\rangle^0 + R_B\langle x - L\rangle^0$$

$$M(x) = -\frac{w_0}{2}\langle x - 0\rangle^2 + \frac{w_0}{2}\langle x - a\rangle^2 + M\langle x - b\rangle^0 + R_A\langle x - 0\rangle^1 + R_B\langle x - L\rangle^1$$

8.47

$$R_A = P\left(1 - \frac{a}{b}\right) + w_0L\left(1 - \frac{L}{2b}\right)$$

$$R_B = \frac{a}{b}P + \frac{w_0}{2}(L^2/b)$$

$$V(x) = -w_0\langle x - 0\rangle^1 - P\langle x - a\rangle^0 + R_A\langle x - 0\rangle^0 + R_B\langle x - b\rangle^0$$

$$M(x) = -\frac{w_0}{2}\langle x - 0\rangle^2 - P\langle x - a\rangle^1 + R_A\langle x - 0\rangle^1 + R_B\langle x - b\rangle^1$$

8.49 $R_A = \frac{w_0}{24}L$

$$R_B = \frac{5}{24}w_0L$$

$$V(x) = -\frac{w_0}{L}\langle x - L/2\rangle^2 + R_A\langle x - 0\rangle^0 + R_B\langle x - L\rangle^0$$

$$M(x) = -\frac{w_0}{3L}\langle x - L/2\rangle^3 + R_A\langle x - 0\rangle^1 + R_B\langle x - L\rangle^1$$

8.51

$L = 4a$ (from loading geometry)

$$R_L = \frac{3P}{4} - \frac{2w_0a}{3}$$

$$R_R = -\frac{P}{4} + \frac{10w_0a}{3}$$

$$V(x) = R_L\langle x - 0\rangle^0 - w_0\langle x - 0\rangle^1 - P\langle x - a\rangle^0 + \frac{w_0}{2a}\langle x - 0\rangle^2 - R_R\langle x - 4a\rangle^0$$

$$M(x) = R_L\langle x - 0\rangle^1 - \frac{w_0}{2}\langle x - 0\rangle^2 - P\langle x - a\rangle^1 + \frac{w_0}{6a}\langle x - 0\rangle^3 - R_R\langle x - 4a\rangle^1$$

8.53

$$M_0 = \frac{\sqrt{2}L^2}{2\pi} - \frac{PL}{4}$$

$$V_0 = P - \frac{\sqrt{2}L}{\pi}$$

$$w(x) = M_0\langle x - 0\rangle_{-2} + V_0\langle x - 0\rangle_{-1} + \sin\left(\frac{\pi x}{L}\right)\left\langle x - \frac{L}{4}\right\rangle^0 - \sin\left(\frac{\pi x}{L}\right)\left\langle x - \frac{3L}{4}\right\rangle^0$$

$$V(x) = M_0\langle x - 0\rangle_{-1} + V_0\langle x - 0\rangle^0 - P\left\langle x - \frac{L}{4}\right\rangle^0 + \frac{L}{\pi}\left[\frac{\sqrt{2}}{2} - \cos\left(\frac{\pi x}{L}\right)\right]\left\langle x - \frac{L}{4}\right\rangle^0 + \frac{L}{\pi}\left[\frac{\sqrt{2}}{2} - \cos\left(\frac{\pi x}{L}\right)\right]\left\langle x - \frac{3L}{4}\right\rangle^0$$

$$M(x) = M_0\langle x - 0\rangle^0 + V_0\langle x - 0\rangle^1 - P\left\langle x - \frac{L}{4}\right\rangle^1 + \frac{\sqrt{2}L}{2\pi}\left\langle x - \frac{L}{4}\right\rangle^1 + \frac{\sqrt{2}L}{2\pi}\left\langle x - \frac{3L}{4}\right\rangle^1 - \frac{L^2}{\pi^2}\left[\sin\left(\frac{\pi x}{L}\right) - \frac{\sqrt{2}}{2}\right]\left\langle x - \frac{L}{4}\right\rangle^0 + \frac{L^2}{\pi^2}\left[\sin\left(\frac{\pi x}{L}\right) + \frac{\sqrt{2}}{2}\right]\left\langle x - \frac{3L}{4}\right\rangle^0$$

8.55 $\theta = 44.05°$, $\beta = 67.98°$, $T_A = 809$ N, $T_B = 1501$ N

8.57 **a.** $T_{AB} = 4.767$ kN, $T_{BC} = 4.413$ kN, $T_{CD} = 4.706$ kN, $\alpha = 22.6°$, $\beta = 4.32°$, $\gamma = 20.75°$

b. $A_x = -4.4$ kN
$A_y = 1.83$ kN
$D_x = 4.4$ kN
$D_y = 1.67$ kN
The sag at B is 1.15 m and at C is 1.42 m.

8.59

$T_{AB} = 1349$ N, $T_{BC} = 1355$ N,

$T_{CD} = 1486$ N, $\alpha = 9.96°$,

$\beta = 11.35°$, $\gamma = 26.64°$, $\ell = 2.0$ m

8.61 $T = 22.36$ kN

8.63 $T = 1000\sqrt{1 + (0.1)^2x^2}$,
$-20 \leq x \leq 20$

8.65

$T(x) = 2.5 \times 10^5\sqrt{1 + (0.008)^2x^2}$,
$-50 < x < 50$

8.67 Reaction at $A = 14.4$ kN

$T(x) = 1197.5 \cosh(0.0417x)$,
$-15 < x < 15$

8.69

$T(x) = 87.88 \cosh(0.5689x)$

$y = [\cosh(0.5689x) - 1]/0.5689$

Chapter 9

9.1 See 3.47
9.3 See 3.49
9.5 See 3.51
9.7 See 3.53
9.9 See 3.55
9.11 See 3.57

9.13 $M = r\mu_s W\left(\frac{1 + \mu_s}{1 + \mu_s^2}\right)$

9.15 See 3.59
9.17 See 3.61
9.19 $f = 1888.43$ N
9.21 $F = 302.78$ N
9.23 $F = 777.75$ N
9.25 **a.** $f(\alpha) = 75 \cos\alpha$
b. $N(\alpha) = 100 - 75 \sin\alpha$
$f < \mu_s N(\alpha)$ for all values of α so the block never moves.
9.27 $\theta = 30.96°$
9.31 $80.578 \text{ kg} < m < 203.65 \text{ kg}$
9.33 The crate will tip down the plane.
9.35 The crate is not in equilibrium and will tip down the incline.
9.37 $P = 367.875$ N applied at any value of $h < 0.5$ m.
9.39 **a.** $P = 220.725$ applied anywhere.
b. $P = 220.725$ at $h < 0.5$ m.
9.41 The box slides.
9.43 The blocks slide together.
9.45 $P_{max} < \mu_2(W_1 + W_2)$
9.47 The boxes are stationary.

9.49 $P = \frac{\mu R}{\mu R + r}W$

9.51 $P = 4441$ N
The crate tips.
9.53 $\mu_1 = 0.0893$, $\mu_3 = 0.268$
9.55 $\mu N_2 = 5062$ N
9.57 $\mathbf{r} \times \mathbf{F} = 1203$ N m
9.59 $d = 0.217$ m
9.61 $P = 55.56$ N
9.63 $P = 203.4$ N
9.65 $P = 1883.57$ N
9.67 $P = 2268$ N
9.69 $P = 1973$ N
9.71 $P = 1725.21$ N
9.73 $P = 276.6$ N
Slips on both wall and wedge.
9.75 $P = 80.43$ N,
$\mu_{smallest} = 0.364$
9.77 No solution for slip at all surfaces.
9.79 $P = 2143$ N
9.81 $P = 2335$ N
9.83 The system is self locking.
$F = 79.4$ N
9.85 $F = 102$ N
9.87 $M = 10.02$ m N
9.89 $M = 5.05$ m N
9.93 $M = 0.11W \cot\theta$ m N
9.95 $M = 3.5$ m N
9.97 1.6 times.
9.99 $F = 2236$ N
9.101 The tension of 2500 N m cannot be transmitted by the center wheel.

9.103 $\frac{L}{L + r} = \mu e^{\frac{3\pi\mu}{2}}$

9.105 $\mu_k = 0.098$
9.107 $T_A = 966.25$ N cm
9.109 $M = 1125$ N mm
9.111 $F = 132.52$ N
9.113 $M_f = 0.943$ N m
9.115 $P = 72$ N
9.117 $P = 19.62$ N

Chapter 10

10.1 $\bar{x} = 80$ mm, $\bar{y} = 50.66$ mm
10.3 $I_{yy} = 1.2375 \times 10^8$ mm^4
10.5 $I_{xx} + I_{yy} = \frac{\pi}{2}\left(R_0^4 - R_1^4\right)$

10.7 $I_{yy} = \dfrac{h\ell^3}{3}$

10.9 $I_{yy} = \dfrac{2}{15}\text{ m}^4$

10.11 $I_{xx} = 320 \times 10^3\text{ mm}^4,$
$I_{yy} = 3840 \times 10^3\text{ mm}^4,$
$I_0 = 4160 \times 10^3\text{ mm}^4$

10.13 $I_{yy} = \dfrac{\pi a^3 b}{4} = \dfrac{\pi(3)^3 2}{4} = \pi\left(\dfrac{27}{2}\right)$

10.15 $I_{xx} = \dfrac{\pi b^3 a}{8}$

10.17 $I_0 = \dfrac{\pi ab}{8}(a^2 + b^2)$

10.19 $I_y = \dfrac{h\ell^3}{\pi}\left(1 - \dfrac{4}{\pi^2}\right)$

10.21 $\bar{y} = \dfrac{\pi}{4}$
$I_{yyc} = \pi^2 - 8$

10.23

Origin at A: $I_x = \dfrac{1}{9}\ell h^3,\ I_y = \dfrac{1}{9}h\ell^3$

Origin at B: $I_x = \dfrac{7}{12}\ell h^3,\ I_y = \dfrac{7}{12}h\ell^3$

10.25 $I_{xx_c} = 0.0457\text{ m}^4$
$I_{yy_c} = 0.0393\text{ m}^4$

10.27 $K_{xx_c} = 0.521$
$K_{yy_c} = 0.684$

10.29 $I'_{xx} = \dfrac{\pi r^4}{2}$

10.31 $I_{xx_c} = \dfrac{bh^3}{36},$

$I_{yy_c} = \dfrac{b^3 h}{48}$

10.33 $I_{xx} = \dfrac{a^4}{12} - \dfrac{\pi r^4}{4},$

$I_{yy} = \dfrac{a^4}{12} - \dfrac{\pi r^4}{4}$

10.35

$$I_{xx} = \frac{1}{6}(b - t)t^3 + (b - t)t\left(\frac{h}{2} - \frac{t}{2}\right)^2 + \frac{1}{12}th^3,$$

$$I_{yy} = \frac{1}{6}(b - t)^3 t + (b - t)t\left(\frac{b - t}{2} + \frac{t}{2}\right)^2 + \frac{1}{12}t^3 h$$

10.37

$$I_{xx_c} = \frac{1}{6}(b - t)t^3 + (b - t)t\left[\left(h - \frac{t}{2} - \bar{y}_c\right)^2 + \left(\bar{y}_c - \frac{t}{2}\right)^2\right] + \frac{1}{12}th^3 + ht\left(\frac{h}{2} - \bar{y}_c\right)^2,$$

$$I_{yy_c} = \frac{1}{6}t(b - t)^3 + 2t(b - t)\left(\frac{b + t}{2} - \bar{x}_c\right)^2 + \frac{1}{12}ht^3 + ht\left(\bar{x}_c - \frac{t}{2}\right)^2$$

10.39 $I_x = I_y = 0.702r^4$

$I_z = 1.404r^4$

10.41 11.68% increase in bending stiffness.
11.68% increase in polar moment of inertia.

10.43 $I_{xx_c} = 1432\text{ cm}^4$
$I_{yyc} = 1495\text{ cm}^4$

10.45 $I_{xx_c} = 0.374 \times 10^{-6}\text{ m}^4$
$I_{yy_c} = 0.4639 \times 10^{-6}\text{ m}^4$

10.47 $\beta = -5.52°$
$I_{x'x'} = 26.94 \times 10^{-6}\text{ m}^4$
$I_{y'y'} = 13.36 \times 10^{-6}\text{ m}^4$

10.49 $\beta = 42.82°$
$I_{x'x'} = 114.375\text{ cm}^4$
$I_{y'y'} = 68.175\text{ cm}^4$

10.51 $\bar{x}_c = 59.99\text{ mm}$
$\bar{y}_c = 26.66\text{ mm}$
$\beta = -40.83°$
$I_{x'x'} = 2.0392 \times 10^{-6}\text{ m}^4$
$I_{y'y'} = 0.9578 \times 10^{-6}\text{ m}^4$

10.53 $I_{xy_c} = 0$ and $I_{xx_c} = I_{yy_c}$

10.55 $\bar{x}_c = 18.84\text{ mm}$
$\bar{y}_c = 20\text{ mm}$
$\beta = -35.33°$
$I_{x'x'} = 0.4359 \times 10^{-6}\text{ m}^4$
$I_{y'y'} = 0.1731 \times 10^{-6}\text{ m}^4$

10.57 $\bar{x}_c = \dfrac{1}{2}\dfrac{ht + l^2 - t^2}{h + l - t},$

$\bar{y}_c = \dfrac{1}{2}\dfrac{h^2 + lt - t^2}{h + l - t},$

$$I_{xx} = \frac{1}{12}th^3 + \frac{1}{12}(l - t)t^3 + \frac{1}{4}\frac{ht(lt - t^2 - hl + ht)^2}{(h + l - t)^2} - \frac{1}{4}\frac{(-l + t)\,th^2(h - t)^2}{(h + l - t)^2},$$

$$I_{yy} = \frac{1}{12}t^3 h + \frac{1}{12}t(l - t)^3 + \frac{1}{4}\frac{htl^2(-l + t)^2}{(h + l - t)^2} - \frac{1}{4}\frac{(-l + t)t\,h^2 l^2}{(h + l - t)^2},$$

$$I_{xy} = -\frac{1}{4}\frac{htl(-l + t)(h - t)}{h + l - t}$$

Use (10.36) and (10.37).

10.59 $\bar{x}_c = 18.3$ mm, $\bar{y}_c = 38.8$ mm, $\beta = -16.07°$, $I_{x'x'} = 1.41 \times 10^{-6}\text{ m}^4$, $I_{y'y'} = 0.25 \times 10^{-6}\text{ m}^4$

10.61 See 10.47

10.63 See 10.49

10.65 See 10.51

10.67 See 10.53

10.69 See 10.55

10.71 See 10.57

10.73 See 10.59

10.75 See 10.47

10.77 See 10.49

10.79 See 10.51

10.81 See 10.53

10.83 See 10.55

10.85 See 10.57

10.87 See 10.59

10.89 $I_{xx_A} = \frac{1}{3}m(h^2 + t^2)$

$I_{yy_A} = \frac{1}{3}m(t^2 + \ell^2)$

$I_{zz_A} = \frac{1}{3}m(\ell^2 + h^2)$

10.91 $I_{zz} = 1867\text{ kg m}^2$

10.93 $\bar{x} = 400$ mm, $\bar{y} = 230.94$ mm, $I_{zz0} = 3.547 \times 10^6\text{ kg mm}^2$

10.95 $I_{xx} = 0.45\text{ kg m}^2$

10.97 $I_{zz} = 0.02081\text{ kg m}^2$ (about the center of mass)

10.99 $I_{z_0} = 0.511\text{ kg m}^2$

10.101 $I_{z_0} = \frac{5}{24}m\ell^2 + \frac{1}{8}m\ell^2\cos\theta$

Chapter 11

11.1 $\theta = 2\tan^{-1}\frac{2F}{mg}$

11.3 $M = \frac{25mg\ell}{18}\sin\theta$

11.5 $F = mg\cot\theta$

11.7 $F = 2\tan\theta(P - mg)$

11.9 $F = 15{,}288$ N

11.11 1274.32 N

11.13 $P(\theta) = 625\frac{1}{\sin\theta} + 22.5$

11.15 $W = 1204.2$ N

11.17

$$F(\theta) = \frac{-W(a\sin\theta - b\cos\theta)}{-c\sin\theta + \dfrac{c^2\cos\theta\sin\theta}{\sqrt{d^2 - c^2\sin^2\theta}}}$$

11.19 $M = mgR\sin\theta$

11.21 $W = 100\cot\theta$ N

11.23 $\theta_1 = 73.3°$, $\theta_2 = 84.3°$

11.25 $\theta_1 = 73.3°$, $\theta_2 = 0°$

11.27 $\theta_1 = 2.3°$, $\theta_2 = 3.89°$, $\theta_3 = 11.52°$

11.29 $U = -10$ Joules

11.31 $U = 20$ Joules

11.33 $U = 10$ Joules

11.35 **a.** $V(x) = 750x^2 - 98.1x$ Joules

b. $V'(x) = kx - mg = 0$

So, $x = \frac{mg}{k}$, $\sum F = kx = mg$ is zero at $x = \frac{mg}{k}$.

11.37 $\theta = \sin^{-1}\frac{mg}{2kL}$, stable and $\theta = \pm\pi/2, \pm 3\pi/2, \cdots$, neutral.

11.39 $k = \frac{294.3}{(1 - \cos\theta)\tan\theta}$, $\theta = 22°$, stable.

11.41 $x = \frac{mg}{k}$

11.43 $k = 101.09$ kN/m

11.45 $\theta = -27.0°$, stable.

11.47 $\theta = 39.97°$
No.

11.49 $\theta = \cos^{-1}\left[\frac{W}{kL}\right]$, unstable.

$\theta = 0$, stable if $kL > W$, neutral if $kL = W$, unstable if $kL < W$.

11.51 $\theta = 0°$, stable if $2kL > W$.

$\theta = \cos^{-1}\left(\frac{W}{2kL}\right)$, unstable.

11.53 No.

11.55 Yes.

11.57 $\mathbf{F} = -2x\hat{\mathbf{i}} - 2y\hat{\mathbf{j}} - 2z\hat{\mathbf{k}}$

11.59 $\mathbf{F} = -2(x - 1)\hat{\mathbf{i}} - 2y\hat{\mathbf{j}} - 2z\hat{\mathbf{k}}$
Yes.

INDEX

A

B

G

H

I

S

T

U

V

W

Z

Vector Algebra

Components of a vector: $\mathbf{A} = A_x\hat{\mathbf{i}} + A_y\hat{\mathbf{j}} + A_z\hat{\mathbf{k}}$

Matrix notation: $\mathbf{A} = \begin{bmatrix} A_x \\ A_y \\ A_z \end{bmatrix}$

Magnitude or the absolute value of a vector: $|\mathbf{A}| = \sqrt{A_x^2 + A_y^2 + A_z^2}$

Unit vector along **A**: $\hat{\mathbf{a}} = \dfrac{\mathbf{A}}{|\mathbf{A}|}$ that is, the vector divided by the magnitude of the vector. Therefore, a vector can always be written as the product of its magnitude and the unit vector.

$$\mathbf{A} = |\mathbf{A}|\hat{\mathbf{a}}$$

Base vectors of a coordinate system: Three non-coplanar unit vectors along the coordinates $\hat{\mathbf{i}}$, $\hat{\mathbf{j}}$, and $\hat{\mathbf{k}}$ in the x, y, and z, respectively.

Vector addition: $\mathbf{A} + \mathbf{B} = \mathbf{B} + \mathbf{A} = (A_x + B_x)\hat{\mathbf{i}} + (A_y + B_y)\hat{\mathbf{j}} + (A_z + B_z)\hat{\mathbf{k}}$

Vector equality: Two vectors are equal if and only if each of their components are equal.

$$\mathbf{A} = \mathbf{B} \rightarrow A_x = B_x \; A_y = B_y \; A_z = B_z$$

Multiplication of a vector by a scalar: $\alpha\mathbf{A} = \alpha A_x\hat{\mathbf{i}} + \alpha A_y\hat{\mathbf{j}} + \alpha A_z\hat{\mathbf{k}}$

Direction cosines of a vector: $\lambda_x = \dfrac{A_x}{|\mathbf{A}|} \quad \lambda_y = \dfrac{A_y}{|\mathbf{A}|} \quad \lambda_z = \dfrac{A_z}{|\mathbf{A}|}$

The direction cosine is the cosine of the angle between the vector and the coordinate axis. The components of the unit vector along A are the direction cosines.

$$\hat{\mathbf{a}} = \lambda_x\hat{\mathbf{i}} + \lambda_y\hat{\mathbf{j}} + \lambda_z\hat{\mathbf{k}}$$

Scalar or dot product between two vectors:

$$\mathbf{A}\cdot\mathbf{B} = |\mathbf{A}||\mathbf{B}|\cos\theta \qquad \text{where } \theta \text{ is the angle between the vectors}$$

$$\mathbf{A}\cdot\mathbf{B} = A_xB_x + A_yB_y + A_zB_z$$

Vector or cross product between two vectors:

$$\mathbf{A} \times \mathbf{B} = |\mathbf{A}||\mathbf{B}| \sin \theta \hat{\mathbf{n}}$$

where θ is the angle between the vectors and $\hat{\mathbf{n}}$ is a unit vector perpendicular to the plane formed by the vectors with its direction determined by the right-hand rule.

$$\mathbf{A} \times \mathbf{B} = (A_yB_z - A_zB_y)\hat{\mathbf{i}} + (A_zB_x - A_xB_z)\hat{\mathbf{j}} + (A_xB_y - A_yB_x)\hat{\mathbf{k}}$$

$$\mathbf{A} \times \mathbf{B} = \begin{vmatrix} \hat{\mathbf{i}} & \hat{\mathbf{j}} & \hat{\mathbf{k}} \\ A_x & A_y & A_z \\ B_x & B_y & B_z \end{vmatrix}$$

Properties of the dot and cross products between unit vectors:

$$\begin{array}{llllll}
\hat{\mathbf{i}}\cdot\hat{\mathbf{i}} = 1 & \hat{\mathbf{i}}\cdot\hat{\mathbf{j}} = 0 & \hat{\mathbf{i}}\cdot\hat{\mathbf{k}} = 0 & \hat{\mathbf{i}}\cdot\hat{\mathbf{i}} = 0 & \hat{\mathbf{i}} \times \hat{\mathbf{j}} = \hat{\mathbf{k}} & \hat{\mathbf{i}} \times \hat{\mathbf{k}} = -\hat{\mathbf{j}} \\
\hat{\mathbf{j}}\cdot\hat{\mathbf{i}} = 0 & \hat{\mathbf{j}}\cdot\hat{\mathbf{j}} = 1 & \hat{\mathbf{j}}\cdot\hat{\mathbf{k}} = 0 & \hat{\mathbf{j}}\cdot\hat{\mathbf{i}} = -\hat{\mathbf{k}} & \hat{\mathbf{j}} \times \hat{\mathbf{j}} = 0 & \hat{\mathbf{j}} \times \hat{\mathbf{k}} = \hat{\mathbf{i}} \\
\hat{\mathbf{k}}\cdot\hat{\mathbf{i}} = 0 & \hat{\mathbf{k}}\cdot\hat{\mathbf{j}} = 0 & \hat{\mathbf{k}}\cdot\hat{\mathbf{k}} = 1 & \hat{\mathbf{k}} \times \hat{\mathbf{i}} = \hat{\mathbf{j}} & \hat{\mathbf{k}} \times \hat{\mathbf{j}} = -\hat{\mathbf{i}} & \hat{\mathbf{k}} \times \hat{\mathbf{k}} = 0
\end{array}$$

Vector identities:

$$\begin{aligned}
&\mathbf{A}\cdot(\mathbf{B} \times \mathbf{C}) = \mathbf{B}\cdot(\mathbf{C} \times \mathbf{A}) = \mathbf{C}\cdot(\mathbf{A} \times \mathbf{B}) \\
&\mathbf{A} \times (\mathbf{B} \times \mathbf{C}) = \mathbf{B}(\mathbf{C}\cdot\mathbf{A}) - \mathbf{C}(\mathbf{A}\cdot\mathbf{B}) \\
&(\mathbf{A} \times \mathbf{B}) \times (\mathbf{C} \times \mathbf{D}) = \mathbf{B}[\mathbf{A}\cdot(\mathbf{C} \times \mathbf{D})] - \mathbf{A}[\mathbf{B}\cdot(\mathbf{C} \times \mathbf{D})] \\
&(\mathbf{A} \times \mathbf{B}) \times (\mathbf{C} \times \mathbf{D}) = \mathbf{C}[(\mathbf{A} \times \mathbf{B})\cdot\mathbf{D}] - \mathbf{D}[(\mathbf{A} \times \mathbf{B})\cdot\mathbf{C}] \\
&(\mathbf{A} \times \mathbf{B}) \times (\mathbf{A} \times \mathbf{C}) = \mathbf{A}[(\mathbf{A} \times \mathbf{B})\cdot\mathbf{C}]
\end{aligned}$$

Components of a vector parallel and perpendicular to a line designated by a unit vector $\hat{\mathbf{n}}$:

$$\begin{aligned}
\mathbf{A}_{\|} &= (\mathbf{A}\cdot\hat{\mathbf{n}})\hat{\mathbf{n}} \\
\mathbf{A}_{\perp} &= \hat{\mathbf{n}} \times (\mathbf{A} \times \hat{\mathbf{n}}) = \mathbf{A} - (\mathbf{A}\cdot\hat{\mathbf{n}})\hat{\mathbf{n}}
\end{aligned}$$

Solution for the components of a vector in a nonorthogonal coordinate system:

$$\mathbf{F} = F_a\hat{\mathbf{a}} + F_b\hat{\mathbf{b}} + F_c\hat{\mathbf{c}}$$

Form a biorthogonal set of base vectors:

$$\mathbf{r} = \frac{\hat{\mathbf{b}} \times \hat{\mathbf{c}}}{|\hat{\mathbf{b}} \times \hat{\mathbf{c}}|} \quad \hat{\mathbf{s}} = \frac{\hat{\mathbf{c}} \times \hat{\mathbf{a}}}{|\hat{\mathbf{c}} \times \hat{\mathbf{a}}|} \quad \hat{\mathbf{t}} = \frac{\hat{\mathbf{a}} \times \hat{\mathbf{b}}}{|\hat{\mathbf{a}} \times \hat{\mathbf{b}}|}$$

The components are:

$$F_a = \frac{\mathbf{F}\cdot\hat{\mathbf{r}}}{\hat{\mathbf{a}}\cdot\hat{\mathbf{r}}} \quad F_b = \frac{\mathbf{F}\cdot\hat{\mathbf{s}}}{\hat{\mathbf{b}}\cdot\hat{\mathbf{s}}} \quad F_c = \frac{\mathbf{F}\cdot\hat{\mathbf{t}}}{\hat{\mathbf{c}}\cdot\hat{\mathbf{t}}}$$

Linear Algebra

A matrix $[A]$ is an ordered rectangular array containing elements a_{ij} occupying the ith row and the jth column.

$$[A] = \begin{bmatrix} a_{11} & a_{12} & \cdots & \cdots & a_{1n} \\ a_{21} & a_{22} & \cdots & \cdots & a_{2n} \\ \vdots & \vdots & \vdots & \vdots & \vdots \\ \vdots & \vdots & \vdots & \vdots & \vdots \\ a_{m1} & a_{m2} & \cdots & \cdots & a_{mn} \end{bmatrix}$$

Such a matrix is said to be of order $m \times n$ and to have m rows and n columns.

Matrix addition or subtraction: Two matrices may be added or subtracted if they are of the same size (say $m \times n$) by the addition or subtraction of their corresponding elements.

$$[C] = [A] \pm [B]$$

$$c_{ij} = a_{ij} \pm b_{ij}$$

Multiplication of a matrix by a scalar α is defined on a per element basis (i.e., the product of the scalar α and the matrix $[A]$, denoted $\alpha[A]$, has elements αa_{ij}).

Product of two matrices: The product of two matrices is defined only if the matrices are of compatible size. In particular, the matrix product $[C] = [A][B]$ is defined only if $[A]$ is an $(m \times p)$ matrix and $[B]$ is a $(p \times n)$ matrix where p is the compatible size parameter. The resulting matrix $[C]$ is of size $(m \times n)$.

$$[C] = [A]\,[B]$$

$$c_{ij} = \sum_{k=1}^{p} a_{ik} b_{kj}$$

The determinant of a square $(n \times n)$ matrix:

$$|A| = \sum_{j=1}^{n} (-1)^{1+j} a_{1j} |A_{1j}|$$

where $[A_{1j}]$ is the $[(n-1) \times (n-1)]$ matrix formed from the matrix A by deleting the first row and the jth column of the matrix A. The matrix $[A_{ij}]$ formed by deleting the ith row and the jth column of the matrix $[A]$ is called a minor of $[A]$.

The adjoint of $[A]$ is denoted by adj$[A]$ and is the matrix defined by forming the elements α_{ij} into a matrix where $\alpha_{ij} = (-1)^{i+j}\,|A_{ij}|$

The inverse of a matrix $[A]^{-1}$ is defined such that

$$[A]^{-1}[A] = [A][A]^{-1} = [I]$$

where $[I]$ is the unity matrix having the diagonal elements equal to 1 and all the off-diagonal elements equal to 0.

$$[A]^{-1} = \frac{adj\,[A]}{|A|}$$

If the determinant of the matrix $[A]$ is zero, the inverse of the matrix does not exist and the matrix is said to be singular.

System of linear equations: $[A][x] = [C]$ is a system of n equations for the n unknowns $[x]$.

$$[x] = [A]^{-1}[C]$$

Transpose of a matrix $[A]^{\mathrm{T}}$ is obtained by interchanging the elements of the rows and columns.

MATHEMATICS WINDOW

Properties of Transpose, Inverse, and Determinants of Matrices

$$([\mathbf{a}] + [\mathbf{b}])^{\mathrm{T}} = [\mathbf{a}]^{\mathrm{T}} + [\mathbf{b}]^{\mathrm{T}}$$

$$(\alpha\,[\mathbf{a}])^{\mathrm{T}} = \alpha\,[\mathbf{a}]^{\mathrm{T}}$$

$$([\mathbf{a}][\mathbf{b}])^{\mathrm{T}} = [\mathbf{b}]^{\mathrm{T}}[\mathbf{a}]^{\mathrm{T}}$$

$$([\mathbf{a}]^{\mathrm{T}})^{\mathrm{T}} = [\mathbf{a}]$$

$$([\mathbf{a}]^{-1})^{\mathrm{T}} = ([\mathbf{a}]^{\mathrm{T}})^{-1}$$

$$([\mathbf{a}][\mathbf{b}])^{-1} = [\mathbf{b}]^{-1}[\mathbf{a}]^{-1}$$

determinant: $\det([\mathbf{a}][\mathbf{b}]) = \det([\mathbf{a}])\det([\mathbf{b}])$

$\det([\mathbf{a}]^{\mathrm{T}}) = \det([\mathbf{a}])$

$\det(\alpha\,[\mathbf{a}]) = \alpha\det([\mathbf{a}])$ where α is a scalar.